AF732230

ESTATES GAZETTE LAW REPORTS

Estates Gazette Law Reports

1992

Volume 1

Edited by
Barry Denyer-Green LLM PhD FRICS
of the Middle Temple, barrister

Audrey Boyle BA
Publisher's editor

Estates Gazette

Routledge
Taylor & Francis Group
LONDON AND NEW YORK

EG Books is an imprint of Routledge
2 Park Square, Milton Park, Abingdon, Oxon OX14 4RN
711 Third Avenue, New York, NY 10017, USA
Routledge is an imprint of the Taylor & Francis Group, an informa business

ISBN for complete set of 2 volumes: 0 7282 0180 1
ISBN for this volume 0 7282 0181 X
ISSN 0951-9289

For information on all EG Books publications
visit our website www.routledge.com

CONTENTS

SUPPLEMENT

Cases not reported in "Estates Gazette"

TABLE OF CASES

Titles of cases shown in bold type
Names of parties reversed shown in ordinary type

INDEX OF SUBJECT-MATTER

AGRICULTURE

Court of Appeal

October 11 1991

(Before Lord Justice PARKER, Lord Justice FARQUHARSON and Lord Justice SCOTT)

LAND v SYKES AND ANOTHER

Estates Gazette January 25 1992

[1992] 03 EG 115

Agricultural Holdings Act 1986 — Claim by landlord for possession — Severance of reversion — Case E notice to quit served — Effect of severance of reversion on validity of notice — Second notice to quit under Case G — Whether first notice to quit valid — Whether implication of new tenancy after determination of previous tenancy — Appeal by tenant dismissed

By a lease the respondent landlord granted the late Kenneth Sykes an agricultural tenancy of some 291.24 acres — The tenancy was for a fixed term of one year from April 1 1985 and thereafter from year to year at an annual rent of £12,900 payable half yearly in arrears on October 11 and April 6 in each year — By a conveyance dated October 24 1986 the landlord conveyed by way of gift a small strip of land which was comprised in the tenancy — The strip of land was a rough track — There was no apportionment of the rent — Following allegations that Mr Sykes in breach of covenant had cut down trees, on March 17 1987 the landlord served a notice to quit pursuant to Case E, Part I of Schedule 3 to the Agricultural Holdings Act 1986 — The solicitors acting for Mr Sykes failed to seek arbitration to prevent the Case E notice from taking effect but contended that by reason of the severance of the reversion the notice was invalid — On March 5 1988 Mr Sykes died — The appellants in this appeal were appointed by his will as his executors — On March 21 1988 the landlord's solicitors served a notice to quit under Case G, a letter accompanying the notice stating that it was served without prejudice to the contention that the tenancy had already been determined by the previous notice — The Case G notice to quit was drafted having regard to the severance of the reversion — The Case E notice expired on April 1 1988, but notwithstanding this the landlord's solicitors served a Case D form of notice to pay rent — No application was made by Mr Sykes' son for a succession tenancy of the holding within the period of three months after Mr Sykes' death

In January 1989 the appellants' solicitors sent a cheque for £6,450 stating that it was the rent due from April 6 to October 11 1988 — By April 1989 the landlord became in urgent need of money and the cheque for £6,450 was presented for payment, the landlord's solicitors writing to the executors' solicitors asking that the executors would not take the point that the money was tendered either as rent or as mesne profits — Proceedings for possession were commenced on June 1 1989

In the court below the judge accepted the submissions on behalf of the landlord that the Case E notice was invalid — The Case G notice had been effective to terminate the tenancy and the executors were ordered to give up possession — On appeal the executors submitted that the Case E notice was valid and that the tenancy created by the lease terminated on April 1 1988 — The Case G notice was irrelevant — It was submitted further, however, that in the period between April 1 1988 and the commencement of proceedings on April 10 1989 the landlord granted the executors a new tenancy of the farm or, alternatively, a licence that was enlarged into a tenancy to be implied from conduct, namely the notice to pay rent of April 26 1988 and the presentation of the cheque sent in January 1989 expressly as rent

***Held*: The Case E notice to quit was valid — The law as to the validity of a notice to quit is summarised in *Carradine Properties Ltd* v *Aslam,* subject to the qualification that the reasonable tenant reading the notice is to be taken to have had the knowledge of the surrounding facts and circumstances which the actual landlord and tenant enjoyed — Following the expiration of the Case E notice the executors became tenants on sufferance — No implied grant of a tenancy had been established — The notice to pay rent and the acceptance of the cheque were inadequate taken singly or together to justify imputing to the landlord an intention to grant a new tenancy or licence to occupy to the executors**

The following cases are referred to in this report.

Carradine Properties Ltd v *Aslam* [1976] 1 WLR 442; [1976] 1 All ER 573; (1975) 32 P&CR 12
Doe d Cheny v *Batten* (1775) 1 Cowp 243
Frankland v *Capstick* [1959] 1 All ER 209; [1959] 1 WLR 204, CA

This was an appeal by the defendants, the executors of the will of the late Kenneth Sykes, from a decision of Mr Timothy Lloyd QC, sitting as a deputy judge of the Chancery Division (see [1991] 1 EGLR 18; [1991] 16 EG 125), in an action by the respondent, Mrs Edna Lydia Land, for possession of Grange Farm and other land in Wyverstone, Suffolk.

Paul Morgan (instructed by Reynolds Porter Chamberlain) appeared on behalf of the appellant executors; Jonathan Brock (instructed by Burges Salmon, of Bristol) represented the respondent landlord.

Giving the first judgment at the invitation of Parker LJ, **Scott LJ** said: This is an appeal from the judgment of Mr Timothy Lloyd QC, sitting as a deputy judge of the High Court. It raises a point regarding the efficacy of a notice to quit an agricultural holding given by the plaintiff, the landlord, to the defendants, the executors of the deceased tenant, and a point regarding the alleged implied grant of a new tenancy to the defendants.

The facts relevant to the issues that have been argued on the appeal can be quite shortly stated.

The plaintiff, the landlord, is Mrs Land. The deceased tenant was Mr Kenneth William Sykes. By a lease dated December 17 1985 Mrs Land granted Mr Sykes an agricultural tenancy of "all those farms farmhouse buildings and land containing 291.240 acres or thereabouts situate in the parish of Wyverstone in the County of Suffolk and known as Grange Farm aforesaid Sudbourne Farm Plough Farm Crooksells Hall Farm and adjoining land".

Particulars of the premises were specified in the first schedule to the lease. The premises included a 1.105-acre plot, part of OS no 227. The tenancy was granted for a fixed term of one year from April 1

A 1985 and thereafter from year to year. The lease reserved an annual rent of £12,900 payable half yearly in arrears on October 11 and April 6 in each year. The deputy judge held, as a matter of construction, that the instalments of rent were payable in respect of half yearly periods up to April 1 and October 1 in each year. It has not been suggested that this was wrong. The terms of the lease did not permit the tenancy to be brought to an end by notice to quit until April 1 1987 at the earliest.

The lease was granted as part of the working out of a court order made in contentious litigation between Mrs Land and Mr Sykes. They had at one time been in partnership farming the land comprised in the lease and other land as partners. Mrs Land terminated the

B partnership and sought to obtain possession of the land from Mr Sykes. She brought possession proceedings in the High Court. Her action failed. She appealed. The Court of Appeal dismissed her appeal*.

By a conveyance dated April 4 1986 Mrs Land conveyed to her grandson, Mr Warner, by way of gift a small plot of land adjacent to the part of OS no 227 that was comprised in the lease. The plot conveyed was not itself comprised in the lease.

By a later conveyance dated October 24 1986 Mrs Land conveyed to Mr Warner, again by way of gift, a small strip of land, roughly 9 ft wide and 200 ft in length lying on the boundary of the plot she had already given him. This strip of land was part of OS no 227 and was comprised in the lease. It was, apparently, a rough track leading from

C the public road. The conveyance of the strip of land to Mr Warner was expressed to be subject to the lease "but without any part of the rent payable under the said lease".

It was found by the deputy judge, and was, I think, common ground, that Mr Sykes and his family, which would include both defendants, were aware at all material times that Mrs Land had conveyed this strip of land. The conveyance made no real difference at all to Mr Sykes. He still had to pay the whole rent to Mrs Land. There had been no apportionment of rent. He could, of course, continue to use the track for the purposes of his farm as he had previously done.

In September 1986 complaints were made by Mrs Land's solicitors that Mr Sykes, in breach of covenant, had cut down trees on the land

D comprised in the lease. And on March 17 1987 Mrs Land's solicitors served on Mr Sykes and his solicitors a notice to quit expressed to be given under Case E: see Part I of Schedule 3 to the Agricultural Holdings Act 1986. The cutting down of the trees was relied on as justifying the service of the notice.

The notice was headed:

> Re: the Holding known as Grange Farm Sudbourne Farm Plough Farm Crooksells Hall Farm and all land held therewith

and commenced as follows:

> As Solicitors for and on behalf of your Landlord Edna Lydia Land we hereby give you notice to quit and deliver up to her possession of all that holding and premises known as Grange Farm Sudbourne Farm Plough Farm Crooksells
> E Hall Farm and all land held therewith and situate at Wyverstone Stowmarket aforesaid which you hold of her as Tenant thereof on the 1st day of April 1988 or at the expiration of the year of your tenancy which shall expire next after the end of twelve months from the date of service of this Notice . . .

The solicitors acting for Mr Sykes failed to take the steps available to be taken under the terms of the 1986 Act in order to prevent the Case E notice from taking effect. They did, however, contend that the notice was invalid. By letter dated April 8 1987 they said:

> We understand that your client has disposed of part of the freehold reversion and if that be the case we believe that the Notices as served are invalid. We invite your comments on this point.

The point taken was based on the conveyance to Mr Warner of the 9 ft by 200 ft strip of land. Mrs Land's solicitors responded by saying

F they would inquire into the point. The point thereafter remained an open one.

Subject to the point that had been taken and in view of the failure of Mr Sykes' solicitors to make use of the relevant procedures provided by the 1986 Act, it is accepted that the effect of the Case E notice would be to bring the tenancy to an end on April 1 1988.

On October 20 1987 Mrs Land's solicitors served a notice on Mr Sykes requiring him to pay the £6,450 rent that had fallen due on October 11 1987. This notice was given by means of the printed form usual for such notices. Failure to pay in accordance with the notice would have given Mrs Land the right to serve a Case D notice to quit.

---

*Editor's note: Reported at (1984) 271 EG 1264.

It must be remembered that the validity of the Case E notice had been challenged. G

On March 5 1988 Mr Sykes died. He left a will by which he appointed his widow and his son, the defendants in this action, his executors. They proved the will on April 10 1989. They are the appellants before us.

Mr Sykes' interest under the lease vested by operation of law in his executors. It was not, of course, known to Mrs Land or her solicitors whether Mr Sykes had died leaving a will or whether he had died intestate.

On March 21 1988 Mrs Land's solicitors served a second notice to quit. This notice was served under Case G. Their letter to Mrs Sykes, the first defendant, enclosing the notice said:

> We are sorry to hear of the death of your late husband. We regret that it is necessary for us to serve a notice to quit following your husband's death. H
>
> This Notice is served without prejudice to the contention that the tenancy has already been determined as a result of a previous notice served on behalf of our client.

That, of course, was a reference to the Case E notice that had been served.

A copy of the letter and of the Case G notice were sent to Mrs Sykes' solicitor, who had previously been Mr Sykes' solicitor.

The Case G notice to quit was so drafted as to avoid the point on which the alleged invalidity of the Case E notice had been based. It was headed:

> Re: So much of the Holding known as Grange Farm, Sudbourne Farm, etc. J

and described the premises to be delivered up as follows:

> ALL THE holding and premises described above which was held of her by the late Kenneth William Sykes as tenant thereof until his death on or about the 5th day of March 1988 . . .

Delivery up of the land was sought "on the 1st day of April 1989 or at the expiration of the year of the said tenancy which shall expire next after the end of twelve months from the date of service of this Notice".

It is easy to understand why this second notice was served. The validity of the Case E notice had been challenged on a technical point of form. The point had not been resolved. If the point were a good one the Case E notice would be invalid. The service of the Case G K notice, expressed to be given without prejudice to the validity of the Case E notice, represented an adoption of the belt and braces principle.

On April 1 1988 the Case E notice expired. But the defendants, Mr Sykes' widow and son, continued in possession of the land comprised in the lease. They showed no signs of vacating possession voluntarily.

On April 13 1988 Mrs Land's solicitors wrote to the defendants' solicitors about the rent for the half year which had ended on April 1 1988. The letter said:

> We are concerned to note that the rent due at the end of the half year of the tenancy which has just expired has not been paid. Our client does not wish unnecessarily to serve a further Notice to Pay but we should be glad if you can L arrange for the executors to pay the rent through your firm as usual as soon as possible.

The rent was not paid and on April 26 1988 Mrs Land's solicitors served a Case D form of notice to pay the rent. The form was the same as that which had been used on the previous occasion to which I have referred. The form of this notice makes no sense where the tenancy under which the unpaid rent had become due has expired. The form is consistent only with a continuing tenancy. Accordingly, Mr Morgan, for the defendants, has contended that by sending this notice to pay Mrs Land, through her solicitors, was implicitly recognising that the defendants were in possession of the land under a continuing tenancy.

If, as the solicitors for Mr Sykes had originally contended, the Case M E notice had been invalid, Mrs Land would have had to rely, if she were to succeed in recovering possession, on the Case G notice. The weakness of the Case G notice, however, was that for a period of three months after Mr Sykes' death on March 5 1988 it would have been open to the second defendant, Mr Sykes' son, to apply for a succession tenancy of the holding. No such application was, however, made. By June 5 1988 it was too late for such an application to be made. The result of this state of affairs was somewhat paradoxical. If the Case E notice were invalid, the defendants would have no possible answer to the Case G notice requiring them to deliver up the premises on April 1 1989. So it was no longer in the interests of the defendants to argue that the Case E notice was invalid. Instead, it

A was necessary for them to find some means of seeking to avoid the Case G notice taking effect.

On January 23 1989 the defendants' solicitors sent to Mrs Land's solicitors a cheque for the sum of £6,450. The cheque was enclosed with a letter which said:

> The Personal Representatives of the above named deceased have asked us to forward to you the enclosed cheque for £6,450 being the rent due from the 6th April to 11th October 1988.

This was fairly remarkable. It had not been the custom for Mr Sykes or for the defendants to pay rent that had fallen due without first being pressed to pay. The cheque sent on January 23 1986 was not sent as a result of any pressure. Mrs Land's solicitors, not
B surprisingly, regarded the cheque as a Trojan horse, as indeed I think it was intended to be. If Mrs Land were to accept rent for a period after April 1 1988, there might be an implication that she was recognising that the defendants were tenants in respect of that period. If the Case E notice was valid, Mr Sykes' tenancy would have come to an end on April 1 1988. So, any tenancy thereafter enjoyed by the defendants could be attributable only to a new tenancy granted by Mrs Land to the defendants. It could not be attributable to the tenancy created by the lease. That, on this hypothesis, had come to an end on April 1 1988.

It is, I think, easy enough, with the advantage of hindsight, to follow the tactics being employed, entirely properly, on behalf of the defendants. They would assert the validity (not the invalidity) of the
C Case E notice. It would then be sought to establish that since April 1 1988 the defendants had acquired a new tenancy of the holding. The Case G notice, served on March 21 1988, and directed to bringing to an end the original tenancy, in case the Case E notice should turn out to be ineffective, would have no bearing on the new tenancy. That was the plan.

Mrs Land's solicitors did not, however, accept the Trojan horse. They replied by letter of January 25 1989 in these terms:

> Thank you for your letter of 23rd January, enclosing cheque for £6,450. We regret that so long as the notice to quit, which was served in 1987 and which expired in 1988, remains outstanding, our clients cannot accept any rent for this holding. We shall, therefore, hold your cheque unpresented.

D There for a while the matter rested.

Then, on April 10 1989, after, be it noted, the expiry of the Case G notice but before any proceedings for possession had been commenced, a family tragedy struck Mrs Land and she became urgently in need of money. Her solicitors still held the cheque for £6,450 which had been sent as rent on January 23 1989. Her solicitors explained the circumstances that had arisen to the defendants' solicitors and pressed them to agree that, if the cheque were presented, the point that there was thereby an implied acceptance that the defendants were tenants would not be taken. The defendants' solicitors refused to bind the defendants not to take the point. But on April 11 1989 the cheque was presented none the less. Mrs Land's solicitors wrote a letter to the defendants' solicitors on
E that date in these terms:

> We refer to our telephone conversation on 10th April, and write to confirm that we have presented the cheque in the sum of £6,450 dated 19th January.
>
> For the record, we have explained to you the somewhat unusual circumstances in which we have been instructed to present this cheque, and we would ask that if the matter comes to the point, your client will not take the point that the money was tendered either as rent or as mesne profits. We fully appreciate, however, that circumstances may mean that your client will not be able to comply with this request.

Proceedings for possession were commenced by writ issued on June 1 1989. It is important to notice that these possession proceedings had been consistently threatened ever since June the previous year. On June 1 1988 Mrs Land's solicitors had asked the
F defendants' solicitors if they had instructions to accept service of proceedings. During the 12 months between then and the issue of the writ there was never a time when the defendants could have been in any doubt but that possession proceedings would be instituted in due course. The correspondence, through which we were taken by Mr Brock, for Mrs Land, makes that clear. Those are, to my mind, the facts relevant to this appeal.

The first issue, albeit a minor issue, that I should deal with relates to some correspondence. The correspondence is not marked "without prejudice". It was argued before the deputy judge that the correspondence was, by implication, without prejudice and should not be admitted into evidence. The deputy judge rejected that submission and allowed the correspondence into evidence but gave leave to appeal against his decision. Mr Morgan, for the appellants, G
has argued that the correspondence followed on from, and formed part of, earlier correspondence consisting of a number of letters which were expressly marked "without prejudice" and in which the parties had been negotiating a settlement. In my judgment, however, the judge was quite right. The letters in question contain no reference to or flavour of negotiations. They are simply letters in which the defendants' solicitors are giving certain information to the plaintiff's solicitors. The author of the letters did not choose to mark these letters "without prejudice". I can see no reason why he should have done so, and I can see no reason why the court should assume to correct the author on this point. I would dismiss the interlocutory appeal.

The next issue concerns the validity of the Case E notice. Although H
initially the defendants' solicitors contended that the notice was invalid in form and, presumably, Mrs Land's solicitors thought it was valid, the positions are now reversed. If the notice is invalid, then the Case G notice rules the field and the *indicia* relied on by the defendants as establishing the grant of a new tenancy are no more than *indicia* of the continuing tenancy granted by the lease and due, by reason of the Case G notice, to expire on April 1 1989. If the Case E notice is invalid it is accepted that the plaintiff's case must succeed and the defendants' appeal must fail. So it is Mrs Land who is arguing for invalidity and the defendants who are arguing for the validity of the Case E notice.

The case for invalidity is based upon the alleged lack of certainty as J
to whether the premises described in the Case E notice and of which delivery-up was sought consisted only of the premises of which Mrs Land was landlord, ie excluding the 9 ft by 200 ft strip, or whether the notice, properly construed, purported to call for the delivery-up of all the premises comprised in the lease, ie including the strip.

The effect in law of the conveyance of the strip to Mr Warner was that Mrs Land could serve a notice to quit only in respect of the premises that remained vested in her. She had no power to serve a notice to quit in respect of the strip: see section 140(2) of the Law of Property Act 1925. So, if the notice, properly construed, did include the strip, Mrs Land was, to that extent, acting outside her power. The notice was, to that extent, calling upon the tenant to do that which he should not have been called upon to do. K

The requirements of certainty that, at common law, a notice to quit must have if it is to be valid are dealt with in two authorities to which we were referred.

In *Frankland* v *Capstick* [1959] 1 WLR 209 the notice under review had misstated the name of the landlord. But the Court of Appeal rejected the proposition that that error invalidated the notice. Sellers LJ said at p 208:

> there is no indication in the case . . . that the widow was in any way deceived by this or was under any misapprehension as to who the landlord was.

and on the same page, he said:

> The intervening negotiations for the dilapidations arising out of the termination of the tenancy had taken place with the son, who she knew to be L
> the agent acting on behalf of his father, and the solicitors throughout were solicitors acting on behalf of the landlord. There cannot have been any mistake about it. It is a technical error: it is indeed a slip which I think in no way affected the widow, her rights, or her understanding of what the proceedings were and her obligations thereunder.

In *Carradine Properties Ltd* v *Aslam* [1976] 1 WLR 442 Goulding J summarised the law in this way at p 444:

> I would put the test generally applicable as being this: "Is the notice quite clear to a reasonable tenant reading it? It is plain that he cannot be misled by it?"

At p 10 of his judgment the deputy judge said*:

> It is a basic principle that a notice to quit must be clear and certain, so as to bind the party who gives it and to enable the party to whom it is given to act upon it at the time when it is given. It may not be ambiguous. On the other M
> hand the court will not strain to find doubts, difficulties or ambiguities but will construe the notice as a whole in the light of the relevant surrounding circumstances and by reference to its effect on a reasonable tenant reading it.

In my judgment, subject perhaps to the qualifications that the reasonable tenant reading the notice is to be taken to have had the knowledge of the surrounding facts and circumstances which the actual landlord and tenant enjoyed, that passage, in my view, accurately states the law. But while I agree with the direction that the judge thereby gave himself, I do not, I regret, agree with his

---

*Editor's note: Reported at [1991] 1 EGLR 18 at p 22 J.

A application of the law to the facts of this case. At p 13 of his judgment he said*:

> . . . is the notice quite clear to a reasonable tenant reading it? Is is plain that he cannot be misled by it? In my view a reasonable tenant, following the severance, would be confused or misled or left in doubt by this notice.

The construction of the notice should, I agree, in general be an objective one. But the objective approach must, in my opinion, take into account the surrounding facts and circumstances that were common knowledge to this landlord and to this tenant. Both Mrs Land and Mr Sykes knew that Mrs Land had divested herself of the strip. Both knew that the strip was not part of the land "which you [ie Mr Sykes] hold of her as tenant . . .".

B The learned deputy judge thought there was a contradiction between the heading "the holding known as Grange Farm . . ." etc on the one hand and the reference to the land "which you hold of him as tenant" on the other hand. I do not think there is any contradiction. The removal of the strip from the 291-acre holding would not make in the least inaccurate the description of the land of which Mrs Land remained the owner as "the holding known as Grange Farm" etc. Of what materiality, so far as the description of "the holding" is concerned, is a 9 ft by 200 ft strip in the context of this 291-acre farm? The common-sense answer and the answer I would give is "None".

I do not think the description in the Case E notice of the land to which it applied would have been in the least confusing to a
C reasonable tenant with knowledge that Mrs Land was not the owner of the strip. For my part, even if the tenant had not had that knowledge, I would have upheld the notice. The proposition that a notice to quit this 291-acre farm was confusing and invalid because it purported to include the 9 ft by 200 ft strip that the landlord no longer owned is not one that I am prepared to accept. The maxim *de minimis non curat lex* is one that I would, if necessary, apply. In my judgment, the Case E notice was valid.

On this part of the case I would add that, in my opinion, if a notice to quit in fact communicates the correct information to the recipient tenant, I do not think that it is any business of the courts, or that there is any requirement of the law of landlord and tenant that operates, to
D deprive the notice of validity by reference to what some hypothetical reasonable tenant might have thought the notice meant. I agree that, if objectively construed in the light of the facts known to both landlord and tenant the meaning of the notice is clear, that is an end of the matter. But if, in fact, the tenant was not confused, I for my part think that that would suffice. I regard the approach of Sellers LJ in *Frankland* v *Capstick* as authority.

It follows from the validity of the Case E notice that the tenancy created by the lease terminated on April 1 1988. It was conceded by Mr Morgan that, if that were so, then, immediately after April 1 1988, the defendants' continued occupation and possession of the farm was a trespass. It may be that the better view is that the defendants became tenants on sufferance rather than trespassers: see
E *Woodfall Landlord and Tenant* vol 1 p 269. Their position as tenants on sufferance would, however, have been no more favourable, for the purpose of the points argued on this appeal, than their position as trespassers.

It also follows from the validity of the Case E notice that the Case G notice, expressed to be given without prejudice to the Case E notice, becomes irrelevant, except perhaps for the purpose of explaining why certain things were done or not done in the ensuing months.

Mr Morgan contends that in the period between April 1 1988 and the commencement of proceedings on April 10 1989 Mrs Land granted the defendants a new tenancy of the farm or alternatively a licence that was enlarged into a tenancy by the provisions of the 1986 Act. An express grant is not suggested. What is contended is that the
F grant is to be implied from conduct. So, on the footing that immediately after April 1 1988 the defendants were trespassers, it is necessary to ask whether anything happened thereafter to enable them to claim an agricultural tenancy or licence for the purposes of section 1 of the 1986 Act.

Mr Morgan relies on two particular matters. He relies on the terms of the notice to pay rent of April 26 1988 and he relies on the presentation of the cheque that had been sent on January 23 1989 expressly as rent for the period April 1 1988 to October 1 1988.

He did, in addition, rely on certain other matters that I should briefly mention. He drew attention to the failure of Mrs Land's solicitors to serve any notice under section 83(2) of the 1986 Act G
within two months of April 1 1988. This failure, he said, was indicative of the acceptance of the defendants as tenants. For my part, I do not think that inaction, where there was no obligation to take any action, gives any support at all to a case of implied grant of tenancy.

He relied on the service of the Case G notice. But this notice was given expressly without prejudice to the Case E notice. It catered for the possibility that Mr Sykes' old tenancy might continue beyond April 1 1988. It is inconsistent with, rather than supportive of, an implied intention on Mrs Land's part to grant *a new* tenancy after April 1 1988.

And finally Mr Morgan relied on certain paragraphs in the H
statement of claim that was served in June 1989. These paragraphs do not, in my judgment, assist Mr Morgan in establishing the alleged implied grant of a tenancy.

So I return to the two matters of substance on which he relied.

Does the service of the notice to pay on April 26 1988, coupled with the acceptance by Mrs Land in April 1989 of the £6,450 that had been tendered as rent in January justify imputing to her an intention to grant the defendants a new tenancy?

The learned deputy judge held that an implied grant of a tenancy had not been established. But he appears to have taken into account subjective evidence of the lack on Mrs Land's part of any intention to grant the alleged tenancy. He referred to a number of authorities and concluded that evidence of subjective intention was admissible. One J
of Mr Morgan's grounds of appeal is that the judge was in error in this respect and that the test to be applied to the conduct relied on in order to determine whether or not an intention to grant the tenancy (or for that matter licence) should be implied therefrom was an objective one.

My instinctive reaction is that Mr Morgan must be right. I do not see any reason why the approach to formation of contract in landlord and tenant cases should be any different from the approach to formation of contracts generally. The judge cited an observation of Lord Mansfield in *Doe d Cheny* v *Batten* (1775) 1 Cowp 243 at p 245 that:

> The question therefore is, *quo animo* the rent was received, and what the real intention of both parties was? K

This observation does not deal with the manner in which the intention is to be ascertained and, for my part, I think it should be ascertained by an objective approach to what has been said and done.

In this court, however, we have not been referred to any subjective evidence of Mrs Land's intention. Argument has proceeded on the basis of what intention can or cannot properly be inferred from the matters on which Mr Morgan relies, considered in the context of the surrounding facts and circumstances known to both sides. This is the objective approach. So it is not necessary for us to decide whether the learned deputy judge was right in concluding that subjective evidence of Mrs Land's intention was, on authority, admissible. I will leave the point with my deep reservations already expressed. L

As to the two matters on which Mr Morgan relied, they are, in my judgment, inadequate whether taken singly or together to justify imputing to Mrs Land an intention to grant a new tenancy or a licence to occupy to the defendants.

The notice to pay rent related to rent that had become due for the period October 1 1987 to April 1 1988. The fact of the demand was not indicative of an intention to continue the tenancy or grant any licence beyond April 1 1988. The language of the form that was used was, I agree, inept in relation to a tenancy that had expired and indicative of a continuing tenancy. But in the context of the correspondence that had been passing between the parties' respective solicitors and in view of the persistent attempts that were being made
on behalf of Mrs Land to regain possession of the farm, an M
implication from the formal language of parts of the form of an intention on the part of Mrs Land either to grant a new tenancy or to authorise the defendants to remain in occupation seems to me unreasonable. It has not been suggested that the defendants read the notice in that sense. If the notice constituted an offer it has not been suggested that there was any overt acceptance of that offer.

From June 1988 onwards there were consistent references in the correspondence between the solicitors to the imminence of possession proceedings. This was the context in which the communications between the solicitors regarding the cheque for £6,450 tendered as rent in January 1989 is to be considered. Mr Morgan submitted that the cheque was tendered as rent and accepted

*Editor's note: Reported at [1991] 1 EGLR 18 at p 23 J.

A as rent. I disagree. The cheque was certainly tendered as rent. But it was not accepted as rent. It was simply accepted as money to be set against the undoubted indebtedness of the defendants to Mrs Land. The defendants had been in occupation and possession of the farm since April 1 1988 initially, the Case E notice being valid, as trespassers, or, perhaps, as tenants on sufferance. Either they owed the £6,450 as rent, if the Case E notice had not been valid, or they owed at least that sum as mesne profits. The circumstances in which the money was accepted by Mrs Land and the language used by her solicitor when communicating with the defendants' solicitors about the cheque could not, in my judgment, have led the defendants, or any objective observer, to suppose that Mrs Land was accepting the defendants' status as tenants or licensees or was intending to grant B them a new tenancy or licence.

On this part of the case, I am in complete agreement with the learned deputy judge. At p 17* he said:

> To suggest that the landlord could be supposed, by the tenant, to have intended to create a new tenancy in those circumstances by the acceptance of rent, on April 10 1989, just as the issue of proceedings was becoming, at last, very close, given the progress on the probate side, seems to me fanciful. I am reminded of the trenchant words of Henn Collins J in *Maconochie Brothers* v *Brand* [1946] 2 All ER 778 as follows at p 779:
> "That they did receive and pay into their bank a cheque paid in respect of a quarter's rent in advance, that quarter being after the expiration of the notice to quit, is true, but it is equally true that the question is: Was a new tenancy thereby created? That is, not a tenancy to be discovered by some act of law, C but a tenancy created by two assenting minds. It is obvious that the one thing which the landlords wanted was possession of these premises, and to tell me that they assented to the creation of a new tenancy is to tell me what I do not believe. Therefore, if I have to find that there was a new tenancy, it would have to be because of some legal ingenuity and in face of what my mind informs me is not the fact."

The learned deputy judge said at p 18 that he "would not, therefore, have held that the defendants were entitled to a new tenancy by virtue of anything done after April 1 1988". Nor would I. The same applies to the argument regarding licence. I would, therefore, dismiss the appeal.

Agreeing, **Farquharson LJ** said: The two main issues in this D appeal are, first, whether the Case E notice to quit, which was expressed to expire on March 17 1988, was valid, and, second, if so, whether a new tenancy was agreed between the parties on or after April 6 1988.

In my judgment, the Case E notice was valid on the terms of the document itself when it stated that the solicitors on behalf of the landlord "hereby give you notice to quit and deliver up to her possession of all that holding and premises which you hold of her as tenant". That phrase could not include the strip of land which had been severed from the landlord's reversionary interest because the tenant did not hold the strip as her tenant but as Mr and Mrs Chittenden's tenant. In the particular circumstances of this appeal, there was in any case no room for ambiguity in the terms of the notice, E because the tenant knew perfectly well that the strip of land had been conveyed to the Chittendens.

Mr Brock, for the respondent, submits that the tenants' knowledge of the situation is irrelevant when construing the terms of the notice. One has to consider, he says, on an objective basis what a reasonable tenant would understand by the service of the notice upon him. In my judgment, that imports an air of unreality into the situation. In deciding whether the notice is void for ambiguity one has to have regard to the circumstances that existed at the time when it was served. Otherwise the clear understanding of the parties as to their rights is frustrated by a strict application of the rules of construction.

So far as the second point is concerned, it is apparent from all the F letters and documents passing between the parties from April 1988 to April 1989 that the landlord had in this context a single purpose and that was to issue proceedings for possession of the land. Looking at the position as a whole, there can be no room for any suggestion that the parties agreed upon the creation of a new tenancy. I agree with the observations of Scott LJ in this connection concerning the notice to pay of April 6 1988 and the treatment of the cheque tendered as rent on January 23 1989.

I, too, would dismiss the appeal.

Parker LJ agreed and did not add anything.

Appeal dismissed with costs.

---

*Editor's note: Reported at [1991] 1 EGLR 18 at p 25 B.

Queen's Bench Division G

June 19 1991

(Before Mr E C EVANS-LOMBE QC, sitting as a deputy judge of the division)

ROPER v PRUDENTIAL ASSURANCE CO LTD

Estates Gazette March 7 1992

[1992] 09 EG 141

Agricultural holding — Tenant's repairing covenant — Agriculture (Maintenance, Repair and Insurance of Fixed H Equipment) Regulations 1973 — Tenant's covenant to repair included *inter alia* "electrical wiring" — Whether distinction between repair and replacement

The plaintiff tenant of an agricultural holding issued an originating summons seeking the court's determination as to whether he was liable for the replacement of the electrical wiring in the farmhouse — The work had been completed at a cost of £1,908.75 — It had necessitated the replacement of the whole electrical installation in the farmhouse — There was no fault on the part of the tenant, the replacement was due simply to age — It was accepted that, although in principle the Agriculture (Maintenance, Repair and Insurance of Fixed J Equipment) Regulations 1973 as amended applied to the tenancy, in fact the terms of the relevant clauses in the tenancy agreement prevailed over the "model clauses" in the 1973 regulations in regard to the particular matter in issue in this case

"Electrical Wiring" was included in a clause in the agreement in which the tenant covenanted to keep in good and substantial repair, order and condition the farmhouse, cottages and farmbuildings together with fittings and fixtures and a number of other specified items — In another clause the landlord covenanted to execute all repairs and replacements to K parts of the farmhouse, cottages and farmbuildings, such as (with certain exclusions) main walls and exterior walls and a number of other items bearing some resemblance to, but by no means identical with, para 1 of the Schedule to the 1973 regulations mentioned above — The tenant made a number of submissions with a view to displacing his *prima facie* liability under the repairing covenant which mentioned "electrical wiring"

The plaintiff tenant submitted that "repair" in the context of the agreement must be interpreted as excluding replacement; that the total replacement of an electrical L installation could not be a repair; and that it was permissible to look at the "model clauses" in the regulations to see where they would put liability for replacement

***Held*: In the tenant's covenant, electrical wiring was to be treated as part of the general obligation to repair the farmhouse, cottages and farmbuildings — In that context a long line of well-known authorities supported the view that the replacement of the electrical installation in the farmhouse fell within the tenant's covenant to repair — The answer to the first question in the summons, as to whether the cost of the electrical wiring fell on the plaintiff tenant, must accordingly be answered in the affirmative and the other questions did not M arise**

The following case is referred to in this report.

Burden v *Hannaford* [1956] 1 QB 142; [1955] 3 WLR 606; [1955] 3 All ER 401, CA

The plaintiff, George Lawson Roper, tenant of the farm known as Lower House Farm, Tyberton, Herefordshire, sought by this summons a declaration on the issue of his liability for the cost of the electrical rewiring of the farmhouse. The defendants to the summons were the landlords, Prudential Assurance Co Ltd.

Michael King (instructed by Badhams) appeared on behalf of the

A plaintiff; Jonathan Brock (instructed by the solicitor to the Prudential Assurance Co Ltd) represented the defendants.

Giving judgment, **Mr Evans-Lombe QC** said: The issue in this case is who must pay the cost of rewiring the farmhouse comprised in the lease of Lower House Farm to the plaintiff, Mr Roper, dated December 31 1963. The landlords, the Prudential Assurance Co Ltd, are successors in title to the reversion on the lease of the tenant, Mr Roper.

The work has been completed at a total cost, inclusive of VAT, of £1,908.75. This is not a large sum but it is said that this is a matter of some importance generally in this field of the law. In the result, however, it seems to me that it is a matter which turns on a simple
B construction of the terms of the lease in question, the terms of which are somewhat idiosyncratic.

The farm in question's electrical installations may be described in the following way: they are divided in two at a junction box from which a three-phase service of electricity departs to the farmbuildings and the general area of the farm and is used for the purpose of the farm operations. Also from the junction box departs a one-phase service to the farmhouse. The work in question required the entire installation at the farmhouse to be removed and replaced, leaving in place only two sockets and the fixed central-heating pump and immersion heater. It is not suggested that the work was necessary for any reasons caused by the tenant, but simply because the installation
C was an old one and had to be replaced in full.

It is accepted that, notwithstanding their date, the Agriculture (Maintenance, Repair and Insurance of Fixed Equipment) Regulations 1973, (SI 1973, no 1473) as amended by regulations of 1988 in an immaterial respect, applied to this lease. Para 3 of those regulations provides as follows:

> The provisions set forth in the Schedule hereto relating to the maintenance, repair and insurance of fixed equipment shall be deemed to be incorporated in every contract of tenancy of an agricultural holding, whether made before or after the commencement of the Agricultural Holdings Act 1948, except in so far as they would impose on one of the parties to an agreement in writing a liability which under the agreement is imposed on the other.

There then follow provisions which do not concern this case.

D It is accepted, following on the case of *Burden* v *Hannaford* [1956] 1 QB 142, that the effect of that provision is that where there is a conflict between the agreement and the regulation as to where liability lies, the agreement prevails.

I now turn to the provisions of the lease itself. The material provision is contained at clause 4(iii)(a) and reads as follows, under the heading "Repairs" and having been preceded by the words "the tenant hereby agrees with the landlord as follows":

> (a) To repair maintain and keep in good and substantial repair order and condition throughout the tenancy using the best and most suitable materials for the same (except in so far as the landlord is responsible under clause 5(ii)(a) hereof) the farmhouse cottages and farm buildings together with all fixtures and fittings including sash cords glass locks and window fastening electrical
> E wiring boilers ranges grates and fixed inside ladders drains sewers including manholes and inspection chambers and their covers and taps water supplies including all pipes above ground and fittings taps tanks cisterns sanitary fittings drinking troughs and underground pipes pumping equipment live and dead hedges gates field gates field walls posts stiles bridges cattlegrids culverts ponds water courses ditches roads and yards in or upon the holding or which during the tenancy may be erected or provided thereon and to put and keep clean and in good working order all roof valleys eaves-guttering and down-pipes gulleys and grease-traps septic tanks cesspools including covers and tops filter beds and all sewerage disposal apparatus and chimneys properly swept and also to use carefully so as to protect from wilful reckless or negligent damage all items for the repair of which the landlord is responsible under clause 5ii(a) hereof and also to report in writing immediately to the landlord any damage however caused to items for repair of which the landlord is
> F responsible.

That subclause shows all the signs of a clause constructed from a number of re-draftings.

Clause 5(ii)(a), referred to in that subclause, provides, under the heading of "Agreements by the Landlord", as follows:

> The landlord hereby agrees with the tenant as follows:
>
> (ii)(a) To execute all repairs and replacements to the undermentioned parts of the farmhouse cottages and farm buildings . . .

The rest of that clause should be read into this judgment*.

It is accepted by counsel for the plaintiff that if clause 4(iii)(a) stood alone it would comprehend the work in question in this case, namely the wiring of the house so as to place upon the tenant the G
obligation to pay for that rewiring. The plaintiff, however, makes a number of submissions which I have produced under five headings.

The first submission is that in the context of this lease a special meaning must be given to the word "repair" in contradistinction to the word "replace" and that this is apparent from both the subclause in question and other provisions of the lease. It is accepted that there is no special meaning to be applied to the word "replace" and it is also accepted that there is a special meaning, applied by authority, to the word "repair".

The second submission is that clause 4(iii)(a) is to be construed so that each item contained in it — and I have read those out and they are many and various — is to be treated as a separate item capable, H
individually, of being repaired. Accordingly, it is submitted, in construing this clause and applying it to the facts of this case, the court should look at that which is being repaired — and it is submitted the court should look simply at the words "electrical wiring" on their own — and to say that because the electrical wiring was almost totally replaced, following the authorities, that cannot be a repair but must be taken to have been a replacement; accordingly the cost of the rewiring does not fall within the provisions of the subclause.

The third submission is, as I have just said, that on the facts of the case the work in question involved replacement and not repair of the wiring for the reasons which I have just summarised.

The fourth submission is that, in consequence, the agreement contained no provisions ascribing liability for replacement of the J
wiring as opposed to repairing it and accordingly, there being no such provision, one is entitled to look at the regulations to see where the regulations would place liability for the cost of the rewiring.

The fifth submission is that, looking at those regulations, the regulations place responsibility for the cost of rewiring the farmhouse upon the landlord. Reference is made, first, to para 1(3) of the Schedule to the regulations which reads as follows:

> Except as provided by paragraph 8, to replace

— and this is an obligation of the landlord —

> anything mentioned in paragraph 5(1) which has worn out or otherwise become incapable of further repair unless the tenant is himself liable to replace it under paragraph 6. K

It is not suggested that either para 6 or para 8 of these regulations impacts on the facts of this case. Accordingly, if one looks at para 5(1) — and this is an obligation upon the tenant — it reads:

> To repair and to keep and leave clean and in good tenantable repair, order and condition the farmhouse, cottages and farm buildings together with all fixtures and fittings boilers ranges and grates drains sewers gulleys grease-traps, manholes and inspection chambers, electrical supply systems and fittings . . .

Then it goes on to describe a number of other items. Those being matters for which the regulations place the duty of repair upon the tenant by para 5(1), it is said that by para (1)(3), where the item is incapable of repair and has worn out, the obligation is then shifted to the landlord to replace it. L

It is further submitted that wiring of the farmhouse would be included in the words "electrical supply systems and fittings" as one of those electrical supply systems. It follows, so the submission goes, that the regulations place upon the landlord the obligation to pay for work such as has taken place in this case, namely the substantial replacement of that wiring.

In my judgment, assuming the first and second submissions to be correct, it follows that submissions three to five are also correct and Mr Brock only faintly contended to the contrary. However, it is necessary to examine the first two submissions in greater detail.

The first submission was that a special meaning is to be given to the word "repair" under clause 4(iii)(a). The second submission was M
that, if such a special meaning can be given to those words, the electrical wiring is a separate item in clause 4(iii)(a) which has to be repaired or replaced and, if it has to be replaced, then the lease does not expressly deal with that circumstance and the regulations step in and their provisions, as I have found, would place the obligation to pay for that replacement upon the landlord.

I will deal with the second submission first. It seems to me that clause 4(iii)(a) must be read so that electrical wiring is treated as a fixture or fitting of the farmhouse, cottages and farmbuildings which it is the duty of the tenant to repair. Notwithstanding that the clause confusingly includes a number of items such as "live and dead hedges" and "ditches roads and yards" — all items which could not

---

*Editor's note: The provision is reproduced at the end of this judgment.

A form part of a house or buildings — it seems to me that the earlier part of the clause or the earlier lines of the clause and the words immediately succeeding the words "farmhouse cottages and farm-buildings" together with all fixtures and fittings including sash cords and various items down to electrical wiring, must require to have notionally inserted a comma so that they are read as forming part of the buildings and not forming part of the separate items the obligation to repair which items the clause also disposed of. It seems to me that the earliest place at which one could properly put that comma is in the [sixth] line between the words "ladders" and "drains" but that the more likely place where the comma should be placed is immediately before the words "drinking troughs" in the
B [eighth] line.

It follows that, in my judgment, clause 4(iii) is to be construed so that electrical wiring is to be treated as part of the farmhouse, cottages and farm-buildings and the clause provides an obligation upon the tenant to repair the farmhouse, cottages and buildings. It would follow from that that replacing the electrical wiring would merely be, in my judgment, a part of the general obligation "to repair maintain and keep in good and substantial repair" those buildings and it would follow that the word "repair", the second word in clause 4(iii)(a), would comprehend the work in question in this case. That is sufficient to decide the issue between the parties and produces an answer in accordance with sentence (a) of para 1 of the summons. However, I would also conclude that the provisions of clause 4(iii)(a)
C and later provisions of the lease — and in particular clause 4(iii)(b), (d) and (g) and clause 5(ii)(b) — do not require a special meaning to be given to the word "repair". When read in context and notwithstanding that the word "replacement" is used in those other subclauses in at least apparent contradistinction to the word "repair", those subclauses do not require a conclusion that a special meaning, other than that ascribed by a now long line of authorities, be given to the word "repair" when viewed in the context of this clause.

It follows that I answer the question in para 1(a) of the summons in the affirmative, or in accordance with sense (a), and the remainder of the summons therefore falls away.

D There must, therefore, be judgment for the defendant.

Costs to be costs in the action.

Editor's note: The provision referred to on p 6 *ante* reads as follows:

(ii)(a) To execute all repairs and replacements to the undermentioned parts of the farmhouse cottages and farm buildings (except such farmhouse cottages and buildings as are hereinafter specified in the Third Schedule hereto as redundant to the proper requirements of the holding or as shall hereafter be agreed between the parties hereto or decided by the award of an arbitrator under Clause 6(vii) hereof to be redundant to the proper requirements of the holding) namely: main walls and exterior walls excluding chimney stacks and pots outside steps and staircases not built of timber walls of open and covered yards and garden walls (whether constructed of brick stone timber or other material but excluding the interior covering of all walls
E and ceilings save where such interior covering is affected by structural defect in the wall) roofs and roof timbers including eaves-guttering and downpipes (the Tenant supplying straw and reed or thatch) and floors interior staircases floor joists doors door frames windows window frames including skylights and their frames (excepting sashcords glass locks and fastenings) but so that in the case of repairs and replacements to floor boards interior staircases doors door frames windows window frames including skylights and their frames outside timber built ladders steps and staircases eaves-guttering and downpipes the Landlord shall be entitled to recover one-half of the cost thereof from the Tenant.

F Queen's Bench Divisional Court

November 27 1991

(Before Lord Justice MANN and Mr Justice ROCH)

R v MINISTER OF AGRICULTURE, FISHERIES AND FOOD, EX PARTE LOMAX

Estates Gazette March 28 1992

[1992] 12 EG 146

Judicial review — Milk quota — Whether Dairy Produce Quotas (Amendment) (No 2) Regulations 1990 valid — Whether regulations flawed by reason of illegality in not giving effect to European Commission's approval in regard to new entrants to milk production G

The milk quota system came into effect in the United Kingdom on April 1 1984 in consequence of EEC Regulations 856/84 and 857/84 — In the UK a producer was entitled to a "primary quota" in proportion to the quantity of milk delivered by him during 1983 — The applicant, Mr Lomax, received both a primary quota and a development quota — In EEC Regulation 3880/89 the European Commission increased the amount of quota available to mitigate hardship in specified circumstances and sought from member states national measures for implementation of the scheme — The H **Minister of Agriculture forwarded to the EEC measures for the benefit of four categories of persons including a category for new entrants — Following discussions between the minister and the EEC, the commission gave approval to modified proposals of the ministry — Following that approval the ministry decided to allocate 22.5m litres of quota originally earmarked for a new entrants scheme to family-type farms, one of the original four categories put before the EEC for approval — The Dairy Produce Quotas (Amendment) (No 2) Regulations 1990 contain nothing in regard to new entrants and the 22.5m litres have otherwise been allocated — The** J **applicant contended that the regulations were flawed by reason of illegality in that they did not give effect to the commission's approval in regard to new entrants**

***Held*: A member state does not have to implement an approved proposal in full — The approval was of measures for implementation which relate to categories selected by the member state — No perceptible purpose would be served by insisting on implementation in full — The allocation or reallocation as between categories is a matter for a member state**

No cases are referred to in this report. K

This was an application by Derek Lomax, the tenant of Hollyoven Beck Farm, Great Eccleston, Preston, Lancashire, seeking to impugn the validity of the Dairy Produce Quotas (Amendment) (No 2) Regulations 1990 made by the respondent, the Minister of Agriculture, Fisheries and Food.

Richard Gordon (instructed by Cartmell Shepherd, of Carlisle) appeared for the applicant; Stephen Richards (instructed by the Solicitor to the Ministry of Agriculture, Fisheries and Food) represented the respondent.

Giving judgment, **MANN LJ** said: Roch J is unable to be present today but has authorised me to say that he agrees with the judgment which I am about to deliver and with the proposal which I make at the L end of it.

There is before the court an application for judicial review for which leave to move was given by Otton J on July 24 1990. The applicant is Derek Lomax, who is a dairy farmer and the tenant of Hollyoven Beck Farm, Great Eccleston, Preston, Lancashire. Mr Lomax sells milk wholesale to the Milk Marketing Board and his operation commenced under an agricultural and horticultural development scheme which covered a four-year period from May 12 1983. The respondent is the Minister of Agriculture, Fisheries and Food (hereafter "the minister"). Mr Lomax impugns the validity of the Dairy Produce Quotas (Amendment) (No 2) Regulations 1990 (SI 1990 No 664 and hereafter "the regulations"). The regulations were made jointly by the minister and the Secretary of State for Scotland M on March 19 1990 in the exercise of powers conferred by section 2(2) of the European Communities Act 1972. They were laid before Parliament on March 20 and came into force the next day.

The regulations are about the grant of additional dairy produce quota to certain producers for wholesale. The quota system came into effect in the United Kingdom on April 1 1984. This was in consequence of the adoption by the council of ministers of Council Regulations (EEC) Nos 856/84 and 857/84, which established that system. Under it each member state was allocated a total reference quantity of milk, which was to be apportioned among its producers. A producer who produces more than his own annual reference quantity has to pay a levy in respect of the excess. The quantities or

A "quotas", as they are called, were determined by applying the terms of articles 2, 3 and 4 of Regulation (EEC) No 857/84. The application in national law was, and is, effected by a number of statutory instruments with the generic title of Dairy Produce Quotas Regulations. The national measures have to comply with the requirements of community law and, subject to the present dispute, we were not informed that they do not. In general terms a UK producer was entitled to quota in proportion to the quantity of milk delivered by him during 1983. This was called "primary quota". Producers with expansion plans could receive what was called "development quota" (in European terms, a "special reference quantity") in response to what are called a development claim or an exceptional hardship claim. The UK's total annual reference B quantity was insufficient to satisfy all the claimants as well as the established producers and accordingly primary quota allocations had to be set at a percentage of the base-year production in order to allow for the claims. Notwithstanding this adjustment, established development and hardship claims themselves had to be subject to a percentage reduction. I should add that the UK was enabled under article 3 of Regulation (EEC) No 857/84 to grant a special reference quantity to "young farmers setting up after 31st December 1980", but this power was not and has not been exercised. At one stage I thought there might have been some complaint that it should have been, but no such complaint was, or, in my view, could have been, pursued in these proceedings.

C Mr Lomax received both a primary quota, which was not large, for he had little base-year production, and a development quota in response to a claim based on his AHDS. His primary quota was set at the percentage of his computed base-year production, which was applied to all producers, and his development quota was awarded after percentage deductions from the claim, which were applied to all claimants. I find it unnecessary to discuss the figures but Mr Lomax thinks that he has been disadvantaged by a system which does not enable him to have a quota properly reflecting the value of his investment. He has displayed a lively interest in the alterations to the system which I shall describe and indeed claims some responsibility for persuading the European Commission to make those changes.

The European Commission perceived that certain producers may D have sustained hardship under the system of Regulation (EEC) No 857/84, and in order to mitigate that hardship they proposed that each member state should have its total reference quantity increased by 1% and that there might be general rules determining how allocations of that amount to individuals should be made by the member states. The proposals were considered by the Council of Ministers, who on December 11 1989 adopted Council Regulation (EEC) No 3880/89. That regulation was inserted into the 1984 regulations as article 3b in these terms:

> 1. For the purposes of determining the reference quantities referred to in Article 2, the Member State may grant additional or special reference quantities within a limit of 1% of the guaranteed total quantity fixed in the E second subparagraph of Article 5c(3) of Regulation (EEC) No 804/68:
> — to producers as referred to in Articles 3 and 4(1)(b),
> — to producers determined in particular on the basis of one or more of the following criteria:
>   — producers who are new entrants,
>   — producers whose individual reference quantities are less than or equal to 60,000 kilograms,
>   — the siting of milk production in one of the zones as defined in Article 3(3)(4) and (5) of Directive 75/268/EEC, as last amended by Regulation (EEC) No 797/85,
>   — or to all the producers of the Member State if, on its territory, the implementation of Articles 3 and 4 has been carried out on quantities coming from the national reserve.
>
> Member States shall forward to the Commission, for prior approval, their F draft national measures for the implementation of the first subparagraph. The Commission shall decide on these measures taking into account the criteria mentioned in the first subparagraph. These criteria may be adapted and/or supplemented, if justified by the diversity in national situations.

The UK was under no obligation to exercise the power conferred by article 3b, but the minister decided that it should be exercised. On February 12 1990 he wrote to Mr Ray MacSharry, who is the European Commissioner responsible for agriculture, seeking the commission's approval "for the method I propose to adopt for distributing . . . the additional 1% of milk quota". The method was for the benefit of four categories of person. The first was that of producers who had been given development awards in accordance with article 3 of Regulation (EEC) No 857/84 together with some producers in Northern Ireland who had received awards on the grounds of exceptional hardship (paras 2 to 4 of the letter). The G second was that of the smaller producers who occupy a family-type holding and the promotion of this category required an adaptation of one of the criteria in article 3b which defines small producers in terms which have no relevance to the UK (para 5). The third category was that of producers in remote parts of less favoured areas (para 6). The fourth category has become important and was described in para 7 as "the category of new entrants". After describing the difficulties confronted by "those wishing to enter dairying" the minister wrote:

> I am convinced that now is the time to take action on this, albeit on a modest scale, and I propose to use 22.5 million litres of quota to make 150,000 litres available to each of 150 new entrants. These would be people who are not already milk producers. If you take the view that they would not be covered by the wording of Article 3b of Regulation 857/84, which refers to "producers H who are new entrants", then I would suggest my proposal should be regarded as an adaptation to the criteria mentioned fully justified by the nature of our situation in the UK as provided for in the regulation. Alternatively, my approach could be regarded as a supplementary category or criterion as also provided for in the regulation.

I doubt whether the proposal could properly be described as an "adaptation to the criteria" because the criteria are criteria in regard to "producers", which those wishing to enter dairying by hypothesis are not. However, the minister went on in para 8 to explain his particular wish that the quotas for would-be dairymen should be issued on loan, free of charge and for a period of perhaps six years. The minister undertook that detailed national measures in draft concerning the distribution of the 1% would be sent so soon as they J were available.

The minister's proposals were then discussed between his own and the commissioner's officials. They resulted in a ministry letter dated February 23 1990 and telefaxed to Brussels on that day. After dealing with points in regard to developers and the smaller producers, the author wrote:

> 6. I come now to the question of new entrants (paragraphs 7 and 8 of Mr Gummer's letter). My authorities have not yet decided how they wish to proceed in this matter. I cannot rule out the possibility that Mr Gummer may wish to make a further approach to Mr MacSharry. Whether or not he does so — and regardless of the outcome — it will take a little time for us to work out the details of a new entrants scheme. This part of our proposals will not K therefore be covered by the draft national measures which we shall be submitting. We envisage seeking a further Commission Decision in respect of our proposals on new entrants at a later date, and I understand that this is acceptable to the Commission.
>
> 7. There remains to be settled the treatment for the quota year 1989/90 of the 22.5 million litres of quota which we intend should eventually be issued to new entrants. I should be grateful for your urgent advice as to whether it is permissible for us to issue this to the generality of producers — ie across-the-board — this year and then withdraw it from them next year for issue to new entrants.

Para 6 was written for an informed audience and may have nuances which escape other readers. That said, it does seem clear that the draft measures which the minister had promised would not contain proposals concerning "new entrants" and that another L commission decision would be sought in respect of them. If my doubt as to adaptation is well founded, then the appropriate course would perhaps have been to be the seeking of amending regulations from the Council of Ministers. However that may be, paras 7 and 8 of the minister's letter of February 12 were, as Mr Stephen Richards, who appeared for the minister, put it, "effectively withdrawn". Para 7 of the letter of February 23 deals with the consequence of the withdrawal and suggests a course which in its terms is quite unsustainable, for article 3b does not permit an issue of any of the 1% "across the board".

There were further discussions between officials and they resulted in a ministry letter dated February 28 and telefaxed to Brussels on the M same day. The letter dealt with the definition of the category of smaller producers and then continued:

> 4. Turning to paragraph 6 of my letter I confirm that we still wish to allocate quota to new entrants and I understand that the Commission Decision will make appropriate provision. The precise modalities in respect of new entrants remain to be settled. Indeed, as I said in my previous letter, Mr Gummer may wish to make a further approach to Mr MacSharry on the question of loaning quota but this does not affect the definition of categories of eligible producer under the terms of Article 3b of Regulation 857/84, on which we have now reached agreement.
>
> 5. On the question posed in paragraph 7 of my previous letter I understand that this is not a course which the Commission could agree for any Member State.

A That the proposal in para 7 of the previous letter had been given its quietus is unsurprising, but para 4 of this letter has an enigmatic quality for the reader who was not party to the discussions. In the light of the previous letters, what is the appropriate provision which is to be made? I think the answer is that the ministry by now wished to have producers who are "new entrants" as an unadapted category in respect of which "modalities . . . remain to be settled" but for which 22.5m litres would be available.

The commission's decision on the UK proposals was given on March 12 1990. The decision had a recital, which included the following:

> Whereas the draft measures notified by the United Kingdom on 12 February
> B 1990, as amended by telefax no 18 of 23 February 1990 and telefax no 38 of 28 February 1990 should be approved.

The decision itself was in two parts. The first dealt with the first three categories in the notification of March 12 and provided for the second of those in terms of an adaptation of an article 3b category. The second part was effectively in these terms:

> The draft national provisions for implementation in the United Kingdom of Article 3b of Regulation (EEC) No 857/84, which provide . . . for the assignment of a special reference quantity to producers newly installed are hereby approved.

We were told that "producers newly installed" has the same meaning as "producers who are new entrants". There is thus no purported adaptation of an article 3b category and the commission
C must have understood the amending telefax of February 28 as having the meaning which suggested itself to me.

I think that on an analysis of the documents there is no doubt as to what was ultimately proposed and approved. I have taken the matter at length, first, because Mr Lomax ought to know what occurred and, second, because during the hearing I was perplexed by Mr Richards' varying submissions as to what was from time to time proposed and then ultimately approved. His submissions on the second day of the hearing were uncharacteristically inconsistent with those on the first. Thus, on the first day he had submitted that the approval in regard to new entrants was "unnecessary", while on the second day he submitted that the approval was of a category determined on an adapted criteria such as was proposed on March 12.
D Neither submission is tenable. Together with those submissions which he variously advanced in regard to para 7 of the letter of February 23, they may betray an unfortunate confusion in the minds of those instructing Mr Richards as to what they or their predecessors were about in February and March 1990.

After the commission's approval had been obtained, the minister had to consider what to do in regard to the approved category of new entrants and for which the 22.5m litres remained earmarked. What he decided was explained in a written reply to a parliamentary question on March 20, which was the day on which the regulations were laid before Parliament. The minister said:

> The proposals which we originally put to the Commission contained
> E provision for the loan of quota to new entrants. The Commission's view, however, is that no satisfactory basis exists in Community law for the issue of quota to new entrants on temporary and conditional terms. We have considered whether quota should instead be granted outright to new entrants but have decided against this course. In reaching this conclusion we took into account the strong opposition of the National Farmers Union to the gift of quota to new entrants. We have therefore decided to allocate the 22.5 million litres of quota originally earmarked for a new entrants scheme to family-type farms, who will now receive an increase of 4% of their existing wholesale quota, and to holders of development awards, which will be made up to approximately 80% of the original figure after abatement to take account of subsequent across-the-board quota cutbacks.

The regulations contain nothing in regard to new entrants and the
F 22.5m litres has in fact been allocated additionally to the approved categories of development award holders and occupiers of family-type holdings. We were told that the commision, who presumably know of what has been done, have raised no objection to the regulations or to the grants of additional quota in fact made.

Mr Richard Gordon, who appeared for Mr Lomax, made it his primary submission that the regulations were flawed by reason of illegality in that they did not give effect to the commission's approval in regard to new entrants. It is Mr Lomax's case that he is a "new entrant" within the true meaning of that term as employed in the approval and was therefore a potential beneficiary of it. There is a dispute as to whether he is a "new entrant", but it is a dispute which need not be resolved in these proceedings.

The inserted article 3b of Regulation (EEC) No 857/84 gives a G
power to a member state to grant quota to specified categories of producer. Mr Gordon accepted that there is no obligation to exercise the power either at all or in regard to all or any of the categories. However, a proposal to exercise must be submitted to the commission for their prior approval, and it may be that a failure to obtain that approval renders unlawful any implementation of the national measure. I express no opinion on that point, which is a different one from that which we have to decide, which is whether the member state *must* implement the approved proposal in full. I entertain no doubt that they do not. First, there is no express obligation that they should. Second, the argument to the contrary presupposes that the approval is of an indivisible package. It is not. It H
is an approval of measures for implementation which relate to categories selected by the member state and those categories are not interdependent. Third, no perceptible purpose would be served by insisting on implementation in full because the commission's concern is that additional reference quantities within the 1% limit are granted only to approved categories of producer. I do not see that they have any concern that the quantities should be granted to all approved categories of producer. I conclude that an approval does not create an obligation towards a particular category of producer which hitherto was absent. I, accordingly, reject Mr Gordon's argument that the regulations were flawed by reason of illegality.

There was an argument as to whether the reallocation of the 22.5m litres required the commission's approval. If it did, none was given. I J
am of the opinion that no approval was required. This would necessarily follow from the freedom of a member state not to grant to an approved category of producer. If the commission had to approve a reallocation, then by withholding it they would in effect either destroy the freedom or diminish the 1%. The allocation or reallocation of the 1% as between categories is, in my judgment, a matter for the member state and it is of course to be observed that the approval of March 12 is silent upon quantities. I should add that, if I were wrong in my view, then, as Mr Richards pointed out, the absence of approval would not affect the validity of the regulations, for they do not deal with the amount available for each category but provide formulae for the distribution by grant of whatever quantity is allocated by the minister for each category of producer. Any K
assault would thus have to be on what he actually did for the greater benefit of two of the approved categories. Such an assault is quite outside the present application.

I would dismiss this application for judicial review. As I have indicated, Roch J has expressed his agreement. The application is accordingly dismissed.

No order for costs. Legal aid taxation of applicant's costs.

Chancery Division

December 12 1991 L

(Before Mr Justice CHADWICK)

FAULKS v FAULKS

Estates Gazette April 18 1992

[1992] 15 EG 82

Agricultural holding — Partnership — Dissolution of partnership — Whether value of milk quota registered in the names of the partners should be taken into account in valuing the assets of the partnership

Pursuant to an agreement dated March 2 1983 the M
respondent, John Faulks, was granted an agricultural tenancy commencing on March 25 1983 of Hills Farm, Colston Bassett, Nottinghamshire — On February 9 1984 the respondent and his brother, Harry Faulks, entered into a deed of partnership to carry on the business of farming Hills Farm — By clause 13 of the partnership deed the partnership would determine on the death of either partner — By clause 16 of the partnership deed provision was made for an account of the assets of the partnership and for these to be valued — By clause 18 of the partnership deed the respondent was declared to stand possessed of the tenancy of Hills Farm in trust for the

partnership — In or about 1984 the partnership was allocated milk quota and the names of the partners were entered on the register in respect of Hills Farm — Harry Faulks died on June 12 1988 at which date the milk quota in respect of which the partnership was registered was 426,640 litres — The appellant was appointed executrix of Harry Faulks' will — Pursuant to the terms of the partnership deed, an arbitrator was appointed and, by an interim award dated November 2 1990, he determined that the milk quota was not an asset of the partnership and accordingly its value was not to be taken into account for the purpose of valuing the assets of the partnership — The partnership said that milk quota is not a right or interest in property separate from or capable of being separated from the land in respect of which it is held and so is not capable of being an asset — By way of an appeal under section 1 of the Arbitration Act 1979 the appellant contended that the arbitrator had erred in law in determining that the value of the milk quota was not an asset of the partnership

***Held*: The appeal was dismissed — In *Wachauf* v *Bundesamt für Ernährung und Forstwirtschaft* [1991] 1 CMLR 328 the European Court of Justice affirmed the principle that milk quota is attached to the land, while recognising that the rules made under community legislation permit member states to make provision for compensating the outgoing tenant of a holding for loss of milk quota in cases where the strict application of that principle would cause hardship — The concept of "quota leasing" was authorised by Council Regulation (EEC) No 2998/87 and introduced into England and Wales by regulation 6 of the Dairy Produce Quotas (Amendment) Regulations 1988 — Re-enacted by the Dairy Produce Quotas Regulations 1989, a producer is permitted "to make a temporary transfer within one region of part of the wholesale quota registered as his to another producer for a period of one quota year" — The transfer of quota from one producer to another under these regulations, by which there are reciprocal adjustments to their respected entries on the register, is properly to be regarded not as the transfer of an asset by the transferor to the transferee but rather as the transfer of a potential liability on the producer to contribute to additional levy imposed on the wholesale purchaser from the transferee of the quota to the transferor — The fact that the transfer of a potential liability is likely to be matched by a payment to the person assuming that liability may enable it to be said that the quota transferred has an intrinsic economic value, but that does not of itself lead to the conclusion that quota is an asset separate and distinct from the holding in relation to which it was, or becomes, registered — In relation to milk quota, the true position in England and Wales is that a producer does not require a licence to produce milk and will not necessarily incur any liability as a consequence of production, but it is in respect of his potential liability that he enjoys an immunity from Formula B contribution by virtue of, and to the extent of, the register entry of an amount of wholesale quota in his name — Accordingly, the milk quota formerly registered in the name of the partnership could not properly be regarded as property and as an asset of this partnership**

The following cases are referred to in this report.

Ambler v *Bolton* (1872) LR 14 Eq 427
Miles v *Clarke* [1953] 1 WLR 537; [1953] 1 All ER 779
Pawsey v *Armstrong* (1881) 18 Ch D 698
Smith v *Mules* (1851) 9 Hare 556
Wachauf v *Bundesamt für Ernährung und Forstwirtschaft* (Case 5/88) [1991] 1 CMLR 328

This was an appeal, pursuant to leave granted by Knox J on May 7 1991, from an interim award of an arbitrator appointed to determine issues arising upon the dissolution of a partnership between John and Harry Faulks upon the death of Harry Faulks. The appellant was the executrix of the will of Harry Faulks deceased; the respondent was John Faulks.

Peter Cranfield (instructed by Roythorne & Co, of Spalding) appeared for the appellant; Kim Lewison QC (instructed by Burges Salmon, of Bristol) represented the respondent.

Giving judgment, **CHADWICK J** said: Hills Farm, Colston Bassett, in the county of Nottinghamshire, comprises some 170 acres of mixed grass and arable land. By an agreement dated September 7 1925 the farm was let to George Faulks on a tenancy from year to year commencing on March 25 1926. On his death the tenancy became vested in his widow, Florence Maud Faulks. She farmed the land, with her two sons, John Albert Faulks and Harry Alfred Faulks, under the firm name "F M Faulks & Sons". It is clear that by the time of her death on December 9 1982 a dairy enterprise had been established on the farm.

On the death of Florence Maud Faulks it was agreed with the landlord that the elder son, John, should succeed to the tenancy under the provisions of Part II of the Agriculture (Miscellaneous Provisions) Act 1976. That agreement is recorded in a memorandum dated March 2 1983. The new tenancy was to commence on March 25 1983 and, save as to rent, was to be on the same terms and conditions as in the 1925 tenancy. It was expressly provided in the memorandum of March 2 1983 that, in the event of John's death before March 25 1983, his brother Harry should succeed to the tenancy of the farm as first successor under the Act. In the events which happened John became tenant of the farm on March 25 1983 and the tenancy has remained vested in him since that date.

On February 9 1984 the two brothers entered into a deed of partnership under which they agreed that they would, with effect from May 31 1983, carry on the business of farmers at Hills Farm under the existing firm name of "F M Faulks & Sons". The partnership deed required that each of the brothers should devote substantially the whole of his time and attention to the partnership business and it further provided, at clause 19, that Harry Faulks and his wife Pamela should be entitled to occupy and reside in the farmhouse free of rent so long as they wished during the continuance of the partnership. Clauses 12 to 15 contained provisions for the determination of the partnership. In particular, clause 13 provided that the partnership should determine on the death or bankruptcy of either partner. Clauses 16 to 18 are in these terms, so far as material:

FINANCIAL PROVISIONS ON TERMINATION

16. In the event of either of the Partners ceasing to be a Partner by reason of his death or his retirement or expulsion in accordance with the provisions of this Deed the provisions contained in this clause shall have effect and so that reference herein to "outgoing Partner" shall be taken as references to the Partner so ceasing to be a Partner and where appropriate shall include his personal representative receiver or trustee in bankruptcy:

(a) An account shall forthwith be taken up to the day upon which the outgoing partner ceases to be a partner and the assets of the partnership shall be valued for the purposes of such account by agreement between the partners or if the partners are unable to agree as to the value of all or any specific assets such assets shall be valued by a qualified valuer (acting as an expert) appointed by the Partners or failing agreement upon a valuer then either party may request the President for the time being of the Nottinghamshire branch of the Royal Institution of Chartered Surveyors to appoint a valuer. The costs of any valuer appointed as aforesaid shall be shared equally between the parties hereto unless such valuer shall otherwise determine.

(b) The remaining partner may purchase the share of the outgoing partner in the partnership upon giving to the outgoing partner notice in writing to that effect at any time within three months from his ceasing to be a partner. Any such purchase shall take effect and relate back to the date of termination of the partnership and the partner exercising such option shall thereupon become exclusively liable for the partnership debts and liabilities whether incurred or arising before or after such date.

(c) If the said option is not exercised within the time aforesaid or if the remaining partner does not wish to exercise it the partnership business shall be wound up in accordance with the provisions of the Partnership Act 1890.

(d) The purchase money for the share of an outgoing partner shall be the net value thereof as shown in the said account after debiting any tax due from the partnership in respect of the outgoing partner.

— Subparas (e) to (k) are not material —

17. Upon dissolution of the partnership in any event not otherwise herein provided for the affairs thereof shall be wound up and the assets and liabilities dealt with in manner provided by the Partnership Act 1890.

PROPERTIES

18. The tenancy of the said property known as Hills Farm Colston Bassett being now vested in the said John Albert Faulks he shall during the continuance of the partnership stand possessed of the same in trust for the

A partnership and the rent of such premises shall be payable by the partnership and the partnership shall indemnify the said John Albert Faulks in respect of all claims demands and liabilities arising in respect of such tenancy. Provided that in the event of the said John Albert Faulks leaving the partnership for any reason whatsoever he will if requested by the said Harry Alfred Faulks use his best endeavours to procure that such tenancy shall be transferred to Harry Alfred Faulks . . .

The partnership deed continues at clause 24 with provision for arbitration in the event, *inter alia,* of any dispute touching the rights and liabilities of the partners and their representatives under the deed or otherwise in relation to the partnership.

On March 31 1984, shortly after the execution of the partnership deed, the Council of the European Communities adopted Council
B Regulation (EEC) No 856/84 for the purposes of introducing for an initial period of five years an additional levy (that is to say, a levy in addition to the uniform co-responsibility levy already in force) on quantities of milk delivered by producers to purchasers in excess of a predetermined reference quantity. On the same day, by Council Regulation (EEC) No 857/84, general rules were adopted for the application of that additional levy with effect from April 2 1984. More detailed rules were laid down on May 16 1984 by Commission Regulation (EEC) No 1371/84. That community legislation was expressed to be directly applicable in all member states. It was implemented within the United Kingdom by the Dairy Produce Quotas Regulations 1984 (SI 1047) which came into operation on July 24 1984. Those regulations, as amended from time to time, were
C themselves replaced by the Dairy Produce Quotas Regulations 1986 (SI 470).

It will be necessary to examine the effect of the community legislation as implemented by the United Kingdom regulations in force at the relevant time in some detail later in this judgment; but, for present purposes, it is sufficient to indicate that that legislation led to the allocation in England and Wales of wholesale quota to each milk producer in respect of the holding which he farmed and the establishment of a register for the purpose of recording and identifying each producer, the amount of his quota and the relevant holding. "Quota" for these purposes means the quantity of dairy produce which may be delivered in a quota year without the producer
D becoming liable to make a payment in respect of additional levy.

In due course, in or about 1984, the partnership was allocated quota under the 1984 regulations, and the names of the partners were entered on the register in respect of the holding which comprised Hills Farm. The brothers continued to farm in partnership at Hills Farm until the death of Harry Faulks on June 12 1988. At that date the quota in respect of which the partnership was registered was 426,640 litres.

On the death of Harry Faulks the partnership determined in accordance with clause 13 of the partnership deed and the financial provisions contained in clause 16 took effect. In particular, it became necessary to take an account between the partners and to value the
E assets of the partnership for the purpose of that account. John Faulks and his brother's executrix were unable to agree whether the tenancy of Hills Farm and the milk quota should be taken into the partnership account as assets of the partnership and, if so, what value should be ascribed to those assets. By an appointment in writing dated May 23 1990 these questions were referred to an arbitrator. In the light of the provisions of clause 16(a) of the partnership deed, which refer questions of value to a qualified valuer acting as an expert, the better view is, I think, that this appointment was made pursuant to an *ad hoc* arbitration agreement and was not a reference under clause 24 of the deed. But it has not been contended that anything turns on this point.

On November 2 1990 the arbitrator made an interim award in the
F following terms:

(1) The tenancy of Hills Farm, Colston Bassett, in the County of Nottingham (*sic*) held in the name of the Respondent is not an asset of the partnership within clause 16 of the Deed of Partnership dated 9 February 1984 and accordingly any value which the said tenancy may have is not to be taken into account for the purposes of valuing the assets of the partnership as at 12 June 1988 pursuant to clause 16 of the Deed of Partnership;

(2) The milk quota which attaches to the land comprised in the said tenancy is not an asset of the partnership within clause 16 of the Deed of Partnership and accordingly any value which such milk quota may have is not to be taken into account for the purposes of valuing the assets of the partnership as at 12 June 1988 pursuant to clause 16 of the said Deed of Partnership . . .

In these circumstances the question of what value should be ascribed to these assets did not arise.

The arbitrator gave a full statement of his reasons for the award. In G
short he held:

(i) that to treat the tenancy as partnership property or as an asset of the partnership would be to disregard the express limitation in clause 18 of the partnership deed that the tenancy is held on trust for the partners only during the continuance of the partnership; and

(ii) that milk quota is not a right or interest in property separate from, or capable of being separated from, the land in respect of which it is held and so is not capable of being an asset within clause 16 of the Partnership Deed.

By an originating summons issued on December 10 1990 the executrix of Harry Faulks, being the disappointed claimant in the arbitration, sought leave pursuant to section 1(3)(*b*) of the Arbitration Act 1979 to appeal to the High Court on the questions of H
law arising under both paras (1) and (2) of the award. That application was duly made to the Commercial Court pursuant to Ord 73, r6(1) of the Rules of the Supreme Court 1991, but it was, in due course, transferred to this division. On May 7 1991 Knox J gave leave to appeal but on condition that the applicant did not seek to set aside para (1) of the award. Accordingly, the matter comes before me on the basis that the arbitrator was correct in holding that by virtue of the express limitation in clause 18 of the partnership deed the tenancy of Hills Farm ceased to be an asset of the partnership on the death of Harry Faulks on June 12 1988; and that thereafter the tenancy was held by John Faulks in equity as well as at law. In these circumstances, the sole question of law which I have to determine is whether the value of the milk quota formerly registered in the names J
of the partners in respect of the land comprised in the tenancy of Hills Farm now held by John Faulks should be taken into account in valuing the assets of the partnership as at June 12 1988 pursuant to clause 16 of the partnership deed.

It is necessary to keep in mind the context in which that question arises. The primary requirement in clause 16(a) of the partnership deed is that an account be taken up to the day on which the outgoing partner ceases to be a partner. The requirement that the assets of the partnership are to be valued is for the purposes of that account. The account is taken in order to establish the net value of the share of the outgoing partner in the partnership — see clause 16(d) of the partnership deed — that being the amount which is to be paid to him
by the remaining partner if the remaining partner chooses to continue K
the partnership business. If the remaining partner does not choose to continue the partnership business, then the partnership must be wound up under the Partnership Act 1890 — see clause 16(c) of the partnership deed. It is, I think, correct to approach clause 16 of the partnership deed on the basis that the partners are likely to have intended that the payment to be made to the outgoing partner should be substantially the same whether or not the remaining partner chooses to continue the business.

Section 39 of the Partnership Act 1890 is in these terms (so far as material):

On the dissolution of a partnership every partner is entitled, as against the other partners in the firm, and all persons claiming through them in respect of L
their interests as partners, to have the property of the partnership applied in payment of the debts and liabilities of the firm, and to have the surplus assets after such payment applied in payment of what is due to the partners respectively after deducting what may be due from them as partners to the firm . . .

Partnership property in this context means:

All property and rights and interests in property originally brought into the partnership stock or acquired, whether by purchase or otherwise, on account of the firm, or for the purposes and in the course of the partnership business . . .

See section 20(*f*) of the 1890 Act.

Adopting the approach which I have indicated, the "assets of the partnership" which are to be valued for the purposes of clause 16(a) M
of the partnership deed ought to be the same as the property, rights and interests which comprise the partnership property and which would have been available, in a dissolution under the Act, to pay the debts and liabilities of the firm and as to the surplus for distribution to the partners. One test, therefore, as it appears to me, is whether the milk quota formerly registered in the name of the partnership would have been available, in a dissolution under the Act, to meet the debts and liabilities of the firm. If so, it should be taken into account in valuing the assets of the partnership under clause 16(a). But it does not necessarily follow that if the nature of milk quota is such that the milk quota formerly registered in the name of the partnership could not have been realised by a sale in the course of a dissolution under the

A Act, then it cannot be taken into account in valuing the assets of the partnership under clause 16(a).

I was referred in the course of argument to *Smith* v *Mules* (1851) 9 Hare 556 and *Ambler* v *Bolton* (1872) LR 14 Eq 427. These cases illustrate that an asset which is vested in the name of one of the partners, and which is not alienable by him, may none the less be treated as a partnership asset if the partners so intend; and, if so, the partner in whose name the asset is vested may be charged with its value in the partnership accounts. The question in each case is whether the partners did intend that that asset should be treated as partnership property.

In the present case it is impossible to attribute to the partners any B specific intention in relation to milk quota at the time that they made the partnership deed. The concept had not then been introduced. The relevant question must be whether, at the date when the partnership was allocated quota under the 1984 regulations and the names of the partners were entered on the register in respect of the holding which comprised Hills Farm, the quota was properly to be regarded as property, or a right or interest in property, separate and distinct from the holding. If not, it seems to me impossible to attribute to the partners any intention other than that the quota should be treated in the same way as the tenancy of the farm; that is to say that, upon the determination of the partnership, it should be held with the tenancy under the provisions of clause 18 of the partnership deed. The intention under that clause was that, so far as possible, the tenancy C should be held by, or transferred to, the partner who was remaining in the business.

In order to answer that question it is necessary to examine the community legislation, as implemented by the United Kingdom regulations, under which milk quota has been introduced and by which it is now regulated. Council Regulation (EEC) No 856/84 introduces the additional levy payable by producers or purchasers of milk or other milk products by inserting into Regulation (EEC) No 804/68 a new article 5(c). That article provides that the levy system shall be implemented in each region of the territory of a member state in accordance with one of the following formulae:

> **Formula A**
>
> D — A levy shall be payable by every milk producer on the quantities of milk and/or milk equivalent which he has delivered to a purchaser and which for the twelve months concerned exceed a reference quantity to be determined.
>
> **Formula B**
>
> — A levy shall be payable by every purchaser of milk or other milk products on the quantities of milk or milk equivalent which have been delivered to him by a producer and which, during the twelve months concerned, exceed a reference quantity to be determined.
>
> — The purchaser liable to the levy shall pass on the burden in the price paid to those producers who have increased their deliveries, in proportion to their contribution to the purchaser's reference quantity being exceeded . . .

It can be seen that under Formula A the levy is imposed directly on each milk producer and becomes payable by each milk producer when he delivers milk in excess of his own reference quantity, E irrespective of the quantity delivered by other producers in the same region. Under Formula B, on the other hand, levy is imposed directly on the wholesale purchaser and becomes payable by the purchaser only when the total quantity of milk delivered to the purchaser exceeds the purchaser's reference quantity. The United Kingdom has adopted Formula B and, in the region constituted by England and Wales, at all material times there has been, in effect, a single monopoly purchaser — namely the Milk Marketing Board. The effect is that levy becomes payable in England and Wales only if the total production of milk exceeds the purchaser's reference quantity for England and Wales; so that overproduction on the part of some producers may be set against underproduction on the part of others. But, if levy does become payable under Formula B, then it is passed F on by the purchaser to those producers who have overproduced.

It is implicit in the direction under Formula B — that the burden suffered by the purchaser by whom the levy is payable shall be passed on "to those producers who have increased their deliveries, in proportion to their contribution to the purchaser's reference quantity being exceeded" — that, under Formula B as well as under Formula A, it will be necessary to determine a quota or quantity in respect of each producer by reference to which an "increase" can be identified and a proportionate contribution can be ascertained.

Council Regulation (EEC) No 857/84 was adopted for the purpose of setting out general rules for the application of the additional levy introduced by the new article 5(c) of Regulation (EEC) No 804/68. Article 2 determines the reference quantity. It is in these terms:

> (1) The reference quantity referred to in Article 5c(1) of the previously G mentioned Regulation [that is, 804/68] shall be equal to the quantity of milk or milk equivalent delivered by the producer during the 1981 calendar year (Formula A), or to the quantity of milk or milk equivalent purchased by a purchaser during the 1981 calendar year (Formula B), plus 1% . . .

Article 9 provides for the collection of levy. In particular, in cases where Formula B is applied, article 9(1)(b) requires that the levy shall be collected by a final account calculated after the end of the 12-month period concerned, on the basis of the actual excess during this same period beyond the purchaser's annual reference quantity. Article 10 is in these terms:

> Where Formula B is applied: . . .
>
> 2. After the end of the twelve-month period concerned the purchaser shall, H where appropriate, on the basis of the final account referred to in Article 9(1)(b), make the necessary adjustments on the basis of the milk or milk equivalent quantity by which each of the producers has, for the relevant period, exceeded an annual quantity corresponding to that taken for the purposes of establishing the purchaser's reference quantity . . .

Article 8 provides that, except as provided for in article 7(1), where Formula B is applied, member states may take the necessary steps making it possible for purchasers to manage the reference quantities allocated to them "including the allocation and re-allocation of the quantities mentioned in Article 10 . . .". The "quantities mentioned in Article 10" include a quantity to be established in respect of each producer "corresponding to that taken for the purposes of establishing the purchaser's reference quantity". This again gives J recognition to the need under Formula B as well as under Formula A to determine a quantity or quota in respect of each producer.

"Producer" is defined in article 12(c) of Regulation (EEC) No 857/84. For present purposes the expression means:

> A natural or legal person or group of natural or legal persons farming a holding located within the geographical territory of the Community: — supplying the purchaser . . .

And in this context "holding" means:

> All the production units operated by the producer and located within the geographical territory of the Community . . .

"Purchaser" is defined as:

> An undertaking or grouping which purchases milk or other milk products: K
> — to treat or process them, or
> — to sell them to one or more undertakings treating or processing milk or other milk products . . .

Article 7(1) of Council Regulation (EEC) No 857/84, as originally adopted, was in these terms:

> (1) Where an undertaking is sold, leased or transferred by inheritance, all or part of the corresponding reference quantity shall be transferred to the purchaser, tenant or heir according to procedures to be determined . . .

Detailed rules for the purposes of applying that article were laid down by article 5 of Commission Regulation (EEC) No 1371/84 in these terms (so far as material):

> . . . the following rules shall apply to the transfer of reference quantities granted to producers and purchasers in application of Formulas A and B and L of reference quantities granted to producers selling for direct consumption:
>
> 1. Where an entire holding is sold, leased or transferred by inheritance, the corresponding reference quantity shall be transferred in full to the producer who takes over the holding.
>
> 2. Where one or several parts of a holding is sold, leased or transferred by inheritance, the corresponding reference quantity shall be distributed among the producers operating the holding in proportion to the areas used for milk production or according to other objective criteria laid down by Member States . . .
>
> 3. The provisions of subparagraphs 1 and 2 above shall also be applicable in other cases of transfer which, under the various national rules, have comparable legal effects as far as producers are concerned . . .

In the circumstances that the concept of the "holding" is linked M directly to the producer — see paras (c) and (d) of article 12 of Regulation (EEC) No 857/84, which are set out above — and only indirectly to a purchaser (in that a purchaser is an undertaking which purchases milk from a producer) the rules laid down in article 5 of Regulation (EEC) No 1371/84 must be taken to apply not only to the producer's reference quantity determined under Formula A but also to the "[producer's] quantity corresponding to that taken for the purposes of establishing the purchaser's reference quantity" referred to in article 10 of Regulation (EEC) No 857/84. So understood, it is clear that article 7 of Regulation (EEC) No 857/84, as amplified by article 5 of Regulation (EEC) No 1371/84, is laying down the general principle that, in both Formula A and Formula B cases, the quantity or quota determined in respect of each producer

A will follow the land upon a transfer of all or part of the holding which is farmed by that producer.

The community legislation to which I have referred was implemented in the United Kingdom by the Dairy Produce Quotas Regulations 1984 (SI 1984 No 1047). In those regulations the expressions "holding", "producer" and "purchaser" have the same meanings as in article 12 of Regulation (EEC) No 857/84. "Purchaser quota" is defined as:

> . . . the quantity of dairy produce which may be delivered by wholesale delivery to a purchaser, from holdings in a region, during a quota year without that purchaser being liable to pay levy.

"Formula B contribution" means the sums to be recovered under B article 10 of Regulation (EEC) No 857/84 by a purchaser from a producer in a region where Formula B is implemented; and "wholesale quota" is defined (so far as material) as:

> (b) The quantity of dairy produce which may be delivered by wholesale delivery to a purchaser . . . from a holding in a region in which Formula B is implemented, in a quota year without the producer in occupation of that holding being liable to pay Formula B contribution . . .

This definition is consistent, and only consistent, with the view that under the community legislation it is necessary, in Formula B cases as well as those under Formula A, to determine a quantity or quota in respect of each producer by reference to which his contribution to the purchaser's liability to pay levy under Formula B can be ascertained.

C The Dairy Produce Quotas Regulations 1984 were drawn on the basis that no decisions had then been taken either as to the regions in which the United Kingdom was to be divided for the purpose of Regulation (EEC) No 857/84 or whether, in respect of any given region, Formula A or Formula B was to be adopted — see regulation 5(2) and (3). Schedule 2 to the regulations applied in respect of the allocation, transfer and surrender of wholesale quota in any region — see regulation 5(6). Para 14(1) of Schedule 2 required the minister to prepare in respect of each producer a wholesale register entry setting out: (a) his name; (b) his wholesale quota; (c) his cumulative quarterly wholesale quota; and (d) identification of his holding. Para 15 required the minister to maintain that register. Para 16 provided for the creation of a running regional wholesale reserve and for the D addition to that reserve of such quantities of wholesale quota as might be surrendered to it by any producer on such terms as might be agreed between the producer and the minister.

Part III of Schedule 2 to the Dairy Produce Quotas Regulations 1984 contained provisions applicable to the transfer of wholesale quota. These provisions are consistent, and only consistent, with the principle that, upon the change in occupation of a holding or part of a holding, the quota follows the land.

In the light of the relevant community legislation and the United Kingdom regulations the position which arose in England on the introduction in 1984 of the additional levy on milk and milk products and the implementation by the United Kingdom of Formula B in E respect of a region constituted by England and Wales may be summarised as follows:

(a) A milk producer is liable to pay Formula B contribution in respect of the milk which he delivers to the Milk Marketing Board if, but only if, two conditions are satisfied: namely (i) the total of wholesale deliveries of milk to the board from holdings in England and Wales exceeds the "purchaser quota" which has been allocated to the board; and (ii) the wholesale deliveries of milk to the board by that producer from his holding exceeds his own "wholesale quota".

(b) There is nothing in the community legislation or in the United Kingdom regulations which prohibits the production or the wholesale delivery of milk by a producer. Registration of wholesale quota in the name of a producer confers on that producer immunity F to the extent of his quota from liability to pay Formula B contributions. Delivery of milk in excess of his wholesale quota may, but not necessarily will, expose the producer to liability to pay Formula B contributions.

(c) Wholesale quota may be surrendered to the wholesale reserve, but, save to that extent, it cannot exist independently of a producer and a holding, both of which must be identified in the register entry. This conclusion necessarily follows from the terms in which "wholesale quota", "producer" and "holding" are defined for the purposes of the United Kingdom regulations.

(d) Upon a change in the occupation of all or a material part of the producer's holding, there will be a corresponding change in the register entry relating to that holding. The change will reflect the principle that quota follows the land. The register entry must be G amended to show the new occupier of the whole, or a part, of the holding as the person entitled to the whole, or a corresponding part, of the quota.

In these circumstances, if the question posed in this appeal were to be answered on the basis of the community legislation and the United Kingdom regulations as adopted and made in 1984, I should have no doubt that the conclusion which the arbitrator has reached was correct. Hills Farm was the relevant holding. For so long as the partnership was in occupation of that holding the partners were — together — the producer in relation to that holding and so were — together — the persons who were potentially liable to pay Formula B contribution in respect of the wholesale delivery of milk from that H holding and the persons who were able to, and did, benefit from the immunity from that liability which was conferred by the registration of wholesale quota in their names. Upon the death of Harry Faulks the equitable interest of the partnership under the trust of the tenancy declared by clause 18 of the partnership deed — by virtue of which, alone, the partnership was entitled to occupy Hills Farm — determined. John Faulks became the sole occupier of the holding, having the whole legal and beneficial interest in the tenancy, and so became the producer in relation to that holding. He became the person who was potentially liable to pay Formula B contribution in respect of the wholesale delivery of milk from that holding and so the only person who could benefit from the immunity from that liability conferred by the wholesale quota registered in respect of that J holding.

Under the United Kingdom regulations the minister was bound to give effect to the change of occupation by amending the register entry. This was consistent with the principle in article 7(1) of Regulation (EEC) No 857/84 as amplified by article 5 of Regulation (EEC) No 1371/84; treating the determination of a beneficial interest under a trust as being one of those "other cases of transfer which, under . . . national rules, have comparable legal effects as far as producers are concerned" to a transfer by inheritance — see, in particular, article 5(3) of the Commission regulations. The answer to the question whether the milk quota formerly registered in the name of the partnership could properly be regarded as property or a right or interest in property, separate and distinct from the holding, would be K a clear negative. It is also clear that the milk quota would not have been available in a dissolution under the 1890 Act to meet the debts and liabilities of the partnership.

It was urged upon me that, whatever had been the intention of those who adopted the community legislation and made the United Kingdom regulations 1984, circumstances had altered by 1988. It was said on behalf of the appellant that, by 1988, milk quota had become recognised as an independent asset having an economic value. I do not doubt that this perception of milk quota is widely held. Nevertheless, I do not think it accords with the community legislation and the United Kingdom regulations as they were at the death of Harry Faulks in June 1988. It is necessary to trace the material L developments.

Within a year of the adoption by the Council of the European Communities of Regulation (EEC) No 857/84 it was recognised that the application of article 7 of that regulation might "in certain cases result in difficult situations at an economic and social level". In particular, it was thought appropriate by the council

> in order to enable a lessee whose lease on a holding is due to expire to continue milk producing elsewhere, that Member States be authorised to place at the disposal of such lessee all or part of the reference quantity corresponding to the holding which he is leaving.

To this end article 7 was replaced. The new article was adopted on February 26 1985, together with a number of other amendments, in M Council Regulation (EEC) No 590/85. So far as material the new article 7 is in these terms:

> (1) Where a holding is sold, leased or transferred by inheritance, all or part of the corresponding reference quantity shall be transferred to the purchaser, tenant or heir according to procedures to be determined.
>
> Where land is transferred to the public authorities and/or for public use . . . Member States may provide that all or part of the reference quantity corresponding to the holding or to the part of the holding transferred shall be put at the disposal of the departing producer if he intends to continue milk production . . .

I need not read subparas (2) and (3); but subpara (4) is in these terms:

> (4) In the case of rural leases due to expire, where the lessee is not entitled to an extension of the lease on similar terms, Member States may provide that all

A or part of the reference quantity corresponding to the holding which forms the subject of the lease shall be put at the disposal of the departing lessee if he intends to continue milk production . . .

As I have already indicated, detailed rules for the purposes of applying article 7(1) of Regulation (EEC) No 857/84 had been set out in article 5 of Commission Regulation (EEC) No 1371/84. By June 1988 that regulation had been amended on a number of occasions. It was decided, in the interests of clarity, to consolidate the rules. This was done by Commission Regulation (EEC) No 1546/88, adopted on June 3 1988. The recitals to that regulation include the following:

> Whereas pursuant to the second subparagraph of paragraph 1 and paragraph 4 of Article 7 of Regulation (EEC) No 857/84 [meaning the new article
> B introduced in 1985] Member States may, where land is transferred to the public authorities or where a non-renewable lease expires and cannot be extended on similar terms, provide that all or part of the reference quantity corresponding to the holding or to the part of the holding concerned is to be put at the disposal of the departing producer; whereas that right constitutes a derogation from the principle laid down in the first subparagraph of Article 7(1) whereby a reference quantity may not be transferred without the land; whereas the said derogation was introduced in order to overcome certain economic and social problems and allow producers to continue milk production; whereas the extent of that derogation should therefore be specified . . .

For this purpose Regulation (EEC) No 1546/88 repealed Regulation (EEC) No 1371/84 and replaced article 5 of that repealed
C regulation by a new article 7. Subparas (1), (2) and (3) of article 7 of Regulation (EEC) No 1546/88 are in substantially the same terms as those of article 5 of Regulation (EEC) No 1371/84. Article 7 subpara (4) of the new article is in these terms:

> (4) In the event of the application of the second subparagraph of Article 7(1) of Regulation (EEC) No 857/84, concerning the transfer of land to the public authorities and/or for public use, and Article 7(4) of the said Regulation, concerning rural leases which are due to expire and which cannot be extended on similar terms, all or part of the reference quantity corresponding to the holding or to the part of the holding which is the subject of the transfer or of the said lease shall be put at the disposal of the producer concerned if he intends to continue milk production, provided that the sum of the reference quantity thus made available to him and the quantity corresponding to the holding which he takes over or on which he continues milk production does
> D not exceed the reference quantity which was available to him before the land was transferred or before the lease expired.

The new article 7(4) of Regulation (EEC) No 857/84 is permissive. It allows provision to be made for the benefit of outgoing tenants who intend to continue in milk production. It was adopted for the purpose of relieving the hardship which would otherwise flow from the application of the principle in the first subpara of article 7(1). That this is so is recognised by the recital in Commission Regulation (EEC) No 1546/88, which I have set out above. The introduction of the new article 7(4) in the commission regulation is expressed in that recital to be for the purpose, *inter alia*, of specifying, or defining, the extent of the derogation from that principle. The principle itself — namely
E that the reference quantity or quota corresponding to a producer's holding follows the land — is affirmed.

Provision for the benefit of outgoing tenants in the United Kingdom was made by section 13 of and Schedule 1 to the Agriculture Act 1986. But that provision does not take the form of an allocation to the outgoing tenant of any part of the quota held in respect of the land formerly subject to his tenancy. Rather, the tenant is entitled, by way of compensation, to a payment from his landlord. The amount of the payment is equal to the value of so much of the relevant quota (defined in para 1(2) of Schedule 1) as is apportioned, or attributable, to the tenant under the complex provisions in the Schedule. But the apportionment or attribution is for the purpose of measuring the compensatory payment and not for the purpose of any transfer of
F quota. There is no doubt that, as counsel for the appellant contended, analysis of the provisions in Schedule 1 to the Agriculture Act 1986 must lead to the conclusion that Parliament has recognised that milk quota can be valued — see, in particular, para 9 of that Schedule — but these provisions do not, in my view, lead to the conclusion that quota is an asset which is capable of being enjoyed, or transferred, independently of the land in respect of which it is registered.

It was further argued on behalf of the appellant that, whatever view I might form of the effect of the community legislation, the relevant provisions had been considered by the Court of Justice for the European Communities and that the conclusions of that court were binding on me. The case to which I was referred, *Wachauf* v *Bundesamt für Ernährung und Forstwirtschaft* (Case 5/88) [1991] 1 G
CMLR 328, was decided by the European Court (Third Chamber) on July 13 1989, on a reference from Germany by the Administrative Court at Frankfurt. The questions referred to the European Court were these:

> (1) Is an agricultural production unit having neither dairy cattle nor facilities (such as milking parlours) capable of being used exclusively for milk production a holding within the meaning of Article 12(d) of Council Regulation 857/84 of 31 March 1984?
>
> (2) Is the surrender of leased property upon the expiry of the lease a case having "comparable legal effects" within the meaning of Article 5(3) of Commission Regulation 1371/84 of 15 May 1984, if the leased property is an agricultural holding without dairy cattle and without any facilities capable of being used only for milk production (for example, milking parlours), and H
> where the lease provided for no obligation on the part of the lessee to engage in milk production?

Both questions were answered in the affirmative. Those are not questions which arise in the appeal before me. But there are observations in the judgment of the European Court and, more particularly, in the opinion of the Advocate-General on which the appellant relies.

The context in which the questions referred to the European Court in *Wachauf* had arisen was a dispute between a tenant farmer and the German authorities regarding the farmer's participation in a scheme granting compensation for the definitive discontinuance of milk production ("an outgoers scheme"). Article 4(1) of Regulation (EEC) No 857/84 had empowered member states, with a view to the J
restructuring of milk production, to grant compensation to producers undertaking to discontinue milk production definitively. The German authorities had adopted a scheme ("the German outgoers scheme") for this purpose. Under that scheme, a claimant who was the tenant of a holding within the meaning of article 12(d) of Regulation (EEC) No 857/84, was required not only to undertake to discontinue production but also to submit written consent from his landlord.

The claimant, Hubert Wachauf, was tenant of a farm which had not previously been, but which had become under his occupation, a dairy farm. In due course the claimant had been allocated a reference quantity or milk quota. His tenancy expired and he vacated the farm. K
He applied for compensation under the German outgoers scheme as a producer who was discontinuing milk production. The landlord refused consent on the grounds that compensation under the scheme would result in the loss to the farm of the quota which had been allocated. The German court was uncertain whether the farm, as leased by the claimant, was a "holding" within the meaning of article 12(d) — hence the first question. That court had also taken the view that the surrender, on the expiry of the lease, of a farm which, as leased, was not a milk-producing holding, could not be regarded as a case comparable to the grant of a lease of a milk-producing holding to which quota already attached so as to fall within article 5(3) of Regulation (EEC) No 1371/84 — hence the second question.

The Advocate-General advised that both questions be answered in L
the affirmative. He then went on to discuss what he described as "the underlying question". He put that question in these terms, at [1991] 1 CMLR 328 at p 340:

> The National Court takes the view that, if the questions referred for a preliminary ruling are answered in the affirmative, then an issue arises as to the compatibility with constitutional guarantees of equality and of respect for private property of the rule requiring the landlord's consent to a tenant's participation in the German outgoers' scheme and the rule that, on expiry of the tenancy, the quota reverts with the land to the landlord. The National Court's concern with constitutional guarantees appears to stem from its conviction that there may be cases in which it is the tenant, rather than the landlord, who through his efforts has attracted a quota to the holding, and that in such cases it would be inequitable if the landlord could, without more, M
> veto the tenant's participation in an outgoers' scheme and if the landlord, on the expiry of the tenancy, were to obtain all the benefit of the quota to the exclusion of the tenant . . .

The Advocate-General then considered the two principles — non-discrimination and respect of private property — and, in relation to the latter, came to this conclusion at p 342:

> That analysis [in *Haur und Land Rheinlandpfalz* [1980] 3 CMLR 42] can be applied, in my view, to the intangible asset constituted by a milk quota, which can properly be regarded as having an independent economic value; and in accordance with that analysis, I would suggest that there may well be cases where the permanent loss to the tenant of the use and value of the quota on expiry of a tenancy can be viewed as a measure of expropriation.
>
> In their written observations in this case, both the Commission and the

A United Kingdom Government have sought to argue that a quota is nothing more than an instrument of market management and cannot be considered as a kind of intangible asset in which property rights can arise. In my view, while this might correspond with the intention of the Community legislation, it does not reflect economic reality. If one considers the nature of the quota from the point of view of the producer, then it is plain that what the quota amounts to is a form of licence to produce a given quantity of a commodity (milk) at a more or less guaranteed price without incurring a penalty (the additional levy). In a market which has been effectively ossified by the introduction of quotas, such a "licence" is bound to acquire an economic value. This value will primarily translate into higher rental and capital values for dairy holdings. But that a quota can also have an intrinsic value is shown by the practice of "quota leasing", ie the temporary transfer without land of unused quota from one producer to another, a practice authorised by Article 5c(1a) of Regulation
B 804/68, as amended by Council Regulation 2998/87. It is also indicated, though more indirectly, by Article 7(4) of Regulation 857/84, which can be seen as designed to protect the tenant's interest in the economic value represented by the quota.

Community legislation has not resolved the issue of ownership of quota, possibly because it was not considered desirable to admit — for fear of creating a market in quota — that quota could be owned at all. The issue is not an easy one. On the one hand, the fact that the transfer rules in principle require the quota to follow the land suggests that these attach to the land and therefore be regarded as the property of the landowner. On the other hand, the existence of Article 7(4) of Regulation 857/84 and the recent authorisation of "quota leasing" indicate that attachment to the land is not absolute. Moreover the quotas are allocated to a person, the individual producer, who may of course be a tenant, on the basis of his production in a given reference
C year, rather than to a holding. These considerations, in my view, suggest that it is possible for either a landlord or a tenant to have a proprietary interest in a quota.

It will be clear, from earlier passages in this judgment, that I think that there is no true dichotomy, as the Advocate-General appears to suggest, between the allocation of quota to a person and allocation to a holding. Reference quantity, or quota, is allocated to a producer who must, reasonably, be a person or group of persons farming a holding — see article 12(c) and (d) of Regulation (EEC) No 857/84. Further, under the United Kingdom regulations, a wholesale register entry must, necessarily, identify both the producer and the holding — see para 14(1) in Schedule 2 to the Dairy Produce Quotas Regulations 1984 and its successor, regulation 30(2) of the Dairy
D Produce Quotas Regulations 1986. Further, I do not think it is correct to describe quota as "a form of licence to produce a given quantity of a commodity (milk)". The true position, at least in England and Wales, is that a producer does not require a licence to produce milk and will not necessarily incur any liability as a consequence of production. It is in respect of his potential liability that he enjoys an immunity from Formula B contribution by virtue of, and to the extent of, the register entry of an amount of wholesale quota in his name. Nevertheless, the appellant is entitled to point to the Advocate-General's view that, as a matter of economic reality, milk quota is to be regarded as an intangible asset having an intrinsic value.

E The European Court did not, I think, find it necessary to endorse the Advocate-General's view. The effect of article 7 of Regulation (EEC) No 857/84 is considered at para 13 of the judgment of that court. The conclusion at [1991] 1 CMLR 328 at p 347 in the report is that:

It is apparent from the provisions quoted, considered as a whole, that the Community legislature intended that at the end of the lease the reference quantity should in principle return to the lessor who retakes possession of the holding, subject, however, to the Member State's power to allocate all or part of the reference quantity to the departing lessee . . .

As I have already indicated, that power has not been exercised by the United Kingdom Government. The outgoing tenant in England and Wales is compensated by a payment of money, not by the allocation
F of quota.

The European Court in the *Wachauf* case accepted at para 19 of its judgment:

. . . that Community rules which, upon the expiry of the lease, had the effect of depriving the lessee, without compensation, of the fruits of his labour and of his investments in the tenanted holding would be incompatible with the requirements of the protection of fundamental rights in the Community legal order . . .

But it reached the conclusion at para 22 that:

The Community regulations in question accordingly leave the competent national authorities a sufficiently wide margin of appreciation to enable them to apply those rules in a manner consistent with the requirements of the protection of fundamental rights either by giving the lessee the opportunity of keeping all or part of the reference quantity if he intends to continue milk G
production, or by compensating him if he undertakes to abandon such production definitively.

In my view, the effect of the judgment of the European Court is to affirm the principle that quota is attached to the land, while recognising that the rules made under the community legislation permit member states to make provision for compensating the outgoing tenant in cases where the strict application of that principle would cause hardship. I do not find anything in that judgment which lends support to the Advocate-General's view that quota can properly be regarded as an intangible asset, distinct from the holding to which it relates.

The provisions comparable to the German outgoers scheme which H
were current in England and Wales on June 12 1988 (when the partnership determined) are found in the Milk (Cessation of Production) (England and Wales) Scheme 1987 — a scheme made under the Milk (Cessation of Production) Act 1985. I can find nothing in the Act or in the scheme which supports the contention that milk quota is to be regarded as an intangible asset in which distinct property rights — that is to say, distinct from rights in respect of the holding to which the quota relates — can arise. Indeed, given the observations which the United Kingdom Government put before the European Court in the *Wachauf* case — see [1991] 1 CMLR 328 at p 342 — it would be surprising if there were support for that contention in the domestic scheme. The position under the scheme is that, on a transfer of milk quota to the reserve by a tenant in J
occupation of a holding, the compensatory payment may be apportioned between landlord and tenant under the provisions of regulations 12 and 13. There is nothing in the scheme which entitles the tenant who has gone out of occupation and so ceased milk production to claim compensation from his landlord. That right is conferred by the Agriculture Act 1986, to which I have already referred.

Counsel for the appellant directed me to the definition of "milk quota" in section 1(8) of the 1985 Act; that is to say:

"milk quota" means —

. . .

(*b*) in the case of a person registered in the wholesale register maintained K
under [the 1984] Regulations, a wholesale quota registered as his in that register . . .

There are similar expressions of a proprietary nature in the scheme — see in regulation 3(2):

. . . an applicant's quota shall be the amount of quota to which he is entitled at the date of the application . . .

And in regulation 4(1):

. . . any person who is or has been a registered milk producer and who —
(a) undertakes . . . to surrender all his milk quota to the reserve . . .

In my view, given the terms in which "wholesale quota" is defined by the Dairy Produce Quotas Regulations 1984 and the requirement in those regulations (and in the 1986 regulations) that a wholesale L
register entry must identify not only the producer but also his holding, these expressions do not indicate any departure from the principle that milk quota attaches to land. They are, I think, no more than a convenient way of referring to the quota shown in the register entry to which the applicant for compensation is named as the producer.

The concept of "quota leasing" to which the Advocate-General refers in the *Wachauf* case was authorised by Council Regulation (EEC) No 2998/87. It was introduced in England and Wales with effect from March 31 1988 by regulation 6 of the Dairy Produce Quotas (Amendment) Regulations 1988 (SI 1988 No 534). That regulation, now re-enacted as regulation 15 of the Dairy Produce Quotas Regulations 1989 (SI 1989 No 380), permits a producer "to M
make a temporary transfer within one region of part of the wholesale quota registered as his to another producer for a period of one quota year". The transferor will retain the untransferred part of his wholesale quota and so will remain on the register in respect of his holding. The transferee must be a producer within article 12(c) of Regulation (EEC) No 857/84; that is to say, he must be a person farming a holding within the same region and so a person who is already the subject of a register entry. In these circumstances, the effect of the "temporary transfer" is that, in respect of a given quota year, the wholesale quota in respect of which the transferor is registered is diminished, and the quota in respect of which the transferee is registered is increased, by a corresponding amount.

That is to say, the extent of the transferor's potential liability to one year's Formula B contribution is increased, and the transferee's potential liability is diminished, by the amount of the transfer.

This does not lead to the conclusion that what has been transferred under a regulation 6 transfer is an asset or some form of property right. It must be kept in mind that liability to formula B contribution is a liability on the producer to contribute to additional levy imposed on the wholesale purchaser. It is analogous to the liability of one of several co-sureties (each in respect of part of a debt) to contribute a limited proportion of the debt owed by the principal debtor to the creditor. An agreement between two co-sureties and the creditor that the proportions in respect of which the sureties shall be liable to contribute shall be varied — so that the potential liability of one is increased and the potential liability of the other correspondingly diminished — would not, on an ordinary use of language, be regarded as a transfer of an asset or property right by the one surety to the other. Rather, it is the transfer of a potential liability from the latter to the former. In the same way, as it seems to me, the transfer of quota from one producer to another — effected, as it must be, by reciprocal adjustment to their respective entries on the register — is properly to be regarded not as the transfer of an asset by the transferor to the transferee but rather as the transfer of a potential liability from the transferee of the quota to the transferor. The fact that the transfer of a potential liability is likely to be matched by a payment to the person assuming that liability may enable it to be said that the quota transferred has an intrinsic economic value, but that does not of itself lead to the conclusion that quota is an asset separate and distinct from the holding in relation to which it was, or becomes, registered.

Counsel for the appellant also placed reliance on the way in which milk quota has been treated for the purpose of the taxation of capital gains; and, in particular, on the provision, introduced by section 12 of the Finance Act 1988, in relation to roll-over relief under the Capital Gains Tax Act 1979. In summary, the position is:

(a) Tax is charged in accordance with the 1979 Act in respect of capital gains accruing to a person on the disposal of assets — see section 1(1) of that Act.

(b) Chapter 1 of Part II of the Act (sections 19 to 27) contains detailed provisions for identifying what shall be regarded for the purposes of the Act as a disposal of assets; in particular, section 20(1) provides that there is a disposal of assets by their owner where any capital sum is derived from assets notwithstanding that no asset is acquired by the person paying the capital sum.

(c) Section 115 of the Act provides, in effect, that roll-over relief shall be available where the consideration which a person carrying on a trade obtains for the disposal of assets used for the purposes of the trade is applied by him in acquiring other assets which, on acquisition, are taken into use for the purpose of the trade, and both the old assets and the new assets are within the classes listed in section 118.

(d) Section 118 of the Act, as originally enacted, listed three classes of assets — namely class 1, assets within Heads A and B; class 2, ships, aircraft and hovercraft; and class 3, goodwill. Head A of class 1 included any land occupied (as well as used) only for the purpose of the trade.

(e) Section 112 of the Finance Act 1988 added two additional classes to the original three: class 2A, satellites, space stations and spacecraft; and class 4, which was in these terms:

> Milk quotas (that is, rights to sell dairy produce without being liable to pay milk levy or to deliver dairy produce without being liable to pay a contribution to milk levy) and potato quotas (that is, rights to produce potatoes without being liable to pay more than the ordinary contribution to the Potato Marketing Board's fund . . .)

It appears from an Inland Revenue press release dated October 29 1987 (reported in [1987] *Simon's Tax Intelligence* at p 806) that the relief afforded by section 112 of the Act was introduced with effect from that date on the basis that "the land itself already qualifies for roll-over relief but under present rules the quota — the right to produce milk or potatoes without incurring a levy — does not qualify".

In the ordinary transaction of a transfer on sale of the whole of the land comprised in a holding in respect of which there is a register entry of an amount of wholesale milk quota, there will necessarily be an amendment to the register entry into the name of the transferee. This will follow from the provisions in Part III of Schedule 2 to the Dairy Produce Quotas Regulations 1984 and (subsequently) regulation 8 of the Dairy Produce Quotas Regulations 1986 (as originally made and as subsequently amended by regulation 5 of the Dairy Produce Quotas (Amendment) Regulations 1988). I find it difficult to understand how, in such a case, it could have been argued that the consideration for the disposal of the land — for the purposes of roll-over relief under section 115 of the 1979 Act — did not include any element of that consideration attributable to the milk quota in respect of which the transferee would become registered as a consequence of the transaction.

If the understanding of the law which gave rise to the comment in the Revenue's press release of October 29 1987, which I have set out above, was to the contrary, then I think that understanding may have been based on a false appreciation of the true position. Further, I would have thought that the surrender of milk quota to the reserve could well have been regarded as a partial disposal of the land — by virtue of section 20(1) of the 1979 Act — and so qualified for roll-over relief under Head A of class 1. These are not questions which I need to decide. There can be no doubt that it was thought by those responsible for the press release dated October 29 1987 and for the relief subsequently enacted that there were circumstances in which a producer might receive a capital sum in respect of milk quota which would not qualify for roll-over relief under the 1979 Act in its original form.

The question for me is whether the inclusion of milk quota within the classes of assets in section 118 of the 1979 Act requires me to alter my view — formed on the basis of an analysis of the community legislation and the United Kingdom regulations — that the milk quota formerly registered in the name of the partnership could not properly be regarded as property, or a right or interest in property, separate and distinct from the holding in respect of which it was registered; and that it would not have been available in a dissolution under the 1890 Act to meet the debts and liabilities of the partnership. In my judgment, it does not. I do not think that it could be right to allow a relieving section in a taxing statute to override what I regard as the clear effect of the primary legislation. I find some comfort in the fact that, whatever may have been the views of those responsible for its treatment under the capital gains tax legislation, the concept of milk quota as an asset independent of the holding to which it related was expressly refuted by the United Kingdom Government in its submissions to the European Court: see the *Wachauf* case [1991] 1 CMLR 328 at p 342.

It follows, therefore, that I do not find anything in the developments which have taken place since the introduction of milk quota in 1984 which causes me to alter the view which I have already expressed — namely that the milk quota formerly registered in the name of the partnership should not be taken into account as an asset of the partnership for the purposes of valuing the assets as at June 12 1988 pursuant to clause 16 of the partnership deed. Accordingly, I dismiss this appeal.

There are two further points which I should mention. First, although it may seem harsh that those entitled to the estate of Harry Alfred Faulks will be deprived of a share in what may well be a substantial part of the economic value of the dairy business which he carried on in partnership with his brother, this is a necessary consequence of the bargain which the brothers made in relation to the tenancy of the farm. It is clear from the provisions of clause 18 of the partnership deed (including, in particular, the proviso to that clause) that the brothers intended that the survivor — or, in the case of a dissolution *inter vivos*, the continuing partner — should succeed to the tenancy.

If John Faulks had predeceased his brother Harry, it is likely that effect would have been given to this intention by the provisions of Part II of the Agriculture (Miscellaneous Provisions) Act 1976. But if John had decided to determine the partnership, say, by notice under clause 12 of the partnership deed during his brother's lifetime, the succession could not have been guaranteed. The landlord might have refused consent to the assignment of the tenancy to Harry which is contemplated by the proviso to clause 18. The tenancy would have remained vested in John, who would have been entitled to continue farming the holding as a dairy enterprise and to be registered in respect of the milk quota by virtue of regulation 8 of the Dairy Produce Quotas Regulations 1986. But Harry would have been entitled, as the remaining or continuing partner, to take all the partnership assets and farm elsewhere. In that situation it would have been bizarre to value the milk quota as an asset of the partnership for the purposes of setting the price which Harry was to pay for John's

A share of the partnership on a purchase under clause 16.

Second, in dismissing this appeal, I am not to be taken as having decided that there are no circumstances, either in this or in a similar case, in which some adjustment may need to be made in taking the accounts of a farming partnership to reflect the fact that the land which, *ex hypothesi*, the partnership has ceased to occupy may have acquired an enhanced value as a result of a dairy enterprise which the partnership has carried on. I was referred, in the course of argument, to *Pawsey* v *Armstrong* (1881) 18 ChD 698 and *Miles* v *Clarke* [1953] 1 WLR 537. The principle in those cases appears to me to be this: that where partnership money has been expended in maintaining or enhancing the value of land which is the exclusive property of one partner, then in taking partnership accounts on a dissolution it may B be fair and right to make some allowance to the partnership against that partner for the money which has been so expended or, perhaps, for a portion of the enhanced value. In making such an allowance the court does not, I think, treat the land or anything on it as an asset of the partnership which is to be valued as such; rather, it adjusts the accounts between the partners by debiting the account of the partner who benefited from the expenditure in order to do what is just and equitable. It is not difficult to imagine circumstances in which that principle might apply in relation to milk quota formerly registered in the name of a partnership — for example, where the dairy enterprise was commenced by, or was wholly attributable to, the efforts of the partnership (*cf* the *Wachauf* case) or where (in so far as that is C possible) some part of the quota had been acquired by the partnership on the back of a transfer of a limited interest in land (as to which see, generally, para 8.4, "Transfer of Quota", in *Essential Law for Landowners and Farmers* (3rd ed, 1990) by Gregory and Sydenham). There is nothing in the facts which I have seen which suggests that the principle applies in the present case and the question is not one which I can properly decide on this appeal.

Appeal dismissed with costs. Certificate issued under section 1(7) of the Arbitration Act 1979. Leave to appeal refused.

D

Queen's Bench Division

June 14 1991

(Before Mr Justice PHILLIPS)

STUBBS v HUNT & WRIGLEY AND ANOTHER

Estates Gazette May 23 1992

[1992] 20 EG 107

Milk quotas — Negligence — Adverse decision by tribunal on plaintiff's application for hardship relief — Whether advice that plaintiff would have failed in challenging tribunal's E decision negligent — Effect of partnership arrangements — Meaning of producer — Meaning and effect of para 17(3)(a) of Schedule 2 to the Dairy Produce Quotas Regulations 1984 — Plaintiff's claim dismissed

In May 1982 the plaintiff and his wife entered into a partnership with the owner of Manor Farm, Thornton Steward, Ripon, Mr A J Crossley, subject to termination on 12 months' notice — Mr and Mrs Stubbs contributed as capital their dairy herd and milking plant — Following notice given by Mr and Mrs Stubbs the partnership was dissolved on April 30 1984 and they moved part of their dairy herd to Broadmires Farm, a farm belonging to Lord Swinton — The F plaintiff was permitted to graze his milking cows on Broadmires Farm, first under two grazing licences and then, from October 11 1984, pursuant to a partnership agreement between the plaintiff and a company belonging to Lord Swinton — Broadmires Farm and adjacent land was leased to the partnership and both partnership agreement and lease were drafted so as to enable Lord Swinton to bring the joint venture to an end by 12 months' notice — The Dairy Produce Quota Tribunal rejected the plaintiff's exceptional hardship claim for milk quota to be determined for himself at Broadmires Farm on the ground, *inter alia*, that the "producer" at the farm was the plaintiff and Lord Swinton, as partners, and that the "producer" had not prior to April 2 G 1984 entered into any transaction and arrangement as fell within para 17 of Schedule 2 to the Dairy Produce Quotas Regulations 1984 — In giving legal advice to the plaintiff, the National Farmers Union stated that the plaintiff could not be advised to challenge the decision of the tribunal as, *inter alia*, the only transaction or arrangement which the plaintiff entered into prior to April 2 1984 was the dissolution of the first partnership and that did not show an intention to go into or remain in business as a producer

The plaintiff claimed that the advice given by the NFU was negligent, that the tribunal erred in law and that, had he H received proper advice, he would successfully have challenged the tribunal's decision and obtained a quota — The plaintiff submitted that the tribunal erred in law in finding that the "producer" was the partnership of the plaintiff and Lord Swinton, failed to consider whether the notice to terminate the first partnership was a transaction or arrangement satisfying para 17(3)(a) of Schedule 2 to the 1984 regulations, failed to consider whether it was fair and reasonable that the wholesale quota be allocated in that it failed to recognise the consequence of a refusal of the exceptional hardship claim was that the plaintiff would have to cease dairy farming, which producers in general would not have to do, and that the J tribunal came to the erroneous view that the "producer" was a partnership, which had not yet commenced in business

***Held*: The NFU had not been negligent in its advice and the plaintiff's claim was dismissed — If the only issue in the way of a successful challenge of the tribunal's decision was whether the plaintiff had established that, when he gave notice terminating the first partnership, he "intended to go into or remain in business as a producer", judicial review should have been sought — On April 2 1984 the plaintiff was not himself a producer but he and his wife were members of a partnership under the first partnership and the partnership was the K producer — Accordingly, the plaintiff was not a person to whom article (1)(c) of Council Regulation 857/84 applied and therefore could not make a claim for exceptional hardship under para 17 of Schedule 2 to the 1984 regulations — The notice to terminate the first partnership was not a transaction or arrangement that had either of the consequences provided for by para 17(3)(a) of Schedule 2 to the 1984 regulations — The plaintiff suffered prejudice not because he had been deprived of quota but because he chose to leave a partnership that occupied land that had a history of use for dairy farming and was thus entitled to quota — The 1984 regulations do not L cater for hardship arising in these circumstances**

No cases are referred to in this report.

This was a claim by the plaintiff, Francis Stubbs, against the second defendant, the National Farmers Union, alleging negligent advice and claiming damages. The claim against the first defendants, Hunt & Wrigley, of Northallerton, solicitors, was discontinued.

Richard Seymour QC (instructed by Ward Hadaway, of Newcastle upon Tyne) appeared for the plaintiff; Stuart Isaacs QC (instructed by Lloyd Cooper) represented the second defendant. The first defendants, Hunt & Wrigley, did not appear and were not represented. M

Giving judgment, **PHILLIPS J** said: On July 24 1984 the United Kingdom brought into force regulations under which milk quotas were allocated to British dairy farmers. Thereafter those who produced milk without, or in excess of, such quotas became liable to pay a levy. The size of this levy soon rendered uneconomic the production of milk which did not fall within a quota. This was in accordance with the intention of the European Community scheme under which quotas were introduced.

Francis Stubbs, the plaintiff, had been a dairy farmer all his life. When milk quotas were introduced he applied for one. His application was rejected by the Dairy Produce Quota Tribunal ("the tribunal"). He sought the advice of the second defendants ("the

A NFU") as to whether he should challenge the tribunal's decision by seeking judicial review. The NFU advised against this course and Mr Stubbs did not pursue it. In this action he claims that the advice given by the NFU was negligent. He contends that the tribunal erred in law and that, had he received proper advice, he would successfully have challenged the tribunal's decision and obtained a quota. He claims, as damages flowing from the NFU's negligence, the loss that he has sustained as a consequence of failure to obtain a milk quota.

The facts

Mr Stubbs devoted himself to dairy farming from the time that he left school at the age of 15. He married when he was 21. His wife shared with him the tasks that dairy farming involved and he treated
B her as an equal partner in his farming business. He owned no land and carried on business by persuading landowners to join with him and his wife as partners in the business of dairy farming. In a succession of partnerships, the landowner provided the land and the plaintiff and his wife provided the livestock that formed the basis of the partnership's dairy herd and took responsibility for the management of the dairy business.

In May 1982 Mr and Mrs Stubbs entered into a partnership with Mr A J C Crossley to farm Manor Farm, Thornton Steward, near Ripon. Mr Crossley had purchased this farm from its previous owners, with whom Mr and Mrs Stubbs had had a similar partnership. The following terms of the partnership are material:

(1) Mr and Mrs Stubbs contributed as capital their dairy herd and
C milking plant ("live and dead stock").

(2) Mr Crossley contributed a sum equal to the value of this live and dead stock.

(3) The management of the partnership business was shared on the basis that Mr Crossley was responsible for matters relating to cropping and grassland and Mr Stubbs for matters relating to the dairy herd.

(4) Profits were shared (broadly) as to one half to Mr Crossley and one quarter each to Mr and Mrs Stubbs.

(5) Any partner was entitled to retire from the partnership on 12 months' notice whereupon (in effect) Mr and Mrs Stubbs were obliged to vacate Manor Farm.

(6) Upon dissolution of the partnership Mr Stubbs had the option
D to purchase all the female dairy livestock and take the same as part of his partnership capital.

Mr and Mrs Stubbs had difficulty in working in partnership with Mr Crossley and, in April 1983, they gave notice to terminate the partnership on April 30 1984.

It was Mr and Mrs Stubbs' intention to take their share of the partnership capital *in specie* and to move their herd to other land. To this end Mr Stubbs had, before giving notice to dissolve the partnership, entered into negotiations with Mr Stanton, the agent of Lord Swinton. Lord Swinton had land on which dairy cattle had been farmed some seven or eight years previously and Mr Stubbs was led to believe that there was a good chance that he and his wife would be
E able to continue dairy farming from Lord Swinton's land. In about February 1984 Mr Stubbs entered into an oral agreement with Mr Stanton under which he was granted a grazing licence on Broadmires Farm on Lord Swinton's estate. This farm extended to some 72 acres and the intention was that, when the Manor Farm partnership terminated, Mr Stubbs would bring at least part of the Manor Farm herd to Broadmires Farm and graze them there, pending the conclusion of a partnership agreement with Lord Swinton in October. In fact, though Mr Stubbs may not have known this at the time, the trustees of Lord Swinton's estate had insisted on advertising Broadmires Farm to let on a partnership basis in order to make sure that there was no alternative tenant more attractive to the estate than Mr Stubbs.

F When the Manor Farm partnership was dissolved at the end of April 1984, Mr and Mrs Stubbs took as their share of the partnership capital, or part of it, 55 milking cows and about 30 followers. The rest of the herd, which by this time had risen in number to 80 milking cows and 90 followers, was sold. Mr Stubbs moved his 55 milking cows to Broadmires Farm. The farm was not large enough to support the followers, and these he put to graze on land belonging to a relative. A formal grazing agreement dated February 15 1984 was drawn up, granting Mr Stubbs grazing rights until September 1. This was then renewed until October 11 1984. As from that date Mr Stubbs went into a partnership on terms that were subsequently incorporated in two contracts dated November 6 1984. The first of those contracts was a partnership agreement between Mr Stubbs and the Healey Farming Co, Lord Swinton's corporate vehicle. The other was a lease G
under which Lord Swinton leased to the partnership both Broadmires Farm and the adjacent Sykes Farm — the two comprising 186 acres. The contracts had been skilfully drafted so as to enable Lord Swinton to bring the joint venture to an end by giving 12 months' notice of determination of the lease. Mr Stubbs was entitled to give 12 months' notice of determination of the partnership.

In August 1984 Mr Stubbs submitted a claim for milk quota to the tribunal. The hearing took place in September and the claim was rejected. Mr Stubbs then joined the NFU with the specific intention of seeking from them advice as to whether he could challenge the tribunal's decision. His case was considered by the NFU legal division and, in particular, a barrister employed by that division, H
Miss Diane Smith. She advised that a challenge was unlikely to succeed and, discouraged by her advice, Mr Stubbs took the matter no further. Without quota, dairy farming on Broadmires and Sykes Farm was unprofitable and in June 1986 the dairy herd was sold for £18,000. The partnership continued but devoted itself to sheep farming and livestock rearing, without financial success. Mr Stubbs' claim is put forward on the basis of his actual financial position as compared with the position he would have enjoyed had he been granted a milk quota of, initially, 300,000 litres pa.

The principal issue

The NFU have accepted that they owed Mr Stubbs a duty to exercise reasonable care when giving him advice. They have based J
their defence essentially upon the contention that the advice given by Miss Smith was correct. They have argued, through Mr Isaacs QC, that the tribunal's decision could not have been successfully attacked on a judicial review and that Miss Smith was correct so to advise. In considering this submission I have, in effect, had to put myself in the position of the court hearing an application for judicial review. This has involved scrutiny of regulations of some complexity.

European Council regulations

The United Kingdom regulations, which governed Mr Stubbs' application for milk quota, were made pursuant to European Council regulations. In order properly to interpret the former it is necessary, first, to consider the latter. K

Council Regulation (EEC) No 856/84

This regulation provided for the introduction of the quota scheme with which this action is concerned. The preamble made it plain that the object of the scheme was to deal with a structural imbalance in the community between supply and demand for milk leading to surpluses and commented:

4. Whereas a careful examination of the different possible ways of re-establishing balance in the milk sector shows that, despite the administrative difficulties which its implementation may involve, the most effective method, and the one having the least drastic effect on the incomes of producers, is the introduction for an initial period of five years of an additional levy on quantities of milk delivered beyond a guarantee threshold. L

The regulation proceeded to insert into an earlier regulation, 804/68, dealing with milk production, article 5c, which provided:

1. During five consecutive periods of 12 months beginning on 1 April 1984, an additional levy payable by producers or purchasers of cows' milk shall be introduced. The objective of the said levy shall be to curb the increase in milk production while at the same time permitting the structural developments and adjustments required, having regard to the diversity of the situations among individual Member States, regions and collection areas in the Community. However, the first period shall start on 2 April 1984.

The levy system shall be implemented in each region of the territory of the Member States in accordance with one of the following formulas:

Formula A

A levy shall be payable by every milk producer on the quantities of milk and/or milk equivalent which he has delivered to a purchaser and which for M
the 12 months concerned exceed a reference quantity to be determined.

Council Regulation (EEC) No 857/84

This regulation, dated March 31 1984, implemented article 5c. The preamble commented:

Whereas the scheme must, as a matter of overwhelming public interest, enter into force on 2 April 1984; whereas, to this end, transitional measures must be adopted so that the levy due from 2 April onwards can be collected reasonably promptly.

The regulation provided as follows:

Article 2

1. The reference quantity referred to in Article 5c(1) of the above mentioned

A Regulation shall be equal to the quantity of milk or milk equivalent delivered by the producer during the 1981 calendar year (formula A) . . .

Article 4

1. In order to complete the restructuring of milk production at national or regional level or at the level of the collecting areas, the Member States may, in connection with the application of formulas A and B:

(a) grant to producers undertaking to discontinue milk production definitively compensation paid in one or more annual payments;

(b) grant an additional reference quantity to producers realising a milk production development plan approved after the entry into force of this Regulation under Directive 72/159/EEC, on condition that this plan meets the criteria referred to in Article 1(2) of Regulation (EEC) No 1946/81;

(c) grant producers undertaking farming as their main occupation an additional reference quantity, whether their herd fulfils the conditions set out B in paragraph (b) or not.

2. The reference quantities freed shall, as necessary, be added to the reserve referred to in Article 5.

. . .

Article 5

For the purpose of applying Articles 3 and 4, additional reference quantities may be granted only within the guaranteed quantity limit . . .

. . .

Article 7

1. Where an undertaking is sold, leased or transferred by inheritance, all or part of the corresponding reference quantity shall be transferred to the purchaser, tenant or heir according to procedures to be determined.

. . .

Article 12

C For the purposes of this Regulation the following meanings shall apply:

. . .

(c) producer: a natural or legal person or group of natural or legal persons farming a holding located within the geographical territory of the Community:

— selling milk or other milk products directly to the consumer, and/or

— supplying the purchaser.

(d) holding: all the production units operated by the producer and located within the geographical territory of the Community;

These definitions were incorporated in the UK regulations, to which I now turn.

The Dairy Produce Quotas Regulations 1984 ("the UK regulations")

The UK regulations [SI 1984 No 1047] were made on July 23 1984 D and came into operation on July 24 1984 ("the operative date"). I have already said that the UK regulations incorporated the definitions in article 12 of Regulation (EEC) No 857/84. The only additional definition that I need refer to is the following:

"wholesale quota" means —

(a) the quantity of dairy produce which may be delivered by wholesale delivery, from a holding in a region in which Formula A is implemented, in a quota year without the producer in occupation of that holding being liable to pay levy.

The regulations governing wholesale quota are set out in Schedule 2. The relevant paragraphs of this schedule provide as follows:

1. (1) Subject to subparagraph (3) each applicant shall no later than the date E five weeks after the operative date (or, in a region for which Formula A is implemented, four weeks after the operative date) submit his application for primary wholesale quota to the Minister on a form (as specified in subparagraph (2)) duly completed by him or under his authority.

. . .

(2) The form referred to in subparagraph (1) shall be provided on request by the Minister and shall require an applicant to provide —

(b) a statement that on 2nd April 1984 the business of the direct sale or wholesale delivery of dairy produce was being carried on from land within the identification of the holding occupied by him on the date of his application.

Paras 2 to 5 make provision for the calculation of primary wholesale quota relating to holdings on the basis of historic production from the holdings in question.

Part II of the Schedule deals with "Registers, Running Regional F Wholesale Reserve and Exceptional Hardship Claims". Its provisions include the following:

14.(1) The Minister shall prepare in respect of each producer a wholesale register entry, dated as at the operative date and setting out —

(a) his name,

(b) his wholesale quota, and

. . .

(d) an identification of his holding,

15.(1) The Minister shall maintain —

(a) a wholesale register (being a register of entries referred to in paragraph 14), and

(b) a register of particulars of wholesale deliveries by each producer.

. . .

16.(1) A running regional wholesale reserve may be created. . .

17.(1) The Minister may make available from the running regional G wholesale reserve an exceptional hardship provision to be used in satisfaction of exceptional hardship claims in accordance with the following provisions of this paragraph.

(2) For the purpose of this paragraph "exceptional hardship claim" shall be construed in accordance with subparagraph (3), and "claimant" shall be construed accordingly.

(3) An exceptional hardship claim shall be a claim by a person to whom Article 4(1)(c) of Council Regulation 857/84 (which deals with producers undertaking farming as their main occupation) applies that —

(a) before 2nd April 1984 he has entered into, or become obliged to enter into, a transaction or made an arrangement —

(i) as a result of which his wholesale quota is substantially less than it would have been had he not entered into or become obliged to enter into that H transaction or made that arrangement, or

(ii) the reasonably expected outcome of which is a level of wholesale delivery of dairy produce in respect of which, or a substantial part of which, wholesale quota is not otherwise capable under these regulations of being allocated to him.

(b) at the time of entering into or becoming obliged to enter into that transaction or making that arrangement, he intended to go into or remain in business as a producer,

(c) he has not received, and will not receive or become entitled to receive, as a result of that transaction or arrangement, benefit reasonably commensurate with the want of wholesale quota to which his exceptional hardship claim relates,

(d) as a result of the matters specified in the preceding paragraphs of this subparagraph, he has suffered or will suffer exceptional hardship in J comparison with producers in general, and

(e) it is fair and reasonable that he should be allocated wholesale quota as a result of his exceptional hardship claim.

. . .

(5) The Tribunal shall determine for any claimant who satisfies them of the matters referred to in subparagraph (3) his gross additional wholesale quota (including cumulative quarterly wholesale quotas), being the quantity of dairy produce which the Tribunal determines is justified by his exceptional hardship claim.

Part III of the Schedule deals with changes of occupation of holdings after the operative date. The effect of its provisions is that the quota remains with the holding and is transferred with it to the new occupier of the holding. K

Mr Stubbs' application for hardship relief

In making his claim Mr Stubbs was assisted by a solicitor and an accountant. The former was a partner in the first defendants, against whom proceedings have been discontinued. Mr Stubbs told me that his solicitor was clearly out of his depth. I have no difficulty in accepting this, for he was navigating in very deep and largely uncharted waters.

Mr Stubbs submitted his claim on a form designed to be used by applicants for one or more of the following: direct sales quota; special case treatment; outgoers scheme.

It does not seem to me that this was an appropriate form for an application for wholesale quota on the ground of exceptional L hardship and it seems plain that Mr Stubbs was in some confusion as to the basis upon which he was seeking relief. To the tribunal that confusion must have been accentuated by the manner in which the application form was completed. In particular:

(1) Mr Stubbs stated that his business title was "F J & L Stubbs", ie Mr Stubbs and his wife.

(2) Mr Stubbs applied in relation to Broadmires Farm.

(3) Mr Stubbs described his status as that of "partner".

(4) Mr Stubbs stated he was a tenant of Broadmires Farm.

(5) Mr Stubbs stated "we rent 186 acres from Swinton Estate".

(6) Mr Stubbs claimed that before March 1 1984 he made significant capital investments aimed at increasing milk production M on his holding. These included the value of his 55 cows and items of equipment purchased both before and after March 1 1984.

(7) Mr Stubbs stated that on April 2 1984 he was carrying on the business of direct sale and/or wholesale sale of dairy produce from Broadmires Farm.

Mr Stubbs included in his statement, however, a narrative which gave a fair picture of events as I have tried to describe them, save that it did not state that Mr Stubbs was farming at Broadmires Farm under a temporary grazing licence, but stated:

In the Spring of 1984 I reduced my herd to 55 cows and followers (about 138 in all) and took the herd to Broadmires Farm, Ilton, Masham, going into partnership with the owner, Lord Swinton. This farm does not have any milk quota.

A He went on to state that a quota had been allocated to the Manor Farm partnership and ended:

In view of the fact that there has been no milk production at Manor Farm since 1st May 1984, which was the day I left, I wish to apply for the whole of the above quota to be transferred to Broadmires Farm where the dairy herd from Urevale Farms is now located and where I am now farming.

The decision

Mr Stubbs' application was heard by a tribunal chaired by Lord Grantchester himself. I must read the tribunal's decision in full:

DECISION

The Tribunal rejects the application made by Mr Stubbs for additional quota to be allocated on grounds of exceptional hardship in respect of the holding
B known as Broadmires Farm and Sykes Farm, Ilton, Masham, Ripon in Yorkshire, which holding is registered under the number 50/719/19.

REASONS

1. Broadmires Farm and Sykes Farm are part of Lord Swinton's estates. No milk was being sold or delivered from these two farms in the period of twelve months ending on the 2nd April 1984, and no quota has been allocated or provisionally allocated in respect of either of them.

2. For some 9 years prior to, and ending on, the 30th April 1984 Mr Stubbs carried on the business of dairy farming at Manor Farm, Thornton Steward, in partnership with the landowner, Mr A J C Crossley. In April 1983 in consequence of a dispute Mr Stubbs gave notice terminating that partnership with effect from 30 April 1984. Accordingly he had to vacate Manor Farm on the 1st May 1984.

3. In February 1984 Mr Stubbs entered into an oral arrangement with Lord
C Swinton under which they agreed to become partners in the business of dairy farming at Broadmires and Sykes Farms with effect from the 1st October 1984. Later that month Mr Stubbs ordered a bulk milk tank and milking equipment, and some minor alteration to be done to the cow byre and milk house on the farms for which he paid sums totalling £5,276 in May 1984.

4. Mr Stubbs moved his herd of 55 cows from Manor Farm at Thornton Steward to Broadmires and Sykes Farms on the 1st May 1984. Since then he has been delivering milk to the Milk Marketing Board from Broadmires Farm. No formal partnership agreement between him and Lord Swinton has been produced to us. We hold that it has been orally agreed that a partnership will commence in October 1984 (that is to say, at a future date). We understand that it has also been agreed that, under the proposed partnership agreement, either party will be able to terminate the partnership on 12 months notice, whereupon Mr Stubbs will be required to vacate Broadmires and
D Sykes Farms with his animals and his belongings. Mr Stubbs has not yet moved into either of the farm buildings as Broadmires Farm building is now being renovated for him to occupy.

5. Mr Stubbs initially applied for the milk quota of 447,776 litres or thereabouts allocated, or provisionally allocated, in respect of Manor Farm to be transferred to Broadmires Farm. We have no power to direct any such transfer on any grounds.

6. Alternatively Mr Stubbs now applies for a quota of 447,776 litres per annum to be determined for himself at Broadmires Farm on the grounds of exceptional hardship. We decline to make any such determination. First, in our judgment, the "producer" at Broadmires Farm for the purposes of the Dairy Produce Quotas Regulations 1984 (SI No 1047) is Mr Stubbs and Lord Swinton, as partners. In our view, such "producer" did not, prior to 2nd April 1984, enter into any transaction or arrangement, or become bound to
E enter into any transaction or arrangement, of the nature specified in provision (a) of sub-paragraph 17(3) of the Second Schedule to such Regulations. Furthermore we have no evidence that either partner individually entered into such a transaction or arrangement prior to that date, excepting the purchase by Mr Stubbs of the bulk tank milking equipment (which remains his property) and the minor alteration done by him after 2nd April 1984 to the cow byre and milk house. In our judgment such matters by themselves are not sufficient to satisfy the requirements of such provision (a).

7. This Tribunal consider that, in its judgment, neither Mr Stubbs nor Lord Swinton either individually or together as partners has or have suffered, or will suffer, exceptional hardship in comparison with producers in general.

8. In our considered judgment, it is neither fair nor reasonable that quota should be determined for a producer in respect of a holding where (a) the holding was not in milk production on the 2nd April 1984, (b) the producer is
F a partnership which has not yet commenced in business, (c) the partnership has no security of tenure, and (d) the partner carrying on the day to day farming operations on the holding with his cows can be required to leave it at 12 months notice or less, and (e) it is alleged that exceptional hardship is being or will be suffered by the partner carrying on the day to day farming operations on the holding, which he began to occupy only after the 2nd April 1984 without then being bound or obliged so to do.

The NFU's advice

Miss Smith set out her conclusions in respect of the tribunal's decision in a memorandum which was passed on to Mr Stubbs.

Miss Smith made a number of criticisms and raised a number of queries in respect of the tribunal's decision, but the most important passages in her advice were as follows:

I do not think that Mr Stubbs would be very well advised to challenge the G
decision of the Dairy Produce Quota Tribunal by an application to the High Court. My reasons for saying this are that, although the Tribunal may have been wrong in some of its conclusions (which I will deal with later) I do not think these errors, if indeed they are errors, have affected the decision it has reached. Furthermore, even were a Court to direct the Tribunal to re-consider the matter the Tribunal could still reach the same conclusion having correctly appreciated the facts, because the determining factors in granting exceptional hardship status are that the producer has suffered exceptional hardship in comparison with producers in general and that it is fair and reasonable that quota should be given. The Tribunal were quite clear in their decision that it is not fair and reasonable that quota should be determined for a producer in respect of that holding (see paragraph 8 of the decision).

It is true that Mr Stubbs did make an arrangement which has caused his quota to be substantially less than it otherwise would have been, that H
arrangement being the giving of notice and termination of the partnership with Mr A J C Crossley. However, in order to rely on this arrangement as a qualification for an exceptional hardship claim Mr Stubbs needs to show an intention to go into or remain in business as a producer, such intention to have existed at the time the arrangement was made. Although this may have been Mr Stubbs' intention there is no evidence that it was. The keeping of cows would not be sufficient as these could always be sold at any time. There is nothing in Mr Stubbs' statement attached to the documents sent to me indicating that there is evidence of such intention.

. . . the only transaction or arrangement which seems to have been entered into prior to 2nd April 1984 is the dissolution of the old partnership and there appears to be no evidence shown to the Tribunal that on that dissolution Mr Stubbs intended to remain in business as a milk producer.

I am sorry I cannot really offer any hope to Mr Stubbs in the application he J
has made for secondary quota but I shall be happy to advise further if you feel this may be helpful.

Mr Stubbs' case

In his helpful written argument Mr Seymour QC set out Mr Stubbs' case for an allocation of quota under para 17(3) of Schedule 2 to the UK regulations as follows:

(i) Mr Stubbs was a person to whom Article 4(1)(c) of Regulation 857/84 applied because he was, as at 24th July 1984 and thereafter until he entered into partnership with Healey, himself undertaking farming as his main occupation;

(ii) Mr Stubbs had, before 2nd April 1984, entered into a transaction or made an arrangement as a result of which his wholesale quota was substantially less than it would have been had he not entered into that K
transaction or made that arrangement in that he had in April 1983 given notice to Mr Crossley to terminate the partnership between them and in consequence had no entitlement to wholesale quota, whereas otherwise he would have been entitled with his partners to the benefit of the wholesale quota allocated to the Second Farm;

(iii) at the time of giving notice to Mr Crossley to terminate the partnership between them Mr Stubbs intended to go into or remain in business as a produer;

(iv) Mr Stubbs received no benefit from the termination of his partnershp with Mr Crossley which was reasonably commensurate with him being left with no allocation of wholesale quota;

(v) it was fair and reasonable that he should be allocated wholesale quota.

Mr Seymour submitted that the following errors of law were apparent on the face of the tribunal's decision: L

(i) The Tribunal, having found as a fact in paragraph 4 that it had been orally agreed that a partnership between Mr Stubbs and Lord Swinton would commence in October 1984, erred in holding, in paragraph 6, that the Exceptional Hardship Claim fell to be determined on the basis that the "producer" at the date of the Exceptional Hardship Claim and/or at the date of the Decision was not Mr Stubbs alone but a partnership of Mr Stubbs and Lord Swinton.

(ii) The Tribunal failed, notwithstanding the facts found at paragraph 2, to consider whether the giving by Mr Stubbs of the notice to terminate the partnership with Mr Crossley was such transaction or arrangement as satisfied the requirements of paragraph 17(3)(a) of Schedule 2 to the UK Regulations and thus failed to hold, as it should have, that the giving of such notice did satisfy such requirements.

(iii) The Tribunal, having made the error set out at (i) above, failed to take M
into account relevant matters and/or took into account irrelevant matters in relation to the issues of exceptional hardship in comparison with producers in general and whether it was fair and reasonable that the wholesale quota be allocated in particular in that it

(a) failed to recognise that the consequence of refusal of the Exceptional Hardship Claim was that Mr Stubbs would have to cease dairy farming, which producers in general would not have to do;

(b) relied in support of its finding as to what was fair and reasonable upon the erroneous view that "the producer is a partnership which has not yet commenced in business".

He submitted that Miss Smith's advice was negligent in that, having identified the relevant transaction or arrangement as being the notice to terminate the partnership, she wrongly discounted this

A on the basis that there was no evidence before the tribunal that Mr Stubbs intended, at the time of giving the notice, "to go into or remain in business as a producer".

Was the tribunal's decision open to challenge?

Miss Smith was called to give evidence for the NFU and sustained prolonged examination and cross-examination. She had been responsible for dealing with all cases referred to the legal division of the NFU in relation to milk quotas, running to hundreds, and I suspect she was accurate in stating that there was no lawyer with more experience of this complex field than herself. At times both counsel and I were tempted to question her more as if she were an expert witness of foreign law than a witness of fact. Miss Smith impressed
B me highly as a competent and conscientious lawyer. In evidence she sought to justify her advice that the tribunal's decision was not vulnerable to challenge both by affirming the reasons given in her memorandum and by advancing other grounds upon which the decision could be supported. I was not persuaded that Miss Smith had all these points in mind when she advised Mr Stubbs — indeed she frankly accepted that it was not possible to recollect or reconstruct her precise thought process at the time.

In my judgment, if the only issue in the way of a successful challenge of the tribunal's decision was whether Mr Stubbs had established that, when he gave notice terminating the Manor Farm partnership, he "intended to go into or remain in business as a
C producer", judicial review should have been sought. This was not an issue on which the tribunal ruled and I consider that there was evidence before the tribunal upon which they could have concluded that Mr Stubbs had satisfied this requirement.

There are, however, a number of questions that have to be considered in order to decide whether, on judicial review, the court would, or might, have remitted Mr Stubbs' application for reconsideration by the tribunal. These are best considered in the context of the requirements of para 17 of Schedule 2 to the UK regulations.

Was Mr Stubbs a person to whom article 4(1)(c) of Council Regulation 857/84 applied?

D In some of the transactions with which this case is concerned Mr and Mrs Stubbs were recorded as being individually concerned as partners. In others Mr Stubbs acted in his own name alone. I doubt if this factor of itself would have prejudiced his claim for quota and I propose to proceed on the basis that Mr Stubbs was at all times acting on behalf of himself and his wife.

A claim under para 17 of Schedule 2 could be made only by "a person to whom article 4(1)(c) of Council Regulation 857/84" applied. That article applied to "producers undertaking farming as their main occupation". A producer was, by definition:

(i) a natural or legal person or group of persons;
(ii) farming a holding; and
(iii) selling milk.

E Mr Isaacs submitted that, under the regulations, a producer qualified as such only if he was producing milk from a holding on April 2 1984, the date when the scheme had to be brought into operation. I am unable to accept that this requirement implicitly qualified the definition of producer wherever that word appeared in the regulations. For example, a purchaser of a dairy farm between April 2 1984 and the operative date would, in my judgment, have been entitled to claim the quota relating to that holding pursuant to para 1 of Schedule 2 and to be registered as a producer pursuant to para 14 of that schedule. In the context of article 4(1)(c), however, I accept Mr Isaacs' submission that the producer had to be producing milk from a holding on April 2 1984. Regulation (EEC) No 857/84 dealt with determining reference quantities for producers with effect
F from that date. Article 4(1)(c) dealt with the grant of "additional" reference quantities to producers "in connection with the application" of the formulas for determining the primary reference quantities. It seems relatively clear that such a grant could be made only to producers entitled to primary quotas as at April 2. Support for this conclusion is afforded by para 17 of Schedule 2 to the UK regulations in that para 17(3)(a)(i) speaks of a claimant whose wholesale quota is "substantially less" than it would have been and para 17(5) provides for the grant of "gross additional wholesale quota".

If this conclusion is correct, Mr Stubbs was entitled to make a claim under para 17 only if he was producing milk from a holding on April 2 1984. At that date he was not himself a producer. He and his wife were members of a partnership together with Mr Crossley, which G
partnership was the producer. It is not possible to apportion all, or any part, of the production to Mr and Mrs Stubbs alone. It follows that Mr Stubbs was not in a position to make a claim under para 17.

Was the termination notice a transaction or arrangement that satisfied para 17(3)(a)?

The fact that Mr and Mrs Stubbs belonged to a partnership which produced milk at Manor Farm places additional difficulties in the way of a claim under para 17:

(1) If one treats as the relevant transaction or arrangement the notice to terminate the partnership given by Mr and Mrs Stubbs, it is not possible to demonstrate that this had the effect that *their* wholesale quota was substantially less than if they had not entered H
into the transaction or made the arrangement. Had they not given the notice they would have had no wholesale quota — they would have remained members of the partnership which would have obtained a quota. The effect of the notice was that they left the partnership and the partner who remained in occupation of the holding obtained the quota. This does not, in my judgment, fall with para 17(3)(a)(i).

(2) I do not see how it can be said that the reasonably expected outcome of the notice to terminate the partnership was "a level of wholesale delivery of dairy produce in respect of which . . . wholesale quota is not otherwise capable . . . of being allocated to" Mr and Mrs Stubbs. That result depended, it seems to me, not upon the negative act of terminating the partnership but on the positive J
actions of:

(i) purchasing some of the cattle from the partnership on termination of the partnership; and
(ii) arranging for those cattle to be moved to a farm which had no quota.

I can summarise my conclusions thus far as follows:

(1) Mr Stubbs was not entitled to invoke article 4(1)(c) because he was not a producer as at April 2 1984 entitled to a primary wholesale quota;

(2) the notice to terminate the partnership was not a transaction or arrangement that had either of the consequences provided for by para 17(3)(a) of Schedule 2.

These conclusions, reached after three days of evidence and K
argument, do not reflect the reasons given by the tribunal, which did not expressly consider the notice of termination as a possible arrangement or transaction within para 17. I do not find this surprising. The tribunal were given an unenviable task of attempting to analyse the confused claim advanced by Mr Stubbs in the light of the complex regulations. They started quite logically by considering the identity of the producer on behalf of whom the quota was sought. They correctly identified that producer as the producer of milk at Broadmires Farm. They then went on to conclude that this producer was not Mr Stubbs alone but Mr Stubbs and Lord Swinton as partners. Mr Seymour has criticised this conclusion, but I do not consider that criticism justified. As I understand the position, Mr Stubbs' application to the tribunal was advanced on the basis that he L
had moved to Broadmires Farm pursuant to the proposed partnership with Lord Swinton. It was not suggested that he enjoyed sole occupation under the grazing licence. Even had this been suggested, it would have constituted no more than a temporary arrangement and the tribunal were plainly correct when considering the overall merits of the application to do so on the basis that the proposed long-term relationship was that of a partnership.

The tribunal then went on to consider whether Mr Stubbs and Lord Swinton, either as partners or individuals, had entered into any transaction or arrangement in relation to milk production on the Broadmires Farm holding prior to April 2 1984 that fell within para 17(3)(a). They concluded that they had not — a conclusion which Mr M
Seymour did not seek to challenge.

Excessive hardship

Mr Seymour submitted that the conclusion that Mr Stubbs had not suffered excessive hardship when compared with producers in general was one which no reasonable tribunal could have reached. Strictly, on the findings made by the tribunal, the question of excessive hardship did not arise. Excessive hardship had to result from deprivation of quota or production not covered by quota attributable to a transaction or arrangement within para 17(3)(a). The tribunal, rightly in my judgment, found that no such transaction or arrangement had been made by Mr Stubbs.

There is no doubt that Mr Stubbs has suffered great hardship as a

A consequence of the introduction of milk quotas, but the reason for this merits analysis. Mr Stubbs has not, in his own right, enjoyed a secure occupancy of farmland in recent times. He has relied upon concluding partnerships with landowners in order to earn his living from dairy farming. The scheme of milk quotas is territorial. The quota attaches to and enhances the value of the land, to the benefit of the landowner. Once quotas were introduced, Mr Stubbs would have been exposed in any event to the risk of prejudice. Had he not himself given notice terminating the partnership he would have been at risk of termination by Mr Crossley on 12 months' notice. Such termination would have left him unable to continue to earn his living by dairy farming unless he could find some other landowner with quota
B prepared to enter into partnership with him. Mr Stubbs has suffered prejudice not because he has been deprived of quota but because he chose to leave a partnership that occupied land that had a history of use for dairy farming and was thus entitled to quota. The regulations do not cater for hardship arising in these circumstances. The finding of the tribunal in relation to excessive hardship was, in my judgment, correct.

Fair and reasonable

In dealing with what was fair and reasonable the tribunal made a short summary of the relevant facts. That summary was an accurate one and included the hardship urged on the part of Mr Stubbs. The tribunal held that it was not fair and reasonable that wholesale quota
C should be allocated in those circumstances. Again Mr Seymour submitted that no reasonable tribunal could have reached this conclusion. I do not agree.

In considering what was fair and reasonable the following considerations were material:

1. The overall quota for distribution between all producers was limited.

2. Quotas were based upon historic production of milk on land holdings.

3. The milk produced by Mr Stubbs' cows when farmed by the Manor Farm partnership had procured quota for the Manor Farm holding.

4. If an additional quota were to be granted to cover the milk
D produced by the self-same cows on Broadmires Farm they would, in effect, have been responsible for the procurement of double quota.

5. The benefit of such a quota would attach to and enhance the value of Broadmires Farm. The principal beneficiary of this would be Lord Swinton, who had suffered no special hardship of any kind as a consequence of the introduction of quotas.

6. Mr Stubbs would benefit from the grant of quota to Broadmires Farm only so long as he was permitted to farm there. At the time of the application he had no vested right to do so (other than a short-term grazing licence of which the tribunal was unaware) and the proposed partnership arrangement was to be subject to termination on 12 months' notice.

Most of those considerations were reflected in the tribunal's
E reasoning and I do not consider their conclusion on what was fair and reasonable to be open to attack.

Had Mr Stubbs sought and obtained leave to seek judicial review of the tribunal's decision the court might have found it as hard to get to grips with the complex regulations that are applicable as I have done. But, for the reasons I have given, I consider that the challenge of the tribunal's decision would have been unsuccessful. The conclusion reached and the advice given by Miss Smith were correct, albeit that I have not been able to concur with every step of her reasoning. She was not negligent.

No one who has heard this story, particularly from his own lips, can fail to sympathise with Mr Stubbs and with his feeling that he has
F been hard done by. He has, however, suffered as a consequence of the application of regulations passed to give effect to community obligations and not as a result of a failure properly to apply those regulations.

Court of Appeal G

November 26 1991

(Before Lord Justice DILLON, Lord Justice GLIDEWELL and Lord Justice STUART-SMITH)

PAHL v TREVOR

Estates Gazette June 27 1992
[1992] 25 EG 130

Agricultural holdings — Fixed-term tenancy — Ministerial consent — Whether tenancy granted on terms consented to by minister H

Mr Malcolm James Cockburn was the owner of and farmed Lower Farm, Hillfield, Dorset, in partnership with the defendant, Mr Trevor, until Mr Cockburn's death in March 1984 — The partnership came to an end on Mr Cockburn's death — His executors included his son, Mr Malcolm Cockburn, and his daughter, the plaintiff — Because the plaintiff had a son of 16 and it was envisaged that he would take over the farming following a period at agricultural college, on September 24 1984 a joint application was made by the defendant and Mr Malcolm Cockburn to the Ministry of Agriculture, Fisheries and Food for consent to the grant of a J
five-year tenancy of the farm within the meaning of section 3B of the Agricultural Holdings Act 1948 — By a letter dated September 28 1984 the minister approved an agreement that section 3 of the 1948 Act should not apply to a tenancy to be granted after the date thereof for a period of not less than two years commencing on October 1 1984 and ending not later than September 31 [*sic* 30] 1989 — The tenancy was executed at the end of November and was expressed to commence on October 1 1984 for a term of five years — Upon the refusal of the defendant to give possession when the contractual term ended on September 30 1989, the plaintiff, who by then was K
solely and beneficially entitled to the reversion, brought proceedings for possession — In Weymouth County Court Mr Recorder Boyle dismissed the plaintiff's claim holding that the agricultural tenancy was protected by the Agricultural Holdings Act 1986 because the fixed-term tenancy commenced at the date of its execution and not the expressed date of October 1 1984 and was accordingly not within the terms of the minister's consent — The plaintiff appealed

***Held*: The appeal was allowed, Dillon LJ dissenting — *Per* Glidewell and Stuart-Smith LJJ, protection given to an agricultural tenant by what is now the Agricultural Holdings** L
Act 1986 does not apply where: the tenancy is for a term of not less than two or more than five years; before the grant of the tenancy the prospective landlord and tenant agree that section 3 of the Act shall not apply to the tenancy; the prospective landlord and tenant make a joint application to the minister for his approval; the minister notifies his approval; and the contract of tenancy must be in writing and must indicate in whatever terms that section 3 of the 1986 Act does not apply to the tenancy — Notwithstanding that the term of the lease (the legal estate) does not commence until the date when the agreement for lease is executed, other obligations, such as payment of rent, may commence from the date stated in the M
lease — It was common ground that the term in this case resulting from the agreement for tenancy entered into at the end of November 1984 commenced at that time and ran until September 30 1989, but the obligation to pay rent was backdated to start from October 1 1984 — All the requirements of what is now section 5 of the 1986 Act were satisfied — The issue turned not so much on any proposition of law but upon the proper construction of the words of the minister's approval — This approval did not specify the date upon which the tenancy was to be granted but made it clear that the minister was consenting to the grant of the tenancy for

A **any period which fell within the ambit of what is now section 5, that is to say not less than two or more than five years — The minister did not require that the term should commence on October 1 1984, three days after his consent — The minister did not use the expression "a *term* of not less than two years" but referred to "a *period* of not less than two years" — There was no reason why the minister should have wished to grant a consent to a tenancy which would commence on a precise date**

Dillon LJ dissenting: The consent of the minister required the term to be for not less than two years commencing on October 1 1984 — The word "period" in the consent was to be
B **read as meaning the same as "term" — "A term of not less than two years" — The wording of the minister's consent had been drawn to follow the wording of the Act**

The following cases are referred to in this report.

Bradshaw v *Pawley* [1980] 1 WLR 10; [1979] 3 All ER 273; (1979) 40 P&CR 496; [1980] EGD 100; 253 EG 693
Cadogan (Earl) v *Guiness* [1936] Ch 515
Colton v *Becollda Property Investments Ltd* [1950] 1 KB 216
Gladstone v *Bower* [1960] 2 QB 384; [1960] 3 WLR 575; [1960] 3 All ER 353; (1960) 58 LGR 313, CA
Javad v *Aqil* [1991] 1 All ER 243; [1990] 2 EGLR 82; [1990] 41 EG 61
Keen v *Holland* [1984] 1 WLR 251; [1984] 1 All ER 75; (1984) 47 P & CR 639; [1984] EGD 9; 269 EG 1043, CA
C *Roberts* v *Church Commissioners for England* [1972] 1 QB 278; [1971] 3 WLR 566; [1971] 3 All ER 703, CA

This was an appeal by the plaintiff, Mrs Janice Pahl, from an order of Mr Recorder Boyle, sitting in Weymouth County Court whereby on March 14 1991 he dismissed the plaintiff's claim for possession of Lower Farm, Hillfield, Dorset, holding that the defendant, Roger Trevor, had a tenancy protected by the Agricultural Holdings Act 1986.

Michael Barnes QC and John Grove (instructed by Batten & Co, of Yeovil) appeared for the appellant; Joanne Moss (instructed by Porter Bartlett & Mayo, of Yeovil) represented the defendant.

D In dissenting, **DILLON LJ** said: This is an appeal by the plaintiff in the action, Mrs Janice Pahl, from an order made by Mr Recorder Boyle sitting as a deputy county court judge in Weymouth County Court on March 14 1991. By that order the learned recorder dismissed with costs the plaintiff's claim for possession of some land known as Lower Farm, Hillfield, Dorset.

The essential question in the proceedings is whether Mrs Pahl is entitled to rely on the provisions for the grant of tenancies of agricultural holdings to which the security of tenure provisions of the Agricultural Holdings Act 1986 will not apply, which were first introduced by section 3B of the Agricultural Holdings Act 1948 (brought in by the Agricultural Holdings Act 1984) and which are now contained in section 5 of the Agricultural Holdings Act 1986.

E The circumstances of the present case are that Lower Farm was owned by a Mr Malcolm James Cockburn of Hillfield Manor, Cerne Abbas, Dorset, who died on March 11 1984. During his lifetime Mr Cockburn had farmed Lower Farm in partnership with Mr Trevor, the defendant, but the partnership came to an end on Mr Cockburn's death. The partnership year ran to September 30. Mr Cockburn's executors were his son, Mr Malcolm Cockburn, and his daughter, Mrs Pahl. They proved his will in May 1984. They had then to consider what should be done with Lower Farm.

Mrs Pahl was entitled to Lower Farm under the terms of the will. She sues now as sole plaintiff as a result of an assent in her favour subsequent to the creation of the tenancy in favour of Mr Trevor to
F which I shall come. She had children and in particular a son of 16 and the idea was that the son would read agriculture at an agricultural college and then take over the farming of Lower Farm, which was a dairy farm. It was therefore thought a good idea to take advantage of the new section 3B, which came into force on September 12 1984. That section provides as follows:

(1) This section shall apply to any tenancy of an agricultural holding for a term of not less than two, and not more than five, years.

(2) Where, before the grant of any such tenancy —

(*a*) the persons who will be the landlord and the tenant in relation to the tenancy agree that section 3 of this Act shall not apply to the tenancy; and

(*b*) those persons make a joint application in writing to the Minister for his approval of that agreement; and

(*c*) the Minister notifies them of his approval; G

section 3 shall not apply to the tenancy if it satisfies the requirements of subsection (3) of this section.

(3) A tenancy satisfies the requirements of this subsection if the contract of tenancy is in writing and it, or a statement endorsed upon it, indicates (in whatever terms) that section 3 of this Act does not apply to the tenancy.

Those being the provisions of the section, Mr Malcolm Cockburn wrote to the divisional executive officer of the Ministry of Agriculture, Fisheries and Food in Taunton a letter which actually bears no date but is believed to have been written on or about September 20 1984. It reads as follows:

Short Term Tenancies: Lower Farm, Hillfield, Dorchester

I am one of the two Executors to the Will of my father, M J Cockburn. He farmed Lower Farm, Hillfield, in partnership with Mr R J Trevor (of Church H Farm [and another village is named]). My sister, Mrs J M Pahl is the beneficiary of Lower Farm under the will and she has children at the late school/University age.

It seems to us, with the agreement of Mr Trevor, that a short term tenancy would be suitable in this situation. We would like the tenancy to last for the maximum, five years. There is some urgency because the financial year for the old partnership finishes at the end of September, but our solicitors

— and he gives the name of a local firm —

say that they will be able to complete the agreements by that date so long as the MAFF give their consent by that date.

I telephoned your department in Taunton and they suggested that I write to you, with our case and two maps of the farm, which are enclosed. I apologise for the pressure on you. J

That letter of course was not a joint application, but the joint application followed. It is dated September 24 1984 and bears a stamp showing its receipt in the ministry offices in Taunton on September 26. It reads as follows:

Agricultural Holdings Act 1948 (1984) Sec 3b Lower Farm, Hillfield, Dorchester, Dorset

This application for a 5 year tenancy on Lower Farm, Hillfield is made by Mr R P Trevor (the tenant) and Mr M D Cockburn (executor to the Will of M J Cockburn). As explained in M D Cockburn's previous letter, Lower Farm, Hillfield was farmed by Mr Trevor in partnership with the late M J Cockburn as a dairy farm with 135 cows. The farm has been left to Mrs J M Pahl (daughter) who has a son aged 16. It is the intention that this boy will take over the farm at the age of 21 after attending college. Meanwhile Mr Trevor is able K and willing to maintain the farm as a dairy farm.

Since the old partnership ends its financial year at the end of September, we would very much appreciate your quick decision.

The ministry obliged with admirable dispatch and granted, as it seems to me, what had been asked for. There is a covering letter of September 28 1984, which as it happens was a Friday. This is addressed to Mr M D Cockburn. It refers to his application with Mr Trevor, which I have just read, and goes on:

I am pleased to tell you that approval has been given and is enclosed, together with a copy of the plan. I would remind you that the tenancy agreement must state, or be endorsed to the effect that Section 3 does not apply.

A copy of this letter has also been sent to Mr Trevor. The consent in its L operative parts is in these terms:

Approval by the Minister of an agreement under section 3B.

In exercise of the powers conferred upon him by section 3B of the Agricultural Holdings Act 1948 as amended, the Minister of Agriculture, Fisheries and Food hereby approves (but only for the purposes of the said section) the agreement between Malcolm D Cockburn and Roger P Trevor that Section 3 of that Act shall not apply to a tenancy to be granted after the date hereof by the said Malcolm D Cockburn to the said Roger P Trevor of the land at Lower Farm

— and it is then described —

for a period of not less than two years commencing on the 1st day of October 1984 and ending not later than the 31st day of September 1989.

In the absence of evidence that the Taunton area lives by a different M calendar to the rest of the world, the last date must be taken to be the day September 30 1989.

The position was, however, contrary to what Mr Malcolm Cockburn had supposed in his original letter to the department, that the parties had not in truth agreed all the terms of their tenancies; they had agreed that there should be a tenancy which should be exempted from the statutory protection under section 3B by the minister's approval, but they had not agreed the rent or some of the other terms.

Having therefore got the approval, they proceeded to negotiate the remaining terms while Mr Trevor no doubt continued farming the land. The tenancy agreement ultimately achieved is dated back so as to bear the date "1st day of October 1984" but was in truth not

A executed until the end of November 1984. It provides that the landlord lets and the tenant agrees to take the holding known as Lower Farm, Hillfield, and it provides in clause 1 that:

> THE TENANCY shall commence on the First of October . . . (1984) and continue for a term of 5 years expiring on the 30th September 1989.

Clause 2 provides that:

> THE RENT is to be seven thousand five hundred POUNDS PER ANNUM payable on March 25 and September 29 in each year the first payment to be made on the twenty fifth of March . . . 1985:

There then follow all the usual detailed provisions that you would expect in an agricultural tenancy and there is the appropriate note to satisfy subsection (3) of section 3B.

B When the contractual term came to an end at September 30 1989, Mrs Pahl, who by then was entitled to the reversion, brought immediate proceedings for possession against Mr Trevor as she was entitled to if Mr Trevor was not protected by the Agricultural Holdings Act 1986 as it is now. If, however, as the recorder held, he was entitled to that protection, then Mrs Pahl's claim for possession was bound to fail because a tenancy for a term of years which is protected by the Agricultural Holdings Act continues as a periodic tenancy after the expiration of the term of years and can be terminated only by a much longer notice to quit, which Mrs Pahl had not given, and by a consent from the agricultural land tribunal if the tenant objects to giving up possession.

C The contention of Mr Trevor, which the learned recorder upheld, was quite simply:

> It is well settled [law] that a lease [or tenancy agreement] cannot retrospectively vest an estate in the lessee. If today a lease is granted for seven years from this day a year ago, no term of seven years is brought into being, but only a term of six years from today.

I read from the judgment of Sir Robert Megarry V-C in *Bradshaw* v *Pawley* [1980] 1 WLR 10* at p 14E. He refers to earlier authority in 1847. There are also many other cases where the same principle has been applied, such as the decision of Clauson J in *Cadogan (Earl)* v *Guinness* [1936] Ch 515 and, even more to the point, the decision of this court in a case in relation to the Agricultural Holdings Act 1948, *Keen* v *Holland* [1984] 1 WLR 251†.

D Mr Trevor's point is, in short, this: that a tenancy agreement made at the end of November 1984 which purports to create a term commencing on October 1 1984 and continuing for a term of five years expiring on September 30 1989 in law creates a term starting only on the execution of the tenancy agreement and continuing until September 30 1989. The term cannot, because of the principle that Sir Robert Megarry enunciated, be made to commence at a date prior to the execution of the tenancy agreement. Therefore, as is accepted by Mr Barnes QC, in his submissions for the appellant, the position of Mr Trevor from October 1 1984 until the tenancy agreement was executed and took effect was that he was not a tenant but a tenant at will of the property: see the recent decision of this court in *Javad* v *Aqil* [1991] 1 All ER 243§.

E Mr Trevor therefore says that the condition of the minister's consent, that the section is not to apply to a tenancy "for a period of not less than two years commencing on the 1st day of October 1984 and ending not later than the 30th day of September 1989", is not satisfied because the tenancy which Mr Trevor was granted was for a term which commenced at some date at the end of November 1984 and not on the first day of October 1984.

Section 3B refers to a "tenancy of an agricultural holding for a term of not less than two, and not more than five, years" and the term actually granted to Mr Trevor by the tenancy agreement at the end of November was indeed for not less than two and not more than five years from the date when the term was granted. That is not the point.

F The point is that the consent of the minister required the term to be for two years commencing on the first day of October 1984. I read the word "period" in the consent "for a period of not less than two years" as meaning the same as "term" — "a term of not less than two years".

As I have said, in *Keen* v *Holland* this principle of construction was held to be applicable to agricultural holdings. The position in that case was that it was an established exception to the Agricultural Holdings Act that there was no protection given where a tenancy was granted for a fixed term which was of more than one year but less than

G two years' duration. That followed from a decision of this court in *Gladstone* v *Bower* [1960] 2 QB 384. That decision was regarded as somewhat anomalous but no legislative step was taken to counter it. Therefore, in *Keen* v *Holland* the parties intended, with the full concurrence of the tenant, to avoid — and I do not use the word pejoratively — the protection conferred by the Act for the tenant by granting him a term for more than one year but less than two years. However, there was delay in the preparation of the necessary agreement while negotiations took place and the tenancy agreement actually granted, though dated back to the date from which it was intended that the period of more than one year but less than two years should run, was actually granted only a few weeks before the end of that period. It was consequently held by this court that the term

H granted was a term for very much less than one year. It was merely a term of some two weeks from the date when the tenancy agreement was executed until the expiration of the intended term and so the tenant was entitled to the protection of the 1948 Act.

Oliver L J, who gave the leading judgment in this court, referred at the foot of p 258 to an alternative argument which was put forward by counsel then appearing for the landlord. He said:

> As an alternative — although it is, we think, merely a different way of expressing the same concept — [counsel] submits that where (as here) the tenant is already in occupation before the formal grant of a tenancy, a tenancy agreement for a term commencing from a prior date is to be treated as clothing [the tenant], as between the parties, with the status of tenant under the agreement as from the agreed commencement date. His "interest" for the purposes of the Act is, therefore, to be determined in accordance with that agreed status. Terms of a tenancy are frequently negotiated well prior to the execution of a formal tenancy and it is a commonplace that the tenancy when formally granted is expressed as creating a term commencing on a date prior to the date of the instrument. Whilst it is no doubt true that no estate is created until the instrument is executed, the agreement as to the duration of the term is effective for all other purposes, for instance, as regards payment of rent and covenants in the lease. J

The court therefore rejected that alternative way of putting the case and held that the tenant was entitled to protection. If one comes to section 3B(1) of the 1948 Act, one finds that that is to apply to any tenancy of an agricultural holding for a term of not less than two and not more than five years. If, therefore, what the parties had purported to do in their agreement of late November 1984 had been to K grant Mr Trevor a tenancy for a term of two years from October 1 1984, that would have been outside the Act and would not have avoided giving him the protection of the Act.

It necessarily follows, to my mind, that the wording of the minister's consent has been drawn to follow the wording of the Act. There is emphasis that the tenancy is to be granted after the date of the approval. It is to be for a period which, as I have said, I read as "term of not less than two years and ending not later than 30th September 1989". It is also, however, to be for a term of not less than two years commencing on the first day of October 1984. As a matter of law it can only be for a term commencing on the first day of October 1984 in that sense, if the agreement was executed on or before October 1 L 1984.

It may well be that the date, October 1 1984, was put into the consent because it had been stressed in Mr Malcolm Cockburn's original letter and in the joint application that expedition was desired because the old partnership ended its financial year at the end of September. But while that may be the reason for having chosen that date and while it may well be the case that the minister might have granted a consent which did not actually specify the date for the commencement of the term, what he has given, and it is the only approval that he gave before the grant of the tenancy, was approval for the grant of a tenancy for a period of not less than two years commencing on the first day of October 1984. In law the term under the tenancy agreement actually entered into commenced not on the M first day of October 1984 but on a day at the end of November. Therefore, the terms of the approval have not been satisfied. What has been done is something outside the scope of the approval granted.

In my judgment, therefore, and speaking for myself alone, the tenant is entitled to the protection of the Act and I would dismiss this appeal.

Mr Barnes raised an alternative submission to the effect that it was, if the consent bears the meaning which, in my judgment, it does bear, unreasonable for the minister to have granted a consent which required the execution of the tenancy agreement on September 29 or 30 or on October 1. But the minister's representative in Taunton had been told by Mr Malcolm Cockburn in his earlier letter,

---

*Editor's note: Also reported at (1979) 253 EG 693.
†Editor's note: Also reported at (1984) 269 EG 1043.
§Editor's note: Also reported at [1990] 2 EGLR 82.

A unfortunately incorrectly, that the solicitors would be able to complete the agreements by the end of September so long as the ministry gave their consent by that date and I cannot regard the minister as having acted unreasonably in a *Wednesbury* sense in acting as he was requested to act in the letters he had received from Mr Cockburn.

I lay aside also any question of whether *Wednesbury* unreasonableness is a concept applicable to the construction of consents given by the minister outside the field of judicial review in the field of ordinary litigation between private people.

As I have said, for my part I would dismiss this appeal.

Giving judgment, **Glidewell LJ** said: Section 3B of the
B Agricultural Holdings Act 1948 as inserted by the Agricultural Holdings Act 1984 and repeated, though not in identical words, in section 5 of the Agricultural Holdings Act 1986 provides that the protection given to an agricultural tenant by section 3 of the 1948 Act shall not apply to a tenancy of an agricultural holding in the following circumstances. First, the tenancy has to be one for a term of not less than two or more than five years. Second, by subsection (2) three conditions which are cumulative have to be fulfilled: that before the grant of the tenancy the prospective landlord and tenant should agree that section 3 shall not apply to the tenancy, that the prospective landlord and tenant must make a joint application to the minister for his approval of that agreement and that the minister must notify his
C approval. Finally, the contract of tenancy must be in writing and must indicate in whatever terms that section 3 of the Act does not apply to the tenancy.

In this case the agreement for a tenancy, which operated as a lease, was expressed to be entered into on October 1 1984 for a term commencing on the same date. However, it is common ground that that agreement was not concluded nor entered into until a date at the end of November 1984. Mr Barnes submits that this agreement had two effects. First, it created a legal estate, that is to say a term of years and, second, it imposed a variety of obligations on the parties in usual form. Mr Barnes accepts that where, as here, a lease is expressed to commence on a date earlier than the date of its execution, the term of the lease (the legal estate) does not commence until the date when the
D agreement or lease is executed. If authority for that proposition be needed, it is to be found in the decision of this court in *Roberts* v *Church Commissioners for England* [1972] 1 QB 278 as repeated in and relied upon by this court in the judgment of the court delivered by Oliver LJ in *Keen* v *Holland* [1984] 1 WLR 251.

However, Mr Barnes argues, and I agree, that other obligations may commence from the date stated in the lease; for instance, it is a commonplace that rent shall start to be payable from some earlier date or that a rent review date shall be at a given period after such earlier date. In *Keen* v *Holland*, at p 259, Oliver LJ quoted a passage from a judgment of Denning LJ (as he then was) in *Colton* v *Becollda Property Investments Ltd* [1950] 1 KB 216, the particular passage being from p 231 in that decision, where Denning LJ said:

E It continually happens that the tenant goes into occupation whilst the negotiations are still pending and before the lease is executed, and then, afterwards, when the lease is drawn up, it is expressed to start as from the date the tenant went into occupation. In such cases, in point of law the legal interest of the tenant starts from the date when the lease was executed: . . . But as between the parties it may start from the date stated in the lease, because the parties, by their agreement, have related its commencement back to the date when the tenant went into occupation. This relation back is not to be regarded as a nullity; it may often mean that the conditions and covenants of the lease relate back to that time also.

Mr Barnes also relied upon a judgment of Sir Robert Megarry V-C, in *Bradshaw* v *Pawley* [1980] 1 WLR 10, the particular passage, which I forbear to read, being from pp 14B-15B in that decision. It is
F to much the same effect as the passage from the judgment of Denning LJ to which I have just referred.

It follows, and indeed it is common ground, that the term in this case resulting from the agreement for tenancy entered into at the end of November 1984 commenced at that time and ran until September 30 1989; in other words, the duration of the term was some four years and 10 months. But the obligation to pay rent was back-dated to start from October 1 1984, and so, though we have not considered them, some of the other obligations in the agreement may have been.

On the face of it, therefore, all the requirements of section 3B were satisfied. This was a tenancy for more than two but less than five years. The landlord and tenant had agreed that section 3 should not apply. They had applied to the minister for his consent for that agreement. On the face of it he had given his consent and that fact was G
recorded in the agreement for lease itself. The argument advanced by Miss Moss, for Mr Trevor, which convinced the learned recorder, was that the minister's approval was subject to a condition which was not satisfied. Since the condition was not satisfied the approval never took effect.

This, therefore, depends not so much upon any proposition of law as upon the meaning of the condition and of the whole of the minister's grant of approval. Before I turn to that wording I would say that, in my view, the decision in *Keen* v *Holland*, though of course binding on us, does not assist us to construe the words of the minister's approval. *Keen* v *Holland* is authority for the proposition that parties who execute an agreement for a tenancy on a date less H
than a year before the expressed end of the term, although they insert in the agreement a commencement date more than a year before the termination, nevertheless create an interest which is "less than a tenancy for year to year", which therefore falls within the protection afforded by section 2(1) of the 1948 Act and, indeed, the same section in the 1986 Act.

The wording of the minister's consent, to which Dillon LJ has referred and to which it is necessary to go back briefly, having referred to section 3B, referred to the agreement between Mr Cockburn and Mr Trevor as follows:

. . . that Section 3 of that Act shall not apply to a tenancy to be granted after the date hereof by the said Malcolm D Cockburn to the said Roger P Trevor of the land . . . [and then defines the land] for a period of not less than two years J
commencing on the 1st day of October 1984 and ending not later than the 31st day of September 1989.

It is noticeable that in the middle of that consent are the words "a tenancy to be granted after the date hereof", not at that point specifying the date upon which the tenancy is to be granted, and it is clear that the minister was consenting to the grant of the tenancy for any period which fell within the ambit of section 3B, that is to say not less than two or more than five years. The terminal date, September 30 (31st as it reads) was obviously designed with the five years in mind.

The question is: did the minister require that the term should commence on October 1 1984, which itself meant, if it was to take effect, that the agreement for the lease had to be executed not later K
than that date, three days after the minister gave his consent, two days after, presumably, it was received by the parties?

Mr Barnes argues that the passage in which the minister has concerned himself with the question, when should the tenancy be granted, is answered by the words "tenancy to be granted after the date hereof" and he points to the fact that the minister has not referred to a *term* of not less than two years but a *period* of not less than two years. We are not of course here construing a statute and, in my view, the same principles do not apply as apply to the construction of a statute. But, nevertheless, I agree with Mr Barnes that we are entitled to assume that, although the minister was here putting into effect a statute which had only just come into force a few weeks L
before the giving of this consent, nevertheless it is of some significance that he chose a different word from that which is contained in section 3B(1), namely "a term of not less than two and not more than five years" and speaks of a "period". Unlike Dillon LJ, I regard that as being of some significance.

The parties here were no doubt concerned that their obligations under the tenancy, particularly the payment of the rent, should commence on October 1 1984. But I cannot see why the minister should have wished to grant a consent to a tenancy which would commence on a precise date. I can see that he would not wish to grant a consent that would be, as it were, left in limbo for a very long period, that he would hope, as was indeed the case, that it would be acted upon in the near future. But as to whether it was part of his M
concern that the term of the agreement should commence on a particular date, I can find no reason to justify that if indeed it is right to consider the practicalities of this matter.

In my judgment, this tenancy, entered into at the end of November 1984, but expressed to run from October 1 1984, did comply with the condition contained in the minister's consent. It follows that I conclude that section 3 of the 1948 Act did not apply to this tenancy. It is not necessary, therefore, for me to express any view about the unreasonableness argument, but had I done so I would on that part of the matter have expressed my entire agreement with Dillon LJ.

I would therefore allow the appeal and order that the plaintiff recover possession of the land.

A Agreeing, **Stuart-Smith LJ** said: In order to avoid the provisions as to security of tenure afforded to tenants of agricultural land under the Agricultural Holdings Acts, section 3B of the 1948 Act (as amended) and now contained in section 5 of the 1986 Act requires that seven conditions are complied with in addition, of course, to the fact that it has to be a tenancy of an agricultural holding: first, that the term must be for not less than two years; second, that it should be for not more than five years; third, that before the grant the parties must agree that section 3 shall not apply; fourth, that a joint application is made to the minister; fifth, that the minister must consent to it; sixth, that the tenancy is in writing and, finally, that the consent is endorsed on the tenancy agreement. All those conditions
B were complied with in this case.

The question is whether, on the true construction of the consent, the minister in fact imposed a further condition specifying that the term of the tenancy had to begin on October 1 1984. It may well be that in an appropriate case the minister can, if he sees fit, impose further conditions other than those required in the statute. But in this particular case, for my part, I can see no point in his having done so, particularly since he expressed the consent in terms of great flexibility, namely that the term should be not less than two years or not more than five. It is very common to find that leases are executed on a particular day but are expressed to run from some prior date. Where this happens there is no tenancy or estate or interest in land prior to the execution of the document. But the parties can relate the
C time of their obligations to such a date. In *Bradshaw* v *Pawley* [1980] 1 WLR 10 at p 14 between D and E, Sir Robert Megarry V-C said:

> First, it is well settled that a lease cannot retrospectively vest an estate in the lessee. If today a lease is granted for seven years from this day a year ago, no term of seven years is brought into being, but only a term of six years from today. A lease, of course, is more than a mere contract, for it operates by way of grant to create an estate or interest in the land; and you cannot grant today that X shall have had a term of years vested in him a year ago. Whatever contractual obligations there may be between the parties, no actual term of years can be created until the lease has been executed and so the grant has been made. Thus where the question is what term has actually been created, as where a statute refers to terms of a particular length, the commencement of the term cannot be earlier than the date of the grant of the lease. There is, of course, no objection to a lease defining the term by reference to some past
> D date, as in the grant of a term of seven years from this day a year ago; but this merely creates a term of six years from today.
>
> Second, a lease for a term from some past date will not, at any rate normally, make into a breach of covenant that which was not a breach when it was committed. An act or omission by a prospective lessee which would be a contravention of the proposed lease is not turned into a breach of covenant by him merely because he subsequently accepts the grant of the lease for a term which is expressed to run from some date anterior to the act or omission.

Then he refers to the authorities for that proposition and at p 15B he says:

> However, when what is in issue is not to title but obligation or liability, I do not see why the parties should not, if they choose, make the obligation or liability enforceable in respect of such anterior date as they wish. Thus if a lease is
> E granted for a term of 21 years from a date two years prior to the grant of the lease, with a provision for determination at the end of the seventh or fourteenth year of the term, then, in the words of Clauson J in *Earl Cadogan* v *Guinness* [1936] Ch 515, 518: "it is perfectly easy as a matter of construction of such a document to say that the seven years according to the obvious intention of the parties is not to run as from the date of the execution of the lease but from the moment spoken of, though inaccurately, as the beginning of the term in the document".

This case, in my judgment, turns solely on the construction of the consent granted by the minister and it is to be noted, as it seems to me, that if the minister had wished that the grant of the lease should be made from October 1 and no other date, he could easily have said so. In fact he did not use that language but he specified that the tenancy
F was to be granted on a date after the date of the consent itself, which was September 28 1984. I can see no good or rational reason why he should want the parties to exchange lease and counterpart on that date in particular, especially as it would be extremely difficult, having regard to the date of the consent, which was Friday September 28, for that to have taken place. As it seems to me, all he was concerned with was that the tenancy should be granted within the full and wide words of the limits of the statute, namely that it should be for not less than two years or more than five. Second, had he wished that the term should commence on October 1, again, it seems to me, he could have said so and thereby have followed the words of the statute. In fact he did not do so. He used the expression "for a period of not less than two years commencing from the first day of October". I do not think that it is necessary to construe that G
expression as the same as the word "term". In my judgment, the minister must be taken to have known the common practice of backdating leases to a date prior to the actual execution of it, and, in my judgment, that is the likely interpretation or it is likely that he had that practice in mind when he used those words at the end of the consent.

I can see nothing in the language of the statute itself or of the decision in *Keen* v *Holland* which compels a different interpretation. It is quite true that if the parties here had simply entered into a lease for a period of two years from October 1 1984, that lease being executed on the date when it was in November, they would have fallen foul of the principle enshrined in *Keen* v *Holland* and would not have avoided the protection of the Act. But that is not what happened and, H
as I have indicated, the term which in fact was created was less than five years and more than two years. I do not think that the interpretation of this consent requires the imposition of what I would regard as an unreasonable and unnecessary condition by the minister and I would so construe the consent. For those reasons I, too, would allow the appeal.

Petition to the House of Lords for leave to appeal was dismissed.

County Court Decision J

SURREY COUNTY COUNCIL v MAIN AND ANOTHER

August 12 1991

(Before Judge SLOT, sitting at Reigate County Court)

Estates Gazette February 15 1992

[1992] 06 EG 159

Compensation for milk quota — Termination of agricultural tenancy — Arbitrator's award — Meaning of para 6 of Schedule 1 to the Agriculture Act 1986 — Dispute concerned calculation of "standard quota" as defined in para 6 of K
Schedule 1 to the 1986 Act — The tenant contended that the "standard quota" is determined by multiplying the relevant number of hectares by the prescribed quota per hectare — The council contended that para 6(2) of Schedule 1 to the 1986 Act applied

The following case is referred to in this report.

Grounds v *Attorney-General of the Duchy of Lancaster* [1989] 1 EGLR 6; [1989] 21 EG 73, CA

Christopher Priday QC (instructed by the solicitor to Surrey County Council) appeared for the applicant council; Andrew L
Densham, of Burges Salmon, of Bristol, represented the first respondent, James Baird Main; the second respondent, the arbitrator Mr Richard Arthur Law, did not appear and was not represented.

Giving judgment, **Judge Slot** said: An arbitration had been conducted by Mr Richard Arthur Law FRICS FAAV under the provisions of para 11 of Schedule 1 to the Agriculture Act 1986 between Surrey County Council as landlords and James Baird Main as tenant, and was concerned with the assessment of the amount of compensation payable by the council to Mr Main for milk quota, as provided by that Act. In the event, the arbitrator accepted the tenant's submissions made in the course of the arbitration and M
awarded Mr Main the sum of £63,290.48 as the statutory compensation due to him under the scheme set out in the Act.

Mr Main became the tenant of the council under a tenancy from year to year from September 29 1980 of an agricultural holding, named Holmlea Farm at Horne, near Horley, Surrey. He surrendered that tenancy in March 1987 by agreement and served a notice claiming compensation for milk quota pursuant to the statute. He claimed a sum in excess of £72,000. The council did not challenge his right to compensation, but the sum could not be agreed, so that the matter went, in accordance with the scheme set out in the statute, to arbitration. Mr Law was appointed the arbitrator in October 1988, but, for reasons with which I am not concerned and about which no

A one has complained, the matter was not heard until December 1990. The award was made on March 21 1991. The council applied to set aside the award by an application to this court dated April 10 1991. The tenant's answer was filed on May 3 1991. The court gave directions on June 18 1991 and the matter was heard on July 30 1991, this judgment being delivered on August 12 1991. It is irrelevant, but satisfactory, that the court, despite the many pressures on it, has been able to deal with this substantial matter a great deal more quickly than the arbitrator was able to do.

Judge Slot said that compensation for milk quota was a comparatively new concept in the law and was the child of statute. In March 1984 a system of milk quotas was agreed, whereby each
B member state of the EEC was allocated, and agreed to produce no more than, an agreed national quota of milk, the plan being that, as the years went by, the national quota would gradually be reduced, so that the quantity of milk produced in the community would become lower.

In England and Wales each milk producer was allocated a quota expressed in litres of milk per base year, the base year being the calendar year 1983: and the quota was tied to the land occupied by the producer, so that, if the occupier of the land changed, the quota, and the right to produce and sell the quantity of milk contained in the quota, was transferred to the newcomer, while the outgoing farmer took no right with him. This was seen as being of little concern to farmers who owned the freehold and could buy and sell as they chose.
C But in 1986 Parliament took the view that a tenant dairy farmer who left his land and his quota behind with his landlord would thus suffer hardship, for which he should be compensated by the landlord. This compensation was first provided by the Agriculture Act 1986, which set out the right to compensation for milk quota and the method by which it was to be calculated.

The method of calculation of the compensation payment in this case is set out in para 5(2)(*a*) of Schedule 1 to the Act and involves mathematical calculations and the use of specific statutory terminology. For this case it provides:

> The amount of the payment . . . shall be an amount equal . . . to the value of the sum of —
> (i) the tenant's fraction of the standard quota, and
> D (ii) the amount of the excess;

The excess is the amount by which "the allocated quota exceeds the standard quota for the land".

The terms "tenant's fraction", "allocated quota" and "standard quota" are all statutory terms with specific statutory meanings and each has to be calculated in accordance with the statute. In this instance the calculation of the "tenant's fraction" creates no difficulty. It has been agreed at 9.5%. Likewise in this case there is no difficulty in calculating the amount of the "allocated quota". For the material year, it is agreed at 379,317 litres.

The problem in this case lay in the calculation of the "standard
E quota": and that term is defined in para 6 of Schedule 1 to the statute. For material purposes that paragraph provides:

> 6(1) . . . the standard quota for any land shall be calculated by multiplying the relevant number of hectares by the prescribed quota per hectare . . .
>
> (2) Where by virtue of the quality of the land in question or climatic conditions in the area the amount of milk which could reasonably be expected to have been produced from one hectare of the land during the relevant period ("the reasonable amount") is greater or less than the prescribed average yield per hectare, then sub-paragraph (1) shall not apply and the standard quota shall be calculated by multiplying the relevant number of hectares by such proportion of the prescribed quota per hectare as the reasonable amount bears to the prescribed average yield per hectare; and the Minister shall by order prescribe the amount of milk to be taken as the average yield per hectare for the purpose of this sub-paragraph.

F As Glidewell LJ has observed in the leading case of *Grounds* v *Attorney-General of the Duchy of Lancaster* [1989] 1 EGLR 6; [1989] 21 EG 73, subpara (2) read at speed does not sound very clear. What is clear, though, is that subpara (2) provides for an entirely different calculation from that set out in subpara (1) when one comes to calculate the amount of the "standard quota", a calculation necessary if the compensation is to be quantified: and whether one applies subpara (1) or subpara (2) the first essential is to ascertain "the relevant number of hectares", which is a multiplicand common to both methods of calculation. In this case, the "relevant number of hectares" has been agreed at 24.

The "prescribed quota per hectare", which is the multiplier in the calculation under subpara (1), is defined in the relevant statutory instrument, in this case the Milk Quota (Calculation of Standard G
Quota) Order 1986, and is there quantified at a figure of 8,000.

The tenant contended that the "standard quota" in this case is simply calculated by muliplying the relevant number of hectares (24) by the prescribed quota per hectare (8,000), which produces a figure of 192,000, from which must be deducted a sum referable to the "development quota awarded but not allocated" of 36,764. On this calculation, which the tenant put forward and the arbitrator found to be correct, the standard quota is 155,236 litres.

The landlord, however, argued that para 6(2) applied, so that the standard quota was at a higher figure of 208,484 litres, which reduced the excess by which the allocated quota exceeded the standard quota and thus reduced the amount qualifying for compensation. It was to H
be noted in this context that the higher the standard quota is, the lower the compensation will be, since the excess (the amount by which the allocated quota exceeds the standard quota), which is the principal element in the compensation calculation in this case, will decrease as the standard quota rises.

The judge said it was therefore necessary to determine whether the standard quota should be calculated in accordance with para 6(1) of the Schedule or para 6(2). The arbitrator made his calculations on the basis of para 6(1) and this is the error of law on the face of the award on which the landlord relied.

The issue turned on the proper construction of para 6(2) and, in his judgment, its meaning was clear. It provides that, if a certain state of affairs exists, the alternative method of calculation, set out in para J
6(2), must be adopted. That state of affairs exists where, by virtue of the quality of the land in question, the amount of milk reasonably expected to be produced from 1 hectare of that land ("the reasonable amount") differs from the statutorily prescribed average yield per hectare. That state of affairs also exists where, by virtue of the climatic conditions in the area, the reasonable amount differs from the prescribed average yield per hectare. It is only, as he read it, if one or other such state of affairs exists (or both) that the para 6(2) calculation comes into play. If neither exists, then the para 6(1) calculation comes into play.

The judge said, in his judgment, for either state of affairs to exist, the difference between the reasonable amount and the prescribed average yield must have been caused either by the quality of the land K
(it may be particularly rich land, or particularly poor land, or something of the kind) or by the climatic conditions in the area (it may be particularly rainy, particularly sunny, or particularly liable to frost or something of the kind). The words "by virtue of" mean "by reason of" or "because of": and unless the difference between the reasonable expected yield and the statutorily prescribed average yield is brought about by one of the two specified causes, then para 6(2) does not come into play.

It was argued for the landlord that the para 6(2) test always comes into play, if for no other reason because one cannot tell whether there is a difference between the reasonable expected yield and the statutorily prescribed average yield, unless one makes the L
calculation: and, unless one knows that there is a difference and what that difference is, one cannot ascertain what its causes may be. The judge disagreed. It may well be open to the arbitrator to make the para 6(2) calculation, see if there is any material difference and then look to see if the difference is due, wholly or in part, to the quality of the land or the climatic conditions. But he need not do so if he is satisfied, on the evidence without more, that the quality of the land and the climatic conditions are not so unorthodox that they would affect the milk yield either way. After all, the skill and pertinacity of the farmer, the nature of the artificial feed which he gives to his stock, and the natural productivity of the cows he places on the land are just as likely, if not more so, to cause the milk yield to rise or fall on ordinary grazing land in ordinary grazing country as anything else: M
and these other factors were, in his view, specifically excluded from consideration, when the choice is made between the para 6(1) and the para 6(2) methods of calculation.

It seemed to the judge, furthermore, that there was sense in this. Whether the compensation system is wise or unwise, fair or unfair, the object of the scheme is to compensate the departing tenant farmer by giving him recompense for the things he leaves behind him when he goes. When he goes he takes with him his farming skills, his ability to work hard, his experience with artificial feeds and all his knowledge as to how to make a dairy farm productive. He also takes with him, in all probability, his cattle, whether they be high-yielding stock or low-yielding stock. He needs and deserves no compensation in respect of these things. He takes them with him. He leaves behind, however, the milk quota. He also leaves behind the land, with its good characteristics and its bad characteristics. He also leaves behind the climatic conditions of the area, be they helpful or unhelpful. These elements, which he leaves behind, redound to the benefit or disadvantage of the landlord and are the elements which the departing tenant farmer has, by departing, lost. If Parliament has decided that his compensation is to be altered from the orthodox para 6(1) computation in certain special circumstances, it seems sensible that those special circumstances should arise from the things of which the farmer's departure has deprived him, rather than from the things which he has taken with him.

There are two further arguments in favour of the tenant, which seemed to him to be irresistible. The first is that, if the landlord's argument is correct, the para 6(1) test will never come into play at all, unless by chance the reasonable expected yield and the statutorily prescribed average yield happen to coincide. There would be no purpose in including para 6(1) in the Schedule at all, let alone in putting it before para 6(2), if this argument were right: and half the figures in the statutory instrument would be nugatory also.

The second argument is that, if the landlord's case is right, the first 17 words of para 6(2) have no effect at all. The landlord's argument, perhaps oversimplified, is that one must always make the para 6(2) calculation, ascertain the difference between the reasonable expected yield and the statutorily prescribed average yield, take into account all the circumstances which have brought that difference about, including the quality of the land and the climatic conditions in the area, and then apply the para 6(2) test to calculate the standard quota. It might be a sensible way to calculate the compensation. It might (he simply did not know) sometimes be a fairer method of doing it. But it is simply not what the statute provides.

Reference had been made in the argument to the decision of the Court of Appeal in *Grounds* v *Attorney-General of the Duchy of Lancaster*: and the judge was comforted by the fact that, in paraphrasing para 6(2), Glidewell LJ considered the words "by virtue of" effectively to be synonymous with "by reason of", a conclusion, in his judgment, incontrovertible as a matter of grammar. But, apart from this, the judge did not consider that decision to be helpful on the issues he had to decide. The Court of Appeal in that case was looking at an entirely different problem: and it would be inappropriate, in his view, to latch on to odd phrases in the leading judgment and apply them to the issues in this case, which are a long way away from the issues which they were considering.

In this case the arbitrator had found, as a fact, that the land was good- to moderate-quality agricultural land and that the climatic conditions in the area were close to the 30-year mean and that this grassland on this farm was classified as average in terms of its soil type and summer rainfall. So far as the judge could see, there was no evidence before the arbitrator which could lead to any other conclusion.

In the judge's judgment, this case turned on the ordinary construction of the ordinary words used in para 6(2): and on that issue the arbitrator and the tenant were right.

It followed that there was, in the judge's view, no error on the face of the award: and the application to set aside the award on that ground was dismissed.

Crown Court Decision

FORESTRY COMMISSION v GRACE AND ANOTHER

March 4 1991

(Before Judge GOWER QC and the Justices of Lewes Crown Court)

Estates Gazette April 18 1992

[1992] 15 EG 115

Forestry — Felling trees — Whether felling licence required — Burden of proof — Whether burden of proof on defendants to show that felling came within exceptions to requirement for a felling licence

The appellants were convicted by Haywards Heath justices of an offence of felling growing trees without the authority of a felling licence contrary to section 17 of the Forestry Act 1967 — On appeal against conviction, the appellants contended that in accordance with established principles the burden of proof fell on the respondent Forestry Commission as prosecution to establish that a felling licence was required for felling the trees in question — The respondent contended that it was for the appellants to prove on a balance of probabilities that the felling of the trees came within one or more exceptions to the requirement for a felling licence set out in section 9 of the 1967 Act

***Held*: The appeal was dismissed — Section 17 of the 1967 Act incorporates section 9 by reference, so that it is an offence to fell growing trees without a licence unless one or other of the exceptions applies — The exceptions to the general criminal burden of proof are shown by authority to apply where the subject-matter of the exception is peculiarly within the knowledge of the defendant, easy for the defendant to establish on a balance of probabilities, but in practical terms exceedingly difficult, if not impossible, for the prosecution to negative — Such was the position with regard to the exceptions in section 9 of the 1967 Act — Accordingly, the case fell within one of the rare exceptions to the general rule relating to the burden of proof in criminal cases and it was for the appellants to establish, on a balance of probabilities, that they fell within one or other of the exceptions in section 9**

The following cases are referred to in this report.

R v *Alath Construction Ltd* [1990] 1 WLR 1255; [1991] 1 EGLR 285
R v *Edwards* [1975] QB 27
R v *Hunt* [1987] AC 352; [1986] 3 WLR 1115; [1987] 1 All ER 1; (1986) 84 Cr App R 163; [1987] Crim LR 263, HL

This was an appeal by Malcolm George Grace and Patricia Margaret Grace against a conviction by Haywards Heath justices of felling trees without a felling licence contrary to section 17 of the Forestry Act 1967 following a prosecution by the respondent Forestry Commission.

Geoffrey Conlin (instructed by Ralph Haeems & Co) appeared for the appellants; Peter Birts QC (instructed by the Ministry of Agriculture, Fisheries and Food) represented the respondents.

Giving the judgment of the court, JUDGE GOWER said: The court has to rule upon a preliminary point on the burden of proof. The appeal is against a conviction, by the justices at Haywards Heath on March 19 1990, of the appellants of an offence of allegedly felling growing trees without the authority of a felling licence issued by the Forestry Commissioners under section 9(1) of the Forestry Act 1967, as amended.

The short point which we have to determine is whether the burden is upon the respondents to this appeal — in other words, the prosecution — to establish that a felling licence was required for the felling of the particular trees in question in this case or whether there is a burden cast upon the appellants, the defendants in other words, to prove on the balance of probabilities that the felling of the trees came within one or more of the exceptions to the requirement for a felling licence set out in the various subsections of section 9 of the Forestry Act 1967.

The way in which Mr Conlin, for the appellants, argues it is first of all to rely upon the principle fundamental to the criminal law in this country, namely that it is for the prosecution to leave the court sure upon the whole of the evidence of all elements essential to the proof of a criminal charge. Starting from that fundamental, and, if I may say so, extremely sound premise, he goes on to argue that, on the authority of a passage cited in the case of *R* v *Alath Construction Ltd* [1990] 1 WLR 1255* from the speech of Lord Griffiths in the case of *R* v *Hunt* [1987] AC 352 at p 374, the language of the relevant sections and subsections of the Forestry Act 1967 is such that it would not be open to this court to depart from the ordinary criminal burden of proof. For these reasons, it is necessary to look very carefully at *R* v *Alath Construction Ltd*. In essence, what the Court of Appeal decided in that case was that where there was an allegation that a

*Editor's note: Also reported at [1991] 1 EGLR 285.

A defendant felled trees contrary to a tree preservation order the burden rested upon the defendant to bring himself within the exceptions specified in the order or within the enabling provisions of the statute under which the order was made. The speech of Lord Griffiths is quoted in *Alath* at p 1258 E-F where he says:

> I would summarise the position thus far by saying that *Woolmington* v *Director of Public Prosecutions* [1935] AC 462

— that is the famous golden thread running through the web of the English criminal law cases —

> did not lay down a rule that the burden of proving a statutory defence only lay upon the defendant if the statute specifically so provided: that a statute can, on its true construction, place a burden of proof on the defendant although it
> B does not do so expressly: that if a burden of proof is placed on the defendant it is the same burden whether the case be tried summarily or on indictment, namely, a burden that has to be discharged on the balance of probabilities. The real difficulty in these cases lies in determining upon whom Parliament intended to place the burden of proof when the statute has not expressly so provided. It presents particularly difficult problems of construction when what might be regarded as a matter of defence appears in a clause creating the offence rather than in some subsequent proviso from which it may more readily be inferred that it was intended to provide for a separate defence which a defendant must set up and prove if he wishes to avail himself of it.

The judgment in *Alath* goes on at p 1258 H:

> Later Lord Griffiths said at p 374:
>
> "However, their Lordships were in agreement that if the linguistic construction of the statute did not clearly indicate upon whom the burden
> C should lie the court should look to other considerations to determine the intention of Parliament such as the mischief at which the Act was aimed and practical considerations affecting the burden of proof and, in particular, the ease or difficulty that the respective parties would encounter in discharging the burden. I regard this last consideration as one of great importance for surely Parliament can never lightly be taken to have intended to impose an onerous duty on a defendant to prove his innocence in a criminal case, and a court should be very slow to draw any such inference from the language of a statute. When all the cases are analysed, those in which the courts have held that the burden lies on the defendant are cases in which the burden can be easily discharged."

Not surprisingly, Mr Conlin relies heavily upon those concluding words as affording strong support to his argument.

D To continue with the judgment from *Alath*, at p 1259 B-C there is a further citation from Lord Griffiths' speech at p 375 of the report in *R* v *Hunt*:

> "In *Reg* v *Edwards* [1975] QB 27, 39-40 the Court of Appeal expressed their conclusion in the form of an exception to what they said was the fundamental rule of our criminal law that the prosecution must prove every element of the offence charged. They said that the exception 'is limited to offences arising under enactments which prohibit the doing of an act save in specified circumstances or by persons of specified classes or with specified qualifications or with the licence or permission of specified authorities'."

In our judgment, those words are of very great importance in the present context. Mr Conlin, in the course of his argument, made the point that the offence in the *Alath* case lay in ignoring the provisions
E of the tree preservation order and not in contravening a statutory provision prohibiting the doing of a particular act. Upon that basis very largely, Mr Conlin founds his argument that the present case is to be distinguished from that of *Alath*. His argument is that we have to look at the wording of the Forestry Act 1967, which is the Act creating the offence. He draws our attention to section 17, which provides:

> Anyone who fells a tree without the authority of a felling licence, the case being one in which section 9(1) of this Act applies so as to require such a licence, shall be guilty of an offence . . . and liable on summary conviction to a fine . . .

His argument is that if we apply the reasoning in Lord Griffiths' speech, as set out at p 1258 F-H, this is a case in which the exceptions
F to the requirements for a felling licence are set out in a section antecedent to the one which creates the offence. His argument, therefore, is that it is not a case in which the construction of the statute presents any linguistic difficulty and that therefore we ought to apply the ordinary criminal law rules as to the burden of proof.

In the judgment of the court, this reasoning is not sound. First of all, if one looks at the words of section 17, their effect in reality is to incorporate within them the whole of section 9 in so far as that section establishes the requirement for a felling licence and sets out exceptions to the requirement. In order to give meaning to section 17 one has to go back to section 9 and incorporate it by reference to section 17, so that, in effect, section 17 is saying that it is an offence to fell growing trees, and one inserts the word "growing", which Mr Birts has invited us to do, that it is an offence to fell growing trees G
without a licence unless one or other of the exceptions applies.

Having said that, it is then necessary to look again at Lord Griffiths' speech cited in the *Alath* case, to which I have already referred: "the exception 'is limited to offences arising under enactments which prohibit the doing of an act save in specified circumstances or by persons of specified classes or with specified qualifications or with the licence or permission of specified authorities'." If one looks at the exceptions, as set out in subsections (2), (3) and (4) of section 9 of the Forestry Act 1967, in many respects although not in every respect, those exceptions fall squarely within the principle enunciated by the Court of Appeal in the *Edwards* case, to which Lord Griffiths had referred. H

One then considers the practicalities of the situation, the purpose of the legislation. The way in which Mr Birts has aptly put it is to argue that the purpose of section 9 is to cast a protective cloak over the whole of forestry in this country by requiring that a felling licence should be obtained before any growing tree is felled, unless the felling of such a tree is within one or other of the exceptions set out in the statute. Going on from there, one looks again at the general principle which applies to exceptions to the general criminal burden of proof. There is a long line of authority, which, in effect, is encapsulated both in the *Edwards* case and in that part of Lord Griffiths' speech cited in the *Alath* case, to the effect that such exceptions are to apply where the subject-matter of the exception is peculiarly within the knowledge of the defendant, easy for the defendant to establish on a balance of J
probabilities, but in practical terms exceedingly difficult, if not impossible, for the prosecution to negative.

If one looks at section 9(2), once trees have been felled and the wood cut up and removed, it is likely to be very difficult to prove that the tree was not at least of the diameter specified, for example, in subsection (2)(*a*). So far as subsection (3) is concerned, so also it would be very difficult, if not impossible in many cases, for the Forestry Commission to prove the cubic content of cut trees. Those matters are well within the knowledge and competence of a defendant to prove on a balance of probabilities. Much the same considerations apply to (*a*) of subsection (4), namely where it is said that the felling is for the prevention of danger or the prevention or abatement of a nuisance. It is not difficult to visualise circumstances K
in which once a tree is felled it would be very difficult, if not impossible, for the Forestry Commission to negative an assertion that the felling of the tree was in the interests of preventing danger, or indeed in some cases for the prevention or abatement of a nuisance, whereas these are matters peculiarly within the knowledge of the owner of the tree or person responsible for the tree, who would have little difficulty in producing evidence to prove such matters upon a balance of probabilities. If one goes on to look, for example, at the remaining exception set out under subsection (4) where no licence is required if the felling is in compliance with any obligation imposed by or under an Act of Parliament, that would be a very simple matter for a defendant to establish; so also would (*c*), namely where the felling is L
carried out by, or at least request of, an electricity board for the reasons there set out.

For all these reasons, without going into further and more tedious detail about it, we are satisfied that this case falls within one of the very rare exceptions to the general rule relating to the burden of proof in criminal cases, and that it is for the appellants to establish, if they can, upon a balance of probabilities, that they fall within one or other of the exceptions set out in section 9 of the Forestry Act 1967.

County Court Decision

WALKER v CROCKER AND OTHERS M

April 4 1991

(Before His Honour Judge PECK at Plymouth County Court)

Estates Gazette June 13 1992

[1992] 23 EG 123

Agricultural holdings — Claim for compensation in respect of, *inter alia*, milk quota on termination of tenancy — Whether letter sufficient notice in writing for the purposes of section 83(2) of the Agricultural Holdings Act 1986 and/or para 11 of Schedule 1 to the Agriculture Act 1986

A The following cases are referred to in this report.

Amalgamated Estates Ltd v *Joystretch Manufacturing Ltd* [1981] EGD 84; (1980) 257 EG 489, CA
BSC Pension Fund Trustees Ltd v *Downing* [1990] 1 EGLR 4; [1990] 19 EG 87
Buckinghamshire County Council v *Moran* [1989] 3 WLR 152; [1989] 2 All ER 225, CA
Deerfield Travel Services Ltd v *Leathersellers' Company* (1982) 43 P&CR 143; 263 EG 254
Hallinan (Lady) v *Jones* (1984) 272 EG 1081
Lees v *Tatchell* (1990) 60 P&CR 228; [1990] 1 EGLR 10: [1990] 23 EG 62
Mountford v *Hodkinson* [1956] 1 WLR 422; [1956] 2 All ER 17, CA
Nunes v *Davies Laing & Dick Ltd* (1985) 51 P&CR 310; [1986] 1 EGLR 106; 277 EG 416
B *Rush & Tompkins Ltd* v *Greater London Council* [1989] AC 1280
Sheridan v *Blaircourt Investments Ltd* [1984] EGD 176; (1984) 270 EG 1290

This was a special case stated by the arbitrator seeking an opinion on a question arising under the Agricultural Holdings Act 1986 and the Agriculture Act 1986.

Martin Rodger (instructed by Stephens & Scown, of Exeter) appeared for the first respondent, John Derek Walker; Andrew Densham, solicitor (of Burges Salmon, of Bristol) represented the second respondents, Ivor William Crocker, Thomas Stanbury Hockin, Richard Henry Perraton, Norman George Preston, Dennis Arthur Robins and Peter Ambrose Duncan Willis, trustees of the C Charity Lands of Diptford.

An arbitration had been conducted by John Wakeham [FRICS] under the provisions of the Agricultural Holdings Act 1986 and the Agriculture Act 1986 between the second respondents as landlords and the first respondent as tenant and was concerned with the assessment of compensation upon the termination of the tenancy held by the first respondent, of Blakewell Farm, Diptford, Devonshire. The arbitrator had stated a special case seeking an opinion as to whether, on its true construction, a letter written by the tenant's solicitors to the landlord's solicitors on November 9 1987 was a sufficient notice for the purpose of section 83(2) of the Agricultural Holdings Act 1986 and/or para 11(1) of Schedule 1 to D the Agriculture Act 1986.

The question posed by the arbitrator did not expressly cover one of the submissions raised in the case and argued before the judge by Mr Densham, namely whether the letter of November 9 1987 was to be regarded as written "without prejudice" and to be treated as privileged, and therefore incapable of being a notice under the Acts, or of being relied upon in the arbitration at all. Logically this issue should be considered before other matters raised in the case.

The requirements which must be met for a notice to be valid under section 83 of the Agricultural Holdings Act 1986 and to entitle a tenant to claim compensation in respect of improvements and tenant rights differ in one respect from the similar requirements of para E 11(1) of Schedule 1 to the Agriculture Act 1986, which enable the tenant to claim in respect of his milk quota. Under the Agriculture Act the notice must: (a) be in writing; (b) be given before the expiry of two months from the termination of the tenancy; and (c) notify the landlord of the tenant's intention to make the claim. Under the Agricultural Holdings Act there is an additional requirement, namely that the notice specify the nature of the tenant's claim. Section 83 of the Act provides, however, that it shall be sufficient if the notice refers to the statutory provision, custom or term of an agreement under which the claim is made. Because of the differing requirements of the two Acts it follows that a notice may be valid for the purposes of the Agriculture Act but fail to comply with the additional requirements of the Agricultural Holdings Act and be F invalid for the purposes of that Act. Conversely, a notice which is invalid under the Agriculture Act will also be invalid under the Agricultural Holdings Act. It is to be observed that under earlier legislation either party giving notice of a claim for compensation was required to include particulars of the claim in the notice: see section 16 of the Agricultural Holdings Act 1927. Since 1948, however, the relevant statutes have not been similarly worded.

Mr Rodger submitted that the test of whether a notice under the Agriculture Act 1986 is valid is whether the terms of the purported notice are sufficiently clear to bring home to the ordinary landlord and/or his professional advisers that the tenant intends to make a claim for compensation: see *Nunes* v *Davies Laing & Dick Ltd* [1986] 1 EGLR 106 *per* Sir Nicolas Browne-Wilkinson V-C at p 107 and the question in the present case is whether the letter of November 9 1987 G written by Stephens & Scown, the first respondent's solicitors, to Michelmores, the solicitors then acting for the second respondents, satisfies that test. The crucial words of the letter are: "if your clients are successful on November 20 then our client expects compensation for all his input into the farm including the value of the (milk) quota". November 20 1987 was, in fact, the date upon which the second respondent's claim for possession of the holding was due to be heard in the county court. On that date, however, the first respondent was given leave to file a defence to the claim and the action was adjourned generally. It was eventually heard on May 6 1988, when an order was made by consent that, the tenancy having terminated on September 29 1987, there should be an immediate order for H possession of the holding suspended, upon punctual compliance by the first respondent with certain terms agreed, until March 25 1989.

Mr Rodger submitted that the letter of November 9 1987 fully satisfied the requirements of both section 83(2) of the Agricultural Holdings Act 1986 and para 11(1) of Schedule 1 to the Agriculture Act 1986 and passed the test enunciated by the Vice-Chancellor in the *Nunes* case by making clear that the first respondent not merely hoped for but expected compensation both for his "input" into the farm and for his milk quota and by specifying the nature of the input he relied upon. Mr Rodger referred the judge to *Deerfield Travel Services Ltd* v *Leathersellers' Company* (1982) 263 EG 254, a decision of the Court of Appeal on the provisions of the Landlord and Tenant Act 1927, in which the court held that it was sufficient J compliance with section 3(1) of that Act if the notice required by the section notified to the landlord the tenant's intention to make improvements merely by implication.

Mr Densham submitted that the letter of November 9 1987 contained no words indicating an intention by the first respondent to make any claim and, in any event, did not specify the nature of any such claim, leaving the second respondents with no indication whether the compensation sought was for dilapidations only or for tenant's rights or as to the extent of the milk quota referred to. He drew the judge's attention to the fact that the letter relied upon in the *Nunes* case, in contrast to that relied upon in the instant case, was stated expressly to be a "formal notice".

He submitted further that the correspondence between the parties K indicated that neither party treated the letter of November 9 1987 as a notice under the Acts, as shown by the subsequent letter from Gordon Vick & Partners (the first respondent's agents) to Luscombe Maye (the second respondents' agents) dated January 25 1989 enclosing a "formal statutory notice" under sections 10 and 83 of the Agricultural Holdings Act 1986. Mr Densham further argued that any claim made in the letter of November 9 1987 was conditional upon the second respondents' action for possession being successful on November 20 1987. On that date, he says, the second respondents were not successful, since the first respondent obtained an adjournment and a possession order was not obtained until May 6 1988. L

In the course of his argument, Mr Densham cited a number of cases, namely *Sheridan* v *Blaircourt Investments Ltd* (1984) 270 EG 1290, *Lady Hallinan* v *Jones* (1984) 272 EG 1081, *Lees* v *Tatchell* [1990] 1 EGLR 10, *BSC Pension Fund Trustees Ltd* v *Downing* [1990] 1 EGLR 4 and *Mountford* v *Hodkinson* [1956] 2 All ER 17. In *Sheridan* v *Blaircourt Investments Ltd* the words relied upon as constituting a notice to apply for the appointment of an independent valuer under a rent review clause in a lease were: "I suggest it would be appropriate to make application to the RICS for an independent valuer. . . ".

Nicholls J (as he then was) held that a form of words "which is limited to the tenant's agents making a suggestion" is not language appropriate to the exercise by a tenant of an option. In the judgment M of Judge Peck, however, the language of the letter of November 9 1987 in the instant case was very far removed from that used in the case cited. In *Lady Hallinan's* case — a county court decision — the words relied upon were: "We give you prior warning that it is our present intention to serve a notice under section 70 of the Act (Agricultural Holdings 1948) in due course". The learned assistant recorder held that this wording did not constitute a sufficient notice under the section. The judge thought this was clearly right; the words used in that case were totally inconsistent with an intention to give actual notice. In *Lees* v *Tatchell,* a decision of the Court of Appeal, Parker LJ referred with approval to the test applied in *Amalgamated Estates Ltd* v *Joystretch Manufacturing Ltd* (1980) 257 EG 489,

A namely that a notice should be in terms which are sufficiently clear to bring home to the ordinary landlord that the tenant is purporting to exercise his right under the relevant section of the Act in question. In *BSC Pension Fund Trustees Ltd* v *Downing* — Mr Andrew Blackett-Ord (sitting as a deputy High Court judge in the Chancery Division) applied the same test. Finally, in *Mountford* v *Hodkinson* — a tenant who had received from his landlord a notice to quit replied: "I don't intend to go. I shall appeal against it and take the matter up with the AEC". The Court of Appeal held that, although a counternotice was not to be construed with great strictness, it must be so expressed as to indicate a clear intention to invoke the right given by the relevant Act and the tenant's letter in that case did not show such an intention.

The judge thought it right, as Mr Densham suggested, to construe B the letter of November 9 1987 in the light of the surrounding circumstances and the contemporary correspondence. Nevertheless, it seemed to him, that letter did, on its face, comply with the requirements of the Agriculture Act and satisfied both the test adopted in the *Nunes* case and that applied in *Amalgamated Estates* and similar cases. The words "our client expects compensation" must have made clear to the second respondents' advisers that the first respondent was claiming such compensation and that claim was on the basis that the second respondents would obtain possession upon the termination of the tenancy. The judge did not think that the second respondents would have been misled into thinking that the claim depended upon possession being obtained by an order made on C a particular date. Nor did he consider that the subsequent formal notice given by the first respondent's agents on January 25 1989 affected the matter. The particulars then provided were required for the arbitration to proceed and the fact that they were said to be a formal statutory notice was, he thought, irrelevant. No doubt when agricultural agents are dealing with compensation claims they normally make use of *pro forma* notices but the Agriculture Act does not require formality provided the landlord is informed in writing of the tenant's intention to make a claim. While it is true that the notice in the *Nunes* case was stated in terms to be a "formal notice", it did not appear that this was itself the *ratio decidendi* of that case.

In Judge Peck's judgment, the letter of November 9 1987 made clear to the second respondents that the first respondent intended to D make a claim both under the Agricultural Holdings Act and under the Agriculture Act in respect of his milk quota and, in his judgment, therefore, the letter constituted a valid notice under the latter statute. The judge had rather more difficulty in deciding whether it complied with the requirement of section 83(2) of the Agricultural Holdings Act that it specify the nature of the claim but on balance he thought that, despite Mr Densham's arguments to the contrary, the letter gave a sufficient indication to the second respondents of the matters in respect of which compensation was sought and he held, therefore, that the requirements of this Act also were satisfied.

Finally, the judge turned to consider whether the letter of November 9 1987 was privileged. It was, on its face, an open letter but E Mr Densham argues that it forms part of a chain of privileged correspondence containing negotiations for a settlement of the dispute between the parties.

The principles governing the "without prejudice" rule were considered by the House of Lords in *Rush & Tompkins Ltd* v *Greater London Council* [1989] AC 1280. Lord Griffiths, giving the opinion of the House, stated, at p 1299, that the rule rests, at least in part, upon public policy, namely that parties should be encouraged as far as possible to settle their dispute without resort to litigation and should not be discouraged by the knowledge that anything that is said in the course of such negotiations may be used to their prejudice in the course of the proceedings. Later in the opinion, Lord Griffiths emphasised that application of the rule is not dependent upon the use

F

of the phrase "without prejudice". A letter genuinely aimed at G settlement is therefore privileged even if it is not described as being written "without prejudice". Conversely, however, a letter headed "without prejudice" is not privileged, merely because it is so described, if it is not genuinely intended to be part of negotiations. If headed "without prejudice" there is a presumption that it is so intended; if not so headed it seems that the presumption is the other way; either presumption may, however, be rebutted by the nature of the contents of the letter. Once privilege is established, a letter covered by the privilege is inadmissible in the current or any subsequent litigation about the same subject-matter.

The letter of November 9 1987 was written by the first respondent's solicitors in reply to two letters from the second respondents' solicitors both dated November 6 1987, one of them being an open H letter and one marked "without prejudice". The first respondent's solicitors did not state specifically to which letter their reply was addressed; it merely thanked the second respondents' solicitors for their two letters and noted their instructions. Therefore, it was important, in the light of Mr Densham's submission, to consider carefully the contents of the letter of November 9 1987 and to have regard to the context in which it was written and its place in the continuing correspondence. That correspondence, so far as it is relevant, began on October 22 1987 with two letters from Stephens & Scown to Michelmores, one of them an open letter and the other, the second, headed "without prejudice"; the first of these letters merely indicated an intention by their client to apply for legal aid and asked J for agreement to an adjournment of the hearing due on November 20 1987; the second contained a proposal for settlement. Michelmores replied with an open letter on October 26 1987 without, however, indicating which of Stephens & Scown's letters they were answering. They stated that they would be discussing Stephens & Scown's letter (in the singular) with their clients. Stephens & Scown wrote again on October 29 1987 by an open letter confirming that legal aid had been applied for, "so that we can advise our client fully and protect (his) interests in the best possible manner". On November 6 1987 Michelmores wrote two letters to Stephens & Scown; the first, an open letter refused agreement to an adjournment of the possession action and expressed the intention to proceed to obtain possession as quickly as possible; the second, "without prejudice", effectively K rejected out of hand the offer of settlement put forward in Stephens & Scown's "without prejudice" letter of October 22 1987. It seemed to the judge that Michelmores' letter brought negotiations to an end. The vital letter of November 9 1987 contained no new offer; it merely set out the facts alleged by the tenant and asserted his rights to compensation: see *Buckinghamshire County Council* v *Moran* [1989] 3 WLR 152 *per* Slade LJ. The correspondence ended there. On November 29 1987 the second respondents' agents sent to the first respondent the landlords' notice under section 83 of the Agricultural Holdings Act 1986. The correspondence between the parties' solicitors was not resumed until August 24 1989 and the subsequent letters are not relevant to the matter being considered. In the L circumstances being outlined, the judge did not regard the letter of November 9 1987 as forming part of the negotiations between the parties designed to achieve a settlement and he saw no reason why, as a matter of public policy, it should be treated as privileged. If, as he believed to be the case, the letter constituted a good notice, the landlords were not prejudiced by its admission in evidence, *a fortiori*, if he had been mistaken in this view. The judge held: (1) that the letter dated November 9 1987 written by the tenant's solicitors to the landlords' solicitors was admissible in evidence; and (2) that the letter was a notice for the purpose of section 83(2) of the Agricultural Holdings Act 1986 and also of para 11(1) of Schedule 1 to the Agriculture Act 1986.

M

For a further case on this subject see p 261

ARBITRATION

A **Queen's Bench Division**

September 4 1991
(Before Mr Justice BLOFELD)

MOORE STEPHENS & CO v LOCAL AUTHORITIES' MUTUAL INVESTMENT TRUST AND ANOTHER

Estates Gazette February 1 1992
[1992] 04 EG 135

Arbitration — Misconduct — Rent review — Request by
B **applicant tenant that arbitrator stand down — Respondent landlord wishing to rely on a comparable — The comparable was determination of the arbitrator acting as an independent expert — Applicant wishing to call arbitrator as witness — Applications under section 23 and 25 of the Arbitration Act 1950 dismissed**

The applicant holds a lease of premises at St Paul's House, Warwick Lane, in the City of London — The first respondent is the landlord — The second respondent was appointed an arbitrator in May 1991 to determine the rent at review — In relation to the rent review of premises at 7-10 Old Bailey, also
C **in the City, the arbitrator had been appointed an independent expert — As independent expert he had determined the rent of 7-10 Old Bailey at £1.4m — Landlord in present case wished to rely upon that determination as a comparable — Tenant objected to comparable on grounds that determination too high — Tenant requested arbitrator to stand down — On landlord's valuer including 7-10 Old Bailey transaction as comparable in submissions, tenant applied for order that arbitrator misconducted himself and that he stand down**

Tenant submitted that arbitrator cannot do justice in considering comparable transaction because of interest in comparable — Arbitrator's integrity and impartiality raises
D **doubt as to whether justice can be seen to be done — Tenant submitted it wished to call arbitrator as a witness in the arbitration to ask him factually how he arrived at his rental determination of 7-10 Old Bailey — If it cannot do this it will be prejudiced**

***Held*: Argument that justice will not be seen to be done unattractive — No reason why tenant should have any fears — No reason why arbitrator should be biased — On third point, the whole basis of calling arbitrator as witness would be to undermine his opinion evidence — In those circumstances arbitrator would be called as an expert witness — Matters of**
E **expert opinion should be addressed to tenant's valuer in arbitration — Questions of fact can be asked of the arbitrator or from other sources — No reason why tenant would be prejudiced by inability to call the arbitrator — Test in "*The Ellisar*" applied**

The following case is referred to in this report.

Haigh and London North Western and Great Western Railway Co, In re [1896] 1 QB 649

This was an application by an originating notice of motion, pursuant to sections 23 and 25 of the Arbitration Act 1950, for the F
removal of Mr A H P Gillett [FRICS], an arbitrator appointed to determine the rent of premises at St Paul's House, Warwick Lane, City of London, and for the appointment of a fresh arbitrator. The applicant tenant was Moore Stephens & Co and the respondent landlord was the Local Authorities' Mutual Investment Trust.

Miss Hazel Williamson QC (instructed by Alsop Wilkinson) appeared for the applicant; Guy Fetherstonhaugh (instructed by Lovell White Durrant) represented the first respondent; and Barry Denyer-Green (instructed by Taylor Joynson Garrett) represented the second respondent arbitrator.

Giving judgment, **BLOFELD J** said: In this case the applicant tenant G
asked the court to decide that Mr Alan Gillett [FRICS], who has been appointed arbitrator, stand down for a number of reasons. The background is that there is a property known as St Paul's House, Warwick Lane, in the City of London, and there is now an arbitration for a new rent to be fixed and Mr Gillett was appointed by the president of the Royal Institution of Chartered Surveyors to be the arbitrator. He was appointed in May 1991. At that time it would appear that it was known generally, in the way that these things get about, that he had, in fact, been used as an expert in the valuation of 7-10 Old Bailey, also in the City. The parties were not aware that the Old Bailey determination in which he, Mr Gillett, was the expert involved would be used as one of the comparables. That happened in June or maybe later. When it was discovered that Mr Gillett was H
prepared to allow this to be used as a comparable, the applicant made this application early in August. At that time, as Mr Gillett had not actually looked at the comparables, because he had deliberately allowed a period of three days to elapse in case objections were taken, he was not aware of the precise reason for this application, but he has since put in an affidavit and the matter is now clear as it comes before this court today.

The applicant put its case forward on three grounds. It started by making it perfectly clear that it did not impugn Mr Gillett's integrity, but it said that when he was acting as an expert in relation to 7-10 Old Bailey the figure that he put in was a figure of £1,400,000. That, it said, was too high and therefore as, it said, the premises with which J
these proceedings are concerned, that is St Paul's House, are very comparable to 7-10 Old Bailey, that that figure will inevitably be lurking in the back of his mind and will influence him in any decision to which he comes.

Without further ado I reject that submission. I am quite certain that somebody of Mr Gillett's standing and expertise would have absolutely no difficulty in being entirely unbiased about the whole matter.

It is further said, as a result of that, that the applicant can reasonably fear that justice will not seem to be done. I find that argument as unattractive as I find the first argument put forward by the applicant. I do not think it does credit to Mr Gillett and all other people who act in this capacity as arbitrators. I see absolutely no K
reason why they should have any fears. I have listened with care to what Miss Williamson has said; she has taken me through the principles which are admirably set out in *Mustill and Boyd on Commercial Arbitration,* and she particularly referred me to pp 529, 530, 289 and 290. I do not propose to read them out because this is in chambers, but I have them all fully in mind. So, on those two grounds I am against the applicant without further ado.

Before going further I should say that the arbitrator is, in fact, represented here today. He has set out his views in an affidavit and his counsel has addressed me and has indicated that his instructions are that he is perfectly willing to continue as arbitrator.

I have called the first two points really 1(a) and 1(b); that is the integrity point and the impartiality point. I now come to the third point although it is probably, if one looks at Wright J's judgment in the case of *In re Haigh and London North Western and Great Western Railway* [1896] 1 QB 649, a bias point, too, because that is how it starts.

The applicant says that it wishes to call Mr Gillett as a witness in this arbitration. It is perfectly clear that it is not possible for it to call the arbitrator itself. It says that it wishes to call him as a witness of fact, to ask him factually how he arrived at the £1,400,000 that he did. It said that it cannot do that if he is the arbitrator and therefore it will be prejudiced. It is said on behalf of the first respondent first of all, and indeed on behalf of the arbitrator, that this is an extremely unusual proceeding to which the applicant says in reply that as the figures get bigger so the investigations get wider and so the number of witnesses that may be called get enlarged, and although it may be that nobody has experience of a witness being called to deal precisely with a comparable, there is always a first time and this is a proper occasion for it to be a first time. Against that, again, the first respondent says that, of course, it will require a *subpoena*, anticipating that Mr Gillett would be extremely reluctant to give evidence without a *subpoena*, and a *subpoena* may well not be granted in the circumstances. Furthermore, the first respondent, echoed by the counsel on behalf of Mr Gillett, said that in fact it really would be seeking to call him as an expert not as a witness to fact. To that Miss Williamson, on behalf of the applicant, strongly disagrees; she says he would be called as a witness of fact. But as the whole basis of calling him would be to undermine his opinion evidence that the proper value of 7-10 Old Bailey was £1,400,000 however it is dressed up in the end, I have come firmly to the view he would be called as an expert witness. I therefore accept the first respondent's argument that the questions that the applicant should wish to ask should be addressed to Mr Arthur Spooner [FRICS], who is to be called as an expert for the applicant tenant. If they want precise questions of fact I have absolutely no doubt, as has been indicated by the arbitrator's counsel, that Mr Gillett would be only too happy to supply them, and furthermore they can get matters of fact from other sources rather than Mr Gillett. So, I cannot for the moment see that they would be in any way prejudiced by their inability to call Mr Gillett.

I further bear in mind, having paid close attention to the decision in *In re Haigh and The London North Western and Great Western Railway,* that arbitrators in these sort of circumstances inevitably are men or women who not only have great experience of commercial rents, they inevitably must have come down firmly in comparable situations. If they had no experience of comparable situations, they would hardly be in a position to do justice as arbitrators. I take the view that the court has to apply the test which is set out concisely in *"The Ellisar"* [1984] 2 LLR 84 and I cite at p 89 from Ackner LJ, as he then was:

> Do there exist grounds from which a reasonable person would think that there was a real likelihood that Mr Clark [Mr Gillett] could not, or would not, fairly determine the safe port issue . . .

or the issue here —

> on the basis of evidence and arguments to be adduced before him? It seems to me that that is a satisfactory way of expressing the objective test which, to my mind, is the appropriate approach. To suggest that the mere lack of confidence which no reasonable person would, in the relevant circumstances, experience, should be a basis for setting aside an award, seems to me quite unacceptable.

I have absolutely no doubt that Mr Gillett can fairly determine this issue on the basis of the evidence and arguments to be adduced before him. I am equally satisfied that the evidence and arguments to be put before him can be perfectly properly and fully put before him if I had seen it in any way necessary to call him as a witness.

So for those short reasons I refuse this motion.

Motion dismissed with costs to the first and second respondents.

COMPULSORY PURCHASE AND COMPENSATION

A

Court of Session

Inner House

July 25 1991

(Before Lord HOPE (Lord President), Lord MAYFIELD and Lord KINCRAIG)

CITY OF GLASGOW DISTRICT COUNCIL AND ANOTHER v MACKIE

Estates Gazette May 23 1992

[1992] 20 EG 114

B

Land compensation — Discretionary disturbance payment — Whether Scottish Lands Tribunal has jurisdiction to determine amount of discretionary disturbance payment under section 34(4) of the Land Compensation (Scotland) Act 1973

The respondent, Duncan G Mackie, held a lease of a shop in Glasgow which was due to terminate in 1992 — Because the shop was in an area the subject of a comprehensive improvement scheme, the respondent was obliged to give up possession in 1987 — Although it was originally envisaged that the respondent's loss of occupation would be only C **temporary, because the improvement works were delayed he became permanently displaced and negotiated a renunciation of his lease with his landlords by which the lease was determined in 1989 — The improvement scheme was carried out by the second appellants, Yoker Housing Association Ltd, as agents for the first appellants, City of Glasgow District Council — On a reference to the Lands Tribunal to have determined the amount of a discretionary disturbance payment under section 34(4) of the Land Compensation (Scotland) Act 1973, the tribunal rejected the second appellants' argument that since a discretionary payment was** D **not a "disturbance payment" within the meaning of section 35(4) of the Act it had no jurisdiction to determine the amount in dispute — The appellants appealed from the decision of the tribunal to repel their pleas to a competency on the grounds that it had jurisdiction to determine the amount in dispute**

***Held*: The appeal was allowed — The jurisdiction of the tribunal is defined by section 35(4) of the 1973 Act and refers to "any dispute as to the amount of a disturbance payment" — A discretionary payment under section 34(4) applies only where a person who is displaced is *not* entitled to receive a disturbance payment — The payment in this case was not a disturbance payment but was for the discretion of the** E **compensating authority — There is no indication in section 34(4) that the amount of the discretionary payment, if in dispute, is to be referred to and determined by the tribunal — *Gozra* v *Hackney London Borough Council* relating to the same provisions in the Land Compensation Act 1973 (section 37(5)) not followed**

The following cases are referred to in this report.

Glasgow Corporation v *Anderson* 1976 SLT 225
Gozra v *Hackney London Borough* [1988] 2 EGLR 20; [1988] 46 EG 87
Lanarkshire and Dunbartonshire Railway Co v *Main* (1985) 22 R 912; 21 R 1018; 31 SC LR 239; 2 SLT 258 F
McMonagle v *Westminster City Council* [1990] 2 AC 716; [1990] 2 WLR 823; [1990] 1 All ER 993, HL
R v *Federal Steam Navigation Co Ltd* [1974] 1 WLR 505; [1974] 2 All ER 97; [1974] 1 Lloyd's Rep 520, HL

This was an appeal by City of Glasgow District Council and Yoker Housing Association Ltd by way of a case stated by the Lands Tribunal for Scotland under the Tribunals and Inquiries Act 1971 in respect of a determination that the tribunal had jurisdiction to determine the amount of a discretionary disturbance payment under section 34(4) of the Land Compensation (Scotland) Act 1973 claimed by the respondent, Duncan G Mackie. G

James McNeill QC (instructed by Campbell Smith & Co, for the first appellants, and MacRoberts for the second appellants) appeared for the appellants; Gerard Moynihan (instructed by McClure, Naismith, Anderson & Gardiner) represented the respondent.

Giving his opinion, **LORD HOPE (LORD PRESIDENT)** said: This is an appeal by case stated by the Lands Tribunal for Scotland under the Tribunals and Inquiries Act 1971 and Rule of Court 291. The respondent was the lessee of a shop in Glasgow. His lease was due to terminate in 1992, but the tenement in which the shop was located was the subject of a comprehensive improvement scheme as a result of which he was obliged to move out of the property in 1987. It was H originally envisaged that his loss of occupation of the premises would be only temporary, but the contractor went into liquidation and the work was delayed. The respondent alleges that he was not able to return to the premises and that he was permanently displaced. He negotiated a renunciation of his lease with his landlords, as a result of which it was determined in 1989. The improvement scheme was being carried out by Yoker Housing Association Ltd, who allege that they were acting as agents for City of Glasgow District Council.

This case is concerned only with the appellants' pleas to the competency of the respondent's application to the tribunal for determination of his claim for compensation for legal costs and loss of profits resulting from these events.

The respondent's application to the tribunal was in terms of J section 35(4) of the Land Compensation (Scotland) Act 1973, which provides that:

> Any dispute as to the amount of disturbance payment shall be referred to and determined by the Lands Tribunal.

His claim was originally presented as being one for a disturbance payment in terms of section 34(1)(*d*) of the Act, by which it was provided for the first time that a payment could be recovered as of right by a person displaced from any land, and also for a discretionary payment in terms of section 34(4), which replaced the power to make an allowance to persons displaced in section 38 of the Land Compensation Act 1963. But his solicitor accepted in the hearing before the tribunal on the preliminary pleas that, as the housing association were not the owners of the land, section 34(1)(*d*) K did not apply. He confined himself to the respondent's claim for the determination of the amount of a discretionary payment under section 34(4), which is in these terms:

> (4) Where a person is displaced from any land as mentioned in subsection (1) but is not entitled, as against the authority there mentioned, to a disturbance payment or to compensation for disturbance under any other enactment, the authority may, if they think fit, make a payment to him determined in accordance with section 35(1) to (3) below.

The appellants' submission was that this was not a question which was open for determination by the tribunal, since a discretionary payment under that subsection was not a "disturbance payment" within the meaning of section 35(4). The tribunal rejected that argument and repelled the pleas to competency on the ground that it had jurisdiction under the subsection to determine the amount which is in dispute.

The tribunal noted that the respondent had no statutory entitlement to any payment under section 34(4). But they pointed out that the housing association had already exercised their discretion by agreeing to make a payment under that subsection to the respondent at the rate of £240 per week for the period of 20 weeks originally envisaged for the completion of the works. The subsection requires that the amount of any payment under section 34(4) is to be determined by the authority in accordance with section 35(1) to (3). They considered these provisions to be mandatory and not directory and said that it would be surprising if Parliament had given no statutory remedy to a displaced person if an authority did not determine the payment in accordance with what section 35(1) to (3) requires. They held that a payment, once it had been assessed under section 35, became a discretionary displacement payment and thus that it was a disturbance payment within the meaning of section 35(4) which could be referred to and determined by the tribunal if there were a dispute as to its amount.

In our opinion, this approach is not in accordance with the words used by the legislature. The jurisdiction of the tribunal is defined by section 35(4), which refers to "any dispute as to the amount of a disturbance payment". But the only dispute which remains in this case relates to a payment under section 34(4), which is to be made at the discretion of the authority. The phrase "disturbance payment" is used in section 34(1) to define the payment which a person who is displaced from land is entitled to receive as of right in the circumstances to which that subsection refers. Section 34(4) applies in terms only where the person who is displaced is *not* entitled to receive a disturbance payment. The phrase which is used to describe the discretionary payment which may be made under that subsection is: "a payment . . . determined in accordance with section 35(1) to (3) below". Thus, whether one starts with section 35(4) or with section 34(4) the result is the same. Section 35(4) is concerned only with disputes about the amount of a disturbance payment which, applying the language used in section 34, the payment in dispute in this case is not. And section 34(4) says that the amount payable under that subsection is to be determined in accordance with section 35(1) to (3), not in accordance with section 35(1) to (4). There is no indication in section 34(4) that the amount of the discretionary payment, if in dispute, is to be referred to and determined by the tribunal under section 35(4).

Mr Moynihan submitted that there was an ambiguity as to the scope of the phrase "discretionary payment" as used in section 35(4). He pointed out that section 34(1), when describing the payment with which that subsection is concerned, says that it is "hereafter referred to as 'a disturbance payment'". It was open to question whether the word "hereafter" meant simply "hereafter within this section" or whether the definition was intended to apply also to section 35. Moreover, section 35(1), which is among the group of subsections referred to in section 35(4) and which sets out the formula to be applied, describes the amount which is to be arrived at as the "amount of a disturbance payment". This showed that a payment under section 34(4) could reasonably be described as a "discretionary disturbance payment" and that section 35(4) applied because it referred to any dispute about the amount of a disturbance payment. He invited us to examine the sidenotes and the heading to these two sections in support of this approach. His point was that the phrase "disturbance payments" was used in the plural both in the heading and in the sidenote to section 34 in order to describe the subject-matter. So it was appropriate to regard section 34 as being concerned with two different types of disturbance payment, namely a "discretionary disturbance payment", which is what a payment under section 34(4) amounted to, as well as the "disturbance payment" which is defined in section 34(1).

We are not persuaded that there is any such ambiguity and, in our opinion, the use of the plural in the heading and in the sidenote to section 34 is of no significance in view of the clarity with which section 34 is expressed. A sidenote is at best a poor guide to the construction of an enactment and such guidance as we may derive from the heading is outweighed by the definition in section 34(1). This confines the expression "disturbance payment" to a payment to which there is an entitlement in terms of the Act. This approach is confirmed when one comes to section 34(5), which provides that a disturbance payment shall carry interest from the date of displacement until payment. That provision would be inappropriate to the discretionary payment to which subsection (4) refers. And the word "hereafter" in section 34(1) is not qualified in such a way as to restrict the meaning which it gives to the words "disturbance payment" to their use in that section only. So the same meaning should be given to these words in any following sections in which they appear, as being a reference back to what section 34(1) provides. In short, there is no statutory basis for the use of the expression "discretionary disturbance payment" as a guide to the proper construction of section 35(4).

We were invited, nevertheless, by Mr Moynihan to construe section 35(4) in what he described as a purposive manner with regard to the consequences of the competing arguments. It should be noted, as background to this part of his submissions, that Mr McNeill for the appellants accepted that the provision in section 34(4) that a payment under that subsection was to be determined in accordance with section 35(1) to (3) was mandatory. There was a discretion as to whether to make the payment but, once it had been decided to make it, the amount of it had to be determined in accordance with those subsections. Thus, there was no discretion as to how its amount was to be calculated, in contrast, for example, to the wide terms of section 83(1) of the Local Government (Scotland) Act 1973, which gives power to local authorities to incur expenditure which, in their opinion, is in the interests of their area or any part of it or all or some of its inhabitants. That being so, said Mr Moynihan, there would be an anomaly if a dispute as to the amount of a disturbance payment within the meaning of section 34(1) could be referred to the tribunal and a dispute as to the amount of a discretionary payment under section 34(4) could not. The provisions of section 35(1) to (3) applied to them both and in both cases they were mandatory. It was not to be thought that Parliament intended that a different result should be arrived at in each case. Yet that could well happen if, in the case of a discretionary payment, the only method open to resolve any dispute was that of judicial review.

This argument reflects the comment by the tribunal in its opinion in this case that it would be surprising if Parliament gave no statutory remedy to a displaced person if an authority did not determine the payment in accordance with the provisions to which section 34(4) refers. But it is not obvious that the intention of Parliament was that the remedy in each case should be the same. It makes good sense, in the case of a disturbance payment, to make express provision for the determination of a dispute as to its amount. There is an entitlement to this payment under the statute in those cases to which section 34(1) applies. It is an amount which can be sued for, if necessary, if it is not paid. The effect of section 35(4) is to exclude all questions as to the amount of a disturbance payment from the courts, because these questions are to be referred to and determined by the tribunal. This is in accordance with a long tradition, in cases where compensation is due as of right in cases of compulsory acquisition by bodies with statutory powers, that determination of the amount to be paid is to be by arbitration and is not a matter for the court: see Lord McLaren's comments in *Lanarkshire & Dunbartonshire Railway Co* v *Main* (1895) 22 R 912 at p 920. But a discretionary payment is in a quite different category. It is not something which can be sued for because it is in the discretion of the authority as to whether it is to be paid. The amount of the payment is prescribed by section 34(4) under reference to section 35(1) to (3). But it seems not unreasonable to leave it to the authority to resolve all questions of fact and degree — or of circumstances and degree, as they were described in *Glasgow Corporation* v *Anderson* 1976 SLT 225 — so long as the particular exercises which are described in section 35(1) to (3) are carried out. Questions as to what was or was not reasonable for the express purposes of these subsections would thus be left to them to decide. Their decision as to what was or was not reasonable might well be relevant to a decision as to whether or not the payment should be made. It might well be thought to be anomalous for an authority, having decided to exercise their discretion to make a payment under section 34(4) and having proceeded to do so upon their own view of what was reasonable entirely in accordance with section 35(1) to (3), to be at risk of the amount of the payment being reviewed in every detail by the tribunal. They might then find that the amount to be paid was greater than had been assumed when the discretion was exercised simply because the tribunal took a different view of the facts. Assuming that a petition for judicial review would be competent in those cases where the payment is at the discretion of the

A authority — and Mr McNeill conceded this point, at least in those cases where the authority had already agreed to make the payment — this would still leave all questions which were pure questions of fact and degree to the authority. That seems to us to be consistent with the nature of a discretionary payment under section 34(4), as to which the displaced person has no right of action under the statute. It is also consistent with the wording of section 34(4), which says that the amount of the payment is to be "determined" in accordance with section 35(1) to (3) — determined, that is to say, by the authority, since the absence of a reference here to section 35(4) seems to us to exclude its determination under that subsection by the tribunal.

In our opinion, this is not one of those cases where it can be said B that to apply the words used by the legislature according to their ordinary meaning would produce a result which cannot have been intended by Parliament. It has been recognised that it may be permissible for a word to be struck out and another substituted for it where without such substitution the provision is unintelligible or absurd or totally unworkable or where by giving effect to the words used in it the operation of the statute would be rendered insensible, absurd or ineffective to achieve its evident purpose: see *R* v *Federal Steam Navigation Co Ltd* [1974] 1 WLR 505 *per* Lord Reid at p 509B and *McMonagle* v *Westminster City Council* [1990] 2 AC 716 at p 726E. But that is not the situation in the present case and we do not think that we would be justified in departing from the clear language of the statute which we would require to do if we were to give effect to C Mr Moynihan's argument.

We were referred to a decision of the Court of Appeal in *Gozra* v *Hackney London Borough* [1988] 2 EGLR 20 in which, contrary to the opinion which we have formed in this case, it was held that the Lands Tribunal had jurisdiction to determine a dispute as to the amount of a discretionary payment. That was a case where the claimant was a weekly tenant of a dwelling-house who, on being rehoused, claimed a disturbance payment. It was agreed that the payment could be made only under section 37(5) of the Land Compensation Act 1973, which is in the same terms as section 34(4) of the Scottish Act of 1973 except for the section numbers. It provides that the amount of the payment under that subsection is to be determined in accordance with section 38(1) to (3) of the English Act. D A preliminary question was raised as to the jurisdiction of the tribunal to determine the amount of a payment made or decided upon under section 37(5) which was in dispute. All three members of the Court of Appeal were of the opinion that the disputed question was referable to the tribunal.

That decision, which is not binding on us, is entitled to respect and we would be disposed to follow it if we could, since the legislation is, E for all practical purposes, identical. But we regret that we are unable to do so because we disagree with the opinions which were expressed. Under reference to the construction of the statute which we favour, Nourse LJ said at p 21:

> As a matter of cool statutory interpretation, that submission has some sterile appeal to it. But I do not think that it can be the function of the court to put such a construction on the Act unless the material provisions absolutely require it. As the Lands Tribunal pointed out, it would produce an incongruous difference of remedy for settling disputes as to the amount of section 37(1) payments on the one hand and section 37(5) payments on the other, even though they have to be determined according to precisely the same criteria. Not only would the remedy be different. In the case of a section 37(5) payment the remedy would be that of judicial review, a remedy far inferior tc F an absolute right to have the amount of the payment determined by the Lands Tribunal under section 38(4).

Having identified the problem in this way, he found the solution to it, with which Mann LJ and Lord Donaldson of Lymington MR both agreed, in holding that, since section 37(5) required there to be determined the equivalent of an amount equal to a disturbance payment, a dispute as to that amount was referable to the tribunal as being a dispute about the amount of a disturbance payment. We find ourselves unable to agree with this approach. It seems to us that the decision overlooks the significance of the point that the question whether to make the payment is entirely at the discretion of the authority and that, in contrast to a disturbance payment, there is no G entitlement to sue for this amount. The reference to the remedy of judicial review as a far inferior remedy seems to recognise, in our view correctly, that the amount of a discretionary payment could not be the subject of a dispute in the courts, since no action could be raised to recover it. But it does not seem to us to be incongruous in these circumstances that the only remedy in the event of a dispute should be that of judicial review, leaving all matters of fact and degree to the determination of the authority. Nor does it seem to be unreasonable that disputes about a disturbance payment, for which there is a right of action, are to be referred to the tribunal for determination there rather than in the exercise of its ordinary jurisdiction by the court.

For these reasons we consider that the decision reached by the H tribunal in this case was unsound and we shall allow this appeal. We answer the single question of law for City of Glasgow District Council and the first question of law for Yoker Housing Association both in the affirmative. We find it unnecessary to answer questions two to five for Yoker Housing Association, since these questions were not argued and are in any event superseded by our decision.

For a further case on this subject see p 265

HOUSING ACTS

A **Court of Appeal**

December 4 1991

(Before Lord Justice NEILL, Lord Justice WOOLF and Lord Justice SCOTT)

TOWER HAMLETS LONDON BOROUGH COUNCIL v MIAH

Estates Gazette May 9 1992

[1992] 18 EG 147

B **Housing Act 1985 — Secure tenancies — Local authority interest in dwelling-house held under licence — Whether agreement with occupying tenant excluded from secure tenancies by para 6, Schedule 1 to the Housing Act 1985 — Local authority's appeal allowed**

By an agreement dated March 1 1990 Lawlodge Property Services permitted the plaintiff council to use premises at 41A Hermitage Road, London N4, to provide temporary housing accommodation for homeless persons — The agreement was described as a licence — On the same date the defendant and his family were given temporary accommodation in the premises by the council — Following the council's offer of a
C **secure tenancy of permanent accommodation elsewhere, and the defendant's declining the offer, the plaintiff council served notice to quit on March 20 1991 and commenced proceedings seeking possession — Judge Tibber gave judgment for the defendant holding that he had a secure tenancy of the premises by reason of being a secure licensee, his licence not being excluded from being a secure licence by para 6 of Schedule 1 to the 1985 Act because the interest of the council was held by way of a licence and not a lease — The council appealed**

***Held*: The appeal was allowed — The terms of the**
D **agreement between the council and Lawlodge plainly demonstrated that it was a licence as so described — The terms "leased" and "lessor" in the context of para 6 of Schedule 1 to the Act not only include a lease in a technical sense and a lessor of such a lease but also cover persons who hold lesser interests under a licence — The word "leased" is used to indicate that the interest of the local authority landlord is to be a lesser interest than the freehold — That interpretation is in accord with what is the policy to be discerned from the schedule which was intended by Parliament — The defendant did not have a secure tenancy**

E The following case is referred to in this report.

Street v *Mountford* [1985] AC 809; [1985] 2 WLR 877; [1985] 2 All ER 289; [1985] 1 EGLR 128; (1985) 274 EG 821, HL

This was an appeal by the plaintiffs, Tower Hamlets London Borough Council, from an order dated October 22 1991 of Judge Tibber in Edmonton County Court dismissing a claim by the plaintiffs against the defendant, Mr Sabru Miah, for possession of 41A Hermitage Road, London N4.

Ashley Underwood (instructed by the acting head of legal services to Tower Hamlets London Borough Council) appeared for the appellants; Martin Seaward (instructed by T V Edwards) represented F
the respondent.

Giving judgment, **Neill LJ** said: This is an appeal by Tower Hamlets London Borough Council against the order dated October 22 1991 of Judge Tibber in Edmonton County Court dismissing a claim by the council against Mr Sabru Miah for possession of 41A Hermitage Road, London N4. The judge held that Mr Miah was a secure licensee of the premises and was therefore entitled to the protection afforded by Part IV of the Housing Act 1985.

Between about November 7 1986 and March 1 1990 the council provided accommodation to Mr Miah and his family at premises other than 41A Hermitage Road. The accommodation was provided on the basis that Mr Miah and his family were homeless. G

On March 1 1990 the council made an agreement with a firm called Lawlodge Property Services permitting the council to use the premises at 41A Hermitage Road for providing temporary housing accommodation for homeless persons. By clause 4 of the agreement the council were given the right to nominate persons, being homeless persons as defined by Part III of the Housing Act 1985, to occupy the premises. The agreement was described as a licence, and the council were described as the licensee. Clause 6 of the additional clauses provided that the licence might be terminated by either party giving not less than seven days' notice in writing. The agreement also contained a provision whereby the licensee (that is the council) agreed:

> H
> to use the Premises for no other purpose than as temporary housing accommodation in accordance with the Licensee's powers under the provisions of the Housing Act 1985 Part 3 and to permit the Grantor [that is the firm] to inspect any agreement licence or other documentation under or by virtue of which the Nominees are temporarily accommodated in the Premises.

In the further clause 3 the licensee agreed "not to do anything whereby the Licensee creates as between itself and any nominee any secure or other tenancy".

On the same day, March 1 1990, Mr Miah and his family moved to 41A Hermitage Road. Six weeks or so later, on May 16 1990, the council made an offer to Mr Miah of a secure tenancy at 33 Oban House, Oban Street, London E14. Mr Miah declined the offer. The council said that when he declined it he declined it unreasonably. On J
March 20 1991, some 10 months later, the council served a notice to quit. The present proceedings were started by the issue of the plaint on August 15 1991.

It is necessary to refer next to some of the provisions of the Housing Act 1985. The Act contains a number of Parts. Part II is concerned with the provision of housing accommodation, Part III is concerned with housing the homeless and Part IV is concerned with secure tenancies and the rights of secure tenants. A number of sections in Part III dealing with housing the homeless impose duties on local housing authorities with regard to homeless persons. Specific duties are imposed requiring the authority to secure that accommodation is made available to such persons.

In the present case Mr Miah relies on the provisions of section 79 of K
the Act, which is the first section in Part IV. That section provides as follows:

> **79.**—(1) A tenancy under which a dwelling-house is let as a separate dwelling is a secure tenancy at any time when the conditions described in sections 80 and 81 as the landlord condition and the tenant condition are satisfied.
>
> (2) Subsection (1) has effect subject to —
>
> (*a*) the exceptions in Schedule 1 (tenancies which are not secure tenancies) . . .

It is not necessary to refer to the other paragraphs of subsection (2). I come next to subsection (3), which is in these terms:

The provisions of this Part apply in relation to a licence to occupy a dwelling-house (whether or not granted for a consideration) as they apply in relation to a tenancy.

It is submitted on behalf of Mr Miah that he was a licensee of this property, no 41A and that (unless Schedule 1 applied) he was therefore a secure licensee by reason of the combined effect of section 79(1) and section 79(3), because it is common ground that the landlord condition described in section 80 is satisfied in this case, as is the tenant condition in section 81. Accordingly, the question in issue is whether the licence under which Mr Miah occupies no 41A is excluded from being a secure licence by the terms of para 6 of Schedule 1.

Schedule 1 to the 1985 Act sets out in 12 paragraphs a list of circumstances in which tenancies are not secure tenancies. It is not necessary to refer to more than three of those paragraphs. We are particularly concerned with para 6, but I must come first to para 4, which is headed *"Accommodation for homeless persons"* and provides as follows:

4.—(1) A tenancy granted in pursuance of —
- *(a)* section 63 (duty to house pending inquiries in case of apparent priority need),
- (*b*) section 65(3) (duty to house temporarily person found to have priority need but to have become homeless intentionally), or
- (*c*) section 68(1) (duty to house pending determination whether conditions for referral of application are satisfied),

is not a secure tenancy before the expiry of the period of twelve months beginning with the date specified in sub-paragraph (2), unless before the expiry of that period the tenant is notified by the landlord that the tenancy is to be regarded as a secure tenancy.

It is unnecessary to read subpara (2) of para 4.

Para 6 is headed *"Short-term arrangements"*. That is in these terms:

6. A tenancy is not a secure tenancy if —
- (*a*) the dwelling-house has been leased to the landlord with vacant possession for use as temporary housing accommodation,
- (*b*) the terms on which it has been leased include provision for the lessor to obtain vacant possession from the landlord on the expiry of a specified period or when required by the lessor,
- (*c*) the lessor is not a body which is capable of granting secure tenancies, and
- (*d*) the landlord has no interest in the dwelling-house other than under the lease in question or as a mortgagee.

Then, finally, para 12 — it is not necessary to refer to all its terms — provides that in the case of almshouses a licence to occupy a dwelling-house is not a secure tenancy if the licence was granted by an almshouse charity, and then there are certain other conditions to which it is unnecessary to refer.

It is accepted on behalf of Mr Miah that if the agreement between the council and the company dated March 1 1990 had been in the form of a lease para 6 would apply and Mr Miah's licence would not have been a secure licence. It is argued, however, that as the agreement between the council and the company was both in form and in substance a licence rather than a lease, therefore the exception in para 6 does not apply.

The judge held first that though the description of a document as a licence or a lease is not conclusive, bearing in mind the decision of the House of Lords in *Street* v *Mountford* [1985] AC 809*, nevertheless looking at the facts of this case the terms were consistent with a licence and were wholly inconsistent with a lease. He further held, however, that the word "leased" in para 6 of Schedule 1 meant a lease rather than a licence and was incapable of meaning anything less than a lease. In his judgment, having dealt with the question of whether the document was a licence or a lease, he put the matter as follows:

That being so, does para 6(a) apply to it?

— then he read the paragraph —

The landlords apparently acquired it with vacant possession for use as temporary housing accommodation, but how did they acquire it, not by lease but by licence. I think it would be wrong to go beyond the terms of the statute and say that the same deprivation of security follows a licence as follows a lease. It would have been perfectly open to the legislature to use some term wider than being "leased" to the landlord but has not chosen to do so. In those circumstances, it seems to me that Mr Miah is a secure [licensee]

— I think that has been amended —

by virtue of section 79 of the Housing Act, 1985.

In this court on behalf of the council Mr Underwood has sought to argue two points. First, he submits that the judge was wrong in coming to the conclusion that this document was a licence rather than a lease. He has drawn our attention to provisions in it which he says are consistent with exclusive possession being given to the local authority, to the fact that there was an obligation on the local authority to make certain that vacant possession was given up at the end of the licence, and to the fact that there was a specific provision in the agreement which required the local authority to ensure that no secure tenancy came into existence. He says these provisions and indeed all the other terms of the agreement were perfectly consistent with a lease and that this should be construed as being a lease.

I find it impossible to accept that argument. It seems to me that if one reads the terms of the agreement they quite plainly demonstrate that this was a licence as it was described. The judge went through its terms. I find it unnecessary to add to what the judge stated as his reasons. He pointed, among other things, to the fact that the weekly fee which was payable in respect of it was payable in respect of a household and that during a period when no one was in occupation as a homeless person no fee became payable. He said that looking at all the provisions of this document they were wholly consistent with the normal terms of a lease. I agree with him and I do not think it is necessary to say any more about it.

I therefore turn to what is the more difficult part of this case. Mr Underwood as his second argument, adopting and elaborating an argument which was originally put to him by the court, makes the submission that in the context of para 6 the terms "leased" and "lessor" not only must mean or include a lease in a technical sense and a lessor of such a lease but also cover persons who hold lesser interests under a licence. In the particular context of this case "lease" can include "licence" and "lessor" can include "licensor".

I see the force of the argument which was urged upon us by counsel for Mr Miah that section 621 of the Act contains a definition of "lease", that the word "lease" is used in this paragraph of Schedule 1, and that if Parliament had wished to say "lease" or "licence" they could quite easily have done so. Mr Seaward quite rightly points out that in para 12 of Schedule 1, when dealing with almshouses, Parliament has in fact referred to a licence to occupy a dwelling-house and it would have been a simple matter to have included some similar provision in para 6.

In the course of the argument we have considered other provisions of the Housing Act, provisions of the Local Government Act (section 111), and most recently the provisions of the Local Government and Housing Act 1989, and have looked to see whether there is any guidance as to the meaning of "lease" or any reason which there may be which would lead one to the conclusion that the word "lease" must be construed in this particular context in a narrow sense. Having had the benefit of that argument, I remain of the opinion, which I formed when I studied these papers at the outset, that in para 6 of Schedule 1 the words "leased" and "lessor" must be construed in their context as including "licensed" and "licensor". At the same time I should make it clear that I accept that section 79(3) is concerned with licences to occupy granted by the local authority and that it does not apply directly to the contract between the local authority and the owner of the property.

Nevertheless, it is necessary, as I see it, to look at section 79 and the exceptions set out in Schedule 1 together. By section 79(1) provision is made for the creation of secure tenancies in the circumstances there described. By section 79(3) a similar protection is conferred on licences to occupy. But from the class of secure tenancies certain tenancies which would otherwise qualify as secure tenancies are excluded by section 79(2)(a) and by the Schedule. I have already read para 4 of Schedule 1. That excludes tenancies granted to homeless persons under some, but not all, of the provisions relating to the housing of homeless persons in Part III of the Act.

Para 6 is concerned with persons who are housed under short-term arrangements where the dwelling-house is provided by a third person so that the local authority or some other body specified in section 80 of the Act can use it to provide *temporary housing accommodation*.

Paras 4 and 6 demonstrate to my mind that where the occupation of the dwelling-house is to be on a short-term basis the special status of a secure tenant is not conferred. The argument on behalf of Mr Miah involves the proposition that someone who occupies the house as a licensee from the local authority has a security which is not available to a tenant in precisely similar circumstances, though it is right to say that Mr Seaward was prepared to accept that in a sense it is an illusory security because, although as between himself and the

*Editor's note: Also reported at [1985] 1 EGLR 128.

A local authority the licensee may be secure, if the head licence between the local authority and the owner is determined then the owner of the property can recover possession against the licensee (that is the homeless person) as a trespasser. That may be so, though I have an anxiety in my mind as to whether it is consistent with section 82 of the Housing Act 1985 for the local authority themselves to determine such a licence and therefore bring to an end indirectly a secure tenancy or licence which they have themselves created.

But it is unnecessary to explore that matter. It seems to me that it is unsatisfactory, when one is considering the proper meaning of para 6, to see whether there is some other way in which the matter can be dealt with. I find it impossible to accept that Parliament intended that B a person who was a tenant might be prevented from becoming a secure tenant by reason of the provisions of para 6 but that someone who held the lesser status of a licensee should remain a secure licensee. It seems to me that though the word "leased" is used, the exception in para 6 must have been intended by Parliament to cover circumstances where, provided the other conditions in the paragraph are satisfied, the owner of the house has made it available to the local authority under either a lease or a licence agreement. As I see it, the word "leased" is used to indicate that the interest of the landlord local authority is to be a lesser interest than the freehold. But a leased dwelling-house includes a dwelling-house held under a lesser interest C such as a licence.

For these reasons, though I fully understand the arguments which led the learned judge in his careful judgment to reach a contrary conclusion, I for my part would allow this appeal.

Agreeing, **WOOLF LJ** said: I would allow the appeal for the reasons that Neill LJ gives. It seems to me that that is in accord with what is the policy to be discerned from the Schedule which was intended by Parliament, that policy being to make better use of existing houses so as to enable them to be used for the accommodation of the homeless.

SCOTT LJ also agreed and said: It simply makes no sense to hold that a secure licence obtained from a local authority holding a licence D gives a greater security than a secure licence obtained from a local authority that hold a lease strictly so-called. It is, in my judgment, a permissible construction of para 6 of Schedule 1 to the 1985 Act to hold that the greater includes the less, that lease includes licence and that para 6 should be construed accordingly.

Appeal allowed; no order as to costs; legal aid taxation of respondent's costs; application for leave to appeal to the House of Lords refused.

LANDLORD AND TENANT
GENERAL

A **House of Lords**

December 21 1991

(Before Lord KEITH OF KINKEL, Lord TEMPLEMAN, Lord OLIVER OF AYLMERTON, Lord GOFF OF CHIEVELEY and Lord JAUNCEY OF TULLICHETTLE)

BILLSON AND OTHERS v RESIDENTIAL APARTMENTS LTD

Estates Gazette January 11 1992

[1992] 01 EG 91

B **Landlord and tenant — Forfeiture — Whether tenant entitled to relief from forfeiture after landlord has peaceably re-entered — Held that tenant so entitled**

By a lease dated July 15 1964 the appellant, Residential Apartments Ltd, held premises at 17 Gledhow Gardens, London SW5, for a term expiring on June 25 1997 — The respondent trustees held the reversion expectant upon the term — By the lease the appellant covenanted not to alter or add to the premises without the lessors' consent — The lease contained a provision entitling the lessors to re-enter upon failure by the lessee to comply with any covenant — By a
C **notice dated July 4 1989 given pursuant to section 146 of the Law of Property Act 1925, the respondents alleged that the appellant had carried out works of alteration without consent, and required the breach to be remedied — On July 18 1989 the respondents effected peaceable re-entry by entering the premises and changing the locks — Within four hours the appellant had recovered possession**

The respondents issued proceedings claiming possession, damages for breach of covenant and damages for trespass — The appellant counterclaimed seeking relief from forfeiture — The trial judge and the Court of Appeal both held that they
D **had no jurisdiction to grant relief — The appellant appealed**

***Held*: The appeal was allowed — The proposition that a landlord who has peaceably re-entered premises is no longer "proceeding" to enforce his rights within the meaning of section 146(2) of the Law of Property Act 1925 is unsound — Section 146(2) enables a tenant to apply for relief whenever and however the landlord claims that the lease has been determined for breach of covenant — *Rogers* v *Rice* distinguished — A tenant cannot apply for relief from forfeiture after a landlord has forfeited a lease by issue and service of proceedings, has recovered judgment and obtained**
E **possession pursuant to that judgment; if such a judgment is set aside or successfully appealed, the tenant will be able to apply for relief in the landlord's action**

The trial judge's order for costs on an indemnity basis against the appellant should not have been made — The practice of ordering indemnity costs as a condition of granting relief from forfeiture is ripe for reconsideration

The following cases are referred to in this report.

Canas Property Co Ltd v *K L Television Services Ltd* [1970] 2 QB 433; [1970] 2 WLR 1113; [1970] 2 All ER 795; (1970) 21 P & CR 601; [1970] EGD 445; 214 EG 1494, CA

Egerton v *Jones* [1939] 2 KB 702 F

Jacques v *Harrison* (1884) 12 QBD 165

Jones v *Carter* (1846) 15 M&W 718

Pakwood Transport Ltd v *15 Beauchamp Place Ltd* (1977) 36 P&CR 112; [1978] EGD 233; 245 EG 309, CA

Quilter v *Mapleson* (1882) 9 QBD 672

Rexdale Investments Ltd & Gibson, Re [1967] 1 Ont LR 251

Rogers v *Rice* [1892] 2 Ch 170; (1892) 8 TLR 511

This was an appeal by the tenant, Residential Apartments Ltd, from the Court of Appeal (see [1991] 1 EGLR 70), which had dismissed an appeal from a decision of Mummery J, who had held that the respondent trustees were entitled to possession of premises at 17 Gledhow Gardens, Earl's Court, London SW5. G

Michael Driscoll (instructed by Gouldens) appeared for the appellant; and Robert Reid QC and Gordon Bennett (instructed by Frere Cholmeley) represented the respondents.

In his speech, **Lord Templeman** said: By a lease dated July 15 1964 freeholders demised the premises, 17 Gledhow Gardens, South Kensington, for a term due to expire by effluxion of time on June 25 1997. On May 3 1989 the lease was assigned to the appellant, Residential Apartments Ltd. The reversion is vested in the respondent trustees.

The provisions of the lease included a covenant by the tenant:

> (f) . . . not to make any alteration in or addition to the demised premises or any part thereof without the consent in writing of the Lessors first had and obtained H

The lease also contained the following proviso for re-entry:

> (i) If and whenever any rent hereby reserved shall be in arrear for twenty-one days after becoming due (whether legally demanded or not) or if and whenever any covenant by the Lessee hereinbefore contained (other than the covenant to pay rent) shall not be performed or observed then and in any such case the Lessors may by re-entry or (at their option) by notice in writing in that behalf given to the Lessees determine this demise and thereupon this demise shall absolutely determine but without prejudice to any right of action or remedy of the Lessors in respect of any breach of covenant by the Lessees.

In breach of covenant the appellant embarked on alterations to the demised premises without the consent of the trustees. Section 146(1) of the Law of Property Act 1925 provides that: J

> A right of re-entry or forfeiture under any proviso or stipulation in a lease for a breach of any covenant or condition in the lease shall not be enforceable, by action or otherwise, unless and until the lessor serves on the lessee a notice —
> (*a*) specifying the particular breach complained of; and
> (*b*) if the breach is capable of remedy, requiring the lessee to remedy the breach; and
> (*c*) in any case, requiring the lessee to make compensation in money for the breach;
> and the lessee fails, within a reasonable time thereafter, to remedy the breach, if it is capable of remedy, and to make reasonable compensation in money, to the satisfaction of the lessor, for the breach. K

The trustees served on the appellant a notice dated July 4 1989 complying with section 146(1) but the appellant did not remedy its breach of covenant within a reasonable time.

Section 146(2) of the Law of Property Act 1925 provides that:

> Where a lessor is proceeding, by action or otherwise, to enforce such a right of re-entry or forfeiture, the lessee may, in the lessor's action, if any, or in any action brought by himself, apply to the court for relief; and the court may grant or refuse relief, as the court, having regard to the proceedings and conduct of the parties under the foregoing provisions of this section, and to all the other circumstances, thinks fit; and in case of relief may grant it on such terms, if any, as to costs, expenses, damages, compensation, penalty, or

otherwise, including the granting of an injunction to restrain any like breach in the future, as the court, in the circumstances of each case, thinks fit.

On July 18 1989 at 6 am agents for the trustees peaceably re-entered the demised premises, which were vacant, and changed the locks. By 10 am on the same day workmen engaged by the appellant had retaken possession of the property, as the trustees say, unlawfully.

By the writ in this action dated July 19 1989 the trustees claim possession, damages for breach of covenant and damages for trespass. By its defence and counterclaim the appellant counterclaims for relief against forfeiture. By their reply the trustees claim that the court has no jurisdiction to grant the appellant relief from forfeiture. The trial judge, Mummery J, and the Court of Appeal (Sir Nicolas Browne-Wilkinson V-C, and Parker and Nicholls LJJ) considered that they were constrained by authority to hold that the court had no jurisdiction to grant the appellant relief against forfeiture pursuant to section 146(2) because the appellant had not applied to the court for relief prior to the re-entry into possession by the trustees on July 18 1989. The appellant now appeals.

By the common law, when a tenant commits a breach of covenant and the lease contains a proviso for forfeiture, the landlord at his option may either waive the breach or determine the lease. In order to exercise his option to determine the lease the landlord must either re-enter the premises in conformity with the proviso or must issue and serve a writ claiming possession. The bringing of an action to recover possession is equivalent to an entry for the forfeiture. Thus in *Jones* v *Carter* (1846) 15 M & W 718 Parke B said at p 726:

> . . . the bringing of an ejectment for a forfeiture, and serving it on the lessee in possession, must be considered as the exercise of the lessor's option to determine the lease; and the option must be exercised once for all . . . for after such an act, by which the lessor treats the lessee as a trespasser, the lessee would know that he was no longer to consider himself as holding under the lease, and bound to perform the covenants contained in it . . .

This observation was cited and applied by Lord Denning MR in *Canas Property Co Ltd* v *K L Television Services Ltd* [1970] 2 QB 433 at p 440.*

Before the intervention of Parliament, if a landlord forfeited by entering into possession or by issuing and serving a writ for possession, equity could relieve the tenant against forfeiture but only in cases under the general principles of equity whereby a party may be relieved from the consequences of fraud, accident or mistake or in cases where the breach of covenant entitling the landlord to forfeit was a breach of the covenant for payment of rent.

Mr Reid QC, who appeared for the trustees, conceded that, where equity claimed power to relieve against forfeiture, the tenant could apply for relief irrespective of the method by which the landlord had exercised his option to determine the lease. Relief could be granted whether the landlord had forfeited by entering into possession or had forfeited by issuing and serving a writ claiming possession.

In 1881 Parliament interfered to supplement equity and to enable any tenant to be relieved from forfeiture. The need for such intervention was and is manifest because otherwise a tenant who had paid a large premium for a 999-year lease at a low rent could lose his asset by a breach of covenant which was remediable or which caused the landlord no damage. The forfeiture of any lease, however short, may unjustly enrich the landlord at the expense of the tenant. In creating a power to relieve against forfeiture for breach of covenant Parliament protected the landlord by conferring on the court a wide discretion to grant relief on terms or to refuse relief altogether. In practice this discretion is exercised with the object of ensuring that the landlord is not substantially prejudiced or damaged by the revival of the lease.

Section 14(1) and (2) of the Conveyancing and Law of Property Act 1881 were provisions which conferred on the court power to relieve against forfeiture and those provisions were reproduced in section 146(1) and (2) of the Law of Property Act 1925 in identical terms. In referring to a section 146 notice I shall therefore mean and include a notice served under section 14(1) of the Act of 1881, and in referring to section 146(1) and (2) I shall mean and include section 14(1) and (2) of the Act of 1881 where appropriate.

Section 146(1) prevents the landlord from enforcing a right of re-entry or forfeiture by action or otherwise so that the landlord cannot determine the lease by issuing and serving a writ or by re-entering the premises until the tenant has failed within a reasonable time to remedy the breach and make reasonable compensation. Section 146(2) enables the tenant to apply to the court for relief where the landlord "is proceeding, by action or otherwise," to enforce his right of re-entry or forfeiture. If the landlord "is proceeding" to determine the lease by issuing and serving a writ, the tenant may apply for relief after the writ has been served. If the landlord "is proceeding" to determine the lease by re-entering into possession, the tenant may apply for relief after the landlord has re-entered.

Mr Reid submitted and referred to authority for the proposition that on the true construction of section 146(2) a tenant cannot apply for relief against forfeiture after the landlord has re-entered without obtaining a court order. Thereafter the landlord is no longer "proceeding" to enforce his rights; he has succeeded in enforcing them. The proposition is, in my opinion, historically unsound because the effect of issuing and serving a writ is precisely the same as the effect of re-entry; in each case the lease is determined. The landlord is entitled to remain in possession if he has re-entered and he is entitled to possession if he has issued and served a writ because the lease no longer exists. In each case the tenant seeks relief because the lease has been forfeited. The proposition is also inconsistent with the language of section 146(2). The tenant may apply for relief where the landlord is "proceeding, by action or otherwise" to enforce his rights. The tenant may apply for relief where the landlord is "proceeding" by action and also where the landlord is proceeding "otherwise" than by action. This can mean only that the tenant may apply for relief where the landlord is proceeding to forfeit by re-entry after the expiry of a section 146 notice. If re-entry bars relief, the right of the tenant to apply for relief where the landlord is proceeding otherwise than by action is substantially inoperative and the words "or otherwise" in section 146(2) have no application. In my opinion, those words must have been included because Parliament intended that a tenant should be able to obtain relief against a landlord whether the landlord has asserted his rights by a writ or by re-entering. It is said that a tenant served with a section 146 notice could during and after the expiration of the notice apply for relief under section 146(2) but if he fails to do so he is at the mercy of the landlord who decides to re-enter and whose rights are therefore, it is said, quite unaffected by the provisions of section 146(2) designed to relieve tenants from the consequences of breach of covenant. In my opinion, the ambiguous words "is proceeding" can mean "proceeds" and should not be construed so as to produce the result that a tenant served with a section 146 notice can ensure that he will be able to apply for relief only if he does so before he knows whether the landlord intends to proceed at all or whether, if the landlord decides to proceed, he will issue and serve a writ or will attempt to re-enter.

When a tenant receives a section 146 notice he will not know whether the landlord can be persuaded that there is no breach or persuaded to accept in due course that any breach has been remedied and that he has been offered adequate and satisfactory compensation or whether the landlord will seek to determine the lease by issuing and serving a writ or will seek to determine the lease by re-entering the premises. The tenant will not wish to institute proceedings seeking relief from forfeiture if those proceedings will be aggressive and hostile and may be premature and unnecessary. Parliament cannot have intended that if the landlord employs the civilised method of determining the lease by issuing and serving a writ, then the tenant will be entitled to apply for relief, but if the landlord employs the dubious and dangerous method of determining the lease by re-entering the premises, then the tenant will be debarred from applying for relief.

Mr Reid concedes that re-entry can avail the landlord only if the entry is lawful. Re-entry is unlawful where the premises are occupied by the tenant but not unlawful where the premises are occupied by the tenant's goods. If the argument of the trustees is correct, section 146 provides a method by which a landlord can sneak up on a shop at night, break into the shop, and install new locks so that the tenant loses his lease and can only press his nose against the shop window, being unable to obtain the assistance of the court because he has become a trespasser entitled to no rights and to no relief. The farce in the present case when the landlord occupied the premises for four hours should not be allowed to defeat the statutory rights of the tenant.

The right conferred by section 146(2) on a tenant to apply for relief against forfeiture may, without violence to the language, be construed as a right to apply "where a lessor *proceeds,* by action or otherwise," to enforce a right of re-entry. So construed, section

*Editor's note: Also reported at (1970) 214 EG 1494.

146(2) enables the tenant to apply for relief whenever and however the landlord claims that the lease has been determined for breach of covenant. I have no doubt that this was the object and intention and is the effect of section 146.

In *Quilter* v *Mapleson* (1882) 9 QBD 672 a landlord forfeited a lease before the Act of 1881 came into force by issuing and serving a writ for possession. He recovered judgment, the tenant appealed and the Act of 1881 came into force before the appeal was heard. The Court of Appeal held that the Act was retrospective and granted relief to the tenant. Lindley LJ at p 676 decided that section 14(2) was applicable:

> The action was brought by the landlord on the ground of breaches committed before the Act, and he obtained judgment before the Act came into operation, but execution was stayed, so that he has never obtained possession. The original action then is not yet at an end . . . so long as the tenant has not been turned out of possession he is within the terms of the enactment, for the lessor is "proceeding to enforce" his right of re-entry. The enactment then being in terms retrospective must be construed according to its terms as being retrospective.

The judgments of Sir George Jessel MR and Bowen LJ were to the like effect and it is now settled law that where a landlord forfeits a lease by issuing and serving a writ for possession the tenant may apply for relief before but not after the landlord has recovered judgment and re-entered. But although the court limited the time during which a tenant could apply for relief against forfeiture constituted by the issue and service of the writ, the court had no power and, in my opinion, did not intend to deprive a tenant of any right to apply for relief after a forfeiture constituted by re-entry without judgment. *Quilter* v *Mapleson* is authority for a case where the landlord forfeits by issue and service of a writ but is not authority for a case where the landlord forfeits by re-entry.

In *Rogers* v *Rice* [1892] 2 Ch 170 a landlord forfeited by the issue and service of a writ recovered judgment and re-entered pursuant to the writ of possession then issued and was held to be no longer "proceeding by action" within section 146(2). The tenant sought and was refused leave to set aside the verdict and the judgment. The tenant later issued an originating summons seeking relief from forfeiture under section 146(2). Lord Coleridge at pp 171-172 said that a section 146 notice had been given and ignored, and continued:

> The action proceeded to judgment, the judgment was executed, so far as possession was concerned, and at the time when the present proceeding was commenced the lessor was in possession. The action then, so far as related to enforcing the right of re-entry, was at end, and it cannot be said that the landlord was "proceeding" to enforce his right of re-entry. The case is clear on the terms of the Act, but I cannot omit to notice that the same view was taken by the judges of the Court of Appeal in *Quilter* v *Mapleson* 9 QBD 672, 677, where all three judges gave their opinion to this effect, though that was not the point on which their decision turned.

The decision can be supported on the grounds that no court could properly exercise its discretion to relieve against forfeiture after the landlord had issued and served a writ, recovered judgment in the action and entered into possession pursuant to that judgment. The decision can also be supported on the grounds set out in the speech of my noble and learned friend Lord Oliver of Aylmerton. But the court had no power and, in my opinion, did not intend to deprive a tenant of any right to apply for relief after a forfeiture constituted by re-entry without judgment.

In *Pakwood Transport Ltd* v *15 Beauchamp Place Ltd* (1977) 36 P&CR 112* the Court of Appeal rejected an argument by a landlord who had served a section 146 notice that the tenant could not apply for relief from forfeiture until proceedings for forfeiture had been instituted by the landlord. All three lord justices derived from the cases of *Quilter* v *Mapleson* (1882) 9 QBD 672 and *Rogers* v *Rice* [1892] 2 Ch 170 the proposition that in the words of Orr LJ at p 117:

> . . . a lessee could not apply for relief against re-entry or forfeiture after the landlord had obtained a judgment of the court entitling him to re-enter on a forfeiture; and it is claimed, and in my judgment rightly claimed, that the same principle must apply where the landlord has peacefully recovered possession. In other words, once he has either recovered possession or obtained an order for possession he can no longer be said to be "proceeding by action or otherwise to enforce a right of re-entry or forfeiture".

My lords, I accept that it is now settled law that a tenant cannot apply for relief after the landlord has recovered judgment for possession and has re-entered in reliance on that judgment. But I do not accept that any court has deprived or is entitled to deprive a tenant of any right to apply for relief if the landlord proceeds to forfeit otherwise than by an action instituted for that purpose.

Orr LJ continued:

> On this basis the argument for the lessor appears to me to involve an absurdity, in that if the landlord has done no more than serve a section 146 notice, it is too early for the tenant to apply for relief; but if the landlord's next step is peaceably to recover possession, it is then too late for the tenant to apply. For my part, I am not prepared to accept an argument which leads to this absurdity, and I have no hesitation in holding that a landlord who serves a section 146 notice is at that stage "proceeding to enforce a right of re-entry or forfeiture" in that the service of such a notice is a step which the law requires him to take in order to re-enter or forfeit.

My lords, I accept the conclusion that a landlord who serves a notice under section 146(1) can be said, for the purposes of section 146(2), to be proceeding to enforce his rights under the lease. A tenant authorised by section 146(2) to apply to the court for relief against forfeiture if he fails to comply with a section 146 notice may make that application after service of the notice for the purpose of elucidating the issues raised by the notice, ascertaining the intentions of the landlord, and setting in train the machinery by which the dispute between the landlord and the tenant can be determined by negotiation or by the court. But the fact that the tenant may apply to the court for relief after service of the section 146 notice does not mean that if he does not do so he loses the right conferred on him by section 146(2) to apply for relief if and when the landlord proceeds, not by action but "otherwise" by exercising a right of re-entry. No absurdity follows from a construction which allows the tenant to apply for relief before and after a landlord re-enters without first obtaining a court order.

In the words of Laskin JA in *Re Rexdale Investments Ltd & Gibson* [1967] 1 Ont LR 251 at p 259 dealing with provisions in the Ontario legislation indistinguishable from section 146(2), the argument that a tenant cannot apply for relief after a landlord has determined the lease by re-entry:

> depends on a detached grammatical reading of the phrase "is proceeding" . . . which makes nonsense of the phrase "or otherwise" (as covering physical re-entry) by making ineffective, in any practical sense, the provision for relief from forfeiture applicable to such re-entry. We do not construe statutes, especially when they are remedial . . . to the point of self-contradiction. In my opinion, the phrase "is proceeding" is more properly read in the sense of "has proceeded", and I am fortified in this view by the fact that the exercise of the power of termination is manifested effectively by the mere taking of proceedings as well as by physical re-entry. What [section 146(2)] means, therefore, is that when the landlord has terminated the lease by action or by actual re-entry without action, the tenant may seek relief from forfeiture in the pending action, if any, or, if none, by proceedings initiated by him. In the latter case, one would expect prompt reaction by the tenant . . . The English cases relied on . . . [*Rogers* v *Rice* [1892] 2 Ch 170; *Locke* v *Pearce* [1893] 2 Ch 271 and *Quilter* v *Mapleson* (1882) 9 QBD 672], are distinguishable, if need be . . . by the fact . . . that they relate to re-entry in pursuance of a judgment for possession.

These observations by a distinguished Canadian judge, who subsequently became Chief Justice of the Supreme Court of Canada, support the views which I have formed concerning the construction of section 146 and the ambit and effect of the earlier decisions.

Mr Reid argued that your lordships should not interfere with 19th-century decisions and for my part I do not intend to do so on this occasion or to question the result of the decision of the Court of Appeal in *Pakwood Transport Ltd* v *15 Beauchamp Place Ltd* (1977) 36 P&CR 112. The authorities were never directed to the point now in issue and certainly never decided that issue.

It was suggested that Parliament in 1925 accepted the views expressed in the 19th-century cases. I agree that Parliament accepted that a tenant cannot apply for relief under section 146(2) after the landlord has forfeited the lease by issuing and serving a writ for possession and in his action has recovered and enforced judgment. The 19th-century cases were not directed to the problem which has now emerged.

We were informed that the researches of counsel had not disclosed any reported case in which a landlord has forfeited by re-ɜntry and then successfully denied the right of the tenant to apply for relief.

The trustees or their advisers, perhaps incensed by the activities of the appellant in the present case, conceived and carried out a dawn raid which fortunately did not result in bloodshed. Since the decision of the Court of Appeal in the instant case there has been a proliferation of section 146 notices followed by pressure on tenants to surrender on terms favourable to the landlord. If this appeal were not allowed, the only safe advice for a tenant would be to issue proceedings for relief against forfeiture as soon as a section 146

*Editor's note: Also reported at (1977) 245 EG 309.

A notice is received at a time when the tenant cannot know whether relief will be necessary. A tenant ignorant of the development in the law pioneered by the trustees in the present case will be at the mercy of an aggressive landlord. The conclusions which I have reached will not entail these consequences and will not again involve Parliament in correcting judicial constructions of statute by further legislation.

The results of section 146 and the authorities are as follows. A tenant may apply for appropriate declarations and for relief from forfeiture under section 146(2) after the issue of a section 146 notice but he is not prejudiced if he does not do so. A tenant cannot apply for relief after a landlord has forfeited a lease by issuing and serving a writ, has recovered judgment and has entered into possession
B pursuant to that judgment. If the judgment is set aside or successfully appealed the tenant will be able to apply for relief in the landlord's action but the court in deciding whether to grant relief will take into account any consequences of the original order and repossession and the delay of the tenant. A tenant may apply for relief after a landlord has forfeited by re-entry without first obtaining a court order for that purpose, but the court in deciding whether to grant relief will take into account all the circumstances including delay on the part of the tenant. Any past judicial observations which might suggest that a tenant is debarred from applying for relief after the landlord has re-entered without first obtaining a court order for that purpose are not to be so construed.

I would therefore allow the appeal and set aside the orders of the
C trial judge and the Court of Appeal. The trustees must pay the costs of the appellant in this House. There will be no order for costs in the Court of Appeal because a number of issues were decided against the appellant and there is no appeal from any decision of the Court of Appeal other than the decision on the right of the appellant to appeal for relief against forfeiture. The application of the appellant for relief against forfeiture will be remitted to the High Court. The judge would have been prepared to grant relief against forfeiture except for the fact that he considered that he had no jurisdiction to do so. On the renewal of the appellant's application for relief the parties will be at liberty to file and adduce further evidence in order that the judge may consider the propriety of granting relief and the terms of relief on the basis of the facts as at the date of the hearing of the renewed
D application for relief. The costs of the original hearing of all the issues by the trial judge and the costs of the renewed application before the trial judge will be in the discretion of the judge who hears the renewed application.

When the judge refused relief to the appellant on the grounds of jurisdiction he made an order for costs against the appellant on an indemnity basis. In my opinion, such an order should not be made. There is no reason why an unsuccessful applicant for relief should be in any worse case than any other unsuccessful litigant. Moreover, if a landlord is entitled to costs on an indemnity basis, whether relief is obtained or not, there will be no inducement to the landlord to compromise his dispute with the tenant.

E For my part I consider that the practice of ordering indemnity costs as a condition of granting relief is ripe for reconsideration. Indemnity costs to a landlord encourage lawyers and surveyors and other advisers to charge large fees. I make no criticism of the advisers of the trustees because I do not know what fees and charges they intend. But it seems to me that, in principle, a tenant should not be at the mercy of an order made by a judge who has no means of knowing the effect of the order and imposes no impartial criterion by which costs can be taxed down.

In his speech, **Lord Oliver of Aylmerton** said: I have had the advantage of reading in draft the speech delivered by my noble and learned friend, Lord Templeman, and I agree that the appeal should
F be allowed for the reasons which he has given. I add a few words of my own only in deference to the very clear and helpful argument presented to your lordships by Mr Reid QC on behalf of the respondents, which, at one point during the hearing, appeared to me to present an insurmountable obstacle in the appellant's path.

Broadly the argument was to this effect. *Rogers* v *Rice* [1892] 2 Ch 170 established that once possession had actually been resumed by the landlord under the forfeiture clause in a lease it was too late for a successful application for relief to be mounted under section 146(2) of the Law of Property Act 1925 (then section 14(2) of the Conveyancing Act 1881) since the landlord was no longer "proceeding" to enforce a forfeiture but had completed the process. That construction of the section, it was urged, had now stood unchallenged for the past 100 years and could be departed from only G
by overruling *Rogers* v *Rice*. Parliament had, in 1925, re-enacted the section in precisely the same terms and, it must be assumed, had done so in the knowledge of the construction accorded to the words "is proceeding" by a strong Court of Appeal consisting of Lord Coleridge CJ and Lindley and Kay LJJ following *obiter dicta* to the same effect from Jessel MR and Lindley and Bowen LJJ in *Quilter* v *Mapleson* (1882) 9 QBD 672. Whatever might have been the position in 1925, when the legislature came to reconsider the legislation, it was now too late for your lordships to adopt a construction which differed from what must have been intended by Parliament when the self-same words were re-enacted. In so submitting, Mr Reid was echoing what was said by Sir Wilfrid Greene MR in *Egerton* v *Jones* [1939] 2 KB 702 at p 707. H

I confess to having been very much troubled by this submission, to which, at the time, I could see no logical answer but which, as it seemed to me, led to a result not only manifestly inconvenient but also opposed to common sense. On further consideration, however, I think that the fallacy of the argument lies in the way in which the essential premise is expressed. Neither *Rogers* v *Rice* nor *Quilter* v *Mapleson* did, in fact, establish that the right to apply for relief under section 146(2) was ousted simply by a landlord's resumption of possession, however that occurred. The expressions used by the judges in both cases were directed only to the position of a landlord who was the plaintiff in legal proceedings against his tenant for possession of the premises comprised in the lease on the ground of J
forfeiture. In *Quilter* v *Mapleson* the point did not arise directly or, indeed, at all, since notice of appeal by the tenant was given before judgment against him was executed and a stay had been granted pending appeal. Bowen LJ's statement (quoted by Lord Coleridge CJ in *Rogers* v *Rice*) that "the tenant may apply for relief until the landlord has re-entered" was made in the context of pending proceedings for possession. It was, in fact, not strictly accurate, since the re-entry dates technically from the service of the writ; but it was in any event clearly directed to the enforcement of a final and unappealed judgment in the proceedings.

In *Rogers* v *Rice*, similarly, the very brief judgments were directed entirely to a situation in which the landlord had obtained a judgment for possession which had been executed and against which there was K
no appeal, either pending or possible. It was pointed out by Lord Coleridge CJ that at the time when the tenant's originating summons for relief was taken out "the action . . . so far as related to enforcing the right of re-entry, was at an end". It could not therefore be said that the landlord was "proceeding" to enforce his right of re-entry. Now although it is true that Lord Coleridge also mentioned that the landlord was in possession, the salient point, as I read the judgment, was not the mere fact that the landlord had resumed possession but the fact that possession was held under an unassailable judgment in proceedings which had been litigated to the end. It is clear, for instance, that where a judgment for possession has been wrongfully obtained because, for instance, no notice was served under section L
146(1), it may, by appropriate procedure, be set aside so as to enable one deriving title under the lessee to defend (see *Jacques* v *Harrison* (1884) 12 QBD 165). What defeats the claim to relief is not the fact of possession simpliciter but possession under a final and unassailable judgment. It is, in fact, worth noting that in *Rogers* v *Rice* there had been, after execution of the writ of possession, an application to set the judgment aside. That had failed and it was therefore unnecessary for the court to advert to it in its judgment; but had it succeeded, I doubt whether the result would have been the same, for even though the landlord was in actual possession, his title to hold the property would have rested upon the forfeiture of the lease and not upon the judgment in the proceedings. He would, therefore, still be "proceeding" to enforce the forfeiture in seeking to restore the M
judgment.

As my noble and learned friend Lord Templeman has said, the court was simply not directing its mind to what the position might be in a case where the landlord sought to enforce the right of re-entry, not by proceedings for possession, but by physical entry upon the demised premises; and I do not think that it necessarily follows at all, as the Court of Appeal seems to have assumed in *Pakwood Transport Ltd* v *15 Beauchamp Place Ltd* (1977) 36 P & CR 112, that because legal proceedings resulting in a final, unappealed and fully executed judgment are then at an end so that the landlord is no longer enforcing his right of re-entry, a landlord who has entered upon the demised premises or, possibly, even upon a portion of the demised

A premises "in the name of the whole" (to use the time-honoured formula), is not, once he has lawfully set foot on the premises, any longer "proceeding . . . to enforce" his right of re-entry. I would, for my part, adopt the analysis suggested by Sir Nicolas Browne-Wilkinson V-C in the Court of Appeal in the instant case. A landlord who re-enters peaceably holds possession of the demised premises as against the tenant and his privies by virtue of the right accorded to him by the proviso for re-entry contained in the lease. If the lease is set up against him, he can justify keeping the tenant out only by pointing to the forfeiture clause and pleading his entry under it; and so long as that remains the position he is "proceeding . . . to enforce" it. Obviously there is going to come a time when a tenant who delays applying for relief will find himself debarred from obtaining it, but
B theoretically the landlord who has elected to enforce his forfeiture in this way remains vulnerable to an application unless and until he obtains a final judgment for possession.

That position is to be contrasted with that of the landlord who adopts the more conventional course of forfeiting by means of an action for possession. Once he has obtained judgment and has been put into possession, any attempt by the tenant to raise the lease against him is met by a simple plea of the judgment. The landlord is no longer proceeding to enforce a right of re-entry, for his possession rests now not upon the exercise of a right under the lease but upon a judgment of the court which, as between the parties and their privies, constitutes *res judicata*. Proceedings for relief under section 146(2)
C cannot therefore be effective unless and until that position is reversed and the judgment set aside, as it might be, for instance, in the case of a default judgment or one obtained without proper service.

This seems to me to be an essential distinction and it is one which is not touched by and was not adverted to in *Rogers* v *Rice*. Accepting, therefore, that Parliament, in re-enacting section 14(2) of the Act of 1881 as section 146(2) of the Act of 1925, was endorsing and perpetuating the Court of Appeal's construction that a landlord holding by virtue of an executed final judgment was no longer "proceeding to enforce" his right of re-entry but was merely relying upon his judgment in proceedings which had conclusively terminated, there is nothing in this construction which governs the quite different case of the landlord who elects to exercise his right
D "otherwise" by physical entry upon the property.

It was, of course, the ratio of the decision of the Court of Appeal in the *Pakwood Transport* case that a peaceable entry by a landlord was to be equiparated with an executed judgment, the court's conclusion in that case that the proceedings were not premature being based on what they saw as the manifest absurdity of the right to relief being destroyed *eo instante* with the forfeiture from which it arises. It does not, however, follow from the rejection of that ratio that the conclusion itself was wrong. "Is proceeding" means, I think, no more than "is taking the necessary steps" or "proceeds", which would include the service of the necessary preliminary notice under subsection (1). Where a landlord has given notice of his intention to
E forfeit, on, for instance, the liquidation of a company tenant, there is obvious good sense in enabling the liquidator to clarify the position without waiting for the landlord's action so that the lease may be sold as soon as possible.

For these reasons and for those more fully expressed in the speech of my noble and learned friend Lord Templeman, I agree that the appeal should be allowed and concur in the order for costs which he proposes.

LORDS KEITH OF KINKEL, GOFF OF CHIEVELEY and JAUNCEY OF TULLICHETTLE agreed with the speeches of Lord Templeman and Lord Oliver of Aylmerton and the reasons given and did not add any observations of their own.

F

G

Court of Appeal

November 1 1991

(Before Lord Justice PARKER, Lord Justice McCOWAN and Lord Justice SCOTT)

PRUDENTIAL ASSURANCE CO LTD v LONDON RESIDUARY BODY AND OTHERS

Estates Gazette February 1 1992

[1992] 06 EG 145

Landlord and tenant — Construction of agreement — H
Agreement providing that tenancy to continue until demised land required by landlord for road widening — Whether notice to quit was effective to determine the tenancy notwithstanding that the land was not required for road-widening purposes — Appeal from decision of Millett J that valid notice to quit had been served allowed

By an agreement dated December 19 1930 the London County Council ("the LCC") as landlord granted a term of land fronting 263-265 Walworth Road, London SE17, to Samuel Nathan — The agreement represented the leaseback element following the sale of the land by Mr Nathan to the J
LCC — By clause 6 of the agreement the tenancy was to continue until the land was required by the LCC for road widening — The reversion to the term devolved first to the Greater London Council and then to the first defendants, the London Residuary Body ("the LRB") — The term of the tenancy is now held by the plaintiff, the Prudential Assurance Co Ltd — The second to fourth defendants are the purchasers of the reversion from the LRB — The proposal to widen the Walworth Road has never been carried into effect — In July 1988 the LRB sold the reversion to the agreement to the second to fourth defendants following the service under section 25 of K
the Landlord and Tenant Act 1954 of a notice purporting to determine the tenancy on December 19 1988 — By reason of the London Electricity Board being in possession, and not the plaintiff, the notice was accepted as not being effective under section 25 but was capable of taking effect as a common law notice to quit if such a notice could be given — In the court below Millett J held that a valid notice to quit had been served determining the tenancy agreement and that the second to fourth defendants were entitled to possession — The plaintiff appealed

***Held*: The appeal was allowed — Clause 6 of the agreement was nothing to do with notices to quit, which are optional** L
notices that a landlord can decide whether or not to give; the clause refers to an obligatory notice where the land was required by the LCC for their statutory road-widening purposes — The effect of the agreement was to grant a tenancy for a term of uncertain duration limited to terminate on the land being required for road-widening purposes — The court was faced with conflicting authorities in *Lace* v *Chantler* and *Ashburn Anstalt* v *Arnold* — Unless the case is one in which the uncertain duration of the term can be controlled by both parties, as in the *Ashburn Anstalt* case, *Lace* v *Chantler* continues to be binding authority that a term of uncertain M
duration is bad for uncertainty as to its maximum duration — However, Mr Nathan entered into occupation and he and his successors paid rent — Mr Nathan became a tenant under a tenancy from year to year on such of the terms of the agreement as were consistent with such a tenancy — There is implied into a tenancy from year to year a term enabling either party to terminate the tenancy by six months' notice — Clause 6 was inconsistent with the nature of a periodic tenancy — However, the fetter imposed by clause 6 remained — It was not open to the LRB to serve the notice determining the tenancy because the tenancy included an implied provision

that a notice to quit may not be served until the land is required for road-widening purposes

The following cases are referred to in this report.

Ashburn Anstalt v *Arnold (No 2)* [1989] Ch 1; [1988] 2 WLR 706; [1988] 2 All 147; (1988) 55 P&CR 137; [1988] 1 EGLR 64; [1988] 23 EG 128, CA
Doe d Rigge v *Bell* (1793) 5 Term Rep 471
Lace v *Chantler* [1944] KB 368
Midland Railway Co's Agreement, Re: Clay (Charles) & Sons Ltd v *British Railways Board* [1971] Ch 725; [1971] 2 WLR 625; [1971] 1 All ER 1007; (1971) 22 P&CR 360, CA

This was an appeal by Prudential Assurance Co Ltd against a decision of Millett J given on January 16 1991 ([1991] 1 EGLR 90; [1991] 25 EG 120) relating to a number of questions on the true construction and effect of a memorandum of agreement relating to land fronting 263-265 Walworth Road, London SE17. The appellant was the tenant of the land the reversion to which had been transferred from the first respondents, the London Residuary Body, to the second, third and fourth respondents, Barron Investments Ltd, Alan Moss Bayes and Joan Estelle Bayes.

David Neuberger QC and Paul de la Piquerie (instructed by the solicitor to the Prudential Corporation plc, holding company for Prudential Assurance Co Ltd) appeared for the appellant; and Robert Reid QC and Stephen Lloyd (instructed by Clifford Watts Compton, for the second to fourth respondents) represented the respondents; the first respondents did not appear and were not represented.

Giving the first judgment at the invitation of Parker LJ, **Scott LJ** said: This is an appeal from the judgment of Millett J, given on January 16 1991. The case raises a short but, in my opinion, difficult point regarding the effect of a tenancy agreement dated December 19 1930.

The parties to the agreement were the London County Council ("the LCC"), as landlord, and a Mr Samuel Nathan, as tenant. The land comprised in the agreement was a small strip of land fronting 263-265 Walworth Road. The circumstances leading up to the agreement and constituting the factual matrix in the context of which the agreement must be construed are described by Millett J in his judgment*:

> It appears that immediately prior to December 19 1930 Mr Nathan owned land fronting the Walworth Road and wished to redevelop the site, or part of it, by putting up a building upon it. The LCC, being the highway authority, contemplated the possible widening of Walworth Road and accordingly, on December 19 1930, acquired part of Mr Nathan's land, consisting of the frontage of the premises to the Walworth Road, and on December 30 1930 leased it back to Mr Nathan, together with a right to put up a temporary building upon it . . . it appears from the internal evidence of the document itself that the intention was that Mr Nathan should have the right to put up a temporary building on the frontage and to occupy it until such time as the LCC should determine to proceed with their proposal to widen the Walworth Road, whereupon he would have to give up possession. The new and permanent building which he intended to erect behind the temporary one would then have a frontage to the widened Walworth Road.
>
> It is not in evidence whether similar transactions were entered into with adjoining premises on either side of the subject premises or further along Walworth Road.

The two critical clauses of the agreement are clauses 1 and 6. But they are, of course, to be construed in the context of the agreement as a whole.

The agreement provided, so far as relevant, as follows:

> 1. The Council hereby let to the Tenant and the Tenant takes from the Council the land (hereinafter called "the said Land") described in the Schedule hereto from 19th December 1930 at the rent of £30 per annum payable quarterly on the usual quarter days until the tenancy shall be determined as hereafter provided the first payment calculated from the date thereof to be made on the 25th day of March next and the last payment if need be to be apportioned up to the date of determination of the tenancy.
>
> 2. On the rebuilding by the Tenant of the premises nos 263 and 265 Walworth Road behind the line of widening of that road the Council will afford to the Tenant all facilities for the erection (subject to the provisions of the London Building Acts and the by-laws and regulations made thereunder) on the said land of temporary one storey shops or buildings of one storey and for the retention of such shops or buildings as temporary structures until the Tenant shall be required to give to the Council vacant possession of such land as hereinafter provided.
>
> 4. The Tenant agrees to keep the said one storey shops or buildings and the fixtures and fittings therein in good and tenantable repair during the tenancy reasonable wear and tear and damage by fire not due to the act default or negligence of the tenant excepted and at the termination of the tenancy as hereinafter provided to deliver to the Council vacant possession of the said land and to leave such land cleared to ground level to the satisfaction of the Council.
>
> 6. The tenancy shall continue until the said Land is required by the Council for the purposes of the widening of Walworth Road and the street paving works rendered necessary thereby and the Council shall give two months' notice to the Tenant at least prior to the day of determination when the said Land is so required and thereupon the Tenant shall give vacant possession to the Council of the said Land as hereinbefore provided.
>
> 7. The Tenant shall (subject to Clause 6 hereof) in the event of holding possession of the said Land after the determination of the tenancy or failing to clear all materials as provided by Clause 4 hereof forfeit and pay to the Council as liquidated and ascertained damages the sum of ten pounds for every day possession shall be held or such material shall remain and such sum shall be recoverable by the Council either as rent in arrear or as a debt due from the Tenant but this condition shall not prejudice any right the Council would otherwise have to recover or take possession of the premises.
>
> 8. The Council may re-enter upon the said land and determine the tenancy on non-payment of rent for fourteen days whether legally demanded or not or on breach by the Tenant of any of the terms of this Agreement.
>
> 9. The Council shall at its own expense when vacant possession of the said land is given to the Council as hereinbefore provided execute all road making and paving works but shall not be called upon to bear any part of the cost of clearing the said land to ground level.
>
> 10. Before any building or rebuilding on the land belonging to the Tenant and adjoining the said land is commenced the Tenant shall give seven clear days' notice in writing to the valuer for the time being of the Council so that the line of widening may be set out and agreed and the tenant shall also give to the said valuer notice when the new building has reached pavement level.

The schedule to the agreement described the land as "ALL THAT piece or plot of land situate on the east side of Walworth Road and being part of nos 263 and 265 Walworth Road . . . as is more particularly delineated and shown by red colour on the plan annexed hereto". The plan shows a line marking the eastern boundary of the proposed widened road. The line constitutes the eastern boundary of the land comprised in the agreement.

As is apparent from the passage in Millett J's judgment which I have read, the tenancy agreement represented the leaseback element of a composite sale-and-leaseback arrangement. The sale of the land by Mr Nathan to the LCC was, it seems, completed on December 19 1930. It appears from the Proprietorship Register at HM Land Registry that the LCC was registered as proprietor of the land on January 29 1931 and that the price paid to Mr Nathan was £2,750.

I can take the subsequent history of the land from the judgment of the learned judge*:

> The interest granted to the tenant by [the agreement] has become vested in the plaintiff, the Prudential Assurance Co Ltd, and in 1975 it sublet the premises subject to the memorandum of agreement, together with the freehold land behind it, to the London Electricity Board, which is still in occupation, at a substantial rent.
>
> The proposal to widen the Walworth Road was never carried into effect and must, I apprehend, have been abandoned many years ago. In the course of time the LCC's rights and obligations became vested in their successor body, the Greater London Council, and a further statutory vesting occurred on the dissolution of the Greater London Council when its property, rights and interests were vested in the first defendant, the London Residuary Body. The LCC had, rightly or wrongly, conceived that it was not open to them to review the rent of £30 a year for the subject premises, which was fixed in 1930, or to determine the tenant's interest thereunder. The LCC and their successor, the GLC, were each the highway authority. The London Residuary Body, however, is not a highway authority. After taking advice they concluded that they could deal with the reversion on the footing that the interest created by the memorandum of agreement was terminable. Accordingly, on July 21 1988 the London Residuary Body sold the subject land to the second, third and fourth defendants, and the land was transferred to them by a transfer dated August 25 1988. Since that date the reversionary interest expectant on the determination of the interest created by the memorandum of agreement has been vested in the second, third and fourth defendants, and of course, ever since it became vested in the London Residuary Body it has been vested in a party other than a highway authority.
>
> . . . Prior to the auction sale, the London Residuary Body purported to serve a notice under section 25 of the Landlord and Tenant Act 1954 determining the interest on December 19 1988. The notice was served on March 31 1988. It would have been a perfectly good notice if Part II of the Landlord and Tenant Act 1954 applied to the interest. However, since the plaintiff had sublet the whole of the land to the London Electricity Board and was not in possession of any part of it, it is plain that the land was not within

*Editor's note: Reported at [1991] 1 EGLR 90 at p 90M.

*Editor's note: Reported at [1991] 1 EGLR 90 at p 91C.

A Part II of the Landlord and Tenant Act 1954 at all. It is now common ground that the notice was not effective as a section 25 notice . . .

It is also common ground that the section 25 notice is capable of taking effect as a common law notice to quit and was apt to terminate an ordinary yearly tenancy if such a notice could be given by the London Residuary Body.

The issue in this case is whether the section 25 notice, treated as a common law notice to quit, was effective to determine the tenancy created under, or as a result of, the agreement of December 19 1930.

The plaintiff, the appellant before us and the tenant under the agreement, is the Prudential Assurance Co Ltd. It relies on clause 6 of the agreement under which the tenancy was expressed to continue "until the said Land is required by the Council for the purposes of the widening of Walworth Road . . ." etc. It is common ground that the B event contemplated, namely the land being required for road-widening purposes, has not happened. So the tenancy continues.

The effective defendants, the respondents before us, are the purchasers of the property from the London Residuary Body. Their answer to the plaintiff's simple reliance on clause 6 of the agreement is a complex one.

First, it is contended that the agreement purported to create a tenancy for a fixed term of uncertain duration. If it did, it is submitted that the agreement cannot take effect as a grant of that tenancy. *Lace* v *Chantler* [1944] KB 368, where the grant of a tenancy "for the duration of the war" was held bad, is relied on.

Second, it is contended that since Mr Nathan entered into C occupation as tenant under the December 19 1930 agreement and paid rent at the rate of £30 pa he must, if the agreement cannot take effect as a grant for the period provided for by clause 6, be treated as holding under a periodic yearly tenancy on so much of the terms of the agreement as are applicable to such a tenancy.

Third, it is contended that in so far as clause 6 restricts the respective ability of the landlord and the tenant to determine the tenancy by notice to quit, the restriction is repugnant to the inherent nature of a periodic yearly tenancy and is void.

Fourth, it is contended, as an alternative to the repugnancy point, that the restrictions imposed by clause 6 on the landlord's right to serve a notice determining the tenancy cannot, as a matter of construction, or alternatively implied term, survive the transmission D of the reversion to a landlord that is not the highway authority.

On these contentions the learned judge concluded, first, that the agreement properly construed granted the tenant a periodic yearly tenancy and not a tenancy for a fixed but uncertain term. He held, second, that clause 6 imposed a restriction on the landlord's right to serve a notice to quit terminating the periodic tenancy, but that the restriction was not repugnant to the nature of a yearly tenancy and was valid. He relied on and applied *Re Midland Railway Co's Agreement* [1971] Ch 725 and *Ashburn Anstalt* v *Arnold* [1989] Ch 1*, both Court of Appeal decisions. He held also that the agreement placed no fetter on the right of the tenant to terminate the yearly tenancy by a notice to quit. But, finally, the learned judge held that the clause 6 restriction on the landlord's right to serve a notice E terminating the yearly tenancy fell away once the landlord and the highway authority were different bodies. When that state of affairs arose, clause 6, he held, ceased to have any effect. So the defendants succeeded below. The learned judge upheld the notice to quit that had been served by the London Residuary Body.

In considering the various matters that have been argued before us, the starting point must, in my opinion, be the nature of the tenancy brought into being by, or as a result of, the agreement of December 19 1930.

The learned judge concluded, as I have said, that, properly construed, the agreement granted a periodic yearly tenancy. I do not agree with that conclusion. Clause 1 of the agreement specified the F date of commencement of the tenancy, December 19 1930, but not the date of termination. The letting was to continue "until the tenancy shall be determined as hereafter provided". Clause 6 contained the termination provisions and can, for construction purposes, be broken down into two parts. First, clause 6 provided that the tenancy "shall continue until the said Land is required by the Council for the purposes of the widening of Walworth Road . . .". Second, clause 6 provided that "the Council shall give two months' notice to the Tenant at least prior to the day of determination when the said Land is so required . . .". The first of these provides, in my judgment, the intended termination date of the tenancy. The second imposes on the council the obligation of giving two months' prior notice of the day when the land is so required, ie two months' prior G notice of the day when the tenancy will come to an end.

The learned judge construed clause 6 as modifying the landlord's right to serve a notice to quit. In my opinion, however, clause 6 is nothing to do with notices to quit. Notices to quit are optional notices that the landlord can decide whether or not to give. The notice which clause 6 is contemplating is not an optional notice. The landlord is obliged to give the notice — "the Council shall give . . ." is the language used. Clause 6 was, in my opinion, dealing with the duration of the tenancy. Once the land was required by the council for their statutory road-widening purposes, the tenancy was to come to an end. But, as a matter of machinery, the council were placed under the obligation of giving the tenant two months' prior notice H specifying, in effect, the day when the land would be so required.

This construction of clause 6 is consistent with the absence of any mention in the clause, or anywhere else in the agreement for that matter, of notices to be served by the tenant terminating the tenancy. If the clause, properly construed, is dealing with the termination of a tenancy granted for a term of uncertain duration, there is no room for notices to quit. If, on the other hand, the tenancy was a periodic yearly tenancy, one would expect to find that the tenant could terminate the tenancy by a notice to quit. The learned judge thought that a provision to that effect could be implied. But an implied term enabling the tenant to serve a notice to quit would contradict the express terms of clause 6 — "The tenancy shall continue . . ." etc. A term cannot be implied into a contract if it would contradict an J express term of the contract: see *Chitty on Contracts* 26th ed p 557. The judge's approach was to conclude from clause 1 that a periodic tenancy was granted and then to hold that a right for the tenant to serve a notice to quit must be treated as included notwithstanding clause 6. In my respectful opinion, this approach puts the cart before the horse. I think one should read clauses 1 and 6 in the context of the whole agreement and ask whether the clauses taken together are more consistent with an intention to grant a tenancy for a term, albeit one of uncertain duration, or to grant a periodic yearly tenancy. If that approach is adopted, I do not, I confess, see how any conclusion is possible other than that clauses 1 and 6 purported to grant a tenancy for a term of uncertain duration.

The next question is whether the grant of a tenancy for a term K limited to terminate on the land being required by the council for road-widening purposes is a grant that the law will recognise. There are three Court of Appeal authorities to which I should at this point refer.

The first is *Lace* v *Chantler,* to which I have already referred. The case concerned the letting of a dwelling-house at a rent of 16s 5d a week. In the rent book it was stated, in effect although not in terms, that the tenancy was to continue for the duration of the war. A notice to quit was served by the landlord before the war had come to an end. The question was whether that notice was valid. The Court of Appeal held that the grant of a tenancy for the duration of the war was bad. But since the tenant was in occupation and paying rent on a weekly L basis he was a weekly tenant and so the notice to quit was good.

At p 370 Lord Greene MR, with whose judgment MacKinnon LJ and Luxmoore LJ agreed, said:

> The question immediately arises whether a tenancy for the duration of the war creates a good leasehold interest. In my opinion, it does not. A term created by a leasehold tenancy agreement must be expressed either with certainty and specifically or by reference to something which can, at the time when the lease takes effect, be looked to as a certain ascertainment of what the term is meant to be. In the present case, when this tenancy agreement took effect, the term was completely uncertain. It was impossible to say how long the tenancy would last. Mr Sturge in his argument has maintained that such a lease would be valid, and that, even if the term is uncertain at its beginning when the lease takes effect, the fact that at some future time it will be rendered certain is M sufficient to make it a good lease. In my opinion, that argument is not to be sustained.
>
> I do not propose to go into the authorities on the matter, but in *Foa's Landlord and Tenant*, 6th ed, p 115, the law is stated in this way, and, in my view, correctly: "The habendum in a lease must point out the period during which the enjoyment of the premises is to be had; so that the duration, as well as the commencement of the term, must be stated. The certainty of a lease as to its continuance must be ascertainable either by the express limitation of the parties at the time the lease is made, or by reference to some collateral act which may, with equal certainty, measure the continuance of it otherwise it is void. If the term be fixed by reference to some collateral matter, such matter must either be itself certain (eg a demise to hold for 'as many years as A has in the manor of B') or capable before the lease takes effect of being rendered so (eg for 'as many years as C shall name'). The important words to observe in

*Editor's note: Also reported at [1988] 1 EGLR 64.

A that last phrase are the words "before the lease takes effect". Then it goes on: "Consequently, a lease to endure for 'as many years as A shall live' or 'as the coverture between B and C shall continue', would not be good as a lease for years, although the same results may be achieved in another way by making the demise for a fixed number (ninety-nine for instance) of years determinable upon A's death, or the dissolution of the coverture between B and C." In the present case, in my opinion, this agreement cannot take effect as a good tenancy for duration of the war.

The authority of *Lace* v *Chantler* seems to me to make inescapable the conclusion that a grant of a tenancy to continue until the land is required for road-widening purposes cannot take effect as a good grant.

But Mr Neuberger QC submitted that later Court of Appeal B decisions required a different conclusion.

In *Re Midland Railway Co's Agreement* there was an express grant of a periodic tenancy from half-year to half-year. Clause 2 of the agreement provided, so far as relevant, as follows:

> The agreement may be determined by either party on giving to the other three months' notice, such notice to be in writing and to expire at any time thereafter without reference to the commencement of the tenancy . . . Provided that this agreement shall not be so terminated by the company [the Midland Railway Co] until they shall require the said premises for the purposes of their undertaking.

British Railways Board, the successor in title to the original landlord, purported to terminate the tenancy by serving a notice to quit. They did not require the land for the purposes of their undertaking.

C Two points were relied on by the board. They contended that the proviso to clause 2 conflicted with the principle that for a tenancy to be valid there must be certainty as to the maximum duration of the estate. Alternatively, they contended that the proviso, being a fetter for a period of uncertain and potentially unlimited duration on the landlord's right to terminate the periodic tenancy, was repugnant to the nature of the periodic tenancy and must be rejected. The Court of Appeal held, on the first point, that the principle regarding terms of uncertain duration that was applied in *Lace* v *Chantler* did not apply to periodic tenancies.

At p 732D Russell LJ (as he then was), giving the judgment of the court, said:

> D Now it appears to us that that decision is confined to a case in which that which was purported to be done was simply to create a leasehold interest for a single and uncertain period. The applicability of this matter of certainty to a periodic tenancy was not under consideration. If *Lace* v *Chantler* had been a case in which there was simply a periodic tenancy with a proviso that the landlord would not give notice during the continuance of the war, this court might not have concluded that such an agreement, which would of course have left the tenant free to determine on notice at any time, was inoperative to create a leasehold. There is nothing in the reasoning of the judgments to lead to the necessary conclusion that such must have been so.
>
> If you have an ordinary case of a periodic tenancy (for example, a yearly tenancy), it is plain that in one sense at least it is uncertain at the outset what will be the maximum duration of the term created, which term grows year by year as a single term springing from the original grant. It cannot be predicated E that in no circumstances will it exceed, for example, 50 years; there is no previously ascertained maximum duration for the term; its duration will depend upon the time that will elapse before either party gives notice of determination. The simple statement of the law that the maximum duration of a term must be certainly known in advance of its taking effect cannot therefore have direct reference to periodic tenancies.

He declined to extend the uncertainty of term principle to periodic tenancies. He said, at p 733C:

> . . . we are persuaded that, there being no authority to prevent us, it is preferable as a matter of justice to hold parties to their clearly expressed bargain rather than to introduce for the first time in 1971 an extension of a doctrine of land law so as to deny the efficacy of that bargain.

He then turned to the second point, the repugnancy point, and said at F p 733F:

> Our instinct, as previously indicated, is to give effect if possible to the bargain made by the parties. It may well be that if in a periodic tenancy an attempt was made to prevent the lessor ever determining the tenancy, that would be so inconsistent with the stated bargain that either a greater estate must be found to have been constituted or the attempt must be rejected as repugnant. But short of that we see no reason why an express curb on the power to determine which the common law would confer upon the lessor should be rejected as repugnant to the nature of the leasehold interest granted. In *Breams Property Investment Co Ltd* v *Stroulger* [1948] 2 KB 1, a curb on the lessors for three years unless they required the premises for their own use was upheld in this court, notwithstanding the earlier cases of *Warner* v *Browne* (1807) 8 East 165 and *Cheshire Lines Committee* v *Lewis & Co* (1880) 50 LJ QB 121. It follows that in a periodic tenancy a similar curb for 10, 20 or 50 years should not be rejected as repugnant to the concept of a periodic tenancy: and once the G argument based on uncertainty is rejected we see no distinction in the present case.

Russell LJ ended his judgment with the following reservation: "We say nothing as to the situation which might arise in law should the defendants sell the reversion to another, when it might be arguable that the proviso would be no longer relevant."

The *Midland Railway Co's* case is authority for two propositions. First, the uncertainty of term principle that was applied in *Lace* v *Chantler* does not apply to periodic tenancies. Second, a fetter in a periodic tenancy of the right of one or other party to serve a notice determining the tenancy is not to be rejected as repugnant to the nature of the periodic tenancy if the fetter falls short of preventing the party from ever determining the tenancy. H The fact that the fetter is of uncertain and of potentially unlimited duration is not enough for it to be rejected on the ground of repugnancy. These propositions are, in my opinion, binding on us. If they are unsatisfactory or are wrong, the correction must come from the House of Lords or from Parliament.

Finally, there is *Ashburn Anstalt* v *Arnold*. That case concerned an agreement for the sale of a lease. The agreement dated February 28 1973 provided under clause 5 that:

> From and after completion Arnold & Co [the purchaser] shall be at liberty to remain at the property as licensee . . . until 29th September 1973 without payment of rent or any other fee to Matlodge Ltd [the vendor] save that Arnold & Co shall pay all outgoings as long as it is in occupation of the J property. From and after 29th September 1973 Arnold & Co shall be entitled as licensee to remain at the property . . . on the like terms save that it can be required by Matlodge Ltd to give possession on not less than one quarter's notice in writing upon Matlodge Ltd certifying that it is ready at the expiration of such notice forthwith to proceed with the development of the property . . . involving, *inter alia*, the demolition of the property.

Ashburn Anstalt was the successor in title of Matlodge Ltd to the freehold and in October 1985 served on Arnold & Co a notice to quit. The issue for decision was whether the 1973 agreement under which Arnold & Co remained in occupation was binding upon Ashburn Anstalt. The judgment of the court was given by Fox LJ and he dealt first with the question whether Arnold & Co was a mere licensee or a leaseholder. He referred to the authorities and concluded that the K reservation of a rent was not necessary for the creation of a tenancy. He then said at p 10:

> There remains the question of the existence of a term. It is the plaintiff's case that clause 5 created no term sufficiently identifiable to be capable of recognition by the law and that no tenancy was created. For that, the plaintiff relies on *Lace* v *Chantler* . . .

Fox LJ cited the passage from Lord Greene's judgment which I have cited, and commented that:

> The ambit of the decision in *Lace* v *Chantler* was limited by the further decision of this court in *Re Midland Railway Co's Agreement*

and gave his attention to that case. He then, after citing passages from the judgment of Russell LJ, said at p 11: L

> So far as *Lace* v *Chantler* is concerned, the present case, it seems to us, is distinguishable. In *Lace* v *Chantler* the duration of the war could not be predicted and there was no provision for either party to bring the tenancy to an end before the war ended and that event might be very hard to pinpoint. In the present case the arrangement, so far as Matlodge was concerned, would continue until Matlodge determined it by giving not less than a quarter's notice upon Matlodge giving the required certificate. The event entitling Matlodge to give the certificate might not, of course, occur. But the same applies to the qualifying event for the giving of the landlord's notice in the *Midland Railway Co's Agreement*.

Fox LJ then noted that Arnold & Co were free under the terms of clause 5 of the agreement to bring the arrangement to an end and continued at p 12: M

> The result, in our opinion, is that the arrangement could be brought to an end by both parties in circumstances which are free from uncertainty in the sense that there would be no doubt whether the determining event had happened. The vice of uncertainty in relation to the duration of a term is that the parties do not know where they stand. Put another way the court does not know what to enforce. That is not the position here. It seems to us, therefore, that as in the *Midland Railway Co's Agreement* there is no reason why the court should not hold the parties to their agreement. That is so even though the tenancy is (or may not be) an ordinary periodic tenancy. The rights of the parties are no more subject to uncertainty than those in the *Midland Railway* case. We do not see why the mere absence of a formula referring to a periodic tenancy or occupancy should alter the position.

I have cited extensively from the *Ashburn Anstalt* case for two

A reasons: first, because Mr Neuberger has submitted that the termination provisions in clause 6 of the agreement in the present case are indistinguishable from the termination provisions in clause 5 of the *Ashburn Anstalt* case; second, because I find very great difficulty in following how the conclusions in the *Ashburn Anstalt* case can be reconciled with *Lace* v *Chantler*, an authority binding on the court in *Ashburn Anstalt* as on us.

The uncertainty that Lord Greene regarded as fatal in *Lace* v *Chantler* was an uncertainty as to the *maximum* duration of a term. His citation from *Foa's Landlord and Tenant* makes that clear. Thus a grant for a term to continue until Britain wins the Davis Cup would be bad. But there would be nothing the matter with a grant for 99
B years terminable if within that period Britain wins the Davis Cup. This principle of law was applied by Lord Greene in *Lace* v *Chantler* and was recognised by Russell LJ in the *Midland Railway Co's* case: see pp 731A to 732B. But in the *Midland Railway Co's* case it was held that the principle did not apply to periodic tenancies.

In the *Ashburn Anstalt* case it was held that clause 5 of the agreement in question produced a tenancy and not merely a licence. It is difficult to see how the tenancy could have been a periodic tenancy. It is true, as Fox LJ noted, that a term of years need not be accompanied by a reservation of rent. It is, however, very difficult to conceive of a periodic tenancy without a reservation of rent. Even if such an interest could be created by express words, there were no words in clause 5 that could have been treated as creating a periodic
C tenancy. The right of occupancy was simply to continue until terminated in the manner and on the event specified in the clause. Under clause 5 the right of occupancy was, in effect, to continue until two events occurred. First, Matlodge had to certify that it was ready to proceed with the development of the property; second, Matlodge had to serve a quarter's notice in writing. The term would not come to an end until both these events had happened. Fox LJ recognised (at p 11H) that the first might never happen. The fact that the parties would know whether the second event had or had not happened could not, I would have thought, remedy the inherent uncertainty of duration of a term that depended upon both events happening. And while the *Midland Railway Co's* case was authority for the
D proposition that an uncertainty as to when a party to a periodic tenancy could serve a notice terminating the tenancy was not fatal to the efficacy of the provision creating that uncertainty, the case could not, in view of *Lace* v *Chantler*, be treated as authority, except in periodic tenancy cases, for the proposition that an uncertainty as to the maximum duration of a term was acceptable. The answer, I would respectfully have thought, to the question posed by Fox LJ in the last sentence of the passage from his judgment which I have cited is that, in the absence of a periodic tenancy, the principle recognised and applied in *Lace* v *Chantler* strikes down the uncertain term.

We are, in my opinion, faced with conflicting authorities in *Lace* v *Chantler* and the *Ashburn Anstalt* case. Both bind us. I have already
E commented that I can see no solid ground of distinction between the present case and *Lace* v *Chantler*. In both, the maximum duration of the tenancy purported to be granted was uncertain. In neither was a periodic tenancy granted. Is there a valid ground of distinction between the present case and the *Ashburn Anstalt* case? I think there is, although I do not pretend to find it very satisfactory. In clause 5 of the *Ashburn Anstalt* agreement the tenancy (as the court found there to be) was expressed to be terminable "on not less than one quarter's notice upon Matlodge certifying that it is ready at the expiration of such notice forthwith to proceed with the development . . ." etc. This double event could be brought about or not brought about by Matlodge as it might choose. It need consult no interests but its own. If it wanted to get rid of the tenant (as the court found Arnold & Co to
F be) it could prepare to proceed with the development, certify accordingly and serve the quarter's notice on the tenant. Or it could choose to leave the tenant and the tenancy in place. Clause 6 of the December 19 1930 agreement in the present case does not permit the termination of the tenancy to be subject to the will of the landlord in the same way. The tenancy is to continue until the land is required by the council for road-widening purposes. Whether the land is or is not so required and the time at which it is so required are matters which depend upon the proper performance by the council of their statutory duties as highway authority. The arrival of "the day of determination" cannot be governed and controlled by the council simply according to their private interests and wishes. I repeat what I have already said regarding the second part of clause 6. The service by the council of the two months' prior notice is a matter of obligation G
— "shall give . . ." — and not a matter of choice.

In *Lace* v *Chantler* the bringing to the end of the uncertainty of duration of the term was not within the control of either landlord or tenant. In the *Ashburn Anstalt* case the tenancy, as it was held to be, could have been brought to an end by the tenant at any time and could have been brought to an end by the landlord upon the occurrence of an event that it was within the power of the landlord to control. In the present case, the tenancy granted by clauses 1 and 6 of the agreement could not, on my construction, have been brought to an end by the tenant (it is not to the point that the tenant would be unlikely to want to do so) and, upon the occurrence of an event over which the landlord had, *qua* landlord, no control, would come to an end H
whether or not the landlord so wished.

In my judgment, the present case is governed by the authority of *Lace* v *Chantler*. Unless the case is one in which the uncertain duration of the term can be controlled by both parties, as in the *Ashburn Anstalt* case, *Lace* v *Chantler* continues, in my opinion, to be a binding authority.

For the reasons I have given I conclude that the tenancy purported to be granted by clauses 1 and 6 of the agreement is bad for uncertainty as to its maximum duration.

Mr Nathan entered into occupation as tenant under the terms of the agreement. He and his successors have paid the rent required by the agreement. The rent has been accepted by the successive landlords. In these circumstances it is accepted that, the tenancy J
which the agreement purported to grant being bad, Mr Nathan became a tenant under a tenancy from year to year on such of the terms of the agreement as were consistent with such a tenancy: see *Doe d Rigge* v *Bell* (1793) 5 Term Rep 471 where Lord Kenyon CJ said:

> Though the agreement be void by the Statute of Frauds as to the duration of the lease, it must regulate the terms on which the tenancy subsists in other respects as to the rent, the time of the year when the tenant is to quit etc.

In the ordinary way, if the parties have not agreed, expressly or impliedly, anything to the contrary, there will be implied into a tenancy from year to year a term enabling either party to terminate the tenancy by six months' notice expiring at the end of a year of the K
tenancy. Mr Reid submits that such a term should be implied into the tenancy from year to year to which Mr Nathan became entitled. If that is right then the notice to quit served by the LRB was valid.

Mr Neuberger, on the other hand, contends that the substance of clause 6, or so much thereof as is consistent with a tenancy from year to year, should be implied as a fetter on the right of the landlord to serve notice terminating the tenancy. He submits that a fetter preventing the landlord from serving a notice to quit until such time as the land is required by the highway authority for road-widening purposes should be an implied term of the tenancy from year to year.

In my judgment, the approach to what terms should or should not be implied into the tenancy from year to year is no different from the approach to what terms should be implied into contracts generally. L
The tenancy from year to year is implied by law from the circumstances that the tenant has entered into occupation and paid rent under an agreement for a tenancy that the law will not recognise. But the law does not, in my judgment, presume to add to the tenancy terms to which the parties would never at the time have agreed or to prevent the addition to the tenancy of terms to which, if the parties had directed their minds to the matter, they would plainly have agreed.

The approach to what should or should not be implied must be an objective one. The intentions of the parties must be objectively ascertained from the words that they have used and the surrounding circumstances. In *Chitty on Contracts*, 26th ed p 554 the test is stated M
thus:

> The court will be prepared to imply a term if there arises from the language of the contract itself, and the circumstances under which it is entered into, an inference that the parties must have intended the stipulation in question. An implication of this nature may be made in two situations: first, where it is necessary to give business efficacy to the contract, and secondly, where the term implied represents the obvious, but unexpressed, intention of the parties. These two criteria overlap and, in many cases, have been applied cumulatively. Both, however, depend upon the presumed intention of the parties.

That passage from *Chitty*, in my opinion, succinctly and correctly expresses the approach that should be adopted in considering what terms should be implied into Mr Nathan's tenancy from year to year.

Clause 6 of the agreement of December 19 1930 cannot, as it stands, be treated as a term of the tenancy from year to year. It is, in my opinion, inconsistent with the nature of a periodic tenancy in that it precludes both the landlord and the tenant from terminating the tenancy by notice to quit. If the parties, the LCC and Mr Nathan, had been asked in December 1990 whether Mr Nathan should be free to terminate the tenancy by an ordinary six months' notice to quit, they would have been astonished by the question but would, I am sure, have answered that if he wanted to do so, of course he could. But if the parties had been asked whether it should be open to the landlord to terminate the tenancy by a six months' notice to quit, they would have been equally astonished but would, in my opinion, at once have replied that the landlord could do no such thing. If asked in what circumstances, if at all, the landlord should be able to serve a notice to quit, the obvious answer they would have given would have been that the landlord could serve notice to quit if the land were required for road-widening purposes.

The background to the December 19 1930 agreement as well as its actual terms make it certain, in my opinion, that these are the answers that would have been given. Prior to the sale-and-leaseback transaction, Mr Nathan had owned a commercial site with a frontage to Walworth Road. The LCC, the highway authority, was contemplating the widening of Walworth Road. To facilitate its arrangements in that regard it agreed with Mr Nathan to purchase from him the strip of his land that it would need for the road widening but to lease the land back to him until the land should be required for that purpose. The LCC would have been astonished and Mr Nathan would have been horrified by the suggestion that the LCC or a successor in title could serve notice on Mr Nathan terminating his tenancy, depriving his site of its road frontage and leaving access to the site to be obtained only by means of a back alley. Clause 6 of the agreement shows the intention of the parties that Mr Nathan should not have to give up his tenancy until the land was required for the widening of Walworth Road. The implied tenancy from year to year should, in my judgment, incorporate a corresponding fetter on the right of the landlord to terminate that tenancy.

The *Re Midland Railway Co's* case is authority, binding on us, for the proposition that the fetter to which I have referred would not be repugnant to the inherent nature of a periodic tenancy and would be valid.

The parties did not contemplate in December 1930 that the time might come when the landlord under the tenancy was not the highway authority. Otherwise they would in the agreement have catered for that contingency. If they had been asked what, in that event, the position would be, their answer would, in my opinion, have been that the fetter would continue to bind the successor landlord until the land should be required by the highway authority for road-widening purposes.

Accordingly, I would treat the tenancy from year to year as incorporating an implied term that the tenancy would be determinable by a six months' tenant's notice to quit expiring at the end of a year of the tenancy or by a two months' landlord's notice to quit in the event that the land was required by the highway authority for the road-widening purposes described in clause 6.

The learned judge, who, as I have said, treated the agreement as granting a tenancy from year to year and treated clause 6 as a contractual fetter on the power of the LCC to serve a notice to quit determining the tenancy, concluded that the clause 6 fetter was limited to the period during which the landlord and highway authority were one and the same. I think this conclusion must have been reached via an implied term route rather than as a matter of construction of the actual language of clause 6, but, whichever the route, I would respectfully dissent from the conclusion. I think the "obvious but unexpressed" intention of the parties, in the event that the landlord and highway authority were no longer one and the same, would have been that the fetter could continue until the highway authority required the land for the road-widening purposes.

A final question debated before us has been whether the event, on which the landlord's right to serve a notice to quit depends, ie the land being required for road-widening purposes, can never now happen and, if it can never happen, what is the effect on that state of affairs?

If the event can never happen, the fetter becomes not simply of uncertain duration, which the court in the *Re Midland Railway Co's* case held did not matter, but absolute. As at present advised, I think that at that point the fetter would become repugnant to the nature of the periodic tenancy and would cease to be effective. I can see no objection in principle to a fetter being initially valid but becoming, by subsequent events, repugnant and ineffective. But this point need not be finally decided in the present case because there is no evidence that justifies the conclusion that the land will never be required for road-widening purposes.

In my judgment, it was not open to the LRB to serve notice determining the tenancy under which the Prudential, as successor in title to Mr Nathan, held the land comprised in the December 19 1930 agreement. The tenancy from year to year under which the land is held includes, in my judgment, an implied provision that notice to quit may not be served by the landlord until the land is required for road-widening purposes. That provision, on the authority of the *Re Midland Railway Co's* case, is valid and effective.

I would allow the appeal.

Agreeing, **PARKER LJ** said: I add a contribution of my own only because we are differing from the learned judge and I do so with some diffidence.

I begin with the judgment of the court in the *Re Midland Railway Co's* case which made clear that, in the absence of authority to prevent us from so doing, we should if possible give effect to the bargain between the parties.

I, therefore, begin by a consideration of the terms of, and surrounding circumstances at the time of, the 1930 agreement.

It is clear from its terms that the intention of the parties was (1) that Mr Nathan should, or at least should be able to, build on his retained land a new building which would have a frontage on to the widened Walworth Road as then contemplated; (2) that, pending such widening, he should be entitled to build on the leased-back land one-storey shops or buildings which would front on to the then existing Walworth Road; (3) that he should retain them until that land was required for road widening; (4) that he was to have two months' notice of the date when it was so required and in that two months was to remove the building and clear the site to ground level. If he failed to do so within that time he was to pay £10 per day until he had done so. All this is clear from the terms of the agreement itself and is reinforced by the physical conditions of the site. What was clearly *not* contemplated was that the tenant should ever be left with a new building with no frontage on to the existing Walworth Road and with no interest in the land between that building and such road.

In my view, there is no possible way of so construing the agreement as constituting a periodic tenancy subject to six months' notice on either side, or so as to produce the result that the council could either (1) abandon any plan of road widening and give notice of either six months or two months or (2) by alienating the land to a third party, whether a highway authority or not, thereby give their successors a right to give either a six or a two months' notice. In any of such cases the tenant, having built on the retained land so as to have a frontage on to the proposed widened road and built shops on the tenanted land with a frontage on to the existing road, would find himself in a position quite clearly contrary to the common intention of both parties. He would have no frontage on to any road.

If the above is right, it does not, in my view, matter whether the original agreement was bad, on the basis of *Lace* v *Chantler* and there must be implied a yearly tenancy, or was good. In either case the landlord could not, in my view, give notice until the widening of Walworth Road became impossible, as for example if it ceased to exist. To hold otherwise would defeat the common intention of the parties. I agree, however, with Scott LJ that the original lease was bad for the reasons he gives. For the reasons given by him and for the additional reasons which I have set out I, too, would, as I have said, allow this appeal.

McCOWAN LJ agreed and did not add anything.

Appeal allowed with costs.

A
Court of Appeal

November 12 1991
(Before Lord Justice MANN and Sir George WALLER)

KING v SOUTH NORTHAMPTONSHIRE DISTRICT COUNCIL

Estates Gazette February 1 1992
[1992] 06 EG 152

Landlord and tenant — Implied term — Whether implied
B **term that landlord liable to repair access path to demised premises — Whether implied term necessary to fulfil the purpose of the demise — Appeal by defendant landlord dismissed**

Since 1967 the plaintiff has been the weekly tenant of a house owned by the defendant council — There was no written tenancy agreement between the parties, and the council's standard conditions of tenancy impose no obligations upon the landlord — The premises have both a front and a rear access, the rear access from a footpath — The plaintiff, who for the past five years has used a wheelchair, has had to use the
C **footpath as her principal means of access to the premises — She claimed a right of way over it and asserted that it was in a bad state of repair — In her action she claimed "a mandatory injunction and damages for pain, suffering, inconvenience and distress" — In the court below the judge found that the plaintiff had a right of way over the footpath — That it was an implied term of the tenancy, or a term implied by section 11 of the Landlord and Tenant Act 1985, that the council would repair the path — There had been a breach of those implied terms — The council appealed, contending that the judge was wrong in holding that they were under an obligation to repair**
D **the path**

***Held*: Bearing in mind the layout of the estate and the houses within it, the houses could not be enjoyed or function in accord with their design without the rear-footpath access — In that circumstance there was an implication of an obligation to maintain the rear-footpath access to fulfil the purpose of the demise — The demise being of a dwelling-house designed to function with the mechanism of rear service — The demise was not efficacious unless the rear-footpath surface was adequately maintained — Although it is a question of fact and degree as to what is included in the exterior of the dwelling-house for the purposes of section 11 of the Landlord and**
E **Tenant Act 1985, it would be an extravagant and unsupportable conclusion to say that the footpath was as a matter of fact and degree part of the exterior of the demised house — Section 4(4) of the Defective Premises Act 1972 had no application because the footpath was not part of the demise**

The following cases are referred to in this report.

Hopwood v *Cannock Chase District Council* [1975] 1 WLR 373; [1975] 1 All ER 796; (1974) 73 LGR 137, CA
Liverpool City Council v *Irwin* [1977] AC 239; [1976] 2 WLR 562; [1976] 2 All ER 39; (1976) 74 LGR 392; [1976] EGD 282; 238 EG 879, HL
Miller v *Hancock* [1893] 2 QB 177
F *Westminster (Duke of)* v *Guild* [1985] QB 688; [1984] 3 WLR 630; [1984] 3 All ER 144; (1983) 48 P&CR 42; [1983] EGD 541; 267 EG 762, CA

This was an appeal by South Northamptonshire District Council from a decision of Judge Wilson given in the Northampton County Court on October 4 1990. By his decision the judge had ordered the council to repair the rear access path to 69 Forest Road, Piddington, and to pay Mrs Irene King, the weekly tenant of the dwelling-house, £750 as general damages.

Andrew Gordon-Saker (instructed by Shoosmiths & Harrison, of Northampton) appeared for the appellants; and Julian Matthews (instructed by Smith Chamberlain, of Wellingborough) represented the respondent.

Giving judgment, MANN LJ said: This is an appeal against a G
decision of Judge Wilson given in the Northampton County Court on October 4 1990. The appellants are South Northamptonshire District Council, who were the defendants in an action brought against them by the respondent, who is Mrs Irene King. By his decision the judge ordered the appellants to repair a certain rear access path and to pay to the respondent £750 as general damages for breach of covenant to repair that path. He also made a consequent order as to costs.

There is no dispute as to the primary facts. Mrs King has since 1967 been the weekly tenant of a house at 69 Forest Road, Piddington. That house is owned by the appellants in their capacity as housing authority. As is commonplace with council lettings there is no written tenancy agreement, but both parties have assumed, and rightly so in H
my judgment, that the council's standard conditions of tenancy apply. Those conditions, again as is commonplace in relation to council lettings, impose no obligations upon the landlord, but they do impose obligations upon the tenant. However, there is not among those imposed obligations any repairing covenant by the tenant. The tenant does, however, covenant to provide access to the landlord for the purpose of repair.

The house is upon an estate. That estate is laid out in terraces of two and four houses. No 69, when viewed from the front, is the third house in a terrace of four, numbered 65 to 71. No 69 has, as do the others, a rear door. A path, which can be reached from the rear door, runs from the rear of no 69 across the rear of no 71, to join at right-angles a path to Forest Road running between nos 71 and 73, which is J
the first house in the next terrace.

Nos 71 and 73 were sold by the council to the sitting tenants in 1981, no doubt under the right-to-buy legislation. Under the transfers, the ownership of the path across the rear of no 71 passed to the new owner and the ownership of the path between nos 71 and 73 was divided laterally between the two new owners, each enjoying a right of way over the half not in his ownership.

Mrs King in her action claimed a right of way over the paths, which she said she had used since 1967. I should say, although I do not think it of significance in law, that the path and its condition have become of importance to Mrs King because she has had the misfortune to become a victim of rheumatoid arthritis and for the past five years K
has had to use a wheelchair. This is difficult to use through the front door of no 69, partly because of the internal arrangement of the house and partly because of an installed stair lift. Accordingly, she has had to use the rear access to reach Forest Road by means of the path. The path, in so far as it comprises the length running between nos 71 and 73, is in a bad state of repair and the plaintiff has found that to use that path in her wheelchair is both painful and dangerous.

In her action, Mrs King claimed a mandatory injunction and damages for pain, suffering, inconvenience and distress. The learned judge found that there was granted to her upon the grant of the tenancy a right of way over the path. He based his conclusion upon section 62 of the Law of Property Act 1925, which provides that on a L
demise there shall pass appurtenant rights and easements. The judge also found that it was an implied term of the tenancy that the council would repair the path, that they were in breach of that implied term and that they were also in breach of a term implied by statute. The statutory term is the term now contained in section 11 of the Landlord and Tenant Act 1985, which re-enacts section 32(1) of the Housing Act 1961. The judge made no finding on an alternative claim by reference to section 4(4) of the Defective Premises Act 1972. In regard to that there is no respondent's notice, but I shall comment upon it later.

The judge's conclusion that a right of way over the path passed with the demise to the plaintiff in 1967 is not now challenged. Nor is it challenged that the path is out of repair. The issue on this appeal is, M
narrowly, as to whether the judge was right in holding that the appellants were under an obligation to repair the path. The appellants' case is that they were not. They say that it was wrong to imply a covenant for repair in principle and that there is no applicable exception to that principle. They say, further, that there was no breach of the statutory covenant because the circumstances in which that covenant can be broken did not here arise.

I approach the problem from the basis that the expressed terms of the tenancy, as derived from the conditions of tenancy, were incomplete. They were incomplete because they lacked any obligation upon the landlord at all. In order to give the arrangement a necessary bilateral character, the landlord's obligations have to be derived by implication of law. That exercise is, of course, an exercise for the purpose of completing an agreement as of 1967. It is in regard to affairs as they then stood that one must look. What has happened since that time, whether it be the alienation of the path or the misfortune which has befallen the plaintiff, is of no materiality.

As a matter of general law, it is not to be implied that there is an obligation upon a landlord to maintain retained land over which a right has been granted to the tenant. That is not in dispute. To that general proposition there is an exception, the germ of which can be found for present purposes in the decision of this court in *Miller* v *Hancock* [1893] 2 QB 177, in particular in the judgment of Lord Esher MR at p 179. I do not find it necessary to rehearse what the Master of the Rolls there said, and find it sufficient to go straight to the decision of the House of Lords in *Liverpool City Council* v *Irwin* [1977] AC 239*. The headnote, so far as is material, states:

> The tenants of a council maisonette on the ninth and tenth floors of a 15 storey tower block withheld their rent as a protest against conditions in the building and in their maisonette. In an action by the council for possession the tenants counterclaimed nominal damages, alleging, *inter alia,* that the council were in breach of their duty to repair and maintain the common parts of the building of which they retained control, including lifts, staircases, rubbish chutes and passages, and were also, in relation to the demised maisonette, in breach of their covenant for quiet enjoyment and of the implied covenant under section 32(1) of the Housing Act 1961 . . .

The terms of the tenancy were in that case incomplete, as were the terms of the instant tenancy. The question thus was, was there to be implied an obligation upon the landlord to repair the lifts and the rubbish chute? Lord Wilberforce at p 254D said:

> There can be no doubt that there must be implied (i) an easement for the tenants and their licensees to use the stairs, (ii) a right in the nature of an easement to use the lifts, (iii) an easement to use rubbish chutes.
>
> But are these easements to be accompanied by any obligation upon the landlord, and what obligation? There seem to be two alternatives. The first, for which the council contends, is for an easement coupled with no legal obligation, except such as may arise under the Occupiers' Liability Act 1957 as regards the safety of those using the facilities, and possibly such other liability as might exist under the ordinary law of tort. The alternative is for easements coupled with some obligation on the part of the landlords as regards the maintenance of the subject of them, so that they are available for use.
>
> My Lords, in order to be able to choose between these, it is necessary to define what test is to be applied, and I do not find this difficult. In my opinion such obligation should be read into the contract as the nature of the contract itself implicitly requires, no more, no less: a test, in other words, of necessity. The relationship accepted by the corporation is that of landlord and tenant: the tenant accepts obligations accordingly, in relation *inter alia* to the stairs, the lifts and the chutes. All these are not just facilities, or conveniences provided at discretion: they are essentials of the tenancy without which life in the dwellings, as a tenant, is not possible.

Lord Salmon at p 263A said:

> I find it difficult to think of any term which it could be more necessary to imply than one without which the whole transaction would become futile, inefficacious and absurd as it would do if in a 15 storey block of flats or maisonettes, such as the present, the landlords were under no legal duty to take reasonable care to keep the lifts in working order and the staircases lit.

The speeches of their lordships were considered in this court in the *Duke of Westminster* v *Guild* [1985] QB 688†, where at p 699C Slade LJ was at pains to point out that *Irwin* was a case where "the essential means of access to the unit are retained in the landlord's occupation".

Mr Gordon-Saker, in his argument for the appellants, emphasised the use of the word "necessary", the use of the word "essential" and the use of the word "solely", all of which are to be found in the authorities to which I have referred or in others which are referred to in those cases. Here, he said, there was a front door and in that circumstance the rear access was not essential or necessary. Accordingly, the implication of an obligation is not the implication of a necessary obligation. It was at this point that the difference between the parties occurred.

The argument of Mr Matthews for the respondent was that the rear access was essential. He pointed out that in the *Irwin* case there were, in addition to the lift, stairs. That the stairs in *Irwin* presented difficulties to those on the higher floors was only a difference in degree, said Mr Matthews, and not in kind, from the difficulty of receiving coal and depositing household refuse through the front door of the house.

In my judgment, it is important, in resolving the dispute between the appellants and the respondent, to bear in mind the layout of this estate and of the terraces within it. The houses within the terrace were designed with front and rear access. The rear access was plainly for the removal of refuse and the delivery of coal and the like, uses to which this rear access was in fact put. The houses could not be enjoyed or function in accord with their design without the rear access. In that circumstance, I, for my part, find no difficulty in concluding that the implication of an obligation to maintain the rear access is a necessary one to fulfil the purpose of the demise. The demise is of a dwelling-house, which is designed to function as such with the mechanism of rear service. In my judgment, the demise of a house which is designed to function in that way is not efficacious unless the rear surface is adequately maintained.

I think that the judge was right, in the particular circumstances of this design and layout, to conclude as he did, that is to say that there was here by law to be implied an obligation to maintain the path. I observe, but do not rely upon, first, that the council did until some 14 or 15 years ago in fact maintain the path and, second, that in a book published in 1982 by the appellants' housing department it is at least a matter of inference that "paths are maintained by the Council".

The other two ways in which the matter had been put I can deal with shortly. The judge himself had dealt very shortly with the breach of statutory covenant. After his careful analysis of the major point, he added:

> I am also of the view that the path forms part of the exterior of the premises for the purposes of section 11 of the Landlord and Tenant Act 1985, and therefore there is a breach also of the statutory covenant.

Section 11, which would have been section 32(1) of the 1961 Act at the time of this demise, provides:

> In a lease to which this section applies . . . there is implied a covenant by the lessor —
>
> (*a*) to keep in repair the structure and exterior of the dwelling-house.

The lease of no 69 is a lease to which the section applies, but I for my part do not see how the path running between nos 71 and 73 could be said to be the exterior of no 69. No doubt, as decisions such as *Hopwood* v *Cannock Chase District Council* [1975] 1 WLR 373 show, it is a question of fact and degree as to what is included in the exterior of the dwelling-house. However, in my judgment, it would be an extravagant and unsupportable conclusion to say that the path was as a matter of fact and degree part of the exterior of no 69.

The final way in which the matter was put, although on it the judge expressed no conclusion, is with reference to section 4(4) of the Defective Premises Act 1972, which provides that:

> Where premises are let under a tenancy which expressly or impliedly gives the landlord the right to enter the premises to carry out any description of maintenance or repair of the premises, then, as from
>
> — a certain time —
>
> the landlord shall be treated for the purposes of subsections (1) to (3) above (but for no other purpose) as if he were under an obligation to the tenant for that description of maintenance or repair of the premises.

The difficulty confronted by the plaintiff in relation to that subsection is that the premises must mean the premises which are let and let the path was not. Accordingly, I think there was nothing in the point under the statute, and in fairness to Mr Matthews I must record that he canvassed it with a degree of diffidence. However, for the reasons which I have given I would dismiss this appeal.

Sir George Waller agreed and did not add anything.

Appeal dismissed with costs.

---

*Editor's note: Also reported at (1976) 238 EG 879.
†Editor's note: Also reported at (1983) 267 EG 762.

Court of Appeal

June 27 1991

(Before Lord Justice PARKER, Lord Justice NOURSE and Lord Justice BUTLER-SLOSS)

MONTROSS ASSOCIATED INVESTMENTS SA v MOUSSAIEFF AND ANOTHER

Estates Gazette February 8 1992

[1992] 05 EG 160

Landlord and tenant — User covenant in lease — Construction — Covenant not to use premises for certain specified trades "but will use the demised premises" for other stated purposes — Whether this was a positive obligation to use the premises for such purposes or merely an emphatic statement of the prohibition of uses other than those specified — Whether use "for the business of high-class retailers of jewellery" covered use for storage, processing of paperwork and negotiations with suppliers to the business — Relevance of factual matrix — Appeal from decision in favour of lessees dismissed

The defendants (respondents to this appeal) carried on a high-class retail jewellery business in the foyer of two properties in Park Lane, London W1, nos 35 and 37, near the Hilton Hotel — The two properties were let under one lease — No 37 had never been occupied by the respondents, while no 35 had been used not as a shop but for purposes connected with their jewellery business at the hotel, such as storage of some stock, processing of paperwork and negotiations with suppliers — The respondents covenanted that they were not to carry on a number of specified unacceptable businesses and activities and that they "will use the demised premises either for the business of high-class retailers of jewellery" or for certain other specified businesses — Mr Thomas Morison QC, sitting as a deputy judge of the Chancery Division, held that the respondent lessees had not been guilty of the breach of covenant alleged — The lessors appealed

***Held*: The appeal was dismissed — Two questions arose — (1) It was one of impression whether the use of no 35 for the supportive activities described was a use "for the business of high-class reatailers of jewellery" within the meaning of the covenant — The second question was whether, accepting that such use was not in itself in breach of covenant, there was a breach involved in the use of no 35 only, no use having been made of no 37**

The business of a high-class retailer did not begin and end at the point of sale to the customer — It necessarily included the storage of stock used in the business, the processing of the paperwork and meetings and negotiations with suppliers — Premises used for these necessary purposes were used "for" the business of high-class retailers within the covenant — The judge below concluded that the words in question were in the nature of an emphatic negative rather than a positive instruction — (2) It was considered neither necessary nor desirable to express a view as to whether the mandatory form of the words "will use the demised premises" indicated the imposition of a positive obligation to use the premises for the purposes stated — The omission to make use of no 37 did not constitute a breach of the covenant even if these words were construed to be positive in substance — Taking no 35 and no 37 together, as two premises which were let as a whole and must be viewed as a whole, the use of half was in all the circumstances a sufficient use of the whole

The following cases are referred to in this report.

Basildon Development Corporation v *Mactro Ltd* [1986] 1 EGLR 137; (1985) 278 EG 406

Levermore v *Jobey* [1956] 1 WLR 697; [1956] 2 All ER 362, CA

Tea Trade Properties Ltd v *CIN Properties Ltd* [1990] 1 EGLR 155; [1990] 22 EG 67

Westminster City Council v *Duke of Westminster* [1990] 23 HLR 174

This was an appeal by the lessors, Montross Associated Investments SA, a Panamanian company, from the decision of Mr Thomas Morison QC, sitting as a deputy judge of the Chancery Division, dismissing a claim by the appellants that Shlomo Moussaieff and his wife, Alisa Moussaieff, lessees of premises at 35 and 37 Park Lane, London W1, had been in breach of a covenant in the lease of the premises. The decision of Mr Thomas Morison QC was reported at [1990] 2 EGLR 61, [1990] 45 EG 109.

David Neuberger QC and Miss Judith Jackson (instructed by Lewis Silkin & Co) appeared on behalf of the appellants; Nicholas Dowding (instructed by Rabin Leacock Lipman) represented the respondents.

Giving the first judgment at the invitation of Parker LJ, NOURSE LJ said: This appeal raises two short questions of construction on a user covenant in a lease of commercial premises at 35 and 37 Park Lane, London W1. The lease was dated September 12 1979 and was made between the freehold owners, Romulus Films Ltd ("Romulus"), as lessors of the one part and the defendants, Mr and Mrs Shlomo Moussaieff, as lessees of the other part. It was for a term of 21 years from December 25 1978 at a fixed rent of £16,485 pa. On or about January 28 1987 the freehold reversion was acquired from Romulus by the plaintiffs, a Panamanian company called Montross Associated Investments SA.

The background to the dispute between the plaintiffs and the defendants and the course which it has taken are fully recounted in the careful judgment in the court below, which is reported at [1990] 2 EGLR 61. The Vice-Chancellor had ordered the trial of the following question:

> Whether there has been a breach of clause 2(17) of the lease dated September 12 1979.

On June 25 1990 Mr Thomas Morison QC, sitting as a deputy judge of the Chancery Division, decided that there had not. The plaintiffs now appeal to this court.

Among the facts admitted or found by the learned judge were these. The defendants have for many years been carrying on business as high-class retailers of jewellery in shop premises in the foyer of the Hilton Hotel, a short distance from the demised premises, through a private company of which they are the sole directors and shareholders, called London Hilton Jewellers Ltd. Between 1971 and 1977 the demised premises were let to other tenants under separate leases for use as a restaurant and coffee shop respectively. In the latter year, after disputes with the chairman of Romulus, Sir John Woolf, the tenants indicated that they would like to assign their leases. Sir John, who was on friendly terms with the defendants and a reasonably frequent visitor to their shop in the Hilton Hotel, introduced them to the premises. In due course planning permissions for changes of use were obtained; deeds of licence and variation giving Romulus' consent to the assignment of the leases and the changes of use were executed; and the leases were assigned to another private company owned and controlled by the defendants. The lease of no 35 was assigned on October 31 1977 and the lease of no 37 on December 5 1978. Thereafter Romulus and the defendants negotiated for a new, single lease of the two premises together, those negotiations resulting in the lease of September 12 1979 with which we are concerned. It is agreed that both premises were then unoccupied and, moreover, that they had been unoccupied since the dates on which the previous leases had respectively been assigned. There was also evidence, which appears to have been accepted by the judge, that the cost of making the premises into retail outlets would amount to over £200,000. We have been told that the floor areas of the two premises are exactly the same, being 671 sq ft in each case.

The material provisions of the lease are these. By clause 1 the lessor demised unto the lessees:

> ALL THOSE the lock up shops situate at and known as 35 and 37 Park Lane on the lower ground floor of the building known as 3 Curzon Place TOGETHER with the basement stores immediately below the same and the storerooms at the foot of the staircases to the rear of the said shops . . .

The lessees' covenants are contained in clause 2, subclause (17) of which is in these terms:

> NOT during the continuance of the term hereby granted to carry on or to permit or suffer to be carried on upon the demised premises or any part

thereof the trade business or occupation of a retailer of beer wines or spirits or retailer of fried fish or onions or use or permit to be used the same premises or any part thereof as offices or rooms for betting or wagering or carrying on or permit to be carried on in the premises any other trade business or calling whatsoever that shall be adjudged by the Lessors to be noisy noisome noxious or offensive or that shall be hurtful or prejudicial or constitute a nuisance or annoyance to any other tenant or tenants of the Lessors or a public nuisance and will without prejudice to the generality of the foregoing not suffer or permit the escape of cooking smells from the premises or use or allow the premises or any part thereof to be used for frying onions but will use the demised premises either for the business of high-class retailers of jewellery and/or antiques and/or luxury goods and/or a travel agency or a recognised Bank the authorised name of which includes the word "Bank".

It will be observed that that subclause falls into two parts, the first of which is negative in form and the second, which starts with the words "but will use", positive in form. Clause 2(13) contains an absolute covenant by the lessees against assigning, underletting, parting with or sharing possession of the whole or any part of the demised premises.

The use which the defendants have made of the demised premises since the grant of the lease was described by the judge at p 4G of the transcript:*

The defendants have never occupied any part of no 37. No 35 has never been used as a retail shop as such, but it has been used for the benefit of the jewellery business carried on by the defendants through Hilton Jewellers at the hotel shop, in a number of different ways. In the first place some of Hilton Jewellers' stock is kept at no 35, thus easing a space problem at the hotel shop; second the defendants have put a desk and filing cabinets in no 35 so that the paperwork generated by Hilton Jewellers can be processed there, either by one of the two defendants themselves or by one of their staff from the hotel shop; third, where the occasion demands, suppliers of items for the hotel shop are sometimes directed to no 35, it being thought prudent not to carry on negotiations with sellers while engaged in selling items to customers at the hotel shop.

Thus no 37 has never been used at all and no 35, although it has been used, has not been used as a retail shop. It has been used for the storage of some of the stock and for the processing of the paperwork of the business of the London Hilton Jewellers Ltd and sometimes for negotiations with suppliers to that business. The first question which we have to decide is whether that use is use "for the business of high-class retailers of jewellery" within the second part of clause 2(17) of the lease.

In both courts there has been some debate as to the extent of the factual matrix which can be taken into account for the purpose of construing the lease. On a broad view of the matter, I think it possible that all the facts to which I have referred as having been admitted, found or agreed would be admissible for this purpose, including the terms of the two deeds of licence and variation, being deeds to which both Romulus and the defendants' other private company were parties. However, the judge said that he had put aside from consideration any of those facts, save that the defendants had a retail jewellery business not far from the demised premises which were empty and had been empty for some months at the date when the lease was granted. He made it clear that his construction of the lease would have been the same whatever the extent of the admissible factual matrix might have been.

As to the first question, the judge was of the view that the premises could be used either as shop premises for the business specified in the second part of clause 2(17) or for the purposes of those specified businesses carried on elsewhere. In seeking to dissuade us from that view, Mr Neuberger, for the plaintiffs, had relied on a number of matters, including the reference in the parcels to "lock up shops" and the facts, first, that the premises both had shop fronts, second, that use as shops was the natural use to make of them and, third, that you would not expect commercial premises in this part of Park Lane to be used for storage and administration as opposed to luxury shops. He has also pointed to the oddity, if the defendants' construction is correct, of permitting the premises to be used for storage and processing of paperwork in connection with a retail business, while prohibiting those uses in connection with a wholesale business. Turning more closely to the wording of clause 2(17), he argues that various of the expressions used in the first part confirm his construction of the second part.

Mr Dowding, for the defendants, argues that the business of a high-class retailer of jewellery does not begin and end at the point of sale to the customer. As is demonstrated by this case, it can include the storage of stock to be used in the business, the processing of the paperwork of the business and negotiations with suppliers to the business. If premises are used for those purposes, they are used "for" the business of high-class retailers of jewellery within the second part of clause 2(17).

The question is one of impression. In my judgment, the argument of Mr Dowding is correct. None of the other provisions of the lease or the other matters on which Mr Neuberger relies can overcome the clear effect of the words which the parties have used. Like the judge, in arriving at that conclusion I do not rely on the factual matrix, although I observe that whatever its admissible extent may be, it cannot detract from, and can only confirm, the result which follows from a true construction of the words themselves.

The second question arises in this way. Given that the defendants' use of the demised premises has not, of its nature, been in breach of clause 2(17), has such a breach nevertheless been caused by the limited extent of their use, that is to say by the use of no 35 only? It has throughout been accepted that in order to answer that question it is helpful to start by asking whether the second part of clause 2(17), which opens with the words "but will use the demised premises etc" is positive in substance as well as in form.

As to that, the judge was of the view that the second part of clause 2(17) was not sufficiently clear as to create a positive obligation to use the premises for one of the specified businesses. He thought, adopting the words of Hoffmann J in *Tea Trade Properties Ltd* v *CIN Properties Ltd* [1990] 1 EGLR 155 at p 157J, that it was "simply an emphatic statement of the prohibition on uses other than that specified". In support of his argument to the contrary, Mr Neuberger relied on the decisions of this court in *Levermore* v *Jobey* [1956] 1 WLR 697 and *Basildon Development Corporation* v *Mactro Ltd* [1986] 1 EGLR 137 and more especially on the very recent decision of Harman J in *Westminster City Council* v *Duke of Westminster* [1990] 23 HLR 174.

Although each of those three decisions is for one reason or another distinguishable from the present case, it is certainly possible that the second part of clause 2(17) is positive in substance as well as in form. For my part, I think it neither necessary nor desirable to express a view on that point. That is because the omission to use no 37 cannot, in my view, constitute a breach of clause 2(17) even if the second part of it is positive in substance.

At this stage Mr Neuberger relies primarily on the general proposition that to hold that a use of half only of the premises is a sufficient compliance with a positive obligation to use them in a specified manner is to stultify the purpose of the obligation. He also relies on the three references in the first part of clause 2(17) to the demised premises "or any part thereof" in contrast to the reference to "the demised premises" in the second part, which shows, he says, that the positive obligation extends to the whole and not merely to half.

Having taking full account of Mr Neuberger's submissions, I can express my own view of the second question by adopting what the judge himself said when dealing with it at p 12E-F:*

. . . If the clause is positive, then I do not consider that clear enough words have been used to require a user of the whole of the premises. In my judgment, even on this basis, the user of no 35 constitutes a sufficiently substantial user of the whole, so that the bystander would say that the demised premises were being used by the defendants for the business of a high-class retailer of jewellery being carried on at the Hilton Hotel. If the remaining part of the premises were used for the sale of doughnuts, then one might readily conclude that the premises were not being used for the business of retail jewellers. A relatively insignificant other user would not alter the character of the prime use. The line would have to be drawn on the facts of each case.

I would emphasise that the two premises, having been let as a whole, must be viewed as a whole without regard to their physical separation. Viewing them thus and taking account of the nature of the use which may be made of them, I agree with the judge that a use of half is, in all the circumstances of this case, a sufficient use of the whole.

For these reasons I would dismiss this appeal and affirm the decision of the judge.

Butler-Sloss LJ agreed with the decision and reasons given by Nourse LJ and did not add anything.

Also agreeing, **Parker LJ** said: Although the drafting of clause

*Editor's note: Also reported at [1990] 2 EGLR 61 at p 63F.

*Editor's note: [1990] 2 EGLR 61 at p 65C-D.

[A] 2(17) is far from happy, the meaning of the expression "for the business of high-class retailers of jewellery" etc becomes clear when one considers the subject-matter of the lease, namely the lock-up shops themselves, the basement storerooms beneath and the storerooms at the foot of the staircases to the rear of the shops. These together form "the demised premises". Suppose that the tenants had in fact used both of the shops as retail shops, one set of storerooms for stock, another for commercial office work and packaging and the third for interviewing and negotiating with suppliers, could it possibly be said that any part of the premises was not being used "for the business of high-class retailers of jewellery"? In my judgment, the answer must plainly be "no". The entirety of the premises would be being used for the purposes of that business. If for any reason the [B] tenants found that it would be more profitable to run the business by using one only of the shops for retail selling and the other for increased storage space the same would apply. It would also apply if for, say, security reasons they decided that retail selling should be done in the storerooms and the remaining operations necessary for the running of the business in the shops. Could it then make any difference if the tenants already had other premises, perhaps no 33, in which they were conducting retail selling together with the ancillary operations but wished to devote the whole of such other premises to retail selling and conduct the ancillary operations in nos 35 and 37? I think not. It would be a single business and nos 33, 35 and 37 would all be being used *for* that business.

[C] As to the question whether the second part of the clause is positive in substance, I find it hard to suppose that, when a series of negative covenants are followed by the words "but will use", the intention was merely to create an emphatic negative. The change of language is marked, it is positive in form and I am not convinced that any good reason exists for holding that the substance is different from the form. Like Nourse LJ, however, I find it unnecessary to reach any conclusion on the point. It is, in my view, clear that, even if positive, the obligation cannot have been to use every part of the demised premises for one of the specified businesses. If, for example, having for a time used the whole of the demised premises for one of the specified businesses, the tenants were forced owing to a fall in trade to contract into part only it could not, as it seems to me, be said that they [D] had ceased to use the premises for that business any more than someone ceases to use a large house as a dwelling-house if he locks up all of the rooms except one bedroom, the kitchen, one of the three reception rooms and the bath and toilet.

In the present case, assuming a positive covenant, I accept that there must be use for one of the specified businesses of such part of the premises as would enable an ordinary person to answer the question "For what are the premises being used?" with the reply "For the business of a high-class retailer". There may be borderline cases, but on the facts of this case I am satisfied that a sufficient part of the premises was and is being used for the specified business to require an affirmative answer to such an inquiry.

[E] For these additional reasons, I, too, would dismiss this appeal.

The appeal was dismissed with costs.

Chancery Division

July 19 1991

(Before Mr Justice HARMAN)

HAGEE (LONDON) LTD v CO-OPERATIVE INSURANCE SOCIETY LTD

[F] Estates Gazette February 22 1992

[1992] 07 EG 122

Forfeiture — Alleged breach of covenant — Notice under section 146 of the Law of Property Act 1925 — Alleged unlawful installation of air-conditioning equipment — Whether tenant knew of works — Whether breach of covenant — Whether section 146 notice effective — Whether landlord entitled peaceably to re-enter — Judgment for tenant

By a lease dated November 6 1984 the plaintiff was the tenant of three shops in the basement and ground floors of 120-122 New Bond Street, London — Lease granted for term [G] of 25 years from June 24 1984 — Defendant was the grantor of the lease and the immediate landlord to the plaintiff — Two of the three subject buildings are listed — First and second floors have been empty for some years — Plaintiff had rights over property outside the demise in connection with easements benefiting the demised premises — Plaintiff covenanting not to make structural alterations or additions to the demised premises — Covenant not to cut any main walls or timbers without the defendant landlord's licence — In 1989 plaintiff decided to replace air-conditioning system — Work involved replacement of condensers on roof of building and outside the demised premises — Landlord's managing agent on learning [H] of works on the roof and outside the demised premises instructed that no further rent be accepted but did not take any further step to prevent the work continuing — Contract placed on behalf of plaintiff with air-conditioning contractor to replace condensers and pipework — Air-conditioning contractor an independent contractor — No reference in contract to precise positioning of units or pipework — Independent contractor commencing work — Following inspection by landlord's surveyors, independent contractor placing pipework in notches in joists in upper floors above the demised premises — Plaintiff, not aware of change in detail of installation, paid independent contractor's bill — On [J] December 20 1989, some three months after commencement of work, landlord served notice under section 146 of the Law of Property Act 1925 alleging breaches of covenant — Defendant landlord effected peaceable re-entry at 3 am on February 20 1990 — Plaintiff readmitted to premises by order of Knox J on February 20 1990 — On following day the plaintiff issued the writ in the present action

Issues for determination of the court — What breaches of covenant were established? — Was there a reasonable time for remedy between service of the section 146 notice and the re-entry? — Was there an effective re-entry? — Was there any [K] jurisdiction in the court to grant relief from forfeiture once a complete re-entry had been effected? — If yes, upon what terms should relief be granted? — Defendants conceding that if there were jurisdiction to grant relief, this would be a case where, upon terms to be discussed, the court would grant relief

***Held*: Defendant landlord had not given consent to the works of replacing the air-conditioning condensers and pipework — Works of notching for joists, the subject of most of the alleged breaches of covenant, were not done by or on the instructions of the plaintiff but were contrary to its instructions and carried out by an independent contractor and [L] its subcontractor without the plaintiff knowing of the acts at the time they were done — Plaintiff did not ratify unauthorised work of its independent contractor by simply paying the contract price in ignorance of how the work was carried out — It followed that the plaintiff had committed no breach of its covenant with the landlord by reason of the fact that works were carried out in the building — Notwithstanding the plaintiff's right to maintain easements, assuming that the pipework and supports to the air-conditioning plant were part of the easements, the plaintiff was not entitled to override a covenant — The acts of cutting [M] the joists, which are main timbers, if done by the plaintiff or at its direction, would have amounted to a serious breach of covenant — No breach of covenant not to make any alteration to the architectural appearance of the demised premises by substitution of new condensers — No breach of covenant not to do work which contravened the Planning Acts — No breach of covenant that landlord not provided with all plans and documents — Landlord's conduct not reasonable in merely placing rent stop to avoid charge of waiver of breach — Replacement of existing permitted air-conditioning plant no breach of covenant not to make any addition or alteration**

without landlord's consent — A replacement cannot be an addition or an alteration — Plaintiff tenant would have committed a serious breach of cutting main timbers had it authorised the work — Whether landlord entitled to re-enter peaceably following breach of a negative covenant being an irremediable breach not decided — Judgment for the plaintiff

The following cases are referred to in this report.

Lister v *Lane and Nesham* [1893] 2 QB 212, CA
Pearlman v *Keepers and Governors of Harrow School* [1979] QB 56; [1978] 3 WLR 736; [1979] 1 All ER 365; (1978) 38 P&CR 136; [1978] EGD 721; 247 EG 1173; [1978] JPL 829, CA

This was a claim by the plaintiff for declarations and for damages arising out of a purported peaceable re-entry by the defendant landlord of premises at 120-122 New Bond Street, London W1. The plaintiff was Hagee (London) Ltd, a subsidiary company of Moss Bros. The defendant was the grantor of a lease dated November 6 1984 in respect of shop premises at 120-122 New Bond Street, London W1, and the plaintiff's immediate landlord.

John McDonnell QC and David Parry (instructed by Jaques & Lewis) appeared for the plaintiff tenant; Michael Barnes QC and John Male (instructed by Nabarro Nathanson) represented the defendant landlord.

Giving judgment, **HARMAN J** said: This is an action by a subsidiary company of the well-known outfitters Moss Bros. The plaintiff (hereafter Hagee) was the tenant of three shops in the basement and ground floors of 120, 121 and 122 New Bond Street under a lease dated November 6 1984, which granted a term of 25 years from June 24 1984. The term granted therefore runs for another 18 years from the present time. The defendant to the action is the Co-operative Insurance Society Ltd, which was the grantor of the lease and is the immediate landlord of the premises. The defendant is itself a tenant of part of the premises under long leases but nothing therein affects the issues in this action. The defendant has, after many years' endeavour, obtained from the Secretary of State for the Environment on March 8 1989 all the multitude of permissions nowadays needed to enable it to carry out extensive development works on an area of land in the City of Westminster bounded by New Bond Street, Brook Street, Avery Row and Lancaster Court. The three shops let to the plaintiff form part of the eastern edge of the area of the projected redevelopment.

Two of the buildings in which the plaintiff has shops were originally early 18th-century dwelling-houses of the typical London-terrace type. Both buildings are listed as Grade II. The third building is of early 20th-century date. The first and second floors of all three buildings have been empty, save for the second floor of no 120, for many years. The plaintiff alleged in its statement of claim that the upper floors of the buildings were "derelict". The defendant admits they were empty and unused but denied the allegation that they were "derelict". At the invitation of the parties I had a view of the buildings and, in my view, the word "derelict" can aptly be applied to them. No modernisation appeared to have been carried out on the upper floors and the structures remain those of an unimproved 18th-century dwelling-house. The ground-floor shops occupied by Hagee have modern shop fronts and internally the ground floor and basement are well-fitted modern men's clothes shops.

The plaintiff was the tenant of the whole of no 122 from March 1970. It is notable that that grant, 21 years ago, contemplated redevelopment in the near future. The plaintiff became the underlessee of the shop premises in nos 121 in March 1973. In 1978 the tenancies were revised and the plaintiff became the tenant of the three shops in nos 120, 121 and 122. The 1984 lease granted to Hagee 25-year terms for the basement and ground-floor shops, excluding therefrom the entrance door, hallway and staircase leading to the upper floors of no 120. It also granted a right to enter other parts of the building (although limited to entry by prior appointment) for the purpose of connecting with and inspecting, maintaining and renewing "sewers, drains, pipes, wires and cables serving the demised premises". Thus Hagee had rights over property not let to it in connection with, broadly, easements benefiting the demised premises. The 1984 lease contained extensive covenants by the plaintiff to this action. By clause 4(5)(a) the plaintiff covenanted not to make structural alterations or additions to the demised premises, and by clause 4(5)(b) not to cut any main walls or timbers of the building without the licence of the defendant. It is important to notice and remember the distinction between "the demised premises" and "the building". By clause 4(8) the plaintiff covenants not to do anything to the demised premises which contravened the Planning Acts (as defined in the lease). By clause 4(9)(c) the plaintiff covenanted not to install on the demised premises machinery which was noisy or which caused dangerous vibrations. The lease also contained a fairly usual covenant for re-entry on breach of covenant.

In 1972 the plaintiff had originally installed air-conditioning for the three shops in no 122. Late in 1976 further air-conditioning apparatus was installed with the knowledge of the defendant's then agents, Bernard Thorpe. The apparatus involved air-dispensing units suspended from the ceiling of each shop, which were connected to six condenser units installed outside the demised premises on the flat roof of the back extension to no 121. The connecting pipes and wires ran through the building and out to the condensers. These condensers were large, rectangular box-like shapes of varying colours, pink, orange and yellow. In 1989 the air-conditioning in the shops was not working adequately and Moss Bros decided to replace the system. No leave was sought from the defendant as landlord, or from any of the other bodies — English Heritage and the local planning authority, which is the Westminster City London Borough Council — which might have an interest in the appearance of the building for these works, which were plainly thought to be unimportant as merely the replacement of existing approved apparatus. I say that, although I have not had evidence from the then property director of Moss Bros who was responsible for the plaintiff's activities, since the inference as to his attitude is, in my judgment, inevitable. The grant of new leases in 1978, after the installation of the air-conditioning condensers on the roof of part of the building outside the demised premises, followed by the new lease granted in 1984 plainly amounted to waivers of any breach of covenant down to that date and to a licence to continue to use and enjoy the apparatus as it stood.

The start of the installation of new air-conditioning units, both condensers on the roof outside the demised premises and units delivering cool air inside the three shops, was observed by two professional men employed by the landlord's advisers in the design and carrying out of the major redevelopment scheme on September 1 1989. The fact of work being done was reported at once to the defendant landlord's managing agents — a wholly-owned subsidiary of the defendant, not an independent agent — and a conscious decision was taken to do nothing about the works being carried out. I had evidence from a Mr Keith Chappell [ARICS], a chartered surveyor, who was the employee of the managing agents in day-to-day charge of the development scheme. His office is in St James's Street, as he said to me about 200 yds away from New Bond Street. On learning that works were being done in the building but outside the demised premises Mr Chappell placed an instruction that no further rent was to be accepted from the plaintiff. He did not walk the short distance to look at the building or to inspect the works then being started. He took no step to prevent the work continuing. I formed the opinion that Mr Chappell, who said that the development scheme would be easier to carry out if the plaintiff's lease were at an end, deliberately did nothing in the hope that a breach of covenant might occur and forfeiture be possible to the advantage of the defendant landlord. Had he wished, he could have stopped the work without difficulty on Monday, September 4 1989 and this action would never have become necessary.

I now turn to find the facts as to the actual events which are relevant to the issue in this action. I had evidence from the following witnesses: Mr Male, a director in BNA Engineering Services, the contractor for supplying and fitting the new air-conditioning units, whose evidence was not challenged; Mr Williams, who was an installer of air-conditioning units who worked on the installation at nos 120, 121 and 122; Mr Joiner, an independent subcontractor engaged by the air-conditioning engineers and in charge of the installation at nos 120, 121 and 122; Mr Moss, a director of the plaintiff and of the eponymous parent company; Mr Barter, the other director of BNA Engineering Services, who actually visited the shop premises, obtained instructions from Mr Michael Gee, then property director of Moss Bros, about the replacement air-conditioning, gave instructions to Mr Joiner who was his company's subcontractor and later altered those instructions; all those were for the plaintiff. I am satisfied that each of them gave me his honest recollection of events to the best of his ability.

A For the defendant I had evidence from Mr William Assheton [RIBA], a chartered surveyor employed by the consulting architects to the landlord for the redevelopment scheme; and Mr Bonano, an Australian structural engineer employed by Kirk Wilson Scott Patrick, the landlord's consulting engineers, who was present with Mr Assheton on September 1 1989 when both young men were in the building at 120, 121 and 122 New Bond Street on behalf of the defendant, although not visiting the demised premises and without any contact with the plaintiff. Both these young professional men gave me, in my judgment, honest accounts to the best of their recollection of what they saw and said that day. Both were intelligent and straightforward and I accept both of them as reliable witnesses with a good recollection of events. In addition, both of them had
B made affidavits in March 1990, which was only some six months after the event and was therefore at a time fairly close to the relevant date.

On the basis of all that evidence I find the following facts. Mr Male, as a director of BNA Engineering Services Ltd, gave a quotation as an independent contractor to the plaintiff to supply and install new air-conditioning units at 120, 121 and 122 New Bond Street. No reference was made to the precise positioning of the units or the pipework. Mr Barter, as a director of BNA Engineering Services Ltd, visited the premises on August 24 1989 with the quotations and there met Mr Michael Gee, the property director of the group. He was instructed by Mr Gee in August 1989 to install new air-conditioning units, both condensers and air dispensers, in place
C of the existing units. Mr Barter was instructed by Mr Gee and agreed that the pipework and the steel channels required to support the new air-dispenser units in the shops should be laid above the floorboards in the upstairs part of the building. Two new condenser units were to be supplied and sited on the rear flat roof of the building near to but in a different position from the existing condensers. New air dispensers inside the demised premises were also to be supplied and installed in place of the existing dispensers. Thus the major part of the work, all of which was the replacement of existing services, was to take place outside the demised premises but in or on the building. As BNA Engineering Services were an independent contractor the general instructions were agreed, notably that the pipework and electric
D wiring should follow one common route and should be above the floorboards of the first floor of the building, but no discussion of the detail of the work to be done took place, as was to be expected. Mr Barter returned to Birmingham and instructed workmen; one was an employee and the other two independent subcontractors of whom one was Mr Joiner, who was to be in charge of the actual carrying out of the work.

Mr Barter instructed Mr Joiner and the others to lay the pipework and the channels on the floor in the first floor of the building and to bolt the suspended air dispensers to the channels above the floor. Mr Barter knew that the upper floors of the building were not let to the plaintiff. He did not know that nos 120 and 121 are listed buildings.
E Mr Joiner received his instructions in Birmingham from Mr Barter. He went to London and to New Bond Street and began work. He had no contact with or any instructions from anyone with authority to deal with property matters in the employ of either the plaintiff or Moss Bros. He was to install the new condensers on the flat roof at the rear of the building and to run all pipework above the floorboards on the most convenient route. Mr Joiner believed that the upper floors of the building were intended to be demolished. On Friday September 1, while Mr Joiner, who was in charge, and the other two men were working on installing above the floorboards the channels for the support for the first air dispenser, Mr Assheton and Mr Bonano entered the building at about 11 o'clock in the morning. They saw but did not speak at once to Mr Joiner and the others and
F went upstairs to the second floor of the building. After some 10 minutes or so they returned to the first floor and Mr Assheton asked what the men were doing. He was told "Installing air-conditioning". Mr Assheton expressed the view that his employer — not identified but understood by Mr Joiner to be the landlord — would not like the pipework and channels being installed above the floorboards. I am satisfied that Mr Assheton in the course of the conversation said he had no authority on behalf of the landlord to make any decisions. Mr Assheton was keen for his own good reasons not to get involved in any argument and had been instructed not to give anything away. He told the group of three workmen that he could not tell them what to do but that he would be reporting to his client. I am satisfied that Mr Assheton wished to and did say as little as possible. I am also satisfied that Mr Bonano added nothing to the conversation between Mr G
Assheton and the workmen.

There followed two telephone conversations between Mr Joiner and another of the workmen in London and Mr Barter in Birmingham. Mr Joiner gave me little evidence other than that he understood that the landlord would not approve of what he was doing. Mr Barter gained the impression that surveyors on behalf of the landlord had instructed Mr Joiner not to lay pipework and channels above the floor. Mr Barter told the workmen to lift the floorboards and inspect the joists. Later that day a further telephone conversation took place and Mr Barter instructed Mr Joiner and the others to lay the channels and the pipework in notches in the joists below the floorboards. He had inquired about the condition of the H
joists and was told that the wood was sound. He gave general instructions as to the number of joists to be spanned, and therefore notched, by the channels. The pipework was half an inch or five-eighths of an inch in diameter so did not require large cuts in the joists for it to be accommodated. I find that no one at Moss Bros or the plaintiff knew of this change of plan which was contrary to the express instructions given to Mr Barter when the tender was accepted.

Mr Barter informed an employee at the head office of Moss Bros that the new air-conditioning would not be working until a week or so later than the original date but he did not given any reason or explanation for the extra time to be taken. Mr Joiner and the workmen did not speak to any person in Hagee or Moss Bros who had J
any authority to do with the works. Mr Barter, as I find, regarded his company as entitled to carry out its contract at its discretion as to method of performance and, since he was not raising his price to Moss Bros, did not consider he needed any further authority to vary the method of installation of the new air-conditioning. I accept that Mr Barter believed that the landlord by its surveyors had required the channels and pipework to be laid under the floor which must involve notching of joints where they cross the line but, as I have held above, I am satisfied that no such requirement had been made, let alone instructions given to lay pipes below the floor.

Mr Joiner and the other workmen then proceeded to install the pipework and channels by lifting the floorboards and notching the K
joists. They also placed the two new condenser units on the flat roof at the back of the building. They completed the work and Moss Bros paid the bill without realising how the installation had been carried out. On September 8 1989 Nabarro Nathanson for the landlord wrote to Jaques & Lewis for the plaintiff asking for details of the works. A most unsatisfactory correspondence continued until late November with no clear answer ever being given to the landlord's inquiries. On December 20 1989 a notice under section 146 of the Law of Property Act 1925 was served by Nabarro on Jaques & Lewis alleging breaches of covenant. This was unfortunately treated as frivolous by Mr Gee as property director of Moss Bros themselves, who wrote on January 5 dated 1989 but obviously 1990, so saying. The service of the section L
146 notice was reported to the Moss Bros board but in a manner which gave the impression that it was not of importance and that surveyors and solicitors would deal with it. In fact very little was done by the plaintiff or Moss Bros.

The next event was, to me, a startling one. At about 3 am on February 20 1990 the defendant landlord by its agents effected what can only be called a forcible entry into the shops. The defendant instructed a well-known firm of certificated bailiffs to effect an entry. A locksmith drilled out the lock, enabled the security grill to be lifted and another specialist disconnected the burglar alarm. When the security to the premises had been forced the bailiffs entered and purported to take lawful possession of the demised premises. Thus the demised premises were effectively repossessed by the defendant M
exactly two months after the service of its section 146 notice. There followed various telephone conversations and an application *ex parte* to Knox J in the afternoon of February 20. On undertakings in damages and otherwise the learned judge made an order readmitting the plaintiffs to the premises. There followed the issue of the writ on February 21, various *inter partes* hearings of the motion and the actual pleadings in the action.

I had a great deal of evidence tendered as to the effect upon the joists as load-bearing parts of the building. In the end I am not satisfied that any serious impairment of the load-bearing capacity was caused by the notches cut by Mr Joiner and his fellow workmen.

Mr Barnes provided a helpful outline of his argument for the

A defendant landlord in which he submitted that there are five issues in this action which he listed as follows:

(1) What breaches of covenant were established?

(2) Was there a reasonable time for remedy between service of the section 146 notice on December 20 1989 and the re-entry of February 20 1990?

(3) Was there an effective re-entry?

(4) Was there any jurisdiction in the court to grant relief from forfeiture once a complete re-entry had been effected?

(5) If yes, upon what terms should relief be granted?

He submitted, and Mr McDonnell for the plaintiff did not demur, that questions (4) and (5) should not be decided in this judgment since
B a pending decision in the Court of Appeal* would affect, and probably determine, the question of jurisdiction to grant relief. Mr Barnes had acknowledged after the mid-day adjournment on the third day of the trial that if there were jurisdiction to grant relief then this would be a case where, upon terms to be discussed, the court would grant relief.

However, question (1) does not raise simple issues of fact as to what has been done in breach of covenant. The plaintiff's first contention in the affidavits on the *ex parte* application and in the pleadings is that the acts were all done in pursuance of permission granted by the landlord's agents, Mr Assheton and Mr Bonano. To that were raised two defences: first, that no permission, express or implied, was in fact given by Mr Assheton and Mr Bonano or either
C of them and, second, that neither of them in fact had or was held out as having authority to grant permission on behalf of the defendant landlord. I am satisfied that no permission was in fact granted. I am also satisfied that neither Mr Assheton nor Mr Bonano had or was held out as having authority on behalf of the landlord. Indeed I am satisfied that Mr Assheton expressly disclaimed any authority. That defence must therefore fail. The *bona fide* belief of Mr Barter, based on some misunderstanding by Mr Joiner, who seemed to me of limited general intelligence however competent a fitter of air-conditioning plant he may be, of what had been said to him, cannot affect the matter.

The second contention of the plaintiff is that it cannot be in breach of covenant in a case where none of the acts complained of was done
D by the plaintiff or any servant or agent of it. Further, the plaintiff argues, the acts alleged to constitute breaches of covenant were not authorised by the plaintiff to be done by an independent contractor but were expressly forbidden by the plaintiff in its instructions to its independent contractor. The plaintiff contends, and I have held, that it had no knowledge of the doing of the acts until long after they were all completed. The plaintiff contends by counsel that it cannot be held to be liable for breach of covenant by acts which it never authorised, which were not done by its officers, servants or agents, and of which it had no knowledge at any relevant time.

Mr Barnes argued that the evidence on behalf of the plaintiff was inadequate in that the director who gave Mr Barter his instructions
E was not called before me, there was no written record of the instructions and it was reasonable to hold that Mr Barter was given authority by Moss Bros on behalf of Hagee to vary the mode of carrying out the work. In my judgment, that is not the correct view of the matter. I am satisfied that Mr Barter was specifically instructed to lay the pipes and channels above the floorboards in the building. Had he done so almost all the alleged breaches of covenant would not have occurred. I hold that Mr Barter believed, on the erroneous statement of Mr Joiner, that the laying of the pipes and channels under the floorboards by notching the joists was authorised and indeed required on behalf of the landlord. Mr Barter did not, in my judgment, consider that he needed authority from the plaintiff or Moss Bros to vary the performance of the contract. Had he expressly
F considered the point he would, in my judgment, have thought it necessary to seek further instructions, since he was proceeding in direct contravention of a term of his instructions. I therefore hold that the works done which are alleged to constitute breaches of covenant were not done by or on the instructions of the plaintiff but contrary to its instructions by an independent contractor and its subcontractor without the plaintiff knowing of the acts at the time they were done.

Mr Barnes argued that even if the plaintiff did not do or authorise the doing of the works in the building in the manner which involved notching joists yet the plaintiff had adopted the works and so was liable to forfeiture. In my judgment, that argument is unsound. G
Hagee had no power to undo the works that had been done; indeed Mr Barnes' contention that notching joists was an irremediable breach of covenant supports the view that Hagee cannot turn the clock back. Mr Barnes argued that a breach of a negative covenant by an act that was complete and not continuing must in logic constitute a breach that is irremediable and I accept the force of that argument. But Mr Barnes' reference to Hagee "ratifying and adopting" the acts of BNA Engineering, which he described inaccurately as Hagee's agent, is not, in my judgment, a proper use of those words. Once the work was done, and the new air conditioning installed, there was nothing Hagee could do which would aptly be described as ratifying the work. No owner who employs an independent contractor to do some work can be said to "ratify" that work by simply paying the H
contract price in ignorance of how the work was carried out. Nor can an owner who uses plant installed by an independent contractor thereby be said to "adopt" the method of installation

No authority on this point was cited to me, no doubt because the facts are extremely unusual in that the works in this case were not done in or to the demised premises or with the knowledge, which might imply approval, even if tacit, of the tenant whose lease is sought to be forfeited for breach of covenant by the tenant. In my judgment, a tenant cannot be held responsible for work not done by it, not authorised by it or known of by it. I hold that Hagee committed no breach of its covenant with the landlord by reason of the fact that these works were carried out to the building. J

Had I been of opinion that Hagee was responsible for the acts done in the building I would have held that some of the acts were breaches of covenant but that some of the acts alleged by the section 146 notice to amount to breaches were not. Mr Barnes, in his final submissions, which I found most lucid and helpful, relied upon five breaches. He abandoned reliance upon any of the other breaches alleged in the section 146 notice. The first was clause 4(5)(b) of the lease, by which Hagee covenanted that it would not without licence in writing cut or remove the main walls or timbers or structural members of the building. I accept Mr Barnes' argument that the adjective "main" qualifies both "walls" and "timbers" but not structural members. I accept that the joists are main timbers and also structural members. Mr McDonnell argued that the rights to maintain easements granted K
with the term by the lease entitled Hagee to enter the building for the purpose of maintaining the easements (which may well include the pipes and wires serving the air-conditioning plant), which is plainly correct, subject to there being prior appointment, which there was not. But the right to maintain easements, assuming that the pipework and supports to the air-conditioning plant were part of the easements, cannot entitle Hagee to override a covenant in the lease and cut a main timber. In my judgment, the acts of cutting, if done by Hagee or at its direction, would have amounted to serious breaches of covenant. The claim that Hagee was performing the obligation in clause 4(4)(a) to repair pipes outside the demised premises which served only the demised premises is not, in my judgment, made out — L
see *Lister* v *Lane and Nesham* [1893] 2 QB 212 especially *per* Lord Esher MR at p 216.

The second breach relied upon was of clause 4(5)(a), which binds Hagee not to make any alteration to the architectural appearance of the demised premises. Mr Barnes submitted that the flat roof at the rear of no 121 formed part of the demised premises and that siting the new condensers thereon constituted a structural alteration and an alteration to the structural appearance of the demised premises. He cited *Pearlman* v *Keepers and Governors of Harrow School* [1979] QB 56* and in particular the judgment of Lane LJ (as he then was) at pp 72D to 73A. Mr Barnes submitted that any change beyond *de minimis* matters amounted to an alteration of architectural appearance. As I have observed, I had a view of those buildings and M
the demised premises. The rear of the buildings gives on to an area which can be described only as a hideous jumble of backs of buildings, old fire escapes and much debris. In my judgment, such alteration as was made by substituting two new condensers for those that were there must have been *de minimis* to the architectural appearance of the building and the demised premises. Further I am of opinion that the flat roof itself, as opposed to its inner skin, was not part of the demised premises. I am of opinion that no breach of this covenant was made out.

The third breach alleged was of clause 4(8)(a) of the lease whereby

*Editor's note: This refers to *Billson* v *Residential Apartments Ltd* now reported at p 43 *ante*.

*Editor's note: Also reported at (1978) 247 EG 1173.

A Hagee covenanted not to do or suffer to be done on or in connection with the demised premises anything which contravened the Planning Acts (as defined). Mr Barnes argued that the cutting of notches in the joists (and the other works) were done in connection with the demised premises because (a) physically the cool air delivered inside the shops was connected with the works in the building in that the pipes and condensers provided the means of cooling the air and (b) the works were done to serve the demised premises. In my judgment, this is to read the covenant altogether too literally. The words are written for a purpose and upon their true construction were never intended to apply to works such as those done here outside the demised premises and in the building. Mr Barnes took me on a learned tour of the Town B and Country Planning Act 1971. In my judgment, the claim that cutting notches in joists which are then covered by floorboards falls within section 55 of that Act as amounting to an alteration to a listed building which will affect its character is fanciful and playing with words. I accept that the physical removal of wood from the notches is an alteration, although trivial, to the building. But the character of both nos 121 and 122 does not derive from details of their construction and the alterations by notching of joists are so trivial as far as character is concerned as to be *de minimis*. I am of opinion that this breach was not made out.

The fourth breach alleged was clause 4(8)(d) in that Hagee was bound upon request to provide the landlord with all plans and documents it may reasonably require to satisfy itself that the earlier C parts of clause 4(8) had been complied with. Mr Barnes referred to the requests for information in Nabarro's letters of September 6 and September 20. The first produced a reasoned response but no documents or plans. The second, unhappily, was dealt with very tardily and incompletely. In my judgment, the landlord fails to make good its claim by reason of the fact that its agents so carefully and deliberately refrained from the obvious and easy step of inspecting the premises at any time from September 4 onwards. It is not reasonable conduct when activities are known of simply to place a stop upon receipt of rent (so as to avoid any charge of waiver of breach of covenant) and instruct solicitors to write letters. The landlord's conduct, by its agents, was not reasonable and the landlord did not reasonably require plans or documents for the D purpose specified when an easily made inspection was deliberately not made. It follows that the breach was not made out.

Finally Mr Barnes alleged a breach of clause 4(12)(a) whereby Hagee covenanted not to make any addition or alteration without the landlord's consent. In my judgment, the replacement of permitted air-conditioning plant (condensers, pipes and wires and the air-dispensing units) by new items to perform exactly the same functions cannot properly be called an alteration or addition. A replacement is not an addition, nor can it fairly be called an alteration in any other but the most literal-minded way. I am of opinion that no breach of this covenant has been made out.

I thus would hold, were it the act of the tenant, that the landlord E has proved one substantial breach of covenant. Mr Barnes submitted that such a breach, being a completed (that is a once and for all) act in breach of a negative covenant, would be an irremediable breach. Upon that basis the landlord was entitled to serve its section 146 notice and to re-enter in pursuance of it. I find this a difficult point. Mr Barnes' argument follows logically step by step to its conclusion, but the result is so surprising that absent clear authority I would not expect the law to be to such effect. Since I do not have to decide the point, because I have held that Hagee has not itself done or by its direction caused any act which is a breach of its covenants, I shall leave the point open for future debate.

In the result I hold that the plaintiff succeeds in its claim that the section 146 notice served on December 20 1989 by the landlord was F and is of no effect. It follows that the declaration sought in para 2 of the prayer ought to be granted. I will hear counsel as to what orders, if any, other than an inquiry as to damages, should be made at this time.

Official Referees' Business G

November 26 1991

(Before Mr Recorder BARRY GREEN QC)

CROWN ESTATE COMMISSIONERS v TOWN INVESTMENTS LTD (NATIONAL WESTMINSTER BANK PLC, THIRD PARTY)

Estates Gazette February 29 1992

[1992] 08 EG 111

H Landlord and tenant — Breach of repairing covenant — Failure to yield up in repair — Landlord incurring costs of repair — Subtenants granted new tenancies under Part II of the Landlord and Tenant Act 1954 — Rents determined on assumption that premises in repair — Whether landlord entitled to recover damages for breach of repairing covenant — Whether *Family Management* v *Gray* distinguishable — Held, subject to proof at trial of damage to reversion by reason of non-correspondence between covenants of headlease and sublease — Landlord's claim for damages extinguished by section 18 of the Landlord and Tenant Act 1927

J The plaintiffs are the owners of 36 St James's Street, London SW1, a building consisting of a basement, ground and six upper floors — The defendant, Town Investments Ltd, acquired the term of the headlease of the whole premises, this lease expiring on April 5 1985 — The whole of the property above the first floor (together with a half-room on the first floor and certain common parts) was demised by an underlease dated April 27 1961 for a term expiring on April 1 1985 ("the first underlease") — The term of the first underlease was severed and National Westminster Bank plc ("the third party") took by assignment the severed term in K respect of the second floor and half the rooms on the first floor — The remaining part of the severed term in respect of the third to sixth floors (and some of the common parts to the lower floors) was assigned to Maris Oil Studies Ltd on April 6 1984 — By a subunderlease dated January 4 1985 Maris Oil subunderlet the third floor to Market Rule Ltd for a term expiring on March 25 1985 — By an underlease ("the second underlease") dated January 23 1976 the defendant sublet the basement, ground and part of the first floor to the third party for a term expiring on April 1 1985

Following a notice served under section 26 of the Landlord and Tenant Act 1954 the plaintiffs granted the third party a L term of the premises, previously occupied by the third party under the first and second underleases, of 25 years from April 1985 at a rent fixed by agreement having regard to the provisions of section 34 of the Act — Following a notice under section 25 of the 1954 Act, the plaintiffs granted Market Rule Ltd a new tenancy of the third floor of the premises for a term of 15 years from April 5 1985 at a rent fixed by negotiation

The headlease held by the defendant contained a covenant to repair both the exterior and the interior of the premises — The first underlease contained a covenant to repair the the interior only — The second underlease contained a covenant M to repair both the exterior and interior — The subunderlease held by Market Rule Ltd contained a covenant to repair the interior of the premises only though there was a liability to pay a service charge in respect of the landlords' repairs to the exterior

The plaintiffs carried out repairs to the premises following the expiration of the headlease for a sum assumed for the purposes of the determination of a preliminary question to be £250,000 — The preliminary issue for the determination of the court was whether the defendant had a defence to the claim for dilapidations by reason of the fact that it left in possession subtenants with rights to new tenancies under the 1954 Act,

and in particular whether *Family Management* v *Gray* could be distinguished on the facts — The plaintiffs contended that that case could be distinguished on the grounds that the landlords had, here, paid for the repairs, the repairing covenants in the new tenancies did not enable the landlord to obtain damages equal to the cost of repairs, there was a lack of correspondence between the repairing covenants in the old headlease and those in the old underleases and part of the premises were unoccupied at the expiry of the headlease

***Held*: It is open to the plaintiffs to call evidence at full trial to show diminution in the value of the reversion (a) based upon the non-correspondence of the repairs obligations in the old headlease and underleases and (b) by reason of the existence of any disrepair in the unoccupied portions of the premises ("the non-correspondence claims") — The plaintiffs' landlords cannot bring any evidence to show that their common law damages are not extinguished by section 18 of the 1927 Act over and above the non-correspondence claims — In respect of the third-floor premises of which Market Rule Ltd were granted a new tenancy, the burden of proof is upon the plaintiffs to assert that the new rent was not referable to the terms of the Act and they had not proved that fact**

The following cases are referred to in this report.

Cunliffe v *Goodman* [1950] 2 KB 237; [1950] 1 All ER 720, CA
Family Management v *Gray* [1980] EGD 77; (1979) 253 EG 369
Fawke v *Viscount Chelsea* [1980] QB 441; [1979] 3 WLR 508; [1979] 3 All ER 568; (1979) 38 P&CR 504; [1979] EGD 146; 250 EG 855, CA
Hanson v *Newman* [1934] Ch 298
Jones v *Herxheimer* [1950] 2 KB 106; [1950] 1 All ER 323, CA
Joyner v *Weeks* [1891] 2 QB 31
Keats v *Graham* [1960] 1 WLR 30; [1959] 3 All ER 919, CA
Salisbury (Marquess of) v *Gilmore* [1942] 2 KB 38
Smiley v *Townshend* [1950] 2 KB 311; [1950] 1 All ER 530; [1950] EGD 139; (1950) 155 EG 110, CA

This was an application to seek determination of certain preliminary questions in proceedings by the plaintiffs, the Crown Estate Commissioners, seeking damages against the defendant, Town Investments Ltd, the tenant of 36 St James's Street, London SW1, under a lease which expired on April 5 1985. National Westminster Bank plc, who had occupied part of the premises under two underleases, were made first third party by the defendant.

Anthony Connerty (instructed by Radcliffes & Co) appeared for the plaintiffs; Edwin Johnson (instructed by Clifford Chance) represented the defendant; and James Barker (instructed by Edwin Coe) represented the first third party.

Giving judgment, **Mr Recorder Barry Green QC** said: The plaintiffs are landlords of premises at 36 St James's Street, London SW1. The defendant was tenant of the premises under a lease which expired on April 5 1985. At that date the major part of the premises (basement, ground, first and second floors) was occupied by National Westminster Bank plc ("the bank"), the third floor by Market Rule Ltd and the remainder (fourth, fifth and sixth floors) was empty. The bank and Market Rule were respectively underlessees and subunderlessees but can be described loosely as subtenants for the purposes of this judgment. At the expiry of the defendant's lease the premises were in disrepair. Subsequently the plaintiffs carried out those repairs for a sum of the order of £250,000. This preliminary issue proceeds upon the *assumption*, though not the admission, that that sum was so expended. The sum remains to be agreed by the parties or determined if necessary at a later hearing. This case is a preliminary issue to determine upon agreed facts whether the defendant has a defence to the claim for dilapidations by reason of the fact that it left in possession subtenants with rights to new tenancies under Part II of the Landlord and Tenant Act 1954. The bank seek to establish a similar defence against the defendant in the third party proceedings. Some time before the expiry of the defendant's headlease the requisite notices had been served under the 1954 Act so that new leases would be granted without opposition by the court. In fact, those leases were granted by negotiation without recourse to the court on terms referable to the provisions of the Act.

In summary the dispute is this. In *Family Management* v *Gray* (1979) 253 EG 369 it was decided in broadly similar circumstances that a landlord had suffered no damage to his reversion because of the operation of section 18 of the Landlord and Tenant Act 1927. The dispute is as to whether that case is distinguishable here on the facts.

Conveyancing history

I said above that the bank and Market Rule can be regarded loosely for the purposes of this judgment as subtenants. But the precise facts, though nothing turns upon the detail, are complicated as follows. On September 29 1905 the Crown Estate Commissioners let the whole premises for a term, expiring on April 5 1985 to tenants who subsequently assigned their interest to Town Investments Ltd, the defendant. By an underlease dated April 27 1961 ("the first underlease") the defendant sublet to Spark Holdings Ltd the whole of the property above the first floor together with a half-room on the first floor and the common parts on the ground floor and basement for a term expiring on April 1 1985. By an assignment dated June 25 1973 Spark Holdings assigned the whole of its interest under the first underlease to a Mr Lawson and a Mr Graham as joint tenants. Mr Lawson died on February 28 1974 and I believe that his interest passed to Mr Graham by survivorship. By an assignment dated June 26 1974 Mr Graham assigned his interest in the second floor and half-room in the first floor to the bank. The bank's immediate landlord in respect of this part of the premises was accordingly the defendant and the bank's interest was due to expire on April 1 1985. There was a four-day fag end between the term date of the bank's interest and the expiry of the headlease on April 5 1985. Mr Graham retained his interest in the remainder of that part of the premises which had been the subject of the first underlease until April 6 1984. On that date he assigned his interest in that remaining part (the third, fourth, fifth and sixth floors and the common parts of the lower floors) to Maris Oil Studies Ltd.

By a subunderlease dated January 4 1985 Maris Oil subunderlet the third floor to Market Rule for a term expiring on March 25 1985. Market Rule's immediate landlord was accordingly Maris Oil. There was a seven-day fag end between the term date of Market Rule's interest under the subunderlease and the expiry of Maris Oil's interest under the first underlease on April 1 1985. On the expiry of the first underlease there was then the defendant's four-day reversion until the headlease fell in on April 5 1985.

The only part of the premises originally let to the defendant under the headlease which did not form part of the first underlease was the basement, ground and first floors excluding the half-room. By an underlease dated January 23 1976 ("the second underlease") the defendant sublet this part of the premises to the bank. The expiry date of the second underlease was April 1 1985 and the legal position was as described above in relation to the bank's interest in the first underlease.

The new leases

Although the details in relation to the new leases are set out in the agreed statement of facts it is convenient that I repeat them in this judgment as follows:

1. The bank

A notice under section 26 of the 1954 Act dated May 21 1984 was served by the bank on the plaintiffs on May 24 1984 and an originating summons requesting the grant of a new tenancy issued on September 14 1984. The bank (the first third party) were subsequently granted by the plaintiffs a lease of the basement, ground, first and second floors. These were the same premises as the bank had previously occupied under the underlease dated April 27 1961 and the underlease dated January 23 1976. That lease is dated April 3 1989 and is for a term of 25 years from April 5 1985. The rent was fixed by agreement between the parties and with reference to the provisions of section 34 of the Act.

2. Market Rule Ltd

A notice under section 25 of the Act dated August 23 1984 was served by the plaintiffs on Market Rule on August 24 1984. A counternotice was served by Market Rule by a letter dated September 10 1984 stating that they were unwilling to give up possession, and on December 19 1984 an originating application requesting the grant of a new tenancy was filed. Market Rule was subsequently granted by the plaintiffs a lease of the third floor of the premises. That lease is dated March 1 1990 and is for a term of 15 years from April 5 1985.

It is to be observed that in the case of Market Rule the agreed facts contain no statement corresponding to that relating to the bank as to the basis upon which the rent was fixed by negotiation. That is to say

A it does not state that the rent was fixed with reference to the provisions of the Act. That omission does not alter my decision below but somewhat complicates my reasoning. I can foreshadow this in summary as follows. As at the date of expiry the authorities show that the parties were entitled to expect (on the facts in this case) that both the bank's new lease and Market Rule's new lease would contain rents fixed either by the court or by negotiation in a sum referable to the provisions of the Act.

The repairs covenants

It is unnecessary to do more than to describe in general terms the relevant covenants, which are as follows:

1. The 1905 headlease contained a covenant to repair both the B exterior and interior of the premises.

2. The Spark Holdings lease (later acquired by the bank) contained a covenant to repair the interior only.

3. The bank's direct underlease contained a covenant to repair both the exterior and the interior.

4. Market Rule's subunderlease contained a covenant to repair the interior of the premises only, though there was a liability to pay, via a service charge, some small part of the landlords' costs, *inter alia*, of repairing the exterior. For the purposes of this judgment this small charge can, I think, be ignored.

5. There were also covenants to paint and similar obligations, which I shall not describe in detail because they do not affect the C principles with which this judgment is concerned.

Thus, in summary, the headlease contained covenants which required the defendant to repair the exterior and interior, but the obligations of the bank and Market Rule, under their interests which expired on or shortly before April 5 1985, did not entirely correspond with those obligations in that between them they were not required to repair the whole of the exterior. It is this lack of correspondence upon which the plaintiffs rely, *inter alia*, as described below.

The law

In *Family Management* v *Gray* the facts were these. The plaintiffs had acquired the reversions in respect of two properties on two leases granted in January 1887 for terms ending in December 1974. The residue of the terms under the leases was vested in Mr Gray. The D premises consisted of shops with living accommodation above, the shops being the subject of subleases to a dry cleaner and a grocer respectively. Those subleases expired in December 1974. There were full repairing covenants in both the headleases and subleases. When the leases expired the shops were in disrepair. Six months before expiry the landlord served notices under the 1954 Act upon the subtenants indicating that he would not oppose the grant of new leases. Subsequently the new leases were granted following negotiations. The landlord brought proceedings against Mr Gray for dilapidations. The common law measure of damages was £6,000 (ie the cost of repairs). The trial judge found that if the premises had not been in disrepair the rents of the new leases would have been higher E and so there was damage to the reversion consisting of a multiplier of the lost rents. The Court of Appeal reversed the decision. It held that the Act required the new rents to be granted upon the assumption that the premises were in repair, for otherwise the subtenants would obtain an advantage from their own wrong. Accordingly, the landlord had suffered no loss of rent and so no damage to his reversion. The common law damages were therefore irrecoverable because of the limit set by section 18.

In the case before me the plaintiffs accept that the principles found in the *Family Management* case apply to the facts of this case but seek to distinguish it upon four grounds:

1. The plaintiffs here have actually paid for the repairs.

2. The repairing obligations in the old underleases do not F correspond (as they seem to have done in *Family Management*) with those in the headlease.

3. The plaintiffs contend that they will be unable to sue the bank and Market Rule under their new leases for the cost of these repairs as could have been done in *Family Management*.

4. In *Family Management* no question arose as to the residential portion of the premises. But here it is common ground that on the assumption that the unoccupied part of the premises was in disrepair the defendant has no defence to that part of the claim.

The defendant's case is that the payment by the plaintiffs for the repairs is irrelevant to the operation of section 18. If, which it denies, the plaintiffs have suffered any injustice, their remedy may be to bring an action for unjust enrichment. But that has not been pleaded or argued and is not before me. In any event, I am concerned with the G law and not the merits, whatever they might be.

Before considering *Family Management* in more detail I observe that certain propositions of law are clearly established on the authorities as to the operation of section 18 as follows.

1. At common law the measure of damages for breach of covenants to repair at the end of a lease was established to be the cost of complying with the covenants together with any consequential loss of rent: *Joyner* v *Weeks* [1891] 2 QB 31. In this case no loss of rent is claimed. It is common ground that subject to liability the plaintiffs are entitled to include in that cost their professional fees and VAT.

2. Section 18 imposes a limit on the common law measure but does not alter the method of its assessment: *Hanson* v *Newman* [1934] Ch H 298.

3. The common law measure of damages is generally *prima facie* evidence of the damage to the reversion, but it is not conclusive evidence thereof: *Jones* v *Herxheimer* [1950] 2 KB 106 and *Smiley* v *Townshend* [1950] 2 KB 311. It will frequently be the case that unless there are special circumstances the court can coalesce the two stages of ascertaining the common law measure and considering whether that is capped by section 18 in one single process. Whether it can do so must be a question of fact in each case.

4. Events subsequent to the expiry of the lease are in general irrelevant to the inquiry under section 18 and are indeed of doubtful admissibility: *Marquess of Salisbury* v *Gilmore* [1942] 2 KB 38, *Smiley* v *Townshend* [1950] 2 KB 311 and *Keats* v *Graham* [1959] 3 J All ER 919.

Family Management v *Gray*

It seems to me that this case is authority for four propositions, namely that in valuing a reversion which is not in possession but is subject to continuation tenancies under the 1954 Act:

1. The court must take that fact into account as a reality and not treat the reversion as if it were in possession.

2. The court must take into account all relevant matters relating to these continuation tenancies which may affect the value of the reversion.

3. The court has limited power to admit evidence of matters which arise after the expiry date of the contractual head tenancy. K

4. The court must assume that the rent will be fixed under the 1954 Act on the basis that the premises are in repair, for otherwise the tenant would be taking advantage of his own wrong. The same principle would apply if the landlord's contractual tenancy imposed the repairs burden upon him alone. He could expect a rent under the Act upon an *unrepaired* basis only, for otherwise it would be he who would gain an advantage from his own wrong: see on this point *Fawke* v *Viscount Chelsea* [1980] QB 441*.

The plaintiffs' case under Family Management v *Gray*

As I said above, the plaintiffs' counsel accepts that the principles decided by this authority apply here but says that it is distinguishable on four grounds. L

1. The landlords have paid for the repairs.

2. The repairs covenants in the new leases do not enable the landlords to obtain damages equal to the cost of these repairs because of the particular machinery of the service-charge clauses.

3. There is a lack of correspondence between the repairs covenants in the defendant's old headlease and those in the old underleases. Accordingly, it is open to the plaintiffs to call evidence at full trial (if such can be obtained) that the reversion has been damaged on that account. This argument was drawn to my attention, with the greatest propriety, by Mr Johnson, the defendant's counsel.

4. It is common ground that part of the premises was unoccupied at the expiry of the lease and if it was in disrepair the *Family Management* defence cannot relate to that portion of the premises. M

I find straight away that the third submission is correct and, as I said, the fourth submission is common ground. So the landlords are at liberty at full trial to seek to prove damage to their reversion based upon the non-correspondence of repairing obligations and based upon damage, if any, to the unoccupied part of the reversion. I suspect that the sums involved are comparatively small.

1. The plaintiffs' distinction based upon payment for the repairs.

There is a major flaw in the landlords' submission as follows. For the court to admit evidence of the payment for repairs after expiry of

*Editor's note: Also reported at (1979) 250 EG 855.

the headlease would be to offend against the rule that the reversion must be valued as at that expiry date. The payment for repairs does not come within Shaw LJ's description in *Family Management* v *Gray* of the exceptions to that rule. He said at p 372: "So events which follow upon the determination of the lease but which are independent of any fact or consideration which was either operative or potential at the date when the lease expired cannot affect the determination of the loss to the value of the reversion." In *Smiley* v *Townshend* Denning LJ said these words to similar effect at p 321 (though under the second limb of section 18(1)): ". . . the subsequent event, when it happens, is not in itself sufficient to extinguish the damages. It is only evidence, in retrospect, of the future as it appeared at the end of the lease". The payment for the repairs was a future event. It took place after expiry of the lease and there is nothing in the agreed statement of facts to indicate that it was either "operative or potential" at the expiry of the lease. An example of a future event which might be admissible would be this. Suppose that in *Family Management* there had been firm evidence that one of the shop tenants was in financial difficulty as at the date of expiry and might well not be able to take up a new lease. If by the date of trial that potentiality had become an actuality and he had not taken up the new lease then that subsequent event would be admissible as corroboration of the evidence of what had previously been only a potentiality.

To vary the example: suppose that there was no such evidence of financial trouble as at the date of expiry of the lease but that by the date of trial the tenant had become insolvent. It seems to me that evidence of the insolvency would be inadmissible because it was not "operative or potential" at the expiry of the lease *upon the evidence* as at that date. The fact that the seeds of the insolvency very probably existed as at the date of expiry is irrelevant unless there was evidence that as at that date those seeds then existed.

The plaintiffs' counsel here puts a further example in support of this submission. He points out that under section 36(2) of the 1954 Act the tenant has the right, if he does not like the rent or other terms granted by the court, to back out of accepting a new tenancy. The plaintiffs say that if the subtenants had operated section 36(2) I could have received evidence of that fact. So, by the same token, I can receive evidence of the payment for repairs. The fallacy in the argument is this. If there had been evidence in *Family Management* that a shop tenant at the date of expiry of the lease was in two minds whether to take a new lease and in the event decided not to do so pursuant to section 36(2) then that decision would be admissible because it was a potentiality *on the evidence* as at the date of expiry. But both in *Family Management* and in the instant case there is no evidence other than that the subtenants showed every indication of desiring to take a new lease. For in both cases the machinery under the Act was in train and there was no suggestion that it would not arrive at its destination, namely the grant of new tenancies.

2. The plaintiffs' distinction based on the machinery of the service charges.

The landlords contend that in *Family Management* the repairing obligations in the new leases could be expected to correspond with those in the old subleases, so that if the premises were left in disrepair the landlord could sue the old subtenants under their new leases. But they argue that in the instant new leases the machinery of the service-charge clauses prevents the landlords from recovering their dilapidations claim. I should emphasise that this is a different submission from that upon which I ruled in favour of the landlords above. There was there a lack of correspondence between the repairing obligations of the *old* headleases and the *old* subleases. The submission here, however, contrasts the covenants in the *old* subleases with those in the *new* direct leases.

It is unnecessary for me to set out in detail the alleged deficiencies in the service-charge machinery. Assuming that these defects exist and assuming that the landlords cannot recover their claim for disrepair under the covenants of the new leases, their submissions are still flawed for the following reasons. The defendant submits that, in valuing the reversion at the expiry of the lease, the court does not need to look at the terms of the lease other than the rent. Rent equals value, they say. I do not consider that that is entirely correct. For a valuer would take into account the rent, the length of the lease, the terms and frequency of the rent review clauses if any, the nature of the repairing covenants, the clauses against alienation and other relevant terms. For commercially a lease is a "package deal". A lower rent may be compensated by more onerous repair obligations upon the tenant, and so on and so forth. It is true that in *Family Management* the court referred to the fact that the new leases would contain repair covenants as had the old. But I do not consider that that consideration was part of the ratio of the judgments. The ratio was that the landlord was entitled to expect new leases upon the terms of the 1954 Act, that is to say, new leases which were more or less of the same commercial value as the old, unless there are unusual considerations to the contrary. It is true to say that the Act nowhere expressly says so. But it is quite clearly implied, in an Act which cuts down the landlord's common law rights, that it does so to no greater extent than it expressly provides. The court and practitioners have, in my experience, always approached the machinery of the Act in that manner.

So the answer to the landlords' submission is this. It is an agreed fact that the new rent in the bank's lease is referable to the Act. Though that is not also an agreed fact in the case of Market Rule's new lease, *Family Management* is authority for the proposition that that rent must also have been expected to be fixed on terms referable to the Act. If the negotiations here have in some way failed to achieve that situation it was perhaps open to the landlords to seek agreement of that fact in the statement of facts. They have not done so and I cannot have regard to it. For reasons which will appear clear below, it would have been irrelevant and inadmissible even if that fact had been agreed. To put the matter another way, I consider that the inclusion in the new leases of defective service-charge machinery is not an event which was "operative or potential" at the date of expiry of the lease. Suppose that in *Family Management* the new leases contained repairing covenants which required the tenants to repair the interior only whereas under their old lease they had to repair both interior and exterior. I consider that that fact would have arisen independently, to adopt Shaw LJ's phrase, and so would be irrelevant to the assessment of the diminution. By the same token, if the new leases had been granted by the court, a landlord could not be heard to say in this court that a previous court had got the matter wrong.

The burden of proof

The burden of proof is said to be of great importance in this case as follows. This preliminary issue proceeds upon the assumption that the landlords are entitled to damages at common law in that they have carried out repairs between the date of expiry of the old leases and the grant of the new. Is it for the defendant to prove that section 18 operates as a cap to extinguish or reduce common law damages or is it for the landlords to prove that it does not? For the landlords to succeed they must prove that they are entitled to common law damages and there must also be a finding in their favour that section 18 does not operate as a cap to extinguish or reduce them. So one starts with a presumption that the landlords must prove both elements necessary to succeed. One then construes section 18(1) to see if it displaces that presumption either expressly or by necessary implication. Section 18(1) provides as follows under limb 1:

Provisions as to covenants to repair

(1) Damages for a breach of a covenant or agreement to keep or put premises in repair during the currency of a lease, or to leave or put premises in repair at the termination of a lease, whether such covenant or agreement is expressed or implied, and whether general or specific, shall in no case exceed the amount (if any) by which the value of the reversion (whether immediate or not) in the premises is diminished owing to the breach of such covenant or agreement as aforesaid;

and then under the second limb:

and in particular no damage shall be recovered for a breach of any such covenant or agreement to leave or put premises in repair at the termination of a lease, if it is shown that the premises in whatever state of repair they might be, would at or shortly after the termination of the tenancy have been or be pulled down or such structural alterations made therein as would render valueless the repairs covered by the covenant or agreement.

There are a number of indications in the above subsection, some of which tend one way and some the other, as to the burden of proof. They are these:

1. The subsection cuts down a common law right. That is an indication that the party who asserts that the common law right is cut down assumes the burden of so proving.

2. Under limb 1 the matters of fact which enable the landlord to succeed are primarily within his own knowledge and not the tenant's.

3. There is clear authority that under limb 2 the burden of proof is upon the tenant: see *Marquess of Salisbury* v *Gilmore* [1942] 2 KB 38;

Cunliffe v *Goodman* [1950] 2 KB 237. One would expect that the burden of proof was on the same party in both limbs.

4. Limb 2 is introduced by the words "and in particular". This suggests that limb 2 situations are a subspecies of those in limb 1 and so indicates that the burden of proof is the same in both.

5. But in fact limb 2 situations differ radically from those in limb 1. For limb 2 applies to demolition or structural alteration so as to render the repairs valueless. There is no clear authority, I believe, but it appears that limb 2 operates in an all-or-nothing manner. It can extinguish damages but not reduce them. This radical distinction (or possibly distinctions) makes more room for a different burden of proof in the two limbs.

6. Limb 2 contains the phrase "if it is shown that". This plainly suggests that it is the tenant who must do the showing. The phrase is not mirrored in limb 1 and is therefore an indication that there may be a difference in the burden of proof.

7. *Jones* v *Herxheimer* shows that in certain circumstances it may be appropriate for the landlord to coalesce the process of proving common law damages and establishing that limb 1 does not cap them. That unified operation would hardly be possible if the burden of proof differed in each of its constituent elements.

Taking all these indications one with another and construing the subsection as a whole, I consider that the burden of proof under limb 1 is upon the landlords. I shall deal later with the effect of this finding.

The wording of the preliminary issue

At trial it became clear that the issue which the parties had previously drafted and put to the court for agreement at the summons for directions by consent was not wholly appropriate. For example, there was no reference to the denial of liability in para 6 of the amended defence. I invited counsel to amend the previous wording. The plaintiffs' counsel was unable to accept the version put forward by the other two counsel and I ruled against his version. In doing so I was mindful of the possible unfairness of ordering a new version of the issue after the plaintiffs had already agreed the statement of facts. But I am sure that no injustice has resulted and indeed the plaintiffs' counsel did not argue as much. The revised issue merely restates the intention behind the former version in clearer language and also drops points which are no longer or never were in dispute (the points as to insurance, professional fees and VAT).

Answers to the questions in the preliminary issue

I have already stated that it is open to the landlords to call evidence at full trial to show diminution in the value of the reversion (a) based upon non-correspondence of the old repairs obligations in the headlease and underleases and (b) by reason of the existence of any disrepair in the unoccupied portions of the premises. I shall refer to these as the "non-correspondence claims". I have to decide whether there is any other class of evidence which they can call, for me to make this decision is not to usurp the function of the trial judge. For it is part of my task to decide whether the consequence of my findings is the extinction or diminution of the plaintiffs' claim (apart from the non-correspondence claims). The question can be posed as follows. Can the landlords bring any evidence to show that their common law damages are not extinguished by section 18 (over and above the non-correspondence claims)? In my view, they cannot do so. Nor has it been suggested that there is any such evidence which might be relevant. The only realistic example which anyone could suggest was this. Suppose there was evidence that at the date of expiry of the lease one of the subtenants was very much in two minds whether to take up a new lease and might very well decide not to do so. I have already said that that fact *might* be relevant to the valuation of the reversion. But it is not a fact which is contained in the agreed statement of facts and so cannot be introduced hereafter. The same goes for any other such facts. It is now open to the landlords to seek a full trial upon different facts from those contained in their agreed statement of facts in this preliminary issue. It follows that the defendant has proved affirmatively upon the agreed facts that section 18 operates to extinguish the claim (with the exception of the non-correspondence claims).

That being so, it seems to me that the defendant does not require my finding that the burden of proof is upon the landlords in order to succeed in this part of the preliminary issue.

I now deal separately with the claim which relates to that part of the premises occupied by Market Rule to show that it does not alter my conclusions above in any way. I pointed out above that the agreed statement of facts lacked agreement that the rent in the new Market Rule lease was referable to the terms of the Act. Nevertheless the plaintiffs' claim in relation to disrepair of the third floor which was and is occupied by Market Rule still fails for the following reasons. First, the burden of proof is upon the plaintiffs. If they wished to assert that the new rent was *not* referable to the terms of the Act it was open to them to seek agreement of that alleged fact. No doubt the defendant would have refused agreement because it is not part of its case. But, for whatever reasons, the plaintiffs have not proved that fact and so cannot rely upon it. In order to satisfy the burden of proof which is upon them. If I am wrong about that reason for my decision, I wish to put the matter in an alternative way as follows. The actual terms of the new leases have arisen independently from the facts as they were at the expiry of the lease as I pointed out in detail above. On this basis the burden of proof is neither here nor there. Even if the plaintiffs were able to prove this fact it would not assist them because it is irrelevant. The defendant on this basis has affirmatively proved its case in relation to the third floor (the part occupied by Market Rule).

The third-party proceedings

The bank (the first third party) succeeds entirely in defeating any claim by the defendant against it. For in the third party proceedings there are no remaining issues based upon noncorrespondence between covenants which remain to be quantified at full trial. The defendant is potentially liable to the plaintiffs on this point because its old lease gave the landlords wider covenants than were given by the subtenants. But the bank cannot be similarly liable to the defendant. For *ex hypothesi* the bank gave only the narrower covenants. So breach of these covenants cannot be causative of the defendant's liability to the plaintiffs. The *Family Management* defence absolves the bank altogether from liability.

The issue between the plaintiffs and defendant

My answers to the questions 1(a) to 2(b) inclusive of the preliminary issue are: the liability of the defendant is extinguished save only that it is open to the plaintiffs to seek to prove at trial that the reversion has been damaged by reason of:

(i) the non-correspondence between the covenants of the contractual headlease and the covenants in the subleases of the first third party and Market Rule Ltd at the expiry of the headlease; and

(ii) breaches of repairs obligations in those part of the premises which were not occupied either by the first third party or by Market Rule Ltd as at April 5 1985.

The issue between the defendant and the first third party

My answer to question 1(a) is: the liability of the first third party is extinguished.

House of Lords

December 5 1991

(Before Lord BRIDGE OF HARWICH, Lord BRANDON OF OAKBROOK, Lord ACKNER, Lord JAUNCEY OF TULLICHETTLE and Lord BROWNE-WILKINSON)

HAMMERSMITH AND FULHAM LONDON BOROUGH COUNCIL v MONK

Estates Gazette March 7 1992

[1992] 09 EG 135

Landlord and tenant — Periodic tenancy — Whether periodic tenancy can be determined by notice to quit given by one of the joint tenants without the concurrence of the other — Held periodic joint tenancy effectively determined by one joint tenant

The respondent local authority granted Mr Monk, the appellant, and a Mrs Powell, a weekly tenancy of a flat at 35 Niton Street, London SW6, where they cohabited — By its terms the tenancy was terminable by four weeks' notice to expire on a Monday — In 1988 the appellant and Mrs Powell fell out and Mrs Powell left the flat — The respondents agreed to rehouse her if she terminated the tenancy of the flat — Mrs

A **Powell gave a notice to quit without the appellant's knowledge or consent — The respondents immediately notified the appellant and in due course brought proceedings to recover possession — In the county court it was held that the notice to quit was ineffective and the claim for possession was dismissed — The Court of Appeal allowed the respondents' appeal and made an order for possession — Mr Monk appealed from that decision**

***Held*: The appeal was dismissed — As a matter of contractual principle if A and B contract with C on terms which are to continue in operation for one year in the first**
B **place and thereafter from year to year unless determined by notice at the end of the first or any subsequent year, neither A nor B has bound himself contractually for longer than one year — To hold that A could not determine the contract at the end of any year without the concurrence of B and vice versa would presuppose that each had assumed a potentially irrevocable contractual obligation for the duration of their joint lives — The application of ordinary contractual principles leads one to expect that a periodic tenancy granted to two or more joint tenants must be terminable at common law by an appropriate notice to quit given by any one of them whether or not the others are prepared to concur — There is**
C **support for the proposition that a yearly tenancy continues only so long as each and all joint tenants concur: see *Blackstone's Commentaries*, and *Doe d Aslin* v *Summersett* — Additionally there is Court of Appeal authority in *Leek & Moorlands Building Society* v *Clark* and *Greenwich London Borough Council* v *McGrady* — Accordingly, unless the terms of the tenancy agreement otherwise provide, notice to quit given by one joint tenant without the concurrence of any other joint tenants is effective to determine a periodic tenancy**

The imposition of a statutory trust for sale meant that the appellant and Mrs Powell held the legal estate in the periodic
D **tenancy as trustees — The submission that trustees act unanimously and neither of them individually could give a valid notice to quit failed — It was doubtful whether Mrs Powell was in breach of trust, but even if the giving of a notice by her was a breach, the notice to quit was effective to terminate the tenancy**

The following cases are referred to in this report.

Alford v *Vickery* (1842) Car&M 280
Bull v *Bull* [1955] 1 QB 234; [1955] 2 WLR 78; [1955] 1 All ER 253
Doe d Aslin v *Summersett* (1830) 1B&Ad 135
Doe d Kindersley v *Hughes* (1840) 7 M&W 139
E *Doe d Whayman* v *Chaplin* (1810) 3 Taunt 120
Gandy v *Jubber* (1865) 9 B&S 15; (1864) 5 B&S 78; 33 LJQB 151; 9 LT 800; 10 Jur NS 652; 12 WR 526
Greenwich London Borough Council v *McGrady* (1982) 81 LGR 288; 46 P&CR 223; [1983] EGD 519; 267 EG 515, CA
Howson v *Buxton* (1928) 97 LJKB 749
Leek & Moorlands Building Society v *Clark* [1952] 2 QB 788; [1952] 2 All ER 492; [1952] 2 TLR 401, CA
Smith v *Grayton Estates Ltd* 1960 SC 349; 1961 SLT 38
Warren, Re [1932] 1 Ch 42

This was an appeal by Mr Monk, who with Mrs Powell held a weekly tenancy of a flat at 35 Niton Street, London SW6, granted to them by the respondent local authority, from a decision of the Court
F of Appeal [1991] 1 EGLR 263 , who had allowed the respondents' appeal from a decision of Judge Roger Cooke in the West London County Court, who had dismissed the respondents' claim to possession of the flat.

Robert Reid QC and Paul Staddon (instructed by Oliver Fisher & Co) appeared for the appellant; and Stephen Sedley QC and Beverly Ann Rogers (instructed by the solicitor to Hammersmith and Fulham London Borough Council) represented the respondents.

In his speech, **Lord Bridge of Harwich** said: The issue in this appeal is whether a periodic tenancy held by two or more tenants jointly can be brought to an end by a notice to quit by one of the joint tenants without the consent of the others. It arises for determination in the following circumstances. The appellant, Mr Monk, and Mrs Powell were granted by the respondent local authority a weekly G
tenancy of a flat at 35 Niton Street, London SW6, where they cohabited. The tenancy was terminable by four weeks' notice to expire on a Monday. In 1988 Mr Monk and Mrs Powell fell out and Mrs Powell left the flat. She consulted the respondents, who agreed to rehouse her if she would terminate the tenancy of the flat by giving an appropriate notice, which she did. The notice was given without Mr Monk's knowledge or consent but the respondents immediately notified him that the tenancy had been determined and in due course brought proceedings in the West London County Court to recover possession. Judge Roger Cooke held that Mrs Powell's notice to quit was ineffective to determine the tenancy and dismissed the claim. The Court of Appeal (Slade, Nicholls and Bingham LJJ) (1990) 61 P&CR 414* allowed the respondents' appeal and made an order for H
possession. Mr Monk now appeals by leave of your lordships' House.

In a previous decision of the Court of Appeal, *Greenwich London Borough Council* v *McGrady* (1982) 46 P&CR 223†, it was held that a notice to quit given by one of two joint tenants without the consent of the other was effective to determine the periodic tenancy to which it related. Much of the argument before the Court of Appeal in the present case was directed to the question whether the court was free to reach a conclusion at variance with *McGrady* on the grounds: (1) that an earlier decision of the Court of Appeal, *Howson* v *Buxton* (1928) 97 LJKB 749, was, as the judge had held, binding authority to the opposite effect; or (2) that, in any event, the decision in *McGrady* was J
given *per incuriam*. The judgment of Slade LJ, with which Bingham LJ agreed, examined these issues very thoroughly and reached the conclusion that *Howson* v *Buxton* was not authority for the proposition sought to be derived from it and that *McGrady* was binding on the court. Nicholls LJ approached the issue more radically and held, both on principle and in reliance on a long line of authority prior to the decision in *McGrady*, that a joint periodic tenancy could be determined by a notice to quit given by one joint tenant.

Your lordships are not technically bound by any previous decision and before examining the relevant authorities I think it helpful to consider whether the application of first principles suggests the answer to the question at issue. For a large part of this century there K
have been many categories of tenancy of property occupied for agricultural, residential and commercial purposes where the legislature has intervened to confer upon tenants extra-contractual rights entitling them to continue in occupation without the consent of the landlord, either after the expiry of a contractual lease for a fixed term or after notice to quit given by the landlord to determine a contractual periodic tenancy. It is primarily in relation to joint tenancies in these categories that the question whether or not notice to quit given by one of the joint tenants can determine the tenancy is of practical importance, particularly where, as in the instant case, the effect of the determination will be to deprive the other joint tenant of statutory protection. This may appear an untoward result and may L
consequently provoke a certain reluctance to hold that the law can permit one of two joint tenants unilaterally to deprive his co-tenant of "rights" which both are equally entitled to enjoy. But the statutory consequences are in truth of no relevance to the question which your lordships have to decide. That question is whether, at common law, a contractual periodic tenancy granted to two or more joint tenants is incapable of termination by a tenant's notice to quit unless it is served with the concurrence of all the joint tenants. That is the proposition which the appellant must establish in order to succeed.

As a matter of principle I see no reason why this question should receive any different answer in the context of the contractual relationship of landlord and tenant than that which it would receive M
in any other contractual context. If A and B contract with C on terms which are to continue in operation for one year in the first place and thereafter from year to year unless determined by notice at the end of the first or any subsequent year, neither A nor B has bound himself contractually for longer than one year. To hold that A could not determine the contract at the end of any year without the concurrence of B and vice versa would presuppose that each had assumed a potentially irrevocable contractual obligation for the duration of their joint lives, which, whatever the nature of the contractual

*Editor's note: Also reported at [1991] 1 EGLR 263.
†Editor's note: Also reported at (1982) 267 EG 515.

A obligations undertaken, would be such an improbable intention to impute to the parties that nothing less than the clearest express contractual language would suffice to manifest it. Hence, in any ordinary agreement for an initial term which is to continue for successive terms unless determined by notice, the obvious inference is that the agreement is intended to continue beyond the initial term only if and so long as all parties to the agreement are willing that it should do so. In a common law situation, where parties are free to contract as they wish and are bound only so far as they have agreed to be bound, this leads to the only sensible result.

Thus the application of ordinary contractual principles leads me to expect that a periodic tenancy granted to two or more joint tenants must be terminable at common law by an appropriate notice to quit B given by any one of them whether or not the others are prepared to concur. But I turn now to the authorities to see whether there is any principle of the English law of real property and peculiar to the contractual relationship of landlord and tenant which refutes that expectation or whether the authorities confirm it. A useful starting point is the following passage from *Blackstone's Commentaries*, Book II, chapter 9, pp 145-147, which explains clearly how the law developed the concept of a yearly tenancy from the earlier concept of a tenancy at will, which gave the tenant no security of tenure:

> The second species of estates not freehold are estates at will. An estate at will is where lands and tenements are let by one man to another, to have and to hold at the will of the lessor; and the tenant by force of this lease obtains possession. C Such tenant hath no certain indefeasible estate, nothing that can be assigned by him to any other; for that the lessor may determine his will, and put him out whenever he pleases. But every estate at will is at the will of both parties, landlord and tenant, so that either of them may determine his will, and quit his connexions with the other at his own pleasure. Yet this must be understood with some restriction. For, if the tenant at will sows his land, and the landlord before the corn is ripe, or before it is reaped, puts him out, yet the tenant shall have the emblements, and free ingress, egress, and regress, to cut and carry away the profits. And this for the same reasons, upon which all the cases of emblements turn; *viz* the point of uncertainty: since the tenant could not possibly know when his landlord would determine his will, and therefore could make no provision against it; and having sown the land, which is for the good of the public, upon a reasonable presumption, the law will not suffer him to be a loser by it. But it is otherwise, and upon reason equally good, D where the tenant himself determines the will; for in this case the landlord shall have the profits of the land.
>
> . . .
>
> The law is, however, careful that no sudden determination of the will by one party shall tend to the manifest and unforeseen prejudice of the other. This appears in the case of emblements before-mentioned; and, by a parity of reason, the lessee after the determination of the lessor's will, shall have reasonable ingress and egress to fetch away his goods and utensils. And, if rent be payable quarterly or half-yearly, and the lessee determines the will, the rent shall be paid to the end of the current quarter or half-year. And, upon the same principle, courts of law have of late years leant as much as possible against construing demises, where no certain term is mentioned, to be tenancies at will; but have rather held them to be tenancies *from year to year so long as both parties please*, especially where an annual rent is reserved: in E which case they will not suffer either party to determine the tenancy even at the end of the year, without reasonable notice to the other . . .

I have added emphasis to the phrase "from year to year so long as both parties please" because in its Latin version *de anno in annum quamdiu ambabus partibus placuerit* this same phrase is used repeatedly in a passage from *Bacon's Abridgment*, 7th ed, which has always been treated as of the highest authority, as apt to describe the essential characteristics of a yearly tenancy.

Hence, from the earliest times a yearly tenancy has been an estate which continued only so long as it was the will of both parties that it should continue, albeit that either party could only signify his unwillingness that the tenancy should continue beyond the end of any year by giving the appropriate advance notice to that effect. F Applying this principle to the case of a yearly tenancy where either the lessor's or the lessee's interest is held jointly by two or more parties, logic seems to me to dictate the conclusion that the will of all the joint parties is necessary to the continuance of the interest.

In *Doe d Aslin* v *Summersett* (1830) 1 B&Ad 135, the freehold interest in land let on a yearly tenancy was vested jointly in four executors of a will to whom the land had been jointly devised. Three only of the executors gave notice to the tenant to quit. It was held by the Court of King's Bench that the notice was effective to determine the tenancy. Delivering the judgment, Lord Tenterden CJ said at pp 140-141:

> Upon a joint demise by joint tenants upon a tenancy from year to year, the true character of the tenancy is this, not that the tenant holds of each the share of each so long as he and each shall please, but that he holds the whole of all so G long as he *and all* shall please; and as soon as any one of the joint tenants gives a notice to quit, he effectually puts an end to *that* tenancy; the tenant has a right upon such a notice to give up *the whole*, and unless he comes to a new arrangement with the other joint tenants as to their shares, he is compellable so to do. The hardship upon the tenant, if he were not entitled to treat a notice from one as putting an end to the tenancy as to the whole, is obvious; for however willing a man might be to be sole tenant of an estate, it is not very likely he should be willing to hold undivided shares of it; and if upon such a notice the tenant is entitled to treat it as putting an end to the tenancy as to the whole, the other joint tenants must have the same right. It cannot be optional on one side, and on one side only.

Now it was rightly pointed out in argument that part of the reasoning in this passage was dictated by considerations derived from H the incidents of joint land tenure at law, which were swept away by the reforming legislation of 1925. But this can in no way detract from the validity of the proposition emphasised in the judgment that the yearly tenant of a property let to him by joint freeholders "holds the *whole* of *all* so long as he *and all* shall please". This by itself is a sufficient and independent ground for the conclusion of the court that notice to quit by any one joint freeholder was effective to determine the tenancy. Precisely the same reasoning would apply to the operation of a notice to quit by one of two or more joint yearly tenants.

Summersett's case was followed in *Doe d Kindersley* v *Hughes* (1840) 7 M&W 139 and *Alford* v *Vickery* (1842) Car&M 280, both cases in which the validity of a notice to determine a yearly tenancy J given to the tenant without the concurrence of one or more of the joint landlords was affirmed. It is interesting that throughout the 19th century there is no reported case in the books where the effect of a notice to quit given by one of two or more joint holders of the tenant's interest under a yearly or other periodic tenancy was ever called in question. I do not, however, find this surprising. The law was probably regarded as settled after *Summersett*'s case, but in any event, before the advent of statutory protection of tenants' rights of occupation, in the case of a notice to quit given by one of two or more joint periodic tenants the parties would in most cases have had little incentive to litigate. If the landlord was content that the other tenants should remain, there would have been nothing to litigate about. If the K landlord wished to recover possession, he could do so by giving his own notice to quit.

In this century the English cases directly in point are: *Howson* v *Buxton* (1928) 97 LJKB 749; *Leek & Moorlands Building Society* v *Clark* [1952] 2 QB 788 and *Greenwich London Borough Council* v *McGrady* (1982) 46 P&CR 223. I will defer consideration of *Howson* v *Buxton* until later. In *Leek & Moorlands Building Society* v *Clark* the point directly in issue was whether one of two joint lessees could validly surrender the lease before the full period of the lease had run without the concurrence of the other joint lessee. Delivering the reserved judgment of the court in favour of the defendant lessees, Somervell LJ said at pp 792-793:

> Counsel for the plaintiffs sought to rely on *Doe d Aslin* v *Summersett* as L supporting a submission that Mr Ellison, by what he did, had brought the joint tenancy to an end. That case was dealing with a lessee from year to year of land which he held from two joint lessors. A notice to quit was served signed by one only of the joint lessors. It was argued that the other lessor had adopted the notice, but Lord Tenterden, who delivered the judgment of the Court of King's Bench, held that without any such adoption a notice to quit by one of the joint lessors, who were joint tenants, put an end to the tenancy as to both.
>
> The ratio of the decision is, we think, to be found in the following sentence: "Upon a joint demise by joint tenants" — that is, the lessors in that case — "upon a tenancy from year to year, the true character of the tenancy is this, not that the tenant holds of each the share of each so long as he and each shall please, but that he holds the whole of all so long as he and all shall please; and as soon as any one of the joint tenants" — that is, the lessors in that case — "gives a notice to quit, he effectually puts an end to that tenancy". It is to be M noted that Lord Tenterden was dealing with a notice to quit in respect of a periodic tenancy. He was not dealing with a right to determine a lease for say 21 years at the end, say, of the seventh or fourteenth year. Nor was he dealing with surrender.
>
> There is, we think, force in the submission made on behalf of the plaintiffs, that in the case of a periodic tenancy Lord Tenterden's principle would apply when there were joint lessees. A periodic tenancy continues from period to period unless the notice agreed or implied by law is given. But if one of two joint lessees who "hold the whole" wishes it not to continue beyond the end of a period, it might well be held that it did not continue into a new period. That would happen only if all, that is, the joint lessees, shall please.
>
> If one considers a lease to joint lessees for a term certain with a right of renewal, it would be obvious, we think, that both must join in requiring a renewal. A periodic tenancy renews itself unless either side brings it to an end.

A But if one of two or more joint lessees does not desire it to continue, we would have thought that it was in accordance with Lord Tenterden's principle, and with common sense, that he should be able to make that effective.

The judgment adds at pp 794-795:

> Even if we are wrong in what we have said with regard to a right to determine within the period of the lease as distinct from a right to terminate a periodic tenancy, we would have thought it plain that one of two joint lessees cannot, in the absence of express words or authority, surrender the rights held jointly. If property or rights are held jointly, *prima facie* a transfer must be by or under the authority of all interested. The answer suggested to this is the principle laid down in *Doe d Aslin* v *Summersett*. That case, for reasons which we have given, is not in our view an exception to the rule we have just stated. It is an illustration, in a highly technical field, of the general principle that if a
> B joint enterprise is due to terminate on a particular day, all concerned must agree if it is to be renewed or continued beyond that day. To use Lord Tenterden's phrase, it will only be continued if "all shall please".

In the *Greenwich* case the point at issue was precisely the same as in the present appeal. After citing the judgment of the court in *Leek & Moorlands Building Society* v *Clark*, Sir John Donaldson MR said, at p 224:

> In my judgment, it is clear law that, if there is to be a surrender of a joint tenancy — that is, a surrender before its natural termination — then all must agree to the surrender. If there is to be a renewal, which is the position at the end of each period of a periodic tenancy, then again all must concur. In this case, Mrs McGrady made it quite clear by her notice to quit that she was not content to renew the joint tenancy on and after June 15 1981. That left Mr
> C McGrady without any tenancy at all, although it was faintly argued by Mr Osman that on, as he put it, the severance of a joint tenancy the joint tenant who did not concur was left with a sole tenancy. That cannot be the law, and no authority has been cited in support of it.
>
> The only point that remains is whether Mr McGrady is entitled to the protection of the Act of 1980 on the ground that what was a secure contractual tenancy has been brought to an end. The short answer to that is that the Act of 1980 operated to give security where landlords give notice to quit; it does not give security where tenants give notice to quit.

In the instant case it has not been suggested either that the notice to quit given by Mrs Powell could have had the effect of "severing" the joint tenancy and leaving Mr Monk in possession as sole tenant or that, if Mrs Powell's notice was effective, Mr Monk was entitled to any statutory protection.

D To this formidable body of English authority which supports the conclusion reached by the Court of Appeal there must be added the decision of the Court of Session in *Smith* v *Grayton Estates Ltd* 1960 SC 349, which shows that Scots law, although using different terminology, applies essentially the same principle to give the same answer to the same question. The issue in the case was whether a tenancy continuing from year to year after the expiry of a fixed term by virtue of the Agricultural Holdings (Scotland) Act 1949 was determined by notice given by one of two joint tenants. Lord President Clyde said, at pp 354-355:

> In considering this matter, it is of importance to realise that in the present case the tenants were occupying under tacit relocation, in other words, that the
> E tenancy was being prolonged from year to year beyond the stipulated term in the lease, but that otherwise the conditions in the lease continued to operate — see *Rankine, Law of Leases*, p 601; *Cowe* v *Millar*, reported only in Connell on the Agricultural Holdings (Scotland) Act 1923, p 346, *per* Lord President Clyde at p 355. The question comes to be whether, in that situation, a timeous notice by one of the two joint tenants is invalid to bring the tenancy to an end. The argument for the appellant was that a valid notice must be from both the joint tenants, and this notice, not being a joint one, consequently is bad.
>
> But, as I see it, this argument overlooks the meaning and effect of tacit relocation. Tacit relocation is not an indefinite prolongation of a lease. It is the prolongation each year of the tenancy for a further one year, if the actings of the parties to the lease show that they are consenting to this prolongation. For, as in all contracts, a tacit relocation or reletting must be based on consent. In the case of tacit relocation the law implies that consent if all parties
> F are silent on the matter. Hence, where there are joint tenants, tacit consent by both of them is necessary to secure the prolongation and to enable tacit relocation to operate. Silence by both is necessary to presume that both the tenants wish the tenancy to continue for another year. On the other hand, if both are not silent, and if one gives due notice of termination, the consent necessary for tacit relocation to operate is demonstrably not present, and tacit relocation will not operate beyond the date of termination in the notice. Clearly, in the present case, there is not such tacit consent, and, in my view, a notice by one of the two joint tenants is enough to exclude the further operation of tacit relocation.

Lord Sorn, at p 356, is to the like effect.

These, then, are the principles and the authorities which the appellant seeks to controvert. In the light of the careful analysis in the judgment of Slade LJ of *Howson* v *Buxton*, which I gratefully adopt and need not repeat, it is now rightly accepted that the case affords no G greater support for the appellant than can be derived from the *obiter dictum* of Scrutton LJ, who said, with reference to a notice to determine a yearly tenancy, at p 752:

> I personally take the view that one joint tenant cannot give a notice to terminate the tenancy unless he does so with the authority of the other joint tenant . . .

Despite the eminence of the author of this observation, I do not feel able to give any weight to it in the absence of any indication of the reasoning on which it is based.

There are three principal strands in the argument advanced for the appellant. First, reliance is placed on the judgment in *Gandy* v *Jubber* (1865) 9 B&S 15 for the proposition that a tenancy from year to year, H however long it continues, is a single term, not a series of separate lettings. The case arose out of an action for damages by a plaintiff who had been injured by a defective iron grating which was out of repair so as to amount to a nuisance. The property was occupied by a yearly tenant, but the claim was brought against the reversioner, who was held liable by the Court of Queen's Bench. The defendant appealed to the Court of Exchequer Chamber on the ground that it was not alleged that the defendant knew of the nuisance, nor that it had existed prior to the commencement of the yearly tenancy. The argument is reported at 5 B&S 485. Judgment was reserved, but before it was delivered the case was settled and Erle CJ announced:

> It will not be necessary to deliver the judgment we have prepared.

J The undelivered judgment in the defendant's favour is nevertheless reported at 9 B&S 15 and has always been regarded as authoritative. The passage relied on reads:

> There frequently is an actual demise from year to year so long as both parties please. The nature of this tenancy is discussed in 4 *Bac Abr* tit *Leases and Terms for Years*, pp 838, 839, 7th ed, and this article has always been deemed to be the highest authority being said to be the work of Chief Baron Gilbert. It seems clear that the learned author considered that the true nature of such a tenancy is that it is a lease for two years certain, and that every year after it is a springing interest arising upon the first contract and parcel of it, so that if the lessee occupies for a number of years, these years, by computation from time past, makes an entire lease for so many years, and after the commencement of each new year it becomes an entire lease certain for the years past and also for K the years entered on, and that it is not a reletting at the commencement of the third and subsequent years. We think this is the true nature of a tenancy from year to year created by express words, and that there is not in contemplation of law a recommencing — or reletting at the beginning of each year.

It must follow from this principle, Mr Reid QC (for the appellant) submits, that the determination of a periodic tenancy by notice is in all respects analogous to the determination of a lease for a fixed term in the exercise of a break clause, which in the case of joint lessees clearly requires the concurrence of all. But reference to the passage from *Bacon's Abridgment*, 7th ed, at p 839, on which the reasoning is founded shows that this analogy is not valid. The relevant passage reads: L

> A parol lease was made *de anno in annum, quamdiu ambabus partibus placuerit*; it was adjudged that this was but a lease for a year certain, and that every year after it was a springing interest, arising upon the first contract and parcel of it; so that if the lessee had occupied eight or ten years, or more, these years, by computation from the time past, made an entire lease for so many years; and if rent was in arrear for part of one of those years, and part of another, the lessor might distrain and avow as for so much rent arrear upon one entire lease, and need not avow as for several rents due upon several leases, accounting each year a new lease. It was also adjudged, that after the commencement of each new year, this was become an entire lease certain for the years past, and also for the year so entered upon; so that neither party could determine their wills till that year was run out, according to the opinion of the two judges in the last case. And this seems no way impeached by the M statute of frauds and perjuries, which enacts, that no parol lease for above three years shall be accounted to have any other force or effect than of a lease only at will: for at first, this being a lease certain only for one year, and each accruing year after being a springing interest for that year, it is not a lease for any three years to come, though by a computation backwards, when five or six or more years are past, this may be said a parol lease for so many years; but with this the statute has nothing to do, but only looks forward to parol leases for above three years to come.

Thus the fact that the law regards a tenancy from year to year which has continued for a number of years, considered retrospectively, as a single term in no way affects the principle that continuation beyond the end of each year depends on the will of the

A parties that it should continue or that, considered prospectively, the tenancy continues no further than the parties have already impliedly agreed upon by their omission to serve notice to quit.

The second submission for the appellant is that, whatever the law may have been before the enactment of the Law of Property Act 1925, the effect of that statute, whereby a legal estate in land vested in joint tenants is held on trust for sale for the parties beneficially entitled, coupled with the principle that trustees must act unanimously in dealing with trust property, is to reverse the decision in *Summersett*'s case and to prevent one of two joint tenants determining a periodic tenancy without the concurrence of the other. It is unnecessary to reconsider the position where the parties beneficially entitled are different from those who hold the legal B interest. But where, as here, two joint tenants of a periodic tenancy hold both the legal and the beneficial interest, the existence of a trust for sale can make no difference to the principles applicable to the termination of the tenancy. At any given moment the extent of the interest to which the trust relates extends no further than the end of the period of the tenancy which will next expire on a date for which it is still possible to give notice to quit. If before 1925 the implied consent of both joint tenants, signified by the omission to give notice to quit, was necessary to extend the tenancy from one period to the next, precisely the same applies since 1925 to the extension by the joint trustee beneficiaries of the periodic tenancy which is the subject of the trust.

C Finally, it is said that all positive dealings with a joint tenancy require the concurrence of all joint tenants if they are to be effective. Thus, a single joint tenant cannot exercise a break clause in a lease, surrender the term, make a disclaimer, exercise an option to renew the term or apply for relief from forfeiture. All these positive acts which joint tenants must concur in performing are said to afford analogies with the service of notice to determine a periodic tenancy which is likewise a positive act. But this is to confuse the form with the substance. The action of giving notice to determine a periodic tenancy is in form positive; but both on authority and on the principle so aptly summed up in the pithy Scottish phrase "tacit relocation" the substance of the matter is that it is by his omission to give notice of termination that each party signifies the necessary positive assent to D the extension of the term for a further period.

For all these reasons I agree with the Court of Appeal that, unless the terms of the tenancy agreement otherwise provide, notice to quit given by one joint tenant without the concurrence of any other joint tenant is effective to determine a periodic tenancy.

An alternative ground advanced in support of the appeal relied on the particular terms of the tenancy agreement entered into by Mr Monk and Mrs Powell with the respondents as requiring notice to quit to be given by both joint tenants in order to determine the tenancy. I agree entirely with the reasons given by Slade and Nicholls LJJ for rejecting this contention.

I would accordingly dismiss the appeal.

E LORDS BRANDON OF OAKBROOK, ACKNER and JAUNCEY OF TULLICHETTLE agreed with the speech of Lord Bridge of Harwich and the reasons given in it and did not add any observations of their own.

In his speech, LORD BROWNE-WILKINSON said: There are two instinctive reactions to this case which lead to diametrically opposite conclusions. The first is that the flat in question was the joint home of Mr Monk and Mrs Powell: it therefore cannot be right that one of them unilaterally can join the landlords to put an end to the other's rights in the home. The second is that Mr Monk and Mrs Powell undertook joint liabilities as tenants for the purpose of providing themselves with a joint home and that, once the desire to live together F has ended, it is impossible to require that the one who quits the home should continue indefinitely to be liable for the discharge of the obligations to the landlord under the tenancy agreement.

These two instinctive reactions are mirrored in the legal analysis of the position. In certain cases a contract between two persons can, by itself, give rise to a property interest in one of them. The contract between a landlord and a tenant is a classic example. The contract of tenancy confers on the tenant a legal estate in the land: such legal estate gives rise to rights and duties incapable of being founded in contract alone. The revulsion against Mrs Powell being able unilaterally to terminate Mr Monk's rights in his home is property based: Mr Monk's property rights in the home cannot be destroyed without his consent. The other reaction is contract based: Mrs Powell cannot be held to a tenancy contract which is dependent for its G continuance on the will of the tenant.

The speech of my noble and learned friend Lord Bridge of Harwich traces the development of the periodic tenancy from a tenancy at will. He demonstrates that a periodic tenancy is founded on the continuing will of both landlord and tenant that the tenancy shall persist. Once either the landlord or the tenant indicates, by appropriate notice, that he no longer wishes to continue, the tenancy comes to an end. The problem is to determine who is "the landlord" or "the tenant" when there are joint lessors or joint lessees.

In property law a transfer of land to two or more persons jointly operates so as to make them, *vis-à-vis* the outside world, one single owner. "Although as between themselves joint tenants have separate H rights, as against everyone else they are in the position of a single owner": *Megarry and Wade, The Law of Real Property*, 5th ed, p 417. The law would have developed consistently with this principle if it had been held that where a periodic tenancy has been granted by or to a number of persons jointly, the relevant "will" to discontinue the tenancy has to be the will of all the joint lessors or joint lessees who together constitute *the* owner of the reversion or the term as the case may be.

At one stage the law seems to have flirted with adopting this approach. Thus in *Doe d Whayman* v *Chaplin* (1810) 3 Taunt 120 there was a periodic tenancy and four persons were the joint lessors. Three only of the joint lessors gave notice to quit against the wishes of the fourth. It is reported that at one stage the court inclined to the J view that in order to determine the tenancy all four lessors had to agree. However, after further argument it was held that each of the three who had given notice to quit was entitled to put an end to the tenancy of his share and the three who had given notice to quit were therefore entitled to recover three parts of the land. As a result, the defendant apparently was entitled to stay on the land in right of his tenancy of one part as tenant in common with the three lessors who had given notice. Although the decision is difficult for a modern lawyer to understand fully, one thing is clear: the giving of notice to quit by three out of the four joint lessors was not sufficient to determine the tenancy of the whole land.

Despite this flirtation, the law was, in my judgment, determined in K the opposite sense by *Doe d Aslin* v *Summersett* (1830) 1 B&Ad 135. The contractual, as opposed to the property, approach was adopted. Where there were joint lessors of a periodic tenancy, the continuing "will" had to be the will of all the lessors individually, not the conjoint will of all the lessors collectively. This decision created an exception to the principles of the law of joint ownership: see *Megarry and Wade* (*supra*) pp 421-422.

It was submitted that this House should overrule *Summersett*'s case. But, as my noble and learned friend Lord Bridge of Harwich has demonstrated, the decision was treated throughout the 19th century as laying down the law in relation to the rights of joint lessors. It is not suggested that the position of joint lessees can be different. Since 1925 the law as determined in *Summersett*'s case has been applied to L notices to quit given by one of several joint lessees. In my judgment, no sufficient reason has been shown for changing the basic law which has been established for 160 years unless, as was suggested, the 1925 legislation has altered the position.

Before 1925 property belonging to two or more persons concurrently could be held by them in undivided or divided shares at law. The Law of Property Act 1925 changed this and requires that, even in the case of joint tenants, they hold the legal estate as joint tenants on trust for themselves as joint tenants in equity: section 36(1). It was suggested that the interposition of this statutory trust for sale has altered the position: since Mr Monk and Mrs Powell held the legal estate in the periodic tenancy as trustees and trustees must act unanimously, neither of them individually could give a valid notice to M quit.

In my view, this submission fails. The trust property in question was a periodic tenancy. As between the lessor and the lessees the nature of the contract of tenancy cannot have been altered by the fact that the lessees were trustees. The tenancy came to an end when one of the lessees gave notice to quit. It may be that, as between the lessees, the giving of the notice to quit was a breach of trust, theoretically giving rise to a claim by Mr Monk against Mrs Powell for breach of trust. Even this seems to me very dubious, since the overreaching statutory trusts for sale imposed by the Law of Property Act 1925 do not normally alter the beneficial rights *inter se* of the concurrent owners: see *Re Warren* [1932] 1 Ch 42 *per* Maugham J at p 47; *Bull* v

Bull [1955] 1 QB 234. But even if, contrary to my view, the giving of the notice to quit by Mrs Powell was a breach of trust by her, the notice to quit was not a nullity. It was effective as between the lessor and the lessees to terminate the tenancy. The fact that a trustee acts in breach of trust does not mean that he has no capacity to do the act he wrongly did. The breach of trust as between Mrs Powell and Mr Monk could not affect the lessors unless some case could be mounted that the lessors were parties to the breach, a case which Mr Reid, for Mr Monk, did not seek to advance. Therefore, in my judgment, the 1925 legislation does not affect this case.

For these reasons and those given by my noble and learned friend Lord Bridge, I, too, would dismiss this appeal.

Court of Appeal

November 20 1991

(Before Lord Justice BALCOMBE and Lord Justice MANN)

AMEC PROPERTIES LTD v PLANNING RESEARCH & SYSTEMS PLC AND ANOTHER

Estates Gazette April 4 1992

[1992] 13 EG 109

Landlord and tenant — Agreement to grant a lease — Claim for specific performance — Whether where intended tenant insolvent the court should have regard to the statutory scheme for the distribution of assets of insolvent companies, the interest of the creditors and the relative hardship to them in rejecting a claim for specific performance of an agreement to grant a lease —Appeal by intended lessee dismissed

In 1987 the respondent, AMEC Properties Ltd, agreed with Globebuy Ltd for the sale of its leasehold interest in premises in the Isle of Dogs, London; it guaranteed payment to Globebuy of the underlying rents, including the rental value of unlet parts —The respondent would find tenants and procure Globebuy to grant leases to such tenants and Globebuy undertook to grant leases at the respondent's request — Globebuy's interest has since devolved upon Midland Bank plc — By an agreement dated September 6 1989 the respondent agreed with the appellant, Planning Research & Systems plc, to procure the Midland Bank plc to grant, and the appellant to take, a lease of part of the premises for a term of 25 years from March 25 1989 at a rent of £313,852.50 pa, subject to a reverse premium payable by the respondent to the appellant of £287,500 upon completion — Midland Bank was not obliged or bound to complete the grant unless the appellant provided a bank guarantee signed by National Westminster Bank plc — On August 24 1989 National Westminster Bank provided the guarantee — On September 5 1989 the reverse premium was paid by the respondent to the appellant and by the appellant into its overdrawn bank account with National Westminster Bank; the appellant was allowed into occupation for fitting out and later to trade — On January 30 1990 the respondent gave notice to complete to the appellant — On October 3 1990 National Westminster Bank were appointed administrative receivers of the appellant — The lease and counterpart were executed in escrow, but the appellant failed to complete and its receiver has since refused to give the necessary authority

Before Millett J the appellant submitted that an order for specific performance of the agreement should not be made because the respondent was not the prospective landlord and had no interest in obtaining the benefit of covenants in the lease, that damages were an adequate remedy, that the appellant's apparent inability to pay damages did not make a contract specifically enforceable if it would not otherwise be so, that no decree of specific performance could be granted against a company in receivership and the court's discretion should be exercised against a decree of specific performance because the appellant was insolvent and could not comply with any of the covenants under the lease — On appeal from the decision of Millett J, who rejected those submissions, the appellant contended that in the light of its insolvency the court should have regard to the effect on the statutory scheme for the distribution of the assets of insolvent companies and the interest of its creditors and the relative hardship to them in rejecting a claim for specific performance

***Held*: The appeal was dismissed — The agreement did not remain wholly executory in that the appellant had been in occupation of the premises for some 10 or 11 months — Had there been no question of the appellant's insolvency there would have been no defence to the claim for specific performance — The statutory scheme for the regulation of the assets of an insolvent company takes effect only when the company goes into liquidation — This was not a circumstance admitting of the possibility of an exception to the well-established principle that a contract for the grant of an interest in land will normally be specifically enforced**

The following cases are referred to in this report.

Gray's Inn Construction Co Ltd, In re [1980] 1 WLR 711; [1980] 1 All ER 814 CA

Patel v *Ali* [1984] Ch 283; [1984] 2 WLR 960; [1984] 1 All ER 978; (1984) 48 P&CR 118

This was an appeal from an order of Millett J made on July 5 1991 whereby, in reversing an order of Master Cholmondeley Clarke, he granted the respondent, AMEC Properties Ltd, summary judgment for a decree of specific performance against the appellant, Planning Research & Systems plc.

Anthony Mann (instructed by Wilde Sapte) appeared for the appellant; Benjamin Levy (instructed by Laytons) represented the respondent. The second defendant, National Westminster Bank plc, did not appear and was not represented.

Giving judgment, **Balcombe LJ** said: This is an appeal from an order of Millett J made on July 5 1991 whereby, reversing an order of Master Cholmondeley Clarke, he granted the plaintiff, AMEC Properties Ltd, summary judgment for a decree of specific performance against the defendant, Planning Research & Systems plc to whom I will refer, as did the judge, as PRS. I take the facts from the judgment of Millett J.

By a formal agreement in writing dated September 6 1989 PRS agreed to take a lease of office premises on the third and fourth floors of a building known as 3 Harbour Exchange, Isle of Dogs, London, for a term of 25 years from March 25 1989, at a rent of £313,852.50 a year. But for the unusual context in which the claim arises, it would be a straightforward claim for specific performance by a proposed landlord against a proposed tenant.

However, AMEC was not the proposed landlord. In 1987 AMEC had agreed with a company called Globebuy for the sale of its leasehold interest in the premises and that when the building was complete it would procure tenants for the various parts of the building. It had guaranteed payment to Globebuy of the underlying rents, including the rental value of unlet parts, so that in economic terms it was in effect an agreement for sale and leaseback; but there was no actual leaseback. There was merely a sale under which AMEC would find tenants and procure Globebuy to grant leases to such tenants; and Globebuy undertook to grant leases at AMEC's request

Globebuy's interest has since devolved upon the Midland Bank. The evidence does not explain why or how that occurred. By the agreement of September 6 1989, AMEC agreed with PRS to procure the Midland Bank to grant the lease to PRS. The landlord was to be the Midland Bank. AMEC agreed to pay PRS a reverse premium of £287,500 upon completion, and condition 12 of the agreement provided that the Midland Bank would not be obliged or bound to complete the grant of the lease unless PRS delivered to AMEC's solicitors or to whomever they should direct prior to completion a bank guarantee signed by the National Westminster Bank in the form of a draft annexed to the agreement.

On August 24 1989, the National Westminster Bank provided the guarantee required. It was a guarantee of the rent payable by PRS under the proposed lease; it is common ground between the parties that it will take effect only if the lease is actually granted.

At this point I cease to quote from the judgment at first instance and I refer to the terms of the guarantee.

The guarantee itself is in clause 1 and the bank undertakes:

. . . that if for any reason the Tenant [PRS] shall fail to pay the sum first reserved as rent under the lease on due date the Bank shall within three

A business days of demand by the Landlord unconditionally pay that sum to the Landlord.

Clause 2 provides that:

> The Bank's liability under this Guarantee shall not be discharged in whole or in part or otherwise be affected in any way by reason of . . .
> (iv) any forfeiture or variation of the Lease or alteration in the obligations [undertaken by] the Tenant . . .
> (v) the liquidation of the Tenant or the disclaimer of the Lease by a liquidator of the Tenant or the Tenant ceasing to exist . . .

Clause 3 provides that the "Guarantee shall expire on the earliest of" and then there are various dates given including six months after the expiry or sooner determination of the lease unless the lease is determined by forfeiture and the tenant is granted relief from B forfeiture and, finally, August 22 1994. So, effectively it is a five-year guarantee, but determinable earlier in certain events.

Finally, I should refer to clause 10 of the guarantee:

> The Bank shall not while this Guarantee subsists exercise in respect of any amount paid by the Bank under this Guarantee any right of subrogation or any other right or remedy which the Bank may have against the Tenant.

The evidence for AMEC in the case makes it clear that the existence of the guarantee was a substantive reason for the choice of PRS as a tenant of this particular property. I return to the judgment giving the relevant facts:

> On September 5 1989 the reverse premium was paid by AMEC to PRS and was paid by PRS into its overdrawn bank account with the National Westminster C Bank. PRS was allowed into occupation, originally for fitting out purposes and later to trade.

— I pause there to stress that fact as an important relevant fact in this case —

> In November 1989, AMEC gave a guarantee to the Midland Bank of the full rent and service charges of the premises until they were let.
>
> On January 30 1990 AMEC obtained all the necessary licences to permit the lease to be granted and on the following day it gave notice to PRS requiring completion of the lease. By March, PRS's solicitors had forwarded to AMEC's solicitors a counterpart, executed in escrow, and thereafter there was a desultory exchange of correspondence between the parties' respective solicitors. By the end of August the position was that the lease had been executed in escrow by the Midland Bank ready for exchange with the D counterpart; and the counterpart had been re-executed in escrow by PRS. Both parts were held by the Midland Bank's solicitors, who had the authority of the Midland Bank to complete, but who had not yet obtained authority from PRS's solicitors to release the counterpart by way of completion.
>
> On October 3 1990, the National Westminster Bank appointed administrative receivers of PRS. Shortly thereafter, the Midland Bank's solicitors forwarded the lease and counterpart lease to AMEC's solicitors, who now hold both parts of the lease executed in escrow

— I pause to interpolate: "who now hold" refers to the date of the hearing before the judge. As will become apparent in a moment, events have proceeded since that date —

> together with the irrevocable authority of the Midland Bank to exchange the lease on completion, but who have no authority from PRS or the E administrative receivers to exchange the counterpart. The administrative receivers have refused to authorise AMEC's solicitors to release the counterpart or to complete.
>
> The position immediately before the appointment of the administrative receivers was that PRS was in occupation of the premises to be demised and had been for some 10 or 11 months. PRS had failed to complete by authorising its solicitors to exchange the counterpart and its receiver has since refused to give the necessary authority. PRS has not returned or offered to return the reverse premium, which is now irrecoverable. There is no outstanding conveyancing objection or requisition. Payment of the rent by PRS is guaranteed by the National Westminster Bank under the agreement given on August 24 1989, but the National Westminster Bank's liability does not commence until and unless the lease is actually granted. The administrative receiver's motive in refusing to authorise completion is not hard to discern. The writ was issued on December 4 1990, and the Master refused to make an F order for summary judgment under Ord 86 on February 14. He subsequently ordered the National Westminster Bank to be added as a defendant and AMEC appeals against that order. PRS is hopelessly insolvent as regards unsecured creditors. It has ceased to trade and has now gone out of occupation of the premises.

In relation to the judge's remark that "PRS is hopelessly insolvent as regards unsecured creditors", I refer to the statement of affairs which was put in evidence and which is the only evidence before the court of the financial position of PRS. The estimated total assets available for preferential creditors amount to £56,468. The preferential creditors alone amount to £651,032, so that there is a deficiency as to them of £594,564. There are then debts secured by a floating charge, all of which are owed to the National Westminster Bank, of £1,167,178; and finally there are unsecured creditors of G £1,913,029. There is a note under the debts secured by a floating charge that there is a contingent liability in respect of a rent guarantee, but this has not yet been quantified. So the judge's statement that PRS is hopelessly insolvent as regards unsecured creditors is clearly well founded.

Although Mr Mann, who appeared for PRS before us, submitted that there was a possibility of further assets being recovered — he suggested that there might be possible claims for misfeasance, avoidable preferences or wrongful trading — I have to say that the statement of affairs is the only evidence before us and on that evidence I am quite content to say that there is no realistic prospect of the unsecured creditors of PRS ever receiving anything. As I have H already mentioned, the National Westminster Bank, which is also the guarantor of the rent, is the only secured creditor.

National Westminster Bank had on its own application been added as a defendant to the action and before the judge appeared by the same counsel as PRS — not, I should say, the counsel currently appearing for PRS before us. Before the judge, the argument took a different form than it did before us and again I go back to refer to the judgment. In this case I will not quote it but merely summarise it.

The first point put by PRS before the judge was a submission that there ought to be no decree of specific performance because AMEC was not the prospective landlord and had no interest in obtaining the benefit of the covenants in the lease. The judge pointed out that AMEC plainly had a financial interest in the grant of the lease J because of the benefit of the National Westminster Bank's guarantee; he said he did not understand why the fact that AMEC was not the proposed landlord made the slightest difference and went on to say that he saw no reason why, on that ground, the agreement should not be specifically enforceable against PRS at the suit of AMEC.

The next point taken before the judge was that damages were an adequate remedy; again, as he put it, essentially for the same reason that AMEC was not the prospective landlord. He rejected that argument.

Then it was submitted that the defendant's inability to pay damages did not make a contract specifically enforceable if it would not otherwise be so. As the judge said, that was true but it was not the K case before him. Regardless of PRS' ability to perform its part of the agreement, that agreement would be specifically enforceable.

Then it was submitted that no decree of specific performance could be granted against a company in receivership. The judge referred to certain authorities which had been cited to him and said:

> The fact that the administrative receivers would cause PRS to execute the lease by affixing the company's seal and witnessing the execution would not impose any personal liability upon the receivers, any more than such conduct would have imposed a personal liability upon the directors had the company not been in administrative receivership.
> Consequently, in my judgment a decree of specific performance in this case will not expose the administrative receivers to any personal liability under the L covenants in the lease.

I will return to that point later in this judgment. The judge went on to say that that was, therefore, no objection to the decree on that ground.

Finally, it was submitted that the court's discretion should be exercised against a decree of specific performance because PRS was insolvent and could not comply with any of the covenants under the lease. He rejected that ground, too, and said that this was, in fact, a straightforward claim for specific performance and none of the unusual factors of the case militated against a decree of specific performance. In the end he granted a decree of specific performance and accepted undertakings from AMEC and the Midland Bank that in the event of a successful appeal they would not oppose rescission of M the lease that was then granted.

This appeal has been brought by PRS only. I stress the fact that National Westminster Bank is not an appellant in this court. The argument before us has taken a different form. We allowed Mr Mann to amend his notice of appeal so as to raise a plea of hardship and I refer to para 2A of his amended notice of appeal, which is in the following terms:

> In the light of the insolvency of the First Defendant, the Court should have regard to the following in rejecting a claim for specific performance:
> (a) the effect on the statutory scheme for the distribution of the assets of insolvent companies
> (b) the interest of the creditors of the First Defendant [PRS] and the relative

> hardship to them of granting specific performance when compared with the interests of the Plaintiffs [AMEC].

Mr Mann's main point was that to impose liability upon PRS would distort the statutory scheme for the administration of the assets of an insolvent company. He directed our attention to a number of cases concerning the position of directors where the company was or was potentially insolvent, which cases he attempted to apply by analogy. I intend no disrespect to him when I say that I did not find these cases of any assistance. He did, however, refer us to the judgment of Buckley LJ *In re Gray's Inn Construction Co Ltd* [1980] 1 WLR 711 at p 717:

> It is a basic concept of our law governing the liquidation of insolvent estates, whether in bankruptcy or under the Companies Acts, that the free assets of the insolvent at the commencement of the liquidation shall be distributed rateably amongst the insolvent's unsecured creditors as at that date.

He (Buckley LJ) went on to develop that point at some length. That was said in the context of validating a transaction which had taken place between the presentation of the petition for winding up and the winding up itself under what was then section 227 of the Companies Act 1948. But even in that context I am prepared to accept that statement as one of general validity.

Mr Mann then pointed out two possible scenarios which could happen in this case. First, if the decree of specific performance were upheld, the company, PRS, would continue to be under an obligation to pay a rent in excess of £300,000 a year, reviewable every five years. There was further rent in the sense of insurance, rent and service charge. The landlords could prove for any deficiency and that obligation theoretically continued until the lease came to an end. However, he accepted that if the company went into liquidation the probability was that the liquidator would disclaim the lease, in which case the landlord could then prove for the amount of its loss. Equally, the landlord could go against the National Westminster Bank under the guarantee until 1994 and he submitted — for the purposes of this judgment I am prepared to accept that this is correct although there is a possible argument the other way, but it does not matter — that National Westminster Bank could then go against the company for what it had paid under the guarantee; in other words, there would be a potential burden of £1.2m or more on the company, a burden additional to what existed at the moment.

Second, if specific performance were not granted then AMEC had its claim for a return of its reverse premium of some £280,000 and AMEC and/or the Midland Bank might be entitled to claim damages depending upon the amount of their loss. In answer to a suggestion from the court that the loss which AMEC and Midland might suffer could possibly be the same as that which the landlord would prove for if the lease were disclaimed, he submitted — and I think there was some force in his submission — that AMEC and/or the Midland Bank might be able to exploit the premises more advantageously than could the company, PRS, and National Westminster Bank. In that event the damages which AMEC and/or the Midland Bank would suffer would be less than the burden thrown on PRS by granting specific performance and that, he submitted, was an issue which should go to trial, it being recalled that this was a case for summary judgment. He pointed out that on that basis the unsecured creditors generally might benefit by a trial which ended up in the result of AMEC's being refused specific performance and merely limited to their remedy in damages.

He then developed an argument based on hardship, but this really added nothing to his argument based on insolvency which, as he rightly said, is a specialised form of hardship. I hope that I have not done any injustice to Mr Mann's submissions by summarising them in the way I have. As I said, essentially his argument was that to grant specific performance would distort the statutory scheme for the realisation of assets of an insolvent company and indeed that was the point to which he returned in his reply.

In my judgment, Mr Levy (for AMEC) had the complete answer to these submissions. He reminded us of the background that this transaction did not remain wholly executory, that PRS had been in occupation for some 10 or 11 months, and he started with the proposition that if there had been no question of PRS' insolvency there would have been no defence to the claim for specific performance. It was only the intervention of the receivers which allowed the defence in its present form to be presented; and he pointed out, again it seems to me with force, that the statutory scheme for the regulation of the assets of an insolvent company takes effect only when the company goes into liquidation and the argument presented by Mr Mann was really an attempt to extend that scheme to a time before liquidation. But even assuming that it were permissible to do that, when looking at the question of hardship or the realisation of the assets of an insolvent company for the benefit of unsecured creditors, one should not look at it in a vacuum. One has to look at the facts in this case — and the facts in this case, as I have already said, on the evidence, are that there is absolutely no prospect of the unsecured creditors receiving anything; it is unrealistic to assume that they will. On the other hand, if specific performance is not granted there is no prospect of AMEC's receiving anything by way of proof or damages if specific performance is refused and particularly for the return of the reverse premium paid.

I have to say that, looking at the matter realistically, this essentially is a contest between AMEC and the Midland Bank on the one hand and National Westminster Bank on the other. There is a small potential point which Mr Mann made that if, as a result of the decree of specific performance and the grant of the guarantee, the liquidation of PRS is prolonged for a period longer than it might otherwise have been, then there might be some small extra cost of the liquidation which would be thrown, on the figures as we have them, upon the unsecured creditors. But in balancing the relative hardships, which we are invited to do, it does seem to me that that possibility weighs very lightly in the scale. As I have said, that is the reality in this case.

Mr Levy asked a very pertinent forensic question. Why is PRS not in liquidation already? He suggested that it was in the hope of creating a situation whereby the National Westminster Bank could get out of the guarantee by allowing a liquidation at a point of time when the lease had not been granted and then the liquidator disclaims the contract to grant the lease under section 178 of the Insolvency Act 1986, which it can now do without the leave of the court.

Accordingly, for the reasons which I have attempted to give, subject to one very small point, I can see no reason why we should disturb the order by Millett J. The small point is this. Under section 44(1)(*b*) of the Insolvency Act 1986, it is provided that:

> The administrative receiver of a company is personally liable on any contract entered into by him in the carrying out of his functions except in so far as the contract otherwise provides.

It could be argued that the exchange of the lease and counterpart pursuant to the decree of specific performance amounted to an entry by the receivers into the contractual obligations of the lease. Millett J was not satisfied that it would, but it seems to me unnecessary to go into that because Mr Mann very fairly conceded that, if there were anything in this point, the words in parenthesis in subpara (*b*) of section 44(1) "except in so far as the contract otherwise provides", could meet the point and he invited us, and I am prepared to accept the invitation, to direct that the lease granted pursuant to the court's decree of specific performance should specifically provide that the receivers were not in any circumstances to be personally liable for the performance of the contractual obligations under it, and I would so order.

Subject to that one point, I would dismiss this appeal.

Agreeing, **Mann LJ** said: The remedy of specific performance, although expressed as a discretionary one, is by now granted in accordance with well-established principles. One of those principles is that a contract for the grant of an interest in land will normally be specifically enforced. I use the word "normally" because one must admit of the possibility of exception in, for example, the case of personal hardship as exemplified in *Patel* v *Ali* [1984] Ch 283. What is here suggested is that there should be an exception by reference to the position of a hopelessly insolvent company which may go into liquidation and where it is said that the making of a decree of specific performance would disturb the statutory pattern for the benefit of creditors.

Like Balcombe LJ, I would decline to make any such exception and I cannot help observing that in this case the disturbance would be minimal if indeed there were any at all.

I also would dismiss this appeal.

Appeal dismissed with costs. Application for leave to appeal to the House of Lords refused.

A
Court of Appeal

December 19 1991

(Before Lord Justice PARKER and Lord Justice McCOWAN)

SKILLETER AND OTHERS v CHARLES

Estates Gazette April 4 1992

[1992] 13 EG 113

Landlord and tenant — Service charges — Recovery of bank interest and charges — Expenses of manager — Recovery of
B legal expenses — Appeal against dismissal of originating application

The applicants are the tenants of three of 13 flats in a Victorian block at South Terrace, Littlehampton, Sussex — The respondent landlord occupies one of the flats — On December 12 1989 the applicants, with the tenants of five more of the flats, issued an originating application in Chichester County Court claiming relief in respect of service charges — The relief sought included a declaration and inquiry relating to the relevant costs incurred by the respondent lessor and properly payable by the applicants in respect of the
C maintenance charge provided for by the leases of the flats — The leases, which were in common form, provided that there could be charged to the maintenance fund certain costs and expenses incurred by the respondent, the cost of employing a managing agent and all legal and other proper costs incurred —The trial judge dismissed the application — The applicants appealed seeking a declaration that certain items should not have been included in the accounts, namely interest and bank charges, management fees and legal expenses

***Held*: The appeal was dismissed — The relevant provision in the lease for the recovery of interest and bank charges provided: "Any interest charges payable by the Lessor on his**
D Bank Account or accounts in respect of any for the purposes of the Maintenance Charge" — Notwithstanding that some word had been omitted after the words "in respect of", the plain intention was that interest should be chargeable and to give effect to that intention it was necessary only to insert after the words "in respect of any" the word "incurred" — Alternatively, the same result can be achieved by implying a term that interest should be so chargeable on the basis that it is necessary to give business efficacy to the terms of the lease or, alternatively, on the basis of the officious bystander test — The legal expenses in question were incurred in recovering
E rents and service charges from certain of the tenants, but the account rendered by the solicitors was not separately itemised between the various tenants concerned — There was insufficient material upon which the Court of Appeal could make any declaration in respect of the particular legal fees — The management fees were payable to a company called Rockford Property Management Ltd, a company formed by the respondent and his wife to manage the block and other properties — The company was not a sham — The lease expressly permitted the employment of a manager and for that manager to be paid —There is no reason why a landlord should not employ a company and charge therefor, even if he
F owned that company, provided that it was not a complete sham — Although the respondent's service charge accounts may have been in a muddle, that was not a basis upon which an inquiry should be ordered, as the true position could readily be ascertained as a result of the documents produced in the proceedings

The following case is referred to in this report.

Finchbourne Ltd v *Rodrigues* [1976] 3 All ER 581; (1976) 238 EG 717, CA

This was an appeal by Kerry Skilleter, Paul Denly, Gina Tamburri, Simon Burn, Alison Richardson, Geraldine Harris, Tony Harbach and Diane Hallady, from an order of Judge Michael Baker whereby G
on May 14 1991 he dismissed the applicants' originating application, which sought various relief in relation to the service charges properly incurred and payable in respect of 21/21a South Terrace, Littlehampton, Sussex, owned by the respondent landlord, David Charles.

Colin MacKay QC and Timothy Sisley (instructed by Williams MacDougall & Campbell, of Worthing) appeared for the appellants; Peter Sheridan QC and Robert Deacon (instructed by Bennett Griffin & Partners, of Worthing) represented the respondent.

Giving judgment, **PARKER LJ** said: The appellants are the tenants of three of 13 flats in a Victorian block situate at South Terrace,
Littlehampton, West Sussex. The respondent is their landlord, who H
also occupies one of the 13 flats.

On December 12 1989 the appellants joined with the tenants of five more of those flats in an originating application in Chichester County Court in which they claimed the following relief:

> 1. A declaration as to what relevant costs of maintenance had been lawfully and properly incurred by the Respondent as Landlord of the various premises occupied by the Applicants, and the extent to which such relevant costs had been reasonably incurred or have been done to a reasonable standard in respect of maintenance of all that block of flats situate at and known as 21/21a South Terrace, Littlehampton in the County of West Sussex, in respect of the years 1987/1988/1989.
>
> 2. An enquiry in consequence of the declaration in paragraph 1 above as to
> what sums are due either from the Applicants to the Respondent, or from the J
> Respondent to the Applicants.
>
> 3. A declaration pursuant to Section 20(c) of the Landlord and Tenant Act 1985 that the costs or some part of the costs as the Court shall think fit incurred by the Landlord in connection with these proceedings shall not be regarded as relevant costs to be taken into account in determining the amount of any service charge payable by the Applicants in these proceedings.
>
> 4. An account or an enquiry and an account of the monies representing the maintenance fund for the said flats for the years 1987/1988/1989.
>
> 5. An Order both interlocutory and final for the appointment of a Manager of the said premises pursuant to Section 24(1) (a) and/or (b) of the Landlord and Tenant Act 1987 with such consequent directions as to functions of the said Manager as the Court deems just.
>
> 6. A mandatory injunction requiring the Respondent whether by himself or
> any other person on his behalf to repair or cause to be repaired the said flats in K
> accordance with the terms of the Lease, and in respect of the schedule of works appended to this Application marked "A".
>
> 7. Such further or other relief as may be just.
>
> 8. Costs.

On May 14 1991, after a four-day trial before Judge Michael Baker, that application was dismissed with costs on scale 2. On June 3 1991 the eight tenants gave notice of appeal seeking to set aside the judge's order and obtain either: (1) orders in accordance with paras 1 to 5 of the original application; or (2) a retrial. The appeal is pursued only on behalf of three of the original applicants, all of whom are legally aided. The other of the original applicants, who are not legally
aided, have disappeared owing to the fact that they have failed to L
comply with orders for security for costs.

Mr MacKay QC, who appeared for the three appellants, has rightly limited the conduct of this appeal very considerably, so that of the many matters canvassed before the learned judge and raised in the notice of appeal we have only a few to consider.

At this point I should say that all the matters now raised relate to the maintenance charge provided for by the leases and that although Mr MacKay restricted the matters raised in the notice of appeal which he desired to argue, he also sought to raise additionally one or two further matters.

All the leases are in identical form. The immediately relevant provisions are contained in the fifth and the eighth Schedules, to
which I now turn. M

Para 2 of the fifth Schedule, which contains the lessees' covenants, is in these terms:

> 2. To pay to the Lessor a Maintenance Charge being that fraction specified in Paragraph 5 of the Particulars of the expenses which the Lessor shall in relation to the Property reasonably and properly incur in each Maintenance Year and which are authorised by the Eighth Schedule hereto (including the provision for future expenditure therein mentioned) the amount of such payment to be certified by the Lessor's Managing Agent or Accountant acting as an expert and not as an arbitrator as soon as conveniently possible after the expiry of each Maintenance Year . . .

The remainder of that provision I need not read.

The eighth Schedule is headed "Costs and Expenses charged upon the Maintenance Fund". That part of it which is presently relevant provides as follows:

> There shall be charged upon the Maintenance Fund such of the following costs and expenses as may from time to time be incurred in connection with the Property . . . The said costs and expenses are:
>
> 1. The cost incurred by the Lessor in complying with its obligations in Part 1 of the Sixth Schedule.
>
> . . .
>
> 9. The cost of employing a Managing Agent or Surveyor to manage the property and to collect the maintenance charges in respect of the flats therein and to carry out such other duties as may from time to time be assigned to him by the Lessor or are otherwise imposed on him by the provisions of this Lease.
>
> . . .
>
> 11. All legal and other proper costs incurred by the Lessor:
>
> (a) in the running and management of the Property and in the enforcement of the covenants on the part of the Lessee and of the Lessees of other flats in the Property and the conditions and regulations contained in the leases granted of the flats in the Property insofar as the costs of enforcement are not recovered from the Lessee in breach . . .

It is also necessary to refer to two provisions of the Landlord and Tenant Act 1985. Section 19(1) provides:

> Relevant costs shall be taken into account in determining the amount of a service charge payable for a period —
>
> (*a*) only to the extent that they are reasonably incurred, and
>
> (*b*) where they are incurred on the provision of services or the carrying out of works, only if the services or works are of a reasonable standard;
>
> and the amount payable shall be limited accordingly.

Section 20C:

> (1) A tenant may make an application to the appropriate court for an order that all or any of the costs incurred, or to be incurred, by the landlord in connection with any proceedings are not to be regarded as relevant costs to be taken into account in determining the amount of any service charge payable by the tenant or any other person or persons specified in the application; and the court may make such order on the application as it considers just and equitable in the circumstances.

The applicants now seek, principally, a declaration that certain specified items, which had been included in the accounts for service charges presented to the tenants, should not have been included. These are:

(1) a sum of £490 plus any interest thereon, which sum was not in fact incurred;

(2) interest and bank charges;

(3) legal expenses;

(4) management fees.

There is no dispute about item (1) and no relief is required in respect of it. As to item (2), the interest and charges were incurred because the tenants had not paid the maintenance charges. It was so found by the judge. No question of reasonableness can arise. Mr MacKay submits that no provision has been made in the lease for the recovery of interest charges and accordingly that no interest or bank charges can be included as relevant expenses for the maintenance charge.

The relevant provision is no 16 in the eighth Schedule, which provides as follows:

> Any interest charges payable by the Lessor on his Bank Account or accounts in respect of any for the purposes of the Maintenance Charge.

It is submitted, rightly, that as that clause reads, it makes no sense. It is apparent that some word has been omitted after the words "in respect of". It is submitted that, because the clause as it stands makes no sense, there can be no interest charge recovered.

I cannot accept this. The plain intention of the parties was that interest should be chargeable and to give effect to that intention it is only necessary to insert after the words "in respect of any" the word "incurred", so that the clause would read "Any interest charges payable by the Lessor on his Bank Account or accounts in respect of any incurred for the purposes of the Maintenance Charge". To read in such a word when there has been an obvious typographical error appears to me to be both permissible and necessary. Alternatively, the same result can be achieved by implying a term that interest should be so chargeable. Such a term could, in my judgment, be implied either on the basis that it is necessary to give business efficacy to the terms of the lease or, alternatively, on the basis of the officious bystander test.

As to legal expenses, the lease provides that they may be charged in so far as they are not recovered from the tenants in breach. Therefore, if the question arises of legal proceedings being taken for the purpose of enforcing the tenants' covenant, there must be a corresponding duty on the landlord to endeavour to recover the expenses from the tenants. It is also plain that the wording of the legal expenses' provision in the eighth Schedule covers not only legal proceedings for the purpose of enforcing covenants, but a number of other legal proceedings.

The legal expenses here in question were incurred in recovering rents etc from certain of the tenants, but the account rendered by the solicitors was not separately itemised between the various tenants concerned. They were asked to itemise the account but declined to do so. The matter was not fully explored in evidence before the learned judge. I do not consider that we have before us any material upon which we could make any declaration in respect of the particular legal fees, much less in respect of all legal fees as a general matter, which is what we are asked to do.

I turn to the management fees. Two points are advanced. First, that the fees are not recoverable at all because they are in respect of the services of a company called Rockford Property Management Ltd, which it is said was the alter ego of the landlord and, it is suggested, was put in place in order that the landlord should be able to recover expenses to which he was not entitled. Second, that they are in any event excessive.

The second point can be shortly disposed of. The judge found them not to be excessive. There was material upon which he could so conclude and we could not, in my judgment, conclude to the contrary.

As to the first point, it appears that the landlord and his wife formed the company after the leases had been created. That company, however, manages other properties, some of them outside Littlehampton, and it cannot be suggested that the company is a sham put in place solely for the purpose which it is suggested lay behind it. The lease expressly permits the employment of a manager and for that manager to be paid. I can see no reason why in such circumstances the landlord should not employ a company and charge therefor, even if he owned that company, provided that it was not a complete sham. In this case there is no evidence that it was such a sham and the evidence that it was not is plain. It further appears that the point was not taken in the court below. Had it been taken, the matter might well have been further explored. It was not, however, further explored and, in my view, it is too late to advance the point now.

Finally on this point I should mention that Mr MacKay referred us to *Finchbourne Ltd* v *Rodrigues* [1976] 3 All ER 581, where it was held that the managing agent must be intended to be someone different from the lessor and that in that case the so-called managing agent was a mere nominee of the landlord. Indeed, it was really the landlord trading under a firm's name. I do not, therefore, consider that that case avails the appellants.

The appellants seek also an inquiry based, as I understand it, on the fact that the landlord's accounts were muddled. The judge declined to grant such relief, holding that although the accounts could be improved upon, the true position could readily be ascertained as a result of the documents which had been produced in the proceedings. He was, in my view, both justified and in fact right in taking the view which he did. To hold an inquiry when the matter can be dealt with more simply and cheaply appears to me to be a good reason for rejecting the call for an inquiry.

One further matter raised related to the question whether the costs of this appeal and in the court below could be properly included in the service charge. That question could arise only if the appellants were substantially successful on this appeal. Since I consider that they are not, the matter does not arise. In my judgment, this appeal should be dismissed.

McCowan LJ agreed and did not add anything.

Court of Appeal

December 20 1991

(Before Lord Justice DILLON, Lord Justice STOCKER and Lord Justice BUTLER-SLOSS)

FULLER AND ANOTHER v JUDY PROPERTIES LTD

Estates Gazette April 11 1992

[1992] 14 EG 106

Landlord and tenant — Forfeiture — Irremediable breach — Validity of two notices under section 146 of the Law of Property Act 1925 — Whether notices valid — Whether tenant entitled to relief from forfeiture

The defendant, Judy Properties Ltd, is the owner of premises consisting of a shop with two flats above — By a lease dated November 8 1976 the defendant demised the premises for a term of 20 years from June 24 1976 — The tenant covenanted not to assign, underlet or part with possession of the premises subject to a proviso that the landlord should not unreasonably withhold consent and, further, not to use the ground floor of the premises other than for the retail sale of drapery — In January 1988 the then tenants, J R Graham Ltd ("Graham"), sought the defendant's consent to assign the premises to the plaintiffs and for a change of use as to the ground floor for an estate agency — Following "subject to contract and without prejudice" correspondence and a letter from the defendant's managing agents stating that the defendant's solicitors were being instructed to prepare a draft licence to assign, subject to the managing agents inspecting the premises before they could advise the defendant to complete, Graham assigned the term on April 13 1988 to the plaintiffs for an agreed premium of £30,000 — The deed of assignment wrongly stated that the defendant landlord's consent had been obtained — In September 1988 the plaintiffs' application for planning permission for change of use was refused and they sought consent to assign the term to another firm of estate agents, Andrews, although by the beginning of January 1989 Andrews had withdrawn — The plaintiffs paid the quarterly payments of rent in June, September and December 1988 and these were accepted — On February 9 1989 the defendant's solicitors sent a notice ("the first notice") under section 146 of the Law of Property Act 1925 to Graham alleging breach of covenant against assignment and requiring the breach to be remedied within seven days, a copy of the notice being sent to the plaintiffs' solicitors for information — On February 24 the defendant peaceably re-entered the premises and changed the locks — On February 28 1989 the defendant served a further notice ("the second notice") pursuant to section 146 on the plaintiffs; this notice was in the same terms as the first notice — In response to the second notice the plaintiffs issued the writ in this action claiming that the defendant could not reasonably have withheld consent to the assignment, that it was not entitled to forfeit the lease and alternatively that any forfeiture had been waived by the acceptance of rent — The relief claimed included an injunction restraining the defendant from entering the premises, damages and relief against forfeiture — On July 10 1990 the defendant granted a new lease of the premises to Hirestar Ltd ("Hirestar")

At the trial of the action, Mervyn Davies J held that no licence to assign had been granted, there had been no unreasonable withholding of consent and therefore the assignment of April 13 1988 had been executed in breach of covenant — He further held that the first notice was ineffective, not having been served on the plaintiffs as the tenants and that the second notice was invalid because the defendant had taken possession: therefore the lease had not been forfeited — The defendant appealed

***Held*: The appeal was allowed — The first notice was ineffective, not having been served on the tenant — The second notice was valid and the writ in these proceedings was issued before the defendant could effect a forfeiture in reliance on it, with the consequence that the plaintiffs' application for relief against forfeiture was properly before the court — The defendant had peaceably re-entered to effect the forfeiture, but before the Court of Appeal the defendant conceded that relief against forfeiture should be granted to the plaintiffs on appropriate terms — Hirestar were purchasers of a legal estate without notice of the plaintiffs' equity to seek relief against forfeiture and were therefore not bound by the equity —The plaintiffs were entitled to relief from forfeiture subject to the lease held by Hirestar; accordingly, the plaintiffs were put into the position of immediate reversioners on Hirestar's lease liable to pay rent under their lease to the defendant but entitled to the rent payable by Hirestar — As a further condition of relief the plaintiffs were to pay the costs of the proceedings down to the judgment of the judge**

The following cases are referred to in this report.

Billson v *Residential Apartments Ltd* [1992] 1 EGLR 43; [1992] 01 EG 91

Horsey Estate Ltd v *Steiger* [1899] 2 QB 79; [1895-9] All ER Rep 515; 68 LJQB 743; 80 LT 887; 47 WR 644; 15 TLR 367, CA

Old Grovebury Manor Farm Ltd v *W Seymour Plant Sales & Hire Ltd* [1979] 1 WLR 1397; [1979] 3 All ER 504; (1979) 39 P&CR 99; [1979] EGD 529; 252 EG 1103, CA

Scala House & District Property Co Ltd v *Forbes* [1974] QB 575; [1973] 3 WLR 14; [1973] 3 All ER 308; [1973] EGD 342; (1973) 227 EG 1161, CA

This was an appeal by the defendant, Judy Properties Ltd, from the order of Mervyn Davies J given in October 17 1990 in relation to the claim by the plaintiffs, Mark Peter Fuller and Joy Carolyn Fuller, for damages and relief from forfeiture in relation to the defendant's peaceable re-entry of premises at 534 London Road, North Cheam, Surrey, of which the plaintiffs, who carry on business as estate agents, had been assigned a lease dated November 8 1976. Mervyn Davies J's decision is reported at [1991] 2 EGLR 41.

Graham Platford (instructed by Cohen & Naicker) appeared for the appellant; Christopher Moger and Alison Potter (instructed by Symons, of Croydon) represented the respondents.

Giving judgment, **DILLON LJ** said: This appeal, from a decision of Mervyn Davies J given in the Chancery Division on October 17 1990 after the trial of the action, raises, upon somewhat unusual facts, novel questions as to the service of a forfeiture notice under section 146 of the Law of Property Act 1925 where there has been an irremediable breach of covenant and as to the granting of relief against such forfeiture.

The appellant, Judy Properties Ltd ("Judy"), the defendant in the action, is the owner of premises at 534 London Road, North Cheam, Surrey. These premises consist of a shop, which is at the end of a terrace of shops, with two flats above.

By a lease dated November 8 1976 Judy demised the premises for a term of 20 years from June 24 1976. By the beginning of 1988, that term had for some years been duly vested in a company called J R Graham Ltd ("Graham"). The lease provided for five-yearly rent reviews and, by the beginning of 1988, the rent was at the rate of £10,500 pa payable by quarterly instalments in advance.

The lease contained in clause 3(15) a tenant's covenant not to assign, underlet or part with possession of the premises or any part thereof subject to a proviso that the landlord should not unreasonably withhold consent to an assignment or underletting of the whole of the premises. There was provision that the tenant would procure that any assignee of the premises would enter into a direct covenant with the landlord to pay the rent and observe the tenant's covenants during the residue of the term. There was also provision for notice of any assignment to be given to the landlord within one month after the execution of the assignment and for a registration fee to be paid. In addition, there was in clause 3(16) a negative covenant by the tenant prohibiting any use of the premises other than as to the ground floor (ie the shop) for the retail sale of drapery and as to the upper floor as residential.

At the beginning of 1988 Graham agreed in principle on the sale of the premises to the plaintiffs in this action ("the Fullers"), who are the respondents to this appeal. The Fullers carry on business as estate agents. They wished to use the ground floor as an office for their estate agency. Before they could do that, they needed planning permission for change of use and also the landlord's consent under the lease to a change of use and to the assignment of the premises to them.

Accordingly, on January 15 1988 Graham's solicitors wrote to Judy's managing agents, Willmotts, for a licence to assign the premises to the Fullers and a licence for the change of use.

On February 1 Willmotts stated in a letter to Graham's solicitors that they wished to carry out an inspection of the premises before the matter proceeded any further. In a letter of February 5, marked "Subject to Contract and Without Prejudice", they stated that Judy was reasonably satisfied with the status of the proposed assignees and they suggested that Judy's solicitors might prepare a draft licence to assign and deeds of variation. Inconclusive further letters passed and on April 14 Willmotts wrote to Graham's solicitors, subject to contract and without prejudice as before, confirming that they were instructing Judy's solicitors to prepare a draft licence to assign but pointing out that until they had gained access to inspect the premises they could not advise Judy to complete any licence to assign.

In fact, on April 13 1988 Graham had assigned the premises to the Fullers for the unexpired residue of the term of the lease and the Fullers had paid Graham the agreed premium of £30,000 for the assignment. The deed of assignment recites that the consent in writing of the landlord to the assignment had been made and duly obtained, but that was not so. It would seem that on the execution of that assignment Graham left the ground floor of the premises vacant and handed over the keys to the Fullers. No notice of the assignment was given to Judy or Willmotts.

In the early summer of 1988 Willmotts obtained access to the premises to inspect them and on June 20 1988 they sent Graham's solicitors a schedule of items of repair and decoration, on the basis that in the licence to assign to the Fullers, which was then circulating in draft, the Fullers should covenant to complete the works in the schedule within three months of completion of the assignment. However, in September 1988 the Fullers' application for planning permission for the change of use of the premises was refused by the local planning authority. Consequently, the Fullers changed their plans and sought permission instead to assign the premises to Andrews, another firm of estate agents. Andrews did not face the same planning difficulty as the Fullers, since Andrews could obtain planning permission for the use of these premises as estate agents' offices by surrendering an existing permission to use their existing nearby premises for that purpose. The proposal that there should be an assignment to Andrews and that Andrews should covenant to carry out the works in the schedule was acceptable in principle to Judy, but at the beginning of January 1989 Andrews withdrew.

Accordingly, at January 5 1989 the Fullers' solicitors asked Judy's solicitors for the original formal licence under which the Fullers would undertake to carry out the works in the schedule. It has been the Fullers' position since their own application for planning permission was refused that they accept that they cannot establish their own estate agents' offices in the premises, but they want to sell the premises to some purchaser for a permitted use in order to recoup the premium which they had paid to Graham.

The quarterly instalments of rent under the lease which were due in June, September and December 1988 were in fact paid by the Fullers and the payments were accepted. It does not appear, however, that Judy, Judy's solicitors or Willmotts knew at that stage that the lease had already been assigned to the Fullers.

On February 9 1989, however, Judy's solicitors sent a notice under section 146 of the Law of Property Act 1925 ("the first section 146 notice") to Graham. This notice refers to the covenant against assignment or parting with possession of the premises in clause 3(15) and asserts that that covenant had been broken in so far as Graham had assigned and/or parted with possession of the premises to the Fullers.

The notice continues in paras 3 and 4:

> 3. We require you to remedy the breach set out in paragraph 2 insofar as the same is capable of remedy and to make compensation to Judy Properties Limited in money for such breach.
>
> 4. On your failure to comply with this Notice within 7 days (or within a reasonable time) it is the intention of Judy Properties Limited to commence proceedings to recover possession of the premises.

Although that notice was addressed only to Graham, a copy of it was sent on February 9 by Judy's solicitors to the Fullers' solicitors for information.

There were letters from the Fullers' solicitors objecting strongly to the first section 146 notice. On February 24 Judy re-entered the premises without any court order and changed the locks. The shop remained otherwise unoccupied; one of the flats continued to be occupied by a statutory tenant.

Thereafter, Judy's solicitors served a further notice ("the second section 146 notice") on the Fullers. This notice is dated February 28 1989, but it is common ground that it was received by the Fullers only on March 3 1989. Save for its date and for the fact that it is addressed to the Fullers and not to Graham, it is in substantially the same terms as the first section 146 notice, and includes the same paras 3 and 4 as set out above.

In response to the second section 146 notice, the Fullers issued the writ in this action on March 8 1989. It was served on Judy on the same day. The statement of claim endorsed on the writ pleaded the first and second section 146 notices. It asserted that the defendant could not reasonably have withheld its consent to the assignment to the plaintiffs and accordingly is not entitled to forfeit the lease and, alternatively, that any forfeiture had been waived by the acceptance of rent. The relief claimed was: (1) an order to restrain the defendant from entering into or remaining on the premises; (2) an order on the defendant to supply the plaintiffs with keys to the premises; (3) damages under section 50 of the Supreme Court Act 1981; (4) damages; and (5) relief against forfeiture.

Certain other matters of fact can be set out shortly:

(1) In a letter of March 6 1989 to the Fullers' solicitors, Judy's solicitors, after referring to both of the section 146 notices, asserted categorically:

> Our clients were perfectly entitled to take peaceable possession of the premises following an irremediable breach of covenant having been committed. The lease has not been forfeited.

(2) On April 26 1989 Judy's solicitors told the Fullers' solicitors that Judy intended to relet the premises and they were being placed on the market for that purpose.

(3) At some unknown date, the Fullers changed the locks again so that they alone had keys to the shop. But on May 22 1989 Willmotts, on behalf of Judy, changed the locks yet again and since then the Fullers have had no key to the premises. At that stage the shop still remained unoccupied.

(4) On July 10 1990 Judy granted a new lease of the premises to a third party, Hirestar Ltd ("Hirestar"), for a term of 16 years from June 24 1990 at a yearly rent of £19,500 with four-yearly rent reviews and in consideration of a premium of £5,000. The user clause limits use to use as a retail shop for the sale of double glazing and associate materials. Hirestar duly went into occupation of the shop, though it appears that they are now seeking to find a purchaser. On all the available evidence it appears that Hirestar had no notice at all, when they took their lease and entered into occupation, either of these proceedings or of the former lease which had been assigned to the Fullers, or of the Fullers' claims in respect of the premises; Hirestar understood that they were taking a new lease from Judy as landlords of premises which — save for the flat occupied by the statutory tenant — had become vacant.

The trial of the action took place in October 1990 and Mervyn Davies J gave judgment on October 17. Having analysed the facts and correspondence, he held that no licence to assign to the Fullers had ever been granted by Judy nor had there been any unreasonable withholding of consent. Thus the assignment of April 13 1988 had been executed in breach of covenant. He further held that there had been no licence *ex post facto* or waiver of the breach of covenant by acceptance of rent or otherwise. None of these findings is challenged on this appeal.

In relation to the first section 146 notice, however, the judge held, following the decision of this court in *Old Grovebury Manor Farm Ltd* v *W Seymour Plant Sales & Hire Ltd (No 2)* [1979] 1 WLR 1397*, that the notice was ineffective because it had been served on Graham and not on the Fullers. That is a point which had been taken by the Fullers' solicitors as long ago as March 3 1989.

It was not suggested in the court below or in this court that the sending by Judy's solicitors of a copy of the first section 146 notice to the Fullers' solicitors on the same day, February 9 1989, on which they sent the original to Graham was a sufficient service of that notice

*Editor's note: Also reported at (1979) 252 EG 1103.

A on the Fullers for the purposes of section 146. It was, however, submitted before the judge that on the correspondence the Fullers were estopped from asserting that because of the assignment of April 13 1988 they, and not Graham, were the lessees on whom a section 146 notice fell to be served. It was submitted that the execution of the assignment had been concealed and that thereby and by the correspondence the Fullers had represented that Graham was still the lessee and so still the proper person to serve with the notice.

The judge rejected those submissions, after examining the correspondence, but they are raised as, logically, the first ground of Judy's appeal. I agree with the judge on this point and I, too, would reject these submissions for, essentially, the same reasons as did the B judge, *viz*: (1) there was no clear representation by the Fullers that Graham was still the lessee and (2) there is no basis for supposing that in choosing to serve the first section 146 notice on Graham Judy's solicitors were relying on anything they had been told by the Fullers or the Fullers' solicitors.

Judy then relied, before the judge, on the second section 146 notice. The judge held that that too was invalid; I shall come to his reasons shortly.

Section 146 provides by subsection (1) that:

> (1) A right of re-entry or forfeiture under any proviso or stipulation in a lease for a breach of any covenant or condition in the lease shall not be enforceable, by action or otherwise, unless and until the lessor serves on the lessee a notice . . . and the lessee fails, within a reasonable time thereafter, to C remedy the breach, if it is capable of remedy . . .

Consequently, the effect of the judge's rulings that both the first and the second section 146 notices were invalid was that the lease subsisted and had not been forfeited. Judy as landlord had not succeeded in getting to the point where it would be entitled to enforce any forfeiture.

Accordingly, by his order the judge: (1) granted an injunction restraining Judy from remaining on the premises; (2) ordered Judy to supply the Fullers with keys to the premises; and (3) directed an inquiry as to what damages the Fullers had sustained by reason of illegal entry. He also awarded the Fullers their costs of the action. That order was stayed by this court pending the hearing of this appeal.

D Taking the view he did, the judge did not have to deal with the Fullers' prayer for relief against forfeiture and in his order he did not deal with it. But in his judgment he said that had Judy succeeded in showing that it was entitled to forfeit, he would have been of the opinion that relief ought to be given to the Fullers in the way of allowing the lease to continue to run on condition that the Fullers paid Judy its costs of the proceedings within six months and without any other condition. He plainly regarded it as unfair that, for having acted precipitately in taking the assignment before licence to assign had actually been granted, the Fullers should wholly forfeit, in effect, their premium of £30,000 paid to Graham, while Judy should have had the advantage of the premium and higher rent payable by E Hirestar.

It is a bit difficult to see what the judge had in mind in granting the first injunction restraining Judy from remaining on the premises, since by the time of the trial Hirestar were in possession of the premises under their lease of July 10 1990. It would seem, however, that the judge intended that that order would require Judy to evict Hirestar from the premises; since Judy had taken it upon itself to let Hirestar into occupation of the premises when the trial was pending, it was for Judy, by whatever means, to get Hirestar out again so as to give vacant possession of the shop to the Fullers.

The difficulty about that is that Hirestar have, under their lease, a business tenancy for a 16-year term, which is also protected by Part II of the Landlord and Tenant Act 1954, and Judy has no legal right to F evict Hirestar. If, therefore, the judge's purpose was to require Judy to get Hirestar out of the premises, that would, in effect, have meant ordering Judy to pay Hirestar whatever sum would be enough to persuade Hirestar to agree to surrender their lease. If that was the purpose I do not believe that that injunction was properly granted.

In fact, after the judge's order had been made, the Fullers started proceedings in the Chancery Division against Hirestar for possession of the premises. The basis of those proceedings is that, if the lease assigned to the Fullers has never been forfeited, the Fullers have, and have at all times had, a legal estate in the premises under that lease which had priority to any interest of Hirestar under their lease. The consequence of that would be that the Fullers would obtain possession against Hirestar (subject to any future forfeiture proceedings against the Fullers) and Hirestar would be left with a G claim for damages against Judy for breach of Judy's covenant for quiet enjoyment in the Hirestar lease. Against that background, I turn to what I regard as the first main issue on this appeal, namely the validity of the second section 146 notice.

The judge held that it was invalid, so that Judy could not found a forfeiture on it, because at the time it was served Judy was already in possession, claiming to have forfeited, and Judy has continually throughout the proceedings sought to maintain a forfeiture based on the first section 146 notice. The judge reinforced his conclusion by reference to the fact that section 146 requires a lessee to be allowed a reasonable time to remedy a breach if it is remediable; ordinarily and inferentially, a lessee will require possession of the premises if he is to H be able to remedy a remediable breach such as a breach of a covenant to repair.

Both parties, however, agreed that the breach in the present case — breach of a covenant against assignment — is to be classified as irremediable. It has happened, once and for all, although relief against forfeiture may be granted: see *Scala House & District Property Co Ltd* v *Forbes* [1974] QB 575, a decision of this court.

The reasons for having a section 146 notice, as it now is, are conveniently set out in the judgment of Lord Russell of Killowen CJ, with which the other members of this court agreed, in *Horsey Estate Ltd* v *Steiger* [1899] 2 QB 79 at p 91 as follows:

> The reason is clear: [the tenant] ought to have the opportunity of considering whether he can admit the breach alleged; whether it is capable of remedy; J whether he ought to offer any, and, if so, what, compensation; and, finally, if the case is one for relief, whether he ought or ought not promptly to apply for such relief. In short, the notice is intended to give to the person whose interest it is sought to forfeit the opportunity of considering his position before an action is brought against him.

In the circumstances of the present case, therefore, what the Fullers as the lessees had to consider was whether there was indeed the breach of covenant alleged, or whether there was no breach because consent to assignment had been given or had been unreasonably withheld, or whether any breach had been waived by the acceptance of rent or otherwise or whether they should apply for relief against forfeiture. All this they were well able to consider, and did consider, without having possession or occupation of the shop at the premises; K the outcome was the issue of the writ in these proceedings on March 8 1989.

I have no doubt at all that the writ was issued within a reasonable time after the service of the second section 146 notice on March 3 and it was therefore issued before Judy could effect a forfeiture in reliance on the second section 146 notice. The consequence of that is that the application for relief against forfeiture in this case is properly before the court irrespective of the decision of the House of Lords in *Billson* v *Residential Apartments Ltd**, delivered since the argument on this appeal.

The seven days, or reasonable time, referred to in para 4 of the second section 146 notice must, in view of the wording already L quoted in subsection (1) of section 146, run from the service of the notice and not from whatever date the notice happens to bear — *semble* in the present case when it was typed.

I see no reason, therefore, why it was necessary in the particular circumstances of the present case for Judy to give up possession of the premises and hand over the keys to the new locks to the Fullers before serving the second section 146 notice. Having the keys to the shop would not have assisted the Fullers in any way in dealing with the second section 146 notice. Indeed, there was nothing they had done before or could have done afterwards in the shop, since, in the absence of planning permission, they could not trade there. For Judy to have sent the new keys to the Fullers would have been a mere token formality. Again, I can see no reason why, because Judy was M contending that the first section 146 notice was valid, Judy should be precluded from serving and relying on the second section 146 notice in the alternative; given that a copy of the first section 146 notice was sent to the Fullers' solicitors at the same time as the notice itself was sent to Graham, the invalidity relied on against the first section 146 notice was somewhat technical and I see no reason why Judy should have to wait until the validity of the first notice had been decided by the court before serving the second notice. I would accordingly hold that the second section 146 notice was a valid notice, even if the first were not.

It is, accordingly, necessary to consider the Fullers' application for

*Editor's note: Reported at [1992] 1 EGLR 43 *ante*.

relief against forfeiture. In relation to that, I have no doubt that, contrary to Mr Moger's submissions for the Fullers, Judy had peaceably re-entered to effect the forfeiture pursuant to whichever notice was valid before Mervyn Davies J gave his judgment. The evidence that that was Judy's intention is overwhelming and Judy's acts are unequivocally against treating the previous lease as still on foot; I refer in particular to the final changing of the locks in May 1989, to the putting of the premises on the market and to the granting of the new lease to Hirestar.

In the court below it was submitted for Judy that relief against forfeiture should not be granted to the Fullers because they were the authors of their own difficulties in having taken the assignment of the lease when they did, at such a high premium, before licence to assign had been granted and before they had obtained the planning permission they needed. In this court that stance has been abandoned in view of the importance attached by the judge to the unfairness of the Fullers losing their £30,000. It is therefore conceded by Judy that relief against forfeiture should be granted to the Fullers on appropriate terms.

But the difficulty about terms arises from the position of Hirestar and was touched on only in the last minutes of the argument in this court. It does not seem to have been explored at all in the court below.

But if the second section 146 notice (or the first) was valid and as Judy had clearly re-entered to enforce the forfeiture before Hirestar took their lease, the position was that the Fullers' lease and legal term in the premises were at an end by forfeiture and the Fullers merely had an equity to seek relief against forfeiture when Hirestar entered and took their legal term. But Hirestar were purchasers of a legal estate without notice of that equity and therefore Hirestar are not bound by the equity. Any terms for relief against forfeiture must therefore recognise the priority of Hirestar's term. As already indicated, Judy cannot evict Hirestar and equally the Fullers cannot if their lease has been forfeited and is only now restored by a grant of relief against forfeiture — albeit the restoration of the forfeited lease by the grant of relief is retrospective.

The effect, therefore, of granting relief on the basis favoured by the judge would seem to be to put the Fullers into the position of immediate reversioners on Hirestar's lease. The Fullers would be liable to pay the rent payable under their lease from the date covered by the last payment accepted by Judy, but they would be entitled to the rent payable by Hirestar until June 24 1996, when the Fullers' term will expire and, since relief against forfeiture is retrospective, they would be entitled to recover from Judy, as moneys had and received, the rent heretofore received by Judy from Hirestar and the £5,000 premium paid by Hirestar.

By contrast, Mr Platford submitted for Judy that any grant of relief against forfeiture to the Fullers should be on the condition that the Fullers forthwith surrendered their restored lease to Judy against payment by Judy to the Fullers of the Fullers' premium of £30,000 less the unpaid rent to date under the Fullers' lease and the amount required to carry out the works in the schedule of works of repair and decoration prepared in 1988 as already mentioned. That was, however, put forward on the assumption that Judy would be retaining the rent and premium already paid by Hirestar, to which the rent to be deducted from the £30,000 would be additional. As to the cost of the scheduled works, the fact is that Judy has the benefit of Hirestar's covenant to repair in its lease. Mr Moger submitted that if the £30,000 were to be payable by Judy to the Fullers, it ought to carry interest from February 1989.

Apart altogether, however, from any question of double accounting and excessive benefits to Judy under Mr Platford's proposals, it would, in my judgment, be wrong in principle for the court to grant relief against forfeiture of a lease on condition that the restored lease is immediately surrendered to the landlord. The object of relief against forfeiture is the continuation of a lease and not its extinction.

It may be that the most satisfactory outcome of these proceedings would be for the Fullers to surrender their lease to Judy for the appropriate payment. But that must be a matter for agreement between the parties. It is not for this court, in the exercise of its powers to grant relief against forfeiture, to make a new contract between the parties or to impose a price of the court's choosing on Judy against Judy's will.

I would therefore grant the Fullers relief against forfeiture on the basis indicated by the judge, with the consequences, as I understand them, which I have indicated. I would impose no conditions except, like the judge, a condition as to the payment of costs, namely the costs of these proceedings down to the judgment of the judge.

Accordingly, I would allow this appeal and vary the order of the judge.

I would discharge the injunctions granted by the judge against Judy, since Hirestar will be remaining in possession of the premises. I would leave, for what it is worth, the inquiry directed by the judge as to damages sustained by the Fullers for wrongful entry into the premises and the direction for payment of the sum found due on the making of that inquiry. But I stress that the entry would have ceased to be wrongful when forfeiture took place pursuant to the second section 146 notice.

I would grant the Fullers relief against forfeiture as indicated, subject to the condition that they pay the costs of these proceedings down to the judgment of the judge. I would consequently set aside the judge's order as to costs.

Agreeing, **STOCKER LJ** said: I have no comment except to express my gratitude upon the recitation of the facts contained in the judgment of Dillon LJ, which I have read in draft.

I also agree with his conclusion (and that of the learned judge) that there was nothing in the correspondence to establish the proposition that Graham remained the lessee and this was the appropriate person upon whom to serve the section 146 notice. The first notice was invalid, since it was served on Graham and not on the Fullers, though a copy was sent to the latter. This fact does not render the copy sufficient service of the notice. It did, however, indicate quite clearly to them the grounds upon which Judy was intending to forfeit. Moreover, it does not seem to me that, even paying full regard to the passage cited by Dillon LJ from the judgment of Lord Russell of Killowen CJ in *Horsey Estate Ltd* v *Steiger* [1899] 2 QB 79 at p 91, those considerations could, on the facts of that case, have made any difference at all to the position of the Fullers. In so far as the Fullers either needed to or did consider these matters, they did not require possession in order to do so — *a fortiori* where possession would be useless to them by reason of the refusal of planning permission. I confess to considerable doubts whether a landlord in possession can ever serve a valid section 146 notice, since the latter by its terms must presuppose that the tenants are in possession. None the less, I agree with Dillon LJ that, in all the circumstances of this case, Judy cannot be said to have been a trespasser when it forfeited and entered into possession under one or other of the section 146 notices. The breach was irremediable whichever party was in possession at the time of the notices and no further consideration of that fact by the Fullers can have altered the position. I therefore agree that the grant of an injunction, the effect of which was to remove Judy from possession of the premises, was inappropriate and that this injunction should be set aside. I also agree that the position as it in fact existed at the date of the trial — *viz* that a third party was in possession under a lease protected by Part II of the Landlord and Tenant Act 1954 — was a further factor which rendered the grant of an injunction inappropriate.

I would grant relief from forfeiture on the terms suggested by Dillon LJ, thus creating the situation to which Dillon LJ has referred. This cannot be disadvantageous to the Fullers, who cannot themselves make any meaningful use of the premises having regard to the planning situation. I agree with the order which Dillon LJ has proposed.

BUTLER-SLOSS LJ agreed and did not add anything.

Appeal allowed; plaintiffs to pay one-half of the defendant's costs of the appeal; application for leave to appeal to the House of Lords refused.

Chancery Division

February 17 1992
(Before Judge Paul BAKER QC, sitting as a judge of the High Court)

DEANPLAN LTD v MAHMOUD AND OTHERS

Estates Gazette April 25 1992
[1992] 16 EG 100

Landlord and tenant — Arrears of rent — Liability of original tenant — Agreement with occupying tenant by assignment releasing assignee from all claims under the lease — Whether release by accord and satisfaction releases an intermediate assignee of a lease

By a lease dated May 19 1972 the plaintiff granted a term of a factory unit at Eley's Estate, Eley Road, Enfield, to CWC Tempest Ltd ("the original tenant") for 20 years from March 25 1972 — On February 18 1974 the plaintiff granted licence to the original tenant to assign the term to the second defendant, CT London Ltd, the second defendant covenanting with the plaintiff to pay the rent and observe and perform the covenants and conditions on the part of the lessee during the residue of the term — The second defendant acquired the term by an assignment dated February 18 1974 — On January 8 1986 the plaintiff granted its licence to assign the term to the first defendant, who covenanted with the plaintiff to pay the rent and observe the covenants during the residue of the term — By an assignment made on January 8 1986 the first defendant acquired the term of the lease — On July 7 1987 the present action was commenced to recover arrears of rent of £11,566 from the first and second defendants — On February 26 1988 the plaintiff took out a distress warrant and on March 2 1988, the day the bailiffs entered and took walking possession, the first defendant agreed to surrender the lease "in consideration of you accepting the goods set out in the inventory below in full and final settlement of all claims and demands against me under the terms of the . . . lease" — That document was accepted and ratified by the plaintiff — Following a demand on behalf of the plaintiff, the second defendant compromised with the plaintiff, paying £18,000 by way of arrears of rent and interest, having been given credit for the value of the first defendant's goods which had been sold for the sum of £309 — Upon seeking to recover this sum and costs from the first defendant under the covenant for indemnity in the assignment of January 8 1986, the second defendant became informed, for the first time, of the agreement of March 2 1988 — In the present proceedings the second defendant counterclaimed for a rescission of the compromise agreement and repayment of the £18,000 on the basis of misrepresentation — Following amendments to pleadings and a concession by the plaintiff, the sole issue at trial between the parties was whether the agreement of March 2 1988 amounted to a release by accord and satisfaction releasing the second defendant from its covenant with the plaintiff

***Held*: Judgment was given for the second defendant — Certain conclusions can be drawn from the decided authorities — First, a release of one joint contractor releases the others as there is only one obligation — A release may be under seal or by accord and satisfaction, but a covenant not to sue is not a release — It is a question of construction of the contract between the creditor and the joint debtor in the light of the surrounding circumstances whether the contract amounts to a release or merely a contract not to sue — Second, the same principles apply to a contract between the creditor and one of joint and several debtors — The document of March 2 1988 was a release from the obligations under the lease purchased by an immediate surrender of the term and the goods in the inventory; this was a clear case of accord and satisfaction — The plaintiff could not say that it never intended to release its rights against the second defendant**

The following cases are referred to in this report.

Allied London Investments Ltd v *Hambro Life Assurance Ltd* (1983) 269 EG 41
Armitage, Re, ex parte Good (1877) 5 ChD 46; 6 LJ Bcy 65; 36 LT 338; 25 WR 422, CA
Baynton v *Morgan* (1888) 22 QB 74
EWA, Re, A Debtor [1901] 2 KB 642; 70 LJKB 810; 85 LT 31; 49 WR 642, CA
House Property & Investment Co Ltd v *Bernardout* [1948] 1 KB 314; [1947] 2 All ER 753
Matthey v *Curling* [1922] 2 AC 180
Nicholson v *Revill* (1836) 4 A&E 675; [1835-42] All ER Rep 148; 6 Nev & MKB; 1 Har & W 756; 5 LJKB 129
North v *Wakefield* (1849) 13 QB 536
Selous Street Properties Ltd v *Oronel Fabrics Ltd* [1984] EGD 360; (1984) 270 EG 643 & 743
Warnford Investments Ltd v *Duckworth* [1979] Ch 127; [1978] 2 WLR 741; [1978] 2 All ER 517; (1977) 36 P&CR 295
Watters v *Smith* (1831) 2 B&A 889

This was a claim by the plaintiff landlord, Deanplan Ltd, in respect of arrears of rent, interest and costs arising out of a lease dated May 19 1972 and made between the plaintiff and CWC Tempest Ltd relating to a factory unit at Eley's Estate, Eley Road, Enfield. The claim against the first defendant, Salah Mahmoud, was for breach of the covenant to pay rent under the lease. The second defendant, CT London Ltd, was an immediate assignee holding the term of the lease prior to its assignment to the first defendant.

Constance Whippman (instructed by B Erlich) appeared for the plaintiff; Nicholas Harrison (instructed by Rowe & Maw) represented the second defendant.

Giving judgment, **JUDGE BAKER** said: The issue in this case is whether an original lessee, or an intermediate assignee of a lease, who has given a direct covenant to pay the rent and observe the covenants, is released from liability following an agreement between the lessor and the occupying assignee of the lease, under which the lessor takes a surrender of the lease and some of the assignee's goods in return for releasing the assignee from all claims under the lease. In short, does the release by accord and satisfaction of one covenantor release other covenantors undertaking the same obligation?

The issue would appear to be a novel one so far as the researches of counsel go. Before addressing it, I must set out the facts, which are not in dispute to any significant degree.

By a lease dated May 19 1972 and made between the plaintiff, Deanplan Ltd ("Deanplan") and CWC Tempest Ltd ("Tempest") and some sureties, Deanplan demised a factory unit on its industrial estate, known as Eley's Estate, Eley Road, Enfield, to Tempest for a term of 20 years from March 25 1972 at a rent of £3,850 for the first five years, with reviews at five-yearly intervals thereafter. There were certain additional rents for insuring and servicing the common parts. The lease contained the normal covenants for the payment of rent, full repairing covenants, a user covenant confined to the business of "home improvement contractors", and the usual qualified covenant against assignment, to which was appended the following proviso:

> Provided always that every assignment or underlease or tenancy of any kind relating to the premises shall, if required by the landlord, contain a covenant by the assignee and the lessee or tenant, as the case may be, directly with the landlord to observe and perform the covenants and conditions in this lease contained, and covenants directly with the landlord in the terms of subclauses (a) and (b) hereof, and covenants that any subunderleases, whether mediate or immediate, shall contain covenants with the landlord in the terms of such subclauses.

The lease contained the usual proviso for forfeiture for non-payment of rent and failure to perform other covenants and concluded with a covenant by the sureties in the normal form.

By a licence made on February 18 1974 between Deanplan of the first part, Tempest of the second and the second defendant, CT London Ltd ("CT"), of the third part, Deanplan granted its licence to assign the lease to CT and consented to the user being changed to that of the business of "light engineering works". CT covenanted with Deanplan that:

> as from the date when the estate and interest in the lease shall be assigned to it, pursuant to the licence hereinbefore contained and thenceforth during the residue of the term granted by the lease, the assignee will pay the rent thereby reserved and observe and perform the covenants and conditions on the part of the lessee therein contained. In particular, without prejudice to the foregoing,

A the assignee will not assign or transfer, underlet or part with possession of the premises demised by the lease without the consent in writing of the landlord for that purpose having obtained.

On the same day (February 18 1974) the lease was assigned by Tempest to CT in consideration of the sum of £8,000 paid by CT to Tempest.

By a licence made on January 8 1986 between Deanplan and CT and Mr Salah Mahmoud (the first defendant), Deanplan granted its licence to assign the lease to Mr Mahmoud and Mr Mahmoud covenanted with Deanplan to pay the rent and observe the covenants during the residue of the term. On the same day an assignment was entered into between CT and Mr Mahmoud, and the first clause of
B that assignment is that:

In consideration of the sum of £4,000 paid by the vendor to the purchaser

— that is by CT to Mahmoud —

the vendor as beneficial owner assigned to the purchaser the property to hold to the purchaser for the unexpired residue.

Going down to clause 3:

The purchaser hereby covenants with the vendor by way of indemnity only that the purchaser and the persons deriving title under him will henceforth and at all times duly pay all rents becoming due under the lease and observe and perform all covenants, agreements and conditions therein contained on the part of the tenant or persons deriving title under it to be observed and performed, and also will at all times keep the vendor and his successors in title effectively indemnified against all proceedings, costs, claims, expenses and
C liabilities in respect thereof.

So, Mr Mahmoud was paid £4,000 to take the lease. At the date of this assignment the annual rent payable under the lease was £9,800. The rent as from March 25 1987 (the final review date) was agreed between Deanplan and Mr Mahmoud at £12,600.

Mr Mahmoud was not a satisfactory tenant and soon fell into arrears with the rent. The rent from June 1986 (just six months after taking over) was not paid on time. On January 29 1987 Deanplan recovered judgment in Romford County Court for part of the outstanding rent. Matters did not improve, so on July 7 1987 the present action was commenced, by which time the arrears amounted to £11,566. The claim was simply for arrears of rent and interest.
D There was no claim for possession pursuant to the proviso for forfeiture. CT was joined as second defendant, but the statement of claim disclosed no cause of action against it, as it failed to plead the covenant contained in the licence of February 18 1974. It was not the original lessee, so was not liable under the covenant to pay rent in the lease itself after assigning over.

A summons under Ord 14 was due to be heard on November 23 1987. On November 16 Mr Mahmoud through his solicitors stated that he was selling the property and "it is hoped that contracts will be exchanged prior to the 23rd of November. If so, our client states that all arrears of rent will be paid when the sale takes effect, which we understand will be very shortly". In consequence, with the consent of all parties the Ord 14 summons was adjourned generally. Nothing
E came of the proposed sale. In January 1988 Deanplan wished to gain access to allow its insurers to inspect and affix sprinklers. Mr Mahmoud made difficulties about this, whereupon Deanplan demanded entry under the clause allowing entry to view the condition of the premises. There was no response to this nor was there any payment of rent or news of any proposed assignment.

On February 26 1988 Deanplan took out a distress warrant in respect of the outstanding rent, then said to amount to £17,866. On March 2 1988 two bailiffs attended to execute the warrant. According to Mr Mahmoud, he admitted them because they pretended to have come on behalf of the insurers. Be that as it may, they gained access and proceeded to make an inventory of the goods there, mostly
F supplies of cloth, particularly linings. Mr Joseph Kasner, a director of Deanplan, who later saw the interior of the premises, says that they were four-fifths empty, the goods occupying only about one-fifth of the premises. He estimated the goods to be worth about £3,000. Mr Mahmoud thought they were worth something like £60,000, although he admitted that to realise that sum would require careful and skilled marketing. When half of the goods were subsequently put up for sale by auction they realised £309. That, however, is to anticipate.

On the day that the bailiffs entered and took walking possession (March 2 1988), there ensued negotiations by telephone between Mr Mendelson, Deanplan's solicitor, and Mr Mahmoud, whose proposals and comments were relayed to Mr Mendelson by one of the bailiffs. The upshot was a written document signed by Mr Mahmoud and, as I understand it, by the other bailiff. It is addressed to G
Deanplan Ltd and dated March 2 1988 and it reads:

Dear Sirs, Re: Factory premises at Eley's Estate, Eley Road, Edmonton, London. I hereby surrender the lease at the above premises dated 19th of May, 1972 in consideration of you accepting the goods set out in the inventory below in full and final settlement of all claims and demands against me under the terms of the said lease and specified in the Notice of Distress dated the 2nd of March, 1998.

And that is signed by Mr Mahmoud and a Mr D J Bennett. The inventory consists of three pages listing items of cloth. The document was accepted and ratified by Deanplan, as one can see in para 6 of the reply:

Whereby, the Plaintiff admits for the purposes of this action, that on the 2nd of March, 1988 it agreed to accept a surrender of the lease from the First H
Defendant, and that by its agents, D Moore, Certified Bailiffs, and as advised by the Third Party, it recorded this surrender in a Memorandum of the 2nd of March, 1988.

Mr Kasner's explanation for thus acting and accepting the proposal was the following in his evidence to me:

I had a phone call from Mr Mendelson [the solicitor]. He asked me if I would agree to accept half of the stock of Mr Mahmoud on condition that I would not go to court against him. I said I agreed. I didn't care very much whether it was a half or a seventh. The goods were of no value to me. I wanted to show him I wanted the premises. If I did not take the goods out, he would stay there. I did not intend to discharge CT. I gave the solicitors instructions to pursue CT after the surrender. I acted in good faith.

In consequence of the agreement, Mr Mahmoud gave up possession J
to Deanplan. Within a day or so the goods had been divided between Deanplan and Mahmoud and removed from the premises. CT had no part in these negotiations and did not immediately become aware of their outcome. On April 22 1988 Deanplan, through its solicitors, demanded payment of the balance of the rent, giving credit for the £309. In their letter, the solicitors did not disclose that there had been a release of all claims against Mr Mahmoud. On the basis of that letter, CT came to a compromise with Deanplan by which it paid £18,000. It then sought to recover this sum and costs from Mr Mahmoud under the covenant for indemnity contained in the assignment of January 8 1986. Thereupon, it was informed of the agreement of March 2 1988. In these proceedings CT counterclaimed K
for a rescission of the compromise and repayment of the £18,000 on the basis of misrepresentation. During the first day of the hearing and following amendments of the counterclaim, which I allowed, Deanplan conceded that the compromise agreement between it and CT should be set aside, leaving as the sole issue whether the agreement reached on March 2 1988 between Mr Mahmoud and Deanplan, and its performance by Mr Mahmoud, is a defence to the claim of Deanplan against CT. Accordingly, the issue arises as stated at the outset of this judgment.

Before passing to it, I should say that I heard oral evidence from Mr Mahmoud, called on behalf of CT, and Mr Joseph Kasner, the director of Deanplan. I have already made some reference to it. I was L
much more impressed by Mr Kasner's evidence than that of Mr Mahmoud, but I do not think that they differ on any material point. It is obvious that Mr Mahmoud was impecunious.

Mrs Whippman's submissions on behalf of the plaintiff landlord are simple and straightforward. In the 1974 licence to assign, CT, then itself about to take an assignment of the lease, entered into a direct covenant with the landlord to pay the rent and observe and perform the covenants and conditions on the part of the lessee contained in the lease during the residue of the term. That amounted to a separate and independent cause of action. The only defence available to CT would be that the covenants had been performed, either by itself or some other covenantor or assignee. If it had to perform them, it has a remedy over and against its assignee by an M
express covenant for indemnity contained in the assignment to Mr Mahmoud in 1986. She put the case as high as this: that a discharge by accord and satisfaction is not a performance, so that CT would be liable for the whole of the unpaid rent. The plaintiff was not bound to give credit for the value of the goods sold (£309), although it was proposing to do so.

Mr Harrison's careful and closely argued submissions opened by pointing out what he described as the "commercially absurd consequences of the landlord's case". Suppose a landlord claims against an assignee for rent and is met by a counterclaim for damages for breach of the landlord's covenants: suppose for that that the claim and counterclaim are compromised on the footing that neither

A party pays the other anything. According to the landlord's case, so said Mr Harrison, the landlord would be entitled to sue the original lessee for the rent, notwithstanding the compromise. The original lessee could then recover from the assignee under the indemnity. This possibility demands a careful analysis of the basis of the liability of the original lessee and of an intermediate assignee who enters into a direct covenant with the landlord.

For the purposes of the analysis, there is no difference between those two cases. Mr Harrison accepts that the liability of the original lessee and other direct covenantees is a primary liability and not a collateral or secondary liability, eg by way of surety. He further accepts that the obligations of the original lessee and an assignee are not joint. However — and this is the crucial point — while the liability B of each is several only, it is not cumulative; payment by one discharges the other. Normally, one meets a non-cumulative several liability only where it is annexed to joint liability. But there is no reason why it should not exist independently of joint liability.

In any event, it is instructive to consider the effect of an accord and satisfaction reached with one joint and several contractor on the liability of the other. Mr Harrison submits that the discharge of one joint and several debtor by accord and satisfaction discharges all the others, unless there is an intention to the contrary which may require express reservation of rights against the others. If this were not done, the benefit to the released debtor would be nullified by the operation of the indemnity covenant. Moreover, the covenant to perform on C the part of the original lessee is merely a covenant to perform in so far as another party, similarly liable, has not performed or otherwise discharged the obligation. These arguments apply just as much where there is a non-cumulative several liability, not annexed to a joint liability, as they do where it is so annexed.

Both counsel, in support of their respective submissions, referred me to a number of authorities, to which I must now turn. I will deal first with those concerned with covenants in leases and then to those concerned with other joint and several obligations. I was also referred to passages in Professor Glanville Williams' monograph on *Joint Obligations*, published in 1949, and to the treatment of the same subject in *Chitty on Contracts*, 26th ed (1989), paras 1301 to 1314, especially para 1307.

D The first case in the former group (landlord and tenant cases) is the decision of the Court of Appeal in *Baynton* v *Morgan* [1888] 22 QB 74. The plaintiff landlord had demised premises to the defendant for 21 years. The premises had been assigned. The assignee surrendered a small part of the premises to the landlord and then assigned the remainder of the premises to an assignee who became bankrupt, hence this claim for rent against the original assignee. It was held that the liability of the original lessee was not extinguished by the partial surrender. It had been argued that after assignment the liability of the original lessee became one of suretyship which was discharged following an alteration of the original contract without the consent of the guarantor. The court rejected this argument. The liability of E the original lessee was, and remained, a primary liability. It was not affected by the surrender of part of the premises, however large, although it would be terminated by a surrender of the whole as then the term would have come to an end. The surrender of part did not amount to an eviction of the assignee, it resulted from an agreement between the assignee and the landlord — and one can see details of that in the report of the divisional court at 21 QBD pp 101 and 102. By assigning, the defendant, the original lessee, had given the assignee power to assign again, which would include the power to surrender, in whole or in part, to the landlord. There was an unresolved discussion in the case as to whether the lessee continued to be liable for the entire rent or merely for an apportioned part of it, which is all that the plaintiff claimed. There was some discussion as to F hardship, but that could cut both ways according to whether a large or small part was surrendered, so that hardship was not a matter which could be taken into account. Those were the main points discussed. There was nothing bearing on the present question, whether a release by accord and satisfaction of a claim for rent against an assignee would release the original lessee and other covenantors.

The next case, *Matthey* v *Curling* [1922] 2 AC 180, does contain some discussion which is material. The decision itself is not of assistance. During the currency of a lease a house was requisitioned by the military authorities. Shortly before the end of the term and while it was still requisitioned it was destroyed by fire. The issue concerned the liability of the lessee under the repairing and insuring covenants. Defences of eviction by title paramount, frustration and G impossibility of performance were raised by the lessee and rejected. The lease had been assigned and the relationship as between the original lessee and the assignee was also discussed, though it did not play a very important part in the case. However, in the judgment of Younger LJ in the Court of Appeal, this passage occurs, at p 208 of the report:

> And here it is convenient to refer pointedly to the fact that it is Mr Matthey, the original lessee, and not the Richardson executors, his assignees, who is defendant to the action and the respondent on this appeal, as well as to the further fact that it was these executors, and not Mr Matthey, who were the real and only actors in the matters to which I have just been referring. In my judgment, however, this is not of any importance as touching the liability of the defendant under his covenants. The transactions just referred to of the H executors, his assignees, in relation to the lease bind him, in my judgment, in a question with the plaintiff as if they had been carried out by himself. The position of a lessee after assignment towards his lessor in relation to the covenants in his lease is concluded in this Court, and I doubt not in all Courts, by the decision of *Baynton* v *Morgan*. The lessee is not a surety for the assignee, any more than is the assignee the agent of the lessee. The lessee in a lease — the defendant so covenanted in this case — covenants in effect that he or his assignee will perform the covenants and observe the conditions contained in it; and when, as in the present instance, the defendant sets up no performance by himself of the covenants sued on, his defence must be either that his assignee has performed them or that his assignee is in some way as between himself and the lessor absolved from performance. *Baynton* v *Morgan* shows, amongst other things, that this last contention may not always be relied on by the lessee; and accordingly the position of the defendant J is not put otherwise than favourably to himself if it is said that the measure of his liability here, that of a lessee after an assignment, is the performance or non-performance of the covenants in question by his assignees when he cannot and does not aver any performance by himself.

In this passage, Younger LJ recognises that an original lessee may have a defence if the assignee not performing has in some way been absolved from performance. As he goes on to point out, the contention may not always be relied on. For example, the assignee may be absolved through the operation of the bankruptcy laws; the classic case where an original lessee may have to pay. But there may be cases where the landlord absolves his tenant from performance in ways which release the other covenantors.

Lord Atkinson in the House of Lords also discussed *Baynton* v K *Morgan* but not in a way which assists in the present case.

The next case I mention is *House Property & Investment Co Ltd* v *Bernardout* [1948] 1 KB 314. The assignee of a lease fell into arrears with the rent and the landlords brought an action against him in 1941. He counterclaimed for damages for breach of the covenant of quiet enjoyment and another claim relating to the drains. Before the action came on for trial the defendant applied for an adjustment of his liabilities under the Liabilities (Wartime Adjustment) Act 1941. I need not go into the procedure under the Act. The upshot was an arrangement endorsed by the court under which the rent was reduced, the defendant paid a fixed sum and abandoned his counterclaim and the lease was disclaimed. The effect of such an L order under the Act is the same as a discharge in bankruptcy. The present action was against the original lessee for the balance of the outstanding rent. It was held that he was not absolved as a consequence of the order made in the suit against the assignee. One of the points taken on behalf of the lessee was that the proceedings before the county court amounted to a compromise. Singleton J disposed of it as follows, at p 319:

> The next point taken by Mr O'Connor

— he was for the lessee who was being sued —

> was that in any event there was a compromise before the county court judge. He reminded me there was a counterclaim for £2,444 and that by the order the plaintiffs in this action were released from any possible liability on the counterclaim. He submitted that the order itself had made a compromise M because it contained the words "counterclaim withdrawn" and because it dealt with the rent so that the plaintiffs had no further claim in respect of it. In reply to that argument, Mr Quass submitted that there was no evidence of any compromise at all. I take the view that, having regard to the recitals and to the terms of the order, one can read into the order itself the element of compromise, but again I do not think it is important to the decision of this case; because the compromise, if compromise it was, was between the plaintiffs and Jack Bernardout, and I do not think it enures for the benefit of Joseph Bernardout, the defendant in these proceedings.

I would not regard those remarks as indicating that a compromise with an assignee could never result in relieving the lessee from liability. It will be noted in that case that the assignee, having obtained relief under the statute, was in the position of one who had

A the protection of the bankruptcy laws. That would give no relief to co-covenantors; quite the contrary.

In *Warnford Investments Ltd* v *Duckworth* [1979] Ch 127, Sir Robert Megarry V-C held that the disclaimer of a lease by the liquidator of a company to which the lease had been assigned did not relieve the original lessee from liability to pay the rent throughout the remainder of the term. The original lessee was not merely a guarantor or surety. The lease was determined only as between the company and the landlord. The case has no direct bearing on the issues before me, but the Vice-Chancellor's judgment contains a full analysis of the original lessee's liability, admirably summarised, if I may say so, in the following paragraph at p 137 G:

> B Generally speaking, an original lessee prima facie remains liable to the reversioner for the payment of the rent throughout the term, even after it has been assigned. This is a direct and primary liability, though after an assignment it is normally accompanied by rights of indemnity against the first assignee, and also, where there have been further assignments, against the assignee in whom the lease is for the time being vested.

The decision of Walton J in *Allied London Investments Ltd* v *Hambro Life Assurance Ltd* (1983) 269 EG 41 is more germane. In a licence to assign a lease the potential assignee, a company, had covenanted with the landlord to pay the rent and perform the covenants in the lease and that covenant had been reinforced by a covenant given by a surety. The assignment duly took place. The following passages from the judgment of Walton J describe what C happened thereafter. I am reading from the first page of the report, the second column:

> In due course, the assignee underlet the premises comprised in the lease in two portions. Thereafter, there were allegations of non-payment of rent, wants of repair, and so forth, and ultimately there were no fewer than eight actions on foot: three by the plaintiff against the assignee; one by the assignee against the plaintiff; two actions by the assignee against its underlessees — one against each — and two actions, one by each of its underlessees, against the assignee.
>
> This legal bonanza was obviously too good to last and ultimately matters were brought to a head in a series of compromises, as a result of which the lease was forfeited but the plaintiff abandoned all its money claims in the relevant action, other than for arrears of rent, and — and for the present purposes this is the crucial matter — the plaintiff released the surety unconditionally from his personal guarantee contained in the licence of December 3 1973.

D In the action before Walton J, the landlord was suing the original lessee for the arrears of rent. The lessee claimed that the release of the surety also released them from liability. This argument was rejected by Walton J on the now familiar ground that the lessee's covenant was a primary liability, not that of a surety. In the course of his review of the authorities, Walton J made extensive citations from *Baynton* v *Morgan*, a case he described as "completely decisive of the present case", and also cited the passage from the judgment of Younger LJ in *Matthey* v *Curling*, which I have already read, and he concluded as follows:

> These two cases, then, make it perfectly clear that when he is sued, as in the present case, on the covenants contained in the original lease, the original E tenant's only possible defences are either, (1) that he has performed the covenants, or (2) that the relevant assignee has performed the covenants, or (3) there has been some other operation conducted upon a lease — for example, surrender of the whole, which has put a complete end to the liability to pay rent. Short of these three defences, there is no defence.

Although the plaintiffs in that case had reached a compromise with the assignee and the surety, it specifically did not involve releasing the assignee from the liability to pay the arrears of rent. It was that liability for which the original lessee was concurrently liable. Hence, that case is clearly distinguishable from the case before me, where the liability of the assignee has been discharged. The decision of Walton J went to the Court of Appeal, but solely on a question of the date of the commencement of the interest on the arrears. The appeal was F dismissed: see [1985] 1 EGLR 45.

Finally in this group of cases, I was referred to the decision of Hutchison J in *Selous Street Properties Ltd* v *Oronel Fabrics Ltd* (1984) 270 EG 643, described in the headnote, justifiably, as "an important and wide-ranging decision", the report of which extended over two issues of *Estates Gazette*. For my purposes, it suffices if I say that one of the issues was whether an original lessee is liable to pay a rent which has been increased by a variation in the terms of the lease reached between the assignee and the landlord. Hutchison J held that he was so liable, relying, *inter alia*, on *Baynton* v *Morgan* and *Allied London Investments Ltd* v *Hambro Life Assurance Ltd*. He founded himself on the principle that an alteration of the terms of the lease agreed between the landlord and the assignee binds the original lessee because the assignee has been put into the shoes of the original lessee G and can do all such things as the original lessee could have done. That does not affect the issue with which I have to deal.

That concludes the review of the cases concerned with covenants in leases, and none deal with the issue here of a release of a covenantor by accord and satisfaction operating as the release of another covenantor. The dictum of Younger LJ in *Matthey* v *Curling* recognises the possibility.

I now turn to the cases on joint and several debts in other branches of the law to which I was referred.

The earliest is *Watters* v *Smith* (1831) 2 B&A 889. The plaintiff had sued one member, Hunter, of a committee of the management of a hospital which had engaged him. Before the pleadings had got beyond the declaration the case was compromised. Under it, the H member paid part of the debt and costs to be taken for the debt and costs in the action. The plaintiff then sued another member of the committee for the balance. It was held that the other members of the committee were not discharged. Lord Tenterden CJ saw the question thus:

> Whether the plaintiff intended to discharge the whole debt or only relieve the party proposing to pay a proportion of the debt.

Lord Tenterden CJ went on to point out that the first defendant would not have been relieved from having to contribute to the defendant in the second case if the latter had been compelled to pay a larger portion than his share. As might be expected, the judgment of Parke J sets out the principles with precision. At p 894 of the report: J

> It is not necessary to consider what would have been the effect of the payment of £20 if it had been made in full satisfaction of the demand against Hunter. If the case rested upon this, it might be a bar to the action, on the authority of *Longridge* v *Dorville*. But I think that question does not arise here from the facts. Looking at the terms of the agreement as contained in the letter from Hunter's attorney, and the receipt, it is manifest that the payment was not made in discharge of the plaintiff's rights against all other parties; and the result of the whole is, that it does not operate as a release, or matter which could have been pleaded as an accord and satisfaction, but amounts merely to an engagement not to sue Hunter, which can only be pleaded by himself; if the action, therefore, had been brought against two parties, it would not have been a discharge of both. Or, it may be considered as a release to one, qualified by a reservation of the plaintiff's rights against the other, as in *Solly* v *Forbes*.

Taunton J adds the point that a release given to one of two joint K contractors enures to the benefit of both. I emphasise the reference to "accord and satisfaction" in the judgment of Parke J, and it is also referred to in that of Taunton J. This case, however, was one of joint contractors.

The next case concerned joint and several contractors. It is *Nicholson* v *Revill* (1836) A & E 675. Mr Harrison relies on it for the statement in the judgment of Lord Denman CJ that:

> The debtee's

— an old word for the creditor or payee —

> discharge of one joint and several debtor is a discharge of all.

It is necessary, in my judgment, to look at the facts somewhat L closely to see what amounts to a discharge and how far the reasoning can be applied to the case not of joint and several debtors but of several debtors alone, albeit all liable for the same debt. The case was tried on special demurrer, that is a demurrer based on specific objections to the form of the pleadings. The pleadings are somewhat convoluted but must be referred to because they contain the facts, as is normal in a demurrer. They open with a declaration based on two promissory notes made on January 1 1832 by the defendant, John Revill, the younger, promising to pay £110 13s 4d 25 months after date, and £56 13s 8d 31 months after date. The plea or defence to this was that the notes were made jointly and severally by the defendant's father, John Revill, and Samuel Revill, as well as by the defendant and that before the action the plaintiff, without the defendant's M knowledge or consent, struck out the name of Samuel Revill on the notes and wholly discharged him from liability. In the judgment, Lord Denman states that the court need not inquire what the effect of a demurrer to that plea would have been, thus hinting that if the plaintiff had then demurred, the defence might not have been found sufficient. At all events, the plaintiff pleaded over and in the replication or reply went on to state many additional facts. From this it appeared that the father was the principal debtor in the sum of £299. Initially, a note was given by the three parties as security for payment of the sum. The note was unpaid and an action brought against the three was compromised by the giving of three notes, one of January 1 1832 for 13 months in the sum of £52 18s 8d, and the

A other two now sued on. The first note was not paid when it fell due. On January 28 1834, before the payment dates of the second and third notes, Samuel Revill offered and the plaintiff accepted £100 in discharge of his liability on the three notes. It will be seen that the plaintiff obtained part payment on the second note, the difference between £52 18s 8d and £100, ie £47 1s 4d, before it was due. Hence, there is an element of consideration and the transaction can be regarded as an accord and satisfaction. It was not simply a payment of part of the debt. Having received that payment the plaintiff handed over to Samuel Revill the first note and endorsed on the second (the note of £110 13s 4d) "received on account of Samuel Revill £47 1s 4d", this sum, with the amount of the first note, makes £100. He struck out Samuel Revill's name on the second note. The
B rejoinder by the defendant to this pleading alleged that the notes were given by Samuel Revill and himself as sureties and were given for the securing of part of the father's debt and for no other consideration or purpose.

Pausing there and turning to the judgment, Lord Denman stated that the court did not proceed on any doctrine as to relation of principal and surety. Hence, I take it that they were laying down principles in relation to joint and several debtors generally.

Going back to the rejoinder, it further alleged that the discharge of Samuel Revill was without the knowledge, privity or consent of the defendant. The rejoinder concluded with the verification (that is the normal way of ending a pleading alleging new facts by way of
C confession and avoidance). It amounts to an undertaking to prove those facts before the jury in due course. The plaintiff's surrejoinder purported to be a traverse, as is evidenced from its conclusion to the contrary, the way of ending a pleading which raises no few facts and puts the other party to proof. The pleading was, however, objectionable on various formal grounds which formed the basis of the special demurrer. In accordance, however, with the normal practice, the court considered the whole case and did not confine itself to the objections in point of form to the surrejoinder. Judgment was given on the substance of the case in the following terms — and picking up the point that I have already mentioned:

But we do not proceed on some of the grounds mentioned at the bar, such as the effect of the plaintiff's alteration of the instrument as making it void, or
D that the defendant thereby lost his right to contribution from the joint makers of the note; nor on any doctrine as to the relation of principal and surety. We give our judgment merely on the principle laid down by Lord Chief Justice Eyre in *Cheetham* v *Ward*, as sanctioned by unquestionable authority, that the debtee's discharge of one joint and several debtor is a discharge of all. For we think it clear that the new agreement made by the plaintiff with Samuel Revill, to receive from him £100 in full payment of one of the three notes and in part payment of the other two, before they became due, accompanied with the erasure of his name from those two notes, and followed by the actual receipt of £100, was in law a discharge of Samuel Revill.

I take this to be a decision that a release by accord and satisfaction reached with one joint and several debtor discharges all. I should refer to the final paragraph in the judgment:

E This view cannot perhaps be made entirely consistent with all that is said by Lord Eldon in the case of *Ex parte Gifford*, where his Lordship dismissed a petition to expunge the proof of a surety against the estate of a co-surety. But the principle to which we have adverted was not presented to his mind in its simple form; and the point certainly did not undergo much consideration. For some of the expressions employed would seem to lay it down that a joint debtee might release one of his debtors, and yet, by using some language of reservation in the agreement between himself and such debtor, keep his remedy entire against the others, even without consulting them. If Lord Eldon used any language which could be so interpreted, we must conclude that he either did not guard himself so cautiously as he intended, or that he did not lend that degree of attention to the legal doctrine connected with the case before him, which he was accustomed to afford. We do not find that any other authority clashes with our present judgment, which must be in favour of the
F defendant.

The view in that last paragraph, that the creditor could not reserve his rights against others, has not prevailed. Lord Eldon's views have proved more acceptable, as is evident from the next case: *North* v *Wakefield* (1849) 13 QB 536. This also concerned a joint and several promissory note, to which the defendant pleaded that the co-debtor had been released by deed on making a composition with the creditors, but the deed contained a proviso that it should not operate to prejudice the rights of others. It was argued for the defendant, on the authority of *Nicholson* v *Revill*, that a creditor could not release one joint and several debtor and hold another liable by reserving remedies. Patteson J, for the court, said at p 540:

Now the deed contained an express clause that the release to Goddard should not operate to discharge any one jointly or otherwise liable to the plaintiff for G
the same debts. It is plain, therefore, that it did not release the defendant. The reason why a release to one debtor releases all jointly liable is, because, unless it was held to do so, the co-debtor, after paying the debt, might sue him who was released for contribution, and so in effect he would not be released; but that reason does not apply where the debtor released agrees to such a qualification of the release as will leave him liable to any rights of the co-debtor.

Re Armitage, ex parte Good (1877) 5 Ch D 46 is a case where, on the somewhat special facts, a receipt given to one party did not release the other party. The Court of Appeal held that the receipt could not be construed as a release. The interest of the case for the present purposes is the comment in the court below of Sir James Bacon, the chief judge in bankruptcy, on *Nicholson* v *Revill*. I read from p 55: H

Now, in point of law, it is not to be disputed that a release to one of several joint debtors releases them all. The case of *Nicholson* v *Revill* is one of the many instances in which this rule of law has been applied, and the reason upon which the rule is founded is as clear and plain as the rule itself. It is to prevent circuity of action, inasmuch as if the released debtor were still liable to an action for contribution by his co-debtor (who had been sued and had paid), he would not in fact be released: *North* v *Wakefield*. But this principle has no application to the present case; for Armitage, being by the resolution of the creditors released from all his debts, could neither sue himself nor could his trustee sue on his behalf, even if the supposed partnership had existed, although the rights of the joint creditors against the joint debtor would be unaffected by the bankruptcy, by sect 50 of the Bankruptcy Act, 1869.

That reasoning applies as much to several debtors each liable for J
the same debt as to debtors jointly and severally liable for such a debt.

Finally, I come to what is probably the most important case of all, the decision of the Court of Appeal in *Re EWA, a Debtor* [1901] 2 KB 642. A and B were liable on a joint and several guarantee to a bank for the account of a company. The guarantee was limited to £6,000. Subsequently, the bank obtained judgment for that sum against A and B jointly and severally. The judgment not being satisfied, the bank presented a bankruptcy petition against B alone for the whole debt. Terms were arranged on which the petition might be withdrawn upon payment of £3,000. A document signed by the bank's solicitors acknowledged receipt of the sum of £3,000 in full discharge of all claims by the bank against B in connection with the company and all guarantees given by him to the bank in connection with that company K
and in settlement of any outstanding questions as to the amount due to the bank. There was no reservation of rights against A in that letter. The petition was withdrawn or dismissed. The bank then served a bankruptcy petition on A for the other half. The Court of Appeal held that he had been discharged by the release given to B. The sole-reasoned judgment is given by Collins LJ, with whom Rigby LJ agreed. Collins LJ posed the question as follows at p 648:

The question really turns on this, whether or not this document has the effect of accord and satisfaction in getting rid of the joint and several liability of B under the judgment. If it has that effect, it is not disputed that the rule of law applies, namely, that the release of one of two joint debtors has the effect of releasing the other.

After dismissing an argument that the doctrine as to release did not L
apply to a judgment debt, he went on at p 648:

But now I come to what was really the main point argued before us. It is clear that, though a document in terms purports to release one of two joint debtors, yet it may contain in terms a reservation of rights against the other joint debtor. Where you find these two provisions you construe the document, not as a release, but merely as an undertaking not to sue a particular individual, and the result is that the right to proceed against the co-debtor is reserved and can be put in force against him. Whenever you can find from the terms of the document an agreement for the reservation of rights against the co-debtor, then, I agree, the document cannot be construed as an accord and satisfaction of the joint debt, and, therefore, as a release of the co-debtor. But it appears to me that, on the face of this document, there is no intention shown so to limit its effect, and that it is framed in the widest possible terms so as to cover not only M
this particular debt, but all other claims by the bank in connection with the Professional & Trade Papers Ltd, for it is admitted that the foundation of the judgment was the guarantee, and at the time this document was drawn up there was this joint liability on the judgment to the extent of £6,000. And there is this circumstance to be observed, that this document expressly states that the cash and bills are accepted "in full discharge of all claims by the Capital & Counties Bank Ltd, against Mr B in connection with the Professional & Trade Papers Ltd".

Having regard to some of the evidence before me, I should notice the following towards the bottom of p 649:

Why, then, should any limitation be placed upon the effect of the discharge of "all claims"? If there is no limitation, the effect of the document is that it releases the claim against both co-debtors. If this legal consequence had been

A pointed out to the bank at the time, they might have said that that was not what they intended, but that is a factor common probably in all cases to which a release is given to one of two joint debtors. The person giving it may not realise the full legal consequences of it as regards the release of a co-debtor; but that is not, in my opinion, a sufficient ground for reading into the document something that is not expressed in it; and unless you find in it something qualifying the general words, it appears to me that the legal consequences in the general words of discharge must follow, notwithstanding that those consequences may go beyond what the person giving the document would have intended if they had been pointed out to him at the time, and he had had an opportunity of addressing his mind to them.

Collins LJ also rejected the argument that surrounding circumstances qualified the *prima facie* meaning of the document.
B He distinguished *ex parte Good*, that there the other debtor had already been made bankrupt and hence no claim could be made against him. Romer LJ did not dissent but expressed a doubt.

This decision has been roundly criticised as "extraordinary" by Professor Glanville Williams in his *Joint Obligations*, at p 117, para 55. His point is that there was simply an agreement to accept a lesser sum in satisfaction of the whole amount and so could not amount to accord and satisfaction. I, of course, have to accept the decision of the Court of Appeal as binding on me, whatever I or others may think of it. I would, however, make the following comments. First, it does not appear that any point other than the construction of the document was argued (and I refer to the argument at pp 645 and 646 of the report). Second, the proceedings were by way of bankruptcy
C petition. It by no means follows that if B had been made bankrupt the dividend would have amounted to the sum obtained, so that the bank might have got less.

From this long review of the cases, I draw the following conclusions. First, a release of one *joint* contractor releases the others. There is only one obligation. A release may be under seal or by accord and satisfaction. A covenant not to sue is not a release. It is merely a contract between the creditor and the joint debtor which does not affect the liabilities of the other joint contractors or their rights of contribution or indemnity against their co-contractor. It is a question of construction of the contract between the creditor and joint debtor in the light of the surrounding circumstances whether the contract amounts to a release or merely a contract not to sue. Second,
D the same principles apply to a contract between the creditor and one of joint and *several* debtors. If one joint and several covenantor is released by accord and satisfaction, all are released. Some have seen this as illogical, and so it would be if the only reason for the rule that the release of one joint contractor releases the other is that there is only one obligation. Professor Glanville Williams sees the reason for the extended rule to have been an early uncertainty as to the nature of a joint and several obligation (I refer to his book at p 135). Two other reasons can be adduced. First, that where the obligations are non-cumulative, ie the obligation of each is to perform in so far as it had not been performed by any other party, the acceptance of some other performance in lieu of the promised performance relieves the others.
E The covenantee cannot have both the promised performance and some other performance which he agrees to accept. Second, unless the co-covenantors were released following an accord and satisfaction, they could claim a right of contribution or indemnity. Thus, by suing the co-contractor the creditor commits a breach of the contract with the released covenantor, for such an action will inevitably lead to the very claim from which the release has been purchased by accord and satisfaction. Third, the reasoning in the preceding paragraph applies equally to a number of separate covenantors each liable to perform the same obligation as in the case before me. Indeed, the dictum of Younger LJ in *Matthey* v *Curling* already adverted to is consistent with this.

Returning now to the facts of this case, we start with a document
F signed by Mr Mahmoud and accepted by the plaintiff. The release from the obligations under the lease is purchased by an immediate surrender of the term and his handing over of one half of his goods. So, it would seem to be a clear case of accord and satisfaction. Furthermore, there are no words reserving rights against any other parties, nor can the document be read as a covenant not to sue. Is there anything in the surrounding circumstances which would rebut the *prima facie* meaning of the agreement? In some cases it has been found in the exemption from liability of the released party to be sued by his co-contractors, as where the released party is already bankrupt (*ex parte Good* is an example of that). In the present case, Mr Mahmoud was not bankrupt when he signed the documents and was open to a claim for an indemnity, which has in fact been made. It might be said that at the time of the signing of the document he was in G
no position to resist both the seizure of *all* his goods under a distress and the forfeiture of his lease with no practical possibility of relief. If the landlords had chosen to act in that way, the liability of CT for the balance of the rent could not be resisted. The landlord, however, did not choose to press his rights in that way but chose to enter into a bargain with Mr Mahmoud so as to get possession speedily without the delay that further proceedings would entail. He has, in my judgment, to accept the consequences of that. It is certainly no good his saying that he never intended to release his rights against CT.

Therefore, the result is that I dismiss the claim as against the second defendant.

Plaintiff's claim against the second defendant dismissed. H

Court of Appeal

November 21 1991

(Before Sir Donald NICHOLLS V-C, Lord Justice RUSSELL and Lord Justice LEGGATT)

CRAGO v JULIAN

Estates Gazette May 2 1992

[1992] 17 EG 108

J

Landlord and tenant — Assignment — Whether a deed necessary for an effective assignment of an oral periodic tenancy — Held, a deed was necessary

From 1966 the appellant, Mrs Julian, and her husband lived in a flat which they rented on an oral weekly tenancy from the plaintiff landlord — The tenancy was originally held in the sole name of Mr Julian — In 1982 Mr and Mrs Julian were divorced but Mr Julian, although he left the flat, took no steps to transfer the tenancy — In their matrimonial proceedings Mr Julian gave an undertaking to do all acts and things as may be necessary to transfer to Mrs Julian the tenancy of the flat — K
No steps were taken as they assumed that the effect of that undertaking was to transfer Mr Julian's interest in the flat to his wife — Mrs Julian remained in the flat and paid the rent — In October 1987, following a request by Mrs Julian, the landlord's agents refused to transfer the tenancy into her name and refused to accept further payments of rent — In February 1988 the landlord served notice to quit on Mr Julian terminating his tenancy and, because he was no longer in possession, he did not become entitled to any statutory protection — In May 1990, following proceedings issued by the landlord, Judge MacNair (in Lambeth County Court) made an order for possession — Mrs Julian appealed L

***Held*: Section 53(1)(*a*) of the Law of Property Act 1925 provides an insuperable obstacle for a person in the position of Mrs Julian — The tenancy had not been disposed of in writing — The statutory language leaves no room for reaching a different conclusion — The exception in section 52(2)(*d*) of the 1925 Act that no deed is required for the creation of leases or tenancies not required by law to be made in writing does not apply to an assignment of a lease — Accordingly, to be effectual as a legal assignment, a transfer of Mr Julian's tenancy should have been by way of deed**

The following cases are referred to in this report. M

Botting v *Martin* (1808) 1 Camp 317
Pocklington (Thomas) Gift Trustees v *Hill* [1989] 2 EGLR 97; [1989] 36 EG 87; (1989) 21 HLR 391
Rye v *Rye* [1962] AC 496; [1962] 2 WLR 361; [1962] 1 All ER 146, HL

This was an appeal by the defendant, Mrs Julian, who lived in a first-floor flat at 55 Hollingbourne Road, London SE24. The flat had been let on a weekly tenancy by the plaintiff, Mrs Crago, to Mrs Julian's former husband, Mr Julian.

David Phillips (instructed by Philcox Gray & Co) appeared for the appellant; Guy Fetherstonhaugh (instructed by Stafford Young Jones) represented the respondent.

Giving judgment, **Sir Donald Nicholls V-C** said: Mrs Julian has lived in a rented flat on the first floor, 55 Hollingbourne Road, London SE24, since early 1966. She lived there with her husband, and subsequently with their children, from the time of their marriage in March 1966. The flat was rented on a weekly tenancy from the plaintiff, Mrs Crago. The flat was the Julians' matrimonial home and, although it was in the sole name of Mr Julian, they assumed that the tenancy belonged to them jointly.

In 1982 Mr and Mrs Julian were divorced. Mr Julian left the flat and had nothing further to do with it. In the matrimonial proceedings he gave an undertaking:

> . . . within 14 days of the granting of the Decree Absolute herein to do all acts and things as may be necessary to transfer to the Petitioner the tenancy of the flat situate at 55 Hollingbourne Road, London SE24 and which Tenancy is presently vested in his sole name.

Nothing was done to transfer the tenancy. Mr and Mrs Julian assumed that nothing further needed to be done. They assumed that the effect of the document was to transfer Mr Julian's interest in the flat to his wife.

Mrs Julian stayed in the flat. The two children grew up and eventually left. Mrs Julian paid the rent. She alone dealt with the managing agents. She spent substantial sums of money on the flat in the belief that she was a protected tenant. New rent books were issued by the agents, one in February 1983 and another in November 1986, both naming Mr Julian as the sole tenant. In 1987 Mrs Julian experienced some difficulty with housing benefit. She wrote to the managing agents and asked them to change the name of the tenant shown in the rent book from her former husband to herself. She thought there would be no difficulty about this. She wrote:

> Could I now have the tenancy (of 55 Hollingbourne Rd) changed over to my name please. If you still need a letter from my ex-husband, it could be arranged. . . . P.S. Here is a copy of the divorce paper stating transferring the tenancy to me. Thanking you.

That was on October 8 1987.

Unfortunately for Mrs Julian, the agents were not willing to do this. They refused to accept further payments of rent from her. Until they received this letter the agents did not know that Mr Julian had moved away and given up all interest in the tenancy. A few months later they served on Mr Julian a notice to quit dated February 18 1988, terminating his tenancy from March 14 1988. Because he was no longer in possession, he did not become entitled to any statutory protection. Proceedings against Mrs Julian, seeking possession of the flat, were begun by Mrs Crago in Lambeth County Court on March 28 1988. The trial took place before Judge MacNair and he gave judgment in favour of the lessor, Mrs Crago, on May 17 1990.

There were several issues before the judge. He held, first, that the contractual tenant was Mr Julian alone; second, that the tenancy was never assigned to Mrs Julian; third, that the tenancy was not determined until the expiry of the notice to quit in March 1988; and, fourth, that the plaintiff was not estopped, by the conduct of herself or her agents, from claiming possession from Mrs Julian. On this appeal, brought by Mrs Julian, only the second of those four determinations is being challenged. That is the only issue before us in this court. Counsel for Mrs Julian accepted that for him to succeed on this appeal he must establish that, before the notice to quit expired, Mr Julian had assigned the tenancy to his former wife.

On the issue of assignment the judge found that the tenancy did not include a term prohibiting assignment. But he decided against Mrs Julian on two grounds: first, that as a matter of law, the tenancy could be assigned only by deed and, second, that even if an assignment could be made orally, there were here no words or conduct which could give rise to an assignment. The judge reached his overall conclusion with reluctance. He said that the consequences for Mrs Julian, if she lost, were very serious. No reason had been given by Mrs Crago for wishing to get Mrs Julian out of the flat which for so long had been her home and Mrs Julian's lack of statutory protection stemmed simply from these parties, inexperienced in legal matters, not having executed a formal deed of assignment of the tenancy. Had either Mrs Julian or her former husband realised a deed was necessary, there would have been no difficulty in getting one drawn up by a solicitor and executed by them.

I turn, first, to the question of law: could this tenancy be assigned effectively only by deed? I can start with the summary of the law set out in *Megarry and Wade on the Law of Real Property*, 5th ed, p665:

> A legal lease, once created, can be transferred *inter vivos* only by deed, in accordance with the general rule. This applies to all legal leases, even those created orally, eg a yearly tenancy. However, on principles similar to those applicable to the creation of leases, an oral or written assignment will be effective in equity as between the assignor and the assignee as a contract to assign, if sufficiently evidenced by writing or part performance.

In the present case nothing less than an assignment at law will assist Mrs Julian. The view stated in *Megarry and Wade* is also expressed in all the other leading textbooks on this subject. Undaunted, Mr Phillips contended that these views are erroneous. The general rule mentioned by *Megarry and Wade* is a reference to certain provisions in the Law of Property Act 1925. Mr Phillips submitted that, properly interpreted, these provisions do not preclude a valid assignment otherwise than by deed of a lease which has been created orally. The statutory provisions are to be found in Part II of the Law of Property Act 1925. The material parts of the relevant sections, sections 52 to 54, provide:

> **52.**—(1) All conveyances of land or of any interest therein are void for the purpose of conveying or creating a legal estate unless made by deed.
> (2) This section does not apply to—
> (*a*) assents by a personal representative;
> (*b*) disclaimers made in accordance with section 54 of the Bankruptcy Act 1914 or not required to be evidenced in writing;
> (*c*) surrenders by operation of law, including surrenders which may, by law, be effected without writing;
> (*d*) leases or tenancies or other assurances not required by law to be made in writing;
> (*e*) receipts not required by law to be under seal;
> (*f*) vesting orders of the court or other competent authority;
> (*g*) conveyances taking effect by operation of law.
>
> **53.**—(1) Subject to the provisions hereinafter contained with respect to the creation of interests in land by parol —
> (*a*) no interest in land can be created or disposed of except by writing signed by the person creating or conveying the same, or by his agent thereunto lawfully authorised in writing, or by will, or by operation of law;
>
> **54.**—(1) All interests in land created by parol and not put in writing and signed by the persons so creating the same, or by their agents thereunto lawfully authorised in writing, have, notwithstanding any consideration having been given for the same, the force and effect of interests at will only.
> (2) Nothing in the foregoing provisions of this Part of this Act shall affect the creation by parol of leases taking effect in possession for a term not exceeding three years (whether or not the lessee is given power to extend the term) at the best rent which can be reasonably obtained without taking a fine.

This is a hotchpot of sections reproducing, with some amendments, provisions which before 1925 were to be found in the Statute of Frauds 1677 and the Real Property Act 1845. I consider section 53 first. The grant of a tenancy is the creation of an interest in land. An assignment of a tenancy of land is the disposal of an interest in land. Thus, both the grant of a tenancy and the assignment of a tenancy fall four-square within section 53(1)(*a*). A tenancy of land cannot be created or assigned save by writing, or by will, or by operation of law. Para (*a*), however, is expressed to be "subject to the provisions hereinafter contained with respect to the *creation* of interests in land by parol" (emphasis supplied). Section 54(2) is such a provision. Section 54 (2) provides that the statutory requirements of Part II of the Act do not affect "the creation" by parol of leases taking effect in possession for a term not exceeding three years at a full market rent. I pause to observe that, thus far, there can be no doubt as to the meaning of the statutory provisions: interests in land cannot be created or disposed of except in writing, but a lease in possession at a market rent for up to three years can be validly created orally. Hence, and this is to be noted, the effect of these statutory provisions is indubitably to draw a distinction between the manner in which a short lease may be created and the manner in which it may be assigned.

The point is not wholly free from authority. Section 53(1)(*a*) derives from section 3 of the Statute of Frauds and section 54 derives from sections 1 and 2. In *Botting* v *Martin* (1808) 1 Camp 317, Serjeant Best argued that by the leases mentioned in section 3, as requiring to be assigned in writing, must be intended such leases as are required by sections 1 and 2 to be created in writing, viz, leases conveying a larger interest to the party than a term for three years. He submitted:

> As a lease from year to year could be originally made by parol, there was no reason why it might not be assigned by parol; and the words of the statute would bear this interpretation, which was clearly consistent with its general import.

McDonald CB rejected the submission.

A In my view, section 53 provides an insuperable obstacle for Mrs Julian in the present case. Like McDonald CB, I consider the statutory language really leaves no room for reaching a different conclusion, even though the result is the curious distinction I have mentioned. Even if there were room for doubt I would be slow to upset an interpretation of statutory provisions which has been accepted by conveyancers for a very long time.

This conclusion is sufficient to dispose of this appeal, because here there was no assignment in writing. However, I do not think I should leave the arguments of law there, given the existence also of section 52. One of the arguments advanced by Mr Phillips, if correct, would have the effect that, although section 53(1)(*a*) requires writing for the
B assignment of an orally created tenancy, section 52 does not apply. Such an assignment does not require to be made by deed.

This submission raises a point on the interpretation of section 52, which seems not to have been the subject of decision. I see force, as a matter of common sense and practicality, in the arguments of Serjeant Best in *Botting* v *Martin*. But, as I have already said, the language of sections 53 and 54 does not admit of that interpretation. Quite apart from section 52, the law imposes a formal requirement on the assignment of an orally created lease which does not apply to its creation. With that in mind, I can see no justification for interpreting the exception in section 52(2)(*d*) as applicable to the assignment of leases not required to be in writing as well as to their creation. Section 52 (1) must be read with the definition of "conveyance" in section
C 205. It bites on particular types of transactions effected by instruments in writing (see *Rye* v *Rye* [1962] AC 496): conveyances, mortgages, leases and so forth. They are void, for the purpose of conveying or creating a legal estate, unless made by deed. Section 52(2) then excepts from the requirement of a deed certain instruments: for example, assents by a personal representative, certain disclaimers, surrenders by operation of law and (para (*d*)) leases or tenancies not required by law to be made in writing. That is the ambit of the exception. In the case of leases, the exception does not touch anything other than the lease itself. Subsequent dispositions of the lease, whether by way of mortgage, charge, assignment or otherwise, are not within para (*d*), unless they themselves qualify as "other assurances not required by law to be
D made in writing". An assignment of a lease does not so qualify.

I am confirmed in this construction of section 52 by the terms of its ancestor section, section III of the Real Property Act 1845. So far as material, this provided as follows (for ease of reading I have numbered the separate limbs):

> . . . [1] a lease, required by Law to be in Writing, of any Tenements or Hereditaments, and [2] an Assignment of a Chattel Interest . . . in any Tenements or Hereditaments, and [3] a Surrender in Writing of an Interest in any Tenements or Hereditaments . . . not being an Interest which might by Law have been created without Writing . . . shall also be void at Law, unless made by Deed . . .

A "chattel interest" in land is a reference, in modern terminology,
E to a leasehold interest. This harks back to the old classification whereunder an interest in land for a term of years was classed as a chattel: a chattel real. Again, the wording of this section leaves no room for doubt. A clear distinction is drawn between leases and assignments of leases. In the former case, but not the latter, there is an exception for leases not required by law to be in writing.

In my view, therefore, the law concerning the need for a deed is correctly stated in *Megarry and Wade*. To be effectual as a legal assignment, a transfer of Mr Julian's tenancy required to be by way of deed.

Mr Phillips sought to place some reliance on the decision of this court in *Thomas Pocklington's Gift Trustees* v *Hill* (1989) 21 HLR 391.* There the court based its decision on a finding of an oral
F assignment of a periodic tenancy. But, so far as one can see, the attention of the court was not drawn to the provisions of the Law of Property Act. No argument seems to have been addressed to the court on sections 52 to 54, nor is the need for a deed mentioned or alluded to in the judgments. This case, therefore, is in no sense an authority on the interpretation of those sections. All that can be said is that it seems to have been assumed on all sides that a deed was not necessary. But that is not a decision by the court on the point.

This appeal, therefore, must fail. The judge's decision was correct. Accordingly, I shall not prolong this judgment by considering whether there was a purported oral assignment in the present case, beyond observing that I should not be taken as necessarily agreeing with the judge's conclusion on this issue. As it is, however, I would G dismiss this appeal. Mrs Julian never became the tenant. Of course, a deed would not have been necessary if Mrs Crago or her agents had expressly or impliedly *agreed* to Mrs Julian's becoming the tenant and had accepted her as the tenant in place of her former husband. There could then have been a new tenancy, which could have been created orally. Unfortunately for Mrs Julian, that never happened.

*Editor's note: Also reported at [1989] 2 EGLR 97.

Russell and **Leggatt LJJ** agreed and did not add anything.

Court of Appeal

December 5 1991 H

(Before Lord Justice GLIDEWELL, Lord Justice BALCOMBE and Lord Justice BELDAM)

NORWICH UNION LIFE INSURANCE SOCIETY v LOW PROFILE FASHIONS LTD AND OTHERS

Estates Gazette May 30 1992
[1992] 21 EG 104

Landlord and tenant — Liability of original tenant following assignment — Whether term can be implied into lease that landlord should take care in consenting to assignment — J Whether landlord should exhaust all other remedies before pursuing his remedy against original tenant

By an underlease dated February 6 1981 the respondent landlord granted the appellant a term of 20 years from September 29 1980 of the first floor of Moray House, 23-31 Great Titchfield Street, W1, at an annual rent of £8,550 — From September 29 1985 the rent was increased to £10,000 pa — In addition to the rent the appellant covenanted to pay a yearly sum of £1,350, representing a proportion of insurance and service charges — The appellant covenanted not to assign the premises without the written consent of the landlord, such K consent not to be unreasonably withheld with the proviso that the landlord may require the proposed assignee to enter into direct covenants for the performance of the lessee's covenants — On February 12 1982 the appellant, with the consent of the landlord, assigned the underlease to the first assignee and on July 13 1983 the first assignee assigned the underlease to the second assignee, again with the consent of the landlord — In each case Mr Ronald Swinson, the second defendant, a director of the first and second assignees, acted as surety and entered into direct covenants with the landlord — Following non-payment of the rent and service charges on April 10 1987 the respondent issued proceedings against the second assignee L to recover rent and other payments due — On May 12 1988 the second assignee assigned the underlease to a third assignee, with the landlord's consent, in consideration of the payment of a sum which discharged in part the outstanding liability to the landlord — Following the compulsory winding up of the second assignee in October 1988, the respondent claimed the balance of arrears of rent of £27,351 from the appellant as original tenant — On appeal from a decision of Mr Patrick Bennett QC (sitting as a deputy judge of the Queen's Bench Division), who had dismissed an appeal from an order of deputy Master Foster giving judgment under Ord 14 of the M Rules of the Supreme Court for the respondent, the appellant contended: (1) that there should be implied into the underlease the term that, in relation to its dealings with the second or subsequent assignees of the lease, the landlord would take reasonable care to ensure that such an assignee was financially able to meet the rental and service payments due under the underlease; and (2) that, since the tenants' continuing liability to the landlord arose from privity of contract, the court should apply an equitable doctrine restraining the landlord from pursuing its remedy against the tenant when, because of the existence of alternative remedies against the second assignee

A **and the surety, it was wholly unreasonable for the landlord to pursue its remedy against the tenant**

***Held*: The appeal was dismissed — In seeking to imply the term contended for, the tenant was seeking to equate business efficacy with the elimination of the hardship which could arise from its being called upon to comply with the express covenants after it has parted with the lease — There was no ground for implying such a term — In a rare case in which it can be shown that a contracting party has no legitimate interest, financial or otherwise, in insisting that he has a right to perform the contract rather than to claim damages, equity**
B **can intervene if the ordinary rules requiring a party to mitigate his loss do not protect the party in breach from having to pay more than the proper measure of damage; but any such principle is of no relevance in the present case when the remedies available to the landlord are against different persons and are cumulative and not alternative**

The following cases are referred to in this report.

Baynton v *Morgan* (1888) 22 QB 74
Clea Shipping Corp v *Bulk Oil International Ltd (No 2) "the Alaskan Trader"* [1984] 1 All ER 129; [1983] 2 Lloyd's Rep 645
Liverpool City Council v *Irwin* [1977] AC 239; [1976] 2 WLR 562; [1976] 2 All ER 39; (1976) 74 LGR 392; [1976] EGD 282; 238 EG 879, HL
C *White & Carter (Councils) Ltd* v *McGregor* [1962] AC 413; [1962] 2 WLR 17; [1961] 3 All ER 1178, HL

This was an appeal against a decision of Mr Patrick Bennett QC (sitting as a deputy judge of the Queen's Bench Division) on November 6 1990, who had dismissed the appellant's appeal against the order of deputy Master Foster of September 24 1990 that judgment under Ord 14 of the Rules of the Supreme Court be entered for the plaintiff respondent, Norwich Union Life Insurance Society, against the appellant defendant, Low Profile Fashions Ltd.

Karen Troy-Davies (instructed by Philip Ross & Co) appeared for the appellant; Robert Pearce (instructed by the solicitor to Norwich
D Union Life Insurance Society) represented the respondent.

Giving the first judgment at the invitation of Glidewell LJ, **Beldam LJ** said: The appellant, the first defendant in the action, appeals from the judgment of Mr Patrick Bennett QC, sitting as a deputy judge of the Queen's Bench Division on November 6 1990, dismissing the appellant's appeal against the order of deputy Master Foster of September 24 1990, that judgment under Ord 14 of the Rules of the Supreme Court be entered for the plaintiff respondent against the appellant, with damages to be assessed, and that the first defendant pay to the plaintiff within 14 days the sum of £10,000 and from an order that the appellant pay the respondent's costs of the
E appeal and before the Master.

In this appeal the appellant seeks unconditional leave to defend the action and an order that its costs of appeal and below be paid by the respondent. The learned deputy judge dismissed the appellant's appeal on the ground that it had shown no arguable defence to the respondent's claim. That claim was for arrears of rent and interest due under the terms of a lease entered into by the appellant on February 6 1981.

The respondent (to whom I shall refer as "the landlord") was entitled to possession of the first floor of Moray House, 23-31 Great Titchfield Street, W1, under the terms of a lease dated May 1 1928. By an underlease of February 6 1981, the landlord demised the premises to the appellant (to whom I shall refer as "the tenant") for the term of
F 20 years from September 29 1980 at an annual rent of £8,550. From September 29 1985 the rent was to be increased to a sum of £10,000 pa. The rent was payable by equal quarterly instalments in advance, on the usual quarter days in every year. In addition, the tenant covenanted to pay additional rent as and when demanded in the yearly sum of £1,350, plus VAT, representing a proper proportion of insurance and service charges to the common benefit of tenants in the building. Further, the tenant covenanted not to assign the premises without the written consent of the landlord (such consent not to be unreasonably withheld) with the proviso that the landlord "may require the proposed assignee to enter into direct covenants with the landlord to perform all the covenants herein contained and on the tenant's part to be performed and observed".

The underlease provided for a right of re-entry and forfeiture in the G
event of breach of covenant by the tenant and, if the rent reserved was more than 14 days in arrears, for the payment of interest at the rate of 5% above minimum lending rate by the tenant.

On February 12 1982 the tenant, with the consent of the landlord, assigned the underlease to a limited company, Al-Awad Swinson International Ltd, to whom I shall refer as "the first assignee".

On July 13 1983 the first assignee, again with the consent of the landlord, assigned the underlease to a second assignee, Al-Mehry Swinson International Ltd. Mr Ronald Swinson, the second defendant in the action, was a director of both the assignee companies and acted as surety on the assignments to them, entering into direct covenants with the landlord. I shall refer to Mr Swinson as H
"the surety".

After July 1983 the second assignee continued to pay the rent and other outgoings until March 25 1986, when the rent due of £2,500 on that quarter day was unpaid. On September 10 1987 the landlord made a demand for the additional rent, the proportion of insurance and other charges in respect of earlier periods. Neither these nor any further payments of rent were made by the second assignees. Accordingly, on April 10 1987 the landlord issued proceedings against the second assignee to recover the rent and other payments due.

By May 1988 the second assignee had apparently found a purchaser for the residue of the term of the sublease and on May 12
1988, with the landlord's consent, it assigned the lease to a third J
assignee in consideration of a payment of £22,500. That payment enabled the second assignee to discharge part of its outstanding liability to the landlord by a payment of £20,000, thus reducing the amount of the rent and other payments outstanding. It is not clear whether the proceedings against the second assignee were thus compromised or were simply not further pursued. Nor is it clear whether the surety was thereby relieved of further liability, but for the purpose of these proceedings I assume that any payments made by the second assignee did not affect the outstanding liability either of the second assignee or of the surety.

On October 5 1988 the second assignee was wound up compulsorily and on November 7 the landlord, by letter, claimed the
balance of arrears of £27,351 from the tenant. The writ in the action K
against the tenant was issued on August 7 1989 and the pleadings closed in December 1989. The tenant delivered further and better particulars of its defence in April 1990. On May 14 the landlord issued its summons seeking judgment under Ord 14.

By its defence the tenant contended that the landlord, when granting the licence to the first assignee to assign to the second, owed the tenant a duty of care properly to assess the credit-worthiness of the second assignee and to take proper and timely measures to enforce the covenants of the underlease against the second assignee. The tenant alleged that the landlord had failed to do so and therefore the tenant had suffered damage. It was pleaded that to allow the
landlord's action to proceed would be to permit circuity of action. It L
was conceded before us that the form of the defence was inadequate and, further, that the grounds upon which the tenant sought leave to defend were not properly reflected in its defence. The court was asked to consider the grounds of defence now put forward on their merits and if, as was contended, an arguable defence in law was disclosed, to grant leave to the tenant to amend the defence.

Before us the tenant argued:

(1) That there should be implied in the underlease a term that, in relation to its dealings with the second or subsequent assignees of the lease, the landlord would take reasonable care to ensure that such an assignee was financially able to meet the rental and service payments due under the underlease.

(2) That, since the tenant's continuing liability to the landlord M
arose from privity of contract, the court should apply an equitable doctrine restraining the landlord from pursuing its remedy against the tenant when, because of the existence of alternative remedies against the second assignee and the surety, it was wholly unreasonable for the landlord to pursue its remedy against the tenant.

It was conceded by the tenant that there was no decided case in which either of these defences had been held to be available to a tenant who, it was accepted, would otherwise be liable on the covenants in the lease.

In support of the first ground, Mrs Troy-Davies, counsel for the appellant, referred us to the well-known passage in the judgment of

A Lord Wilberforce in *Liverpool City Council* v *Irwin* [1977] AC 239 at p 253. In that case the agreement between the parties had only partially been stated in writing and it was necessary to complete the agreement and in particular to give it a bi-lateral character to imply some terms. As Lord Wilberforce stated at p 254A:

> The court here is simply concerned to establish what the contract is, the parties not having themselves fully stated the terms. In this sense the court is searching for what must be implied.

By contrast, the lease in the present case was a comprehensive document clearly intended by the parties to include all the terms agreed and to constitute the entire agreement. The term the appellant seeks to imply is one said to be necessary to give business efficacy to
B the contract. But, quite apart from the difficulty of implying any term into this detailed lease, it is not, in my view, established that the term proposed is in any way necessary for that purpose. Leases in similar form have for many years been created and their covenants performed notwithstanding the absence of such a term, which would operate in only very limited circumstances. It would not serve to modify the tenant's obligations in any way but at best would give the tenant the opportunity to intervene on the granting of licence to assign or subsequently when an assignee had defaulted. In practice, no steps have been suggested which it would be open to the landlord to take that he would not already have taken in his own interest. To avoid subsequent criticism of his steps a landlord might feel constrained to seek the consent of the tenant for any subsequent
C assignment and this would clearly lead to greater delay in the granting of consent. It would be likely to undermine the purpose of the Landlord and Tenant Act 1988, which was to speed up transactions of this kind. In the case of long leases when several assignments may have taken place, including some of the landlord's right to the reversion, even further difficulties can be envisaged in the operation of such a term.

It was argued in *Baynton* v *Morgan* (1888) 22 QB 74 that the contingent liability of the tenant on his personal covenant should be construed as a contract of suretyship, but in the light of the terms in which the obligation in that case was expressed, Lord Esher said it could be contended to be only a contract of suretyship by virtue of
D some implication, and he added at p 77:

> I see nothing that authorises us to make any such implication. It was suggested that, unless the contract were treated as one of suretyship, some hardship might arise. I do not think that such considerations afford any ground for altering the effect of the plain words of the covenant. If there were any hardship, it would be one arising out of the terms of the contract which the parties have chosen to make.

The same is equally true of the term we are asked to imply into the lease in the present case and I see no more reason to do so than did Lord Esher. It is to be observed that, in the proviso to the covenant against assignment, the tenant agreed to leave it to the discretion of the landlord whether to require that an assignee should enter into a direct covenant with the landlord, whereas the tenant could have
E stipulated for a requirement that the landlord must do so and thus have ensured at least the continuing liability of the first assignee whose financial strength the tenant had itself been able to verify. In truth the tenant seeks to equate business efficacy with the elimination of the hardship which could arise from its being called upon to comply with the expressed covenants after it has parted with the lease — no doubt for a suitable consideration. The hardship, if any, is considerably less than in the instances given by the Law Commission in 1986 in their Working Paper no 95, para 2.18. I would therefore reject the implication of the term suggested.

The second arguable defence for which the tenant contended was based upon the observations of Lord Reid in *White & Carter (Councils) Ltd* v *McGregor* [1962] AC 413, at pp 427-431. After
F stating the general rule that an innocent party to a contract is not bound in law to accept a repudiation of it and sue for damages for breach (whether or not the time for performance had arrived) Lord Reid considered a case in which the innocent party could completely fulfil the contract without any co-operation from the party in breach. At p 431 he said:

> It may well be that, if it can be shown that a person has no legitimate interest, financial or otherwise, in performing the contract rather than claiming damages, he ought not to be allowed to saddle the other party with an additional burden with no benefit to himself.

We were referred to the decision of Lloyd LJ in *The "Alaskan Trader"* [1984] 1 All ER 129 in which he reviewed the way in which the principle referred to by Lord Reid has on occasion been applied and we are asked to say that in the present case the landlord ought not G
in equity to be allowed to pursue his remedy against the tenant before he has exhausted other remedies available to him.

No doubt in the rare case in which it can be shown that a contracting party has no legitimate interest, financial or otherwise, in insisting that he has a right to perform the contract rather than to claim damages, equity can intervene if the ordinary rules requiring a party to mitigate his loss do not protect the party in breach from having to pay more than the proper measure of damage. I cannot myself see that any such principle is of relevance in the present case, when the remedies available to the landlord are against different persons and are cumulative and not alternative. I can see no equity in requiring the landlord to pursue one rather than another means of seeking to recover his loss or requiring him to run the risk that by H
pursuing one party he may jeopardise his chances of recovering against another. It has long been the law that the landlord may sue either the original lessee or the assignee, or both at the same time. I can see no justification for trying to import into the law of landlord and tenant long-since-settled concepts developed to meet a specific difficulty in a different class of case altogether.

I would thus hold that neither defence is arguable in law and that it is unnecessary to consider the evidence before the court. I would say, however, that after reading the evidence I entertained no doubt that on the facts disclosed there was no conceivable basis for saying that the respondents had either failed to exercise proper care in granting a licence to assign to the first or any subsequent assignee or that they J
failed to take all proper steps to mitigate the amount for which the tenant might be liable under its covenants.

The answer to the tenant's complaint of perceived injustice lies, in my view, in reform by legislation. The manner in which that should be done was recommended by the Law Commission in their report, *Landlord and Tenant Law, Privity of Contract and Estate* (Law Com no 174). The first principle on which the recommendations were based was that a landlord or a tenant of property should not continue to enjoy rights, or be under any obligation arising from a lease, once he has parted with all interest in the property. A considered recommendation has the advantage over attempted development on a case-by-case basis, that it is made after widespread consultation with parties who may be affected by changes in the law concerned — K
an advantage which is denied to this or any other court. Implementation of the commission's recommendation in this field would certainly obviate the need for the kind of elaborate argument developed in the present case for seeking to avoid a tenant's continuing liability.

I hope the present case will provide added impetus for the implementation of the recommendation for reform.

I would dismiss the appeal.

Balcombe and **Glidewell LJJ** agreed and did not add anything.

Appeal dismissed with costs. Leave to appeal to the House of Lords refused.

L

Chancery Division

March 2 1992

(Before Mr Justice CHADWICK)

DALLHOLD ESTATES (UK) PTY LTD v LINDSEY TRADING PROPERTIES INC

Estates Gazette June 13 1992

[1992] 23 EG 112 M

Landlord and tenant — Rent — Whether section 48 of the Landlord and Tenant Act 1987 applies to an agricultural holding — Whether any notice given pursuant to that section — Whether effective notice to pay served pursuant to Case D under the Agricultural Holdings Act 1986

The plaintiff, Dallhold Estates (UK) Pty Ltd, is the tenant of Upp Hall, a 940-acre agricultural holding in Hertfordshire, holding under a five-year lease from July 30 1987 at a yearly rent of £50,000 payable by equal quarterly payments in advance on the four usual quarter days — The defendant,

Lindsey Trading Properties Inc, is the owner of the freehold reversion to the lease — The plaintiff failed to pay rent in respect of the quarter days June 24 1990 to and including June 24 1991 — By a notice dated August 7 1991 the defendant landlord's solicitors gave notice pursuant to Case D, in Part I of Schedule 3 to the Agricultural Holdings Act 1986 requiring the tenant to pay the arrears of rent within two months — On October 9 1991 the defendant's agents gave to the plaintiff a notice to quit the holding on July 30 1993 on the ground that the plaintiff had failed to comply with the notice to pay rent — A provisional liquidator appointed in respect of the plaintiff challenged the service of the notice to pay rent and requested the landlord's agents to state whether any notice under section 48 of the Landlord and Tenant Act 1987 had been served — By a further notice dated December 3 1991 the landlord's agents confirmed for the purposes of section 48 their client's address for service of all notices, including notices in proceedings, and enclosed a further notice to pay rent

By an originating summons the plaintiff tenant sought declarations (1) that by reason of section 48 of the Landlord and Tenant Act 1987 no rent was due from the plaintiff on August 7 1991 and December 3 1991; and (2) that the purported notices to pay rent dated August 7 1991 and December 3 1991 were invalid and of no effect

***Held*: Section 48 of the 1987 Act applies to premises which consist of or include a dwelling — The Upp Hall agricultural holding included a dwelling — The expression "premises" is intended to have a wider meaning than dwelling and means the subject-matter of a letting and therefore applies to an agricultural holding if it includes a dwelling — On the evidence, the defendant failed to give any notice pursuant to section 48 of the 1987 Act until the letter of December 3 1991 — In respect of an agricultural holding a notice to pay rent must comply with the Agricultural Holdings (Forms of Notice to Pay Rent or to Remedy) Regulations 1987 and require the tenant to pay within two months of the date of service of the notice the *rent due* — The letter of December 3 1991 was a valid notice under section 48 of the 1987 Act and on that date, and not before, rent in respect of earlier quarters became due — Accordingly the notice to pay rent of December 3 1991 failed to comply with the 1987 regulations in stating that it was in respect of rent due on the quarter days from June 24 1990 to September 29 1991**

The following cases are referred to in this report.

Bracey v *Read* [1963] Ch 88; [1962] 3 WLR 1194; [1962] 3 All ER 472
Dickinson v *Boucher* [1984] EGD 19; (1983) 269 EG 1159, CA
Maunsell v *Olins* [1975] AC 373; [1974] 3 WLR 835; [1975] 1 All ER 16, HL
Official Solicitor v *Thomas* [1986] 2 EGLR 1; (1986) 279 EG 407
Pickard v *Bishop* (1975) 31 P&CR 108; [1975] EGD 24; 235 EG 133, CA
Whitley v *Stumbles* [1930] AC 544; 46 TLR 555

This was a hearing of an originating summons issued by the plaintiff, Dallhold Estates (UK) Pty Ltd, against the defendant, Lindsey Trading Properties Inc.

Edward Bannister QC and Martin Rodger (instructed by Mills & Reeve) appeared for the plaintiff; David Neuberger QC and Joanne Moss (instructed by Franks Charlesly & Co) represented the defendant.

Giving judgment, **CHADWICK J** said: The plaintiff, Dallhold Estates (UK) Pty Ltd, is a company incorporated under the Companies (Western Australia) Code 1981. On August 9 1991 a provisional liquidator of the plaintiff company was appointed by an order of the Federal Court of Australia. On August 14 a provisional liquidator was appointed in England by an order made in this court. On January 24 1992 I made an administration order for the purposes of achieving a more advantageous realisation of the plaintiff company's assets than would be effected on a winding up. The circumstances in which that order was made appear from a judgment which I delivered on that day. They form part of the background to the present application.

The plaintiff is the tenant under a lease dated July 30 1987 of an agricultural and sporting estate comprising some 940 acres in Hertfordshire. The estate includes a large Elizabethan manor house known as Upp Hall and a number of cottages, some of which are occupied by estate workers under service tenancies. Upp Hall itself is occupied by Miss Suzanne Bond, the daughter of Mr Alan Bond, who was until August 9 1991 a director of the plaintiff company. It is common ground that the estate is an agricultural holding for the purposes of the Agricultural Holdings Act 1986.

The defendant, Lindsey Trading Properties Inc, is a company incorporated and having its registered office in Panama. Lindsey is owned and controlled by Mr Jorg Bolag, who is resident in Switzerland. Lindsey is entitled to the freehold reversion expectant upon the determination of the lease.

The lease is expressed to be for a term of five years from July 30 1987 at a yearly rent of £50,000 payable by equal quarterly payments in advance on the four usual quarter days to Franks Charlesly & Co, of Hulton House, London, or to such other agent as the landlord shall from time to time notify to the tenant.

No rent was paid by the tenant under the lease on June 24, September 29 or December 25 1990, or on March 25 and June 24 1991. On August 7 1991 Franks Charlesly, acting as solicitors and authorised agents for Lindsey, gave notice to the tenant requiring that rent be paid within two months of that date. The notice was expressed to be given in accordance with Case D in Part I of Schedule 3 to the 1986 Act. The notice dated August 7 1991 was served at the registered office of the tenant at 201 Adelaide Terrace, Perth, Western Australia. Payment of rent was not made in response to that notice or at all. No rent was paid by the tenant on September 29 1991.

On October 9 1991, a little over two months after the date of the notice to pay rent, Franks Charlesly, as agents for the landlord, gave notice to the tenant to quit the estate on July 30 1993. The notice to quit was expressed to be given on the grounds that the tenant had failed to comply with the notice to pay rent dated August 7 1991.

Section 26(1) of the Agricultural Holdings Act 1986 imposes a restriction on the operation of a notice to quit an agricultural holding. Provided that the tenant gives the appropriate counternotice within one month, the notice to quit shall not have effect unless on an application by the landlord consent to its operation is given by an agricultural land tribunal. That restriction does not apply in any of the cases set out in Part I of Schedule 3 to the Act. In particular, it does not apply in Case D; that is to say where, at the date of the giving of the notice to quit, the tenant has failed to comply, *inter alia,* with a notice to pay rent which satisfies the conditions in para (a) of that case. The effect is that, on a failure to pay rent within two months of the service of a valid notice to pay, the tenant will lose the protection of the Agricultural Holdings Act 1986: see *Pickard* v *Bishop* (1975) 31 P&CR 108* and, in particular, the remarks of Lord Denning MR at p 109.

The tenant's initial response, through its provisional liquidator, was to challenge service of the notice to pay rent. Subsequently, by letter dated October 31 1991, the tenant's solicitors asked Franks Charlesly whether any notice under section 48 of the Landlord and Tenant Act 1987 had been served and, if so, for a copy of that notice. On the same day the tenant's solicitors served a counternotice under section 26(1) of the Agricultural Holdings Act 1986 in respect of the notice to quit dated October 9. Neither party has sought to argue before me that the service of that counternotice prevents this court from considering the questions raised in these proceedings.

Section 48 of the Landlord and Tenant Act 1987 is in these terms so far as material:

> (1) A landlord of premises to which this Part applies shall by notice furnish a tenant with an address in England and Wales at which notices (including notices in proceedings) may be served on him by the tenant.
>
> (2) Where a landlord of any such premises fails to comply with subsection (1), any rent or service charge otherwise due from the tenant to the landlord shall, (subject to subsection (3)) be treated for all purposes as not being due from the tenant to the landlord at any time before the landlord does comply with that subsection.

Subsection (3) has no application to the present case.

Failure to serve a notice under section 48 of the 1987 Act in cases to which that section applies has the effect that rent which would otherwise be due is to be treated for all purposes as not being due. It would follow in such a case that the landlord would be unable to rely upon a notice to pay rent for the reason that the court would be

*Editor's note: Reported also at (1975) 235 EG 133.

A required to hold that the rent demanded in the notice to pay was not due.

Faced with the tenant's request for a copy of the notice under section 48 of the 1987 Act upon which the landlord relied, and without admitting that no such notice had been served, the landlord's solicitors served a further notice to pay rent dated December 3 1991. The notice was served under cover of a letter of that date. The final paragraph of that letter is in these terms:

> We would confirm on behalf of our client and for the purposes of section 48 aforesaid that our client's address for service of all notices, including notices in proceedings, is care of this firm.

It is common ground that that letter does itself constitute an
B effective notice for the purposes of section 48 of the 1987 Act. It is not, however, accepted by the tenant that the notice to pay rent served on the same day is in a form upon which the landlord can rely for the purposes of Case D in Part I of Schedule 3 to the Agricultural Holdings Act 1986.

In these circumstances, shortly after the making of the administration order, on January 24 1987, the tenant commenced proceedings by originating summons for a declaration that no notice under section 48 of the Landlord and Tenant Act 1987 had been served by the landlord prior to the service of the notice to pay rent dated December 3 1991. The originating summons was subsequently amended and the declarations now sought are in these terms:

> (1) a declaration that on a true construction of section 48 of the Landlord
> C and Tenant Act 1987 no rent was due from the plaintiff to the defendant under the lease dated 30th July 1987 . . . on (a) 7th August 1991 and (b) 3rd December 1991.
>
> (2) a declaration that on a true construction of section 48 of the Landlord and Tenant Act 1987 and of Case D of Schedule 3 to the Agricultural Holdings Act 1986 and/or of the Agricultural Holdings (Forms of Notices to pay Rent or to Remedy) Regulations 1987, notices dated 7th August 1991 and 3rd December 1991 purporting to be notices to pay rent served on the plaintiff by the defendant were invalid and of no effect.

Section 48 of the Landlord and Tenant Act 1987 is contained in the group of six sections which forms Part VI to that Act. The first of those sections, section 46, is in these terms:

> (1) This Part applies to premises which consist of or include a dwelling and
> D are not held under a tenancy to which Part II of the Landlord and Tenant Act 1954 applies.
>
> (2) In this Part "service charge" has the meaning given by section 18(1) of the 1985 Act.

In this context the 1985 Act means the Landlord and Tenant Act 1985.

The primary question in these proceedings is whether the word "premises" in section 46(1) of the 1987 Act is to be given what Lord Diplock and Lord Simon described, in their dissenting judgment in *Maunsell* v *Olins* [1975] AC 373 at p 392, as the meaning which it bears in ordinary legal parlance, namely the subject-matter of a letting, or whether it is to be given a more limited meaning which would not extend to the agricultural holding let under the lease dated
E July 30 1987.

It is I think clear that the word "premises" is capable of a wide range of meanings. In strict conveyancing language it means ". . . everything in a deed which precedes the habendum . . ." — see *per* Lord Wilberforce in *Maunsell* v *Olins* at p 386B; or ". . . the subject-matter of the habendum in a lease . . ." *per* Viscount Hailsham in *Whitley* v *Stumbles* [1930] AC 544 at p 546; *per* Cross J in *Bracey* v *Read* [1963] Ch 88 at p 92; and *per* Lord Diplock and Lord Simon in *Maunsell* v *Olins* at p 392B.

In a popular sense, in the appropriate context, the word may mean "buildings" or "dwelling-houses", including in the latter expression houses let with a garden or yard, garage or paddock — see *per*
F Viscount Dilhorne in *Maunsell* v *Olins* at p 384B-D and *per* Lord Wilberforce at p 389A-B. Both of those members of the House of Lords had no doubt that the word would never be used in a popular sense to mean or include farmland — see at p 383H and p 386G-H. Cross J took a similar view in *Bracey* v *Read* at p 92, where he said:

> I do not think that in the popular sense anybody would say that some gallops on a downland with no building on or near them could be described as premises.

Nevertheless, Cross J held in *Bracey* v *Read* that property of that nature was within the meaning of "premises" for the purposes of section 23(1) of the Landlord and Tenant Act 1954.

The question which I have to decide is what, within the wide spectrum of meanings which the word "premises" can bear, is the meaning which it was intended to have for the purposes of Part VI of G the Landlord and Tenant Act 1987. That question must be answered by construing the language which Parliament has used in that Act. It is wrong to start from the position that the word "premises" has some *prima facie* meaning which must prevail in the absence of sufficient reason to the contrary: see *per* Lord Wilberforce in *Maunsell* v *Olins* at p 386. It is necessary, first, to look at the Act as a whole, in order to see whether some overall statutory purpose can be identified, from which the word "premises" in section 46(1) may take its meaning.

The long title to the Act of 1987 describes its purpose in these terms:

> An Act to confer on tenants of flats rights with respect to the acquisition by H them of their landlord's reversion; to make provision for the appointment of a manager at the instance of such tenants and for the variation of long leases held by such tenants; to make further provision with respect to service charges payable by tenants of flats and other dwellings; to make other provision with respect to such tenants; to make further provision with respect to the permissible purposes and objects of registered housing associations as regards the management of leasehold property; and for connected purposes.

As that long title suggests, Parts I, II and III apply only to flats or to premises which consist of the whole or part of a building and which contain flats. In Parts I, II and III of the Act premises has a defined meaning which is clearly limited in this way. For example, in section 1(2) it is provided that:

> . . . this Part applies to premises if — J
> (a) they consist of the whole or part of a building; and
> (b) they contain two or more flats held by qualifying tenants;
> . . .

Sections 21(2) and 25(2) contain similar provisions in relation to Parts II and III of the Act.

Part IV of the Act (variation of leases) applies primarily to flats but extends to dwellings other than flats: see section 40 of the Act.

Part V applies not only to flats but also to other dwellings. The first section in that Part (section 41) is in these terms so far as material:

> (1) Sections 18 to 30 of the 1985 Act (regulation of service charges payable by tenants) shall have effect subject to the amendments specified in Schedule 2 (which include amendments —
> (a) extending the provisions of those sections to dwellings other than flats, K
> . . .)

Sections 18 to 30 of the Landlord and Tenant Act 1985 do regulate service charges payable by tenants of flats. Section 18(1) of the 1985 Act defines service charge (so far as material) to mean an amount payable by a tenant of a flat as part of or in addition to the rent which is payable directly or indirectly for services, repairs, maintenance or insurance for the landlord's cost of management.

The effect of Schedule 2 to the 1987 Act is to substitute the word "dwelling" for "flat" in section 18(1) and in each of the other relevant provisions in sections 18 to 30 of the 1985 Act. "Dwelling" in this context means a building or part of a building occupied or intended to be occupied as a separate dwelling, together with any yard, garden, outhouses and appurtenances belonging to it or usually L enjoyed with it: see section 38 of the 1985 Act. "Dwelling" has the same meaning in the 1987 Act: see section 60(1).

The effect, therefore, of Part V of the 1987 Act is to extend the regulation of service charges introduced by the 1985 Act in respect of flat-tenants to a wider class of tenant, including tenants of buildings let with land which fall within the definition of "dwelling".

Part VI of the 1987 Act is concerned, in part, also with tenants who are liable to pay service charges, as section 46(2) itself indicates. In that subsection "the meaning given by section 18(1) of the 1985 Act" must be the meaning in section 18(1) as amended by section 41 of, and para 1 of Schedule 2 to, the 1987 Act, that is to say "service charge" includes amounts payable by tenants of dwellings. M

Sections 47 and 48 of the 1987 Act are directed towards ensuring that the tenant shall know the identity of his landlord and shall have a definitive address in England and Wales for the service of notices, including notices in proceedings, on his landlord. The sanction against the landlord who fails to comply with his obligation to provide that information is, in the case of failure to comply with section 47, that the tenant will not have to pay the service-charge element of any amount demanded under the terms of the tenancy; and, in the case of failure to comply with section 48, that the tenant will not have to pay rent or service charge — at least until the landlord does provide the information which the section requires him to provide.

Section 49 of the 1987 Act has the effect that an address furnished by the landlord under sections 48 or 47 shall be treated as the last-known place of abode or business of the landlord for the purposes of section 196(3) and (4) of the Law of Property Act 1925 (regulations respecting notices required or authorised to be served under that Act).

Section 50 of the 1987 Act extends section 3 of the 1985 Act. Section 3(1) of the 1985 Act required (so far as material) that, if the interest of the landlord under a tenancy of premises which consist of or include a dwelling is assigned, the new landlord shall give notice in writing of the assignment and of his name and address to the tenant. The effect of the new sections 3(A) and 3(B), introduced by section 50 of the 1987 Act, is that the old landlord remains liable in certain cases to the tenant in respect of breaches of landlord's covenants under the tenancy.

The language of section 3(1) of the 1985 Act — namely "premises which consist of or include a dwelling" — has been adopted in section 46(1) of the 1987 Act. When enacting the 1985 Act, Parliament was careful to distinguish between the concepts "premises occupied as a dwelling" in section 1(1) and "premises which consist of or include a dwelling" in section 3(1). It is the latter concept that has been adopted in section 46(1) of the 1987 Act.

Section 51 of the 1987 Act, which is the last of the group of sections in Part VI, introduced a new section, section 112C, into the Land Registration Act 1925. The new section gave to a tenant of premises to which the section applied the right to inspect the register relating to those premises. The premises to which the new section applied were described in the same terms as those used in section 46(1) of the 1987 Act.

The definition of statutory tenant in section 112C(4) includes a statutory tenant within the meaning of the Rent (Agriculture) Act 1976. So the premises to which the new section 112C applied might include premises occupied by agricultural workers. Examination of section 9(3) of the Rent (Agriculture) Act 1976 shows that a dwelling-house which is itself subject to a statutory tenancy within the meaning of that Act may form part of premises which are let as an agricultural holding. This suggests, at the least, that Parliament was aware, when it enacted Part VI of the 1987 Act, that the phrase "premises which consist of or include a dwelling" could refer to an agricultural holding on which there were cottages occupied by agricultural workers.

I conclude from this examination of the Landlord and Tenant Act 1987 that it is impossible to regard that Act as a whole as having a single theme or purpose which should colour the court's approach to the word "premises" in section 46(1). Parts I to III of the 1987 Act apply only to flats, but this is made clear by the limited terms in which "premises" is defined for the purposes of those Parts. Parts IV and V extend to dwellings other than flats. Part VI, notwithstanding that the long title to the Act might suggest otherwise, extends beyond flats and other dwellings; it extends to "premises which consist of or include a dwelling". In my view, the words "or include" in section 46(1) of the 1987 Act must have been used with the intent that "premises" for the purposes of Part VI should have a wider meaning than simply "dwelling".

Further, I am unable to identify any context in Part VI itself which points to a particular meaning of the word "premises" other than the definition in section 46(1). The purpose of Part VI is to enable a tenant of premises which consist of or include a dwelling to take practical steps — and, in particular, to serve notices including notices in proceedings — for the purposes of enforcing his landlord's obligations. It may well be that these provisions are of particular benefit to tenants holding undertenancies which require the landlord to provide services — for example, repairs, maintenance and insurance — in that it is in respect of tenancies of this nature that the landlord may have a particular incentive to conceal his identity or his address so as to avoid enforcement of his obligations. But, although the provisions are likely to be of particular benefit to those holding undertenancies of this nature, I can see nothing in Part VI of the Act which restricts its application to tenancies under which a service charge is payable. The requirement is, simply, that the premises consist of or include a dwelling. Further, I can see no reason of policy apparent in Part VI of the Act which suggests why some tenants holding undertenancies which include a dwelling and not others should have the benefit of knowing the identity and address of their landlord. There is, after all, nothing in section 46(1) which restricts the operation of Part VI to cases in which the tenant is in occupation of the dwelling which comprises or forms part of the premises or, indeed, is himself in occupation of the premises at all.

It seems to me, therefore, that the proper approach is to construe the words which appear in section 46(1) of the 1987 Act on the basis that there is nothing elsewhere in Part VI, or in the Act as a whole, which points to those words having some particular meaning which they might not otherwise bear.

Adopting this approach, I conclude that premises are intended to have a wider meaning than dwelling, that is to say, that the word "premises" is intended to mean more than a building or part of a building together with any yard, garden, outhouses and appurtenances belonging to it or usually enjoyed with it. So, "premises" can include land which is the subject-matter of the same letting. The question is — how much land? The difficulty which I see in giving the premises a meaning which is narrower than the subject-matter of the letting is that there is no guide in section 46(1) which would enable the court to say where the limit is to be drawn.

Section 46(1) of the 1987 Act does, of course, contain an express exclusion in relation to business tenancies. It must have been thought that, but for the express exclusion, the word "premises" could include premises which were held under a tenancy to which Part II of the Landlord and Tenant Act 1954 applies. Part II of the 1954 Act does not apply to tenancies of an agricultural holding: see section 43(1) of that Act. So, as it appears to me, Parliament must have appreciated that, if tenancies of an agricultural holding were to be excluded from the definition in section 46(1) of the 1987 Act, an express provision would be required to put the matter beyond doubt. Some indication of Parliament's awareness in this area can be derived from section 9(3) of the Rent (Agriculture) Act 1976, which contains an express provision reversing the effect of the House of Lords decision in *Maunsell* v *Olins* and section 137(3) of the Rent Act 1977 as amended by para 60 of Schedule 14 to the Agricultural Holdings Act 1986. No provision excluding agricultural holdings has been introduced into section 46(1) of the 1987 Act and I do not think that it is appropriate to construe "premises" in a way which introduces such a restriction by implication.

Accordingly, I construe section 46(1) of the 1987 Act in the sense that premises means the subject-matter of a letting. I do not think that I am precluded from reaching this conclusion by the decision of the House of Lords in *Maunsell* v *Olins* [1975] AC 373. That was a decision on the meaning of the word "premises" in section 18(5) of the Rent Act 1968; a statute dealing with a particular subject-matter. There is no reason in principle why the application of the process of reasoning indicated by Lord Wilberforce — which I have sought to follow — will necessarily lead to the same construction of the word "premises" where it appears in a different statute dealing with different subject-matter. As I have said, the fact that the word "premises" would not normally be used in popular parlance to include a large sporting and agricultural estate is not, of itself, a bar to giving to the word the meaning which I have indicated in the context of a statute dealing with landlord and tenant matters: see *Bracey* v *Read* [1963] Ch 88.

In construing section 46(1) of the 1987 Act as I do, I take some comfort from the fact that sections 47 and 48 are said to apply to agricultural holdings in one of the standard textbooks on that subject: see *Scammell & Densham's Law of Agricultural Holdings*, 7th ed (1989) at pp 176 and 177. A decision, therefore, that those sections do apply to agricultural holdings, if they include a dwelling, ought not to cause difficulty in practice. It has been foreseen.

The second question, then, is whether a notice under section 48(1) of the 1987 Act was given prior to December 3 1991. To comply with the subsection, the notice must furnish the tenant with an address in England and Wales at which notices, including notices in proceedings, may be served on him by the landlord. The notice must be in writing: see section 54(1) of the 1987 Act.

With those two requirements in mind I look at the candidates which have been suggested by the landlord. They are summarised in a letter dated December 2 1991 from Franks Charlesly.

First, reliance is placed on the notice to pay dated August 7 1991. That notice is signed over the text "Solicitors and authorised agents for Lindsey Trading Properties Inc of Hulton House, 161/166 Fleet Street, London EC4A 2DY". There is nothing in that notice which indicates that the address of the solicitors is the address at which notices, including notices in proceedings, may be served on the landlord by the tenant.

Second, reliance is placed on a statutory demand dated August 6

A 1991 made by or on behalf of the landlord under the Insolvency Act 1986. That asserts that the demand is served by the creditor, Lindsey Trading Properties Inc. It gives the registered office in Panama and includes an address "care of" Franks Charlesly at Hulton House in London. The statutory demand, at Part A, indicates that any communication regarding that demand is to be addressed to Franks Charlesly. It does not anywhere contain notice that Franks Charlesly are authorised to receive notices which may be served by the tenant on the landlord in connection with matters under the lease other than the rent which is the subject-matter of the statutory demand.

Third, reliance is placed on section 93 of the Agricultural Holdings Act 1986. Subsection (1) provides that any notice, request, demand
B or other instrument under that Act shall be duly given to or served on the person to or on whom it is to be given or served if it is delivered to him or left at his proper address or sent to him by post in a registered letter or by the recorded delivery service. Subsection (3) provides that any such instrument to be given or served on a landlord or a tenant shall, where an agent or a servant is responsible for the control of the management or farming as the case may be of the agricultural holding, be duly given or served if given to or served on that agent or servant.

In my judgment, section 93 of the Agricultural Holdings Act 1986 does not assist the landlord. That section, as it indicates, applies only to notices given under that Act. Section 48 of the 1987 Act requires a notice to be furnished which gives an address for all purposes.

C Finally, reliance was placed on an affidavit sworn on November 26 1991 by Delores Caboshe, the secretary of the plaintiff company in Australia. She deposed on November 26 1991 that on or about the beginning of August 1987 she was contacted by telephone by Mr Bolag on behalf of Lindsey and told that, although the provisions of the plaintiff company's lease required payment of the rent to Franks Charlesly, the rents payable under the lease should in fact be paid to Strutt & Parker at Salisbury, whom Lindsey had appointed to act as its managing agents for the property. She goes on to say that, on the same occasion, Mr Bolag told her that all formal communications, including any and all notices which might have to be given from time to time by the plaintiff company to Lindsey in relation to the property
D — not being in the nature of day-to-day management issues — should be addressed to Lindsey c/o Franks Charlesly.

It is not necessary to decide whether section 48 requires a single address capable of covering both day-to-day management issues and formal communications, because it is clear from Miss Caboshe's affidavit that she was given no notice in writing as required by section 54 of the 1987 Act. Accordingly, her affidavit takes the landlord no further.

In those circumstances, I hold that on the true construction of section 48 of the Landlord and Tenant Act 1987 no notice giving the tenant an address in England and Wales at which notices, including notices in proceedings, may be served on him by a tenant was furnished by the landlord to the tenant prior to the letter of December
E 3 1991.

The effect of that failure to give a notice under section 48(1) must be that, in accordance with section 48(2), rent which would otherwise have been due from the tenant to the landlord must be treated for all purposes as not being due from the tenant to the landlord at any time before the receipt by the tenant of the letter dated December 3 1991. There is, I think, no reason to give a more extensive construction to section 48(2). In particular, I reject the argument, put forward (albeit faintly) on behalf of the tenant, that failure to give a notice under section 48(1) had the effect that rent which would otherwise have been due was irrecoverable for all time.

That leads to the third question, namely whether there has been
F service of an effective notice to pay under Case D in Part I of Schedule 3 to the Agricultural Holdings Act 1986.

The Agricultural Holdings (Forms of Notice to Pay Rent or to Remedy) Regulations 1987 (SI 1987 No 711) require that a notice to pay rent for the purposes of Case D in the 1986 Act shall be in Form 1 to those regulations. Form 1 appears in the Schedule to the 1987 regulations. Para 1 is in these terms:

> I hereby give you notice that I require you to pay within two months from the date of service of this Notice the rent due in respect of the above holding as set out below:

Then there is a note which reads:

> This Notice may not be served before the rent is due.

Then appear the words "Particulars of rent not paid" under which there are two columns: "*Date when due*" and "*Amount due*". The G
effect of the 1987 regulations read in conjunction with para 10 of Part II of Schedule 3 to the Agricultural Holdings Act 1986 is that the prescribed form must be used. There will be no compliance with the regulations if the information which is required to be included in the prescribed form is inaccurate.

In the notice dated August 7 1991, the dates when rent due was said to be due were June 24 1990, September 29 1990, December 25 1990, March 25 1991 and June 24 1991. I have held that no notice under section 48(1) of the 1987 Act had been served at any time before those five dates. Section 48(2) requires that the rent which would otherwise have been due from the tenant on those dates is to be treated for all purposes as not being due from the tenant to the landlord. In those
circumstances it seems to me that the notice served on August 7 1991 H
could not be valid, on the ground that, when served, no rent was due from the tenant.

The position in relation to the notice dated December 3 1991 is different. The notice to pay rent was served under cover of a letter of that date which has itself been accepted as a valid notice under section 48 of the 1987 Act. It follows that on December 3 1991 the landlord did comply with section 48(1).

An interesting argument was advanced on behalf of the tenant as to the exact time at which the suspension imposed by section 48(2) ceased to have effect. I can see no reason why it should continue beyond the time when the letter dated December 3 1991 was actually received by the tenant. But it is not necessary to decide the point, J
because, in my view, examination of the notice to pay shows that it fails to comply with the 1987 regulations. The provisions of section 48(2) require only that rent be treated for all purposes as not being due from the tenant to the landlord before receipt of the letter dated December 3 1991.

In the notice of December 3 1991, under the heading "Particulars of rent not paid", the rent is said to have been due on the quarter days from June 24 1990 to September 29 1991. I have already held that the language of section 48(2) requires me to hold that the rent was not due on those quarter days. It did not become due until receipt of the letter dated December 3 1991. Accordingly, the information which is required to be included in the prescribed form is inaccurate and the notice to pay rent of December 3 1991 does not comply with the 1987 K
regulations.

That there is need for strict compliance with regulations in this field was emphasised by the Court of Appeal in *Pickard* v *Bishop* (1975) 31 P&CR 108, to which I have already referred. In that case the defect in the notice to pay was that it did not state that it was served on behalf of the persons who were the landlord.

A similar point arose in *Dickinson* v *Boucher* (1983) 269 EG 1159. There the misstatement in the notice to pay was as to the amount of the rent due; £650 was demanded when only £625 was due. The Court of Appeal followed *Pickard* v *Bishop* and held the notice to be invalid. The court held that the notice to pay was, in effect, the foundation of a forfeiture of an agricultural tenancy and that, in L
connection with forfeiture, strict compliance with the requirements of statute was required.

The Court of Appeal considered the matter again in *Official Solicitor* v *Thomas* [1986] 2 EGLR 1. The defect was in a notice to remedy, under para (*b*) in Case D of what was then section 2(3) of the Agricultural Holdings (Notice to Quit) Act 1977. The court affirmed the need for strict compliance with the statutory requirements in respect of notices to pay under this legislation but held that, on the particular facts of that case, the notice did specify with sufficient accuracy the source of the obligation which was said to require remedy. Sufficiency of description was a question of degree and the description given in the notice was such that the tenant could not reasonably have been misled. M

In the present case, the defect in the notice to pay is not in the nature of a misdescription; it is a statement as to the date when rent was due, which — in the light of the provisions of section 48(2) of the 1987 Act — is simply inaccurate. It is, I think, immaterial that the tenant would not, in fact, have been misled. In my judgment, the notice to pay rent of December 3 1991 does not satisfy the requirement in the 1987 regulations and, therefore, must be treated as invalid.

Accordingly, I answer the questions raised in the originating summons in the following terms:

I declare that on the true construction of section 48 of the Landlord and Tenant Act 1987 no rent was due from the plaintiff to the defendant under the lease on August 7 1991.

A I declare that on the true construction of that section no rent was due from the plaintiff to the defendant at any time before receipt of the letter dated December 3 1991.

And I declare that on the true construction of section 48 and of Case D in Part I of Schedule 3 to the 1986 Act and of the 1987 B regulations the notices of August 7 1991 and December 3 1991 are invalid and of no effect.

Declaration accordingly. Plaintiff entitled to its costs.

For further cases on this subject see p 271

LANDLORD AND TENANT

BUSINESS TENANCIES

A

Chancery Division

July 10 1991

(Before Mr Desmond PERRETT QC, sitting as a deputy judge of the Chancery Division)

CITY OF LONDON CORPORATION v FELL AND OTHERS

Estates Gazette January 18 1992

[1992] 02 EG 172

Landlord and tenant — Arrears of rent — Assignment of lease B — Expiration of contractual term — Tenancy continued by Part II of the Landlord and Tenant Act 1954 — Landlords claiming rent from original tenants — Whether original tenants liable for arrears after expiration of contractual term

Plaintiff lessors granted 10-year term to defendant lessees from March 25 1976 — Defendants assigned term to Grove Bell Group plc in June 1979 — Initial yearly rent of £27,500 reviewed and increased to £38,500 with effect from March 25 1981 — Rent payable quarterly in advance — Contractual term expired on March 24 1986 — Assignees continued in occupation of the premises pursuant to Part II of the Landlord C and Tenant Act 1954 — Assignees surrendered premises to plaintiff lessors on January 23 1987 — Assignees were compulsorily wound up on December 1 1986 — No rent was paid by assignees after Christmas quarter day 1985

Plaintiff lessors claimed rent from defendants as original tenants — Defendants denied any liability for arrears of rent arising after expiration of the contractual term — Dispute as to whether assignees remained in occupation after expiration of contractual term — Plaintiffs contended that the term of the lease was continued by the 1954 Act — Defendants submitted that their covenant as original tenants to pay rent for the term did not extend their liability to any continuation D tenancy under the 1954 Act

***Held*: Assignees did remain in occupation after the expiration of the contractual term — The words "tenancy" and "term" must bear their precise meaning — There is no reference to the extension of "the term" in the 1954 Act — No provisions in the 1954 Act which may be taken to perpetuate the obligations of anyone other than the lessor and the business occupant at the time of the expiry of the lease — Proposition of Nourse J in *GMS Syndicate Ltd* v *Gary Elliott Ltd* [1982] Ch 1 that the term granted by a tenancy continues by way of statutory extension not followed — The contract of E lease should be strictly construed in so far as it defines the duration of the original tenant's liability to pay rent — It is not possible to bring within the ambit of the object of the 1954 Act an imposition and continuation of the contractual position which has no relevance to enabling a tenant occupying business premises to obtain a new tenancy — The defendants' liability to pay rent upon the assignees' default came to an end at the expiration of the term as defined in the lease**

The following cases are referred to in this report.

Associated Dairies Ltd v *Pierce* [1983] EGD 383; (1982) 265 EG 127, CA
Baker v *Merckel* [1960] 1 QB 657; [1960] 2 WLR 492; [1960] 1 All ER 668, CA F
Baynton v *Morgan* [1888] 2 QB 74
Bolton (HL) (Engineering) Co Ltd v *Graham (TJ) & Sons Ltd* [1957] 1 QB 159; [1956] 3 WLR 804; [1956] 3 All ER 624, CA
Centrovincial Estates plc v *Bulk Storage Ltd* (1983) 46 P&CR 393; [1983] EGD 556; 268 EG 59
Cornish v *Brook Green Laundry Ltd* [1959] 1 QB 394; [1959] 2 WLR 215; [1959] 1 All ER 373; [1959] EGD 116; (1959) 173 EG 307, CA
GMS Syndicate Ltd v *Gary Elliott Ltd* [1982] Ch 1; [1981] 2 WLR 478; [1981] 1 All ER 619; (1980) 41 P&CR 124; [1981] EGD 331; 258 EG 251
Green v *Bowes-Lyon* [1963] AC 420; [1961] 3 WLR 1044; [1961] 3 All ER 843; [1961] EGD 301; (1961) 180 EG 429, HL
Junction Estates Ltd v *Cope* (1974) 27 P&CR 482; 232 EG 335
Plesser (A) & Co Ltd v *Davis* (1983) 267 EG 1039
Skelton (William) & Son Ltd v *Harrison & Pinder Ltd* [1975] QB 361; [1975] 2 WLR 238; [1975] 1 All ER 182; (1974) 29 P&CR 113 G
Weinbergs Weatherproofs Ltd v *Radcliffe Paper Mill Co Ltd* [1958] Ch 437; [1958] 2 WLR 1; [1957] 3 All ER 663

This was a claim by the plaintiff lessors, the City of London Corporation, seeking £33,460.64 as arrears of rent from the defendants, John Fell, John Hayward and Edward Sturmer partners in Wilde Sapte, the original lessees under a lease dated July 15 1977 of Boston House, 63-64 New Broad Street, London EC2.

Mark Pawlowski (instructed by the solicitor to the City of London) appeared on behalf of the plaintiffs; Miss Erica Foggin (instructed by Wilde Sapte) represented the defendants.

H Giving judgment, **MR DESMOND PERRETT QC** said: This is an action brought by the plaintiffs, the lessors, against the defendants, the lessees, for recovery of rent payable but unpaid in respect of business premises forming part of a building known as Boston House, 63-64 New Broad Street, in the City of London, which are business premises. By a lease dated July 15 1977 and made between the parties to this action, the premises were let by the plaintiff lessors to the defendant lessees for a term of 10 years from March 25 1976. The lease provides for a yearly rent of £27,500 or more as agreed to be paid during the term of 10 years and specifies the dates upon which and the manner in which the yearly rent is to be paid.

The defendants are and were a distinguished firm of solicitors and by 1979 they required other premises. J In June 1979 the lease of the demised premises was assigned to Grove Bell Group plc, the assignees, for the residue of the unexpired term of years, subject to the payment of the rent and observance of the covenants and conditions which on the part of the lessees fell by the lease to be observed. At that time the term of the lease still had six years and nine months unexpired.

By provision in the lease, the rent was to be paid quarterly in advance upon specified quarter days. In June 1981, in accordance with the provisions in the lease, the assignees by then having been in occupation of the business premises for some two years, the yearly rent was reviewed and increased to £38,500 with effect from March 25 1981.

K Pursuant to clause 1 of the lease, the final payment of rent was due at Christmas 1985 in respect of the final quarter year of the 10-year term. I am told that the final quarter's rent was paid to the plaintiffs by the assignees at or about Christmas 1985. At the expiration of the 10-year term on March 24 1986, the assignees continued in occupation of the premises pursuant to the provisions of Part II of the Landlord and Tenant Act 1954, as tenants of the lessors, until January 23 1987, when they surrendered the demised premises to the lessors. On December 1 1986 the assignees were compulsorily wound up by an order made in the companies court and on March 2 1987 W H Cork Gully & Co ("Cork Gully") were appointed joint liquidators

A of the assignees. Neither on nor after March 25 1986 had the assignees paid any rent to the lessors in respect of their tenancy of the business premises, and at the date of the demised premises being surrendered to the lessors the sum of £33,460.64 was owed to the lessors for rent and other outgoings covenanted for in the lease.

Inquiries were made by the plaintiffs of the official receiver as to whether there was any prospect of a dividend being paid to them as unsecured creditors of the assignees. Cork Gully, as liquidator, in letters to the lessors of April 28 1987 and November 27 1987 expressed their view that the prospect of a dividend was, in April 1987, "uncertain", and in November 1987 "remote". Time passed, and it was not until September 7 1989 that the lessors, by letter of that date,
B stated their intention of looking to the original lessees for payment of the outstanding charges. The defendant lessees denied the lessors' entitlement to look to them for arrears of rent arising after expiration of the contractual term contained in the lease and, accordingly, on July 6 1990 the lessors issued their writ against the lessees.

In the proceedings the only question of fact which required determination was whether the assignees, either by themselves or by their subsidiary companies, had remained in occupation of the demised premises after the expiration of the contractual term until the alleged date of surrender to the lessors. In that regard, I heard evidence from Mr Kenneth Gardner to the effect that he, as company secretary of the assignees, Grove Bell Group plc, and as company secretary of Grove Bell Group plc's several subsidiaries, maintained
C a presence with other people at the premises, conducting the affairs of the various subsidiaries to a greater or lesser degree right up to the date of surrender. There was some support for this from Mr Ferguson of Cork Gully who gave evidence, and, as I indicated after short but not strenuous argument from Miss Erica Foggin, counsel for the lessees, my finding is that either by themselves or their subsidiaries the assignees remained in occupation of the premises for business purposes until January 23 1987, upon which date the lessors regained possession.

By clause 1 of the lease the lessors demised the premises to the lessees for the lessees to hold from March 25 1976 for a term of years (hereinafter called "the term") yielding and paying therefor during the term unto the lessor the agreed rent or as later reviewed in the
D course of the term. It is the defendants' contention that clause 1 of the lease defines clearly and without equivocation the extent in time to which, as original lessee, they will remain liable for payment of rent, and that the lessors, having received the agreed rent for the full duration of the term, are entitled to no more from them than was payable in respect of the occupancy of the premises during the term. The plaintiffs, however, allege that this being a business tenancy to which the provisions of Part II of the Landlord and Tenant Act 1954 apply, the privity of contract between the original parties to the contractual lease continues during the statutory extension of the contractual lease. By section 24(1) of the Act, a business tenancy is not to come to an end unless determined in accordance with the
E provisions of the Act, and the effect of the Act is that the contractual tenancy continues with a statutory variation as to the mode of determination and thus remains, throughout, one and the same tenancy. For these reasons, it is contended, the lessees' obligation to pay rent, despite the unequivocal agreement with the lessors as to the duration of the obligation contained in the contractual lease, is extended beyond "the term" and into the period of the statutory continuation of the tenancy.

There is apparently no case in which the point at issue has been decided as between the lessor and lessee and it appears to be the opinion of the author of the current edition of *Woodfall* (at p 22/31) that the point should, by analogy with the decision in *GMS Syndicate Ltd* v *Gary Elliott Ltd* [1982] Ch 1*, be resolved in favour of the
F lessor. The broad effect of Part II of the Landlord and Tenant Act 1954 was alluded to in general terms by Denning LJ (as he then was) in *H L Bolton (Engineering) Co Ltd* v *T J Graham & Sons Ltd* [1957] 1 QB 159 at p 168 where he said, when discussing the effect of a notice to quit on a business tenancy for a term of years certain:

> The right view is that the common law tenancy subsisted with a statutory variation as to the mode of determination.

In *Weinbergs Weatherproofs Ltd* v *Radcliffe Paper Mill Co Ltd* [1958] Ch 437, Harman J at p 445, when deciding whether a notice purporting to terminate a contractual business tenancy in accordance with the terms of the lease was effective, even though no notice had

*Editor's note: Also reported at [1981] EGD 331; (1980) 258 EG 251.

been served under section 25 of the Landlord and Tenant Act 1954, G
said:

> The term must be thought of as continuing by way of a statutory extension.

He cites the passage from Denning LJ's judgment in *Bolton Engineering*.

Mr Mark Pawlowski, counsel for the lessors in this case, places reliance on the apparent view expressed by Harman J that it is "the term" that must be thought of as continuing, and I can see the force of that when he is attempting to extend the obligations contained in the covenants of the contractual lease in this case beyond the term as therein defined. But it appears from Denning LJ's judgment in *Bolton Engineering* that at no time was he referring to anything other than "the tenancy"; that is to say, the right to occupy pursuant to the H
lease, and that it was this tenancy or right to occupy which subsisted with the statutory variation by reason of the Act. Denning LJ made no mention of "the term" in its strict sense as continuing. It was in no way essential to his decision in *Weinbergs* that Harman J should have an eye to any difference in meaning between the words "tenancy" and "term", but, in my view, if he was basing himself upon Denning LJ's view as expressed in *Bolton Engineering*, he was using the word "term" as if it were synonymous with "tenancy". In my view, and for the purposes of the instant case, the words "tenancy" and "term" must bear their precise meaning and are not to be regarded as interchangeable, as clearly they were in the context of their use by Harman J.

In *Cornish* v *Brook Green Laundry Ltd* [1959] 1 QB 394, Romer LJ J
in the Court of Appeal, in deciding that the rights and obligations created by a lease of business premises were still in force during the period of statutory extension, cited (at p 409) Harman J's proposition that "the term" must be thought of as continuing. The Court of Appeal in that case were not concerned to differentiate between the words "term" and "tenancy" in their task of ascertaining whether the landlord was a competent landlord within the meaning of Part II of the 1954 Act, and the precision in interpretation which seems to me to be essential in this case was unnecessary to the decision in *Cornish*.

A reference to the long title of the Landlord and Tenant Act 1954 shows that the object of Part II was: K

> To enable tenants occupying property for business purposes to obtain new tenancies in certain cases.

There is no reference to the extension of "the term" of an existing lease. It is clearly the law (see *Cornish*) that when the contractual tenancy is by reason of Part II to be thought of as continuing by way of a statutory extension, the tenancy continues upon the same terms as were contained in the contractual lease before its expiry. That is to say, the landlord and the party entitled to the tenancy as continued by statute are bound by the same covenants as were contained in the original contractual lease. The contractual lease, after expiry of the contractual term, defines the rights and obligations of the parties to the statutory extension of the tenancy. That statutory arrangement L
holds no benefit to the original lessee. If he is not in possession at the date of expiry of the contractual term, he is afforded no right to possession. He is not privy to the statutory consequences in any way. Whether an assignee of the original lease elects to hold over after the expiry of the contractual term is a matter entirely beyond the control of the original lessee. There appears to be no provision in the Act which may be taken to perpetuate the obligations of anyone other than the lessor on the one hand and the business occupant at the time of the expiry of the lease on the other.

In *GMS Syndicate* v *Gary Elliott Ltd* Nourse J held that the effect of section 24(1) of the Landlord and Tenant Act 1954 was to extend "the term" of a business lease so that, although by the instrument itself a tenant's covenant was expressed to be given "during the M
residue of the term granted", those words were to be construed as extended by the section as "the residue of the term so extended by section 24(1)", so as to require observance of the covenant after the "contractual term" and during the statutory subsistence of the tenancy. Again, Nourse J adopts the use of the word "term" as being that which Part II of the Act is extending. Although I do not doubt that the learned judge would have reached the same conclusion had he referred to the extended term as "the common law tenancy subsisting on the same terms", it is in my view not correct to hold that Part II extends "the term" if by that is meant the original contractual period, for the original lessee has no tenancy or right of occupation after the expiry of the term. It is the assignee in occupation whose

rights are extended by the statute, albeit upon the same conditions and subject to the same covenants as governed the rights of the original parties to the lease Thus, I do not differ from Nourse J's conclusion that a landlord may successfully sue an assignee in possession, or sublessee in possession, in respect of breaches of covenant during the period of statutory extension of the tenancy. But I respectfully decline to follow the proposition on p 10 at B:

It is now well established that the term granted by the tenancy continues by way of statutory extension.

It is my conclusion upon the wording of Part II of the 1954 Act that it is the tenure of the business premises that is sustained by operation of the Act and not the original term. His Honour Judge Edgar Fay QC in *William Skelton & Son Ltd* v *Harrison & Pinder Ltd* [1975] QB 361 at p 367, when referring to the Court of Appeal's decision in *H L Bolton (Engineering) Co Ltd* v *T J Graham & Sons Ltd*, correctly recited Denning LJ's dictum that under Part II the common law tenancy subsists, and he made no reference to the Act extending the term. That it may be important for certain purposes under Part II of the Act to distinguish between "term" and "tenancy" is apparent from the speech of Lord Reid in *Green* v *Bowes-Lyon* [1963] AC 420, at p 434. There Lord Reid records his agreement with the expressed wording of Denning LJ in *Bolton* and, by implication, of the conclusion of Harman J in *Weinbergs*, and makes it clear that it is the tenancy and not the term that is carried on by Part II of the Act.

Mr Pawlowski further relied on the judgments in *Baker* v *Merckel* [1960] 1 QB 657 and *Centrovincial Estates plc* v *Bulk Storage Ltd* (1983) 46 P&CR 393*. Both those cases were concerned with the liability of original tenants during the currency of either the original term or the term as extended pursuant to an option in the original lease which was exercised by the assignee. In both cases the default of which the landlords complained occurred during the term which was within the contemplation of the parties to the original leases, as defined by the leases themselves. I do not find much assistance from those authorities in the present case, for there was no question in either case of liabilities being explored beyond the scope of the express agreements.

In *Junction Estates Ltd* v *Cope* (1974) 27 P&CR 482† MacKenna J considered the position of a guarantor of a lessee's obligation to pay rent, the reference to arrears of rent arising in the period of a Part II statutory extension of the tenancy. He held that the guarantor's liability fell to be determined by construction of the contract of guarantee contained in the lease and that the guarantee was in respect of payment of rent reserved by the lease during its seven-year term. Within that term the original lessee whose obligation to pay rent was the subject of the guarantee assigned his interest in the lease to a third party who fell into arrears during the statutory extension of the tenancy. The learned judge recites the guarantee (clause 3) verbatim at p 483. He concludes at p 484:

In my opinion the question between the plaintiffs and the defendants depends on the true construction of clause 3. I construe it as guaranteeing the performance of the obligation created by clauses 1 and 2 of the lease, namely, in the case of rent, the payment of the rent reserved by the lease during its seven years' term. I do not read clause 3 as covering any obligation of the tenant to pay his rent during any statutory extension of the term. If the plaintiffs had wanted a guarantee which would impose that indefinite liability on the defendants, that is, a liability continuing until either the plaintiffs had served a notice of termination, or the tenants had served a notice to quit, or a new tenancy had been applied for by the tenant and granted by the court, or the landlord had forfeited the tenancy, they should have made that intention clear in the language of the guarantee. They have not done so. I think that the language of clause 3 makes it clear that the guarantors are to be liable only for the seven years' rent if any part of that should be unpaid. If clause 3 is ambiguous about the guarantor's obligation, it should, on well-established principles, be construed against the plaintiffs, and I would so construe it.

So also in *A Plesser & Co Ltd* v *Davis* (1983) 267 EG 1039 on facts which he found indistinguishable from those in *Junction Estates Ltd* v *Cope*, French J arrived at a similar conclusion and found that in effect the duration of the guarantor's liability was defined by strict construction of the contract and was thus coterminous with the contractual term, adding that it was clear that a contractual term and a statutory continuation are distinguishable as concepts. He continues:

A document may be couched in language which without doubt or equivocation points to a guarantee of liability in respect of contractual term obligations or in respect of statutory continuation obligations or it may be of both. An example is to be found in the documents which the court had before it in *Associated Dairies Ltd* v *Pierce*. Clause 31 of each of the leases under consideration in that case started with the words: "To yield up at the expiration or sooner determination of the term *or any statutory continuation thereof* . . .

— the judge's emphasis —

. . . as the case may be". Some such expression could have been used here and, in my judgment, should have been used here in order to produce the result for which the plaintiffs contend.

It has been urged upon me that the position of a guarantor is distinguishable from that of an original lessee in these circumstances, but I can see no practical difference between the two situations and can conclude only that, in order to discover the contractual obligations attaching to either position, one must construe the respective contracts to see whether they cover the nature, extent and circumstance of the debt sought to be recovered. It is said that the contract of guarantee should be strictly construed: see *Associated Dairies Ltd* v *Pierce* (1982) 265 EG 127 at p 129; *per* May LJ. I can see no reason why the same degree of precision should not also be brought to bear upon the construction of the contract in this case. It has been submitted that an original lessee would well have in contemplation when he enters upon a lease of business premises for a defined term of years that the tenancy would, by operation of Part II of the 1954 Act, outlast the term. If that is the case, it seems to me to be all the more important that the contract should be strictly construed in so far as it defines the duration of the original tenant's liability to pay rent; and I can see no reason for importing any implied term into the lease when, if it had been the intention of the parties so to agree, a simple form of wording could have catered for the payment of rent during any statutory continuation of the tenancy. Although it may not be the most reliable method of testing a proposition of landlord and tenant law, I suspect it would come as a surprise to a man who had promised in plain terms to be liable to pay rent either by himself or his assigns for a fixed period that the fixed period meant nothing and that the assignee might continue in possession indefinitely, with the landlord looking each quarter day to the original lessee for rent for as long as the landlord chose. I can find nothing in Part II which evinces a desire on the part of Parliament to confer such an open-ended advantage upon a landlord or impose such a burden upon a business tenant for a term certain who assigns his tenancy. Part II of the Act, as I have recited earlier, has as its object to enable tenants occupying the property for business purposes to obtain new tenancies in certain cases, and it is simply not possible, in my view, to bring within the ambit of that object an imposition and continuation of the contractual position which has no relevance to enabling a tenant occupying business premises to obtain a new tenancy. It has been pointed out to me that, since *Baynton* v *Morgan* [1888] 2 QB 74, it is settled that an original lessee is not a surety. That, of course, I accept, but both functions arise out of contract and, in my view, in this case the contract defines the limit of the duration of the defendants' liability to pay rent to the plaintiffs, namely during the term defined in the lease.

So I can find no reason for drawing any distinction, upon the facts of the *Junction Estates* case, between the liability of the guarantor in that case (who, "being connected" with the original business tenants, might, would or should have had in mind, in 1964, the effect upon business tenancies of Part II of the 1954 Act) and the position of the lessee in this case. For those reasons I conclude that the defendants' liability to pay rent upon the assignees' default came to an end at the expiration of the term as defined in the lease and, accordingly, there will be judgment for the defendants.

Judgment for the defendants.

---

*Editor's note: Also reported at (1983) 268 EG 59.
†Editor's note: Also reported at (1974) 232 EG 335.

A
Court of Appeal

October 16 1991

(Before Lord Justice NOURSE, Lord Justice STOCKER and Lord Justice WOOLF)

MIAH v BROMLEY PARK GARDEN ESTATES LTD

Estates Gazette March 14 1992

[1992] 10 EG 91

Landlord and tenant — Application for a new tenancy under
B **Part II of the Landlord and Tenant Act 1954 — Directions for hearing of application — Applicant tenant appearing at hearing without legal representation or valuation expert — Trial judge refused applications for adjournment — Judge proceeding to determine rent on basis of valuation evidence given by landlords' expert — Whether judge wrongly exercised discretion to refuse adjournment — Whether prejudice to applicant tenant to proceed to determine rent on basis of expert evidence given only on behalf of landlords — Tenant's appeal allowed on terms as to costs**

By a lease, which commenced in September 1975, the appellant held a 15-year term of premises at 131 Fortess Road,
C **London NW5, expiring in September 1990 — The appellant used the premises as an Indian take-away restaurant and occupied part as his home — The rent at the last review in September 1985 had been determined at £5,250 pa — Following the service by the respondent landlords of a notice under section 25 of the 1954 Act expiring on September 29 1990, the appellant made application for a new tenancy — The registrar of the county court gave directions for the action and ordered that experts' reports should be exchanged and that those reports should include evidence as to comparables — Five weeks' notice was given to the appellant that his**
D **application was to be heard on December 5 1990 — On the date of the hearing the appellant attended, but his solicitors did not — The appellant's solicitors had informed the respondents' solicitors that they did not have any instructions and had received only a small amount of payment on account of costs — Further, the appellant's solicitors had made no arrangements for their client to be represented by counsel at the hearing — At the hearing the appellant applied for an adjournment on the ground that he was not well — He made a further application on the same day after the luncheon adjournment — The trial judge refused both applications but**
E **caused the respondents' solicitors to contact the appellant's solicitors during the course of the hearing to ascertain the position — In the event the judge had before him the unchallenged evidence of the respondent landlords' valuer supporting a rent for the premises of £14,000 pa — The judge accepted that evidence and made the appropriate order — The appellant appealed contending that the judge had wrongly exercised his discretion to refuse the applications for adjournment and further had misdirected himself in allowing the hearing to proceed when it became apparent that the appellant was not able to represent himself**

***Held*: The grounds raised by the notice of appeal were**
F **singularly for the discretion of the trial judge — The judge's decision not to allow an adjournment was an exercise of a discretion which could not be interfered with — Although other judges may have taken a different view, the trial judge had no clear evidence as to the medical condition of the appellant and had no proper explanation as to why he was not represented — The judge had before him powerful evidence of the respondents' valuer, but it was evidence which was capable of being the subject of cross-examination because there were areas that required exploration — The appellant himself was incapable of subjecting the respondents' valuer's evidence to the sort of investigation which was necessary — It is unusual for the evidence of an expert on a matter of this nature to be** G
accepted in its entirety — There was a substantial and material risk that injustice was done to the appellant — By reason of the order of the registrar that any new tenancy was to commence upon the expiration of the previous term, there would be no prejudice to the respondents because although there would be a delay in the payment of the increased rent it would not affect that *quantum* — Further, the judge could have taken steps to ensure that the respondents were protected with regard to costs if the appellant had been granted an adjournment — The appeal was allowed and the appellant entitled to a rehearing of his substantive application on terms as to costs and as to a H
payment into court

The following case is referred to in this report.

Maxwell v *Keun* [1928] 1 KB 645; 97 LJKB 305; 138 LT 310; 44 TLR 100

This was an appeal by Rupa Miah from the decision of Judge Dobry QC, who on December 5 1990 (in Bloomsbury County Court) made an order relating to an application by the appellant for a new tenancy under Part II of the Landlord and Tenant Act 1954 in respect of 131 Fortess Road, London NW5. The appellant was in occupation of the premises under a continuation tenancy; the respondents, Bromley Park Garden Estates Ltd, were the landlords.

Ali Azhar (instructed by Paul Yanakas & Co) appeared for the J
appellant; William Elland (instructed by Hamlin Slowe) represented the respondents.

Giving the first judgment at the invitation of Nourse LJ, **Woolf LJ** said: This is an appeal from a decision of Judge Dobry QC, on December 5 1990, when sitting in Bloomsbury County Court. The judge had before him an application for a new tenancy under Part II of the Landlord and Tenant Act 1954 in respect of 131 Fortess Road, NW5, of which the applicant was the tenant. The judge had to deal with four issues on the application, the most important of which was the fixing of the initial rent under the new tenancy. However, the judge also had to deal with the pattern of reviews, some subsidiary matters in relation to insurance and whether there should be a K
prohibition on the sale of alcohol at the premises. The rent of the premises had been £5,250 since the date of the last review on September 29 1985.

The previous tenancy had commenced in September 1975 under a lease of some 15 years. That lease had been assigned to the applicant on August 4 1981 and he used the premises as an Indian take-away restaurant. In addition, he occupied the residential part of the premises as his home. The issues which arose on the application were of considerable importance to the applicant because his business livelihood depended on his occupation of the premises and, in addition, the premises provided him with a home.

The previous tenancy came to an end as a result of a notice to L
terminate the tenancy which was given by the respondent landlords on October 11 1989 to expire on September 29 1990. An application for a new tenancy was made by solicitors on behalf of the applicant in proper form on January 17 1990. In that application, which was for a new tenancy for 15 years, there was inserted the proposal that the rent should be £6,500. Quite clearly that was a remarkably optimistic figure.

On July 3 1990 the registrar made orders dealing with interlocutory matters. Among those orders was one that when the new tenancy was granted, it should commence on September 29 1990, which was the same date as the expiry of the notice to terminate. In addition, the registrar ordered that experts' reports should be exchanged and that those reports should include evidence as to comparables. M

The matter was set down for hearing and notice was given to the applicant that the application was to be heard on December 5 1990. The notice was given approximately five weeks before the hearing. At that time the applicant still had solicitors acting for him. Those solicitors remained on the record up to the date of the hearing. On the date of the hearing the applicant attended, but his solicitors did not.

The non-attendance of those solicitors was not entirely unexpected by the landlords (as appears from an attendance note of the landlords' solicitors) since there had been a discussion on December 4, the day prior to the hearing, between the parties' solicitors. The attendance note records that the landlords' solicitors had telephoned the applicant's solicitors on five occasions during that day and had

A eventually been told that the applicant's solicitors did not have any instructions on the further amendments to the lease, that they had received only a small amount of payment on account of costs from the applicant, that the applicant had given them no instructions, and that they understood the situation was the same with regard to the applicant's surveyor. They also told the landlords' solicitors that they had not arranged counsel, that they would again try to speak to their client to impress upon him the urgency of the situation, and that they thought it possible that he might appear alone.

At the commencement of the hearing the applicant made an application for an adjournment. The judge in his note of judgment records that the applicant told him at the outset that he was not very well. The judge goes on to say:

B It was not very easy to understand what [the applicant] said (that he had been sick for a long time and could not talk properly. He said he could not contact his solicitor and could not conduct his own case).

The judge indicates that that application was repeated later in the day, after the luncheon adjournment, at about 2 pm. On that occasion the applicant showed the judge some medicine. The judge says that he encouraged the applicant to stay and that the applicant was good enough to do so. The judge tried to explain to him what was going on, but the applicant did not appear to be interested and that he did not cross-examine. The judge refused the applications for an adjournment, and in the note of his judgment he gives the following reasons:

C His solicitors give one reason for their non-attendance and why he was not represented and the applicant gave another. The applicant had five weeks' notice of the hearing. He produced no medical evidence that he was unwell. I was sorry to go ahead although he was not represented, but the court could not be treated in this way. It was a case that could be finally determined as the plaintiff was going to call an expert witness.

He [the applicant] did not appear to be unwell and did not produce any medical evidence that he was ill, although I asked him for such evidence. I would be surprised if he did not have the necessary understanding of what was going on. An attempt was made to convey to the applicant's solicitor during the short adjournment that in the light of the weighty evidence which was in the process of being heard from the landlords, evidence the court could do no other than accept, expert evidence as none was provided by the tenant, I do D not think the respondents' solicitors managed to get through.

The reason for the applicant's solicitor's absence had been given over the telephone in an open conversation to the landlords' solicitors as being that Mr Miah would not pay their fees or enough money to secure representations. But that would not discharge their duty to the court and was most unhelpful. Either they should have taken their name off the record or appeared to have apologised to the court. But I do not intend to take any action in this respect and it would not be a proper case for me to direct them to appear because no adjournment was granted and they have not by their absence caused a great deal of delay, although they have caused some delay as the landlords' counsel has taken much longer over the facts than would have been necessary otherwise in order to ensure justice was done.

The hearing accordingly continued. It is right to say, looking at the E note we have before us of what occurred at the hearing, that the judge made efforts to ensure that no injustice was done to the applicant. In addition Mr Elland, who has appeared before us today and who appeared before the judge in the court below, went to considerable and commendable lengths to ensure that the matter was put before the judge fully, so that the applicant could understand what was happening and so that the judge could have all the relevant facts placed before him.

In the course of the evidence which was given on behalf of the respondents the judge caused them to put in evidence a report which had been prepared on behalf of the applicant as to the premises, but not to adopt it as part of their own case. That report was prepared by a Mr Ronald Creed FSVA FIAS IRRV, who is employed by the firm of F Drivers & Norris, who are surveyors of Holloway Road, which is in the area of the applicant's premises. In that report the valuer indicates that the current market rental value of the premises is £8,750 pa.

That report was the only evidence which was placed before the judge on the applicant's behalf. The rest consisted of evidence called by the respondents, the most important being that of Mr John Broomfield, a distinguished valuer who has considerable experience in these matters and who, until earlier this year, had managed on behalf of the respondents this property and a number of other properties in the immediate vicinity. In his evidence, which was based upon a report that was put before the judge, Mr Broomfield set out a carefully reasoned argument for supporting a rental for the premises of £14,000 pa. That evidence was supported by reference to a number G of comparable premises in the area including premises which were also owned by the respondents. The most significant evidence concerned the next-door premises, that is 129 Fortess Road. At the time of the hearing a new letting of 129 Fortess Road had been agreed, although no binding contract had been made, at a rent equivalent to £14,000 for the applicant's premises. However, the new letting of no 129 was subject to a reduction of £2,000 pa owing to the condition of the premises, the new tenant being under an obligation to spend a substantial sum in putting them into a better state of repair.

So far as the applicant himself is concerned, it is clear that he had difficulty in making himself understood at the hearing. In fact, he H played an extremely limited role. He was not present throughout, although he stayed longer than he wished, as the judge had indicated. We do not know why the applicant did not take a fuller part in the hearing. We do not have any more evidence before us than had the judge as to his state of health or why he was complaining of being ill on December 5. With regard to the non-attendance of his solicitors, it is accepted by Mr Azhar that he is not in a position to suggest that they had been put in funds to enable them to represent the applicant at the hearing. If they had not been put in funds, they could not be required to represent the applicant, although we strongly endorse the judge's view about the desirability of their attending the hearing on December 5 to explain their situation. They were on the record, and that makes their duty clear. We have not heard from them for the J purposes of this appeal. I do not propose to say any more about their part in the matter apart from this. By the luncheon adjournment the judge was clearly concerned about the future progress of the case. He therefore asked the respondents' solicitors to communicate with the solicitors on the record for the applicant. They did so by faxing a letter to those solicitors in which they notified them that they were acting on the instructions of the judge. They informed them that the hearing had not been adjourned, and that due to the lack of any evidence other than that called on behalf of the respondents the judge felt constrained to accept the respondents' evidence. The faxed letter continued:

The judge said he would endeavour to ensure that your client is fairly treated K but you must, of course, consider your position.

There is also before this court a note prepared by the respondents' solicitor of the exchanges which took place during the hearing that deal with the same matter. The judge has seen that note and has no objection to it. The terms in which the matter appears in that note are slightly different from those contained in the fax, but, in my judgment, not significantly different.

It is clear from that note that the judge considered that his task was to decide whether the rent should be that indicated in the applicant's valuation or the figure which the respondents were arguing was appropriate. At one stage the judge is recorded as saying:

What I got (*sic*) to decide is whether £9,000 right or £14,000.

L At another stage he is recorded as saying:

Subject to being satisfied that Mr Broomfield is a person who can be relied upon how (*sic*) one can substitute anything for figure he gives.

Mr Broomfield was questioned by the judge, and the judge decided that he could act on his evidence. He came to the conclusion that the initial rent under the new tenancy should be £14,000. He also dealt with the pattern of review, and came to the conclusion that the landlords' case was right. He also accepted the landlords' arguments on the other matter which was in dispute. He therefore made the appropriate order and granted the respondents the cost of the hearing of the application on scale 3.

Before this court Mr Azhar has relied almost exclusively on the first two grounds of his notice of appeal. The first is that the learned M judge wrongly exercised his discretion in refusing to allow the application for an adjournment to enable the applicant to obtain legal representation at the hearing. The second is that the learned judge misdirected himself when he allowed the hearing to proceed when it became apparent that the applicant was not able to represent himself.

As I understood Mr Azhar's argument in support of this appeal, those are two separate and distinct grounds. The first relates to the decision of the judge not to adjourn the hearing at the outset so that the applicant could obtain legal representation. The second is suggesting that during the hearing the learned judge misdirected himself because it should have become apparent to him that the

A applicant was not able to represent himself. That being so, the judge should either of his own volition have given him the opportunity of being represented or he should have given him the opportunity of being represented on his renewed application. Thus the judge should have granted the necessary adjournment.

In relation to both those grounds of appeal, it is important to bear in mind that they are matters which are singularly for the discretion of the trial judge. That being so, there are only limited circumstances in which this court can intervene. So far as the first ground of appeal is concerned, there is the additional point that technically there is a requirement to apply for leave to appeal and in this case no such application was made. However, because of the second ground of
B appeal, which, in my judgment, is the important ground, that is not a matter of significance. Speaking for myself, if any point turned on the absence of an application for leave, I would be prepared to grant an extension of time for applying, then grant leave, and treat the hearing of this appeal as both the hearing of the application for leave and of the substantive appeal.

However, returning to what should be the approach of the court, I refer shortly to a decision of this court in the case of *Maxwell* v *Keun* [1928] 1 KB 645 to which Mr Elland helpfully referred us. At p 650 there is a passage from the judgment of Lord Hanworth MR which I regard as particularly helpful. In it the Master of the Rolls said:

> It is obvious that there must not be any attempt to interfere with the discretion of the learned judge in reference to the trial. It is most important that we
> C should uphold that discretion. But in *Sackville West* v *Attorney-General,* to which our attention has been called, gives an illustration that, although it is only on very rare occasions that the discretion of the learned judge will be interfered with, yet that there may be such occasions. It is a decision of the Court of Appeal, and therefore one which is binding upon this court. The statement is made in the judgment, which is the judgment of Cozens-Hardy MR and Moulton and Buckley LJJ, that, although it could not be said that under no circumstances would the Court of Appeal be justified in interfering with the discretion of the learned judge in a Court below as to the proper mode and time of trying an action, yet it would only be in the most extraordinary circumstances that an application to review the decision of the learned judge as to the conduct of the business in his own Court could succeed; that the only case in which the Court of Appeal would so interfere would be if satisfied that the decision was such that, notwithstanding any exercise by the learned judge
> D of the power of control which he would have over the action when it came on for trial, justice did not result, and he had failed to see that such would be the effect of his decision.

There is also a passage in the judgment of Atkin LJ which is relevant. He said at p 653:

> The other point that was made by the defendants was that this was a discretionary order, and that the Court of Appeal ought not to interfere with the discretion of the learned judge. I quite agree the Court of Appeal ought to be very slow indeed to interfere with the discretion of the learned judge on such a question as an adjournment of a trial, and it very seldom does do so; but, on the other hand, if it appears that the result of the order made below is to defeat the rights of the parties altogether, and to do that which the Court of Appeal is satisfied would be an injustice to one or other of the parties, then the
> E Court has power to review such an order, and it is, to my mind, its duty to do so.

It seems to me that although those statements were made at a time when perhaps the court attached more importance to parties adhering to the procedural requirements and more rigid standards were often applied as to the time-limits than are today, they still provide guidance which this court should follow in considering the way in which a judge at first instance exercised his discretion in relation to matters such as adjournments. I, accordingly, propose to adopt the approach indicated in those passages.

Turning, therefore, to the first ground of appeal relied upon by Mr Azhar, I have come to the conclusion that the learned judge's decision cannot be interfered with. It seems to me that, although
F some judges may have taken a different view from that taken by His Honour Judge Dobry, he was perfectly entitled to come to the conclusion at that stage that the trial should proceed. He had no clear evidence as to the medical condition of the applicant; he had no proper explanation as to why he was not represented; and he had no reason to think at that time that the applicant would not be in a position to give the necessary assistance to the court to enable it to do justice between the parties.

However, I do not take the same view with regard to the second ground of appeal. As the hearing proceeded it must have become more and more apparent that the applicant was not capable of making any positive contribution. It is clear that by the luncheon adjournment the judge was concerned about the position in which he found himself. He took the view that, unless he came to the G
conclusion that Mr Broomfield was not a reliable witness, he would be forced to accept his valuation as to rent. I have reservations as to whether the judge was right in taking that view. By that stage he had already come to the conclusion that the written valuation of the applicant was virtually of no relevance, having been influenced by the fact that that valuation was not prepared as a report for the purpose of the hearing. It had a disclaimer which was in very broad terms, and in addition made no reference to comparables. However, although those criticisms could justifiably be made of the report, bearing in mind that it was not supported by oral evidence, it still seems to me that it was something to which the judge could have regard in evaluating the evidence of Mr Broomfield.

H
Being in the state of mind which I have indicated, the judge made an attempt to secure the attendance of the applicant's solicitors at a late stage. We know not what would have happened if they had attended the court. But they did not attend. The situation was one, therefore, where the court had powerful evidence from Mr Broomfield. However, although that evidence was powerful, it was evidence which was capable of being the subject of cross-examination because there were areas which required exploration, although Mr Broomfield may have been able to provide satisfactory explanations.

Some of those matters have been canvassed in the course of argument before this court. With regard to most of them Mr Elland, after taking instructions over the luncheon adjournment, has been in J
a position to provide explanations. Nevertheless, there were areas which could have been explored. In particular, it is right to say that there has recently been an arbitration in relation to a number of properties owned by the respondents and Mr Broomfield's evidence was not entirely accepted. That goes a long way to suggest that Mr Broomfield's valuations, although *bona fide* and genuine, are capable of being considered as overestimations of the proper value.

Quite clearly Mr Miah was incapable of subjecting Mr Broomfield's evidence to the sort of investigation which was necessary. The judge did his best, but the investigation that he could conduct was limited, and, in my judgment, he should have appreciated that. Instead he took the view, which, in my view, was K
wrong, that this was a case where there was no possibility of injustice to Mr Miah because, whatever representation he may have had, he could not have done more to undermine Mr Broomfield's evidence than the judge.

The fact of the matter is that it is very unusual for the evidence of an expert on a matter of this nature to be accepted in its entirety. It is not without precedent for that to happen, but it is unusual. It seems to me that, when one looks at the notes of what took place before the judge, it is apparent that there were areas which were not investigated in the correct way. Because of that, there is a substantial and material risk in this case that injustice was done to the applicant.

That is one side of the scale. What is the situation with regard to the L
other side of the scale? Would any injustice be caused to the respondents if the judge had taken the view that the situation was one where he was required to intervene by offering the applicant an adjournment? In my view, this was a case where it would have been possible for the judge substantially to have protected the position of the respondents. As a result of the order of the registrar, any new rent would start from the date when the previous tenancy expired. If there were delay, the only consequence would be a delay in the payment of the increased rent. It would not affect the *quantum* of that rent. In addition, the judge could have taken steps to ensure that the respondents were protected with regard to costs. The situation was one where, if the applicant were to be granted an adjournment, it would be granted only on the basis that the respondents were M
protected with regard to the costs thrown away. It was a case where there had been no delay. Remarkably perhaps in matters of this sort, the substantive hearing was taking place within three months of the expiry of the former tenancy. It was not a case where there had been other attempts to adjourn the matter on the part of the applicant. While it would be inconvenient for the respondents if the matter were adjourned, the consequences would not be the same for them as for the applicant, who would be very materially affected by the outcome of the application. That being so, it seems to me that not only was there a substantial risk of injustice being done to the applicant but the case was one where it could not be said that there was any countervailing consideration so far as the respondents were

concerned which prevented the judge from taking the course suggested in the grounds of appeal.

Accordingly, I have come to the conclusion that this is a case where this court can and should intervene. It should do so, however, only on the basis that it protects the respondents' position so far as possible, having regard to the order which this court is required to make. I am of the opinion that this appeal should be allowed on the basis that the applicant is entitled to a rehearing of his substantive application on the matters which were in issue previously. This will be subject to his paying into court, within a time specified by this court, an appropriate sum which represents a proportion of the increased rent which would be payable from the end of September 1990 to date, and also a sum which represents an appropriate amount in relation to the costs which he is likely to have to pay in respect of the costs thrown away by the need for a rehearing. So far as those costs are concerned, in my judgment, if the judge had offered the applicant an adjournment, it would have been on the condition that he paid those costs. Before this court he should have his right to achieve the same result but on the basis that he pays those costs.

Accordingly, I would allow the appeal and order a rehearing conditional upon the applicant fulfilling those requirements.

Agreeing, **STOCKER LJ** said: Although we are overruling the judge's exercise of his discretion, there is nothing that I would wish to add to the judgment of my lord. I agree with the conditions and order he has proposed.

NOURSE LJ also agreed and said: Because it is very rare for this court to interfere with a trial judge's refusal of an application to adjourn the trial, I will state my reasons in my own words.

Although other judges might have taken a different course, I think that Judge Dobry's decision to refuse the initial application for an adjournment was within his discretion. He was entitled to take a serious view of the applicant's lack of candour in giving a different reason for the non-attendance of his solicitors from that which had been given by the solicitors themselves. He was entitled to take the view that the applicant was not so unwell that he could not conduct his own case. He was entitled at that stage to take the view that the applicant was capable in other respects of conducting his own case.

However, it seems clear that as the case progressed the judge became concerned that the applicant was not, in fact, able to conduct his own case. That concern was manifested by the judge's observations at the end of the morning session and by his direction that a message should be sent to the applicant's solicitors. About that he said in his judgment:

> An attempt was made to convey to the applicant's solicitors during the short adjournment that in the light of the weighty evidence which was in the process of being heard from the landlord, evidence the court could do no other than accept as expert evidence as none was provided by the tenant. I do not think the Respondents' solicitors managed to get through.

It is also to be noted that in recounting his reasons for refusing an adjournment, the judge said:

> It was a case that could be finally determined as the plaintiff was going to call an expert witness.

Later he said in reference to the evidence of Mr Broomfield:

> Furthermore I entirely accept the answers that he gave to the question I put in order to attempt what might have been done if the tenant had been represented. I tried to get him to reduce the figures. He told me he had been giving expert evidence, he would have given the same evidence for the other side, and that he had not come here to negotiate with the court or anybody else. Having regard to his demeanour and the competent and highly professional way his evidence was prepared it is very unlikely that Mr Miah lost anything by not having another expert. The result would have been the same.

Those passages show that, in spite of his earlier concern, it must have been in the judge's mind when the second application for an adjournment was made that there was no substantial risk of injustice to the applicant in allowing the case to go ahead with only the evidence of Mr Broomfield on the all-important question of the amount of the rent. Although I can understand how it came about that the judge decided to carry on with the trial, I would respectfully disagree with him that there was no substantial risk of injustice to the applicant. I think that his first thoughts were best. It was not simply the lack of another expert. To my mind the more important point, as Woolf LJ has said, was that there was no one there to put careful and skilled questions to Mr Broomfield in cross-examination.

The judge evidently thought that he had done at least something of what was necessary in that respect. As to that, I must say that it is not, except within strict limits, the function of a judge to act as advocate to an unrepresented party and very few judges would regard it as being within their capacity. In the result, there was no proper opportunity for questions to be put to Mr Broomfield which might have had some effect in reducing his rental figure of £14,000.

For these reasons as well as for those given by Woolf LJ with which I entirely agree, I think that the only reasonable course for the judge to have taken on the occasion of the second application for an adjournment was to grant it on terms that the applicant should pay the costs thrown away in any event. The judge would no doubt have made it clear to him that that was the only chance he was going to get and that, if his present solicitors were not prepared to act for him any longer, he should get others to represent him. He could have added that at the adjourned hearing, whatever the state of the applicant's representation, the trial would go ahead. That course not having been taken, this court can and must interfere with the judge's decision and order a retrial of the issues which were in dispute at the commencement of the first trial.

I entirely agree that the appeal should be allowed and a retrial ordered only on condition that the applicant pays into court an appropriate amount within an appropriate period, as to both of which matters we will now hear submissions from counsel.

Appeal allowed. Applicant to pay £10,000 into court within 56 days or appeal dismissed. Terms as to costs of trial and appeal.

Queen's Bench Division

December 20 1991

(Before Mr Justice TUDOR EVANS)

HERBERT DUNCAN LTD v CLUTTONS

Estates Gazette June 6 1992

[1992] 22 EG 110

Landlord and tenant — Liability of original tenant for arrears of rent — Whether original tenant liable for rent during continuation tenancy under Part II of the Landlord and Tenant Act 1954 — Whether original tenant liable for interim rent — Whether tenancy and not term continued by the 1954 Act

By an underlease dated July 9 1985 the plaintiff demised 74 Grosvenor Street, London W1, to the defendant firm for a term commencing December 1 1984 and ending March 25 1990 at a rent of £70,000 pa — By a licence dated September 13 1988 the defendant assigned the term of the lease to Warrington plc — On July 5 1989, the plaintiff served on Warrington plc a notice under section 25 of the Landlord and Tenant Act 1954 terminating the lease on March 25 1990 — Following a counternotice given by Warrington plc to the plaintiff, and an application to the High Court for a new tenancy under Part II of the 1954 Act, the plaintiff issued a summons for the determination of an interim rent pursuant to section 24A of that Act — Warrington plc served notice discontinuing the application for a new lease and its occupation of the premises was terminated on February 5 1991 — The plaintiff and Warrington plc consented to an order in the interim rent proceedings that Warrington plc would pay £200,000 pa from March 25 1990 for so long as the lease continued under the provisions of the Act — Warrington plc paid £60,794.24 towards its liability for interim rent but it then went into receivership, leaving a balance of £113,452.05 unpaid — The plaintiff claimed this sum from the defendant as the original tenant — By clause 5(g) of the lease the expression "the said term" was stated, where the context admits, to include not only the term thereby granted but also the period of any holding over or any extension thereof whether by statute or at common law — The defendant denied liability for any rent during the period of the continuation tenancy, submitting that

a distinction must be drawn between the continuation of the tenancy and the continuation of the contractual term — Alternatively, if the defendant was liable for rent during the continuation tenancy, this was restricted to the rent reserved in the lease itself

***Held*: Judgment was given to the plaintiff — There is a consistent body of judicial opinion to the effect that the contractual term granted by the tenancy continues during the period of holding over by the 1954 Act — It follows that the term of the lease was extended by the 1954 Act and clause 5(g) came into play — The effect of the decided cases is that the words "tenancy" and "term" bear the same meaning and are interchangeable: see *Weinbergs Weatherproofs Ltd* v *Radcliffe Paper Mill Co Ltd, H L Bolton (Engineering) Co Ltd* v *T J Graham & Sons Ltd* and *Cornish* v *Brook Green Laundry* — The opinion of Nourse J in *GMS Syndicate Ltd* v *Gary Elliott Ltd* is to be preferred to that of Mr D Perrett QC (sitting as a deputy judge of the Queen's Bench Division) in *City of London Corporation* v *Fell* — The defendant's liability is not restricted to the original rent of £70,000 because the effect of section 24A is that the court determines the statutory rent during the period of holding over**

The following cases are referred to in this report.

Bolton (HL) (Engineering) Co Ltd v *Graham (TJ) & Sons Ltd* [1957] 1 QB 159; [1956] 3 WLR 804; [1956] 3 All ER 624, CA
Castle Laundry (London) Ltd v *Read* [1955] 1 QB 586; [1955] 2 WLR 943; [1955] 2 All ER 154
City of London Corporation v *Fell* [1992] 1 EGLR 95; [1992] 02 EG 172
Cornish v *Brook Green Laundry Ltd* [1959] 1 QB 394; [1959] 2 WLR 215; [1959] 1 All ER 373; [1959] EGD 116; (1959) 173 EG 307, CA
GMS Syndicate Ltd v *Gary Elliott Ltd* [1982] Ch 1; [1981] 2 WLR 478; [1981] 1 All ER 619; (1980) 41 P&CR 124; [1981] EGD 331; 258 EG 251
Junction Estates Ltd v *Cope* (1974) 27 P&CR 482; 232 EG 335
Plesser (A) & Co Ltd v *Davis* (1983) 267 EG 1039
Weinbergs Weatherproofs Ltd v *Radcliffe Paper Mill Co Ltd* [1958] Ch 437; [1958] 2 WLR 1; [1957] 3 All ER 663

This was a claim for arrears of rent by the plaintiff, Herbert Duncan Ltd, the leasehold owner of 74 Grosvenor Street, London W1, against the defendant, Cluttons, a firm of chartered surveyors.

Jonathan Brock (instructed by Forsythe Kerman) appeared for the plaintiff; Barry Denyer-Green (instructed by Wray Smith & Co) represented the defendant.

Giving judgment, **Tudor Evans J** said: I give judgment in open court at the request of both counsel. This is a summons by the plaintiff for the determination of a point of construction and/or law under Ord 14A of the Rules of the Supreme Court. If the arguments are resolved in the plaintiff's favour, the plaintiff asks for final judgment under Ord 14. There is also a summons for judgment under Ord 14 in respect of a claim for dilapidations.

There is no dispute as to the facts relating to the summons under Ord 14A. The plaintiff is the leasehold owner of premises at 74 Grosvenor Street, London W1. By an underlease ("the lease") dated July 9 1985, the plaintiff demised the premises to the defendant for a term beginning on December 1 1984 and ending on March 25 1990 at a rent of £70,000 pa payable by equal quarterly payments in advance. Pursuant to a licence dated September 13 1988 between the plaintiffs, the defendant and Warrington plc ("the assignee"), the defendant assigned its interest in the lease to the assignee for the residue of the term.

On July 5 1989, the plaintiff served on the assignee a notice under section 25 of the Landlord and Tenant Act 1954 terminating the lease on March 25 1990. The assignee gave notice to the plaintiff that it was not prepared to give up possession of the premises and on November 23 1989 applied to the High Court for a new tenancy under Part II of the Act. On December 18 1989 the plaintiff issued a summons for the court to determine an interim rent pursuant to section 24A of the Act. But the assignee subsequently served notice discontinuing the application for a new lease, thereby bringing its occupation of the premises to an end on February 5 1991. The plaintiff and the assignee then agreed and on February 18 1991 a consent order was made in the proceedings for an interim rent that the assignee would pay £200,000 pa from March 25 1990 for so long as the lease had continued under the provisions of the Act. The assignee paid £60,794.24 towards its liability for interim rent but it then went into receivership, leaving a balance of £113,452.05 outstanding. The plaintiff claims this sum from the defendant as the tenant.

The question I have to determine is whether the defendant, as the original tenant under the lease, is liable, on the true construction of the lease and/or as a matter of law, to pay the interim rent of which the balance remains unpaid.

On behalf of the plaintiff, Mr Brock submitted that since the assignee held over after the termination of the lease and since the defendant by an original covenant in the lease is bound to pay the rent for such period of holding over, it must follow that the defendant is liable for the balance of the interim rent due to the plaintiff. The relevant clauses are these:

(1) By clause 1 of the lease dated July 9 1985, the plaintiff demised the premises to the defendant:

> . . . to hold the demised premises unto the lessee for a term commencing the 1st day of December [1984] and expiring on the 25th day of March [1990] (hereinafter called "the said Term") yielding and paying therefor during the said Term the rent of £70,000 . . .

(2) By clause 2(1)(a), the defendant covenanted:

> . . . to pay the yearly rent hereinbefore reserved and made payable at the times and in the manner at and in which the same is hereinbefore reserved . . .

The words "hereinbefore reserved" in clause 2(1)(a) refer to the words "the said term" in the habendum and that phrase is defined by clause 5(g) of the lease as follows:

> The expression "the said term" shall where the context admits include not only the term hereby granted but also the period of any holding over or any extension thereof whether by statute or at common law.

Part II of the Landlord and Tenant Act 1954 applied to the premises. The assignee held over under the provisions of section 24(1) of the Act, which provides:

> (1) A tenancy to which this Part of this Act applies shall not come to an end unless terminated in accordance with the provisions of this Part of this Act; and, subject to the provisions of section 20 of this Act, the tenant under such a tenancy may apply to the court for a new tenancy —
> (*a*) if the landlord has given notice to terminate the tenancy, or
> (*b*) if the tenant has made a request for a new tenancy in accordance with section 26 of this Act.

It must therefore follow, submitted Mr Brock, that there was a holding over by statute within the meaning of clause 5(g) and the defendant has therefore covenanted to pay the rent during "the said term", which phrase includes the period of holding over.

On behalf of the defendant, Mr Denyer-Green submitted that the defendant was not liable for any rent during the period of holding over. A distinction must be drawn between the continuation of a tenancy and the continuation of the contractual term. Mr Denyer-Green contended that the effect of the language of section 24(1) is merely to continue the relationship of landlord and tenant with a right to occupy but that it does not extend the contractual term and therefore there can be no obligation to pay the rent. The section simply confers a statutory right to continue in occupation, upon such terms as to interim rent as the landlord and tenant may agree or the court may impose, until the tenancy is terminated as the result of a landlord's notice or until a new tenancy is granted. Mr Denyer-Green submitted that since the term cannot be extended into the period of holding over by the Act, then clause 5(g) by its language does not come into play and the defendant cannot be liable for the rent during the period of holding over.

There is a formidable body of judicial opinion against the construction of section 24(1) which was advanced by Mr Denyer-Green. Before I refer to the cases, I should mention that the learned editors of *Woodfall's Law of Landlord and Tenant* (28th ed), para 22.042, express an opinion, citing authority, to some of which I shall refer, which is contrary to the argument for the defendant.

> . . . section 24(1) provides for the old tenancy simply to carry on. This it does quite automatically without any notice or other action needed to be given or taken by either party. In this respect the statutory continuation under section 24 is like a statutory tenancy under the Rent Acts, but there the resemblance ends. Under the Rent Acts the statutory tenancy is not a tenancy at all; it is a mere personal privilege of exemption from eviction. But section 24 of the Landlord and Tenant Act continues the existing tenancy with a statutory variation as to the mode of determination. If the tenant holds over paying rent, which is accepted, after the contractual end of his tenancy, he will be assumed to do so under section 24 and not under a new tenancy at common law. The estate in the land vested in the tenant is by statute prolonged until it is brought to an end in one of the prescribed ways. So all the terms and conditions of the contractual tenancy, save those relating to the termination

A by the landlord of the contractual relationship, will continue to govern the continuation tenancy under the Act.

A little later the editors write:

> It

— the statutory continuation of the tenancy —

> is not a personal privilege of the tenant. It is a piece of property which he can assign or dispose of to a third person provided it was not prohibited by the terms of the contract.

The first case to which it is convenient to refer is *Weinbergs Weatherproofs Ltd* v *Radcliffe Paper Mill Co Ltd* [1958] Ch 437, a decision of Harman J (as he then was). The question which fell to be decided was whether the landlord's notice to quit was effective in
B view of the fact that no notice to quit had been given in accordance with section 25 of the 1954 Act. Harman J said at p 445, referring to the words "a business statutory tenancy" used by Sellers J (as he then was) in *Castle Laundry (London) Ltd* v *Read* [1955] 1 QB 586:

> This is a convenient phrase so long as one does not confuse it with the so-called statutory tenancy under the Rent Restrictions Acts, that, of course, being not a tenancy at all, whereas having regard to the language used in the Act of 1954 the term must be thought of as continuing by way of a statutory extension (see the observations of Denning LJ in *H L Bolton (Engineering) Co Ltd* v *T J Graham & Sons Ltd* where the following appears: "In his Lordship's opinion the right view was that the common law tenancy subsisted with a statutory variation as 'to the mode of termination'"). Sellers J also speaks of the "contractual relationship" between the parties.

C The second case is *GMS Syndicate Ltd* v *Gary Elliott Ltd* [1982] Ch 1, a decision of Nourse J (as he then was). I must refer briefly to some of the facts. The plaintiff was the landlord of premises, the upper part of which was residential. The plaintiff sublet the ground floor and basement to the first defendant who, with the consent of the plaintiff, sublet the basement to the sublessees and they in turn assigned the benefit of the subtenancy to the second and third defendants. The second and third defendants covenanted directly with the plaintiff and the first defendant to observe the covenants "during the residue of the term granted by the underlease as extended by the tenancy agreement". After the expiration of the contractual term, the second and third defendants held over under Part II of the Landlord and Tenant Act 1954. The basement was used for immoral purposes.
D There were covenants against a user for any illegal and immoral purpose. The residents objected to the user. Nourse J at p 7E said:

> One of the questions in these proceedings is whether that direct covenant expired with the contractual term . . . or whether it continues while [the second and third defendants] hold over (as they now do) under Part II of the Landlord and Tenant Act 1954.

Nourse J held at pp 9G and 10D:

> The convenient course will be for me to deal next with the question whether the direct covenant by [the second and third defendants] with the plaintiff contained in the deed of December 23, 1974, expired with the contractual term on June 30, 1977, or whether it continues while [the second and third defendants] hold over under Part II of the Landlord and Tenant Act, 1954. As
> E I have said, the material words are "and henceforth during the residue of the term granted by the underlease as extended by the supplemental tenancy agreement." On one reading of those words it certainly might be said that the covenant expired with the contractual term, on the ground that that, and no more, was the term granted by the agreement of December 29, 1969, as extended by that of January 30, 1974. But it seems to me that such a reading ignores the effect of section 24(1) of the Act of 1954, which provides that a tenancy to which Part II applies shall not come to an end unless terminated in accordance with the provisions of Part II. It is now well established that this means that the term granted by the tenancy continues by way of a statutory extension and with a statutory variation as to the mode of determination: see for example, *Cornish* v *Brook Green Laundry Ltd* [1959] 1 QB 394, 409. In the circumstances, it seems to me that the term granted by the first agreement as extended by the second is still subsisting, albeit that it has been further
> F extended by the Act of 1954. It is still the same term. On that footing, it seems to me that the better reading of the material words in the direct covenant is to construe the obligation as continuing so long as [the second and third defendants] hold over under Part II of the Act of 1954. I am fortified in that conclusion by the knowledge that the rival construction would produce an anomalous state of affairs which cannot have been within the contemplation of the parties to the deed.

The next case to which I should refer is *Cornish* v *Brook Green Laundry Ltd (supra)*, a decision of the Court of Appeal consisting of Jenkins, Romer and Pearce LJJ. It is unnecessary to refer to the facts save to say that both Brook Green, who held a lease from the trustees, and the plaintiff, who was Brook Green's sublessee, held over after the termination of the lease under the protection of Part II of the Landlord and Tenant Act 1954. The court had to decide whether, in the period of holding over, Brook Green was the plaintiff's landlord. G
This required the court to consider the effect of section 24(1) of the Act. The reserved judgment of the court was delivered by Romer LJ. He said at p 408:

> By reason of section 24(1) of the Act of 1954, both Brook Green's tenancy from the trustees and Mrs Cornish's subtenancy from Brook Green continued in operation notwithstanding that, as a matter of contract, they both expired in September 1956; and, as neither tenancy was terminated subsequently in any manner authorised by the Act, the continuation of both tenancies was still effective when, in September 1957, Brook Green gave notice of their intention to oppose Mrs Cornish's application for a new lease. In *H L Bolton (Engineering) Co Ltd* v *T J Graham & Sons Ltd* Denning LJ considered the position which arises from the continuation of a tenancy under the Act of 1954, and said that in his opinion the right view was that the common law H
> tenancy subsisted with a statutory variation as to the mode of determination. In *Weinbergs Weatherproofs Ltd* v *Radcliffe Paper Co Ltd* Harman J adopted the view of Denning LJ and said that "having regard to the language used in the Act of 1954 the term must be thought of as continuing by way of statutory extension".

Although none of the cases to which I have referred was directly concerned with the effect of holding over in circumstances such as exist in the present case, there is a consistent body of judicial opinion to the effect that the contractual term granted by the tenancy continues during the period of holding over by statute. The words of section 24(1) are: "A tenancy to which this Part of this Act applies shall not come to an end . . .". Unless, therefore, it is valid to construe the word "tenancy" in the limited sense for which Mr J
Denyer-Green contends and not as being synonymous with the contractual term save for the question of termination, it must follow that the term of the lease was extended by the Act and then it must follow that clause 5(g) does come into play, contrary to the submission of Mr Denyer-Green.

I have been referred to the decision of Mr Desmond Perrett QC (sitting as a deputy judge of the Chancery Division) in *City of London Corporation* v *Fell* (unreported, July 10 1991)*. The same point arose on the facts of that case as exist in the present, save that the lease in that case did not contain any clause such as clause 5(g). The plaintiff, the lessors, had granted a lease of business premises to the defendant for a period of 10 years from March 25 1976. In 1979 the lessees assigned the demised premises for the residue of the term. At the K
expiration of the term, the assignees held over under the provisions of Part II of the Act of 1954. But in December 1986 the assignees were compulsorily wound up. The plaintiff contended that the contract between the original parties to the lease continued during the statutory extension of the contractual lease and thus the tenant remained liable for the rent since the obligation to make payment continued. The defendant argued that the lease defined the extent of time during which the defendant, as original lessee, remained liable for the payment of rent and that, the rent having been paid by the lessee during the contractual term, the lessor was not entitled to any further payment.

The judge held that it was the tenure of the business premises and L
not the original term which was sustained by the operation of the Act, that is to say, it was the tenancy and not the term that was carried over.

The cases to which I have been referred were cited to the judge and I must relate how he distinguished them. Having referred to the observations of Denning LJ (as he then was) in *H L Bolton (Engineering) Co Ltd* v *T J Graham & Sons Ltd (supra)* and having quoted the passage from the judgment of Harman J in *Weinbergs Weatherproofs Ltd* v *Radcliffe Paper Mill Co Ltd (supra)* the judge continued [at p 174]:

> . . . But it appears from Denning LJ's judgment in *Bolton Engineering* that at no time was he referring to anything other than "the tenancy"; that is to say, the right to occupy pursuant to the lease, and that it was this tenancy or right to M
> occupy which subsisted with the statutory variation by reason of the Act. Denning LJ made no mention of "the term" in its strict sense as continuing. It was in no way essential to his decision in *Weinbergs* that Harman J should have an eye to any difference in meaning between the words "tenancy" and "term", but, in my view, if he was basing himself upon Denning LJ's view as expressed in *Bolton Engineering*, he was using the word "term" as if it were synonymous with "tenancy". In my view, and for the purposes of the instant case, the words "tenancy" and "term" must bear their precise meaning and are not to be regarded as interchangeable, as they clearly were in the context of their use by Harman J.

The judge distinguished the judgment of the court in *Cornish* v *Brook*

*Editor's note: Now reported at [1992] 1 EGLR 95 *ante*.

A *Green Laundry* in which the passage from the judgment of Harman J was cited with approval by the court, saying [at p 174]:

> The Court of Appeal in that case were not concerned to differentiate between the words "term" and "tenancy" in their task of ascertaining whether the landlord was a competent landlord within the meaning of Part II of the 1954 Act, and the precision in interpretation which seems to me to be essential in this case was unnecessary to the decision in *Cornish* . . . It is clearly the law (see *Cornish*) that when the contractual tenancy is by reason of Part II to be thought of as continuing by way of a statutory extension, the tenancy continues upon the same terms as were contained in the contractual lease before its expiry. That is to say, the landlord and the party entitled to the tenancy as continued by statute are bound by the same covenants as were contained in the original contractual lease. The contractual lease, after expiry of the contractual term, defines the rights and obligations of the parties to the
> B statutory extension of the tenancy. That statutory arrangement holds no benefit to the original lessee. If he is not in possession at the date of expiry of the contractual term, he is afforded no right to possession. He is not privy to the statutory consequences in any way. Whether an assignee of the original lease elects to hold over after the expiry of the contractual term is a matter entirely beyond the control of the original lessee.

As to the judgment of Nourse J in *GMS Syndicate* v *Gary Elliott Ltd*, the judge held [at p 175]:

> . . . Nourse J adopts the use of the word "term" as being that which Part II of the Act is extending. Although I do not doubt that the learned judge would have reached the same conclusion had he referred to the extended term as "the common law tenancy subsisting on the same terms" it is in my view not correct to hold that Part II extends "the term", if by that is meant the original
> C contractual period, for the original lessee has no tenancy or right of occupation after the expiry of the term and it is the assignee in occupation whose rights are extended by the statute, albeit upon the same conditions and subject to the same covenants as governed the rights of the original parties to the lease. Thus, I do not differ from Nourse J's conclusion that a landlord may successfully sue an assignee in possession, or sublessee in possession, in respect of breaches of covenant during the period of statutory extension of the tenancy. But I respectfully decline to follow the proposition on p 10 at B: "It is now well established that the term granted by the tenancy continues by way of statutory extension".

In my opinion, it is inescapable from the authorities to which I have referred that it was held in them that the contractual term continues during the period of holding over, subject to the statutory variation
D as to determination. I consider that the effect of the decisions in those cases is that the words "tenancy" and "term" bear the same meaning and are interchangeable. Harman J in *Weinbergs Weatherproofs Ltd* v *Radcliffe Paper Mill Co Ltd* made his observations in direct contemplation of the language of Denning LJ. In the quoted passage from *H L Bolton (Engineering) Co Ltd* v *T J Graham & Sons Ltd*, when Denning LJ said that the common-law tenancy subsisted with a statutory variation as to the mode of termination he was, as I understand the language, stating expressly that all features of the tenancy (that is the contract) subsisted, except that termination could take place only in accordance with the statutory provisions. In *Cornish* v *Brook Green* the Court of Appeal observed that Harman J adopted the language of Denning LJ.

E If the argument for the defendant in this case is correct, the learned editors of *Woodfall's Law of Landlord and Tenant* have also fallen into error because, in a passage I have already cited, they state that all the terms and conditions of the contractual tenancy save termination continue and they use the word "tenancy" in the context of all the terms and conditions of the contract, thus equating the two terms.

I am faced with two persuasive authorities, the decision in the *City of London* case and the observations of Nourse J in the *GMS Syndicate* case, fully backed by the other authorities to which I have referred. I prefer to follow the opinion of Nourse J that it is well established that the term granted by the tenancy does not come to an end but continues into the period of holding over with the statutory variation as to the mode of determination.

F It follows that the terms, save that as to termination, continued during the holding over by the assignee and, specifically, the liability to pay the rent. Mr Denyer-Green's submission would mean, as Mr Brock pointed out, that a defendant or assignee holding over would not, for example, be liable for the rates. In this lease, the lessee covenanted "from time to time and at all times during the said term" — which is a defined phrase by clause 5(g) — "to pay the rates and other outgoings". If Mr Denyer-Green is correct, who will pay the rates in a situation in which a tenant or assignee holds over under the lease, a holding over which the landlord cannot determine? The phrase "the said term" is used in many of the tenant's covenants in this lease and from the nature of these covenants these phrases have to cover a period of holding over; for example, apart from the rates, the covenant to keep in good and substantial repair the whole of the G demised premises.

I was referred to cases in which a landlord had sought to recover rent from a guarantor. In *Junction Estates Ltd* v *Cope* (1974) 27 P&CR 482, guarantors joined in a lease for seven years covenanting that the tenant would pay the rent. The lease was assigned and the assignees remained in possession after the term under the provisions of the Act of 1954. They failed to pay the whole of the rent due. The lease was forfeited and the landlord claimed the unpaid rent from the guarantor. The action failed. McKenna J held that, upon its proper construction, the guarantee clause applied only to the lease during the seven-year term and not to any statutory extension of the term. The case does not, as I understand it, lay down any general principle H applicable in the present case. It turned on the construction of the relevant clause but McKenna J observed that if the landlord wanted to impose a longer liability then he should make it clear in the language of the guarantee.

The question here is whether the plaintiff has done so in the present case. In a second guarantee case, *A Plesser & Co Ltd* v *Davis* (1983) 267 EG 1039, French J followed *Junction Estates Ltd* v *Cope* and doubted the relevance of the judgment of Nourse J in *GMS Syndicate* to the case which was before him. But he did not question the remarks of Nourse J which I have already considered.

I therefore hold that the old tenancy or term (which words are interchangeable in meaning) continues into the period of a holding over under section 24(1). The question is whether clause 5(g) provides J that the defendants shall be liable for the payment of rent during the period of holding over. It is worthwhile recalling the language of clause 5(g):

> The expression "the said term" shall where the context admits include not only the term hereby granted but also the period of any holding over or any extension thereof whether by statute or at common law.

The holding over within the meaning of the clause was pursuant to the Landlord and Tenant Act. The clause provides that "the said term" shall include the period of holding over. The words "the said term" carry one back to the habendum and the obligation by clause 2(1)(a) "to pay the . . . rent hereinbefore reserved", that is, during "the said term". In my judgment, the defendant's liability to pay rent during the period of holding over follows inexorably. K

Mr Denyer-Green submitted that if all his contentions fail, as I fear they have, and if the defendant is liable for the rent during the period of holding over, as I have held it is, then the liability is restricted to the original rent, £70,000 pa. The provision in the habendum is "yielding and paying . . . during the said term the rent of £70,000". It follows, argued Mr Denyer-Green, that the liability must be limited to £70,000 during the said term, that is including the period of holding over. Mr Brock submitted that this approach is wrong because, pursuant to section 24A of the Act of 1954, it is the court that determines what is the statutory rent which it would be reasonable for the tenant to pay during the period of holding over. I was attracted to Mr Denyer-Green's argument but I think that Mr Brock is right and that the court L determines, albeit here by a consent order, recording the agreement between the plaintiff and the assignee, the rent under section 24A. As Mr Brock pointed out, it is in the nature of a statutory rent review.

I therefore answer the question under Ord 14A in favour of the plaintiff. It follows that the plaintiff is entitled to judgment for £113,452.05 under Ord 14. The claim for dilapidations has been agreed between the parties. It is agreed that there will judgment for the plaintiff, the damages to be assessed by an official referee and I make that order accordingly.

Judgment for the plaintiff with costs. Leave to appeal to the Court of Appeal.

M

Chancery Division

January 29 1992

(Before Mr Justice Harman)

SIMONITE v SHEFFIELD CITY COUNCIL
HOLMES v SAME
T G THOMPSON LTD v SAME
SIMONITE v SAME

Estates Gazette June 20 1992

[1992] 24 EG 134

Landlord and tenant — Landlord and Tenant Act 1954 — Determination of rent and interim rent

The defendants, in each of four applications by the plaintiffs for the grant of new tenancies under the Landlord and Tenant Act 1954, are Sheffield City Council and owner of two markets, Sheaf Market and Castle Market in the City of Sheffield — Pursuant to section 25 of the 1954 Act, the defendants gave notice determining some 384 tenancies of stalls in the two markets on March 24 1986 — The defendants did not oppose the grant of a new tenancy to each recipient of a notice — Four applications were selected by the parties for hearing as to the determination of rent and interim rent as samples representative of the 384 tenancies — The parties described the new tenancies as quarterly with a review of rents payable not to be made until three years from the date of grant of each new tenancy — In giving evidence on behalf of the plaintiff tenants, Mr Christopher Tattersall FRICS asserted that the defendants were the owners of a franchise to conduct markets within the manor of Sheffield; they therefore had a monopoly on the supply of stalls and they must be assumed to behave reasonably by offering stalls without the tenant's being subject to pressure because there could be no other supplier — He further asserted that at least 384 stalls were vacant and available to let because they were simultaneously the subject of section 25 notices — Mr Michael Newton FRICS, in giving evidence for the defendants, asserted that inflation would remain a factor

***Held*: Although the defendants, as owners of the franchise to conduct markets, had a right recognised by the law as entitling them to the sole right to conduct markets within the defined area, there was no reason why that property right should not be fully exploited by the defendants — There was no duty to reduce the income available to the defendants from rents and other income derived from their market franchise by increasing the number of stalls available so as to create conditions of oversupply — The supply of stalls is not to be assumed to be large and each stall must be considered in the context that most other stalls are already let — The plaintiffs' valuer had omitted to notice that each section 25 notice also stated that the landlords would not oppose the grant of a new tenancy — The assertion that inflation is likely to remain was dangerously speculative now that the UK had joined the exchange rate mechanism of the European Monetary System — In determining the interim rent as at April 30 1986 the appropriate figure is 75% of the current rent as determined in each case**

No cases are referred to in this report.

These were four adjourned originating summonses and counterclaims in which Messrs Simonite, Holmes, Simonite, and T G Thompson Ltd sought the determination of the rent for the grant of new tenancies pursuant to Part II of the Landlord and Tenant Act 1954 and the defendants in each case, Sheffield City Council, sought the determination of an interim rent.

Keith Rowley (instructed by Dibb Lupton Broomhead, of Sheffield) appeared for the plaintiffs; David Neuberger QC and Martin Rodger (instructed by the solicitor to Sheffield City Council) represented the defendants.

Giving judgment, **Harman J** said: I have before me four applications by plaintiffs for the grant of new tenancies under the Landlord and Tenant Act 1954, Part II. All four applications are by tenants whose tenancies were determined by notices sent pursuant to section 25 of the 1954 Act which, it is agreed, determined the former tenancies on March 24 1986. At the same time a very large number, I was told over 380, similar tenancies were also determined on the same date. The four applications now before the court have been selected by agreement between the plaintiff tenants and the defendant landlords, who are the council of the City of Sheffield, as "sample" applications. I was also told that all the section 25 notices determining all tenancies were in the form which states that the landlords will not oppose the grant of a new tenancy to each recipient of a notice.

The four applications selected by the parties for hearing are intended to be "samples" of the various sorts of holding so as to represent the 380-plus tenancies. All the tenancies are of market stalls in two large covered markets built and owned by the city council. The four applications before me are in respect of stalls nos 219-220 in the Sheaf Market, stall no 18 in the meat and fish section of the Castle Market, stall no 144 in the miscellaneous section on the upper-ground floor of Castle Market and stalls nos 66-67 in the miscellaneous section of the lower-ground floor of Castle Market.

As can be seen from the above short identifications the four stalls are from four differing parts of the two markets. Each is intended to be representative of the general run of stalls in the respective parts of the markets. I am asked to determine the proper amount of rent payable for each of these four stalls with the intent that from those determinations appropriate figures for new rents under the 1954 Act can be worked out without all the remaining 380 or thereabouts applications each having to come before a court for determination. Consequently the evidence before me concentrates very little on the proper rent payable for each of the four particular stalls before the court and much more upon general surveys of rents payable in the two markets, or more precisely in each of the four identified sections of the two markets. The parties' agreement and the emphasis of the evidence are made notwithstanding that the evidence agrees that the actual attractiveness of particular stalls within each section varies according to its position in the section. Both sides apparently accept that it is in each of their interests to have a uniform level of assessment of basic rent payable for a stall within each of the four identified sections.

This agreement is made notwithstanding that the evidence for both sides agrees that not only the individual position of a particular stall within a section will affect its attraction to a tenant and therefore the rent that it would command in a true open market but also that the permitted trade to be carried on at a particular stall will carry a different level of profitability from that of another trade at the next-door stall. It must in common sense follow that in a true open market the particular trade permitted would affect the level of rent which a tenant would be prepared to pay for the tenancy of that stall with that permitted user. Thus, two important factors which the court would have considered in deciding what would be the rent properly payable for a particular stall upon the statutory hypothesis provided by the 1954 Act are deliberately and by agreement of both sides excluded from the court's consideration of the rents which are to be determined upon these four applications. This agreement, and its consequent effect upon the evidence given before me, has made these applications extremely difficult to decide and makes the reasoning to support the decisions more difficult to follow through on a logical path.

The task of reaching a logical decision is made even more difficult by the nature of the tenancies under consideration and of the new tenancies which the city council landlords are prepared to grant. Both landlords and tenants describe the tenancies as quarterly tenancies but which the landlords can determine only by six months' notice. That is, of course, one of the inevitable effects on any business tenancy of the 1954 Act, since no section 25 notice can take effect in less than six months, but it is unusual to find the legal result expressed as the parties' agreement and makes the description "a quarterly tenancy" seriously inapt.

The new tenancies intended to be granted to applicants under the 380-plus section 25 notices are to be quarterly tenancies determinable by the landlords on six months' notice, but in addition a review of rents payable under the new tenancies will not be made until three years from the date of grant of each new tenancy. It seems to me that

A a much more accurate description in law of such a tenancy would be the grant to each tenant of a three-year term but with a proviso enabling the tenant to determine the tenancy by one quarter's notice. As a matter of experience the rent commanded by the grant of a three-year term is greater than the rent commanded by a quarterly tenancy; that remains the fact even when the so-called quarterly tenancy cannot be determined by the landlords on less than two quarters' notice. In my judgment, confused descriptions of the subject-matter for consideration lead to confused formulation of reasoning about the subject-matter.

One further difficulty of obtaining useful evidence upon which to base the determination of rent is caused by the fact that many,
B probably most, of the existing tenancies in the markets are assignable and all four of the plaintiff applicants before me hold tenancies which permit assignment. But more recently the city council have changed the form of their standard tenancy agreement to include in new lettings an absolute prohibition against assignment. Naturally the rent commanded by the grant of a tenancy with the right to assign will be greater than the rent commanded on the grant of a non-assignable tenancy. This obvious fact makes evidence as to levels of rent payable on the grant of new tenancies by the city council substantially less reliable as a comparable from which to assess the rent for these four tenants.

The evidence given before me consisted of three individual witnesses called by the plaintiff tenants. First, they called Mrs
C Holmes, the stallholder of stalls nos 219-220 in Sheaf Market, who carries on a haberdashery business there and is president of the Market Traders Association. She had acquired another stall, no 218, immediately adjoining her present stalls, in March 1990. The assignor from whom she acquired the tenancy had held at least three other stalls as well as no 218. Mrs Holmes paid no premium on the assignment but agreed to indemnify the assignor against rent and rates including any increased rent. Mrs Holmes had also some little time ago acquired a tenancy of stall no 237 by tender to the city council at a rent of £1,000 per quarter but was not granted by the city council the same user for stall no 237, haberdashery, as she normally carried on. She failed to make stall no 237 a success and had determined the tenancy. She was not cross-examined.

D The plaintiffs also called Mr Carter, the tenant of stalls nos 106 and 107. He had traded from those stalls for about six years and had paid a premium of £25,000 for the goodwill and fixtures and fittings of the business carried on by his predecessor in those stalls. The goodwill included the trading name under which the business had been and continued to be conducted. At the same time, Mr Carter assigned the tenancy he then held of stall no 110, together with the goodwill built up by his trading in that stall under his own name, for a premium also of £25,000. He told me that his purchaser continued now to trade at stall no 110 under Mr Carter's name. Mr Carter also told me that in the autumn of 1990 he tried to sell his present stalls with his present handbag business but although he had had two inquiries he had no
E serious offer. He was asked in cross-examination the price he had been asking and answered £20,000 plus stock at valuation. I found his evidence perfectly reliable.

The plaintiffs also called Mr Harrington, who held three stalls in Castle Market, nos 187, 188 and 189. His business was selling men's and boys' clothing. His family had long connection with that market, his father having been one of the first tenants when the market hall was built. Mr Harrington had held several other stalls, all in what was called in the evidence phase V of the markets. He agreed in cross-examination that none of the four stalls with which this case is directly concerned is in phase V and that phase V was regarded by market traders as the least attractive part of all the market halls. He had failed to make a success of stall no 199, which he acquired for a
F premium of £6,000, and eventually, after failing to sell the business, he surrendered that lease to the city council. Mr Harrington told me that his trading conditions, both turnover and profit, had deteriorated in each of the last two years. I found Mr Harrington a convincing witness.

The bulk of the evidence called was expert evidence. The plaintiffs called Mr Christopher Tattersall, a Fellow of the Royal Institution of Chartered Surveyors, qualified in 1967 after a degree in estate management, who has since 1986 been senior partner in his firm. He lives and practises in Sheffield and has had substantial experience of advising on stall rents in these markets and also other markets. He had had long negotiations with the expert witness called by the city council before me in an endeavour to agree figures for the various rents. Both sides' expert witnesses are obviously very well qualified, G
both generally and in respect to these particular holdings, and naturally the professional integrity and personal honesty of neither was challenged.

The expert called on behalf of the city council was Mr Thomas Michael Newton; his paper qualifications were very similar to those of Mr Tattersall and similarly impressive. Mr Newton's experience was a little wider than that of Mr Tattersall, since he spent five years after becoming an ARICS in practice in the City of London and three years in Nottingham. However, since about 1963 Mr Newton has practised in Sheffield as a partner in a local firm. In 1987 eight regional partnerships amalgamated to form the large commercial consultant surveyors under the name Lambert Smith Hampton. Mr
Newton is the third senior listed Fellow of the Royal Institution of H
Chartered Surveyors in the very long list of partners printed on his proof.

I will comment on the parts of the evidence of each surveyor which seemed to me to be challenged where either the challenge seemed to weaken the evidence or seemed to confirm it. Mr Tattersall showed that the city council are the owners of a franchise to conduct markets within the manor of Sheffield by acquisition from one of the Dukes of Norfolk. Mr Tattersall in para 2.3300 onward of his proof characterised the ownership as conferring a monopoly on the city council. He asserted at para 2.3306 that for the city council to retain their monopoly imposed a duty to operate that monopoly to the
benefit of the citizens and at para 2.3309 that the city council as J
landlords must be assumed to behave reasonably by offering stalls without the tenant's being subject to pressure because there could be no other supply of the market stalls than from the city council. He referred (para 2.3310) to "abuse of monopoly position". Finally, at paras 2.3318 and 2.3319 he doubts the value of evidence from actual transactions on the ground that the city council have a monopoly power and use that power to maximise profit.

All this part of Mr Tattersall's evidence is, in my judgment, founded on an unsound premise that in some way the city council, because they owned the market franchise, owed an extra duty as landlords to their market tenants. The city council are the owners of an item of incorporeal property which they acquired by purchase.
That right is recognised by the law as entitling the city council to the K
sole right to conduct markets within the defined area. There is no reason why that property right should not be fully exploited by the city council. Of course the city council will have regard to the interest of all the electors within the corporation's boundaries and in that sense they will consider what is for the greatest benefit of the city of Sheffield. But the city council have no duty to reduce the income available to the city council from rents and other income derived from their market franchise by increasing the number of stalls available so as to create conditions of over-supply. Nor can the city council be under any duty to permit others to infringe the franchise by opening markets within the area where the city council are entitled to a monopoly so as to allow others to increase the supply of stalls and so L
reduce the price in rent of the city council's stalls. The general law imposes no duty on an owner of property (and a market franchise is obviously property) to do other than deal honestly with those who wish to use that property or part of it. I reject Mr Tattersall's evidence on this part of his proof.

Mr Tattersall also asserted in his proof of evidence (para 2.231), first, and with classic correctitude, that the price (he wrote "value", but in my judgment he must have meant "price", since "value" involves other than pure economic questions) of all commodities, including stalls in a market hall, is governed by supply and demand. But he added in the same paragraph that the state of supply was taken by him to be "at least 384 Stalls [were] vacant and to let". In response
to a question from me, Mr Tattersall said that that did not mean that M
he envisaged a huge, echoing, vacant market hall with 384 stalls vacant. In my judgment, it was difficult to understand the words in his proof "vacant and to let" otherwise, but the evidence clearly is that he did not envisage the 384 stalls as vacant and available. Mr Tattersall asserted that he meant that 384 stalls are simultaneously the subject of section 25 notices terminating each of the tenancies. He seemed to omit to notice that each section 25 notice also stated that the landlords would not oppose the grant of a new tenancy. In my judgment, the "supply" of stalls is not to be assumed to be large and each stall must be considered in the context that most other stalls are already let.

Mr Tattersall's proof dealt with tendered rents at para 3.400

A onwards to para 4.130. He regarded such rents as a poor, though perhaps the only, guide to true open market rents. Mr Newton deals more shortly with tendered rents starting in the course of his para 7.1 with an analysis at para 7.8. In my judgment, both valuers are correct in regarding the tendered rents available as evidence in this case as a very unsatisfactory basis upon which to reach a figure for a true open market rent.

In addition to that difficulty it is clear that in putting in a bid on a tender the person making the bid has to consider "what are the premises worth to me?". If for some reason of his own the person making the bid had a particular desire or need for those premises he (or she) may bid well above the true open market price so as to be sure of a good chance of obtaining the premises. I say only "a good B chance" because in the city council's, as in most other, tender processes, the tenders are invited upon the condition that the city council are not bound to accept the highest (or any) tender and the actual examples show several cases where a tender was accepted by the city council which was not the top bid. It is commonly accepted that the true open market price for an object is most objectively assessed at auction. There the successful bidder has to top all other bidders by the relevant sized "step" in the bidding process. But it is notorious that successful bidders at auction frequently say "I would have gone higher if I had had to". The true open market price of the premises or other subject of the auction was below the "value to me" for that bidder.

C In addition, all the tenders for stalls since 1986, which were the only ones referred to before me, were for non-assignable tenancies of market stalls. Thus the difficulty of assessing by how much the value of an assignable tenancy exceeded that of a non-assignable tenancy was added to the natural unreliability of tender bids as true evidence of open market prices. In the result I conclude that the tender evidence before me is of very limited value and I have merely had it in the very back of my mind in trying to assess the figures.

Mr Newton, as expert witness for the city council, asserted (para 8.1.5) that it seemed "inevitable that inflation is likely to remain with us". That view seems to me to be dangerously speculative in the new economic circumstances imposed by the UK having joined the exchange rate mechanism of the European Monetary System. On all D sides the country is told that the purpose of that political step was to prevent the tendency to inflation historically displayed over the past 40 years in this country and to tie this country to the monetary tendencies of the German Bundesbank so as to ensure that in future inflation will be minimal. Whether that objective will be achieved is obviously uncertain, but so also is the prospect of continued inflation. Mr Newton's evidence is based, no doubt, on the universal experience of all of us whose adult lives have covered the past 30 years, but times may be changing.

Each witness gives elaborate reasons at great length showing how he has reached his conclusion as to the proper open market rent for these four stalls. Each is an honest, experienced and skilled valuer. E Since both are human neither is infallible. In the end, as Mr Tattersall F says at para 4.410, the conclusion is a matter of judgment. I have to form an opinion as to whose judgment I believe is more reliable and in the end, having read and reread the very extensive proofs, considered my notes of cross-examination and, above all, formed a judgment from my observations of each of these two gentlemen in the witness box, I consider that Mr Newton's judgment is the more reliable. I do, however, consider that Mr Harrington's evidence as to falling turnover and profits over both the past two years and Mr Carter's evidence of inability to sell his stall, show that the market is suffering. I also consider that no assessment should be based simply on a belief in future inflation. I have therefore reduced Mr Newton's valuation by what, in my judgment, is the appropriate fraction of one-tenth. G

I thus decide that the appropriate figures for the open market rents of the plaintiffs' stalls are as follows:

(a) stalls nos 219-220, Sheaf Market....................................£7,064 pa
(b) stall no 18, meat and fish section, Castle Market................£8,568 pa
(c) stalls nos 66-67, miscellaneous section, lower-ground floor, Castle Market..£13,990 pa
(d) stall no 144, miscellaneous section, upper-ground floor, Castle Market .£7,466 pa

Finally, I have to consider four applications by the city council for an interim rent for each stall. These are to take effect from April 30 1986, as I was told by counsel. Mr Tattersall does not deal in his proof with interim rents, but he was asked questions about his opinion in cross-examination. He asserted that 70% of the figures for rent given H in his proof would represent the 1986 open market rent of each stall if let on an annual tenancy. Mr Newton dealt with interim rents in his proof. He said in cross-examination that if the open market rents today were the same amount as the figures he had provided as the rents obtainable in 1986 then he still stood by his 1986 valuations. On the basis that, in his professional judgment, rents today are higher than in 1986, I found that part of Mr Newton's evidence difficult to accept. Mr Newton's valuation in his proof was that 1986 rents should be fixed at 80% of the current rent. I am not satisfied that that is entirely justified and, in my judgment, the appropriate figure is 75% of the current rent as I have determined it.

I therefore fix the interim rents as follows: J

(a) stalls nos 219-220...£5,305.50 pa
(b) stall no 18...£6,425.25 pa
(c) stalls nos 66-67...£10,492.50 pa
(d) stall no 144...£5,599.50 pa

I realise that in the real world it is most unlikely that open market rents would be fixed at figures involving pence per annum, but these figures are inevitably somewhat arbitrary, since they are entirely retrospective and are based on scanty evidence which in both valuers' evidence was simply based on a proportion of current rents. Such an exercise is likely to produce slightly unreal figures, but, in my judgment, the figures represent a proper determination of this question on the evidence available to me in these four cases. K

For a further case on this subject see p 281

LANDLORD AND TENANT
LEASEHOLD REFORM

Court of Appeal

February 5 1992

(Before Lord Justice DILLON and Lord Justice STEYN)

MALPAS v ST ERMIN'S PROPERTY LTD

Estates Gazette May 2 1992

[1992] 17 EG 112

Leasehold Reform Act 1967 — Meaning of "house" — Whether building can reasonably be called a house, notwithstanding that it has two separate front and back doors, and two floors occupied by separate families as their residences

The respondent, Mrs Malpas, satisfied all the requirements of the Leasehold Reform Act 1967 for acquiring the freehold of 22-22a Clarendon Road, Colliers Wood, London SW19 ("the premises"), provided the building was a house within the meaning of section 2(1) of the Act — The premises were constructed as part of a terrace in 1905 as a pair of adjoining maisonettes — Alternatively, they were converted in the early 1920s from a single residence into a pair of maisonettes — The front of the premises contains two front doors, one for no 22 and the other for no 22a, which are slightly recessed with a sloping slate-covered porch running over the two of them — At the rear of the premises the appearance is of a single Edwardian house except that there are two back doors, one serving the ground floor and the other serving the first floor — Unlike the position in *Sharpe* v *Duke Street Securities NV*, the respondent holds the unexpired residue of the long lease of the premises as a whole and occupies the upper flat or maisonette, no 22a; the ground-floor flat, no 22, had been sublet by her predecessors for the unexpired residue of the term less one day and was occupied by a separate family — The appellant landlords, St Ermin's Property Co Ltd, appealed the decision of Judge Willis that the respondent was entitled to acquire the freehold of the premises

***Held*: The appeal was dismissed — Although the premises could reasonably be called a building divided horizontally into two flats or maisonettes, it was quite plain that it can also reasonably be called a house, notwithstanding that it had two front doors and two back doors — The concept of a house divided into maisonettes is one which the Act recognises — Further, the parcels clause in the lease described the premises as a messuage or dwelling-house**

The following cases are referred to in this report.

Lake v *Bennett* [1970] 1 QB 663; (1969) 213 EG 633

Sharpe v *Duke Street Securities NV* (1987) 55 P&CR 331; [1987] 2 EGLR 106; 283 EG 1558; 19 HLR 506, CA

This was an appeal by St Ermin's Property Co Ltd from a decision of Judge Willis in the Croydon County Court (February 22 1991) whereby he held that the respondent, Mrs Malpas, was entitled to acquire the freehold of 22-22a Clarendon Road, Colliers Wood, London SW19.

Christopher Russell (instructed by Hamlin Slowe) appeared for the appellants; Rabinder Singh (instructed by James O'Neill, of Morden) appeared for the respondent.

Giving judgment, **DILLON LJ** said: This is an appeal by landlords, St Ermin's Property Co Ltd, against a decision of Judge Willis in the Croydon County Court given on February 22 1991, whereby he held that the applicant in the proceedings, the respondent to the appeal, Mrs Malpas, was entitled to acquire the freehold of a property called 22-22a Clarendon Road, Colliers Wood, London SW19, under the Leasehold Reform Act 1967.

The issue raised by the appeal lies in a very small compass. It is whether the building, 22-22a Clarendon Road, is reasonably called a "house". Section 1 of the 1967 Act provides that Part I is to have the effect of conferring:

> on a tenant of a leasehold house, occupying the house as his residence, a right to acquire . . . the freehold or an extended lease of the house and premises. . . .

The meaning of "house" is set out in section 2. Section 2(1) provides that for the purposes of the Act:

> "house" includes any building designed or adapted for living in and reasonably so called, notwithstanding that the building is not structurally detached, or was or is not solely designed or adapted for living in, or is divided horizontally into flats or maisonettes; . . .

The subsection goes on to provide that:

> (*a*) where a building is divided horizontally, the flats or other units into which it is so divided are not separate "houses", though the building as a whole may be; and
> (*b*) where a building is divided vertically the building as a whole is not a "house" though any of the units into which it is divided may be.

The intention seems to have been to ensure that the acquisition of the freehold of a house should not result in there being flying freeholds of parts of buildings above freeholds belonging to someone else. It is expressly provided in section 2(2) that:

> References . . . to a house do not apply to a house which is not structurally detached and of which a material part lies above or below a part of the structure not comprised in the house.

It is not in doubt that the respondent, Mrs Malpas, satisfies all the requirements of the Act for acquiring the freehold if the building is a house and the only question is whether it is a house reasonably so called.

What the building is, we can see from the colour photographs with which we have been helpfully supplied. Nos 22 and 22a were constructed as part of a terrace in 1905. It seems that they were originally constructed as a pair of maisonettes adjoining nos 20 and 20a similarly so constructed as a pair of maisonettes. There are many others in the same road constructed in the same way. It seems probable that they were originally constructed as maisonettes. The only alternative is that they were converted in the early 1920s from a single residence into a pair of maisonettes. But if that is so, it would be odd that so many others were so converted at the same time.

I take the photograph of the front view of nos 22 and 22a. I see what looks like an ordinary Edwardian house with a ground floor and a first floor and a roof above, except for one matter, which is that there are two front doors next to each other, one for no 22 and the other for no 22a, which are slightly recessed with a sloping slate-covered porch running over the two of them. Next door, nos 20 and 20a appear to be the mirror image. The front doors in nos 22-22a are to the right of the facade of the building as you look at it. Immediately next to them are the front doors of nos 20 and 20a under a similar sloping porch, the ends of the two porches being supported apparently by the same pillar. At the back again the appearance is of a single Edwardian house except that there are the two back doors, one serving the

A ground floor and the other serving the first floor. The garden is divided so as to be shared by each of the maisonettes having a separate part. Nothing turns on that. Down the side to the left of nos 22-22a there is a side passage. Nos 22-22a are in fact of the same sort of construction as was considered by Fox LJ in *Sharpe* v *Duke Street Securities NV* (1987) 55 P&CR 331*.

But the crucial factor in that case, which is not present in the present case, is that both maisonettes were occupied by one family as their residence and a connecting door had been opened in the partition wall which, when the maisonettes were separately occupied, divided the entrance hall for the upper maisonette from the entrance hall for the lower maisonette. In the *Sharpe* case it was held that what was occupied in that case by the one family as their residence was B reasonably to be called a house and therefore the tenants were entitled to acquire the freehold. In the present case the respondent, Mrs Malpas, has the unexpired residue of the long lease of the building as a whole, with its front and back gardens, and occupies the upper flat or maisonette, no 22a. TThe ground-floor flat, no 22, has been sublet by her predecessors for the unexpired residue of the term of the whole less a single day and is occupied by a separate family, the subtenants, as their residence. There is not in this case the factor, which there was in some of the authorities to which we have been referred, of mixed commercial and residential use. The use of the upper and lower maisonettes has always been residential.

The question then comes down to a very short one. Can the C building reasonably be called a house notwithstanding that it has the two separate front and back doors and the two floors are occupied by separate families as their residences? It is well established (see, for instance, the judgment of Fox LJ in the *Sharpe* case and the judgment of Salmon LJ and the other members of this court in *Lake* v *Bennett* [1970] 1 QB 663) that if the building can reasonably be called a house, the fact that it might also reasonably be called something else is not D relevant. This could reasonably be called a building divided horizontally into two flats or maisonettes, but to my mind it is quite plain, looking at the photographs, that it can also reasonably be called a house notwithstanding that it has two front doors and two back doors. I do not regard the fact that there are the two front doors as meaning that this is not a house and some longer circumlocution has to be used to describe it. The concept of a house divided into maisonettes is, of course, one which the Act recognises and which is familiar to anyone who sees what has happened to Victorian houses in central London built in those spacious days for occupation by a single family but now subdivided horizontally into maisonettes sharing a single entrance.

I agree, therefore, with the conclusion which the judge reached. E

There is the further factor on which the judge also relied: that in the 1905 lease the property demised is demised as a:

> piece or parcel of land at Merton in the County of Surrey situate on the north side of a road called Devonshire Road West and having a frontage thereto [and the various dimensions are given]. Together with the messuage or dwelling-house and premises erected thereon and known as [and then the description Number 1 Beresford Villas is given].

It seems, therefore, that at that time the conveyancer who was disposing by long lease of these various buildings was content to describe each as a messuage or dwelling-house and saw nothing odd about that. That is a factor which certainly does not assist the landlords in asserting that the building is not a house. F

For these reasons, which are substantially those of the judge in a careful judgment, I would dismiss this appeal.

Steyn LJ agreed and did not add anything.

Appeal dismissed with costs.

*Editor's note: Also reported at [1987] 2 EGLR 106.

LANDLORD AND TENANT
RENT ACTS

A
Court of Appeal

October 28 1991

(Before Lord Justice LLOYD and Lord Justice STUART-SMITH)

LEWIS-GRAHAM AND ANOTHER v CONACHER

Estates Gazette January 18 1992

[1992] 02 EG 171

Rent Act 1977 — Resident landlord — Whether two adjoining dwellings were at all material times one building for purposes B of section 12 of the Rent Act 1977 — Appeal by tenant dismissed

The plaintiff landlords were the owners of a semi-detached house to which a predecessor in title had added an extension by way of a granny flat and garage — In July 1987 the plaintiffs let a room in the extension to the appellant — At the time of the grant of that tenancy there was a common gas, electricity and water supply to the house and extension, although the intercommunicating doors were locked and sealed with sound insulation as the extension had its own front door — A notice to quit was served on the appellant expiring in August 1990 — Prior to the notice to quit expiring the plaintiffs carried out C building works blocking off the two doorways between the properties — The trial judge found that although the plaintiffs initiated work to create two buildings, the work to separate the two properties had not been completed at the date of the expiration of the notice to quit and at the date of the trial the properties still constituted one building — The tenant appealed, contending that effectively all the work to create two separate buildings had occurred

***Held*: The question as to whether there was one building or two was a question of fact for the trial judge — That question in the present case was a borderline one — The trial judge was D entitled to take into account that until the time of the building works the properties were plainly one building and the building works were not so extensive as to make a great difference — The sealing of the doors was a difference in kind rather than in substance — The trial judge was entitled to hold that the works to create two separate buildings had not been completed and that the tenancy was not a protected tenancy, and therefore did not become a statutory tenancy, by virtue of section 12 of the Rent Act 1977**

The following case is referred to in this report.

E *Bardrick* v *Haycock* (1976) 31 P&CR 420; *sub nom Bardrick* v *Vernon* [1976] EGD 321; (1976) 239 EG 427, CA

This was an appeal by Marjorie Conacher from the judgment of Judge Stockdale, sitting at Barnet County Court, when he made an order for possession of 198a Nether Street, Finchley, London N3, in favour of the landlords, Derek and Patricia Lewis-Graham.

Philip Walter (instructed by S Newman & Co) appeared for the appellant; James Simpson (instructed by David D Lewis & Co) represented the respondents.

Giving the first judgment at the invitation of Lloyd LJ, STUART-SMITH LJ said: This is an appeal from the judgment of Judge Stockdale, sitting at Barnet County Court, given on July 10 1991, F when he made an order for possession of premises known as 198a Nether Street, London N3, and an order for mesne profits in the sum of £2,931.75. The defendant tenant now appeals against that judgment.

The learned judge found in favour of the respondent landlords on two grounds: first, that the premises were not within the protection of the Rent Acts by virtue of section 12 of the Rent Act 1977; and, second, that the plaintiffs reasonably required possession of the residence for themselves, that is to say, for the female plaintiff, who was the landlady, and her family, and that it was reasonable to make the order under Case 9 of the Rent Acts. In order to succeed in this appeal the appellant must succeed on both those grounds. We have heard the first point argued by Mr Walter on behalf of the tenant and, G in view of the conclusion which we have formed, it has been unnecessary to consider the second point.

Shortly put, the issue on the first point under section 12 of the Rent Act 1977 is whether 198 and 198a Nether Street are and were at all material times one building or whether, by virtue of some work which was done in the summer of 1990, they had become two buildings.

The history of the premises is as follows. No 198 is part of two semi-detached houses. It is on the left-hand side as you look at it from the street. It was built in about 1920. At some time before the plaintiffs bought the house an extension was added. It seems that the purpose of the extension was to provide a granny flat and to add a garage. The extension was formed by means of building two rooms between the H existing no 198 and the garage. Those rooms were roofed by a flat roof. There was then a room above the garage with a gabled roof to tie in with the other gables of the house. Originally there was a room on the ground floor, a garage next to it, an outside entrance, and two rooms above.

The plaintiff, Mrs Lewis-Graham, and her mother bought the house in 1969. At that time she was not married to Mr Lewis-Graham. She lived there with her son and her mother lived in the granny flat. At that time there were communicating doors at both ground-floor and first-floor levels from the original no 198 to what became no 198a. But it seems that Mrs Lewis-Graham's mother wanted to receive her own mail, and for that reason the entrance door of the granny flat was called 198a so that she should have a separate J address. Towards the end of her life she lived on the ground floor.

Mrs Lewis-Graham's mother died in 1980. Mrs Lewis-Graham then became the sole owner and lived in the premises with her husband. At that time there was one central heating system for the whole building, no 198 and no 198a. There was a common gas supply, a common electricity supply and a common water supply. The house was singly rated and was subject to a single water rate. The judge found — and there is no dispute about this — that at that stage and, indeed, at the time of the grant of the tenancy to the defendant there was no question but that it was one building.

On the mother's death, the plaintiffs decided to let both the rooms in no 198a, and for that purpose the intercommunicating doors K between no 198 and no 198a were locked and sealed with sound insulation in such a way that it was not possible to pass through from no 198 to no 198a. Other than that, everything remained as before. The judge concluded that the sealing up of those doors made no difference; the building was still one building. On July 24 1987 the downstairs room was let to the defendant by an agreement described as "Furnished accommodation 198a Nether Street . . . (Part of 198 Nether Street)", and it was a furnished letting. By 1990 the plaintiffs' other ventures into the property market became unsuccessful and they were in considerable financial difficulties. They decided that they would make use of a planning permission which had already

A been granted and make a separate dwelling of 198a. For that purpose certain building works were done. Notice to quit was served on the defendant, which expired in August 1990.

The learned judge rightly directed himself that the issue with which he had to be concerned was whether at all times during the existence of the tenancy the building was one building or whether by reason of the building works which were carried out in the summer of 1990 what had in fact been one building had become two buildings. The learned judge heard evidence, and his findings of fact are to be found at pp 15-24 of the bundle. He said:

> Mr O'Hara [the builder] was less competent than Mr Gardiner and seemed determined to go about his contract in his own way, disregarding his client's wishes and instructions. Mr and Mrs Lewis-Graham say he was told not to do B the work in the defendant's room at an early stage until the room became available. Yet he chose to block off the two doorways between the properties at an early stage. It is not finished yet but, before the notice to quit had expired, he sealed off with one layer of materials the gaps linking the houses. It is said by Mr Simpson that this is only temporary and can be knocked down. One can always knock holes in walls. People are constantly knocking down walls and when they do so they often have to put an RSJ in but an RSJ would not be called for here. However, it is not difficult to reverse the works where a door has been there before.
>
> I am satisfied that the work to separate the two properties has not yet been completed apart from the central heating. The works in the defendant's bedroom are not yet completed as the defendant's washing and cooking facilities remain and there is painting and decorating to be done. The works in the former garage are completed. What remains to link the two houses? There C is the gas central heating. So far as the electricity is concerned there is no direct evidence either of separate supplies or of separate meters. It may well still come in via 198. I do not know what the situation was regarding 198a — *vis-à-vis* 198 but there has been no reason to cut the supply off. I was reminded of what Mrs Lewis-Graham said but there is nothing to indicate a separate supply. There was no direct evidence about it. There is nothing to enable me to decide definitely. Rating is no longer relevant but in the past rates and water rates were one and title at the Land Registry remains joint. I accept however that there is an intention on the part of the plaintiffs to separate the titles to 198 and 198a and to have two separate buildings. Had that separation been effected by the expiry of the notice to quit? It is possible to complete severance days before expiry of a notice to quit and then the section 12 exemption will go. Counsel are agreed that there must be one building on expiry of the notice to quit for section 12 exemption to prevent the creation of a statutory tenancy. D The landlord to get the benefit of section 12 has to show there was one building not only at the time when the tenancy was granted but also at the end of the tenancy. Had separation been completed between 198 and 198a to a sufficient extent by the time the notice to quit expired? Well, it had got pretty close by August 1990. Had the defendant co-operated it might be that the works would have been completed and title separated. However, the works were not completed by August 1990. The back garden is still as one; the front is still as one. I find as a matter of fact that complete separation had not yet occurred and it had not become two buildings by the end of August 1990. It has still not occurred and they are still one building today. I find the resident landlords remained the resident landlords at all times and therefore it is not a protected tenancy and no statutory tenancy sprang up.

Philip Walter, on behalf of the appellant, has criticised that E finding. He submits that effectively all the work had been done. He concedes that the central heating had not been separated, but he says that that is *de minimis*. He submits that the learned judge was wrong in his conclusion about the electricity, because in the course of her cross-examination Mrs Lewis-Graham did not refer to that matter as being something which was outstanding, and he submits that in order for the kitchen to be installed upstairs there must have been a separate supply installed downstairs. For my part, I do not accept that submission at all. He says that other matters, such as the removal of the shower from the appellant's bedroom, which was something which was due to take place but had not taken place, was something which could not make what was two buildings into one and the fact that there was still redecoration work to be done again did not alter F that position.

The question whether it was one building or two was a question of fact for the county court judge. That was held by this court in *Bardrick* v *Haycock* (1976) 31 P&CR 420*, and I think it would be helpful to read a passage in the judgment of Scarman LJ. He said at p 424:

> The English word "building" covers an immense range of all sorts of structures. It is an ordinary English word, and its meaning must therefore be a question of fact, always assuming that the court directs itself correctly as to the intention and meaning of the statute which uses it. As a matter of law, to give a defined or precise meaning to the word "building" is an impossibility. It is beyond the capacity of even the most consummate master of the English G language to do so. This itself is, in my judgment, an indication that Parliament is leaving the question of fact to the judge. One must, moreover, remember that the county court judge is likely to be in the best possible position to determine within the area of his jurisdiction what are the housing circumstances, what is the local situation, and the way in which he should approach an answer to this question of fact.
>
> I have no doubt that the intention of Parliament in enacting section 5A

— the predecessor of section 12 —

> was to relieve landlords where it was reasonable to do so. One must remember that the Act of 1974 introduced furnished tenancies into Rent Act control. This meant that resident landlords, who had under the earlier legislation counted on being able to require furnished tenants in the same building as that in which they lived to leave, could no longer do so. Therefore, the mischief at which the section was aimed was the mischief of that sort of social H embarrassment arising out of close proximity — close proximity which the landlord had accepted in the belief that he could bring it to an end at any time allowed by the contract of tenancy.

In my judgment, the question of whether this had become two separate buildings was a borderline one, but the learned judge was entitled to take into account that until the time of the building works it was plainly one building. The building works were not so extensive as to make a great difference. The sealing up of the doors was a difference in kind rather than in substance, because previously no one had been able to get through, and it seems to me plain that the learned judge was right in holding that the works had not yet been completed. It was accepted by Mr Walter that the central heating had J not been connected. That, in my judgment, was not a *de minimis* matter. There was still decorative work to be done, although that may be not so significant, and the learned judge found that the gardens had not yet been separated as had been originally intended.

For my part I am not entirely convinced that, even when the building work was finished, this would have amounted to two buildings rather than one. Where something has been built as an extension to an existing house, where the outer walls are continuous and the roofs are contiguous, it may be that it still remains as one building notwithstanding that it was intended to be two separate dwellings. Be that as it may, for my part I would not interfere with the learned judge's finding of fact. I do not think that the criticisms made by Mr Walter affect this decision, and I would dismiss the appeal. K

LLOYD LJ agreed and did not add anything.

Appeal dismissed.

*Editor's note: Also reported at (1976) 239 EG 427.

Queen's Bench Division

June 21 1991

(Before Mr Justice AULD)

N & D (LONDON) LTD v GADSDON L

Estates Gazette January 18 1992
[1992] 02 EG 176

Decision of London Rent Assessment Committee — Applicant landlord had requested a rent of £500 per month — Committee determined a rent of £5 per month — Application to set aside decision of committee on the ground that the low rent failed to take into account the obligations of the respondent tenant's predecessor to redecorate and keep the premises in good repair M

The applicant is the landlord of a residential dwelling-house — The respondent tenant succeeded to the tenancy of the premises previously held by his late father — His late father had let the premises fall into disrepair — The landlord applied to a London Rent Assessment Committee under section 14 of the Housing Act 1988 for the determination of an open market rent for an assured tenancy — Section 14(2)(c) provides that in determining the rent the committee is to disregard "any reduction in the value of the dwelling-house attributable to a failure by the tenant to comply with any terms of the tenancy"

— The present tenant's father had been granted a lease in July 1983 and agreed to "re-decorate the interior of the premises as often as may reasonably be necessary and to keep in good repair the glass in and locks on all doors and windows and the sanitary apparatus waste water and drain pipes throughout the term and in such repair to yield up the same at the determination of the tenancy" — A rent was registered for the premises at £30 per week from August 15 1988 — The present tenant's father died in July 1989 — The landlords had been unable to gain access to the premises or to carry out any repair — Neither party requested a hearing before the committee — The committee visited the premises and concluded that the premises were in a deplorable condition, dangerous to health and were virtually unlettable in the open market — They determined a rent of £5 per month — The committee concluded that the internal disrepair was long standing and could not be attributed, except in a very small measure, to the period of the present tenancy

In seeking to set aside the decision of the committee, it was submitted on behalf of the landlord that section 14(2)(c) of the 1988 Act required the committee to disregard any failure by the present tenant's predecessor to comply with the terms of the tenancy as well as by the present tenant

***Held*: The application was dismissed — Section 14(2)(c) refers to "the tenant" and "the tenancy" and is a clear indication that a rent assessment committee are concerned only with the assured periodic tenancy over which they have jurisdiction — The definitions of those terms in section 45 of the 1988 Act make no reference to a predecessor or statutory tenant or tenancy — The draftsman of the statute could have inserted some formula in section 14(2)(c) if the obligations of a predecessor in title were to be taken into account in applying the disregard — It is quite clear from the provisions of Schedule 1 to the Rent Act 1977 (as amended) and its companion provisions in section 39(5) and (6) of the 1988 Act that there is a new interest created for the successor upon the death of a statutory tenant — It is not simply a continuing interest — It creates a new periodic tenancy as an estate in land which the previous tenant's statutory entitlement under the Rent Act was not — It is clear that the effect of section 14(2)(c) is that the sins of the predecessor of the tenant whose reference is being considered are not to be visited upon him — There was no principle of construction, whether described as purposiveness or fairness, which could override clear words of the statute**

The following case is referred to in this report.

Jones v *Wrotham Park Settled Estates* [1980] AC 74; [1978] 3 WLR 585; [1978] 3 All ER 527; (1978) 37 P&CR 289; 246 EG 223, CA

This was an application by notice of motion by the applicant, N&D (London) Ltd (a subsidiary of the Northumberland & Durham Property Trust Ltd), the landlord of a dwelling-house at 10 Chilham Close, Perivale, Middlesex, to set aside the decision of a London rent assessment committee determining the rent of the property at £5 per month.

Guy Fetherstonhaugh (instructed by Church Adams Tatham & Co, of Reigate) appeared on behalf of the applicant; the respondent did not appear and was not represented; David Elvin appeared as *amicus curiae*.

Giving judgment, **AULD J** said: By a reamended notice of motion dated August 2 1990 the Northumberland & Durham Property Trust Ltd, the landlord of a dwelling-house at 10 Chilham Close, Perivale, Middlesex, seeks to set aside the decision of a London rent assessment committee made on February 20 1990 determining the rent of that property and for an order remitting the matter to the rent assessment committee.

The tenant is a man named John Gadsdon. The applicant had requested a rent of £500 a month, but the rent assessment committee considered that £5 a month was appropriate because of the poor condition of the property. John Gadsdon had "inherited" the tenancy from his late father, who had let the property fall into disrepair.

There is now only one ground of appeal. It is that the committee in determining that low rent had failed to take into account the obligations of John Gadsdon's father to redecorate the interior and to keep it in good repair. John Gadsdon has not appeared on this application; nor have the rent assessment committee in any formal sense, not being parties to the application. However, in addition to hearing counsel for the applicant, Mr Fetherstonhaugh, I have also had the advantage of Mr Elvin as counsel instructed on behalf of the committee but acting, for all practical purposes, as *amicus*.

The issue concerns the construction of section 14(2)(c) of the Housing Act 1988. Section 14 provides for the determination of rent by a rent assessment committee on an open market rent for an assured tenancy. The section appears in Part I of the 1988 Act, a part which the editors of the *Encyclopedia of Housing Law and Practice*, at p 950, describe as introducing what is in substance, if not in name, a new Rent Act, containing assured tenancies in place of protected tenancies, and shortholds, and applicable to private landlords and housing associations.

Provision is made in section 13 of the Act for the increase of rent of assured periodic tenancies, subject to the protection given by section 14(4)(a) to refer a proposed increase to a rent assessment committee. Section 14 sets out the basis upon which a rent assessment committee must approach its task in each case.

So far as material, section 14 provides:

(1) Where, under subsection (4)(a) of section 13 above, a tenant refers to a rent assessment committee a notice under subsection (2) of that section, the committee shall determine the rent at which, subject to subsections (2) and (4) below, the committee consider that the dwelling-house concerned might reasonably be expected to be let in the open market by a willing landlord under an assured tenancy —

(a) which is a periodic tenancy having the same periods as those of the tenancy to which the notice relates;
(b) which begins at the beginning of the new period specified in the notice;
(c) the terms of which (other than relating to the amount of rent) are the same as those of the tenancy to which the notice relates;
(d) in respect of which the same notices, if any, have been given under any of Grounds 1 to 5 of Schedule 2 to this Act, as have been given (or have effect as if given) in relation to the tenancy to which the notice relates.

(2) In making a determination under this section, there shall be disregarded —

(a) any effect on the rent attributable to the granting of a tenancy to a sitting tenant;
(b) any increase in the value of the dwelling-house attributable to a relevant improvement carried out by a person who at the time it was carried out was the tenant, if the improvement —
  (i) was carried out otherwise than in pursuance of an obligation to his immediate landlord, or
  (ii) was carried out pursuant to an obligation to his immediate landlord being an obligation which did not relate to the specific improvement concerned but arose by reference to consent given to the carrying out of that improvement; and
(c) any reduction in the value of the dwelling-house attributable to a failure by the tenant to comply with any terms of the tenancy.

The 1988 Act is directed to the determination of an open market rent rather than a fair rent, as under the regime of the Rent Acts. However, in the context of section 14(2)(c) of the 1988 Act, the clear Parliamentary intention appears to have been to prevent a tenant under an assured periodic tenancy from securing a lower rent than would otherwise be appropriate by causing or permitting the property to fall into disrepair in breach of such terms of the tenancy as require him to keep it in a good condition. But for that provision, the malign or irresponsible tenant could artificially reduce the appropriate figure for open market rent for the property and thus pay less rent than would otherwise be appropriate having regard to the nature of the property and the terms of his tenancy.

The facts here, giving rise to the question whether under section 14(2)(c) John Gadsdon was entitled to have the breaches of obligation by his father disregarded, are as follows.

His father, Arthur Gadsdon, was granted a lease of the property on July 29 1983 for five years at a monthly rent of £22.58. The property is a semi-detached house, consisting of five rooms, kitchen, bathroom and lavatory, and a front and back garden, in a cul-de-sac of modest houses near Perivale tube station. Under the lease the father, as tenant, undertook by clause 2(iii) to "re-decorate the interior of the premises as often as may reasonably be necessary and to keep in good repair the glass in and locks on all doors and windows and the sanitary apparatus waste water and drain pipes throughout the term and in such repair to yield up the same at the determination of the tenancy''. He was also obliged under clause 2(v) ''to permit the landlord [the applicant] and its servants or agents with or without workmen at all reasonable times to enter and examine the premises and upon notice given by the landlord to repair in accordance therewith''.

For its part, the landlord, by clause 3(i) of the lease, undertook to keep the exterior and structure of the premises in good repair.

Both of those provisions in the lease must now be read subject to section 11 of the Landlord and Tenant Act 1985 which, together with sections 32 and 33 of the Housing Act 1961, which it replaced, implied into the lease and the statutory and assured tenancies that followed it a covenant by the landlord to keep in repair and in proper working order the installations in the dwelling-house for the supply of water, gas and electricity, for sanitation and for space heating and heating water. In addition, the landlord's repairing covenant is to be construed as a covenant to repair within a reasonable time of notice of the disrepair being given by the tenant.

Clause 4 of the lease recited that, upon the expiration of the five-year term created by it, the landlord would act in relation to the tenant as if the agreement constituted a regulated tenancy as defined in the Rent Act 1977.

On April 29 1988, shortly before the expiration of the five-year term, the landlord applied for registration under the Rent Acts of a fair rent. The rent proposed was £30 a week to apply from July 29 1988, the expiry date of the lease. There was an inspection of the house on or about June 22 1988, as a result of which, on August 15 1988, the rent officer notified the landlord's agents that he had registered the new rent at the rate sought of £30 a week to take effect from August 15 1988.

On July 7 1989 Arthur Gadsdon died. His son, John Gadsdon, then became entitled to an assured tenancy by succession under amended provisions of the Rent Act 1977 and under provisions in the 1988 Act to which I shall refer. On November 25 1989 the landlord served a notice under section 13(2) of the Housing Act 1988 proposing an increase in rent from £30 a week to £500 a month. John Gadsdon, in a document which is undated, referred the matter to a London rent assessment committee. Pending the determination of that reference, the landlord's agent wrote to the rent assessment committee on December 13 1989, informing them of certain matters, which included the following: first, that in the view of the landlord's agents the landlord was entitled to a market rent for the house and their initial inquiries with a local letting agency indicated a figure of not less than £500 a month; second, that they had not been able to gain access to the premises as the tenant had not kept the gardens and interior in good tenantable order and that, despite several requests and visits to the premises, they had been unable to carry out any repairs that might be required to the structure of the building.

No hearing was requested by either party. An inspection by the members of the committee took place on Wednesday January 31 1990.

The committee made and delivered their determination of rent under section 14 of the 1988 Act on February 20 1990. They determined a rent of £5 a month. Under the heading ''Remarks'' in the determination notice the following summary was given:

House practically uninhabitable internally;
No water to bath, basin or wc;
Cold water only to kitchen sink;
No electricity to rear of house, on either ground or first floor;
WC choked and unusable;
Plaster has fallen off ceiling in rear ground-floor room and off wall of kitchen and is defective around almost all windows;
Damp penetration around almost all windows;
Metal windows rusting and leaking;
External condition fair but gardens run wild.

In detailed reasons prepared by the committee, dated March 12 1990, those matters were amplified, so far as material, as follows:

3. The Committee found Chilham Close to be in a cul-de-sac very near Perivale Underground railway station and line, one of a group of similar modest 1930s semi-detached houses with good front and rear garden space. The garden of No 10 was so overgrown with brambles that it was difficult to negotiate the path to the front door. The roof and the external structural condition of the house seemed to be fair but paintwork was shabby and windows in need of attention.

4. We set down under ''Remarks'' in our Decision the principal defects found to the interior of the property. In sum, every room was suffering from serious plaster deterioration and dampness and the functioning utilities comprised electricity to the front of the house and one cold water tap to the kitchen sink. Except in the front ground-floor room there was no sign of normal tenant-like use and care.

5. Under section 14 of the Housing Act 1988 we are bound to determine a rent at which we consider that the dwelling-house concerned might reasonably be expected to be let in the open market by a willing landlord under the assured tenancy. We are to disregard any effect on the rent attributable to the granting of a tenancy to a sitting tenant (section 14(2)(a)) and any reduction in the value of the dwelling-house attributable to a failure by the tenant to comply with any terms of the tenancy (section 14(2)(c)). According to section 39, the ''successor'' to a previously regulated tenancy becomes entitled to an assured tenancy. That tenancy shall be a periodic tenancy ''arising by virtue of this section'' (section 39(1)(5) and one:
section 39(6)(a) ''taking effect in possession immediately after the death of the protected or statutory tenancy . . .''
Section 39(6)(b) ''deemed to have been granted to the successor by . . . the landlord of the predecessor under his tenancy'', and
section 39(6)(e) ''under which, subject to sections 13-15 above, the other terms are the same as those on which, under his tenancy, the predecessor occupied the dwelling-house immediately before his death''.

6. Our interpretation of these provisions was that, unlike the position of a successor under the unamended Rent Act 1977, the successor under the Housing Act 1988 holds a new tenancy, if on the terms of the old. The most important implication in the present case was that the disregard of defects due to failure by ''the tenant'' applied only to failure by the new successor tenant, not to any default of the predecessor.

7. It was also clear that section 11 of the Landlord and Tenant Act 1985 applied and overrode with statutory landlords' responsibilities the terms of the agreement between the original parties in relation to repair.

8. In our opinion the internal disrepair of this house was of long standing and could not be attributed, except in very small measure, to the period of the present new tenancy. It would not have been reasonable to expect the present tenant to carry out his decorating responsibilities under the agreement terms still applying, given the decay and dysfunction in the structure (internal) and amenities for which the landlords were liable under the agreement or under section 11 of the Landlord and Tenant Act 1985. Another of the terms of the 1983 agreement was that the landlord company was permitted at all reasonable times to enter and examine the premises and upon notice given by the company to repair in accordance therewith. It would seem that for many years the company could have enforced inspection and repair, but had not done so.

9. We considered that in its present deplorable condition this house was dangerous to health and virtually unlettable ''in the open market'', the yardstick required by section 14: a tenant could not reasonably be expected to come forward, except on the basis of an extended ''rent free'' period, because he would require a considerable amount of time, effort and expenditure to bring the house in reasonably habitable condition and probably also to ensure that the landlords carried through their responsibilities in achieving that outcome.

10. We therefore determined a rent of £5 per month, exclusive of rates, effective from December 1 1989, the beginning of the new period specified in the landlords' notice.

The question is whether, if a person becomes an assured tenant by succession by virtue of section 39(5) and (6) of the 1988 Act, a rent assessment committee are required under section 14(2)(c) to disregard any failure by his predecessor to comply with the terms of the tenancy as well as by him when assessing an open market rent.

Here, John Gadsdon, the present tenant, succeeded his father who was protected under the Rent Acts, but John Gadsdon is an assured tenant under the Housing Act 1988. That is achieved by amendment of Schedule 1 to the Rent Act 1977 and by the provisions in section 39(5) and (6) of the 1988 Act.

As to the 1977 Act, para 3 of Schedule 1 (as amended by Schedule 4, Part I, to the Housing Act 1988) reads, so far as material, as follows:

Where a person who was a member of the original tenant's family was residing with him at the time of and for the period of 2 years immediately before his death then, after his death, that person or if there is more than one such person such one of them as may be decided by agreement, or in default of agreement by the county court, shall be entitled to an assured tenancy of the dwelling-house by succession.

Para 6 (as amended) provides:

(1) Where a person who —
(a) was a member of the original tenant's family immediately before that tenant's death, and
(b) was a member of the first successor's family immediately before the first successor's death,

was residing in the dwelling-house with the first successor at the time of, and for the period of 2 years immediately before, the first successor's death, that person or, if there is more than one such person, such one of them as may be decided by agreement or, in default of agreement, by the county court shall be entitled to an assured tenancy of the dwelling-house by succession.

A Those provisions as they now appear are to be compared with their earlier form. The concluding words of para 3 used to read "shall be the statutory tenant if and so long as he occupies the dwelling-house as his residence".

Para 6 was in much shorter form:

> If the first successor was a man who died leaving a widow who was residing with him at his death then, after his death, the widow shall be the statutory tenant if and so long as she occupies the dwelling-house as her residence.

Corresponding provisions are to be found in section 39(5) and (6) of the 1988 Act. So far as material they read as follows:

> (5) In any case where, by virtue of any provisions of —
> (a) Part 1 of Schedule 1 to the Rent Act 1977, as amended in accordance with
> B subsection (2) or subsection (3) above, or
> (b) section 4 of the Rent (Agriculture) Act 1976, as amended in accordance with subsection (4) above,
> a person (in the following provisions of this section referred to as 'the successor') becomes entitled to an assured tenancy of a dwelling-house by succession, that tenancy shall be a periodic tenancy arising by virtue of this section.
>
> 6. Where, by virtue of subsection (5) above, the successor becomes entitled to an assured periodic tenancy, that tenancy is one —
> (a) taking effect in possession immediately after the death of the protected or statutory tenant or protected occupier (in the following provisions of this section referred to as "the predecessor") on whose death the successor became so entitled
> (b) deemed to have been granted to the successor by the person who,
> C immediately before the death of the predecessor, was the landlord of the predecessor under his tenancy;
> (c) under which the premises which are let are the same dwelling-house as, immediately before his death, the predecessor occupied under his tenancy;
> (d) under which the periods of the tenancy are the same as those for which rent was last payable by the predecessor under his tenancy;
> (e) under which, subject to sections 13 to 15 above, the other terms are the same as those on which, under this tenancy, the predecessor occupied the dwelling-house immediately before his death; and
> (f) which, for the purposes of section 13(2) above, is treated as a statutory periodic tenancy;
> and in paragraphs (b) to (e) above 'under his tenancy' in relation to the predecessor, means under his protected tenancy or protected occupancy or in
> D his capacity as a statutory tenant.

Mr Fetherstonhaugh, on behalf of the landlord, argued that section 14(2)(c) of the 1988 Act should be read so as to require the rent assessment committee, in determining an open market rent, not only to disregard any breach of the tenancy agreement by the current tenant but also any breach of the tenancy by his statutory predecessor. He accepted that, for the provision to be so construed, it requires the reading into it after the word "tenant" the words "or the tenant from whom he derived his tenancy". Mr Fetherstonhaugh acknowledged in making that submission that there are formidable obstacles to it, not least of which are the plain words of the provision as it stands. However, he urged that the statute should be construed in a purposive way to give it the effect for which he contended.

E He said that there was nothing in the language of the Act which required the disregard in section 14(2) to be restricted to the tenant's current tenancy. He said that to do so would produce a contrary result, in that where a statutory periodic tenancy arises on the coming to an end of the fixed-term tenancy, the same tenant's failures under the fixed term would be disregarded. He said that if, on the other hand, section 14(2)(c) were to be construed in the way suggested by him, so as to bring within the ambit of the disregard the tenant's defaults under previous tenancies, it would be logical and fair to consider not only those defaults of the tenant but also those of his predecessors in title. I do not use that expression in any particular technical sense.

F At the heart of his submission lay the complaint that if the section were not interpreted in the way he suggested, it would result in a landlord's being penalised for its tenant's predecessor's default. He pointed out that that would be contrary to one of the main purposes of the Housing Act, which was to arrest the decline in housing in the private rental sector by encouraging the landlord into the market to enable him to achieve a reasonable market rent for his rented property. Shorn of legal niceties, he said that there is no reason why the landlord here should suffer a lower rent because of the state of the premises which, at least in part, appears to have been attributable to John Gadsdon's father's neglect. He said, by way of conclusion, that, in so far as a literal construction of the provisions dictate or might dictate such a result, the court ought to adopt a purposive construction to achieve what Mr Fetherstonhaugh described as a G fairer result.

Mr Elvin drew my attention, on the subject of a purposive approach to legislation, to the words of Lord Diplock in *Jones* v *Wrotham Park Settled Estates* [1980] AC 74* at p 105. Lord Diplock identified three criteria that he considered required fulfilment before one could embark upon a purposive interpretation of a statute. This is what he said:

> My Lords, I am not reluctant to adopt a purposive construction where to apply the literal meaning of the legislating language used would lead to results which would clearly defeat the purposes of the Act. But in doing so the task on which a court of justice is engaged remains one of construction; even where this involves reading into the Act words which are not expressly included in it. *Kammins Ballrooms Co Ltd* v *Zenith Investments (Torquay) Ltd* [1971] AC H 850 provides an instance of this; but in that case the three conditions that must be fulfilled in order to justify this course were satisfied. First, it was possible to determine from a consideration of the provisions of the Act read as a whole precisely what the mischief was that it was the purpose of the Act to remedy; secondly, it was apparent that the draftsman and Parliament had by inadvertence overlooked, and so omitted to deal with, an eventuality that required to be dealt with if the purpose of the Act were to be achieved; and thirdly, it was possible to state with certainty what were the additional words that would have been inserted by the draftsman and approved by Parliament had their attention been drawn to the omission before the Bill passed into law. Unless this third condition is fulfilled any attempt by a court of justice to repair the omission in the Act cannot be justified as an exercise of its jurisdiction to determine what is the meaning of a written law which Parliament has passed. Such an attempt crosses the boundary between J construction and legislation. It becomes a usurpation of the function which under the constitution of this country is vested in the legislature to the exclusion of the courts.

Mr Elvin said — and it is common ground — that the purpose of the Act was to regulate open market rent between landlord and tenant and to revitalise the letting market and to attract landlords to it. He referred me in that connection to the Government White Paper, *Housing: The Government's Proposals*, Cmd 214, presented to Parliament in September 1987.

As to the second of Lord Diplock's criteria, Mr Elvin submitted that the Housing Act 1988 had not been carelessly drawn. Mr Fetherstonhaugh agreed with that and so do I. As I shall indicate in a moment, it has clearly been drawn with great care and in such a way as K to distinguish it from comparable problems that had arisen under the regime of the Rent Acts.

In any event, as Mr Elvin pointed out, there is no harm to a landlord in being limited to a disregard of the defaults of a current tenant because he had his remedy to enforce the identical provisions, subject only to section 11 of the Landlord and Tenant Act 1985, against that tenant's predecessor in title. Here, the landlord could have taken steps over a period of years to require John Gadsdon's father to comply with the tenancy so that the property did not deteriorate into that state, but the landlord did not do so. The same would apply if John Gadsdon himself had, under an earlier tenancy, allowed the property to fall into disrepair. L

Accordingly, Mr Elvin submitted, there is no justification for reading into the section words which are not there.

Mr Elvin went on to construe the provision itself and those related to it. First, section 14(2)(c) refers to "the tenant" and "the tenancy", which, Mr Elvin submitted, is a clear indication that a rent assessment committee are concerned only with the assured periodic tenancy over which they have jurisdiction. He drew attention to the fact that the definitions of "tenant" and "tenancy" in section 45 of the 1988 Act make no reference to a predecessor or statutory tenant or tenancy. He emphasised the contrast in that respect between the 1988 Act and the unamended provisions, to which I have already referred, of Schedule 1, Part I to the Rent Act 1977, paras 3 and 6, in their use of the general term "statutory tenant". M

Mr Elvin also contrasted the simple words of section 14(2)(c) with the provisions as to the determination of a successor's rent under section 70(3) of the Rent Act 1977. That provides:

> There shall be disregarded —
> (a) any disrepair or other defect attributable to a failure by the tenant under the regulated tenancy *or any predecessor in title of his* to comply with any terms thereof;
> (b) any improvement carried out, otherwise than in pursuance of the terms of the tenancy, by the tenant under the regulated tenancy *or any predecessor in title of his*; . . .

---

*Editor's note: Also reported at (1978) 246 EG 223.

A (e) If any furniture is provided for use under the regulated tenancy, any improvement to the furniture by the tenant under the regulated tenancy *or any predecessor in title of his* or, as the case may be, any deterioration in the condition of the furniture due to any ill-treatment by the tenant, any person residing or lodging with him, or any sub-tenant of his.

(Emphasis supplied.)

Mr Elvin submitted that it is inconceivable that the draftsman would not have inserted some such formula in section 14(2)(c) of the 1988 Act if it had been intended.

Finally, Mr Elvin and Mr Fetherstonhaugh drew my attention to a contrasting provision within section 14 itself which militates against the construction for which the latter contended. Section 14(2)(b) expressly provides that, in certain circumstances, any increase in the B value of a dwelling-house attributable to an improvement carried out "by a person who at the time it was carried out was the tenant" is to be disregarded. This formula, which is to be contrasted with the simple reference in section 14(2)(c) to "the tenant" was clearly intended to achieve the disregard of a predecessor tenant's improvements works. A similar formula would have been employed in section 14(2)(c) if it had been intended to disregard any deterioration in tenanted property because of a failure by a current tenant's predecessor to comply with any of the terms of the tenancy.

In my judgment, the submissions made by Mr Elvin on those matters of construction are overwhelming. If more is needed, it is quite clear from the provisions of Schedule 1 to the Rent Act 1977 (as C amended) and its companion provisions in section 39(5) and (6) of the 1988 Act carrying them through that there is a new interest here created for the successor. It is not simply a continuing interest. It creates a new periodic tenancy as an estate in land which the father's statutory entitlement under the Rent Act was not. If one runs down the various subparagraphs of section 39(6), it can be seen that most or all of them would have been unnecessary if what had been created by those provisions was a continuing and not a new tenancy.

As a matter of construction, therefore, it is clear that the effect of section 14(2)(c) is that the sins of the predecessor of the tenant whose reference is being considered are not to be visited upon him.

Mr Fetherstonhaugh suggested that that is unfair. He submitted that a family successor should not start with a clean slate so as to fix D the landlord with what may be years of neglect. However, if the landlord is vigilant in enforcing the rights that are available to him under each tenancy agreement, he should not be faced with years of neglect when a statutory "succession" of the tenancy takes place in this way.

As to fairness, Mr Elvin commented first that there is no canon of construction that a statute is to be construed fairly to its plain words. And he asked: how is fairness to be identified and employed as a tool in the construction of a statute such as this? For instance, why should a tenant be held responsible for breaches by a former tenant even though he may be a member of his or her family? He points out here that John Gadsdon is not benefiting from his own breach because the E breaches have been found by the rent assessment committee to be substantially those of his father. He has an entirely new tenancy and a different status.

I should add that, from the reasons given by the committee for their determination, it cannot be said the responsibility for the neglect should have been apportioned in some different way. The committee were amply justified in the conclusion which they reached on the facts, that the internal disrepair was of long standing and could not be attributed, except in very small measure, to John Gadsdon and that he could not reasonably have been expected to do anything about it given the structural defects which it was the landlord's contractual and statutory duty to remedy.

Accordingly, the application is contrary to the plain words of the F statute, and there is no principle of construction, whether described as purposiveness or fairness, which could properly override those words. The application is dismissed.

Queen's Bench Division G

May 24 1991

(Before Mr Justice HUTCHISON)

BTE LTD v MERSEYSIDE AND CHESHIRE RENT ASSESSMENT COMMITTEE AND ANOTHER

Estates Gazette April 25 1992

[1992] 16 EG 111

Landlord and tenant — Rent Act 1977 — Assessment of fair rent by rent assessment committee — Whether committee H failed to take into account evidence of comparable lettings in arriving at fair rent

The appellant, BTE Ltd, is the owner of 11 Cronton Road, Liverpool, a three-bedroomed terraced house with an upstairs bath and lavatory, no wash-basin, two living-rooms and a kitchen let under a regulated tenancy to the second respondent, Mrs K Jones — On appeal from the determination of the rent officer, who had fixed a rent of £13 per week, the respondent rent assessment committee determined the fair rent at £16.50 per week — The appellant contended for a rent of £30 a week putting in evidence lettings of properties supporting the rent contended for — The J appellant appealed the decision of the committee pursuant to the Tribunals and Inquiries Act 1971 contending that the committee having first found there was not any scarcity of comparable properties for letting purposes and that such properties were being let at the market rent, it was wrong for the committee then to find that those lettings did not afford a guidance because they were not at a reasonable rent — The appellant further said the committee's reasons for disregarding open market lettings and their decision that regard should be had to evidence of rents of property occupied by tenants with statutory tenancies, were wrong K

***Held*: The appeal was allowed — In a climate where there was no scarcity of comparable rented accommodation, the issue for a rent assessment committee is to determine the fair market rent — Having found a fair market rent it was not for the committee to decide that fair market rents do not represent reasonable rents — The committee were wrong in the light of section 70 of the 1977 Act to test the market by using properties occupied by tenants with statutory tenancies and not lettings with the benefit of vacant possession — Having accepted the evidence on behalf of the appellant of lettings in the open market, the committee should have taken account of those when arriving at the fair rent — It was not for the L committee to arrive at a fair rent simply by applying their own notions of what was reasonable**

The following cases are referred to in this report.

Mason v *Skilling* [1974] 1 WLR 1437; [1974] 3 All ER 977; (1973) 29 P&CR 88; [1974] EGD 230; 230 EG 1271, HL

Metropolitan Property Holdings Ltd v *Finegold* [1975] 1 WLR 349; [1975] 1 All ER 389; (1974) 29 P&CR 161; 233 EG 843, DC

This was an appeal pursuant to the Tribunals and Inquiries Act 1971 by the appellant, BTE Ltd, against a decision of the first respondent, the Merseyside and Cheshire Rent Assessment M Committee determining a fair rent of 11 Cronton Road, Liverpool, let under a regulated tenancy to the second respondent, Mrs K Jones.

Christopher Nugee (instructed by Gregson & Ashton, of Liverpool) appeared on behalf of the appellant; the first and second respondents did not appear and were not represented.

Giving judgment, **Hutchison J** said: This appeal is brought pursuant to the provisions of the Tribunals and Inquiries Act 1971, which permits an appeal on a point of law from an assessment of a fair rent by a rent assessment committee acting within the scope of Part IV of the Rent Act 1977. It relates to property in Liverpool at 11 Cronton Road. In my judgment, it is possible to deal with the appeal quite shortly. I should say, by way of introduction, that Mr Nugee

A has appeared and has succinctly and attractively presented the case on behalf of the appellant, but neither the tenant nor the rent assessment committee have appeared on this appeal. It is, I am told, quite usual to find that no respondent appears.

The position is this. Under the scheme embodied in that Part of the Act, it is possible to obtain registration of a fair rent on application to the rent officer and, if dissatisfaction is felt with his determination, there is provision for a reference to the rent assessment committee, who hear evidence and determine what should be the fair rent.

In this case, the landlord was contending for a rent in the order of £30 a week. The rent officer had fixed a rent of some £13.

Looking ahead to the conclusion at which the rent assessment committee arrived, they determined the fair rent at a slightly higher B figure, £16.50 a week.

It is unnecessary, because nothing turns on it, for me to refer to the statutory provisions which give rise to the right to have a fair rent registered and the right of review to which I have just referred.

I go straight to the relevant provisions of the 1977 Act, so far as this case is concerned. They are to be found in section 70 of the Act (as amended) which, as material, provides as follows:

(1) In determining, for the purposes of this Part of this Act, what rent is or would be a fair rent under a regulated tenancy of a dwelling-house, regard shall be had to all the circumstances (other than personal circumstances) and in particular to —

(*a*) the age, character, locality and state of repair of the dwelling-house, and

C (*b*) if any furniture is provided for use under the tenancy, the quantity, quality and condition of the furniture, and

(*c*) any premium, or sum in the nature of a premium, which has been or may be lawfully required or received on the grant, renewal, continuance or assignment of the tenancy.

Subsection (2), which is the crucial provision so far as this case is concerned, provides:

For the purposes of the determination it shall be assumed that the number of persons seeking to become tenants of similar dwelling-houses in the locality on the terms (other than those relating to rent) of the regulated tenancy is not substantially greater than the number of such dwelling-houses in the locality which are available for letting on such terms.

D The effect of those two subsections is most conveniently summarised in a passage in the judgment of Lord Widgery CJ in *Metropolitan Property Holdings Ltd* v *Finegold* [1975] 1 WLR 349*. At p 352B Lord Widgery, having set out the provisions of a predecessor Act, which were in the same terms as those of section 70, said:

This is the provision which is intended to eliminate what is popularly called "scarcity value" from the fair rent fixed by the committee, and it is a provision which has given rise to a great deal of difficulty in practice.

I think that before one begins to consider the difficulties

— they were difficulties which arose in that case but do not arise in this case —

and before one begins to consider the section in detail, one must have clearly in E mind what parliament's obvious intention was in including this provision in the Act. It seems to me that what parliament is saying is this. If the house has inherent amenities and advantages, by all means let them be reflected in the rent under subsection (1); but if the market rent would be influenced simply by the fact that in the locality there is a shortage, and in the locality rents are being forced up beyond the market figure, then that element of market rent must not be included when the fair rent is being considered. Parliament, I am sure, is not seeking to deprive the landlord of a proper return on the inherent value and quality of his investment in the house, but parliament is undoubtedly seeking to deprive a landlord of a wholly unmeritorious increase in rent which has come about simply because there is a scarcity of houses in the district and thus an excess of demand over supply.

That passage summarises the object of the two parts of the F sections, and also, by implication, makes it clear that, subject to adjustment in appropriate cases to reflect the scarcity element, the fair rent is to be equated with the market rent. It is said in other authorities that a fair rent must be fair to the landlord as well as the tenant. That supports the view that, scarcity value apart, the fair and proper market rent is the criterion which ordinarily should guide the rent officer and the rent assessment committee. That is not to say that the only way of arriving at a fair rent is by considering comparable lettings, because there are other possible ways of going about it, but it seems to me it is clearly established that, in general and subject to the scarcity value qualification, the object must be to try to arrive at what is the fair market rent of the property.

*Editor's note: Also reported at (1974) 233 EG 843.

With those preliminary observations and with that citation of G authority, I turn to the facts of the case and the decision appealed against. The property is a three-bedroomed terraced house with an upstairs bath and lavatory, though no wash-basin, two living-rooms and a kitchen below. The living-rooms are heated by two gas fires which had been installed by the tenants. There was evidence, the committee said, of considerable damp.

They go on:

The landlord submitted a schedule of market rents achieved on some of his properties, including two in the same road as the subject property. He contrasted these with a list of regulated rents determined by the Rent Officer on similar properties. Mr Bennett — that is the landlord — stated that his firm had carried out considerable work on 11 Cronton Road in 1986. This had included a new roof, replacing window frames in one of the H bedrooms, new joists and replacing a wall between bedroom and bathroom because of an outbreak of dry-rot; part of the kitchen ceiling had also been replastered on.

I omit a paragraph. The decision continues:

The Rent Assessment Committee decided that the state of repair both externally and internally could more accurately be described as fair/good rather than the fair which the rent officer attributed to it. When questioned about the damp, Mr Bennett stated that this was difficult to eradicate and his firm could not afford to carry out such repairs.

I then come to the crucial passage on which this appeal is based. It reads as follows: "Mr Bennett addressed the committee on the effects of recent legislation, adding that scarcity of accommodation no J longer applied." I interpolate that that is not a submission that the law had been changed to eliminate the necessity to have regard to scarcity of accommodation. It is meant to encapsulate Mr Bennett's submission that, in current conditions and given the liberalisation achieved by recent legislation, there was no longer a shortage of property of a comparable nature for letting.

The decision letter continues:

As a result, he contended that market rents should be the appropriate level. The rent assessment committee took careful note of Mr Bennett's comments, but felt that despite more housing becoming available for rental, there was still a paucity of accommodation available at reasonable rents. The properties which Mr Bennett's firm had recently let at market rents were likely to be those with the benefit of vacant possession, some, at least, of which would have been refurbished by the landlord prior to letting. This situation was clearly not K the same as accommodation which in some cases was in need of repair and which was occupied by tenants on statutory tenancies.

It seems to me it is a reasonable construction of those words that the committee were accepting the evidence they had from Mr Bennett, to the effect that there was an abundance of comparable properties for letting. They appear to have accepted that the lettings which he had achieved recently (of which he had given evidence) had been at market rent, because they used those words. But what they are saying is that they do not regard market rents as being appropriate because there is a paucity of accommodation available at what they regard as reasonable rents. As I ventured to suggest in the course of argument, that seems to be begging the question, since the issue L which the committee had to address was, what was a fair rent? The test of that, in a climate where there was no scarcity of comparable rented accommodation, was, what was the fair market rent? It is nothing to the point, having made a finding about the level of fair market rents, to go on to say that fair market rents do not represent reasonable rents.

Of course, tenants will always wish to have as low a rent as they can. From one point of view, if one applies absolute or arbitrary standards, it may be possible to say that the market rates do not represent a reasonable rent, but that is not the approach which, it seems to me, the statute enjoins.

Second, the committee seem to be saying that it is illegitimate to use as the test of the comparable, as the means of determining the M market, lettings with the benefit of vacant possession, and that regard should be had to a different type of property; that is to say, property occupied by tenants with statutory tenancies. That plainly, on the basis of the wording of subsection (1), is illegitimate. I refer to the case of *Mason* v *Skilling* [1974] 1 WLR 1437, to which Mr Nugee drew my attention. It is contrary to authority.

The other thing they seem to be saying, and this is, it seems to me, a legitimate comment, is that there may be a distinction between the properties let by Mr Bennett (of which he gave evidence) which may have been in an apple-pie state of repair, and the instant property, the state of repair of which, though fair to good in their assessment, suffered from the problem of damp. That is, as Mr Nugee conceded,

A a legitimate distinction, if the evidence supported it, because, plainly, property in bad repair, on the wording of the section, should command a lower rent than comparable property in good repair.

Allowing for that legitimate consideration, it does seem to me that in two important respects the committee have approached the matter in an incorrect manner. The way in which Mr Nugee put it was to submit, first of all, that having in effect found that there was not any scarcity of comparable properties for letting purposes and having found that such properties were being let at the market rate the committee found that those lettings did not afford a guidance because they were not at a reasonable rent. They took some arbitrary standards, the basis of which does not appear, as their measure of a B reasonable rent.

Second, Mr Nugee submits that the committee discounted the comparison which was put forward to show what the market rent was, and the reasons they gave, apart, possibly, from the point about the state of repair, were invalid reasons. Of course, a current letting is a letting with vacant possession. That is a statement of the obvious. The consideration of cases where there was a sitting statutory tenant was illegitimate for the reasons I have indicated.

Accordingly, it seems to me that the decision of the committee discloses an error of law in the approach which they adopted. Assuming, as I think their finding indicates, that they accepted the evidence of Mr Bennett and they accepted that the lettings of which he gave evidence were of comparable value, they should have taken C account of those when arriving at the fair rent. In so far as there was evidence on which they could find that the state of repair of those properties was better than the state of repair of the property they were considering, that is a factor which they could have reflected when fixing the fair rent. But they could not arrive at a fair rent simply by D applying their own notions of what was reasonable or by the decision that the lettings adverted to by Mr Bennett were lettings with vacant possession and were to be contrasted with lettings to tenants on statutory tenancies.

That is sufficient, in my judgment, to compel me to allow this appeal. I am asked by Mr Nugee to allow the appeal and to declare that the fair rent should be registered at £30. I am not prepared to do that. It would, in my judgment, not be proper for me to do that, on the basis of such information as I have. The appropriate course, in my judgment, is to allow the appeal, which will involve quashing the decision of the committee and to remit the matter to be reheard by a committee, applying these relevant criteria as laid down by section 70 of the 1977 Act to the determination of the fair rent. E

Mr Nugee tells me that there is more than one rent assessment committee in this area, and invites me to direct that the matter be reheard by a differently constituted committee. He is at pains to make it clear that in making that request he does not intend any reflection on the detachment or integrity of the committee whose decision is under appeal, but merely intends to reflect the recognition that it is often difficult for any tribunal, which has once determined the matter on an incorrect basis, to approach it again, following a successful appeal. It would be preferable that the matter should be dealt with by a differently constituted committee.

Decision of the Merseyside and Cheshire Rent Assessment Committee set aside. F

Editor's note: The fair rent was increased to £19.50 per week on February 13 1992.

LANDLORD AND TENANT

RENT REVIEW

Chancery Division

March 18 1991

(Before Mr Justice MUMMERY)

PRUDENTIAL ASSURANCE CO LTD v 99 BISHOPSGATE LTD

Estates Gazette January 25 1992

[1992] 03 EG 120

Landlord and tenant — Rent review clause in lease — Construction — Application made by landlords for leave to appeal under section 2(3) of the Arbitration Act 1979 and for the necessary qualifying certificate

The landlords sought a declaration as to the true construction of rent review provisions in a lease for a term of 98 years from September 1975 of premises at 99 Bishopsgate, London EC2 — The lease provided for 14 successive rent periods, the initial rent reserved being £1,531,250, which then represented about 50% of the initial yearly rental value of the demised premises — The lease was an "equity sharing lease" — The rent review provisions provided that the rent during each of the subsequent rent periods should be whichever was the greater of the rent payable during the last relevant year or 50.6% of the yearly rental value of the demised premises as at the appropriate date — The yearly rental value of the demised premises was the amount agreed or determined as representing a fair yearly rent, having regard to rental values current for property let without a premium with vacant possession and to the provisions of the actual lease "(other than the rent hereby reserved)" — The issue was whether these words in brackets were intended to exclude all the provisions as to rent reserved in the actual lease, or none of these provisions or only the rent review provisions, and, if only the rent review provisions, what part if not all of them

***Held*: 1. The tenants' primary submission, which was that the words in brackets were intended to exclude from the hypothetical lease only the quantum of rent payable from time to time by the actual tenant under the actual lease was preferred — The effect of this construction was to exclude the reference to the fixed percentage of 50.6% from the hypothetical rent, since it was merely a term as to the quantum of rent — The practice of trying to find answers by looking at decisions on other rent review clauses was to be criticised, but the Vice-Chancellor's guidelines in *British Gas Corporation* v *Universities Superannuation Scheme* was followed — The accepted construction was in accordance both with the natural and ordinary meaning of the words and with the underlying commercial purpose of rent review clauses — The exclusion from the hypothetical lease of the reference to the fixed 50.6% of the yearly rental value was analogous to the exclusion of the £7,500 premium in the case of *Guys 'n' Dolls Ltd* v *Sade Brothers Catering Ltd* — A declaration in favour of the tenants' construction and an order for payment of costs by the landlords**

2. On the landlords' application under section 2(3) of the Arbitration Act 1979 for leave to appeal to the Court of Appeal and for the necessary certificate which was a condition of appeal — The question was not one of general public importance and therefore the landlords had to rely on the existence of a special reason justifying the grant of a certificate — They relied on the large sums involved but this did not constitute a special reason for the grant of a certificate and it was refused; consequently leave to appeal was refused

The following cases are referred to in this report.

Amax International Ltd v *Custodian Holdings Ltd* [1986] 2 EGLR 111; (1986) 279 EG 762

Antaios Compania Naviera SA v *Salen Rederierna AB* [1985] AC 191; [1984] 3 WLR 592; [1984] 3 All ER 229; [1984] 2 Lloyd's Rep 235, HL

Babanaft International Co SA v *Avant Petroleum Inc* [1982] 1 WLR 871; [1982] 3 All ER 244; [1982] 2 Lloyd's Rep 99

Basingstoke and Deane Borough Council v *Host Group Ltd* [1988] 1 WLR 348; [1988] 1 All ER 824; (1987) 56 P&CR 31; [1987] 2 EGLR 147; 284 EG 1587, CA

British Gas Corporation v *Universities Superannuation Scheme Ltd* [1986] 1 WLR 398; [1986] 1 All ER 978; (1986) 52 P&CR 111; [1986] 1 EGLR 120; 277 EG 980

Equity & Law Life Assurance Society plc v *Bodfield Ltd* [1987] 1 EGLR 124; (1987) 281 EG 1448, CA

Guys 'n' Dolls Ltd v *Sade Brothers Catering Ltd* (1983) 269 EG 129, CA

This was an originating notice of motion by the landlords, Prudential Assurance Co Ltd, under section 2 of the Arbitration Act 1979, to determine questions of law arising in the course of an arbitration on the rent review provisions of a lease of 99 Bishopsgate, London EC2. The defendants were the tenants, 99 Bishopsgate Ltd.

Christopher Priday QC and Paul Morgan (instructed by Mr M C Gilbey, solicitor to Prudential Assurance Co Ltd) appeared on behalf of the plaintiffs; Terence Cullen QC and Miss Carolyn Walton (instructed by Stephenson Harwood) represented the defendants.

Giving judgment, **MUMMERY J** said: This is an application by originating notice of motion under section 2 of the Arbitration Act 1979 to determine various questions of law arising in the course of a reference to a chartered surveyor as arbitrator pursuant to the provisions of a rent review clause in a lease. The application is made with the consent of the parties to current arbitration proceedings as to the rental value, as at March 25 1989, of premises at 99 Bishopsgate, London EC2, demised by lease dated September 29 1975. The Prudential Assurance Co Ltd are the landlords. The original lessee was Bishopsgate Developments Ltd. The residue of the term of 98 years from September 29 1975 was assigned to 99 Bishopsgate Ltd on June 24 1976.

The lease provides for 14 successive rent periods. The first rent period ran from the date of the lease until September 29 1982. During that first rent period the reserved rent was the "initial rent" of £1,531,250, that is just over 50% of the "initial yearly rental value" as defined in clause 1(f) of the lease: see also clauses 1(c) and 2(a)(1). There then follow 12 successive periods of seven years each, the first commencing on September 29 1982, and then the balance of the term. It is provided in clause 2(a)(2) that the rent reserved during each of the subsequent rent periods should be a yearly rent of whichever should be the greater of:

(i) The yearly rent payable during the last year of the last preceding rent period, or

(ii) 50.6 per cent . . . of the amount of the yearly rental value of the demised premises as at the last quarter day but one before the end of the last preceding rent period . . .

It is common ground that the lease is a "speculative development" or "equity sharing" lease which, in effect, provides for the yearly rental value of the premises to be shared between the parties.

The expression "the yearly rental value of the demised premises" is defined in clause 1(d) of the lease. The construction of that definition has given rise to a disputed question of law which this court is asked to decide. Omitting immaterial parts, the definition reads:

> (d) "The yearly rental value of the demised premises" as at the last quarter day but one before the end of a rent period shall mean the amount agreed at any time between the landlords and the Lessee as the yearly rental value of the demised premises as at such day on the bases specified by this sub-clause or failing agreement before such day the amount determined as the yearly rental value of the demised premises as at such day on the said bases by an arbitrator such arbitrator to be a Fellow of the Royal Institution of Chartered Surveyors and failing agreement between the parties as to the person to be appointed such arbitrator to be nominated by the President for the time being of that Institution on the application of the Landlords or the Lessee not earlier than the quarter day but one before or at any time after such day and so that in the case of such arbitration the amount to be determined by the arbitrator shall be the amount which shall in his opinion represent a fair yearly rent for the demised premises as at such day having regard to rental values current as at such day for property let without a premium with vacant possession *and to the provisions of this Lease (other than the rent hereby reserved)* . . .

(Emphasis added.)

The question of law which has arisen concerns the meaning of the words "(other than the rent hereby reserved)". For the purposes of the valuation exercise to be conducted by the arbitrator, a hypothetical letting of the premises is assumed to take place and to be subject to the provisions of the actual lease "(other than the rent hereby reserved)". The issue is whether these words exclude all the provisions as to rent reserved in the actual lease, or none of those provisions, or only the rent review provisions, and what part, if not all, of the rent review provisions.

The landlords' primary contention is that the clear effect of the words in brackets is to exclude from the hypothetical letting the reservation of rent and the provisions as to the reservation of rent contained in the lease, ie clause 2(a)(1), which refers to the initial rent reserved during the first rent period, and clause 2(a)(2)(i) and the first six lines of clause 2(a)(2)(ii), as quoted above, referring to the rent reserved during each of the subsequent rent periods.

The landlords' alternative contention is that if the words in brackets do not exclude all those provisions relating to the reservation of rent in clause 2(a)(1) and (2) then *all* of those rent reservation provisions should have full effect in the hypothetical lease, including the provision in clause 2(a)(2)(ii) which refers to the 50.6% of the amount of the yearly rental value of the premises.

The lessee's construction falls somewhere between the two constructions proposed by the landlords. The lessee contends that the words in brackets do not exclude from the hypothetical lease all the rent reservation provisions contained in clause 2(a), nor do they have the effect of including all those provisions in the hypothetical lease. The construction advanced by the lessee is that the words in brackets have the effect of excluding from the hypothetical lease only the quantum of rent payable from time to time by the actual tenant under the actual lease. The effect of this construction is to exclude the reference to the fixed percentage of 50.6% at the beginning of clause 2(a)(2)(ii) from the hypothetical lease, since it is a term as to the quantum of rent.

The lessee has an alternative argument that the landlords are estopped *per rem judicatam* from asserting that the hypothetical lease does not contain the rent review provisions of the actual lease and from asserting that the hypothetical lease provides for the initial rent payable thereafter to be reviewed every seven years to a rent which is 50.6% of the yearly rental value of the premises. Those arguments are based on the conduct of the landlords in relation to an earlier arbitration which led to an award on June 21 1983 in respect of the rent payable for the second rent period from September 29 1982. A point taken by the landlords on the construction of the rent review provisions was unsuccessfully appealed by the landlords to the Court of Appeal in January 1985*.

It was unnecessary for me to hear arguments on the estoppel point at this stage, since it was agreed between leading counsel for the parties that, in the interests of saving costs and time, the construction point should be argued and decided first, and that it would be necessary for the estoppel point to be argued only if the landlords were successful on their primary or alternative contentions.

*Editor's note: Reported at [1985] 1 EGLR 72.

In my judgment, the lessee's construction of the words "(other than the rent hereby reserved)" is to be preferred. I have reached this conclusion in the following way:

1 I approach the question of construction on the basis that the function of the court is to construe these particular rent review provisions in this particular lease and that, in construing the words "(other than the rent hereby reserved)", it is both wrong and likely to lead to confusion and error for this court to rely on decisions of other courts on other documents as an aid to the construction of this lease: see *Equity & Law Life Assurance Society plc* v *Bodfield Ltd* [1987] 1 EGLR 124, at p 125C-D. As was observed in *Basingstoke and Deane Borough Council* v *Host Group Ltd* [1988] 1 WLR 348*, at p 353D, the function of the court is to ascertain the intention of the parties expressed in the language used, having regard to the context provided by the whole document and the matrix of the material surrounding circumstances, including the commercial purpose of the provision in question.

2 It is permissible to follow guidelines for the correct approach to the construction of particular types of provision. "Rough guidelines" for the construction of rent exclusion provisions in rent review clauses were stated by the Vice-Chancellor in his judgment in *British Gas Corporation* v *Universities Superannuation Scheme Ltd* [1986] 1 WLR 398†, at p 403, and were approved by the Court of Appeal in the *Equity & Law* case (*supra*) at p 125D-F. The Vice-Chancellor stated:

> In my judgment the correct approach is as follows: (a) words in a rent exclusion provision which require *all* provisions as to rent to be disregarded produce a result so manifestly contrary to commercial common sense that they cannot be given literal effect; (b) other clear words which require the rent review provision (as opposed to all provisions as to rent) to be disregarded (such as those in the *Pugh* case (1982) 264 EG 823) must be given effect to, however wayward the result; (c) subject to (b), in the absence of special circumstances, it is proper to give effect to the underlying commercial purpose of a rent review clause and to construe the words so as to give effect to that purpose by requiring future rent reviews to be taken into account in fixing the open market rental under the hypothetical letting.

The Vice-Chancellor added that these are not intended to lay down mechanistic rules of construction, as opposed to principles of construction, and as Dillon LJ observed in the *Equity & Law* case (*supra)*, at p 125G, those guidelines do not entitle a court to construe and apply a clause which the parties might have entered into but, on the true construction of the relevant provisions, have not in fact entered into.

3 There is no dispute that the underlying commercial purpose of a rent review clause, to which effect should be given if the relevant words allow, is the protection of a landlord from a situation in which, as a result of increases in property values or falls in the real value of money in an inflationary period, a fixed rent has become outdated and unduly favourable to the tenant. Consistently with this purpose, the lease which is hypothesised in rent review provisions should be on the same terms (other than as to quantum of rent) as those still subsisting between the parties under the actual lease. As was observed in the *Basingstoke* case, at p 354D:

> Of course rent review clauses may, and often do, require a valuer to make his valuation on a basis which departs in one or more respects from the subsisting terms of the actual existing lease. But if and in so far as a rent review clause does not so require, either expressly or by necessary implication, it seems to us that in general, and subject to a special context indicating otherwise in a particular case, the parties are to be taken as having intended that the notional letting postulated by their rent review clause is to be a letting on the same terms (other than as to quantum of rent) as those still subsisting between the parties in the actual existing lease. The parties are to be taken as having so intended, because that would accord with, and give effect to, the general intention underlying the incorporation by them of a rent review clause into their lease.

4 Bearing in mind that approach and the guidelines laid down by the Vice-Chancellor, I turn to consider the natural and ordinary meaning of the words "(other than the rent hereby reserved)" in their context. Like Hoffmann J, who had to consider the identical words in *Amax International Ltd* v *Custodian Holdings Ltd* [1986] 2 EGLR 111, at p 112D-E, I am of the view that the natural meaning of the words in brackets is to exclude from the hypothetical lease the amount of the rent which the tenant actually has to pay under the existing lease. I see no difference between the specific amount

*Editor's note: Also reported at [1987] 2 EGLR 147.
†Editor's note: Also reported at [1986] 1 EGLR 120.

A mentioned in that lease in pounds and pence (ie the initial rent) and the actual amount arrived at for subsequent rent periods. The words in brackets do not, as a matter of construction or of necessary implication, clearly exclude all the provisions relating to the reservation of rent and the machinery for reviewing the amount of rent reserved from time to time. I also agree with Hoffmann J that, at their lowest, the words in brackets are capable of a meaning which excludes from the hypothetical letting only the quantum of rent payable under the actual lease. A construction should be favoured for those words which promotes, rather than frustrates, the overall commercial purpose of the rent review provisions, ie one which does not exclude future rent reviews but only excludes the sums of rent payable under the actual lease.

B 5 On such a construction the reference in clause 2(a)(2)(ii) to the fixed 50.6% should be excluded from the hypothetical lease, and I can see no special context requiring or indicating otherwise. The application of that fixed percentage to the yearly rental value is a mechanical exercise in quantifying from time to time the actual amount of rent which the lessee should pay from time to time under the existing lease. The fixed percentage presupposes that the amount of the yearly rental value of the premises has been arrived at. That percentage is then applied to that value in order to produce a figure which, if greater than the figure specified in clause 2(a)(2)(ii), will be the amount of revised rent payable for the relevant subsequent rent period. The fixed percentage has no more relevance to the valuation C exercise which involves a hypothetical lease than those provisions of the actual lease which specify a fixed amount of rent in pounds and pence, or than terms, such as those considered in *Guys 'n' Dolls Ltd* v *Sade Brothers Catering Ltd* (1983) 269 EG 129, which provide for the addition of a fixed amount of premium over the fair rate rental market value agreed or ascertained on a rent review.

6 Mr Priday QC, for the landlords, placed considerable reliance on the decision of the Court of Appeal in the *Equity & Law* case, where the words "upon the terms of this lease other than as to . . . rent" in relation to a hypothetical lease were held to exclude the complex rent review machinery which included a requirement that a sum equal to the amount by which 85% of the net rental value exceeded the stated sum of £28,500 be paid by way of additional rent D or increased additional rent. The overall effect was to require payment of a rent less than the net rental value, the ascertainment of which was the whole object of the valuation exercise on rent review. The Court of Appeal held that in view of those percentage discount provisions the rent review provisions with which they were bound up could not be imported into the hypothetical letting. What had to be determined on the hypothetical letting was the best rent which the premises might reasonably be expected to fetch on the open market and not 85% or some other percentage of the best rent. The provisions of the lease in that case were so different from the present that I am not tempted to disregard Dillon LJ's warning in the *Equity & Law* case about decisions on different documents in other cases.

E I therefore find in favour of the construction proposed by the lessee. I would add that the same construction will also apply to the reference to 50.6% in clause 6(iii)(c), which sets out the definition of the expression "the initial rent" in any new lease granted in consequence of the lessee's exercise of an option to renew the lease. I will now hear argument as to the terms in which I should make the appropriate declaration, and other matters such as costs.

After a discussion of the form of the judge's order and costs, the judgment continued as follows:

I have given judgment on the question of law raised by the originating notice of motion. I will make a declaration along the lines set out in para 5.4 of the points of defence which have been filed in F the arbitration, and I would ask that a minute of order containing the form of declaration be agreed between junior counsel and lodged with the court.

I will make an order that the costs of this application be paid by the Prudential Assurance Co Ltd to 99 Bishopsgate.

I will now deal with the application which has been made by Mr Priday on behalf of the Prudential Assurance Co. That is an application for leave to appeal to the Court of Appeal and for a certificate. The need for leave and for a certificate of the High Court arises from the provisions of section 2(3) of the Arbitration Act 1979, which is in these terms:

> A decision of the High Court under
>
> — subsection (1) above —
>
> G shall be deemed to be a judgment of the court within the meaning of section 27 of the Supreme Court of Judicature (Consolidation) Act 1925 (appeals to the Court of Appeal), but no appeal shall lie from such a decision unless —
>
> (*a*) the High Court or the Court of Appeal gives leave; and
>
> (*b*) it is certified by the High Court that the question of law to which its decision relates either is one of general public importance or is one which for some other special reason should be considered by the Court of Appeal.

I consider the question of the certificate first. It is accepted by Mr Priday that he cannot argue in this case that the question is one of general public importance. There is no question of the rent review clause in this case being in a common form found in precedent books so that it would be desirable in the public interest to have the validity of a common form rent review clause considered by the Court of H Appeal. Nor does Mr Priday contend that there is any new question of principle involved or any challenge to established principles of construction of rent review clauses, such as the rough guidelines laid down by the Vice-Chancellor and approved by the Court of Appeal in the *Equity & Law* case.

So the question whether I should grant a certificate comes down to whether I am satisfied as to the existence of a special reason for this matter being considered by the Court of Appeal. On that matter I have in mind the remarks which have been made in the Court of Appeal in the case of *Babanaft International Co SA* v *Avant Petroleum Inc* [1982] 1 WLR 871 and by the House of Lords in *Antaios Compania Naviera SA* v *Salen Rederierna AB*, a case in J [1985] AC 191.

The submission of Mr Priday on special reasons really came to this, that there was a special reason for having the question of law in this case considered by the Court of Appeal because of the large sums involved on the rent period, which is under consideration and which will affect subsequent rent periods and which, in turn, will have an effect on the value of the reversion and of the term.

In my judgment, the fact that large sums are involved, as they undoubtedly are, does not constitute in this case a special reason for my granting a certificate under section 2(3) of the Arbitration Act 1979. I therefore refuse a certificate and, consistently with that, I refuse leave to appeal.

K

Chancery Division

October 24 1991

(Before Mr Justice Mervyn DAVIES)

KINGS (ESTATE AGENTS) LTD v ANDERSON AND ANOTHER

Estates Gazette February 8 1992 L

[1992] 05 EG 166

Construction of a rent review clause — Whether time of the essence for an arbitrator's determination of rent — Parties failed to agree rent — Application for declarations by tenant dismissed

The plaintiff tenant holds the term of a lease dated December 15 1975 of premises at 27 Watling Street, Canterbury, Kent — The defendants are the owners of the reversion to the term — The lease made provision for an upward-only rent review to the current market rental value in respect of what was described as the fourth rent period M commencing June 25 1990 — The current market rental value was to be the amount agreed between the landlords and the tenant six months prior to the commencement of the fourth rent period "or in the absence of agreement (time to be of the essence of this provision) as shall be determined by an Arbitrator" — The lease further provided that the arbitrator was to make his determination by a date three months prior to the commencement of the second rent period or the third rent period (as the case may be) — The tenant sought a declaration that the landlords had failed to comply with the time schedule for the rent review and, in applying for the appointment of an

arbitrator on June 29 1990, had no authority to make that application

***Held*: Upon the true construction of the lease the direction in the rent review clause to the arbitrator was to make his determination by a date three months prior to the commencement of the second rent period or the third rent period, but there was no similar direction in relation to the fourth rent period — The decision of Walton J in *C Bradley & Sons Ltd* v *Telefusion Ltd* distinguished — Even if the arbitration provision had included a reference to the fourth period, it was more likely that the words "time to be of the essence" relate back to the provision for agreeing a rent six months prior to the rent review period and do not relate forward to the time-limit in which the arbitrator has to make his award**

The following cases are referred to in this report.

Bradley (C) & Sons Ltd v *Telefusion Ltd* (1981) 259 EG 337
Phipps-Faire Ltd v *Malbern Construction Ltd* [1987] 1 EGLR 129; (1987) 282 EG 460
Power Securities (Manchester) Ltd v *Prudential Assurance Co Ltd* [1987] 1 EGLR 121: (1986) 281 EG 1327
United Scientific Holdings Ltd v *Burnley Borough Council* [1978] AC 904; [1977] 2 WLR 806; [1977] 2 All ER 62; (1977) 33 P&CR 220; [1977] EGD 195; 243 EG 43 & 127, HL
Wickman Machine Tool Sales Ltd v *L Schuler (AG)* [1974] AC 235; [1973] 2 WLR 683; [1973] 2 All ER 39; [1973] 2 Lloyd's Rep 53, HL

The plaintiff, Kings (Estate Agents) Ltd, is the tenant under a lease dated December 15 1975 of premises at 27 Watling Street, Canterbury, Kent, and sought declarations as to the true construction of that lease against the defendant landlords, Christopher Austen Anderson and Michael John Austen Anderson.

Peter Kirby (instructed by Farmar & Shirreff, of Sevenoaks) appeared for the plaintiff; Andrew Goodman (instructed by Colin Hayward & Co, of Ashford) represented the defendants.

Giving judgment, **MERVYN DAVIES J** said: This case concerns the construction of a rent review clause in a lease dated December 15 1975 of premises at 27 Watling Street, Canterbury. There is an originating summons dated October 26 1990. The plaintiff tenant is Kings (Estate Agents) Ltd. The defendant landlords are Christopher Austen Anderson and Michael John Austen Anderson. The plaintiff seeks two declarations:

1. A declaration that on the true construction of the lease the defendants are not entitled to have the rent reviewed for the period from the 25th June, 1990 and during the residue of the term, having failed to comply with the strict time schedule for such a rent review to come into operation as set out in Clause 1(4) of the lease and

2. A declaration that in the events which have happened the defendants had no authority under the lease to apply to the President of the Royal Institution of Chartered Surveyors to appoint an Arbitrator to determine the current market rental value of the premises.

The parties to this action are not the original parties to the lease but it is not necessary to set out any devolution of title. It is common ground that the plaintiff is now the tenant and the defendants are now the landlords. By the lease the premises 27 Watling Street were demised to the tenant for 20 years from June 25 1975. The rent payable is expressed in these words:

YIELDING AND PAYING: therefor unto the Landlord by equal quarterly payments in advance on the usual Quarter Days in every year the first of such payments to be made on the signing hereof

(i) Until the Twenty Fifth day of June One thousand Nine hundred and Eighty the yearly rent of One thousand Three hundred pounds (hereinafter called "the initial rent")

(ii) From the said Twenty Fifth day of June One thousand Nine hundred and Eighty Five (hereinafter called "the second rent period") the yearly rent (hereinafter called "the second rent") of One thousand Six hundred pounds or such amount (whichever be the greater) ascertained as hereinafter provided as representing the current market rental value of the demised premises for the second rent period.

(iii) From the said Twenty Fifth day of June One thousand Nine hundred and Eighty Five and until the Twenty Fifth day of June One thousand Nine hundred and Ninety (hereinafter called "the third rent period") the yearly rent (hereinafter called "the third rent") of the same amount as the second rent or such amount (whichever be the greater) as shall be ascertained as hereinafter provided as representing the current market value of the demised premises for the third rent period.

(iv) From the said Twenty Fifth day of June One thousand Nine hundred and Ninety and during the residue of the said term (hereinafter called "the fourth rent period") the yearly rent of the same amount as the third rent or such amount (whichever be the greater) as shall be ascertained as hereinafter provided as representing the current market rental value of the demised premises for the fourth rent period.

The expression "the current market rental value" as hereinbefore used shall mean such amount as shall be agreed between the Landlord and the Tenant by a date not later than six months prior to the commencement of the second rent period, the third rent period or the fourth rent period (as the case may be) as representing the current market rental value at such date of the demised premises or in the absence of agreement (time to be of the essence of this provision) as shall be determined by an Arbitrator as sole arbitrator acting under the provisions of the Arbitration Act 1950 or any statutory modification or re-enactment thereof for the time being in force to be nominated by the President for the time being of the Royal Institution of Chartered Surveyors on the application of the Landlord and so that in case of any arbitration the current market rental value as at such date at which the same is required to be ascertained shall be of such amount as shall in the opinion of the Arbitrator represent the rent at which the demised premises might reasonably be expected to be let for a term equivalent to the then unexpired residue of the term hereby granted as between a willing landlord and willing tenant with vacant possession and taking no account of any goodwill attributable to the demised premises by reason of any business carried on there by the Tenant or of any improvements made to the premises by the Tenant otherwise than pursuant to any obligation of the Tenant hereunder and in all other respects on the terms and conditions of this Lease and in the case of any arbitration the Arbitrator shall make his determination by a date three months prior to the commencement of the second rent period or the third rent period (as the case may be) and all fees costs and expenses payable in respect of such arbitration shall be borne by the parties hereto in equal shares.

A reading of the above shows that if a new rent was to run from June 25 1990 it was supposed that such a new rent could be agreed between the parties by a date "not later than six months prior to the commencement of . . . the fourth rent period", ie December 25 1989. In the absence of agreement there is to be arbitration. There was no agreement prior to December 25 1989. In a letter dated March 12 1990 the landlords' agent proposed a rent of £19,500. That was not acceptable and so on June 29 1990 the agent informed the tenant that he had written to the Royal Institution of Chartered Surveyors requesting the appointment of an arbitrator, adding:

. . . although I hope that we may be able to reach agreement before the appointment is made.

The plaintiff was not content with that course. The plaintiff took the view that any new rent should have been agreed by December 25 1989 and, since it was not so agreed, it was then (in June 1990) too late to go to arbitration because, time being expressly of the essence of the review provision, any arbitration decision ought to have been made by March 25 1990.

For general guidance I was referred to several authorities and in particular to the words of Lord Diplock in *United Scientific Holdings Ltd* v *Burnley Borough Council* [1978] AC 904* at p 930G. He says:

. . . I would hold that in the absence of any contra-indications in the express words of the lease or interrelation of the rent review clause itself and other clauses or in the surrounding circumstances the presumption is that the time table specified in a rent review clause for completion of the various steps for determining the rent payable in respect of the period following the review date is not of the essence of the contract.

See also *Phipps-Faire Ltd* v *Malbern Construction Ltd* [1987] 1 EGLR 129 at p 131F and *Power Securities (Manchester) Ltd* v *Prudential Assurance Co Ltd* [1987] 1 EGLR 121 at pp 122L to 123J.

Thus, as I understand, while there is a general presumption that time is not of the essence when operating a rent-fixing procedure, the presumption yields to compelling contra-indications. Here there is an express and compelling contra-indication because the material clause includes the words ". . . (time to be of the essence of this provision)". But then the question arises whether, on the true construction of the clause, that phrase (a) relates back to the provision about agreeing "by a date not later than six months prior . . ." that is, the agreeing provision, or (b) relates forward to the provision whereby the arbitration is to be concluded "by a date three months prior to the commencement of . . ." and that is the arbitration provision, or (c) relates to both the agreeing provision and the arbitration provision.

In *C Bradley & Sons Ltd* v *Telefusion Ltd* (1981) 259 EG 337 Walton J had to construe a clause which, apart from some

*Editor's note: Also reported at (1977) 243 EG 43.

A differences which I will mention later, was the same as the clause now before me. There the learned judge rejected — see p 338 — a submission that the time-essence words related to both the agreeing provision and the arbitration provision, ie he rejected (c) above. Before me counsel did not contend for (c). In these circumstances I am left with (a) and (b).

Walton J took the view that the time-essence words related to (b), ie to the arbitration provision. He said at p 339:

> The clause is in two parts. There is first what may be called the "agreement" part and then there is the "arbitration" part, and those are separated by the word "or" — "or in the absence of agreement (time to be of the essence of this provision)". It seems to me that the more natural way of reading "time to be of the essence of this provision" is that it is applying to what is to come, not B what has gone previously. When one adds to the fact that if it did apply to what had gone previously, it would, in my view, be virtually making an absolute nonsense of the provisions. It seems to me beyond all cavil that that is really the way in which this clause ought to be read.

So in that case the landlord was unable to achieve a rent review as being out of time. However, the same result does not emerge in the present case even if one adopts the construction of Walton J. This is because of the surprising wording at the end of the arbitration provision. One sees the words ". . . the arbitrator shall make his determination by a date three months prior to the commencement of the second rent period or the third rent period (as the case may be)". One notes that there is no mention of the fourth rent period. So, time C being of the essence according to the construction of Walton J (and disregarding, on this matter of construction, any relevant provision of the Arbitration Acts 1950-1979) the arbitrator would have to give his decision on rent for the second period by March 25 1980 and for the third period by March 25 1985. But looking again at the words just quoted there is not laid on the arbitrator any time period for the making of his determination of the fourth rent by a date three months prior to the commencement of the fourth rent period, ie March 25 1990. Thus it seems to me that even if time was intended to be of the essence of the arbitration provision it was intended to be of the essence only as respects the arbitrator's determinations for the second and third rent periods. It follows that a determination of rent for the fourth period did not have to be made by March 25 1990 or at D all. The position appears to me to be that with the passing of December 25 1989, with the parties not having agreed the fourth period rent, the landlord was entitled, at any rate within a reasonable time, after December 25 1989, to apply for the appointment of an arbitrator. The application referred to in the letter dated June 29 1990 was made within a reasonable time. The arbitrator appointed pursuant to that application is, for the reasons I have given, able to make his determination after March 25 1990.

One may of course speculate to the effect that the omission of a reference to the fourth period at the end of the arbitration provision has arisen by mistake. But I do not think that there is here any more than a possibility of a mistake. I cannot say that there was probably a E mistake in omitting a reference to the fourth period. I have in mind the words of Lord Simon of Glaisdale in *Wickman Machine Tool Sales Ltd* v *Schuler (AG)* [1974] AC 235 at p 263:

> There is one general principle of law which is relevant to both questions. This has been frequently stated, but it is most pungently expressed in *Norton on Deeds* (1906), p 43, though it applies to all written instruments:
> ". . . the question to be answered always is 'What is the meaning of what the parties have said?' not 'What did the parties mean to say?' . . . it being a presumption juries et de jure . . . that the parties intended to say that which they have said".
> It is, of course, always open to a party to claim rectification of an instrument which has failed to express the common intention of the parties; but, so long as the instrument remains unrectified, the rule of construction is F as stated by *Norton*. It is, indeed, the only workable rule.

It follows that the plaintiff is not entitled to the declaration sought.

Had the arbitration provision included at its end a reference to the fourth period I would have been obliged to decide whether the time-essence words related to (a) the agreeing provision or (b), as Walton J decided, the arbitration provision. I have great difficulty in agreeing with Walton J. It seems to me that the words "time to be of the essence . . ." relate back to the word "agreement" and have the force of indicating that if an agreement is not reached by the relevant date then there is to be no more time allowed so that the landlord can then apply for arbitration. The fact that the parties are still thereafter able to agree does not seem to me to vitiate that construction. So the declarations are not made.

Chancery Division G

June 19 1991

(Before Mr Justice KNOX)

SKILLION PLC v KELTEC INDUSTRIAL RESEARCH LTD

Estates Gazette February 8 1992

[1992] 05 EG 162

Rent review — Determination by independent surveyor — Hypothetical lease on same terms as actual lease — Use of H premises restricted to "research development testing manufacture repair and training in the use of engines motors and components and mechanical devices" — Whether "mechanical devices" should be given a special meaning — Landlord submitted that "mechanical devices" should be given their ordinary meaning — Tenant submitted that the words "mechanical devices" must be construed restrictively having regard to the earlier genus in the words "engines" "motors" and "components" having reference and/or consisting or all or part of an apparatus for providing motive power — Whether *ejusdem generis* rule applied and *contra* J *proferentem* rules applied

***Held*: The words "mechanical devices" are not to be limited by reference to the preceding words "engines motors and components" — The *ejusdem generis* rule does not apply to restrict the meaning of mechanical devices — The *contra proferentem* rule does not mean that one has got to decide the case, if one possibly can, against the landlord's interest in every particular circumstance, but that one construes the clause in its context against the interest of the person who put it forward — The indication derived from the *contra proferentem* rule is that a larger rather than a smaller construction is appropriate to the tenant's activities permitted K by the lease — The meaning of "mechanical devices" is not reduced by the meaning of, or any common feature in, the other permitted uses — The justification for the *ejusdem generis* construction, that one ought to restrict the wide words because otherwise the earlier particular words are of no practical effect, was of singularly little use in the present case for two reasons — If "mechanical devices" were the "general words" the clause would be highly eccentric — The surplusage argument which justifies the *ejusdem generis* rule carried no weight in the present case because of the great deal of overlap between the items listed — The draftsman of the lease included L the phrase "mechanical devices" to include the general run of engineering, consisting of mechanical devices with the intention of its having an effect of extending the list of permitted activities on the part of the tenant**

The following cases are referred to in this report.

Anderson v *Anderson* [1895] 1 QB 749
Chandris v *Isbrandtsen-Moller Co Inc* [1951] 1 KB 240; [1950] 2 All ER 618, CA
Gibbs v *Lawrence* (1860) 30 LJ Ch 170; 3 LT 367; 7 Jur NS 137; 9 WR 93
Tea Trade Properties Ltd v *CIN Properties Ltd* [1990] 1 EGLR 155; [1990] 22 EG 67
Tillmanns v *SS Knutsford Ltd* [1908] 2 KB 385 M

This was an application by way of an originating summons seeking a declaration as to the true meaning and effect of a lease of premises at East Country Yard, Greenland Dock, London SE16, by the plaintiff landlord, Skillion plc. The defendant, Keltec Industrial Research Ltd, was the tenant of the premises.

Jonathan Brock (instructed by Rowe & Maw) appeared for the plaintiff; Wayne Clark (instructed by Titmuss Sainer & Webb) represented the defendant.

Giving judgment, **Knox J** said: By a lease dated June 20 1985, made between the Borough of Southwark as landlord and Keltec Industrial Research Ltd ("Keltec") as tenant, some premises

A described in the lease as the three eastern-most part bays, part of the ground, first, second and third floors of the building erected thereon situate at East Country Yard, Greenland Dock, London SE16, were demised to Keltec, which is the defendant, for a term of 22 years from January 1 1984 at a rent subject to rent review provisions. By clause 2 the lease provided that it incorporated what are described as "the special provisions below" and "the general provisions herein".

The special provisions contained a limited licence for the lessee to install a petrol tank on certain conditions and, in particular, contained a provision for doing away with it if the lessee assigned the benefit of the lease or the lease otherwise expired.

The general provisions contained quite usual tenant's covenants,
B in particular a rather limited repairing covenant to keep the premises in a wind and watertight condition and a user covenant in the following terms:

> Not to carry on or permit or suffer to be carried on in or upon the premises or any part thereof any trade or business whatsoever other than that described in the said schedule of particulars.

The schedule of particulars to which various matters such as the description of the premises, the term, the rent and the review dates were consigned, contained the following regarding the use:

> As to that part coloured red on the same plan for the research development testing manufacture repair and training in the use of engines motors and components and mechanical devices together with ancillary office accommodation and storage

C and other parts of the demised premises coloured blue on the plan were to be used for parking, but nothing turns on that.

The rent review provisions, in fairly common form, provide for a notice to be served by the landlord providing for an increase in the rent payable under the lease and a fairly normal provision for a lessee to serve a counternotice and at the end of the day, failing agreement, there is a provision for the question whether any, and if so what, increase ought to be paid in the rent payable under the lease as from the review date to be referred to an independent surveyor. The independent surveyor is required to make various assumptions in reaching his conclusion acting, as the clause says, as an expert and one of those matters to which he is required to have regard is contained in subclause (2) of the rent review provision in clause 3(e)
D of the lease. It reads as follows:

> Having regard to market rental values current at the commencement of the relevant period for a property let without a premium with vacant possession and subject to the provisions of this lease other than the amount of rent hereby reserved but including all the provisions for rent review for a term equal to the original term of this lease by a willing landlord to a willing tenant.

The issue which has arisen between the parties revolves round the requirement that the surveyor should have regard to market rents subject to the provisions of the lease and, in particular, subject (as is common ground between the parties) to the provisions of the lease regarding the restrictions upon user which are imposed by the combination of the covenant in the general provisions and the
E incorporation in that covenant of the use that I read from the schedule of particulars.

The originating summons does not perhaps shed a great deal of light on what the main issue between the parties has been. What it asks for, and it is issued by the successor in title to the original lessor, that successor being Skillion plc (which I will call the landlord), is a declaration as to the true meaning and effect of clause 1(x) of the general provisions and the use as described in the second schedule to the above-mentioned lease and, in particular, whether, and if so to what extent, the word "mechanical devices" in the said schedule should be given a special, and if so what, meaning.

As can readily be inferred from the perhaps slightly tendentious way in which that question is drafted the plaintiff landlord is
F concerned to establish that the words "mechanical devices" should be given what the landlord would describe as its ordinary meaning, ie a meaning which is not cut down by the context. On the other hand, the tenant has argued that two presumptions or rules of construction or aids to construction (each has a Latin name) apply to restrict the width of the words "mechanical devices" to the earlier genus in the words "engines", "motors" and "components", which is defined on the part of the tenant as consisting of all or part of an apparatus for providing motive power.

Put perhaps slightly more shortly, the question that has been argued before me is whether "mechanical devices" are limited to devices concerned with or as part of apparatus for the provision of motive power or whether they are unlimited otherwise than by the normal meaning of the words. I have been referred to dictionary G
definitions which, although they illuminate one's ideas of such common words as "engines, motors, components and mechanical devices", do not, in my view, advance the arguments of either side to any significant extent.

So far as the law is concerned, I have been referred to three authorities. On behalf of the tenant, reliance is placed on *Tillmanns* v *SS Knutsford Ltd* [1908] 2 KB 385. That was a case where a provision in a bill of lading read as follows:

> Should a port be inaccessible on account of ice, blockade or interdict, or should entry and discharge at a port be deemed by the master unsafe in consequence of war, disturbance or any other cause, it shall be competent for the master to discharge goods intended for such port on the ice or at some other safe port or place, at the risk and expense of the shippers, consignees or H
> owners of the goods; and upon such discharge the ship's responsibility shall cease.

There was such a discharge but the Court of Appeal came to the conclusion that the reliance on ice as the justification for it was factually not established. The headnote reads:

> Inaccessibility was a question of fact, and that the port must be inaccessible not merely at the moment of the ship's arrival off the port but also for a reasonable time after her arrival and first attempt to enter.

So that argument failed. I am not concerned with that. Reliance was also placed on the words "or any other cause" in the passage that contains the phrase "be deemed by the master unsafe in consequence of war, disturbance or any other cause". Reliance was placed J
because the master had formed that view, but, of course, there is the obvious overlap between the specific reference to ice as causing inaccessibility, as giving rise to the right to discharge goods, and to the argument that was being advanced for "any other cause" to be construed literally as including, in particular, ice.

The decision of the Court of Appeal to which I was referred was given by all three members, but one of them went into the authorities on the *ejusdem generis* rule in greater detail than the other two and, in particular, Vaughan Williams LJ went at some length into the question whether there was a presumption of any sort where one found particular words followed by general words. He said at p 398:

> The real contest comes in here. In *Anderson* v *Anderson* [1895] 1 QB 749 there can be no doubt as to what was laid down by the Court of Appeal and in the K
> absence of a conflicting decision of equal authority we are bound by that case unless the facts are different. It was there undoubtedly held that the general words must be treated, even when following the antecedent enumeration of particular cases, as having a wide application and construction and not as being narrowed by the antecedent words. But my difficulty arises thus. In *Maxwell on the Interpretation of Statutes*, 4th edition, page 499 which I cite not as an authority but as a convenient frame on which to hang a series of cases to which I propose to refer, I find this: "But the general word which follows particular and specific words of the same nature as itself takes its meaning from them and is presumed to be restricted to the same genus as those words, or in other words as comprehending only things of the same kind as those designated by them unless, of course, there be something to show that a wider sense was intended."

That is the end of that quotation from *Maxwell.* Vaughan L
Williams LJ goes on to consider the authorities at some length and he says, having done that, at p 401:

> The series of cases to which I have referred are all cases which go to show that the true rule is that the restricted meaning is the one which primarily applies.

He cites a decision of Wood V-C, *Gibbs* v *Lawrence* (1860) 30 LJ Ch 170, where that conclusion was reached in these words:

> I think the cases

— this is the quotation from Wood V-C —

> which are very numerous on this subject have some common principle upon which they all seem to have been decided and which is not difficult of application with regard to gifts in general. M

And of that, Vaughan Williams LJ observes:

> Of course it is more difficult of application when you come to a contract and remember that *Anderson* v *Anderson* was a case of a gift by a husband by a settlement on his wife, the words following a specific enumeration being confined to things *ejusdem generis.*

He goes on to say that under these circumstances and in particular that decision of *Gibbs* v *Lawrence* he, Vaughan Williams LJ, was not prepared to say that the construction which Channell J at first instance had placed upon the words in question was wrong.

In my judgment, there is no doubt but that Vaughan Williams LJ did express a preference for the view that was stated to be the true interpretation of the authorities in the passage in *Maxwell,* then the

A 4th ed, that he cited. But the other two members of the Court of Appeal did not put the case on that footing. They put it on the footing that there was in this case a genus to be found and that it would be right in all the circumstances to apply the *ejusdem generis* rule in relation to that phrase "or any other cause".

There are at least two points to be noted in relation to that. First, there is the point to which I have already adverted, namely that there was a specific provision in relation to inaccessibility owing to ice, *inter alia,* and it was on the face of it rather surprising that the provision should deal with that potential situation in more than one way. That was a very strong argument, as it seems to me, in favour of a limited construction of the words "or any other cause". The other consideration is that one does have the classic case for the application B of the *ejusdem generis* principle in an enumeration of two things in this case, which have a common feature — war and disturbance — followed by some phrase, such as "or any other cause" which is very close in sense to "etc" or some other quite general word which plainly embraces what has gone before. I say "plainly embraces" because the inclusion of the word "other" makes it clear that "war" and "disturbance" are two causes of that nature and it is, as it seems to me, a matter of common sense and language that it is much easier to apply the *ejusdem generis* rule when one has that type of collocation.

As against that authority I was also referred to the decision of Devlin J in *Chandris* v *Isbrandtsen-Moller Co Inc* [1951] 1 KB 240. The clause there in question was one which was contained in a voyage C charterparty which provided that the cargo was to consist of lawful general merchandise "excluding acids, explosives, arms, ammunition or other dangerous cargo". What happened was that there was loaded turpentine which became dangerous and the issue arose whether that was within the phrase "or other dangerous cargo". The learned judge, having stated the question and the two alternative arguments, both of which seemed to proceed on the assumption that the *ejusdem generis* rule was applicable, started by expressing the view that the first question he had to determine was whether the rule in one form or another applied at all. To do that, he had to examine the nature and scope of the rule. He took as his point of departure an earlier decision of the Court of Appeal, *Anderson* v *Anderson* [1895] 1 QB 749. That, of course, was the decision that was D referred to in the passage that I have quoted from Vaughan Williams LJ's decision in the *Tillmanns* case. Devlin J said at p 244:

> *Anderson* v *Anderson* is a decision of the Court of Appeal in relation to a post-nuptial settlement in which all three Lords Justices clearly laid it down that general words were *prima facie* to be considered as having their natural and larger meaning and not to be restricted to things *ejusdem generis* previously enumerated, unless there was something in the deed to show an intention so to restrict them. In *Tillmanns & Co* v *Steamship Knutsford Ltd,* Vaughan Williams LJ, without impeaching the authority of *Anderson* v *Anderson,* set out other authorities to the opposite effect designed to show that it is the restricted meaning which primarily attaches to the general words. It is perhaps important for this case concerns a charterparty. In *Thorman* v *Dowgate Steamship Co Ltd* Hamilton J, also dealing with a charterparty, thought it E unnecessary to embark on this discussion. *Anderson* v *Anderson* is a decision which, unless distinguishable on the ground that I am concerned with the commercial document rather than a settlement, is binding on me. Apart from its authority, if I may respectfully say so, I entirely agree with it.

The learned judge then continues to discuss the nature of the rule in question. He says:

> A rule of construction cannot be more than a guide to enable the court to arrive at the true meaning of the parties. The *ejusdem generis* rule means that there is implied into the language which the parties have used words of restriction which are not there. It cannot be right to approach a document with the presumption that there should be such an implication. To apply the rule automatically in that way would be to make it the master and not the servant of the purpose for which it was designed — namely, to ascertain the meaning F of the parties from the words they have used. The first approach will often plainly show that the words are used far more widely than the parties could have intended;

And he gives examples of that; for example, "all other perils" in a policy. He then continues to deal with the question whether commercial documents should be dealt with in a more stringent way than settlements. I need not, I think, deal with that aspect of the matter. At p 245 he says this, which is significant:

> Moreover, the main argument of construction which justifies the application of the rule does not apply in commercial documents. It is that if the general words have an unrestricted meaning the enumerated items are surplusage. The presumption against surplusage is of little value in ascertaining the intention of the parties to commercial documents, as many great commercial judges have recognised.

G And he gives various examples, which I need not cite, of that type of observation of a commercial judge. I was, I mention in passing, referred to something that my brother Hoffmann LJ said in *Tea Trade Properties Ltd* v *CIN Properties Ltd* [1990] 1 EGLR 155 at p 158A of this argument against surplusage:

> I see the force of that argument, but I have never found the presumption against superfluous language particularly useful in the construction of leases. The draftsmen traditionally employ linguistic overkill and try to obliterate the conceptual target by using a number of words or phrases expressing more or less the same idea. I cannot, therefore, rely upon the language alone but must, as it seems to me, construe the words also by reference to the commercial effect which would be produced by one construction or the other.

Returning to Devlin J's decision, he continued at p 246:

> H I apply the principle laid down in *Anderson* v *Anderson* and so inquire whether there is anything in the text of this charterparty or in the circumstances in which it was made which would lead me to suppose that the parties intended "other dangerous cargo" to have some limited meaning. I can find no such indication.

He therefore rejected the argument that the *ejusdem generis* rule applied at all. He did, however, go on to consider the other point which assumed that the *ejusdem generis* rule — contrary to his decision — did apply, saying on this score:

> But since it was fully argued, and is a point much in dispute, I may permit myself to express shortly my view of it. If the *ejusdem generis* principle is a rule of automatic application, it becomes of the first importance to determine exactly what the rule is. If it is merely, as I think, an aid to ascertaining the intention of the parties, no point of controversy need arise at all. If there is J something to show that the literal meaning of the words is too wide, then they will be given such other meaning as seems best to consort with the intention of the parties. In some cases, it may be that they will seem to indicate a genus; in others, that they perform the simpler office of expanding the meaning of each enumerated item. If a genus cannot be found, doubtless that is one factor indicating that the parties did not intend to restrict the meaning of the words. But I do not take it to be universally true that whenever a genus cannot be found the words must have been intended to have their literal meaning, whatever other indications there may be to the contrary. I see no reason why, if it accords with the apparent intention of the parties, the words should not be treated as suggested by Lord McNaghten in *Thames & Mersey Marine Insurance Co Ltd* v *Hamilton, Fraser & Co* as being "inserted in order to prevent disputes founded on nice distinctions" and "to cover in terms whatever may be within the spirit of the cases previously enumerated". K

I find myself in a position which is not seriously dissimilar from that to which Farwell LJ referred at the outset of his judgment in the *Tillmanns* case, when he said of this point:

> The second point is the question of *ejusdem generis,* upon which question I felt much embarrassed by the superfluity of authority.

There is, in my view, some considerable difficulty in reconciling the general approach of Vaughan Williams LJ, in particular in the *Tillmanns* case with what was said in the *Anderson* v *Anderson* decision, particularly as interpreted and applied by Devlin J in the *Chandris* case. If it were necessary for me to choose between the two I would prefer and follow the approach that Devlin J took. However, I doubt whether it is necessary for me so to decide because, if one L looks at the clause itself, in the context in which one finds it, it seems to me that the answer is reasonably clear without resort to any particular presumption of law.

I have an affidavit before me which describes the condition of the premises at the date when the lease was granted and that can be described compendiously as being, in the words of the deponent of the affidavit, "scruffy". Hence, no doubt, the rather limited repair obligation imposed upon the tenant. I do not myself derive any particular guidance from that state of affairs. Certainly I do not derive any indication that the parties would have intended to limit closely the extent to which the lessee should be entitled to trade, but the point is fairly neutral.

M The next point does seem to me of somewhat more help. That is, that the clause in which one finds the relevant phrase by incorporation by reference is a clause that enumerates the permitted users in which the tenant can indulge in using the premises by way of trade or business. It is an enumeration of permitted activities. In that context I have no doubt but that the second of the two rules mentioned in the argument before me with a Latin name, the *contra proferentem* rule, operates in favour of a large, rather than a small, construction; because there is common ground between the parties that the *contra proferentem* rule, for what it is worth, does apply to leases and the person who is doing the proffering is the landlord and therefore it is against the landlord that one construes the relevant clause.

A Now the relevant clause, as I have said (I fear more than once), is an enumeration of what the tenant may lawfully do on the demised premises and there can, in my view, be no doubt at all but that, in approaching that question of construction, the large construction is the one that permits the tenant to do more rather than less. By a fairly typical irony in connection with rent review clauses it is, of course, in the tenant's interest in a rent review that the scope of his activities should be restricted by the terms of the user covenant because it is at least possible that a narrow user clause will persuade a valuer who is in the position of the expert in this case that the open market rent is lower than it would be if there were no such restriction. That is a matter of very common experience. The *contra proferentem* rule B does not mean that one has got to decide the case, if one possibly can, against the landlord's interest in every particular circumstance. The *contra proferentem* rule is that in approaching the construction of a clause, in this case in a lease, one construes the clause in its context against the interest of the person who put it forward and had it included in the lease. The particular context is, of course, the extent to which the tenant's activities are permitted or forbidden by way of trade or business.

I start, therefore, with some indication, derived from that rule, that a larger rather than a smaller construction is appropriate.

The next consideration is that we have got here an enumeration of the permitted activities and that in such an enumeration one would expect the several forms of activity that are defined to be C cumulatively permitted — A + B + C — and one would not expect *a priori* C to be reduced by the meaning of, or any common feature in, A and B.

Third, the justification for the *ejusdem generis* construction referred to as the principal one intellectually by Devlin J, namely that surplusage indicates that one ought to restrict the wide words because otherwise the earlier, particular, words are of no practical effect, is of singularly little use in this case for two separate reasons. First of all, if one just had what are described as the "general words", ie mechanical devices, the clause would be, in my view, highly eccentric. It would read: "for the research, development, testing, manufacture, repair and training in the use of mechanical devices", D and one can well see that a draftsman who was faced with a grant of a lease which included a temporary permission to put in a petrol tank and, therefore, I think as a matter of necessary implication it must have contemplated a fairly widespread use of the internal combustion engine by the tenant, would have balked at not making it a great deal clearer than that turn of phrase, ie a reference only to mechanical devices, as a description of permitted user by a tenant who was plainly going to be using engines and motors. So that I find a very natural explanation for the specific mention of the particular words before one comes to what is relied upon as the general. Second, the surplusage argument seems to me of no use in this case, in that this is not — on the face of it — a draftsman who avoided surplusage; he E used "engines" and "motors" in immediately close collocation, and it was very rightly expressly accepted before me by Mr Clark, on behalf of the tenant, that there is a very great deal of overlap between those two. The surplusage argument, therefore, does not seem to me to carry any weight at all in this case.

Next, there is the double use of the word "and" which is slightly unusual but I think explicable. It will be recalled that the phrase is "in the use of engines motors and components and mechanical devices". If one uses brackets, it seems to me that one bracket should be put round "(in the use of engines motors and components)" and another round "(mechanical devices)" because it does seem to me that there may very easily be compelling reasons, and indeed were it necessary to decide it I would decide, in favour of limiting components to what F has gone immediately before; in other words, that the components that are referred to are components of engines and motors, and that, it seems to me, justifies the insertion of the word "and" in front of "components" because those three things are meant to go together. True, there is a very substantial degree of repetition in "engines and motors" but, disregarding that for a moment, the components do seem to me to be intended to be the components of the engines or the motors. But when one comes to "mechanical devices" that argument does not seem to me to apply with anything like the same strength, and also the structure of the clause distinguishes it from the very large number of cases of which I was, mercifully, referred to three only, where there is an enumeration with, at the end of the enumeration, either "and" and then a general word or "or" and then a general word, depending frequently on whether it is a positive or a negative G phrase.

Here we do not have that structure. This is not a long enumeration of which "mechanical devices" is at the end without more. It is a phrase which, it seems to me, falls into two parts: one, "engines, motors and their components" and I would effectively include the word "their" to make the meaning of components clear; and, two, "mechanical devices".

That leads me to the final consideration. I ask myself, why did the draftsman include and why did the parties agree to the inclusion of the expression "mechanical devices"? If it had just been methods of propulsion, or — as it is put in argument on behalf of the tenant — apparatus for providing motive power, it seems to me that it would not have been very difficult to find some much narrower phrase than H "mechanical devices". The justification that I think I detect for the inclusion of "mechanical devices" is that the draftsman (one assumes, on the instructions of the parties) was concerned to allow the general run of engineering, consisting of mechanical devices. It is an unusual phrase, in my experience, to find in a lease, but one has to take the clause as one finds it and it does seem to me that it must have been put in with the intention of its having an effect of extending the list of permitted activities on the part of the tenant.

All those reasons added together lead me to the conclusion that this is not a case in which I should apply the *ejusdem generis* rule to restrict the meaning of mechanical devices.

I will hear counsel on the form of the order that I should make to J give effect to that decision. I am minded to declare that the words "mechanical devices" contained in the schedule of particulars in the lease are not to be limited by reference to the preceding words "engines motors and components", but there may be more felicitous ways of expressing that conclusion.

Court of Appeal

December 16 1991

(Before Lord Justice DILLON, Lord Justice STOCKER and Lord Justice BUTLER-SLOSS) K

ICELAND FROZEN FOODS PLC v STARLIGHT INVESTMENTS LTD

Estates Gazette February 22 1992

[1992] 07 EG 117

Landlord and tenant — Rent review — Construction — Whether landlord entitled to assume hypothetical improvements — Whether premises are to be assumed as ready for immediate occupation and use for any purpose and in any condition that the hypothetical tenant would wish them L to be in — Appeal by landlord dismissed

By a lease dated May 7 1985 the respondent tenant, Iceland Frozen Foods plc, holds a term of 25 years from January 25 1985 of premises at 441-447 Brixton Road, London SW9 — The appellant landlord, Starlight Investments Ltd, holds the reversion to the lease — The premises are divided into two units, one occupied by the tenant and the second by a subtenant — The lease provided for five-yearly rent reviews to an open market rent, being the best yearly rack-rent at which the premises might reasonably be expected to be let as a whole or in parts and assuming, if not the fact, that the premises M remain in existence and are ready for immediate occupation and use — Improvements carried out by the tenant with landlord's consent are to be disregarded — The landlord had two principal contentions as to the proper construction of the rent review clause — (1) The ground floor may be regarded as divided or capable of division into seven units and that tenants may be prepared to pay a higher rent taking the chance of getting consent from the landlord for the subdivision, paying a continuous rent but on the basis that the tenant would be carrying out the works of conversion — (2) That it must be assumed that the premises are ready for occupation and use

for any purpose and in any condition that the hypothetical tenant would wish them to be in; therefore it is permissible to take into account hypothetical works to be carried out after the rent review date by the hypothetical tenant

***Held*: Upon the true construction of the rent review provisions: (1) it is not possible to charge rent for actual improvements under the rent review clause; accordingly it would make nonsense of that clause and of the aim that the rent on review should be fixed so as to bear as close a resemblance to reality as possible to charge the actual tenant with rent for hypothetical improvements which have never been carried out — (2) The words "ready for immediate occupation and use" amplify "remain in existence" to indicate that it is not enough that the premises remain in existence as, for instance, a gutted shell**

The following cases are referred to in this report.

Norwich Union Life Insurance Society v *Trustee Savings Banks Central Board* [1986] 1 EGLR 136; (1986) 278 EG 162

Orchid Lodge (UK) Ltd v *Extel Computing Ltd* [1991] 2 EGLR 116; [1991] 32 EG 57

Trust House Forte Albany Hotels Ltd v *Daejan Investments Ltd* (1980) 256 EG 915

This was an appeal by the defendant landlord, Starlight Investments Ltd, from an order of Judge O'Donoghue (sitting as a judge of the High Court) on the hearing of an originating summons seeking a declaration as to the meaning of the rent review provisions in a lease dated May 7 1985 of premises at 441-447 Brixton Road, Brixton, London SW9, the term of which was held by the respondent tenant, Iceland Frozen Foods plc.

Nicholas Dowding (instructed by David Gouldman & Co) appeared for the appellant; David Gilliland QC (instructed by Bullivant & Co, of Liverpool) represented the respondent.

Giving judgment, **Dillon LJ** said: This is an appeal by a defendant landlord, Starlight Investments Ltd, from an order of Judge O'Donoghue made on December 11 1990 when he was sitting as a judge of the High Court on the hearing of proceedings commenced by originating summons relating to a rent review clause under a lease.

The lease is dated May 7 1985, and it is a lease of premises known as 441-447 Brixton Road, Brixton, London SW9. The original landlord was a company called Wade's Departmental Stores Ltd. The original tenant was a company called Calmgrove Ltd. The demise was for a term of 25 years from January 25 1985 with rent reviews every five years. The questions at issue arise out of the first rent review due at January 25 1990. By then the reversion on the lease had become vested in the present defendant, Starlight, and the term of the lease had become vested in the present respondent, Iceland Frozen Foods. The vesting in Iceland came about by virtue of the landlord's consent and licence to assign granted on September 16 1986. The assignment to Iceland was by Calmgrove Ltd, the original tenant, acting by its liquidators. At the time of the assignment there were various variations in the terms of some of the covenants in the lease.

At the time of the lease in 1985 it was expressed to be granted subject to and with the benefit of the provisions in an underlease, which was a lease of October 12 1966 made in favour of the mayor, aldermen and councillors of Lambeth London Borough Council. That has expired and need not now be considered. It related to certain of the floors of the premises (part of the premises only) used as offices which are now unoccupied and do not give rise to the difficulty on this appeal.

The present condition of the premises, which consist of a large Victorian-looking building at a corner of the road junction between Brixton Road and Electric Avenue, is that the upper floors of the Victorian building were used as offices, the ground floor has been built over what were formerly the gardens of the Victorian building, and save in so far as part of the ground floor is naturally required for access to the floors above, the ground floor is divided into two units. One of those units and the basement are occupied by the plaintiff, Iceland, for use as a frozen-food store. The other part of the ground floor and the first floor are occupied by Murray's Meat Markets Ltd. Murray's use their part of the ground floor as a butcher's shop and their part of the first floor as their offices and relaxation rooms for the staff and so forth.

It appears that Murray's came in under a sublease dated September 18 1986, which was the day of the various licences to assign granted by Calmgrove to Murray's for 25 years less 10 days from January 25 1985. The particular question that arises on this appeal is how, on considering the new rents under the rent review, the ground floor is to be looked at: is it to be looked at as divided only into two units, *viz* one occupied by Iceland and one occupied by Murray's, or is it, as the landlord's surveyors suggest, to be regarded as divided or capable of division into seven units (or any number greater than two), which in the professional view of the landlord's surveyors would yield a greater aggregate rent?

One must have regard to certain of the provisions in the lease as varied by the licence to assign to Iceland. Under the lease as so varied the lessee has the right to underlet parts of the premises at rack-rents subject to the landlord's consent, which is not to be unreasonably withheld. Furthermore, there is a provision barring alterations to the premises or the cutting or maiming of main walls, and so forth, without the landlord's consent. Also under the licence and deed of variation of September 16 1986, which gave consent to the assignment to Iceland, a new user clause is substituted for the user clause in the lease. This tenant's covenant (15) provides in subclause (a):

> (a) (i) Not to use the whole or any part of the Premises for any illegal or immoral purpose
>
> (ii) Not to use the whole or any part of the Premises for any offensive noisy or dangerous trade business or manufacture
>
> (iii) Not without the previous written consent of the Landlord and (if applicable) the Superior Landlord (such consent on the part of the Landlord not to be unreasonably withheld or delayed to a change of use of the ground floor of the Premises to a use which is permitted under Class I of the Town and Country Planning (Use Classes) Order 1972):
>
> (1) to use or permit that part of the ground floor of the Premises sub-demised to Murrays Meat Markets Limited to be used otherwise than as a Butchers Shop; and
>
> (2) to use or permit the remainder of the ground floor of the Premises to be used otherwise than as a Freezer Centre; and
>
> (3) to use or permit the upper floors of the Premises to be used otherwise than as offices within the meaning of Class II of the said Town and Country Planning (Use Classes) Order 1972.

The crucial parts of the rent review clause are to be found in a schedule to the lease. The key phrase is the definition of "Open Market Rent", to which I will come. The basis of the review at each review date is that there should be an assessment of the open market rent for the relevant review date and that open market rent would be the rent from the review date. It was expressly provided that, however, at no time should the rent decrease below the amount of the rent payable immediately before the relevant review date.

The scheme again was that there should be a negotiation for agreement on the rent. Failing agreement it was to be determined by a surveyor, and in the absence of agreement between landlord and tenant on the appointment of the surveyor the surveyor was to be appointed by the president for the time being of the Royal Institution of Chartered Surveyors. In fact in the present case a surveyor has been appointed, but the submissions to the surveyor so appointed by the parties as to the basis on which the open market rent should be calculated differ on a point of law, and the parties have accordingly agreed that it would be fruitless for the surveyor to proceed with his assessment of the open market rent until the question has been decided by the court.

I come back, therefore, to the definition of "open market rent". With a minor and, for present purposes, immaterial variation introduced by a deed of variation of September 18 1986; it reads as follows:

> "Open Market Rent" means the best yearly rack rent at which the Premises might reasonably be expected to be let as a whole or in parts

— I stress those words —

> at the Relevant Review Date by a willing landlord to a willing tenant in the open market with vacant possession and without premium or any other consideration for the grant thereof for a term of years equivalent to the period then unexpired of this Lease but in any event not less than ten years assuming if not the fact that the Premises remain in existence and are ready for immediate occupation and use

— I stress those words also —

> and subject to the due performance and observance of the covenants and conditions on the part of the Tenant herein contained (but without prejudice to any rights or remedies of the Landlord in regard thereto) and on a lease which shall otherwise contain the same terms and provisions in all respects as this Lease (other than the amount of rent but including the provisions for

review of rent herein contained) there being disregarded (i) those matters set out in paragraphs (*a*) and (*b*) of Section 34 of the Landlord and Tenant Act 1954 (ii) any effect on rent of any improvements carried out by the Tenant during the term with the Landlord's consent otherwise than in pursuance of an obligation under this Lease.

It is accepted that under that form it would be possible, for instance, for the surveyor to value the premises either as a whole or in parts. The parts could, for instance, treat as one part the first floor occupied as its offices by Murray's Meat Markets and the three upper floors which are vacant offices. Equally, the ground floor could be one unit or it could be two units. But the landlord says that if that would produce the best rent it could be, say, seven units. But the units envisaged in the landlord's proposal are units self-contained and separated off. A sketch plan has been put in evidence by the surveyors which indicates the general layout, but only as a sketch. Some of the units will have entrance from Electric Avenue, others from Brixton Road, and in effect two or three units would be created out of the part of the ground floor occupied by Iceland and four or five out of the part occupied by Murray's.

As I see it, however, the first stage in this is that the assessment of the new rent is to be on the basis of the rental value of the landlord's premises in their actual condition for the time being. That is in accordance with a statement of the law by Fox J in *Trust House Forte Albany Hotels Ltd* v *Daejan Investments Ltd* (1980) 256 EG 915. At p 916 Fox J said:

> . . . As to the protection of the landlord against inflation, I accept that such is the purpose of the clause, but what is it that is being protected against inflation? In the absence of clear language to the contrary, I would assume that it is the rental value of the landlord's premises in their actual condition for the time being.
>
> . . . You protect the landlord against inflation by reassessing every seven years the rent of the actual premises which he owns and not of other premises. I do not see why the landlord should get the benefit of assumed alterations which the landlord has not made.

That passage was approved by this court in *Orchid Lodge (UK) Ltd* v *Extel Computing Ltd* [1991] 32 EG 57*. It was cited by Judge O'Donoghue in the court below in the present case, together with a statement by Hoffmann J in *Norwich Union Life Insurance Society* v *Trustee Savings Banks Central Board* [1986] 1 EGLR 136 at p 137 where he referred to the "presumption that the hypothesis upon which the rent should be fixed upon a review should bear as close a resemblance to reality as possible".

It consequently follows — and this I apprehend is accepted in this court by Mr Dowding for the landlord — that the rent is not to be fixed on the basis that either the landlord or the hypothetical tenant will, before the review date, have carried out the works of conversion necessary to subdivide the premises into the seven self-contained units on the ground floor envisaged in the landlord's surveyor's suggestion.

So, Mr Dowding says, we must consider that there may be tenants who would be prepared to pay a higher rent, or between them pay higher rents collectively, for the ground floor subdivided into seven units if each one takes the chance that he will get consent from the landlord for the subdivision of the premises and for any necessary change of use, that he will pay a continuous rent without any fitting-out period at a lower rent, and will calculate that rent that he is prepared to pay on the basis that he, the tenant, will be carrying out the works of conversion which are necessary and will be putting up with an absence of return from the unit, whether by trading in it or by subletting, until the works of conversion have been carried out.

But the difficulty about that, as it were the Charybdis to the Scylla of the works being deemed executed before the review date, is that if the execution of these works will produce a higher rental value for the premises or an improved return from trading in the premises which will warrant paying a higher rent for the pleasure of converting the premises, the works of conversion will rank as tenant's improvements which are to be disregarded under the express terms of the rent review clause. Mr Dowding says: "But these are merely hypothetical alterations by a hypothetical tenant under a hypothetical lease. Under the rent review clause you are only required to disregard actual improvements carried out by the real tenant." But it seems to me to make nonsense of the rent review clause as drawn and of the aim that the rent on review should be fixed so as to bear as close a resemblance to reality as possible, if it is not possible to charge rent for actual improvements but is possible to charge the actual tenant with rent for hypothetical improvements which have never been carried out.

The further point taken by Mr Dowding is this. In the rent review clause it is to be assumed, if not the fact, that the premises remain in existence and are ready for immediate occupation. Mr Dowding says that the assumption that the premises remain in existence no doubt relates to the possibility that the premises, being a building, have been damaged by fire or an explosion and have had to be demolished, but the words "and are ready for immediate occupation and use" must refer to something more and therefore are to be taken as indicating an assumption that the premises are ready for occupation and use for any purpose and in any condition that the hypothetical tenant would wish them to be in; therefore, it is permissible to take into account hypothetical works to be carried out after the rent review date by the hypothetical tenant.

Again, I am afraid I do not agree. It seems to me that the words "and are ready for immediate occupation and use" amplify "remain in existence" to indicate that it is not enough that the premises remain in existence as, for instance, a gutted shell.

Accordingly, for these reasons, which are substantially those of the judge, I would dismiss the appeal. I agree with the judge's conclusion in his careful judgment.

Stocker and **Butler-Sloss LJJ** agreed and did not add anything.

Appeal dismissed with costs.

* Editor's note: Also reported at [1991] 2 EGLR 116.

Court of Session: Inner House

July 25 1991

(Before Lord HOPE Lord President, Lord ALLANBRIDGE and Lord MAYFIELD)

VISIONHIRE LTD v BRITEL FUND TRUSTEES LTD

Estates Gazette March 14 1992

[1992] 10 EG 95

Rent review — Provisions for initiating review — Time-limit — Whether presumption that time is not of the essence applies in Scotland — Whether upon construction of the lease time of the essence — Appeal by landlord dismissed

The pursuers hold the term of a shop unit under a 24-year lease from August 1 1978 — The defenders are the landlords — The lease made provision for five-yearly rent reviews — Provision for the appointment of an arbiter or an expert valuer — The application for a nomination may be made not more than six months before or at any time after the relevant rent review dates — The lease further providing that if by the first date for the payment of rent for the relevant five-year period the lessors had not before that date made an application for a nomination, the lessee may at any time thereafter state by notice the amount of such revised rent and that amount shall be the revised rent for the relevant period unless the lessors make an application for nomination within three months after the date of service of such notice — The relevant rent review date was August 1 1988 — The parties had not agreed the revised rent and by a letter dated July 29 1988 the lessors gave notice that the rent should be revised to £29,750 pa but did not make any application for nomination before August 1 1988 — By a notice dated August 23 1988 the lessees gave notice to the lessors that the revised rent should be £13,200 — The lessors failed to make any application for nomination within three months after the date of that notice — The lessees sought a declarator that the revised rent proposed by them of £13,200 was the revised rent for the rent review period commencing August 1 1988

In the Outer House the lord ordinary granted decree of declarator as contended for on behalf of the lessees — The lessors sought review of that interlocutor on the ground that

the lord ordinary misdirected himself in holding that in Scots law there is no presumption that time is not of the essence of a rent review clause and that in any event, on a proper construction of the rent review clause in question, if there were such a presumption, it had been rebutted

***Held*: The decision of the House of Lords in *United Scientific Holdings Ltd* v *Burnley Borough Council* can be treated as laying down principles which are applicable equally to Scotland as to England — In Scotland the rules which according to English law are stated as presumptions are really to be seen as rules of construction which take their place along with various other rules in order to ascertain the intention of the parties**

Dealing with the particular wording in the instant lease, the lessees had followed the procedure in the rent review clause and were entitled to rely on what the clause said about the effect of the expiry of the time-limit — The purpose of the clause was to enable the lessee to fix a time-limit which will be strictly enforced — This being an ultimatum procedure which the parties had agreed as part of their contract — This was a clear indication that time was to be of the essence

The following cases are referred to in this report.

Bowes v *Shand* (1877) 2 App Cas 455; [1871-80] All ER Rep 174; 46 LJQB 561; 36 LT 857; 25 WR 730, HL
Gatty v *Maclaine* 1921 SC (HL) 1
Grahame v *Magistrates of Kirkcaldy* (1882) 9 R (HL) 91
HMV Fields Properties Ltd v *Bracken Self Selection Fabrics* 1991 SLT 31
Leeds Permanent Pension Scheme Trustees Ltd v *William Timpson Ltd* 1987 SCLR 571
Legal and Commercial Properties Ltd v *Lothian Regional Council* 1988 SLT 463; 1988 SCLR 201
Mecca Leisure Ltd v *Renown Investments (Holdings) Ltd* (1984) 49 P&CR 12; [1984] EGD 200; 271 EG 989, CA
Phipps-Faire Ltd v *Malbern Construction Ltd* [1987] 1 EGLR 129; (1987) 282 EG 460
Rodger (Builders) Ltd v *Fawdry* 1950 SC 483
Smith's (Henry) Charity Trustees v *A WADA Trading & Promotion Services Ltd* (1983) 47 P&CR 607; [1984] EGD 103; 269 EG 729, CA
United Scientific Holdings Ltd v *Burnley Borough Council* [1978] AC 904; [1977] 2 WLR 806; [1977] 2 All ER 62; (1977) 33 P&CR 220; [1977] EGD 195; 243 EG 43 & 127, HL
Yates, Petitioner 1987 SLT 86

The pursuers and respondents, Visionhire Ltd, held the term of a lease of a shop unit in the Wellgate Shopping Centre, Dundee. The defenders and reclaimers, Britel Fund Trustees Ltd, held the reversion to the lease.

Nimmo Smith QC (instructed by Tods Murray WS) appeared for the defenders and reclaimers; Ronald Mackay QC and J R Doherty (instructed by Bird Semple Fife Ireland WS) represented the pursuers and respondents.

In giving his opinion, the LORD PRESIDENT said: The parties to this case are in dispute about the meaning and effect of a rent review clause in a lease. The lease relates to a shop unit in a shopping centre known as the Wellgate, Dundee. It was entered into between the Prudential Assurance Co Ltd, Caledonian Terminal Securities (Wellgate) Ltd and the pursuers [Visionhire Ltd] dated March 3, August 25 and September 7 1980 and registered in the Books of Council and Session on March 19 1985. The lessors were Prudential Assurance Co Ltd and the lessee was Visionhire Ltd. The landlord's interest in the lease was assigned to the defenders in 1986.

The premises were let to the lessee for 25 years from the date of entry, which was August 1 1978. By clause 3 of the lease the pursuers undertook to pay to the lessors throughout the currency of the lease the respective rents set out in the second schedule, together with a service charge as defined by the third schedule to the lease. We are concerned in this case with the provisions of the second schedule, which stipulated that rent was to be paid at the rate of £7,500 pa for the period from October 1 1978 to the fifth anniversary of the date of entry and that thereafter in respect of each quinquennium from the anniversary of the date of entry it was to be subject to review.

The provisions of the second schedule, so far as material, have been set out in full by the lord ordinary. It is sufficient at this stage simply to summarise the structure of these provisions before quoting the declaration which is in dispute in this case. The opportunity for the rent to be reviewed is to occur in each quinquennium following the initial period of five years. The rent for the ensuing period of five years is to be the highest rent payable during the immediately preceding quinquennium or such revised rent, whichever shall be greater, as shall be agreed between the lessors and the lessee at any time or, if there is no such agreement before the expiry of the immediately preceding quinquennium, as may be determined by an arbiter or, at the option of the lessors, by an expert valuer. The initiative for referring the matter to an arbiter or to an expert valuer in terms of declaration (1) of the second schedule lies with the lessors. Their application for a nomination may be made to the president of the Royal Institution of Chartered Surveyors not more than six months before or at any time after the anniversary of the date of entry in the years in which the revised rent may become payable. It should be noted that there is no fixed time-limit within which such an application may be made after that date, since it is provided that the application may be made "at any time". Declaration (5) provides a mechanism for rolling up the difference between the rent previously payable and the revised rent, which is recoverable from the lessee without any right to claim interest on the first date for the payment of rent after the revised rent has been ascertained. Then there is declaration (6), which is in these terms:

> if and so often as a revised rent in respect of any quinquennium has not been agreed between the Lessors and the Lessee before the first date hereby provided for payment of rent for the relevant quinquennium, and the Lessors have not before that date made any application to the President for the time being of the Royal Institution of Chartered Surveyors as aforesaid, the Lessee may at any time thereafter give to the Lessors notice in writing containing a proposal as to the amount of such revised rent and the amount so proposed shall be the revised rent for the relevant quinquennium unless the Lessors shall make such application as aforesaid within three months after the date of the service of such notice.

The facts relevant to the dispute which has arisen in this case are all agreed. The rent payable by the pursuers to the defenders fell to be reviewed on the terms set out in the second schedule for the quinquennium commencing on August 1 1988. The parties did not agree the revised amount of the rent before the expiry of the preceding quinquennium. By letter dated July 29 1988 agents for the defenders gave notice to the pursuers that the rent should be revised to £29,750 pa, but the defenders did not make any application to the president of the Royal Institution of Chartered Surveyors for the nomination of an arbiter or an expert valuer before August 1 1988. It is agreed, therefore, that the conditions which are necessary to entitle the pursuers to serve a notice in terms of declaration (6) were fulfilled. The pursuers then proceeded to take advantage of this provision by serving a notice on the defenders by recorded delivery letter dated August 23 1988. In this letter they proposed that the revised amount of the rent should be £13,200. It is admitted that the defenders did not make an application to the president within three months after the date of the service of that notice. It was not until December 29 1988 that they took this step, by which time the period of three months had expired. In these circumstances the pursuers seek declarator that the revised rent proposed by them of £13,200 is the revised rent for the quinquennium from August 1 1988. The lord ordinary, having heard argument on the Procedure Roll, held that the provisions of declaration (6) on which the pursuers' action is based must be regarded as mandatory and that, as a result, the pursuers are entitled to succeed in their action. He sustained their pleas in law and granted decree of declarator as concluded for. The defenders now seek review of that interlocutor on the grounds that the lord ordinary misdirected himself in holding that in Scots law there is no presumption that the timetable specified in a rent review clause is not of the essence of the contract and that, in any event, on a proper construction of the provisions of the second schedule, if there is such a presumption it has been rebutted.

It has not been suggested that declaration (6) is in any respect ambiguous or that there is any reason other than that related to the presumption for not giving effect to it according to its terms. Indeed, if one were to look only to the terms of the agreement which the parties have expressed in the second schedule, there would plainly be no escape from the conclusion that, since the defenders did not make an application to the president within three months after the date of the service of the notice by the pursuers, the amount proposed in that notice is to be the revised rent for the new quinquennium. But the stipulation that the defenders' notice required to be served within three months in order to avoid that result is a stipulation as to time and their argument is that time was not of the essence in this contract.

Much of the argument before the lord ordinary and again in the hearing in the reclaiming motion before us was taken up with the question whether the principles laid down by the House of Lords in *United Scientific Holdings Ltd* v *Burnley Borough Council* [1978] AC 904* are applicable in Scots law. It was held in that case that there was a presumption that a timetable specified in a rent review clause in a contract of lease was not of the essence of the contract and that there was nothing in either of the two leases there in question to displace that presumption. It is clear from an examination of the speeches in that case, especially those of Lord Diplock at pp 923-924 and of Lord Fraser of Tullybelton at p 958, that their lordships were concerned particularly with the question whether a failure to adhere strictly to the timetable in a rent review clause deprives the landlord of his right to have the rent reviewed. This was a question which had arisen on a number of occasions in previous years and had appeared to turn on fine distinctions between the wording of particular clauses in each lease. The Court of Appeal had decided that the commercial nature of the contract and the legal nature of the right granted to the landlord by a rent review clause raised a presumption that time specified in such a clause for anything that needed to be done was of the essence. They held that this presumption must prevail unless there were strong contra-indications in the actual wording of the lease. This decision was reversed by the House of Lords after a detailed examination of the development of the law of England in regard to the enforcement of stipulations in contracts as to time. This involved a discussion of the dichotomy between the rules of equity and of common law which plainly has nothing to do with the law of Scotland, which has never recognised this distinction: see, for example, *Grahame* v *Magistrates of Kirkcaldy* (1882) 9 R (HL) 91 *per* Lord Selborne LC at p 97. But the opportunity was taken, in particular by Lord Fraser of Tullybelton at p 958, to approve of a summary of law in *Halsbury's Laws of England*, 4th ed, vol 9, para 481, which is in these terms:

> The modern law, in the case of contracts of all types, may be summarised as follows. Time will not be considered to be of the essence unless: (1) the parties expressly stipulate that conditions as to time must be strictly complied with; or (2) the nature of the subject matter of the contract or the surrounding circumstances show that time should be considered to be of the essence; or (3) a party who has been subject to unreasonable delay gives notice to the party in default making time of the essence.

On the question of principle which was raised in the appeals about the effect of stipulations as to time in rent review clauses, the House of Lords held, in the words of Lord Diplock at p 930 which are very close to those used in the passage which Lord Fraser of Tullybelton quoted from *Halsbury*:

> . . . upon the question of principle which these two appeals were brought to settle, I would hold that in the absence of any contra-indications in the express words of the lease or in the interrelation of the rent review clause itself and other clauses or in the surrounding circumstances the presumption is that the time-table specified in a rent review clause for completion of the various steps for determining the rent payable in respect of the period following the review date is not of the essence of the contract.

An examination of that decision leads, therefore, to two questions which must be examined in order to see to what extent, if at all, it can be accepted as being consistent with the principles of Scots law. The first question is whether *Halsbury*'s statement of the modern law of England in its application to contracts of all types is in accordance with the Scottish approach to the construction of contracts generally. In addressing this question it should be noted that, in para 482, *Halsbury* elaborates upon the point made in the previous paragraph as follows:

> Apart from express agreement or notice making time of the essence, the Court will require precise compliance with stipulations as to time wherever the circumstances of the case indicate that this would fulfil the intention of the parties. Broadly speaking, time will be considered of the essence in "mercantile" contracts and in other cases where the nature of the contract or of the subject matter or the circumstances of the case require precise compliance.

The second question is whether, whatever may be said about contracts of other types, stipulations as to time in rent review clauses should, in the absence of indications to the contrary, be seen as not being of the essence of the contract.

The lord ordinary, after a detailed examination of the speeches in that case and such indications as are available as to the position in Scots law, was not satisfied that there has ever been an equivalent in Scotland to the English rule as stated in *Halsbury* at para 481. He expressed his conclusion on the first question which I have mentioned above in a passage which is in these terms:

> In all the circumstances, although there are some similarities between the laws of Scotland and England with regard to the effect of stipulations as to time, it appears to me there are also material differences. It further appears to me that the instances considered do not provide a basis in Scots law for a presumption, either general or applying to a particular class of contract, that time is not of the essence, if by presumption is meant a rule which must be given effect in a situation in which the actual contract made by the parties, construed naturally according to its particular terms, provides for some different result.

Turning then to the particular point about rent review clauses, he quoted passages from the speech of Lord Fraser of Tullybelton at p 959, but here again he was not persuaded that the English approach was consonant with Scots law:

> With all diffidence, however, I do not think that the considerations set out by Lord Fraser are of such force as to make it necessary, given the different Scottish background, to override, or give an artificial construction to, express provisions agreed by the parties. Similarly, given the difference in background, certainty in the application of the contract terms may be of more importance than is allowed in the English cases.

Mr Nimmo Smith QC, for the defenders, disclaimed any intention to characterise the English rule in these terms. He accepted that any rule of construction, including any presumption as to the effect of provisions as to time, must be subordinated to the express terms of the contract. Having submitted that in this regard the lord ordinary had overstated the effect of the English rule, which recognised that contra-indications may be found in the express words of the contract, he invited us to regard the history of the development which had led to the modern rule as irrelevant to the question whether, in its modern form, that rule could be accepted in Scotland. He pointed out that in his discussion of provisions as to time, *Gloag on Contract* 2nd ed pp 615-618 cites English authorities along with Scottish in a manner which indicates that he regarded the principles to be found in the English cases as applicable also in Scotland. *Gloag*'s statement at p 617 that "Stipulations as to time of payment are not treated as material conditions of the contract except in various special cases" was entirely consistent with the English approach. Counsel then referred us to three recent cases in the Outer House in which, although the contrary had not been argued, the decision in *United Scientific Holdings* had been referred to without any indication that it did not apply in Scotland. These were *Yates, Petitioner* 1987 SLT 86 in which Lord Davidson, without commenting directly on that particular case, made reference at p 91 with approval to dicta from two of the later English cases which apply that decision to particular rent review clauses; *Leeds Permanent Pension Scheme Trustees Ltd* v *William Timpson Ltd* 1987 SCLR 571, in which Lord Cowie at p 573 said that it is clearly established that in general there is a presumption that the timetable in rent review clauses is not of the essence but that this presumption can be rebutted in the manner mentioned in *United Scientific Holdings*; and *Legal and Commercial Properties Ltd* v *Lothian Regional Council* 1988 SLT 463 where, at p 465, Lord Jauncey said that *United Scientific Holdings* could be treated as laying down principles which are applicable equally to Scotland as to England.

The pursuers' counsel, on the other hand, supported the lord ordinary's approach. Their argument was that in all cases the question was simply one of construction. The law of Scotland took a neutral approach, without making any presumptions either one way or the other as to whether time was of the essence of the contract. The proper question to ask, therefore, was not whether time was of the essence but simply to ask what does the contract mean. If the language used is clear on this point, then the consequences of a failure to adhere to the stipulated timetable are those which the contract itself provides. It was contrary to the Scottish approach to ignore the terms of the contract and in effect to rewrite it so as to mean something else. Reference was made to *Gatty* v *Maclaine* 1921 SC (HL) 1 *per* Lord Birkenhead LC at p 6 and Lord Shaw of Dunfermline at p 11. There was no support in the Scottish authorities for any general rule in the terms indicated by *Halsbury*. *Gloag*'s discussion proceeds merely by looking at particular instances, and the approach in this country had always been to see the question when it arose as one about the materiality of the breach of contract. The strict approach to questions of construction as illustrated by the Scottish approach to irritancy clauses was inconsistent with the

*Editor's note: Also reported at (1977) 243 EG 43 & 127.

A English view that it was open to the court to weigh up the prejudices to each party with a view to deciding whether or not a contract should be strictly enforced.

I recognise the force of these arguments and also of the points made by the lord ordinary in his careful analysis of the differences between Scotland and England in regard to stipulations as to time. But, in my opinion, the conclusion which he reached on this matter is mistaken and I prefer Lord Jauncey's view as expressed in *Legal and Commercial Properties Ltd* at p 465 that *United Scientific Holdings* can be treated as laying down principles which are applicable equally to Scotland as to England. It seems to me to be well established in Scotland that whenever a provision as to time is contained in a contract the question is raised whether time is of the essence. There
B has over a long period been a tendency to examine the English cases in order to find a solution to this problem whenever it is raised. *Gloag*'s reference at p 615 to *Bowes* v *Shand* (1877) 2 App Cas 455, which is English authority for the rule that in mercantile contracts time will be considered to be of the essence, provides an example of this approach. Further examples are to be found at p 617 where stipulations as to time of payment are discussed. I do not think that it is necessary to examine the details of these examples. The point seems to me simply to be this, that in Scots law, as in the law of England, one can find cases where stipulations as to time are treated as being of the essence of the contract and other cases where they are not. No Scottish writer has yet sought to rationalise the position in the way in
C which the general rule has been stated in *Halsbury*, but I do not find this surprising nor do I find the absence of any such statement indicative of an incompatibility with the English approach.

We were referred to *Trotter's The Law of Contract in Scotland* (1913), which was written some 16 years before *Gloag*'s 2nd ed, where at p 250 the author states that whether time is of the essence of the contract is a question of construction which must be determined by the particular facts of the case. The pursuers' counsel founded upon this passage as demonstrating that the approach in Scotland was essentially to construe the contract according to its terms. I note that examples are given in *Trotter*'s discussion of this point of cases where time was of the essence and where it was not and that, after a reference to section 10 of the Sale of Goods Act 1893 that stipulations
D as to time of payment are not to be deemed to be of the essence of a contract of sale, he concludes by repeating that whether any stipulation as to time is of the essence of the contract or not depends on the terms of the contract. But there is no attempt in this discussion at any detailed analysis, and since only a very few cases are cited in comparison to the many noted by *Gloag* at pp 615-618, it does not appear that the writer was concerned to examine more closely the various factors which might be relied upon to arrive at a decision as to whether time is of the essence.

The only objection of substance which has been raised to *Halsbury*'s formulation of the modern rule as an acceptable statement in Scotland is that it is inconsistent with the Scottish
E approach to the construction of contracts, which is said to be to look only to what the contract provides. But I am not persuaded that that is an accurate statement of the Scottish position in regard to stipulations as to time. I do not find it helpful to refer by way of analogy to the strict approach which the common law has always taken to contractual irritancies or to notices to quit where the rule as to time-limits is that they must be strictly observed. Stipulations as to time generally are, I think, in a special position because of the hardship which may arise if, for example, a party is held to be in material breach where he is just one day late in doing something which the contract requires. It appears to have been accepted that it is not inconsistent with the general principle that contracts should be enforced according to their terms to ask whether the parties really
F intended that provisions as to time should be enforced in this way.

I see *Gloag*'s recognition at p 617 that stipulations as to time of payment are not treated as material conditions of a contract except in very special cases as illustrating this point. If the approach is essentially to look only to what the contract actually says and then to enforce it according to its terms, I would have expected an equally strict view to be taken in regard to materiality. The point which emerges from *Gloag*'s discussion is that, while stipulations as to time which are expressed in terms which show that strict compliance is required will be enforced accordingly, there is room also for a consideration of the nature of the contract and the circumstances in order to decide whether a stipulation is material and, as such, is of the essence of the contract. That seems to me to be sufficiently close to *Halsbury*'s position for it to be a position which Scots law can accept G
also as a statement of the modern law. So far as the lord ordinary's position is concerned as indicated in the two passages which I have quoted above, I agree with Mr Nimmo Smith that the lord ordinary appears to have overstated the effect of the English rule since, as the passage which I have quoted from para 482 indicates, the court will require precise compliance with stipulations as to time wherever the circumstances of the case indicate that this would fulfil the intention of the parties. It seems to me, therefore, that there is no essential difference between the positions adopted in the two countries and that the rules which according to English law are stated as presumptions are really to be seen as rules of construction which take their place along with various other rules in order to ascertain what
the intention of the parties truly was in order that the contract which H
they have made should be enforced.

Turning to the particular instance of rent review clauses in commercial long leases, I see no good reason why Scots law should take up a different position from that which has been adopted in England for the reasons which were explained by Lord Fraser of Tullybelton in *United Scientific Holdings* at p 958:

> Rent review clauses have only become common in comparatively recent years, certainly since the last war, and their main purpose is to protect the revenues of landlords from the effects of inflation. From the landlord's point of view a rent review clause is an important, almost indispensable, term of the contract if he is to agree to a lease for a long period, during which inflation may well continue. The clause is also in a less direct way of benefit to the tenant, J
> because, without it, he would not normally be able to get the security of tenure which is afforded by a long lease, except perhaps by paying a rent which in the early years of the lease would be far above the current market level. The rent review clause has thus become a convenient device to facilitate the granting of long leases in an inflationary age, and its main purpose is the same whatever the exact machinery specified in a particular clause.

With that introduction he formulated the approach to the construction of these clauses in these terms at p 959:

> As the substance of a review clause is, in my opinion, to provide machinery for ascertaining the market rent from time to time, at the intervals agreed in the interests of both parties, rather than to confer a benefit on the landlord, it seems to me that stipulations as to time ought not to be strictly enforced unless there is something in a particular clause to indicate that time is of the essence in that case. K

I find myself entirely in sympathy with the lord ordinary's concern not to be drawn too easily into an acceptance of what was said in that case, and he was right not to follow the cases in the Outer House where the point was not fully argued without a more careful consideration of the implications of this approach. But it appears to me that it would not be doing violence to the Scottish approach to the construction of contracts to accept Lord Fraser's dictum as a rule of construction which we can adopt. The concluding words of the passage which I have quoted show that careful attention must be paid to the wording of the provision which has to be construed. The function of the rule is to fill the gap where there is an absence of provision or a lack of clarity on the point, not to override express L
provisions in the contract which show either expressly or by necessary implication that the parties intended that the time-limit was to be strictly applied.

It is also worth noting that rent review clauses which have become commonplace in leases of this kind in Scotland have tended to follow English styles both as to their general structure and the detail of their wording. These styles are themselves influenced by the development of the case law in England especially since *United Scientific Holdings* established the principles which were to be applied. There can, of course, be no compromise of the well-established principles of Scots law such as those relating to the use of irritancies. As I said in *HMV Fields Properties Ltd* v *Bracken Self Selection Fabrics* 1991 SLT 31 at p 35, there are significant differences between the law of England M
and the law of Scotland relating to irritancies. Any investor in commercial property in this country must accept that these differences will continue to exist. But rent review clauses seem to me to present a quite different problem, which is simply one relating to the use of language and the meaning to be attached to the wording of each clause. It seems to me that, as in the case of mercantile contracts generally, there are sound practical considerations for adopting a common approach to the construction of these clauses and for taking notice of the case law in England where it may assist towards understanding the effect of what has been provided as the machinery for the rent review.

This brings me to the particular wording which is in issue in this

A case. The lord ordinary was of opinion that the provisions of declaration (6) must be held to be mandatory. He makes it clear that his conclusion on this point would have been the same even if, contrary to his view on the issue of principle, he had to decide whether the terms of the lease displaced the presumption that time is not of the essence. The precise purpose of declaration (6) seemed to him to be to enable the tenant to bring the question of the rent to an issue, and, in his view, the specific provision for what was to happen in the event of the landlord's failure to apply for the appointment of an arbiter or expert showed that the observance of the time-limit was regarded as essential. Mr Nimmo Smith submitted that the lord ordinary misdirected himself as to the proper construction of the declaration. B He invited us to follow the views expressed by Eveleigh and May LJJ in *Mecca Leisure Ltd* v *Renown Investments (Holdings) Ltd* (1984) 49 P&CR 12* and the decision of Warner J in *Phipps-Faire Ltd* v *Malbern Construction Ltd* (1987) 282 EG 460†, where the clause in issue was almost identical to the declaration in the present case. His contention was that it was up to the lessee, if he desired to make time of the essence, to use the ultimatum procedure to bring this about. He could do this by serving a notice on the lessors which stipulated a reasonable time-limit within which the lessors must apply for the appointment or be held to have lost the right to have the rent reviewed by an arbiter or expert. Counsel for the pursuers, on the other hand, referred to dicta in a number of other cases in England, including *Henry Smith's Charity Trustees* v *AWADA Trading & Promotion* C *Services Ltd* (1983) 47 P&CR 607§, in support of the view that this was contrary to a fair reading of the declaration and its effect. It was not appropriate, they said, for the ultimatum procedure to be used in this case because its function was to provide a remedy for the innocent party to a contract who was faced with a breach. That was the situation which Lord Sorn in *Rodger (Builders) Ltd* v *Fawdry* 1950 SC 483 had in mind at p 492 when he described the procedure and said that failure to pay within the reasonable time-limit would be treated as breach of an essential condition entitling the seller to rescind. The position in this case was that the lessors were not in breach of any obligation and the lessee was not entitled to demand performance from them of something which they were not bound to do. In any event, declaration (6) provided its own ultimatum D procedure, and since the pursuers had followed that procedure by giving notice in writing to the defenders as lessors they were entitled to rely on what the declaration said about the effect of the expiry of the time-limit and were not required to do anything else to bring this into effect.

I am in full agreement with the lord ordinary on this branch of the argument, both as to the result and as to the reasoning which he applied. Approaching the matter first without reference to the authorities, the purpose of declaration (6) seems to me to be to enable the lessee to fix a time-limit which will be strictly enforced.

Declaration (1) enables the lessors to apply for a nomination by the president "at any time" after the anniversary date in the years in E which the revised amount of the rent becomes payable. There is no time-limit set by this declaration within which the application must be made if it is to receive effect. The lessors can set a limit to this uncertainty by making their application at any time, but the lessee has no means within the framework of this declaration of achieving this result. That is the context for declaration (6) which enables the lessee, in a situation where the necessary steps to achieve a rent review have not been taken before the first date for the payment of rent for the quinquennium, at any time thereafter to give to the lessors a notice with the consequences which the declaration describes. I agree with the lord ordinary that this is itself a kind of ultimatum procedure which the parties have agreed as part of their contract and that there is no room here for imposing a requirement on the lessee to embark F on a further ultimatum procedure, since the parties have already set out in the declaration what the consequences are to be if the period of three months is allowed to elapse. I find here a clear indication that time was to be of the essence, with a result that the stipulation that the amount proposed in the lessee's notice was to be the revised rent unless the lessors' application was made within three months was intended to be strictly enforced.

I do not agree with the submission for the pursuers that the ultimatum procedure is confined to those cases where a party is in breach of contract and the innocent party wishes to enforce a remedy G against him in respect of their breach. The procedure is available generally in all those cases where there is a provision in the contract about time which is not of the essence but which one of the parties to the contract wishes to be made of the essence for the purposes of the contract. But that additional step is unnecessary in the present case in view of the terms in which declaration (6) is expressed.

This view is in conflict with what Warner J decided in *Phipps-Faire Ltd* on a construction of a rent provision which, as the lord ordinary points out, was indistinguishable from the present case. He described his approach to the matter in these terms:

> The authorities seem to me to show that the presumption that time is not of the essence of a provision in a rent review clause is strong and that it will not be H rebutted by any contra indication in the express terms of the lease unless it is a compelling one.

He did not find the contra-indications compelling and took the view that an extension of the period of three months by serving a notice specifying a reasonable period would not substantially impair the protection intended to be afforded to the lessee. He was attracted by Eveleigh LJ's description of clauses of this type in *Mecca Leisure Ltd* as "a trap" in a passage in his opinion at p 17:

> I read the clause as a whole and bear in mind that there is a presumption to be displaced that the parties are clearly intending to arrive at a fair rent to be determined amicably if possible. It seems to me that to hold otherwise would make the clause a trap and the chosen machinery dangerous. In my opinion, time is not of the essence in this case. J

But the approach to construction which is indicated by these dicta seems to give insufficient weight to what the parties themselves have agreed. It seems to set a standard for the rebuttal of the presumption which departs significantly from the principles which were laid down in the *United Scientific Holdings* case. The only question, in Lord Fraser's words, is whether there is something in the wording of the particular clause which indicates that time is of the essence. As Lord Donaldson of Lymington MR said in *Henry Smith's Charity Trustees* at p 615, it is for the parties to make their own contract. It was held in that case, in regard to a clause which is not very different from that in issue here, that its provisions were inconsistent with the necessity for the tenants to serve a time of the essence notice and that they clearly indicated that the ordinary presumption against time K being of the essence did not apply. As Lord Donaldson put it at p 614: "the parties have not only set out a timetable, but have in terms provided what is to happen in the absence of strict compliance with their timetable." The same point was made by Slade LJ at p 621 when he said that the parties had set out "in definition, specifically and clearly, what is to happen in default of the exercise of the rights given to the respective parties within the permitted periods of time". At p 623 he said that the provisions in question were quite inconsistent with the survival of any right in the landlords to apply for the appointment of a surveyor after the expiration of the period which was specified. To hold otherwise would, as Browne-Wilkinson LJ said in his dissenting opinion in *Mecca Leisure Ltd* at p 24, "involve L not simply extending the time-limits within which the parties' bargain could be performed but an alteration of the parties' bargain itself". In *Yates, Petitioner* at p 91, Lord Davidson described that opinion, which accords with the decision in *Henry Smith's Charity Trustees*, as convincing and I agree both with him and with the lord ordinary in the present case in endorsing this approach.

For these reasons I consider that the pursuers are entitled to decree of declarator as concluded for. I would refuse this reclaiming motion and affirm the interlocutor of the lord ordinary.

Agreeing, **Lord Allanbridge** said: I have read the opinion of your lordship in the chair and am in full agreement with it both as M regards its conclusions and the reasons for such conclusions.

In particular I agree that the principles laid down by the House of Lords in the *United Scientific Holdings* case are applicable in Scots law. This is consistent with the view expressed in *Gloag on Contract* 2nd ed, at p 617, to the effect that stipulations as to time of payment are not to be treated as material conditions of the contract except in various special cases.

I consider that the present case is such a special case. It is made quite clear in the contract that if the lessee is in a position to operate the procedure in declaration (6) then he can fix a time-limit which will be strictly enforced. He is in fact imposing a kind of ultimatum procedure which the parties have agreed as part of their contract. It

*Editor's note: Also reported at (1984) 271 EG 989.
†Editor's note: Also reported at [1987] 1 EGLR 129.
§Editor's note: Also reported at (1983) 269 EG 729.

would be otiose to require the lessee to initiate a further ultimatum procedure as was suggested by counsel for the lessors.

In these circumstances I agree that this reclaiming motion should be refused.

LORD MAYFIELD agreed, and did not add anything.

Court of Appeal

November 19 1991

(Before Lord DONALDSON OF LYMINGTON MR, Lord Justice NOURSE and Lord Justice BUTLER-SLOSS)

SHEERNESS STEEL CO PLC v MEDWAY PORTS AUTHORITY

Estates Gazette March 28 1992

[1992] 12 EG 138

Landlord and tenant — Rent review — Construction of rent review provision — Whether, *inter alia*, improvements carried out by tenants to be disregarded in giving effect to the expression "a fair yearly rent for the Site" — Appeal allowed in part

The appellant landlord, Medway Ports Authority, is the owner of some 53 acres of land — The land originally included part of a defensive moat extending to 13 acres — By one of two leases the respondent tenant held the term of the major part of the land, including the 13 acres of moat and two pieces of land beyond it, for a term of 125 years from September 1 1968 — Under the lease the tenant covenanted to erect a building consisting of a steel-making plant and rolling mill — The tenant was at liberty to drain and fill the moat with suitable solid matter — The evidence showed that the tenant at considerable cost did drain and fill 11 acres of the moat, leaving the balance as cooling ponds, and that it erected the steel-making plant and rolling mill and further buildings — The lease made provision for the rent to be reviewed every 21 years — The rent at review was to represent a "fair yearly rent for the Site" having regard to rental values of property let for 125 years from the date of valuation " . . . for industrial purposes with an obligation for the Tenant to construct thereon such buildings as may at the time of such valuation be situate thereon . . . " no account being taken of the value of any buildings actually standing on the premises — Upon the hearing of an originating summons issued by the tenant, Browne-Wilkinson V-C granted declarations in answer to, *inter alia*, four questions arising out of the rent review provision; two questions were decided in favour of the landlord and two in favour of the tenant and the landlord and tenant appealed and cross-appealed respectively

***Held*: (1) "The Site" means the site in its state and condition on September 1 1989 (the review date) — (2) The site is to be taken to be an unbuilt plot with the moat thereon wholly filled except for that part not filled and used for cooling ponds — (3) In the passage in the rent review clause worded "with an obligation for the Tenant to construct thereon such buildings as may at the time of such valuation be situate thereon", the expression "such buildings" did not include plant and machinery affixed to the realty — (4) Clause 2(14) of the lease contains a covenant by the tenant not to use or occupy the premises other than for the purposes of steel-making, steel-rolling and operations ancillary thereto or for such other purposes as may from time to time be approved by the lessors — The notional lease incorporated this clause but with the user provision modified so as to permit use for industrial purposes or for such other purposes as might from time to time be approved by the landlord — The notional lessee is to be regarded as being entitled to carry out all alterations necessary or appropriate to convert the premises for an industrial use of his choice, notwithstanding the covenant in the actual lease by the tenant not to carry out alterations or improvements save in connection with the use of the premises for steel-making**

The following case is referred to in this report.

Basingstoke and Deane Borough Council v *Host Group Ltd* [1988] 1 WLR 348; [1988] 1 All ER 824; (1987) 56 P&CR 31; [1987] 2 EGLR 147; 284 EG 1587, CA

This was an appeal by the defendant landlord, the Medway Ports Authority, and a cross-appeal by the plaintiff tenant, Sheerness Steel Co plc from an order of Sir Nicolas Browne-Wilkinson V-C, who on October 24 1990 granted declarations in answer to a number of questions arising out of the rent review provisions of a lease dated February 1 1971 of a 51-acre site at Sheerness, Isle of Sheppey.

David Neuberger QC and Timothy Fancourt (instructed by Norton Rose) appeared for the appellant; Robert Reid QC and Nigel Thomas (instructed by Slaughter & May) represented the respondent.

Giving the first judgment at the invitation of Lord Donaldson of Lymington MR, NOURSE LJ said: By an order made on October 24 1990 Sir Nicolas Browne-Wilkinson V-C granted declarations in answer to a number of questions arising out of a rent review provision in a lease of commercial premises. Of the four questions which are still in dispute he decided two in favour of the landlord and two in favour of the tenant, who now appeal and cross-appeal respectively.

The landlord is the defendant, the Medway Ports Authority. The tenant is the plaintiff, Sheerness Steel Co plc. The demised premises are at Sheerness on the Isle of Sheppey. Their total area is between 52 and 53 acres. The great majority of them were acquired by the landlord in 1963 from the War Department which, until shortly before that date, had used them for military purposes. There were only three buildings on the land at that time. The premises included part of a defensive moat by which the entire north-western tip of the Isle of Sheppey had been almost cut off from the rest of the isle. The area of the part of the moat included in the demised premises was 13 acres. Two pieces of land of an aggregate area of 7.62 acres were cut off from the main area by the moat.

The demised premises are held under two leases dated February 1 1971 and October 14 1973, each made between the landlord of the one part and the tenant of the other part. The first of them comprised the major part of the premises, of an area of about 51 acres, including the 13 acres of moat and the two pieces of land beyond it. The term granted was 125 years from September 1 1968. The tenant agreed to erect a steel-making and rolling mill. It is out of the provisions of this main lease that the questions of construction have arisen. The second lease, which added a small area to the west of the main area, was expressed to be supplemental to the main lease and to be made upon the same terms. Hereafter I will refer to the main lease simply as "the lease".

The principal provisions of the lease are these. The parcels were expressed as follows:

> ALL THAT piece or parcel of land situate at Sheerness in the County of Kent shown on the plan annexed hereto and thereon edged red (hereinafter called "the Site") Together with the Buildings erected thereon or on some part or parts thereof (hereinafter called "the said premises").

The rent reserved for the first 21 years of the term was £22,000. The rent reserved for the next 21 years, that is to say from September 1 1989, was whichever should be the greater of £22,000 and "the ascertained rent", with similar provisions during each period of 21 years thereafter. The ascertained rent was defined to mean the rent agreed between the landlord and the tenant or in default determined by an independent expert:

> being the amount which shall represent a fair yearly rent for the Site having regard to the rental values then current for property let for one hundred and twenty five years from the date of valuation without a premium with vacant possession for industrial purposes with an obligation for the Tenant to construct thereon such buildings as may at the time of such valuation be situate thereon and otherwise on the terms and conditions of this Lease (other than the rent hereby reserved) provided that in determining such rent no account shall be taken of
>
> (i) any effect on rent of the fact that the Tenant has been in occupation of the said premises and
>
> (ii) any goodwill attached to the said premises by reason of the business then carried on thereat by the Tenant

(iii) the value attributable to any buildings actually standing on the premises.

The tenant's covenants were contained in clauses 1 and 2 of the lease. By clause 1(1) it covenanted that it would:

proceed expeditiously to erect and complete by the 31st day of December 1973 a new building consisting of a fully equipped steel making plant and rolling mill capable of producing not less than Fifty thousand tons of steel products per annum . . .

By clause 2(10) the tenant covenanted, among other things, not to discharge any solid matter into the moat, with a proviso in these terms:

PROVIDED THAT the Tenant shall after a water-tight wall shall have been constructed between the points marked X and Y on the said plan be at liberty to drain and fill in the moat with suitable solid matter up to the level of the remainder of the said premises at the point or points respectively where the same are contiguous and such filling in shall not constitute a breach of this Clause.

I shall refer to others of the tenant's covenants when dealing with the individual questions.

It will be observed from these provisions that, although the tenant was obliged to erect a steel-making and rolling mill of a specified minimum capacity, it was not obliged to drain and fill in the moat. At the same time it was within the contemplation of the parties that the moat might be drained and filled in, an operation which, as has been pointed out by the tenant, would result in the reclamation not only of the 13 acres of moat but of the 7.62 acres of dry land beyond it. The evidence does not establish whether the specified minimum capacity could or could not have been achieved without that reclamation and I make no assumption either way. What the evidence does show is that the tenant, at what it claims was a considerable cost, has drained and filled in all but approximately two of the 13 acres of moat (now used as cooling ponds); that it duly erected and completed a fully equipped steel-making plant and rolling mill capable of producing not less than 50,000 tons of steel products pa; and that it has carried out further works of improvement, including the erection of further buildings, resulting in a current productive capacity of the order of 850,000 tons pa.

Early in 1989 discussions started between surveyors on both sides with a view to agreeing the rent which should be payable from September 1 1989. The figure of £900,000 pa proposed by the landlord having proved to be approximately four times the figure acceptable to the tenant, it became clear that the rent would have to be determined by an independent expert. A nomination was duly made by the president of the Royal Institution of Chartered Surveyors on December 5 1989. It also became apparent that there were a number of differences of opinion between the landlord and the tenant as to the true construction of various of the provisions of the lease which, if they were not resolved beforehand, would or might present the expert with a very difficult task in making his valuation. And so these proceedings were commenced by an originating summons issued by the tenant in the Chancery Division on January 18 1990.

By the time that the matter came before the Vice-Chancellor several of the questions had been determined by consent and his order so records. Seven questions, of which the first two were in substance one, were argued and decided. Two of them are no longer in dispute; so five, in substance four, remain. Each of them is short. But it seems that the outcome of two of them may have a considerable influence on the rental value of the premises. I will deal with the questions in turn, using the numbering and formulation set out in the list annexed to the Vice-Chancellor's order.

Questions 1 and 2:

1. Whether "the Site" means the Site as defined by the earlier Lease in its state and condition on 1st February 1971 or on 1st September 1989.

2. Whether (a) the Site is to be taken to be an unbuilt plot and (b) the moat thereon to be wholly unfilled.

The Vice-Chancellor answered these questions by declaring that "the Site" meant the site in its state and condition as at September 1 1968 and that it was taken to be an unbuilt plot and the moat thereon wholly unfilled. That decision was favourable to the tenant because "a fair yearly rent for the Site" will clearly be less if it is taken to include an unfilled moat extending to about a quarter of its area. This is one of the questions whose outcome may, so it seems, have a considerable influence on the rental value of the premises.

The Vice-Chancellor prefaced his consideration of the individual questions with some observations on the general approach of the court to the construction of rent review provisions. Those observations were primarily directed towards questions 1 and 2. He said that in general the purpose of a rent review provision was to give the landlord a rent which reflects, as at the review date, changes in the value of money and in the value of property; it was not to confer some additional windfall on the landlord. Having referred to the decision of this court in *Basingstoke and Deane Borough Council* v *Host Group Ltd* [1988] 1 WLR 348*, he continued:

Mr Neuberger [for the landlord] further accepts what is merely a corollary to that underlying approach, namely that one does not expect under a rent review provision that the tenant will be called upon to pay an additional rent for improvements made at his own expense. However, he emphasises — and I fully accept — that it is difficult to call these "principles": they are mere commercial expectations with which one approaches the construction of a particular document. Each lease has to be construed individually according to the words used in that document. If there are clear words showing an intention which conflicts with one's normal expectation, those words must be given effect according to their tenor. But if the words are not clear and if there is a doubt as to the actual intention as expressed in the words used, it is legitimate to construe the words containing such doubt or ambiguity so as to give effect to the assumed intention of the parties which I have mentioned.

He expressed the view that there was nothing in the rent review provision in the main lease to suggest that in general the intention of the parties was any different from that which it was assumed to be.

So far as material to questions 1 and 2, the rent is to be:

The amount which shall represent a fair yearly rent for the Site . . . with an obligation for the Tenant to construct thereon such buildings as may at the time of such valuation be situate thereon . . .

It is clear from the parcels that the expression "the Site" means only the land demised and does not include the buildings erected on it. It is clear from the words in the rent review provision "to construct thereon such buildings as may at the time of such valuation be situate thereon" that reference is there made to the site as at the date of the valuation. Moreover, the double use of the preposition "on" demonstrates that it can be a reference only to the site in its physical condition. And, if that be right, the words "a fair yearly rent for the Site", appearing, as they do, in the very same provision, can refer only to the site in its physical condition at the date of the valuation. This view of the matter is strongly confirmed by para (iii) of the proviso to the rent review provision, which requires there to be left out of account the value attributable to any buildings actually standing on the premises but which is silent as to the value attributable to any improvements to the site.

The reasoning of the Vice-Chancellor and the arguments of Mr Reid QC, for the tenant, having failed to persuade me to the contrary, I would on these short grounds decide questions 1 and 2 in favour of the landlord. To express it in terms of the approach stated by the Vice-Chancellor, I would say that here there are clear words showing an intention which overrides the normal commercial expectation that a tenant will not be called upon to pay an additional rent for improvements made at his own expense.

I do not arrive at this conclusion with any enthusiasm, since the merits, as the Vice-Chancellor so clearly recognised, are all in favour of the tenant. What I suspect may have happened was that when the terms of the lease were being negotiated there was no certainty that it would be necessary for the moat to be filled in; that the lease as originally drafted by the landlord's solicitors made no mention of it; that the proviso to clause 2(10) was added by the tenant's solicitors because they thought that the site lease as it stood might prevent it from filling in the moat, should it be necessary to do so; and that in any event the impact on the rent review provision of its being filled in, understandably enough at the time, was never considered by the parties.

Whether these suspicions be accurate or not, I would, for the reasons stated, discharge the declarations made by the Vice-Chancellor in answer to questions 1 and 2 and declare instead that (1) "the Site" means the site in its state and condition on September 1 1989 and (2) the site is to be taken to be an unbuilt plot with the moat thereon wholly filled except for the cooling ponds.

Question 4 (a):

Whether, in the passage in the Review Provision worded "with an obligation for the Tenant to construct thereon such buildings as may at the time of such valuation be situate thereon", the expression "such buildings" includes plant

* Editor's note: Also reported at [1987] 2 EGLR 147.

A and machinery affixed to the realty, and if so, whether any distinction is to be drawn between trade and other fixtures.

The Vice-Chancellor answered this question by declaring that the expression "such buildings" did not include plant and machinery affixed to the realty; thus the second part of the question did not arise. That decision was favourable to the landlord because it means that the allowance made to the tenant in making the rental valuation is limited to a notional obligation to construct the buildings alone. It does not extend to the installation of plant and machinery affixed to the buildings. This is the other question whose outcome may, so it seems, have a considerable influence on the rental value of the premises.

B The argument of Mr Reid on this question relied principally on the tenant's obligation, under clause 1(1) of the lease, to erect and complete a new building consisting of a *fully equipped* steel-making plant and rolling mill capable of producing not less than 50,000 tons of steel products pa. He submitted that the tenant's notional obligation under the rent review provision should be construed as being in line with the particular obligation contained in clause 1(1). I do not think that Mr Reid's submission is correct. Although, as a matter of law, plant and machinery affixed to realty, not being tenant's fixtures, become part of the realty, as a matter of construction the word "buildings" does not necessarily, or even ordinarily, include such plant and machinery. Indeed, as Mr Neuberger pointed out, there are positive indications in other parts of

C the lease which support an exclusive construction in this case. Thus the tenant's covenant to repair in clause 2(7) refers to "the said premises and all other buildings erected on the said premises . . . the fixtures and fittings and all additions thereto". In like manner the covenant to yield up at the end of the term in clause 2(11) refers to "the said premises so repaired and maintained amended and kept as aforesaid together with all additions and improvements made thereto in the meantime and all fixtures and fittings of every kind . . . except tenants or trade fixtures".

It is unnecessary to decide whether this question could have been determined in favour of the landlord on these indications alone. Perhaps more significant is the requirement of the rent review

D provision that the premises should be notionally let "for industrial purposes" and not merely for the purposes of a steel-making plant and rolling mill. I agree with the Vice-Chancellor that it would be inconsistent with this requirement to hold that the notional obligation to construct the buildings included an obligation to install plant and machinery which could be used only for the purposes of a steel-making plant and rolling mill. I would therefore affirm the declaration which he made in answer to question 4(a).

Question 6:

Whether such notional lease incorporates Clause 2(14) as to user or not and if not, whether and if so what restrictions on user there are in the notional lease.

E Clause 2(14) of the lease contains a covenant by the tenant, the first part of which is in these terms:

Not to use or occupy the said premises other than for the purposes of steelmaking steel rolling and operations ancillary thereto or for such other purposes as may from time to time be approved by the Lessors (such approval not to be unreasonably withheld) . . .

The Vice-Chancellor answered this question by declaring that the notional lease incorporated clause 2(14), but with the user provision therein modified so as to permit use for industrial purposes or for such other purposes as might from time to time be approved by the landlord (such approval not to be unreasonably withheld). That decision was favourable to the landlord because the tenant under the notional lease would be able, at any rate in theory, to use the premises

F as of right for industrial purposes other than those of steel making, steel rolling and operations ancillary thereto.

The Vice-Chancellor thought that the requirement that the rent should be fixed on the hypothesis that the premises could be used for industrial purposes generally made it inappropriate to include in the notional lease a limitation to use for steel-making and steel-rolling purposes only. He continued:

Nor, as a matter of language, is that, to my mind, ridiculous result forced upon me, since all that is incorporated from the terms and conditions of the actual lease is introduced by the words "and otherwise on the terms and conditions of this lease".

I entirely agree. I would therefore affirm the declaration made by the Vice-Chancellor in answer to question 6.

G *Question 9*:

Whether, given the absolute covenant against alterations contained in Clause 2(6), the notional lessee is to be regarded as being entitled (a) to carry out all alterations necessary or appropriate to convert the said premises for an [industrial] use of his choice (b) to make any alterations necessary or appropriate to the demised premises in conjunction with a change of use effected pursuant to Clause 2(14) or actually authorised by the landlord and (c) to make improvements under Part I of the Landlord and Tenant Act 1927.

Clause 2(6) contains a covenant by the tenant in these terms:

Not at any time during the said term to erect make or maintain or suffer to be erected made or maintained any building erection alterations or improvements nor to make or suffer to be made any change or addition whatsoever in or to the said premises save in connection with the use of the said premises for the purposes of steelmaking steel rolling and operations ancillary thereto. H

The Vice-Chancellor answered this question by declaring that the notional lessee was to be regarded as being entitled to make improvements only under Part I of the Landlord and Tenant Act 1927. That decision was favourable to the tenant, at any rate in theory, because the tenant under the notional lease would not be able, as of right, to convert the premises for another industrial use of his choice.

It might have been thought that the reasoning which led the Vice-Chancellor to hold that clause 2(14) should be modified so as to permit user for industrial purposes generally would equally have led him to hold that clause 2(6) should be modified so as to permit the J notional tenant to carry out as of right all alterations necessary or appropriate to convert the premises for an industrial use of his choice. However, he held that, if the provisions of clause 2(6) were to be modified, a repugnancy between those provisions and the obligations of the notional tenant must be demonstrated. He held that there was no repugnancy because, if the tenant were to be refused permission to carry out the alterations, he could apply to the court under section 3 of the 1927 Act. While I agree that a repugnancy must be demonstrated, I respectfully disagree that it is avoided by the remedy available to the notional tenant under section 3. This, like the other questions, is a question of construction arising out of the terms of the lease. Looking only at those terms, I see the same repugnancy in the case of clause 2(6) as has already been seen in the case of clause K 2(14). Moreover, an application under section 3 might not succeed and, if it did, it would be likely to cause the tenant delay and expense. The remedy under that section cannot be regarded as a satisfactory substitute for the right to make the alterations without troubling either the landlord or the court. Thus there remains a repugnancy between clauses 2(6) and the more ample rights of the notional tenant under the rent review provision.

For these reasons I would discharge the declaration made by the Vice-Chancellor in answer to question 9 and declare instead that the notional lessee is to be regarded as being entitled to carry out all alterations necessary or appropriate to convert the premises for an industrial use of his choice. L

Having decided all the outstanding questions in favour of the landlord, I would allow the landlord's appeal, dismiss the tenant's cross-appeal and vary the order of the Vice-Chancellor accordingly.

Lord Donaldson of Lymington MR and **Butler-Sloss LJ** agreed and did not add anything.

Appeal allowed with costs; cross-appeal dismissed with costs; costs order below to remain undisturbed; leave to appeal to the House of Lords refused.

Chancery Division M

October 28 1991

(Before Mr Justice KNOX)

BRITISH GAS PLC v DOLLAR LAND HOLDINGS PLC

Estates Gazette March 28 1992

[1992] 12 EG 141

Construction of rent review clause — Hypothetical lease on same terms as actual lease — Whether hypothetical term is of the same term as the actual lease commencing on the rent

review date or is for a term equal to the residue of the actual lease — Application by tenant to vary or set aside the award of the arbitrator dismissed — Certificate under section 1 of the Arbitration Act 1979 granted

The plaintiff tenant holds the term of a lease dated December 17 1969 of premises at Sydney Road and Essex Road, Enfield, Middlesex, for 35 years from the date of the lease — The defendant owns the reversion — The lease provides for the rent to be reviewed to the full market yearly rent defined as "the yearly rent at which the demised premises might reasonably be expected then to be let in the open market with vacant possession by a willing Lessor by a Lease in the same terms in all other respects as this present Lease (other than the original rents hereby reserved but including the provisions for review of rent herein contained)" — In respect of the rent review date on December 17 1990, the tenant submitted to the arbitrator that the hypothetical term was of 35 years from the review date, thus throwing up a lower rent — The landlord had contended for a hypothetical term of 14 years from the review date, being the residue of a term of 35 years from the original commencement date — The arbitrator rejected the tenant's argument and made alternative awards in relation to the parties' respective contentions, namely a rent of £1,392,200 on the basis of a hypothetical term of 14 years commencing on December 17 1990 and £1,252,980 on the basis of a hypothetical term of 35 years commencing at the review date — The tenant applied to vary or set aside the arbitrator's principal award

***Held*: The application to vary and/or to set aside was dismissed — The hypothetical term was for a period equal to the residue of the term of the actual lease — Adopting an absolutely literal importation of the hypothetical terms of the lease granted by the actual lease one gets a *habendum* by which a lease is held from the commencement of the actual lease for a term of 35 years — The provisions regarding stepped rents in the early years of the lease are excluded in any event in the phrase "in the same terms in all *other* respects as this present Lease" — A statement specifically excluding provisions as to *quantum* of rent is frequently found in the decided authorities on rent reviews — To consider that the hypothetical term is for a term equal to the actual lease commencing on the rent review date is far from reality and involves treating the tenant as enjoying something other than what he actually has — The amount of rent in issue was a factor in granting a certificate under section 1(7) of the Arbitration Act 1979 — Leave to appeal was refused**

The following cases are referred to in this report.

Basingstoke and Deane Borough Council v *Host Group Ltd* [1988] 1 WLR 348; [1988] 1 All ER 824; (1987) 56 P&CR 31; [1987] 2 EGLR 147; 284 EG 1587, CA

Lynnthorpe Enterprises Ltd v *Sidney Smith (Chelsea) Ltd* [1990] 2 EGLR 131; [1990] 40 EG 130

Norwich Union Life Insurance Society v *Trustee Savings Banks Central Board* [1986] 1 EGLR 136; (1986) 278 EG 162

Ponsford v *HMS Aerosols Ltd* [1979] AC 63; [1978] 3 WLR 241; [1978] 2 All ER 837; (1978) 38 P&CR 270; [1978] EGD 137; 247 EG 1171, HL

Ritz Hotel (London) Ltd v *Ritz Casino Ltd* [1989] 2 EGLR 135; [1989] 46 EG 95

Tea Trade Properties Ltd v *CIN Properties Ltd* [1990] 1 EGLR 155; [1990] 22 EG 67

This was an application by an originating motion by the tenant, British Gas plc, under a lease dated December 17 1969 for the award of an arbitrator, dated January 2 1991, to be varied or set aside. The defendant landlord was Dollar Land Holdings plc. Both parties consented to the appeal to the court.

Peter Birts QC (instructed by the Regional Solicitor for British Gas plc, of Potters Bar, Hertfordshire) appeared for the plaintiff; David Neuberger QC (instructed by Saunders Sobell Leigh & Dobin) represented the defendant.

Giving judgment, KNOX J said: I have before me an amended originating motion by the tenant under a lease dated December 17 1969 for the award of an arbitrator, dated January 2 1991, to be varied or set aside in certain limited respects. Both parties consent to the appeal so there is no issue as to leave by the court.

The tenant is British Gas plc, the respondent in the arbitration and the applicant before the court. The landlord is Dollar Land Holdings plc, the claimant in the arbitration and the respondent in this court. (I propose to call them the "tenant" and the "landlord" respectively.) The points in issue are what length of hypothetical term and what commencement date should the arbitrator use in determining full market yearly rent for the purposes of a rent review clause in the lease.

The lease was over land and an office building built in 1966, at Sydney Road and Essex Road, Enfield, Middlesex. The term was for 35 years, from December 17 1969 (the date of the lease). The arbitrator found that the tenant held the premises on conventional full repairing and insuring terms and that, apart from the review provisions, none of the terms of the lease was remarkable in the context of value.

The rents reserved by the lease were as follows:

> (a) for the period from the date hereof until the 10th day of March 1970 the rent of one peppercorn if demanded
> (b) for the period of one year from the 11th day of March 1970 the rent of One hundred and forty five thousand pounds
> (c) for the period of one year from the 11th day of March 1971 the rent of One hundred and forty seven thousand and five hundred pounds
> (d) for the residue of the term hereby created subject nevertheless to revision as hereinafter provided the rent of One hundred and fifty thousand pounds . . .

The review provisions are contained in clause 5 of the lease. Clause 5(1) provides:

> That the Lessor shall be entitled by notice in writing given to the Lessee during the first six months of every seventh year of the said term call for a review of the yearly rent payable under this Lease and if upon any such review it shall be found that the initial rent payable hereunder or by virtue of this clause shall be less than the full market yearly rent of the premises at the date of such review then as from the end of every such seventh year the yearly rent payable under this Lease shall be increased to such full market yearly rent.

It is common ground that the relevant review date is December 17 1990, that is to say, 21 years after the commencement of the term. Clause 5(2) starts as follows:

> For the purposes of this clause the full market yearly rent means the yearly rent at which the demised premises might reasonably be expected then to be let in the open market with vacant possession by a willing Lessor by a Lease in the same terms in all other respects as this present Lease (other than the original rents hereby reserved but including the provisions for review of rent herein contained) . . .

and then there follow certain things that are to be disregarded, which can be summarised as follows. First, the fact of occupation by the lessees; second, goodwill through the carrying on of their business; third, sublettings below full market rent and, fourth, improvements by the lessee. The further machinery in the clause for the conduct of the arbitration can be disregarded, since no issue has arisen with regard to its operation.

The dispute before the court is limited to the questions:

(a) What length of term is contemplated by clause 5(2)? and

(b) What is the commencement of that term?

The evidence regarding the market, which the arbitrator accepted, included the following from his award:

> It is I think generally accepted that the current market is very difficult and I accept Mr Dorkings' view that a letting of a building of this size in this location and of this specification might border on the impossible.

Mr D J Dorkings [FRICS] appeared for the tenant before the arbitrator. It is no doubt this circumstance that contributes to the feature, unusual in comparison with reported cases over the past 10 to 15 years, that it is the tenant who is contending for a long hypothetical term of 35 years from the review date, since that throws up a lower rent, and the landlord who contends for a hypothetical term of 14 years from the review date or, which seems to me much the same thing, a term of 35 years from the original commencement date, December 17 1969, since that throws up a higher rent. The arbitrator made alternative awards in relation to the parties' contentions on these disputed issues. His main or substantive award was that the rent payable on review, with effect from December 17 1990, should be £1,392,200 pa. In reaching that conclusion, he rejected the

A arguments addressed on behalf of the tenant that the hypothetical term should be 35 years and should commence on the review date. It is not entirely clear to me whether his award was on the basis of a term of 35 years, commencing on December 17 1969, or a term of 14 years, commencing on the review date.

In his alternative award, he said:

> In my award, I have found in favour of the landlord's claim that the hypothetical lease is for a term of 14 years.

In his principal award he said:

> The new hypothetical tenancy will start in December 1990 as rightly pointed out by Mr Hughes even though the term is backdated . . . .

Mr H G Hughes [ARICS] was the landlord's surveyor. What matters B more is that the figure in his alternative award, in respect of a hypothetical term of 35 years commencing at the review date, December 17 1990, was £1,252,980, which is exactly 10% less than the principal award of £1,392,200, and is therefore, not surprisingly, the figure which the tenant seeks to support.

Both sides accept, as binding in this court, the general principle stated by Nicholls LJ in *Basingstoke and Deane Borough Council* v *Host Group Ltd* [1988] 1 WLR 348*, where he said, at p 353, letter D:

> The question raised on this appeal is one of construction of a rent review clause in a lease. In answering that question it is axiomatic that what the court is seeking to identify and declare is the intention of the parties to the lease expressed in that clause. Thus, like all points of construction, the meaning of this rent review clause depends on the particular language used interpreted C having regard to the context provided by the whole document and the matrix of the material surrounding circumstances. We recognise, therefore, that the particular language used will always be of paramount importance. Nonetheless it is proper and only sensible, when construing a rent review clause, to have in mind what normally is the commercial purpose of such a clause.
>
> That purpose has been referred to in several recent cases, and is not in doubt. Sir Nicolas Browne-Wilkinson V-C expressed it in these terms in *British Gas Corporation* v *Universities Superannuation Scheme Ltd* [1986] 1WLR 398,401:
>
> "There is really no dispute that the general purpose of a provision for rent review is to enable the landlord to obtain from time to time the market rental which the premises would command if let on the same terms on the open market at the review dates. The purpose is to reflect the changes in the value of D money and real increases in the value of the property during a long term."

At p 354, letter D, Nicholls LJ said:

> Of course rent review clauses may, and often do, require a valuer to make his valuation on a basis which departs in one or more respects from the subsisting terms of the actual existing lease. But if and in so far as a rent review clause does not so require, either expressly or by necessary implication, it seems to us that in general, and subject to a special context indicating otherwise in a particular case, the parties are to be taken as having intended that the notional letting postulated by their rent review clause is to be a letting on the same terms (other than as to quantum of rent) as those still subsisting between the parties in the actual existing lease. The parties are to be taken as having so intended, because that would accord with, and give effect to, the general intention underlying the incorporation by them of a rent review clause E into their lease.
>
> We are fortified in this view by observations made in several cases. First, this view accords with comments made in passing in *Ponsford* v *HMS Aerosols Ltd* [1979] AC 63. The point now in question was not in issue there, but in the context of a rent review clause that made no express direction for the terms of the existing lease to be taken into account on the review, Viscount Dilhorne described the task of the valuer in these terms, at p 76:
>
> "Surely it is to assess what rent the demised premises would command *if let on the terms of the lease* and for the period the assessed rent is to cover at the time the assessment falls to be made."

The *Basingstoke and Deane* case was not a case in which the length of the hypothetical term was in issue, but the principle laid down extends to that type of issue. Moreover, there can, in my view, be no doubt that the residue of the term at the review date constitutes the F term still subsisting between the parties in the actual existing lease. Lord Dilhorne says so in terms in *Ponsford* v *HMS Aerosols Ltd*†. And that is the natural inference from Nicholls LJ's reference to the subsisting terms of the actual existing lease. The question is, therefore, does this lease, either expressly, or by necessary implication, or through some special context, require a different term to be used as a valuation yardstick?

For the tenant, Mr Birts QC submits that it does so require. He submits that in clause 5(2) the word "then" in the phrase "might reasonably be expected *then* to be let in the open market" refers to the review date. That was not disputed. Next, he submitted that the G requirement that the lease should be "in the same terms in all other respects as this present Lease" showed that each and every term or provision of the existing lease, including its term or duration, should be included in the hypothetical lease, which is therefore one of 35 years. Third and critically, he pointed to the exclusion of "the original rents" (in the plural) "hereby reserved" and identified these rents as the original stepped rents in the *reddendum* of the existing lease. From this he deduced that the whole of the 35-year term was intended to go into the hypothetical lease, rendering it necessary to exclude the provision for stepped rents in respect of the initial part of the term originally granted. A hypothetical lease of the residue of the term on this review date, 14 years, would not require such an H exclusion, because the stepped-rent period lay in the past and was not imported, so such an exclusion would be unnecessary.

I am unconvinced that this argument is one which amounts to a necessary implication that the hypothetical lease should be one of 35 years from the review date.

First, if one gives maximum weight to the use of the words "same terms", and adopts an absolutely literal importation of the hypothetical terms of the lease granted by the actual lease, one gets a *habendum* "to hold unto the Lessee from the 17th December 1969 for the term of 35 years". That is what it actually says.

Second, the provisions regarding the stepped rents are excluded already by the word "other" in the phrase "in the same terms in all *other* respects as this present Lease". "Other than what?" can only J be answered "Other than the yearly rents reserved" and that would exclude the stepped rental provision. So the subsequent exclusion of "the original rents hereby reserved" is tautological and that is destructive of the underlying assumption that a separate meaning and purpose needs to be found for each provision.

Third, such a statement of the obvious, in specifically excluding provisions as to *quantum* of rent, is in fact frequently indulged in both in statute law — see the Landlord and Tenant Act 1954, section 34(1) "Having regard to the terms of the tenancy (other than those relating to rent)" — and in judgments — see the passages quoted above from Nicholls LJ in the *Basingstoke and Deane* case, where he expressly excludes terms as to *quantum* of rent on more than one occasion. Many other examples of this could be found. K

In my judgment, the expression "other than the original rents hereby reserved" was both tautologous and unnecessary in excluding the rents reserved by the original lease. But that is not an unprecedented feature in a commercial lease. The plural "rents" is accounted for by the fact that four different rents were indeed originally reserved. I was referred to a number of authorities with different clauses, *Lynnthorpe Enterprises Ltd* v *Sidney Smith (Chelsea) Ltd* [1990] 2 EGLR 131; *Ritz Hotel (London) Ltd* v *Ritz Casino Ltd* [1989] 2 EGLR 135; *Tea Trade Properties Ltd* v *CIN Properties Ltd* [1990] 1 EGLR 155, and *Norwich Union Life Insurance Society* v *Trustee Savings Banks Central Board* [1986] 1 EGLR 136. In none of those cases did an argument in favour of a L hypothetical lease of the same length as the original term, but commencing at the review date, succeed. I accept, however, Mr Birts' submission that they are all distinguishable on the basis that the words used were different, with the possible exception of the *Norwich Union Life Insurance* v *Trustee Savings Banks Central Board* case, where the report does not in terms contain a verbatim quotation of the relevant clause, but Hoffmann J is reported as saying that "the rent review clause provided that the new rent was to be assessed on the review date on the assumption that the premises were being let in the open market by a willing lessor to a willing lessee, taking a lease otherwise on the terms and conditions of the lease". If that is a verbatim quotation, which it may well be, it is quite close to the present clause, and although Hoffmann J was only concerned to M refuse leave to appeal, he did so on the basis that he had no doubt that the arbitrator was right to take as the term of the hypothetical lease the unexpired period of the existing lease. However, although I derive comfort from that decision, in view of the uncertainty as to the exact content of the clause, I base my decision, in dismissing this appeal, upon the general principle laid down in the *Basingstoke and Deane* decision and what seems to me to amount to the absence of an express or necessary implication in favour of the tenant's construction.

I should add that although this is exceptionally, so far as reports go, not a case where it is the landlord who is arguing for the longer hypothetical term, it can, in my view, make no difference whether it is

* Editor's note: Also reported at [1987] 2 EGLR 147.

† Editor's note: Reported at (1978) 247 EG 1171.

the landlord or the tenant who contends for a valuation based on something other than what the parties are actually enjoying at the relevant date. Putting it in another way, reality, so far as length of term is concerned, consists in the term which the tenant enjoys at the review date. It is true that Staughton LJ, in *Lynnthorpe Enterprises* v *Sidney Smith (Chelsea) Ltd (supra)*, at p 135, said that while there was clearly a presumption in favour of reality so far as length of term was concerned, reality pointed in two directions, one in favour of the unexpired residue of the term, because that was what the tenant had, the other in favour of a 15-year term, which was what the original term in that case was. He went on as follows, at p 135:

> . . . I would have given greater weight to the second aspect of reality rather than to the first, because that is more directly concerned with the monetary calculation that has to be done. So I would have wished to reach a solution whereby the valuer had to go into the market-place and find a 15-year rent rather than a 6-year rent.
>
> Of the three possible solutions that have been put forward, the first is that the valuer should consider a hypothetical lease for 15 years from the date of the review. The second is that he should consider a hypothetical lease for 15 years from the date of the original lease; the third is that he should consider a hypothetical lease for the unexpired period of the original term. That, so far as concerns August 1987, is six years.

He rejected the first for two reasons: first, it seemed to have been uniformly rejected by a number of Chancery judges at first instance. The other reason was the existence of specific stepped rents at the outset of the actual term granted, which was irreconcilable with the requirements of the hypothetical lease. They are specifically excluded here, so that reason is not available to the landlord. The first is, but I would add, for my part, that the first of the three possibilities mentioned by Staughton LJ is much the furthest from reality of the three, because unlike the other two it involves treating the tenant as enjoying something other than what he actually has. For these reasons this appeal will be dismissed.

Following an application by Peter Birts QC, on behalf of the plaintiff, for leave to appeal to the Court of Appeal and for a certificate under section 1(7) of the Arbitration Act 1979, **Knox J** continued: The decision that I have made was intended to be founded on the general principle which both parties before me adopted and which was laid down in the *Basingstoke* case (as I can call it for the sake of brevity). It is submitted that there is a question of general public importance in the application of that principle to the terms of this lease.

The arguments that were addressed to me were, very rightly, detailed arguments in connection with the construction of this particular clause. It is not possible to assert conclusively that this particular clause is one which is unique or very widely adopted in leases that were granted some little while ago. The indications are that it is not a form that is currently very widely adopted, but, of course, commercial leases have a way of being for quite a lengthy period of time, more especially if they have rent review clauses in them, and therefore one has to have regard to what has happened in the past as well as to what is being done today.

Essentially, in my view, this decision was concerned with the consideration whether there were particular circumstances in the construction of this particular lease which took it out of the general run of cases, which is quite strongly, as Staughton LJ noted in the *Lynnthorpe* decision, one where there has been a fairly consistent trend among judges at first instance in the Chancery Division, a trend which has been recognised and given effect to, perhaps with something less than unqualified enthusiasm in the case of Staughton LJ, by the Court of Appeal.

In those circumstances, it does seem to me difficult to say, and I am not able to find that there is a question of general public importance in this particular case, because the arguments have really been concerned with the particular rather than the general. By that I mean the particular considerations that impinge on the true construction of this clause and the emphasis that was very properly placed by Mr Birts on the singular (in a numerical sense) framework of this clause in excluding the original rents. That was a very particular argument, quite rightly addressed as of some significance, but nevertheless particular rather than general. Accordingly, I am not persuaded that there is a question of general public importance in correlating this type of clause with what is, I think now, a very well-established general principle, which has been laid down by the Court of Appeal and is binding.

The other possibility is one that there should be some other special reason requiring the question of law to be considered by the Court of Appeal. Questions of construction are, of course, questions of law, so there is no problem on that score. What this comes down to is whether the amount of money involved — because that is the importance — is sufficient to warrant the case being considered by the Court of Appeal. It is not easy to see whether the section contemplated the size of the amount at stake as being a special reason, not one of general public importance. On balance, it does seem to me that it must be possible for the amount at stake to fulfil the criterion that Parliament has adopted. The amount here has not been very precisely calculated, but it is clear that if one takes 10% of the actual award that was made by the arbitrator and multiplies it by seven, one reaches a figure of the order of a million pounds, and there is a possibility of further reviews, where this question could come up again.

On balance, I have reached the conclusion that this is just, because of the very large size of the amount involved, capable of being a special reason which should be considered by the Court of Appeal. I do not regard, as an ingredient in what the subsection requires, any view to be formed as to the intrinsic merits of the appeal. That would be a difficult process for a judge, whose decision is sought to be upset, in which to indulge.

I propose to give a certificate that there is a special reason why the question of law to which my decision related should be considered by the Court of Appeal — if anyone wants to know what it is, it is the amount at stake — but I do not propose to give leave to appeal.

Appeal dismissed with costs. Certificate that there is a special reason why a question of law should be considered by the Court of Appeal granted. Leave to appeal refused.

Chancery Division

October 30 1991

(Before Mr Justice FERRIS)

ART & SOUND LTD v WEST END LITHO LTD
SAME v TAMPO SUPPLIES LTD

Estates Gazette April 11 1992

[1992] 14 EG 110

Landlord and tenant — Construction of a rent review clause — Whether time of the essence of an arbitrator's determination of rent — Parties failed to agree rent — Application for declarations by landlord dismissed

The plaintiff is the owner of the reversion to two separate leases of the first and second floors of premises at 326 Kensal Road, London W10 — By a lease dated September 18 1973 the defendant in the first action holds a 25-year term of the second floor subject to five-yearly rent reviews — By a lease dated November 27 1973 the defendant in the second action holds a 25-year term of the first floor — Subject to differences in dates, and one point, the two leases were in the same terms — In each case the rent was to be reviewed to "the current market rental value" — That amount was to be agreed between the parties not later than three months prior to the commencement of a relevant review period "or in the absence of agreement (time to be of the essence of these provisions) as shall be determined by an Arbitrator . . . [who] shall make his determination by a date three months prior to the commencement of [the relevant rent review period]" — In neither case was there any agreement three months prior to the relevant rent review periods — Although the plaintiff landlord applied for the appointment of arbitrators to determine the rent, the defendant tenants contended that time was of the essence of the date of the arbitrator's awards, and those dates had expired — The plaintiff landlord issued two originating summonses and contended that time was of the essence only of the part of the rent review clause relating to the agreement of the parties and was not of the essence of any part of the machinery for determination of the rent by an arbitrator

A ***Held*: The originating summonses were dismissed — The same question relating to a clause which was for all practical purposes identical to the provisions of the first-floor lease was considered in *C B Bradley & Sons Ltd* v *Telefusion Ltd* — Walton J held that the words "time to be of the essence of this provision" applied to the part of the provision which related to determination by arbitration — There was no room to distinguish that case, and it was impossible to avoid the conclusion, as a matter of language, that either by agreement or by arbitration the rent must be determined not later than three months before the commencement of the relevant rent**
B **review period — The words "these provisions" in the plural in the case of the second-floor lease strengthens the linguistic argument that time was of the essence of the machinery for arbitration**

The following cases are referred to in this report.

Amherst v *James Walker (Goldsmith & Silversmith) Ltd* [1983] Ch 305; [1983] 3 WLR 334; [1983] 2 All ER 1067; (1983) 47 P&CR 85; [1983] EGD 157; 267 EG 163, CA
Bradley (C) & Sons Ltd v *Telefusion Ltd* (1981) 259 EG 337
Essoldo (Bingo) Ltd's Underlease, Re (1971) 23 P&CR 1
Laing Investment Co Ltd v *Dunn (GA) & Co* [1982] EGD 252; (1981) 262 EG 879
C *Touche Ross & Co* v *Secretary of State for the Environment* (1982) 46 P&CR 187; 265 EG 982, CA
United Scientific Holdings Ltd v *Burnley Borough Council* [1978] AC 904; [1977] 2 WLR 806; [1977] 2 All ER 62; (1977) 33 P&CR 220; [1977] EGD 195; 243 EG 43 & 127, HL

The plaintiff lessor issued two originating summonses: West End Litho Ltd was the defendant to the first summons and Tampo Supplies Ltd was the defendant to the second summons. Each summons raised for determination the proper construction of the rent review provisions in two leases concerning the first and second floors respectively of premises at 326 Kensal Road, London W10.

Adrian Jack (instructed by Cornillie & Co) appeared for the plaintiff; Kim Lewison QC (instructed by Alsop Wilkinson)
D represented West End Litho Ltd; and John Male (instructed by Wedlake Bell) represented Tampo Supplies Ltd.

Giving judgment, **Ferris J** said: There are two originating summonses before me, each of which raises what is essentially the same question arising out of the rent review provisions contained in separate leases of the first and second floors of premises known as 326 Kensal Road, London W10.

The earlier of these two leases is that in respect of the second floor. It was dated September 18 1973 and was made between Art & Sound Ltd (which is the plaintiff in both originating summonses and to which I shall refer as "the lessor") of the one part and West End Litho Ltd (which is the defendant in action 1990 A No 10492) of the
E other part. By it the second floor was demised to West End Litho for a term of 25 years from November 1 1973. A separate rent was fixed for each successive five years of the term, the last four of which were referred to respectively as the second, third, fourth and fifth rent periods, the rent reserved during these periods being referred to as the second, third, fourth and fifth rents respectively. The question raised in these proceedings concerns the fourth rent, which is payable from November 1 1988 until November 1 1993. Under para (d) of Part III of the schedule to the lease the fourth rent is to be:

> the same amount as the third rent or such amount (whichever shall be the greater) as shall be ascertained as hereinafter provided as representing the current market rental value of the demised premises for the fourth rent
> F period.

This formula is then repeated as regards the fifth rent period, after which there follows the clause with which I am concerned, which is in the following terms:

> The expression "the current market rental value" as hereinbefore used shall mean such amount as shall be agreed between the Lessor and the Lessee by a date not later than three months prior to the commencement of the second rent period or the third rent period or the fourth rent period or the fifth rent period (as the case may be) as representing the current market rental value at such date of the demised premises or in the absence of agreement (time to be of the essence of these provisions) as shall be determined by an Arbitrator as sole Arbitrator acting under the provisions of the Arbitration Act 1950 or any statutory modification or re-enactment thereof for the time being in force to be nominated by the president for the time being of the Royal Institution of Chartered Surveyors on the application of the Lessor and so that in case of any G
> arbitration the current market rental value as at such date at which the same is required to be ascertained shall be of such amount as shall in the opinion of the Arbitrator represent the rent at which the demised premises might reasonably be expected to be let for a term equivalent to the then unexpired residue of the term hereby granted as between a willing landlord and a willing tenant with vacant possession and taking no account of any goodwill attributable to the demised premises by reason of any business carried on there by the tenant and in all other respects on the terms and conditions of this Lease and in the case of any Arbitration the Arbitrator shall make his determination by a date three months prior to the commencement of the second rent period the third rent period the fourth rent period or the fifth rent period (as the case may be) and all fees costs and expenses payable in respect of such Arbitration shall be borne by the parties hereto in equal shares.

What has happened in the case of the second floor is that there was H
no agreement between the parties as to the amount of the fourth rent, either three months prior to the commencement of the fourth rent period or at all. On November 11 1988 (that is to say more than a week after the commencement of the fourth rent period) the lessor applied to the president of the Royal Institution of Chartered Surveyors ("RICS") for the appointment of an arbitrator to determine the fourth rent. An arbitrator was duly appointed on January 27 1989. Having regard to the date of the application for his appointment there was never any possibility that he could determine the rent by the time prescribed by the words I have read, namely by a date three months prior to the commencement of the fourth rent period. West End Litho, which remains the tenant of the second floor, contends J
that time is of the essence of the requirement that the arbitrator shall make his determination by that date and the lessor is not now entitled to require the arbitrator to determine the rent. Pending the outcome of the proceedings he has not done so. The originating summons in this matter was not issued until October 8 1990. The delay between the arbitrator's appointment and the commencement of these proceedings is over 20 months, which is surprising, excessive and unexplained. However, no special point is sought to be made as a result of this delay.

The lease of the first floor is dated November 27 1973. The parties were the lessor and a company named Panel Printers Ltd. The demise was for 25 years from June 24 1973. Subject to differences in dates and one point which I shall mention in a moment, the lease was, so far K
as material, in the same terms, *mutatis mutandis*, as that of the second floor. The fourth rent period began on June 24 1988 and ends on June 24 1993. The only material difference between the relevant provisions of the two leases is that in the first-floor lease the words in parentheses expressing time to be of the essence are "time to be of the essence of this provision". It will be recalled from the extract that I have read that the equivalent words in the second-floor lease were "time to be of the essence of these provisions".

The relevant events in relation to the first-floor lease are as follows. On or about August 15 1985 Panel Printers Ltd assigned the lease to Tampo Supplies Ltd, which remains the tenant and is the defendant in action 1990 A No 10494. It seems that there was a suggestion at L
about the time of the assignment that it was unlawful by reason of failure to obtain the lessor's consent, but this was disputed and the lessor has long since accepted Tampo Supplies as its tenant. Accordingly nothing turns on this. The fourth rent was not agreed by the prescribed date or at all. On March 25 1988, which was the day after the last date for agreement, the lessor applied to the president of the RICS for the appointment of an arbitrator. The arbitrator was appointed on May 19 1988. He has not determined the fourth rent because of the objections of Tampo Supplies Ltd, which are the same as those of West End Litho. The originating summons was not issued until October 8 1990, 30 months after the arbitrator's appointment. I make the same comments as regards delay as in the case of the second-floor lease. M

The tenants under each lease accept that the effect of *United Scientific Holdings Ltd* v *Burnley Borough Council* [1978] AC 904* is that there is a presumption that, in the absence of contra-indications, time is not of the essence of the machinery established by the rent review provisions. But the parties to a lease may expressly provide that time is to be of the essence and where they do so the court will give effect to their expressed intention, as Lord Diplock specifically stated in the *United Scientific* case at p 923. There is no doubt that there is, in each of the present leases, an express provision that time is to be of the essence. The question is whether time is of the essence of the whole of the procedure provided for by the passage

*Editor's note: Also reported at (1977) 243 EG 43 & 127.

A which I have cited from the leases or of only part of that procedure and, if so, which part.

Both sides recognise that the passage which I have cited provides for a procedure in two parts but there is a dispute whether these parts were envisaged as being successive or concurrent. The first part is that which relates to agreement. The second part is that which relates to the determination of the rent by an arbitrator. The contention of the tenants is that time is of the essence of both parts or, alternatively, of that part relating to determination of the rent by an arbitrator. The contention of the lessor is that time is of the essence only of the part relating to the agreement of the parties and that time is not of the essence of any part of the machinery for determination of the rent by an arbitrator.

B The argument of the tenants begins with the submission that the two parts of the procedure are separated by the words "or in the absence of agreement". The crucial words as to time being of the essence occur after this separation and grammatically they belong to the second part. Alternatively they apply to both parts, this being particularly relied upon on behalf of West End Litho, where the relevant words are "time to be of the essence of these provisions". This linguistic argument is said to be reinforced by a number of considerations. In particular, it is said to be virtually meaningless to say that time shall be of the essence of a stipulation that rent shall be fixed by agreement, because if the parties were to agree the rent after the prescribed time the fact that time was expressed to be of the
C essence would be no impediment to their giving effect to their agreement. It was also pointed out that it is in the tenant's interests that it shall know what rent it has to pay before the commencement of the relevant rent period, because the need to pay considerable arrears after a late determination may give rise to hardship. Accordingly, there is good reason for making time of the essence of the arbitration determination, particularly when, as here, it is only the lessor who can apply for the appointment of an arbitrator, so that if time were not of the essence the lessor could delay the process of initiating a determination.

Against these considerations it is urged on behalf of the lessor that it is grammatically possible to read the words "time to be of the essence of this provision" in the first-floor lease as referring back to
D the provision for determination by agreement rather than forward to the provision for determination by arbitration. This is also said to be possible in the case of the second-floor lease, where the concluding words are "these provisions", because the stipulation for determination by agreement is to operate on four separate occasions, namely in respect of the second, third, fourth and fifth rent periods, and it is as aptly referred to by the plural "these provisions" as by the singular "this provision".

To treat the words as producing this result would, it was said, produce a result which accords with commercial common sense, whereas the tenants' construction flouts commercial common sense. The basis of this submission was that both the language of the
E provision and common sense suggest that determination by arbitration is to take place only after there has been an opportunity for determination by agreement. There was no attempt to resurrect the argument which failed in *Re Essoldo's Underlease* (1971) 23 P&CR 1, which was not cited to me but was referred to in passing in the course of argument. Thus it is accepted that the words "in the absence of agreement" are to be construed literally and do not require that there must be some attempt at agreement or that the matter must be allowed to run until the last day for agreement. Nevertheless, it was submitted that the two parts of the relevant provisions are to operate sequentially and not simultaneously. If it were otherwise the lessor would probably need to initiate an arbitration some 15 months before the commencement of the
F relevant rent period, because the dates in this case show that it takes between two and three months to obtain the appointment of an arbitrator and neither party has any real control over the speed of the arbitration. If the argument of the tenants is right the two parts of the provisions cannot operate sequentially, although that is how one would expect them to operate. They must operate simultaneously, which is extraordinary.

I was referred to *Touche Ross & Co* v *Secretary of State for the Environment* (1982) 46 P&CR 187*, where the Court of Appeal referred to the fact that appointment of an arbitrator and the performance of the arbitration functions were outside the control of the parties as being a pointer against time being of the essence. But the issue in that case was whether it could be implied from the language of G
the rent review machinery as a whole that time was to be of the essence. Here, the question is how words which undoubtedly make time of the essence are to operate. If grammatically they apply to the part of the provisions that relate to determination of the rent by arbitration then the fact that the parties have thereby made time of the essence of a procedure over which they have little or no control does not appear to me to be of any real relevance. This factor may, however, be of some relevance if the words are fairly capable of two different meanings, because it may be that I should then prefer the meaning which does not attribute to the parties an unlikely intention.

The question which I have to consider is not free from authority. In *C Bradley & Sons Ltd* v *Telefusion Ltd* (1981) 259 EG 337 Walton J had to consider the same question in relation to a clause in which for H
all practical purposes were terms identical to the provisions of the first-floor lease in the present case, save that the time by which the parties were to determine the rent by agreement was not later than six months prior to the commencement of the relevant rent period. The time by which the arbitrator was to make his determination was the same as that prescribed by the two leases with which I am concerned, namely a date three months prior to the commencement of the period. Accordingly there was a possibility that an arbitration initiated after a failure to agree by the first prescribed date could lead to a determination by the second prescribed date although the timetable was, as Walton J observed, a very tight one. Walton J held that the words "time to be of the essence of this provision" applied to J
the part of the provision which related to determination by arbitration. Three considerations led him to this conclusion. First, he attached no significance to the tight timetable, having regard to the power of the court under section 13(2) of the Arbitration Act 1950 to extend the time for making an award. Second, he took the view that it would be meaningless to make time of the essence of the provision for the determination of the rent by agreement of the parties, having regard to the fact that if the parties were agreed they could perfectly easily make and give effect to their agreement after the stipulated time. Third, reading the clause as a whole, he found that the vital words applied grammatically to what was to come (namely the determination of the rent by arbitration), not to what had gone previously (namely the provision for determination of rent by K
agreement).

Mr Jack, on behalf of the lessor, sought to persuade me to reach a conclusion different from that reached by Walton J on two grounds. First, he sought to distinguish the *Telefusion* case on the ground that, while in that case there was at least an interval of three months between the latest date for agreement of the rent and the latest date for the determination of the rent by arbitration, in the present case there is no such interval at all. In this case the timetable is not merely tight but impossible. Although at first sight this may appear to be an important distinction, I have come to the conclusion that in this case it points in the opposite direction from that contended for by Mr Jack. In the *Telefusion* case the fact that there was an interval of three L
months between the latest dates for the two operations left scope for the assumption, which I think Walton J made, that determination by arbitration was to be consequent upon, and in succession to, a failure to reach agreement. But in the present case there is no room for the equivalent view. Whether or not time is of the essence, the provision in both the leases which I have to consider prescribes the same date as the latest date for determination of the rent by agreement or by arbitration. I find it impossible to avoid the conclusion, as a matter of language, that either by agreement or by arbitration the rent must be determined not later than three months before the commencement of the relevant rent period. This negatives the view that arbitration was to be consequent only upon a failure to agree within the prescribed period. Both leases with which I am concerned seem to proceed on the M
basis that an attempt to agree and arbitration would proceed contemporaneously. This is somewhat unusual but it is by no means impossible, because the parties could always settle an arbitration by agreement. However this may be, the meaning of the language used is not changed by regarding time as not being of the essence. All that this would do is to mitigate the consequences of failure to act in accordance with the language. But if, as I think, the language does not envisage the two stages as being necessarily consecutive there is nothing in the timetable which gives rise to any need to depart from the ordinary meaning of the words used in respect of time being of the essence. I respectfully agree with Walton J that the more natural reading of the provision as a whole requires that the words "time to

*Editor's note: Also reported at (1982) 265 EG 982.

A be of the essence of this provision" apply to the determination of the rent by arbitration.

The second ground on which Mr Jack invited me to differ from Walton J was that, in the light of subsequent decisions, I ought to accept that it was not as improbable as Walton J had thought it to be to treat the stipulation that time should be of the essence as applying to that part of the provision which relates to the determination of the rent by agreement.

Mr Jack referred me to the decision of Peter Gibson J in *Laing Investment Co Ltd* v *G A Dunn & Co* (1981) 262 EG 879. It appears from the judgment in that case that the judge had initially been strongly inclined to make an implication as to time being of the essence because, if he did not do so, an express stipulation would have B made time of the essence only of a provision for the determination of the rent by agreement which, following Walton J in the *Telefusion* case, he thought would serve little practical purpose. However, after being referred to the decision of the Court of Appeal in *Amherst* v *James Walker (Goldsmith & Silversmith) Ltd* (1980) 254 EG 123, he concluded that it would be wrong to make such an implication. In the result he accepted that the stipulation as to time being of the essence applied only to determination of the rent by agreement. While I accept this as a correct analysis of the *Laing* case I do not find it to be of assistance in the cases which I have to consider. It provides no basis on which I can give anything other than what I have indicated to be the natural meaning to the stipulation as to time in the leases which C are before me.

In the result I find no ground for distinguishing or departing from the decision of Walton J. In my view, the argument presented on behalf of the lessor not only goes against the actual and grammatical meaning of the clause but abandons a timetable under which determination by arbitration was required to be completed three months before the relevant rent period in favour of a timetable under which arbitration may well not be begun until after that time. I think that this involves a rewriting of the clause, not merely the construction of the words used.

This reasoning applies to both the cases before me. In the case of the second-floor lease the fact that the crucial words refer to "these provisions" in the plural strengthens the linguistic argument which, D like Walton J, I accept. I do not think that the use of the plural can be explained as a reference to the repeated operation of a single provision. If it were relevant to do so I would hold that in the case of the second-floor lease time is of the essence of both parts of the clause. However, I do not think that in practical terms this takes the matter any further.

I apprehend that the consequential order which I must make is to dismiss both originating summonses, although I will hear counsel about this.

E Chancery Division

November 14 1991

(Before Mr Justice MUMMERY)

BRITISH AIRWAYS PLC v HEATHROW AIRPORT LTD AND ANOTHER

Estates Gazette May 16 1992

[1992] 19 EG 157

Landlord and tenant — Rent review — Premises at Heathrow and Gatwick Airports — Construction of several leases — F Issues relating to character of demised premises and identity of hypothetical bidder — General principles

The plaintiff, British Airways plc, holds the terms of eight leases of parts of Heathrow Airport and Gatwick Airport — The first defendant, Heathrow Airport Ltd, is the owner of the reversion of the four leases of parts of Heathrow Airport and the second defendant, Gatwick Airport Ltd, is the owner of the reversion of the other four leases of parts of Gatwick Airport — The rent review provisions in the leases are in broadly similar terms and the disagreement between the parties about the true construction of the rent review clauses is typified in the lease dated December 31 1986 of the West Base G No 1 Maintenance Area at Heathrow for a term of years from April 1 1971 to June 30 2055 ("the West Base lease") — The West Base lease was granted in consideration of the plaintiff's erecting and constructing a servicing hangar, computer-training building and other buildings and works — The demise was of plots of land containing 84 acres and the "demised premises" were defined as including the land together with all buildings, works, fixed plant and fixtures for the time being thereon — The rent review provision distinguished between what were defined as the first premises and the second premises, the latter comprising nine individual lettered plots H of land — From April 1 1990 until June 30 2000 the plaintiff lessee covenanted to pay the initial reserved rent or the current market rental value of the plots, whichever was the higher — The current annual market rental value was defined of the plots as if they were let individually on a building lease for a term of years equivalent in length to the actual lease on the same terms and conditions and excluding the first premises — The lease contained covenants by the lessee relating to the carrying out of improvements and user — The nine plots were not undeveloped at the date of the lease but had been equipped with hardstanding and associated drainage — The plaintiff issued an originating summons which, following amendment, J raised nine questions of construction

***Held*: Guidance can be gathered from the relevant authorities on the approach of the court to the construction of rent review clauses — (1) The paramount principle is that stated in *Basingstoke and Deane Borough Council* v *Host Group Ltd* — (2) The purpose of a rent review clause is to ensure that the rent payable during the term of a long lease reflects changes in the value of money and of the property let — (3) The court should not trespass on the territory of the expert valuer — (4) The only relevant hypotheses are those expressly or impliedly agreed by the parties — (5) It is a K question of construction whether the rent to be determined on the hypothetical lease should take account of the buildings actually on the property — (6) An assumption or disregard about the actual state of the property may have to be implied to give commercial efficacy**

Applying these principles, declarations were made as follows: (1) It must be assumed that each of the plots comprised in the second premises is an undeveloped plot excluding all buildings, hardstanding and drainage at the review date — (2) Each hypothetical lease imposed upon the tenant a covenant to build at the tenant's cost what actually exists on the respective plots at the review date — The L hypothetical lease is required to be a "building lease", not just a lease — (3) On the hypothetical letting of any one plot comprised in the second premises none of the other plots are assumed to be available for letting on the open market but rather are assumed to be occupied by the plaintiff on the terms of the actual lease — That assumption reflects the reality of the position — (4) On the hypothetical letting of any plot comprised in the second premises to which access cannot be obtained by virtue of the right granted by the hypothetical lease for the purpose for which the plot is permitted to be used, the valuer is to disregard the lack of such access — (5) Taking M three questions of construction together, it is for the independent surveyor, and not for the court, to decide who is likely to be a bidder in the open market for the hypothetical lease and, if he concludes that the plaintiff is likely to be a bidder, what is likely to be the amount of its bid for such a lease — (6) In relation to a lease of the East Base maintenance area at Heathrow, the tenant is only entitled to use certain areas as a means of access to the demised premises as defined in the actual lease and/or as a taxi track for aircraft — (7) In relation to a lease of land at Gatwick Airport, it was to be assumed that the land identified on the plan annexed to the actual lease was to be let on a single hypothetical lease

The following cases are referred to in this report.

Basingstoke and Deane Borough Council v *Host Group Ltd* [1988] 1 WLR 348; [1988] 1 All ER 824; (1987) 56 P&CR 31; [1987] 2 EGLR 147; 284 EG 1587, CA
City of London v *Nash* (1747) 3 Atk 512; 1 Ves Sen 12
Compton Group Ltd v *Estates Gazette Ltd* (1977) 36 P&CR 148; [1977] EGD 237; 244 EG 799, CA
Cornwall Coast Country Club v *Cardgrange Ltd* [1987] 1 EGLR 146; (1987) 282 EG 1664
Evans (FR) (Leeds) Ltd v *English Electric Co Ltd* (1977) 36 P&CR 185; [1978] EGD 67; 245 EG 657
First Leisure Trading Ltd v *Dorita Properties Ltd* [1991] 1 EGLR 133; [1991] 23 EG 116
Goh Eng Wah v *Yap Phooi Yin* [1988] 2 EGLR 148; [1988] 32 EG 55; [1985] Malayan Law Journal 329, PC
Hallett v *Martin* (1883) 24 Ch D 624
Ipswich Town Football Club Co Ltd v *Ipswich Borough Council* [1988] 2 EGLR 146; [1988] 32 EG 49
Jeffries v *O'Neill* (1983) 46 P&CR 376; 269 EG 131
Jones d Cowper v *Verney* (1739) Willes 169
Law Land Co Ltd v *Consumers' Association Ltd* (1980) 255 EG 617, CA
Ponsford v *HMS Aerosols Ltd* [1979] AC 63; [1978] 3 WLR 241; [1978] 2 All ER 837; (1978) 38 P&CR 270; [1978] EGD 137; 247 EG 1171, HL

By an originating summons the plaintiff lessee, British Airports plc, sought declarations as to a number of questions of construction relating to eight leases of parts of Heathrow and Gatwick Airports, the reversions to which are held by the defendants, Heathrow Airport Ltd and Gatwick Airport Ltd respectively.

David Neuberger QC and Erica Foggin (instructed by Herbert Smith & Co) appeared for the plaintiff; Hazel Williamson QC and Jonathan Gaunt QC (instructed by McKenna & Co) represented the defendants.

Giving judgment, **MUMMERY J** said: Exercises in property valuation sometimes generate difficult demarcation disputes between the real world, occupied by the relevant property, and an abstract world peopled with spectral willing vendors and purchasers, or lessors and lessees, higgling hypothetically in accordance with stipulated assumptions.

The valuation dispute in this case is unusually complex. It arises in the context of rent review clauses contained in eight leases of parts of Heathrow Airport and Gatwick Airport. In its original form the originating summons taken out by the present lessee, British Airways plc ("BA"), on November 30 1989 raised no less than 53 questions of construction. Thanks to the good sense of the parties and their legal advisers the questions have been reduced, to the immense relief of the court, so that there are now only nine identified in an extensively amended summons.

There are four leases of parts of Heathrow Airport. The lessor is Heathrow Airport Ltd, a company owned by the British Airports Authority plc ("BAA"). There are four leases of parts of Gatwick Airport. Gatwick Airport Ltd, also a subsidiary of the British Airports Authority, is the lessor. Together those companies are the freeholders of, and are responsible for operating, respectively, Heathrow Airport and Gatwick Airport. I shall refer to them collectively in the judgment as "the defendants".

The rent review provisions in the leases are in broadly similar terms. It is agreed between the parties that the six questions which arise on the construction of the lease of the West Base No 1 Maintenance Area at Heathrow, to which I shall refer as "the West Base lease", will dispose of most of the questions arising on the other seven leases. I shall initially deal with the construction of the West Base lease. I shall then construe the terms of the three other leases in order to answer the three remaining questions.

The other three leases are the lease of the East Base No 1 Maintenance Area at Heathrow (I shall call that the "East Base lease"); the lease of MT Operations Base at Heathrow, which I shall refer to as the "MT Operations Base lease"; and the lease of the Maintenance Base East No 3 at Gatwick, to which I shall refer as the "No 3 Gatwick lease". BA is the present lessee under all the leases.

The position on the machinery of the rent review clauses is that notices have been served by the defendants proposing increases in rent to which BA objects. BA has served counternotices, but, because of disagreements about the true construction of the rent review clauses, which emerged in the course of discussions between the surveyors advising the parties, no progress has been made with the determination of the new rent.

The West Base lease

I shall now turn to the terms of the West Base lease. The West Base lease was granted by Heathrow Airport Ltd to BA on December 31 1986 for a term of years from April 1 1971 expiring on June 30 2055. The grant of the lease was expressed in clause 1(a) to be:

> In consideration of the expense incurred by the lessee in erecting and constructing a servicing hangar, a computer training building, engine overhaul offices and ancillary works and of the rents hereafter reserved and of the covenants by the lessee and condition hereafter contained . . .

The demise was of:

> All those plots of land (hereinafter called "the said land")

containing about 84 acres delineated on an attached plan verged in different colours and comprising 22 different titles registered at HM Land Registry. The demise was:

> Together with all buildings, works, fixed plant and fixtures for the time being thereon (which together with the said land are collectively called "the demised premises").

It is important to note that the lease draws a distinction between the plots of land and the buildings on those plots. The expression "demised premises" includes *both* the land and the buildings on the plots. This distinction is made in other provisions in the lease and is relevant to one of the important questions for decision.

The demise also includes rights, easements and privileges specified in the first schedule and expressed to be:

> Appurtenant to the said land for all purposes connected with the use, occupation and enjoyment thereof.

Para (a) of the first Schedule refers to the rights granted to the lessee in these terms:

> (a) A right at all times during the said term for the lessee and persons authorised by the lessee in common with the lessor and all other persons authorised by it or entitled thereto to pass and repass with or without vehicles to and from the demised premises over such road or routes on the airport as may from time to time be specified by the lessor.

There are reserved to the lessor rights variously expressed to be in respect of "the said land" or of "any building thereon" or of "the demised premises".

The rent provisions of the lease are in these terms; that the lessee shall pay:

> in respect of those plots of land verged red and verged red hatched black on the plan (hereinafter called "the First Premises") the following rents namely:
> (a) until the thirtieth day of June Two Thousand the clear yearly rent of Forty-Two thousand four hundred and fifty nine pounds (£42,459).
> (b) from and including the first day of July Two thousand and during the remainder of the said term such rent as shall become payable by virtue of the provisions of the second schedule hereof.

I pause to mention that this case is not concerned with the operation of the rent review provisions for the first premises, which do not come into operation until the year 2000. This case is concerned only with what are next defined as "the Second Premises".

The second premises comprise nine individually lettered plots of land on the perimeter of the first premises. The first premises contain the principal buildings required for BA's aircraft maintenance and administration operations and the nine individual plots are peripheral. I shall describe them in more detail later.

The lease refers to the second premises as:

> Those plots of land verged blue and verged blue hatched black on the plan and being individually lettered "A" "B" "C" "D" "E" "F" "G" "H" and "J" respectively (the extent of each individual plot being marked by dividing blue lines on the plan) (hereinafter together called the "Second Premises").

The rents payable in respect of the second premises are:

> (a) until the thirty-first day of March 1990 the clear yearly rent of Thirty three thousand and thirty six pounds (£33,036).
> (b) from and including the first day of April One thousand nine hundred and ninety until the thirtieth day of June Two thousand either the clear yearly rent reserved in sub-clause (a) hereof or the current market rental values (as hereinafter defined) of those plots comprising the Second Premises as at the first day of August 1989 ("the review date") whichever is the higher.
> (c) from and including the first day of July Two thousand and during the remainder of the said term such rent as shall become payable by virtue of the provisions of the second schedule hereto.
> Provided that for the purposes of the sub-clause (b) hereof it is agreed that the following definitions and provisions shall apply.

These definitions and provisions are at the heart of most of the arguments in this case. There is a definition of the expression "current market rental value" as meaning:

A The total of the current annual market rental values of those plots comprising the Second Premises if let individually on a building lease for a term of years equivalent in length to the term of years hereby granted such lease being on the same terms and conditions (other than as to amount of rent but including the provisions as to review of rent and excluding the First Premises and the provisions of this lease solely relating to the First Premises) as this present demise without the payment of any fine or premium.

Detailed provisions are then made for the manner in which the current annual market rental value for the second premises should be determined. I omit those provisions and turn to the final provision as to rent, in which all such rents in respect of both the first premises and the second premises are provided to be paid without any deduction by equal quarterly payments in arrears on the quarter days in every year.

B I read para (b) on p 8 of the lease:

The demise hereby made shall not include or operate so as to confer on the lessee any liberty privilege easement or advantage (except such as are specifically granted by this lease) in through over or upon any land or premises adjoining or near to the said land nor shall the lessee be entitled hereunder to any airport or other facilities over from or upon such land or premises.

Clause 2 contains various covenants by the lessee with the lessor. The relevant covenants are as follows:

(1) To pay the rents hereby reserved at the times and in manner aforesaid.

(2) From time to time and all times during the said term to pay and discharge all rates taxes duties charges assessments and outgoings whatsoever (whether Parliamentary parochial local or of any other description) which are now or may at any time hereafter be assessed charged or imposed upon or payable in
C respect of the said land and all buildings works fixed plant and fixtures for the time being erected constructed placed or standing thereon or therein or in the course of erection or construction thereon the demised premises and the said rights or on the owner or occupier thereof (Income Tax on the said rents excepted) or if the demised premises are not separately assessed for rates to refund to the lessor any payments made by it or on its behalf in respect of rates.

Clause 2(10)(a) provides:

Save as provided in clause 10(b) hereof not any time during the said term to erect or suffer to be erected any new building or erection whatsoever on the said land without the written consent of the lessor (which consent shall not be unreasonably withheld).

— and —

10(b) In respect of the land verged blue hatched black and lettered "A" on
D the plan not at any time during the said term to erect or suffer to be erected any new building or erection whatsoever on such land without previous written consent of the lessor (which consent shall not be unreasonably withheld).

Clause 2(13)(a) is a covenant:

Not to use or occupy the said [land] (with the exception of the land verged blue hatched black to which the provisions of cl.13(b) hereof shall apply) or any part thereof for any purpose other than those required in connection with the lessee's business as an airline operator provided always that such land will not without the consent of the lessor be used for:

(i) Passenger handling
(ii) freight handling or
(iii) in such a manner as to compete with concessionaires of the lessor provided that in this case the lessor's consent shall not be unreasonably
E withheld; and

(b) not to use or occupy the land verged blue hatched black on the plan for any purpose other than as a means of access to other parts of the demised premises and a taxi track for aircraft.

Clause 2(16) provides as follows:

Not to fix or exhibit or permit or suffer to be fixed or exhibited to or upon any part of the said land or to or upon any part of the exterior of the said buildings or of the external railings or fences (if any) of the demised premises any placard poster electric sign or other advertisement or sign whatsoever without the previous consent in writing of the lessor and forthwith to remove at the lessee's cost from the interior of the demised premises any placard poster electric sign or other advertisement or sign which shall be objected to in writing by the lessor.

F There is a provision on p 18 of the lease concerning the eventuality of a surrender of part of the demised premises. It is provided that in that event:

(i) The lessee will undertake to grant to the lessor its lessees tenants and licensees at no cost to the lessor such rights (*inter alia*) of access to the part surrendered (any such part being hereafter called "a part") and to a metered supply of services to the part as the lessor might reasonably require and

(ii) In assessing the sum to the paid by the lessor to the lessee as aforesaid in respect of the lessee's leasehold interest in the part the rent shall be deemed to be payable by the lessee in respect of the part and shall be calculated as follows.

and there follow two paragraphs:

(a) In respect of the First Premises by apportioning the rent payable in respect of the First Premises at the date of the surrender in the proportion which the area of the part bears to the area of the whole of the First Premises; G

(b) In respect of the Second Premises by apportioning the rent payable in respect of the Second Premises at the date of the surrender in the proportion which the area of the part bears to the area of the whole of the Second Premises.

Clause 2(24) contains an insurance covenant in which it is provided that at all times during the term the lessee is to keep the demised premises insured against loss or damage.

Clause 2(25) contains a covenant:

To permit the lessor or its agents with or without workmen and other at all reasonable times to enter the demised premises for the stated purposes which include viewing the state and progress of any building or erection at any time during the said term in course of erection on the said land . . .

Clause 2(28) provides that: H

If the lessee shall at any time make default in the performance of any of the covenants herein contained for or relating to the repair or rebuilding of the demised premises it shall be lawful for the lessor to enter upon the demised premises and repair or rebuild the same at the expense of the lessee in accordance with the covenants and provisions contained.

Clause 2(39) provides that the lessee is:

To make good or pay compensation for any damage loss or injury (including injury resulting in death) to any property or person by reason of anything done or purported to be done or omitted by the lessee or its servants and arising out of the lessee's use and occupation of the demised premises.

The provisions of the Second Schedule complete my reference to the West Base lease. The provisions in the Second Schedule apply to J
the rent review of the first premises and of the second premises as from June 30 2000. They are relevant for present purposes only in so far as they bear on the construction of the rent review provisions which operate in relation to the second premises before that date. I need read only para 1 of that Schedule:

(1) The rent payable under this lease in respect of the demised premises shall be revised after the 30th day of June Two thousand so as to represent the then current annual market rental value of the demised premises (without any buildings erected or other constructions made thereon by the lessee) if let on a building lease from the 30th day of June Two thousand for the residue of the said term then unexpired on terms similar to the terms of this lease but excluding this sub-clause and such revised rent shall thereupon be deemed to have been substituted for the rents received hereunder from and including the K
first day of July Two thousand and this lease shall thenceforth be read and construed accordingly but so nevertheless that until the lessor shall have notified the lessee of the amount of the revised rent the said rents hereby reserved shall continue to be payable subject to such adjustment as may be necessary being made between the parties hereto on the quarterly date for payment of rent next following such notification as aforesaid.

I should next describe the property at the West Base. The second premises let under the West Base lease are nine plots lettered alphabetically A to J and they are geographically peripheral to the first premises. I should say something about the historical and geographical background of these premises.

Leases of part of the demised premises were granted to BA's predecessor, BEA, as long ago as September 1956. Prior to L
September 1956 BEA had erected part of the West Block Building on the West Base and it was demised to it on September 27 1956 until June 2000. At about the same time BEA was granted a lease of the East Block Building on the West Base for a term expiring in the year 2055. BEA covenanted to construct an engineering base comprising hangars and workshops with aprons. The lease provided for a rent review as at June 2000 on the basis of the market value of the land without the buildings.

In 1958 and 1967 further leases were granted to BEA with similar rent review clauses and limited to expire in 2055. BEA covenanted to construct a car park on the premises. By an agreement dated January 5 1970 BEA covenanted to erect on 28 acres of the West Base area a servicing hangar, computer-training building, engine overhaul M
offices and various ancillary works. On completion BEA and BAA were to take a lease in an agreed form to expire on June 30 2055 with rent to be reviewed as at June 30 2000 to the then current annual market value of the land without buildings made thereon by the lessee if let on a building lease.

The agreement also provided for BAA to grant to BEA a "subsidiary lease" of various plots including plots A to J for the same term. Occupation was taken and rent paid in accordance with the agreement, but the leases were not in fact granted until 1986 prior to the imminent privatisation of BAA.

The physical state of the individual plots can be briefly stated as follows:

A Plot A is of an area of 6.92 acres used as a taxi way for aircraft and access to the maintenance base on the first premises.

Plot B is 2.95 acres in area and provides an apron for standing aircraft and access similar to that provided by plot A to the first premises. Both of the plots are covered with pavement-quality concrete or with grass.

Plot C is a concreted area of 0.25 acre of apron and access to the maintenance base. Part was surfaced by the lessee between 1967 and 1970.

Plot D is a tarmac area of 1.8 acres surfaced by the lessee pursuant to a lease in 1967 and used as an open car park.

Plot E is also used as an open car park. It is 1.06 acres and was
B tarmacked by the lessee pursuant to an agreement made in 1969.

Plot F is a two-lane roadway forming part of Viscount Way and is 0.64 acre in area.

Plot G is a small area of 0.15 acre of landscaped grass and flower beds.

Plot H is 1.66 acres in area. It is concreted and used for apron manoeuvring, turning space and parking aircraft.

Plot J is 1.47 acres in area and is of similar surface and use.

In many cases the plots comprised in the second premises were not undeveloped plots at the date of the 1986 leases or when they were previously demised but had been equipped with hardstanding and associated drainage. In most cases the leases did not impose an obligation on the lessee to construct any buildings or works on the
C second premises. This is to be compared to the provisions in relation to the construction of buildings on the first premises.

Approach of the court

I now turn to the approach of the court to the construction of clauses of this kind. I should preface the exercise of construing the leases by stating briefly the guidance which can be gathered from the relevant authorities.

1. The paramount principle in the construction of a rent review provision is that stated by the Court of Appeal in *Basingstoke and Deane Borough Council* v *Host Group Ltd* [1988] 1 WLR 348* at p 353 D-F:

> D What the court is seeking to identify and declare is the intention of the parties to the lease expressed in that clause. Thus, like all points of construction, the meaning of this rent review clause depends on the particular language used interpreted having regard to the context provided by the whole document and the matrix of the material surrounding circumstances. We recognise therefore, that the particular language used will always be of paramount importance. None the less it is proper and only sensible, when construing a rent review clause, to have in mind what normally is the commercial purpose of such a clause.

2. It is common ground that the purpose of a rent review clause is to ensure that the rent payable during the term of a long lease reflects changes in the value of money and of the property let. This purpose is achieved by enabling the lessor to obtain from time to time the market rent which the property would command if let at the rent review date
E on the same terms in the open market. It is a procedure for up-dating the rent to keep pace for inflation during the residue of a long term. Clauses of increasing complexity are negotiated by the parties as part of the original bargain and usually postulate, for valuation by an independent surveyor, a notional lease of the property by a hypothetical lessor to a hypothetical lessee on the same terms as the subsisting lease.

3. In construing such a clause the court should be on its guard against trespassing on the territory of the expert valuer. This is a court of law, not a court of valuation. The construction of the clause is a question of law for decision by the court, so that the independent surveyor can know what is to be valued and what, as a matter of
F construction of the lease, the parties have expressly or impliedly agreed should be the assumptions on which his valuation is to be made. It is not, however, the function of the court to answer questions or give general directions as to how the independent valuer should go about making his valuation, such as the factors to be taken into account or the weight to be attached by him to those factors: see *Compton Group Ltd* v *Estates Gazette Ltd* (1977) 36 P&CR 148† at pp 152, 159 and 161.

4. The court should also be alert to the danger of confusing reality and hypothesis. The only relevant hypotheses are those expressly or impliedly agreed by the parties in the rent review clause. Agreed assumptions and disregards must be adhered to but, subject to that, it G
is the real circumstances affecting the actual property which are relevant to valuation: see *Cornwall Coast Country Club* v *Cardgrange Ltd* [1987] 1 EGLR 146 at pp 149 G - 150 E.

Thus, the definition of the expression "current market rental value" in the West Base lease predicates the letting of the relevant properties on certain terms. The properties are real but the letting is hypothetical; so also are the lessor and lessee who are parties to the hypothetical letting. The hypothetical lessee is not the actual lessee in actual occupation of the property. He does not have the personal characteristics of the actual lessee which might operate to increase or decrease the rent which he would be willing to pay: for example, his actual occupation of the property; his financial position; his H
resources or lack of them; or his monopolistic position in the market: see *F R Evans (Leeds) Ltd* v *English Electric Co Ltd* (1977) 36 P&CR 185* at pp 189-192.

That does not mean, however, that the actual lessee is necessarily excluded from consideration as a potential lessee of the property, unless the rent review clause expressly or impliedly makes that direction to the valuer. As Vinelott J observed in *First Leisure Trading Ltd* v *Dorita Properties Ltd* [1991] 1 EGLR 133 at pp 137M and 138, there is no principle of law which requires the valuer to assume that a lessee is not a potential hypothetical lessee of the relevant property. Depending on the circumstances of the particular case (for example, where the actual lessee is also the lessee of adjoining property), the independent surveyor may be entitled to J
find as a fact that the lessee is a potential hypothetical lessee.

5. Although the property to be let under hypothetical letting is real, the terms of the rent review clause may direct that, for valuation purposes, certain aspects of it be disregarded or that certain assumptions be made about it. For example, the hypothetical lease of the property carries with it a right to vacant possession of the property, even though in reality the property is occupied at the relevant time by the lessee under the terms of the actual lease. It is also a question of construction of the particular rent review clause whether the rent to be determined on the hypothetical lease should take account of the buildings actually on the property or should be fixed on the assumption that the property is free of the buildings which are actually on it: see *Ipswich Town Football Club Co Ltd* v K
Ipswich Borough Council [1988] 2 EGLR 146; compare *Ponsford* v *HMS Aerosols Ltd* [1979] AC 63.†

Prima facie if the demised premises consist of land and buildings the rent should be determined by reference to both: *Goh Eng Wah* v *Yap Phooi Yin* [1988] 2 EGLR 148. But the lessor and the lessee are contractually entitled to direct that, for valuation purposes, the presence of the buildings on the land let should be ignored by the independent surveyor.

6. In some cases it may be necessary, in order to give commercial efficacy to a rent review provision, to imply into that provision an assumption or a disregard about the actual state of the property. In *Jeffries* v *O'Neill* (1983) 46 P&CR 376§ Nourse J implied a disregard L
of any effect on rental value caused by the position of access to the property since, without such an implication of a disregard, no practical effect could be given to the clause requiring the rent to be assessed in the open market. In that case the lessee failed to persuade the court that the rental value should proceed on the basis that, having regard to the access arrangements, there was, in fact, no open market for the property. The rent review clause predicated the existence of an open market for the property. The lessee could not negative that hypothesis by asserting that he was in a very special position and that, in fact, there was no open market for that property.

By reference to that guidance I shall now attempt to construe the particular questions which have been fully argued before me. I shall M
state as briefly as I can the conclusions which I have reached on each question.

Each question on the West Bank lease is prefaced in the originating summons by these words:

> Whether on the true construction of a lease dated 31st December 1986 between the first defendant as lessor of the one part and the plaintiff as lessee of the other part of property at West Base number one maintenance area, Heathrow Airport, London ("West Base"), the current market rental value of the plots of land edged blue and edged blue hatched black on the plan annexed to the actual lease and lettered individually "A", "B", "C", "D",

*Editor's note: Also reported at [1987] 2 EGLR 147.

†Editor's note: Also reported at (1977) 244 EG 799.

*Editor's note: Also reported at (1977) 245 EG 657.

†Editor's note: Also reported at (1978) 247 EG 1171.

§Editor's note: Also reported at (1983) 269 EG 131.

A "E", "F", "G", "H" and "J" respectively and therein referred to as the Second Premises is to be assessed on the review date on the following assumptions.

The questions are then formulated by reference to the various assumptions.

In the amended originating summons the first question arising for decision is under (iii). The alternative constructions are posed in these terms:

(a) that each of the plots comprised in the Second Premises is an undeveloped plot excluding all buildings hardstanding and drainage thereon at the review date;

Alternatively:

B (b) that each of the plots comprised in the Second Premises is in its actual physical state at the review date it being assumed (if not a fact) that the covenants on the part of the tenant in the actual lease have been complied with.

I agree with BA that, on the true construction of the West Base lease, the current market rental value of the plots of land which are part of the second premises are to be assessed on the assumption that each of the plots is an undeveloped plot, excluding all buildings, hardstanding and drainage thereon at the review date. I shall make a declaration in the terms of (iii) (a).

In many parts of the West Base lease the parties have drawn a distinction between (1) the land; (2) the buildings on the land; and (3) the land *together with* the buildings, and they have defined terms C accordingly.

On this question the definition of the expression "current market rental value" is crucial. On p 5 of the lease it is defined as meaning "the total of the current annual market values of those plots comprising the Second Premises". What are the second premises? That is a term defined at the top of the same page as the aggregation of "those *plots of land* verged blue and verged blue hatched black on the plan".

It appears from other provisions in the West Base lease that the expression "plots of land" means land alone without buildings on it. When reference is made to land *and* buildings collectively a different defined term is usually used, namely "the demised premises". The D independent surveyor is not directed to determine the rental value of the demised premises, that is land plus buildings, but of the plots, by which I understand the parties to mean the plots of land lettered A to J respectively.

I am unable to accept the defendants' submission that the reference to plots in the lease and, in particular, in relation to the second premises is not a reference to land as a green field site but is simply a matter of delineation of boundaries of the relevant areas on the attached plan. I am also unable to accept the defendants' submission that the second premises referred to in the definition of current market rental value on p 5 should be regarded, like the first premises, as simply a subdivision of the "demised premises", defined earlier in the lease to include both land and buildings.

E In my judgment, the general rule that what should be valued is the land with the buildings is displaced in this case by the language of the lease, even though there is in the rent review clause no express disregard of buildings of the kind that is found in the Second Schedule on p 30 of the lease. In my judgment, the language of the rent review provisions on p 5 makes a clear distinction for valuation purposes between land which is the subject-matter of the notional lease and the land and buildings which are the subject-matter of the actual lease as "Demised Premises".

The next question in the originating summons is (iv) and there are posed no less than five possible constructions:

(a) That each hypothetical lease imposes upon the tenant a covenant to F build at the tenant's cost what actually exists on the respective plots at the review date.

Alternatively:

(b) That each hypothetical lease is granted in consideration of the tenant having built, at its own cost, what actually exists on the respective plots at the review date.

Alternatively:

(c) That each hypothetical lease is only on the terms and conditions of the actual lease excluding the First Premises and the provisions relating solely to the First Premises and without any covenant to build.

Alternatively:

(d) That each hypothetical lease is only on the terms and conditions of the actual lease excluding the First Premises and the provisions relating solely to the First Premises without any covenant to build but in consideration of such G expense, if any, as may have been incurred in the construction of such works as exist at the review date as have been constructed on the premises by the actual tenant or its predecessors in title.

Alternatively:

(e) That each hypothetical lease is on the terms and conditions of the actual lease excluding the First Premises and the provisions relating solely to the First Premises and is subject to a covenant on the part of the tenant to develop the plot by executing such works as would be appropriate as at April 1990 in order to develop the plot for use by an airline operator in the manner permitted by the terms of the hypothetical lease.

In order to decide which of these is the appropriate construction it is first necessary to decide what is meant by the expression "building H lease" as used in the definition of "current market rental value" on p 5 of the West Base lease. The expression "building lease" is not defined in the West Base lease. It should, therefore, be construed according to its ordinary and natural meaning, as controlled, of course, by the context in which the parties have used that expression.

In its ordinary meaning the expression "building lease" connotes a lease for building purposes granted, usually for a substantial period, in consideration of the lessee's entering into a covenant to build; for example, by erecting new buildings or adding to or improving existing buildings in accordance with agreed specifications. This appears from *Jones d Cowper* v *Verney* (1739) Willes 169 at p 175; *City of London* v *Nash* (1747) 3 Atk 512 and *Hallett* v *Martin* (1883) 24 Ch D 624 at p 631. I also derive this meaning from section 99 (3) J and (9) and section 205(1)(iii) of the Law of Property Act 1925 and similar provisions in the Settled Land Act 1925.

Adopting that approach, I agree with BA that the definition of "current market rental value" on p 5 of the lease requires the independent surveyor to assess the rent on the review date on the assumption that the hypothetical lease imposes on the lessee a covenant to build, at his cost, what actually exists on the respective plots at the review date. I shall make a declaration in the terms of para (a) of question (iv).

The hypothetical lease is required to be a "building lease", not just a lease. What did the parties mean by that expression? As I have said, they did not define it, either on p 5 or in the Second Schedule on p 30 K where the same expression is used in a different context of a rent review due to take place on June 30 2000.

The defendants do not dispute that the adjective "building" requires the lease to have some quality not present in an ordinary lease. The area of dispute is to what that quality is. Various alternatives are set out under question (iv). As I have indicated, in the absence of a contractual definition I take the parties as intending the term "building lease" to be used in its ordinary sense of a lease in which the lessee is under an obligation to build. The crucial question is what is the hypothetical lessee to be under an obligation to build. The building works are not specified and therefore have to be inferred.

Using the touchstone of reality, I conclude that the only realistic L workable assumption to make is that, for the purposes of determining the rental value of the plots of land uncovered by any buildings, the lessee is to be assumed to be under an obligation to build what is already on the plots of land. I reject the defendants' submission that the assumption to be made does not import a covenant to build or that the "building purposes" element of the hypothetical lease is a reference either historically to the grant of a lease in consideration of past building or prospectively to future building in a form appropriate to the land in question at the rent review date. I reject the submission that "building" is referred to simply as a possibility permitted by the terms of the hypothetical lease.

Neither the ordinary meaning of the expression "building lease" M nor the context in which it is used here carries with it any of the connotations submitted by the defendants. The assumption that the lease is to be a building lease in the ordinary sense does not involve a departure from reality in the form of the physical state of the second premises at the date of the review. The independent surveyor must look at the real state of the premises as at that date, but, in determining the rental value of what is in fact there, he must assume that the hypothetical lessee is under an obligation to build what is in fact there. That does not seem to me to be an absurd assumption to make nor does it seem to me to be inherently likely to produce such an unfair result that this construction should be regarded as contrary to the purpose of the rent review clause.

A I turn now to question (v). The alternatives proposed are:

(a) that on the hypothetical letting of any one plot comprised in the Second Premises none of the other plots are assumed to be available for letting on the open market but rather are assumed to be occupied by the plaintiff on the terms of the actual lease

Alternatively:

(b) that on the hypothetical letting of any one plot comprised in the Second Premises each of the other plots are assumed to be available for letting.

I agree with BA's submission that it is for the independent surveyor to value each plot forming part of the second premises on the assumption that the remaining plots are in fact occupied by BA on the terms of the actual lease. I therefore propose to make a declaration in B the terms of (a). That assumption reflects the reality of the position and the language of the clause does not, in my view, require any other assumption to be made to the contrary.

It is true, as the defendants contend, that in the valuation of each plot that particular plot is assumed *not* to be subject to the actual lease and to be available for letting with vacant possession on the terms of a hypothetical lease. Such an assumption does depart from reality but is implicit in the exercise of postulating a lease of that plot to a hypothetical lessee. It is not, however, implicit in that hypothesis that all the other plots to be separately valued should also be assumed for that purpose to be available for letting in the open market.

The defendants' argument that, on the hypothetical letting of any C one plot, each of the other plots is assumed to be available for letting relies on wringing from the language of the rent review clause a meaning which the words do not bear. In my judgment, neither the fact that the rent review of each plot comes up at the same time nor the plural reference to "values" in the rent review provisions nor the reference to the hypothesis that the plots are let "individually" signifies that any of the plots, other than the particular one being valued for rent review purposes, is to be assumed to be available for letting in the open market when the fact is that it is not available.

Question (vi) proposes two alternatives:

(a) that on the hypothetical letting of any plot comprised in the Second Premises if access cannot be obtained by virtue of the rights deemed to be available

D (i) to any such plot or

(ii) to the First Premises

for the purpose or any of the purposes for which they are respectively permitted to be used, (A) the valuer is to disregard the lack of such access when determining the current market rental value of each such plot, alternatively (B) the value is to take into account the lack of such access

Alternatively:

(b) that on the hypothetical letting of any plot comprised in the Second Premises to which access cannot be obtained by virtue of the rights granted by the hypothetical lease for the purpose or any of the purposes for which the plot is permitted to be used, the valuer is to disregard the lack of such access when determining the current market rental value.

On this question I agree with the defendants that on the E hypothetical letting of any plot comprised in the second premises to which access cannot be obtained by virtue of the rights granted by the hypothetical lease for the purpose or any of the purposes for which the plot is permitted to be used, the independent surveyor is to disregard the lack of such access when determining the current market rental values. I shall therefore make a declaration in the terms of (b).

The starting point of this dispute is the provision that the rent review clause predicates a lease "on the same terms and conditions as this present demise". Those terms and conditions include the rights of access expressly set out in para (a) of the First Schedule: that is, access over roads or routes in the airport as specified by the lessor.

On p 8 of the West Base lease there is a provision carried over into F the hypothetical lease, which excludes from the demise easements, except those specifically granted by the lease, in through over or upon land or premises adjoining or near the land let. Under the terms of the hypothetical lease there is therefore a lack of rights of access from one plot to another plot comprised in the second premises or from any plot comprised in the second premises to any part of the first premises. All the plots, except plot J, abut the first premises. Plots F, G, H and J abut the highway and have access to it subject to possible planning permission being obtained.

The point emphasised by BA is that, as a matter of legal right of access to other plots or to the first premises, many of the plots are landlocked and it is therefore wrong to assume that, where the adjoining premises are leased to and occupied by BA, the lessor under the hypothetical lease would be able to grant rights of access to the G hypothetical lessee over the adjoining premises. This lack of legal right is, however, quite different from the position about access in practice, because BA is the lessee and occupier of all the other plots and of the first premises.

BA submits that the independent surveyor should take into account the lack of legal rights of access when determining the current market rental value of each plot. This leads to the conclusion that the only possible bidder for those plots would be BA and that the rent would, therefore, be nominal. On the other hand, the defendants submit that the independent surveyor should disregard the effect of such lack of legal access on valuation on the basis of an implied term of the kind justified on the authority of *Jeffries* v H *O'Neill* (*supra*). Such a term can be implied in order to give practical or commercial efficacy to the rent review clause.

I agree that, without such an implied disregard of the lack of legal access, it is difficult to see here how there could be a market for a hypothetical lease of the property. There should, in my judgment, be an implied disregard of the kind suggested by the defendants as a matter of commercial efficacy.

I also agree with the defendants that, if access cannot be obtained to the first premises by virtue of the rights granted by the actual lease or otherwise for the purpose or any of the purposes for which they are permitted to be used, the valuer should not disregard the lack of such access when he is determining the current market rental value of each plot comprised in the second premises. I shall make a declaration to J that effect.

The parties are agreed that, on the true construction of the West Base lease, the current market rental value of the plots of land comprised in the second premises should be assessed on the basis that:

(a) the first premises are occupied by BA in accordance with the terms of the West Base lease; and

(b) that lease does not confer any rights over any of the second premises in favour of the first premises.

In addition, I have already declared, contrary to BA's submissions, that the valuer should disregard the lack of legal access to any plot comprised in the second premises on the hypothetical letting of any such plot. BA does not seek to go behind that ruling. K Indeed, BA submits that, if lack of access to individual plots should be disregarded by the valuer, so also should lack of access to the first premises. It is in the interests of BA to advance this argument since, if the valuer may take into account lack of legal access to the first premises, that factor may well tend to increase the market rental value of those plots, such as plot A, which do in fact provide access to the first premises. BA submits that, if the valuer is entitled to take into account lack of legal access to the first premises, the defendants may well be able to argue for a very substantial increase in the rent of individual plots based on a "ransom value", since access over some of the plots comprised in the second premises is vital to BA's use in the first premises and without that access the first premises would be L sterilised.

BA does not base its argument on an implied disregard of lack of access such as I have found, in the case of lack of access to the individual plots, should be made as a matter of commercial efficacy in accordance with *Jeffries* v *O'Neill* (*supra*). BA's submission is simpler. It is that, following para 4 of the guidelines summarised above, valuation for rent review purposes should be by reference to the real circumstances affecting the actual property and not by reference to hypothesis or, even worse, to hypothesis piled upon hypothesis. In support of this submission particular reliance was placed on *Cornwall Coast Country Club* v *Cardgrange Ltd* at p 152 and also on *Law Land Co Ltd* v *Consumers' Association Ltd* (1980) 255 EG 617 at p 619. What are the real circumstances on access M affecting the first premises and the second premises? Together they comprise the "demised premises" under the West Base lease. BA is the tenant and occupier of both of the first and second premises, which will be valued together as a single unit for rent review purposes in the year 2000 in accordance with the provisions of the Second Schedule. There are no access problems in reality and it would, therefore, be in accordance with reality to disregard the lack of legal access to the first premises under the terms of the hypothetical lease. To act otherwise would be to embark on a wholly unrealistic valuation exercise. BA submits that the parties cannot have intended that the rental value of individual plots should vary according to BA's needs for access to the first premises.

BA's submissions have been persuasively argued. I find the point an elusive one. After further argument from both sides, I have been persuaded that BA's contentions are not in fact correct.

As I pointed out in the guidelines above, the paramount principle is that the court should act in accordance with what is identified as the agreement of the parties. The parties have agreed that the independent surveyor should determine the total of the rental values of the plots of land comprised in the second premises if let individually on the same terms and conditions as the West Base lease but excluding the first premises. The terms of such a hypothetical lease do not confer any right of access to the first premises. Further, postulating a lease of each individual plot involves the necessary assumption that each plot is vacant and that BA is not in occupation of it. It necessarily follows from what the parties have agreed about the hypothetical lease that the independent surveyor, in determining the rent payable under the hypothetical lease, should not take into account the fact that BA, as tenant in occupation, could use the plot in question as a means of access to the first premises. In other words, the subject-matter of the hypothetical lease of each plot is a vacant plot over which there is no legal right of access to the first premises, which is not in the actual occupation of BA and which cannot therefore afford access in fact to the first premises. To accede to BA's arguments on this point would involve a departure from the assumptions agreed upon by the parties for the purposes of valuing the hypothetical lease of each plot.

I shall deal with questions (vii), (viii) and (x) together.

(vii) that the plaintiff is not a bidder for the hypothetical lease.

(viii) in the alternative to (vii), if the plaintiff is to be assumed to be a bidder for the hypothetical leases of the plots, the plaintiff's bid is assumed to be in such amount as is just sufficient to secure a letting to it in the open market.

(x)(a) that the effect on rent of the existence of the First Premises and/or of the fact that they are occupied by and/or let to the plaintiff in accordance with the terms of the actual lease shall be disregarded.

Alternatively:

(b) that the First Premises are occupied by the plaintiff in accordance with the terms of the actual lease and that the actual lease does not confer any rights over any of the Second Premises in favour of the First Premises.

Prima facie it is for the independent surveyor, and not for the court, to decide who is likely to be a bidder in the open market for the hypothetical lease, whether the likely bidders do or do not include the actual lessee and as to the amount of the bid which is sufficient to secure a letting in the open market for the lessee or for any other likely bidder. The question for the court is whether the lessor and the lessee have agreed in the rent review clause that the independent surveyor is to act on the assumption that the lessee is not a bidder for the hypothetical lease or, if he is a bidder, whether the bid is assumed to be in any particular amount.

For reasons which I now state, I decline to make the declarations sought in relation to questions (vii), (viii) and (x). BA asks the court to hold that, as a matter of construction, the lease requires the independent surveyor either to assume that BA is not a bidder for the hypothetical lease or, if it is regarded as a bidder, it is to be assumed to bid only in such amount as is sufficient to secure a letting to it in the open market. This is what has been described as a "special purchaser" point. If BA is assumed not to be a bidder then a question arises as to whether there would be any other bidder.

I agree with the defendants that there is nothing in the language of the lease which requires the independent surveyor to make any assumptions about the identity of a likely bidder or the amount of the rent to be bid. It is for the independent surveyor to decide who, in fact, is likely to be a bidder in the open market for the hypothetical lease and, if he concludes that BA is likely to be a bidder, what is likely to be the amount of its bid for such a lease.

A similar point arises on question (x). On that I do not find BA's arguments convincing. BA's submissions were based on the words in parenthesis at the top of p 6 of the lease which exclude from the terms and conditions of the hypothetical lease "the First Premises and the provisions of this lease solely relating to the First Premises". It was submitted that those words have the effect of excluding from the determination of the rental value of the second premises any effect attributable to the first premises or any tenancy or occupation of the first premises by BA. Each plot comprised in the second premises should be valued according to its intrinsic value and not be affected by the fact that BA is the tenant in occupation of the adjoining first premises. It was submitted that BA's actual occupation as a tenant of the second premises should be disregarded and so also should its actual tenancy and occupation of the first premises.

In my judgment, this construction is not supported by the language of the lease and is not forced upon the court by the reasoning in *Cornwall Coast Country Club Ltd* v *Cardgrange* (*supra*): compare *First Leisure Trading Ltd* v *Dorita Properties* (*supra*) at pp 135, 136 and 138. It is for the independent surveyor, and not for the court, to decide on the facts in relation to the identity of the bidder in the market, the likely amount of any bid and what effect, if any, on rental values of the second premises will occur by reason of BA's tenancy and occupation of the first premises under the actual lease.

Those conclude the questions on the West Base lease. I turn to the one question on the East Base lease.

Whether on the true construction of a lease dated 31st December 1986 made between the First Defendant as lessor of the one part and the Plaintiff as lessee of the other part of property at East Base, Number One Maintenance Area, Heathrow Airport, London ("East Base"), the current market rental value of the plots edged blue and edged blue hatched black on the plan annexed to the actual lease and lettered individually "A", "B", "C" and "D" respectively (and therein referred to as "the Second Premises") is to be assessed on review on the following assumptions, namely:

(ix)(a) that under the hypothetical lease of the Second Premises the tenant is only entitled to use the plots lettered "A" and "B" on the plan annexed to the lease as a means of access to the plots lettered "C" and "D" on the said plan and/or as a taxi track for aircraft

Alternatively:

(b) that under the hypothetical lease of the Second Premises the tenant is only entitled to use the plots lettered "A" and "B" on the plan annexed to the lease as a means of access to the demised premises as defined in the actual lease and/or as a taxi track for aircraft.

The demised premises include plots of land lettered A, B, C and D respectively on the plan. They are defined as the second premises. There are also other premises defined as the first premises. As under the West Base lease, only the second premises fall within the current rent review clause. The rent review clause and the definition of the "current market rental value" on pp 5 and 6 are in substantially the same terms as those contained in the West Base lease, including the reference to the exclusion from the hypothetical lease of "the First Premises and the provisions of this lease solely relating to the First Premises".

The difference from the West Base lease, according to the submissions of BA, lies in the covenant in clause 2(13) on pp 13 and 14 of the East Base lease. Clause 2(13)(b) contains a covenant by the lessee not to use or occupy the plots lettered A and B on the plan "for any purpose other than a means of access to other parts of the demised premises and a taxi track for aircraft". BA submits that this covenant is a provision "solely" relating to the first premises. It should therefore be excluded from the terms of the hypothetical lease and the independent surveyor should act on the assumption that under the hypothetical lease the lessee is entitled to use the plots lettered A and B on the plan only as a means of access to the plots lettered C and D on the plan and/or as a taxi track for aircraft. BA therefore asks for a declaration in the terms of (a).

I decline to make that declaration. I agree with the defendants that the flaw in BA's argument is that it fails to take account of the word "solely" in the words of exclusion. Clause 2(13)(b) is not a provision which refers solely to the first premises. It refers to plots A and B as a means of access to other parts of "the demised premises" and the demised premises include the lots lettered C and D as well as the first premises. I shall therefore make a declaration in the terms of (b).

The next question concerns the one issue of construction on the MT Operations Base lease. The question is:

Whether on the true construction of a lease dated 10th November 1983 made between British Airports Authority as lessor of the one part and British Airways Board as lessee of the other part of property at MT Operations Base, Heathrow Airport, London ("MT Operations Base"), the open market rental value of MT Operations Base is to be assessed on review on the following assumptions, namely:

(v) that the Plaintiff is not a bidder for the hypothetical lease.

(vi) in the alternative to (v), if the Plaintiff is to be assumed to be a bidder for the hypothetical lease of the premises the Plaintiff's bid is assumed to be in such an amount as is just sufficient to secure a letting to it in the actual market.

These questions are similar to those raised on the West Base lease on which I declined to make two similar declarations claimed by BA. The wording of the rent review clause is, however, different from that in the West Base lease. It is contained in the Third Schedule on pp 25 and 26 of the lease. In clause (b) it is agreed that the open market

A rental should be taken to be the best annual rent of aggregate annual rents at which the land (excluding the buildings or other erections belonging to the lessee thereon) could reasonably be expected to be let on the open market at the review date with vacant possession by a willing lessor to a willing lessee.

It is then provided that certain matters should be expressly disregarded. They include all or any matters referred to in paras (*a*), (*b*), (*c*) and (*d*) of (1) of section 34 of the Landlord and Tenant Act 1954. Para (*a*) of section 34(1) provides for a disregard of "any effect on rent of the fact that the tenant has, or his predecessors in title have, been in occupation of the building". BA relied on that particular provision in support of the general argument that the independent
B surveyor should assume that BA is *not* a bidder for the hypothetical lease.

In my judgment, the express incorporation in section 34(1)(*a*) makes no difference to the general reasons which I have given in relation to the West Base lease for declining to make similar declarations sought by BA on that lease. I agree with the defendants that the only effect of the disregard (*a*) is to apply for a disregard of the tenant's occupation of the relevant premises. It is not directed to the different point which BA attempted to establish, namely a disregard of the presence of the lessee as a potential bidder in the market for the hypothetical lease. I therefore decline to make either declaration sought in (v) or (vi).

I come finally to the No 3 Gatwick lease on which one question
C remains to be decided:

> Whether on the true construction of a Deed of Surrender and Supplemental Lease dated 21st November 1977 made between British Airports Authority as lessor of the one part and British Caledonian Airways Limited as lessee of the other part and the leases and deed therein referred to of property at Maintenance Base East (No 3) Gatwick Airport, London ("Maintenance Base East (No 3)"), the open market rental value of part of the said land shown hatched yellow and hatched brown on the plan annexed to the actual lease is to be assessed for the purpose of the rent reviews on 29th September 1987 and 29th September 1997 on the following assumptions namely:
>
> (i)(a) that the said land shown hatched yellow and hatched brown is let on a single hypothetical lease
>
> Alternatively:
>
> D (b) that the land hatched yellow (the hangar land) and the land hatched brown (the car park land) are to be taken to be let either on one lease or on two separate leases, whichever will produce the best rent.

On p 16 of the Gatwick No 3 lease there is a definition of the expression "open market rental value" to be determined by the independent surveyor. It is to be taken to be "the best annual rent or *aggregate* annual rents at which the relevant land" could reasonably be expected to be let on the open market at the review date in question with vacant possession.

It is provided that, with effect from each review date (being September 29 1987 and September 29 1997), the yearly rent payable by the tenant shall be whichever is the greater of:

> E (i) the rent payable immediately before the relevant review date, or
>
> (ii) £2,680 per annum plus 75% of the open market rental value (as hereinafter defined) at the relevant review date of that part of the said land shown hatched yellow and hatched brown on the said plan (1.72 acres).

Under the lease the land shown hatched yellow can be used only as an aircraft hangar and offices, workshops and ancillary accommodation all associated with the lessees' business as an airline. The land shown hatched brown on the said plan can be used only as an area for parking cars of people visiting the premises and of the tenant or tenant's staff employed there: see clause 4(16)(a) and (b) on p 8 of the Gatwick No 3 lease.

It should be noted that there is a different basis of review on the review date arising on September 29 2007 under clause 2(d)(ii) on
F p 16, which provides that, with effect from that date and the last day of every fifth year of the term thereafter, the yearly rent payable by the tenant shall be whichever is the greater of:

> (i) the rent payable immediately before the relevant review date.
>
> (ii) the open market rental value (as hereinafter defined) at the relevant review date of that part of the said land and any buildings thereon not hatched yellow or brown on the said plan (3.08 acres), plus the open market rental value (as hereinafter defined) at the 29th September 2007 of that part of the said land excluding any buildings thereon shown hatched yellow and brown on the said plan (1.72 acres).

The defendants, relying heavily on the words "or aggregate" in the definition of "open market rental value", submit that the independent surveyor is to assume that the hangar land hatched yellow and the car park land hatched brown are to be taken to be let G either on one lease or on two separate leases, whichever will produce the best rent.

In my judgment, the defendants' submissions involve reading too much into the two words "or aggregate" in that context. I agree with the submissions made on behalf of BA that, in this context, the draftsman has used the expression "or aggregate" with reference to the difference basis of review in and after the year 2000 and that, for the purposes of the current review and the 1997 review, valuation should be on the basis of a single best annual rent for the two areas of land let on a single hypothetical lease. I shall therefore make a declaration in the terms of (a).

Those conclude my judgment on the individual questions. I will now hear argument from counsel as to the form of the order which I H should make and as to costs.

Declaration accordingly.

Chancery Division

March 19 1992

(Before Judge Paul BAKER QC, sitting as a judge of the High Court)

PONTSARN INVESTMENTS LTD v KANSALLIS-OSAKE-PANKKI J

Estates Gazette June 6 1992

[1992] 22 EG 103

Landlord and tenant — Rent review — Meaning of "fit for immediate occupation and use"

By a lease dated October 28 1985 the plaintiff landlord demised to the defendant premises at 80 Bishopsgate, London, for a term of 25 years from March 25 1985 at an initial yearly rent of £1,242,000 subject to five-year rent reviews — The demised premises were defined as excluding K "all tenants' and trade fittings" — At review the open market rent was to be determined, *inter alia*, on the assumption that the premises are "vacant but fit for immediate occupation and use"; there was to be disregarded tenant's improvements — In respect of the March 25 1991 review, a surveyor was appointed under the lease as an "expert" and he made his determination on July 3 1991 — In his decision the expert, who had the benefit of written submissions by which each party relied upon the opinions of leading counsel given in comparable transactions, stated that (1) there should be no increase in rental value on account of the words "fit for immediate occupation and use" and (2) the words were to indicate that no L allowance should be made against rental value to reflect a tenant's fitting-out period — The plaintiff issued an originating summons seeking a declaration that the expert wrongly held that there should be no increase in rent consequent on the assumption that he was directed to make that the premises were "fit for immediate occupation and use" — The defendant, who contended that the expert's decision was final and binding, applied by way of notice of motion to strike out the relevant paragraph of the originating summons, alternatively an order under Ord 14A of the Rules of the Supreme Court to determine as a question of law M whether the expert's determination was a nullity

Held: **1. The expert had considered the submissions of the parties and the question of the interpretation of the contentious words was one he had necessarily to undertake along the way to determinine the defined open market rent; it was within his remit and final and binding: see *Jones* v *Sherwood Computer Services plc***

2. The expression "fit for immediate occupation and use" meant the building was free from defects and ready for the tenant to go in and fit it out for his business and commence trading — Ready for occupation means when the building is

A **ready to be occupied for fitting-out purposes and meant that the tenant could negotiate for a reduction of rent during the fitting-out period (although such a reduction was not claimed by the defendant)**

The following cases are referred to in this report.

Anisminic Ltd v *Foreign Compensation Commission* [1969] 2 AC 147; [1969] 2 WLR 163; [1969] 1 All ER 208, HL
Arenson v *Casson Beckman Rutley & Co* (on appeal from *Arenson* v *Arenson*) [1977] AC 405; [1975] 3 WLR 815; [1975] 3 All ER 901; [1976] 1 Lloyd's Rep 179, HL
Basingstoke and Deane Borough Council v *Host Group Ltd* [1988] 1 WLR 348; [1988] 1 All ER 824; (1987) 56 P&CR 31; [1987] 2 EGLR 147; 284 EG B 1587, CA
Burgess v *Purchase & Sons (Farms) Ltd* [1983] Ch 216; [1983] 2 WLR 361; [1983] 2 All ER 4
Campbell v *Edwards* [1976] 1 WLR 403; [1976] 1 All ER 785; [1976] 1 Lloyd's Rep 522, CA
Dean v *Prince* [1954] Ch 409; [1954] 2 WLR 538; [1954] 1 All ER 749, CA
Iceland Frozen Foods plc v *Starlight Investments Ltd* [1992] 1 EGLR 126; [1992] 07 EG 117
Jones v *Sherwood Computer Services Ltd* [1992] 1 WLR 277
Nikko Hotels (UK) Ltd v *MEPC plc* [1991] 2 EGLR 103; [1991] 28 EG 86
O'Reilly v *Mackman* [1983] 2 AC 237; [1982] 3 WLR 1096; [1982] 3 All ER 1124, HL
Racal Communications Ltd, Re [1981] AC 374; [1980] 3 WLR 181; [1980] 2 All ER 634, HL

C

This was an application by the defendant, Kansallis-Osake-Pankki, by way of a notice of motion to strike out a paragraph in the originating summons issued by the plaintiff, Pontsarn Investments Ltd, by which the plaintiff sought a declaration that the decision of an expert, Mr Ian V Oddy FRICS, appointed to determine a rent dispute between the parties, was a nullity. The demised premises are located at 80 Bishopsgate, London EC2.

Michael Barnes QC (instructed by Nabarro Nathanson) appeared for the plaintiff; and Paul Morgan (instructed by Clifford Chance) represented the defendant.

Giving judgment, **JUDGE PAUL BAKER QC** said: This is yet another D case arising out of a rent review provision in a lease. The review has proceeded to the point where the rent has been determined by a surveyor appointed by the parties. The landlord is dissatisfied with the determination and has issued an originating summons seeking a declaration that it is a nullity on various grounds of which only one is now pursued, namely that the surveyor wrongly held that there should be no increase in the rent consequent on the assumption that he was directed to make that the premises were "fit for immediate occupation and use".

The tenant's response to this was a notice of motion now before me seeking to strike out the relevant paragraph of the originating summons or alternatively seeking an order under the Rules of the Supreme Court Ord 14A to determine as a question of law whether E the surveyor's determination is a nullity.

While the precise question is whether the determination is a nullity, I have heard much argument on whether the determination is indeed wrong. I have been invited to express my views on that question also, as that may be of use in the future, though I do not propose to make any formal declaration about it.

The facts are not in dispute. By a lease dated October 28 1985 the plaintiff as landlord demised to the defendant as tenant for a term of 25 years from March 25 1985 a substantial building at 80 Bishopsgate in the City of London comprising a basement, ground floor and seven upper floors. It is used to accommodate a restaurant and a shop in the basement and part of the ground floor and offices in the whole F of the remainder. This dispute is confined to the office portion. The initial rent was £1,242,000 pa with review dates at five-yearly intervals, the first being March 25 1991. It is what occurred during that review which is the subject of the dispute.

The demised premises are defined so as to include:

> all additions and alterations thereto and all Landlord's fixtures and fittings now or hereafter in or about the same but excluding all tenants' and trade fittings.

The review provisions are contained in the second schedule to the lease. They begin with a definition of open market rent, which for the office portion of the premises is defined as follows:

> the yearly rent for which the Office Portion might reasonably be expected to be let as a whole but forming part of the Demised Premises with vacant G possession for any use within the same class of the Town and Country Planning (Use Classes) Order 1972 as the user of the Office Portion hereby permitted on the relevant Rent Review Date in the open market by a willing lessor to a willing lessee for a term equal to the residue of the term hereby granted (or if greater a term of ten years with a review of rent at the fifth year thereof) on the assumptions that at the Relevant Review Date they are:

The first one is the critical one for this case:

> 1.2.2.1. vacant but fit for immediate occupation and use and
>
> 1.2.2.2. that no work has been carried out thereon by the Tenant or its predecessors in title which has diminished the rental value of the Demised Premises.

There follow a number of further assumptions which are not material to the issue I have to decide. There follow after that three disregards H and the third one is the one that I have to read:

> 1.2.3. disregarding . . .
>
> 1.2.3.3. Any effect on rent of any Voluntary Improvement carried out by the tenant or Permitted Underlessee at any time after the commencement of the term hereby granted. In this sub paragraph "Voluntary Improvement" means an improvement which the Tenant had no obligation to the Landlord herein to carry out and in respect of which the Landlord has not contributed to and has no liability to contribute to the cost thereof.

The next provision in this schedule which I should look at is clause 3:

> If the Landlord and the Tenant shall not have agreed the Open Market Rent by the relevant Rent Review Date or (if earlier) within three months of the service J of the Notice by the Landlord or the Tenant referred to in Clause 2 of this Schedule the Open Market Rent may be determined by a surveyor (hereinafter called "the Appointed Surveyor") who shall be agreed upon in writing by the Landlord and the Tenant . . .

The parties were able to agree on a surveyor so I need not read further as it deals with appointments in default of agreement. By a letter dated December 20 1990 Mr Ian V Oddy [FRICS], who is accepted by all parties as a suitably qualified surveyor for this purpose, was asked by both parties to act as the appointed surveyor.

Now I go to Clause 4. This again is a critical clause in the context of this case:

> The Appointed Surveyor shall be a Chartered Surveyor in practice in the City of London for at least 5 years prior to the date of his appointment and who K shall be an expert in rental values and lettings in the City of London

— as I have said, there is no question about his qualification in that regard —

> and shall be bound to afford to both the Landlord and the Tenant the opportunity to make representations (whether written and/or oral) regarding the Open Market Rent for the Demised Premises. The Appointed Surveyor shall make his determination including his reasons of the Open Market Rent acting as an expert and not as an Arbitrator and his determination shall be final and binding on the Landlord and the Tenant.

The rest of the clause deals with his fees, which are not in issue here.

Both parties made submissions in writing to Mr Oddy which were followed by counter-submissions on each other's submissions. Mr Oddy made his award on July 3 1991. I must, of course, look at it, but L I must also look at the submissions, or some of them, in view of a submission by Mr Morgan, for the tenant, that the interpretation reached by Mr Oddy on the matter in dispute was reached, first, pursuant to the express terms of the schedule or, alternatively, pursuant to the joint invitation of the parties by reason of their surveyor's submissions to Mr Oddy.

The landlord's surveyors in their submissions referred to a comparable where the expert had allowed an uplift as a consequence of the assumption that the premises were fitted out for immediate occupation and use. It appeared in that case that the expert had had an opinion of counsel, Mr Kim Lewison, which supported the view he took. Having referred to that case, the landlord's surveyors M presented calculations which, they said, would produce an uplift of £10 per sq ft for 37,821 sq ft, say £380,000, as an addition to the rent. The surveyors concluded their submissions on this point in the following terms:

> 7.7 As mentioned previously, as you are the Expert in this matter, you will have to decide upon the correct basis of valuation. If, in your opinion, there should be no uplift then I would ask you to determine an open market rent of the Office Portion of £2,335,000.
>
> If, however, you consider that there should be an uplift to reflect the willing lessee being provided premises which are fit for immediate use and occupation at no capital cost to the lessee then I would request you to determine an open market rent of the Office Portion of £2,715,000.

That, of course, includes the uplift of £380,000.

A So it is clearly put to the expert in that paragraph that he has to decide the point.

The tenant's surveyors countered this by contending that the assumption "vacant but fit for immediate occupation and use" merely meant that one disregards any period of time necessary to fit out the premises. If that was wrong, they say that the work needed to fit them out was minimal. They attack the instructions to Mr Lewison and his opinion in the case referred to by the landlord's surveyors and provided a further opinion of leading counsel, Mr David Neuberger QC, in another comparable case, which differed from Mr Lewison and supported the tenant's submissions.

With that I can now go to Mr Oddy's decision on this point. The
B material paragraphs, the only ones which I propose to read out, are paras 3.7 and 3.8:

> 3.7 Mr Neuberger took an opposite view to that of Mr Lewison concerning the meaning and relevance of "fitted out". I entirely accept the arguments put forward by Mr Neuberger. In this subject rent review clause the reference is not to "fitted" but to "fit for immediate occupation and use". The reasons given by Mr Neuberger in respect of the wording in the Triton Court lease, satisfy me entirely with regard to the way in which the wording in the subject rent review clause should also be treated. *Accordingly, I Determine that there shall be no increase in the annual rental value on account of the words included in the Rent Review Clause, "fit for immediate occupation and use".* [Mr Oddy's italics]
>
> 3.8 In my opinion the purpose of including the words "fitted out and available for immediate occupation and use" or "fit for immediate
> C occupation and use" or such other phrases, are to indicate to the parties and to any third party that no allowance is to be made against the rental value to reflect a tenant's fitting out period.

Those are the facts and I can now turn to the question whether the expert's decision is final and binding and cannot be challenged.

The law on this subject is now governed by the decision of the Court of Appeal in *Jones* v *Sherwood Computer Services plc* [1992] 1 WLR 277, a decision which was handed down in December 1989. This decision has set the law on a new course. Before coming to it and having regard to some submissions made to me by Mr Barnes QC, for the landlords, I should refer briefly to the earlier history.

In *Dean* v *Prince* [1954] Ch 409 Denning LJ (as he then was)
D expressed the view that a valuation could be impeached not only for fraud but also for mistake or miscarriage. At p 427 of the report he says:

> For instance, if the expert added up his figures wrongly:

— I should say this was an accounting type of case —

> or took something into account which he ought not to have taken into account, or conversely: or interpreted the agreement wrongly: or proceeded on some erroneous principle. In all these cases the court will interfere. Even if the court cannot point to the actual error, nevertheless, if the figure itself is so extravagantly large or so inadequately small that the only conclusion is that he must have gone wrong somewhere, then the court will interfere in much the same way as the Court of Appeal will interfere with an award of damages if it is a wholly erroneous estimate. These cases about valuers bear some analogy
> E with the cases on domestic tribunals, except of course that there need not be a hearing. On matters of opinion, the courts will not interfere; but for mistake of jurisdiction or of principle, and for mistake of law, including interpretation of documents, and for miscarriage of justice, the courts will interfere: see *Lee* v *Showmen's Guild of Great Britain*.

At that time it was thought that valuers could not be sued in negligence in relation to their valuation. That was changed by the decision of the House of Lords in *Arenson* v *Casson Beckman Rutley & Co* [1977] AC 405. This development led Lord Denning MR (as he had then become) to revise his views. In *Campbell* v *Edwards* [1976] 1 WLR 403 at p 407E he said:

> In former times (when it was thought that the valuer was not liable for
> F negligence) the courts used to look for some way of upsetting a valuation which was shown to be wholly erroneous. They used to say that it could be upset, not only for fraud or collusion, but also on the ground of mistake: see for instance what I said in *Dean* v *Prince* [1954] Ch 409, 427. But those cases have to be reconsidered now. I did reconsider them in the *Arenson* case in this court: [1973] Ch 346, 363. I stand by what I there said. It is simply the law of contract. If two persons agree that the price of property should be fixed by a valuer on whom they agree, and he gives that valuation honestly and in good faith, they are bound by it. Even if he has made a mistake they are still bound by it. The reason is because they have agreed to be bound by it. If there were fraud or collusion, of course, it would be very different. Fraud or collusion unravels everything.
>
> It may be that if a valuer gives a speaking valuation — if he gives his reasons or his calculations — and you can show on the face of them that they are wrong it might be upset. But this is not such a case. Chestertons simply gave the figure. Having given it honestly, it is binding on the parties. It is no good for G either party to say that it is incorrect.

Later at p 408B he said:

> The position of a valuer is very different from that of an arbitrator. If a valuer is negligent in making a valuation, he may be sued by the party — vendor or purchaser — who is injured by his wrong valuation.

The issue of a speaking valuation came up in *Burgess* v *Purchase & Sons (Farms) Ltd* [1983] Ch 216, a decision of Nourse J (as he then was). It was pressed on him that the Court of Appeal would, in its then state of the authorities, make no distinction between a speaking and an unspeaking valuation. In response to that Nourse J said at p 223G:

> It seems to me that I could not tread that path without trespassing upon the H rarest ethers of speculation. The function of a judge of first instance is to apply the law as it stands. He is not to speculate on what some higher court may one day declare it to be. If the law is declared by earlier decisions at first instance he ought to follow them unless he is satisfied that they are wrong. In this case the law is declared by three earlier decisions at first instance. I am certainly not satisfied that they are wrong. I must therefore follow and apply them.

He stated the law then to be in the following terms on p 225B:

> A non-speaking valuation made of the right property by the right man and in good faith cannot be impugned, although it may still be possible, in the case of an uncompleted transaction, for equitable relief — as opposed to damages — to be refused to the party who wishes to sustain the valuation. On the other hand, there are at least three decisions at first instance to the effect that a J speaking valuation which demonstrates that it has been made on a fundamentally erroneous basis can be impugned. In such a case the completion of the transaction does not necessarily defeat the party who wishes to impugn the valuation.

That was how the law stood when the lease in this case was entered into and may explain in part why the valuer was required to give reasons. And so we come to *Jones* v *Sherwood Computer Services*. The defendant company had agreed to purchase the plaintiff's shares in a public limited company for a consideration fixed by reference to the amount of the sales of the products of some subsidiaries of the plc. The amount of the sales was to be reviewed by the accountants representing the plaintiff and the defendant and, if they were unable to agree, as happened, the matter was to be referred to independent K accountants to determine and report the amount of the sales. The accountants were to act as experts and not as arbitrators and their determination was to be "conclusive and final and binding for all purposes". The disagreement which necessitated a reference to the independent accountants was as to the inclusion of two categories of transactions as sales for the purposes of the calculation. The Court of Appeal (which consisted of a two-judge court, Dillon and Balcombe LJJ) decided as follows, and I read from the headnote:

> *Held*, allowing the appeal, that, where parties had agreed to be bound by the report of an expert, the report, whether or not it contained reasons for the conclusion in it, could not be challenged in the courts on the ground that mistakes had been made in its preparation unless it could be shown that the expert had departed from the instructions given to him in a material respect; L and that, since the third firm of accountants had done precisely what they had been asked to do and there was no question of bad faith, their determination was not subject to challenge by the plaintiffs . . .

The first point I would note is that whether the queried transactions were sales involved a question of construction of the share-purchase agreement between the plaintiff and the defendant. On this question Dillon LJ said at p 287E:

> Any number of issues could arise under the various sub-paragraphs of paragraph 2 of appendix 1 as to the application of the wording of those sub-paragraphs to particular facts. All these issues are capable of being described as issues of law or mixed fact and law, in that they all involve issues as to the true meaning or application of wording in paragraph 2. I cannot read the categorical wording of paragraph 7 as meaning that the determination of the M accountants or of the expert shall be conclusive, final and binding for all purposes "unless it involves a determination of an issue of law or mixed fact and law in which case it shall only be binding if the court agrees with it".

Second, the relevance of speaking awards. In *Jones* v *Sherwood Computer Services* the accountants had not given reasons. On this point Dillon LJ had this to say, and I turn back to p 284F:

> . . . it is convenient to say a little at this juncture about the distinction between speaking and non-speaking valuations or certificates, which to my mind is not a relevant distinction. Even speaking valuations may say much or little; they may be voluble or taciturn if not wholly dumb. The real question is whether it is possible to say from all the evidence which is properly before the court, and not only from the valuation or certificate itself, what the valuer or certifier has

A done and why he has done it. The less evidence there is available, the more difficult it will be for a party to mount a challenge to the certificate.

The general principle was thus stated by Dillon LJ at the top of p 287:

> On principle, the first step must be to see what the parties have agreed to remit to the expert, this being, as Lord Denning MR said in *Campbell* v *Edwards* [1976] 1 WLR 403, 407g, a matter of contract. The next step must be to see what the nature of the mistake was, if there is evidence to show that. If the mistake made was that the expert departed from his instructions in a material respect — eg, if he valued the wrong number of shares, or valued shares in the wrong company, or if, as in *Jones (M)* v *Jones (RR)* [1971] 1 WLR 840, the expert had valued machinery himself whereas his instructions were to employ an expert valuer of his choice to do that — either party would be able B to say that the certificate was not binding because the expert had not done what he was appointed to do.

He concluded his judgment by commenting on the *Burgess* case at p 289E:

> It follows from the foregoing that I respectfully differ from much that was said by Nourse J in *Burgess* v *Purchase & Sons (Farms) Ltd* [1983] Ch 216. This stems primarily from a difference in our respective assessments of the decisions in *Campbell* v *Edwards* [1976] 1 WLR 403 and *Baber* v *Kenwood Manufacturing Co Ltd* [1978] 1 Lloyd's Rep 175, which I regard as binding on us, as the starting point of the reassessment of the law in this field after the decisions of the House of Lords which I have mentioned, but which he, I apprehend, regarded as heretical and illogical departures from the established doctrine enunciated in *Dean* v *Prince* [1954] Ch 409 — in effect, Lord Denning's first thoughts in *Dean* v *Prince* ought to be preferred to his second C thoughts in *Arenson* v *Arenson* [1973] Ch 346 and *Campbell* v *Edwards* [1976] 1 WLR 403. However it is unnecessary to encumber this over-long judgment with further analysis of *Burgess* v *Purchase & Sons (Farms) Ltd* [1983] Ch 216 which on any view is not binding on this court.

Balcombe LJ, in a short concurring judgment, picks up that point at the beginning of his judgment at p 289H:

> The principle applicable to a case of this type is that stated by Lord Denning MR in *Campbell* v *Edwards* [1976] 1 WLR 403, 407. [His Lordship set out the passage cited by Dillon LJ, ante, p 284c, and continued:] That principle was expressly approved by this court in *Baber* v *Kenwood Manufacturing Co Ltd* [1978] 1 Lloyd's Rep 175. In my judgment that principle cannot be affected by the chance whether the valuation is or is not a "speaking" valuation. I agree with Dillon LJ that that is not a relevant distinction and it follows that I also D respectfully disagree with the judgment of Nourse J in *Burgess* v *Purchase & Sons (Farms) Ltd* [1983] Ch 216.

That is the governing decision now and it fell to Knox J in *Nikko Hotels (UK) Ltd* v *MEPC plc* [1991] 28 EG 86* to apply this decision to a case concerning a rent review provision. The subject premises were an hotel and the rent was to be regulated according to charges made for accommodation in the hotel. The mechanism involved an elaborate formula linked to changes in the room rate which gave rise to questions of interpretation which the expert had to decide. Knox J expressed the effect of the decision in *Jones* v *Sherwood Computer Services* in this way at p 99 of the report:

> This decision of the Court of Appeal in *Jones* v *Sherwood Computer* E *Services* is, of course, binding on me, as counsel on both sides before me accepted. In my judgment, it provides for a contractual analysis of the task set for the expert to perform and it gives full effect to the parties' agreement regarding with what it was that the expert should be entrusted.
>
> The result, in my judgment, is that if parties agree to refer to the final and conclusive judgment of an expert an issue which either consists of a question of construction or necessarily involves the solution of a question of construction, the expert's decision will be final and conclusive and, therefore, not open to review or treatment by the courts as a nullity on the ground that the expert's decision on construction was erroneous in law, unless it can be shown that the expert has not performed the task assigned to him. If he has answered the right question in the wrong way, his decision will be binding. If he has answered the wrong question, his decision will be a nullity.

I respectfully and gratefully adopt that and turn to the case before F me.

What is remitted to the determination of the expert is the determination of the open market rent as defined in clause 1.2 of the second schedule. The definition involves the making of certain assumptions and disregarding certain effects. The task may involve interpreting the language used, for example the reconciliation between the direction to assume in fixing the rent that the premises are fit for immediate occupation and use, and the direction to disregard any effect on that same rent of improvements carried out by the tenant after the commencement of the term but before the review date.

Mr Oddy indeed saw as one of the matters referred to him the G application of the assumption that the premises were to be regarded as fit for immediate occupation and use. He noted the different interpretations of the landlord and tenant, backed up by the opinions of counsel in comparable cases, and reached his determination accepting the arguments of the tenant. This resulted in no increase of the rent to take account of any notional fitting out at the expense of the landlord, or at all events not at the expense of the tenant, or any decrease of rent to take account of any rent-free period to allow for the time which the tenant might take to fit out the premises at his expense.

The resolution of this question of interpretation is, in my judgment, one which the expert necessarily had to undertake along H the way to determine the open market rent as defined in the second schedule. Accordingly, as I see it, the question was within the remit of the expert and the answer of Mr Oddy is final and binding and not subject to review.

I would add that I would not have accepted the alternative submission of Mr Morgan. I do not think that one can spell out of the submissions of the experts to Mr Oddy an agreement to submit an issue for his decision. They were merely guiding themselves by the rent review clause in the lease.

I must, however, go on to the careful and penetrating argument of Mr Barnes. He began by considering whether the expert had come to a correct view of the requirement that he has to assume that the premises are fit for immediate use and occupation. He submitted that J it meant that one had to assume that at the review date the premises had been fitted out in readiness for the immediate occupation of a hypothetical tenant for the purposes of his business. This meant that such a tenant had two advantages: (a) no period of time when the tenant cannot use the premises while the fitting out was being done and (b) not have the financial burden of paying a capital sum for the fitting out.

The expert declined in principle to allow anything for the second advantage. In so doing, so it is submitted, he failed to carry out his instructions. He was instructed to value the premises fitted out and deliberately valued them as not fitted out. Accordingly, picking up the respective formulations in *Jones* v *Sherwood Computer Services* and *Nikko Hotels* it is said that he (a) departed from his instructions K and (b) asked himself the wrong question.

I am unable to accept this submission. It depends, as I see it, upon the expression "fit for immediate occupation and use" bearing the meaning attributed to it by Mr Barnes as being beyond any question or argument. As I shall show later, other meanings are fairly arguable. Accordingly, the true construction of this expression was a matter with which the expert was required to grapple. The fact that he may have placed upon it a particular construction and sought to justify it with reasoning which is flawed — or even incomprehensible (Mr Barnes went that far) — would simply mean that he had answered the right question wrongly. I would go further. Even if the construction of Mr Barnes were self-evidently right, I would not L consider that the expert departed from his instructions in a material respect. The examples given by Dillion LJ are cases of departure from express instructions. In this case the express instructions are to determine the open market rent as defined and to give reasons. If he had totally ignored a material element, there might be a question, but the expert had made an attempt to deal with the element as part of his instructions. At worst, he had come to a wrong conclusion. The fact that he may be patently wrong does not mean that he has not done what he was appointed to do nor that he has asked himself the wrong question. To take any other view would lead to the sort of refined arguments such as have been deployed here and go a long way to emasculate the requirement that the decision of the expert, as a matter of contract between the parties, be final and binding. Thus, M the advantages of cost, speed and finality would be seriously diminished.

Mr Barnes went on boldly to argue that in any event, if the expert's determination is wrong in law, it necessarily means that he has exceeded his jurisdiction so that the purported jurisdiction or exercise of it is a nullity. In support of this I was taken through a series of decisions at the highest level commencing with *Anisminic Ltd* v *Foreign Compensation Commission* [1969] 2 AC 147, proceeding through *Re Racal Communications Ltd* [1981] AC 374 to *O'Reilly* v *Mackman* [1983] 2 AC 237, which define the modern doctrine of control by the High Court over inferior courts and tribunals. To see what that doctrine is I cannot do better than cite from the speech of

---

*Editor's note: Also reported at [1991] 2 EGLR 103.

one of its main architects, the late Lord Diplock, in the last case at p 278. After citing section 4(4) of the Foreign Compensation Act 1950, which provided:

> The determination by the commission of any application made to them under this Act shall not be called in question in any court of law,

his speech went on at p 278:

> It was this provision that provided the occasion for the landmark decision of this House in *Anisminic Ltd* v *Foreign Compensation Commission* [1969] 2 AC 147, and particularly the leading speech of Lord Reid, which has liberated English public law from the fetters that the courts had theretofore imposed upon themselves so far as determinations of inferior courts and statutory tribunals were concerned, by drawing esoteric distinctions between errors of law committed by such tribunals that went to their jurisdiction, and errors of law committed by them within their jurisdiction. The breakthrough that the *Anisminic* case made was the recognition by the majority of this House that if a tribunal whose jurisdiction was limited by statute or subordinate legislation mistook the law applicable to the facts as it had found them, it must have asked itself the wrong question, ie, one into which it was not empowered to inquire and so had no jurisdiction to determine. Its purported "determination", not being a "determination" within the meaning of the empowering legislation, was accordingly a nullity.

I note the reference there to "asked itself the wrong question". Mr Barnes sought to say that, as experts shared many of the characteristics of inferior tribunals, this doctrine should apply to their determinations, albeit they operated in the field of private law. I am quite unable to accept this submission for a number of reasons.

First, tribunals have been imposed from outside by Parliament or some other authority to adjudicate on disputes between executive authorities and citizens or sometimes between citizens and citizens. Some control over them is necessary. Sir William Wade, in his well-known textbook, *Administrative Law*, 6th ed, p 297, said:

> If administrative tribunals and authorities could trespass uncontrollably outside their proper field, there would no longer be order in the legal system. Order can be preserved only if jurisdictional demarcation disputes can always be carried to the regular courts of law and so be brought within a unified hierarchy of authority.

The position of an expert is quite different. The expert's jurisdiction (so-called) is settled as a matter of contract between the parties. It is not imposed from without.

Second, if it is thought that some question of law requires resolution, either party can seek a declaration from the court.

Third, if the expert is negligent, a party suffering loss as a result of the determination may have a remedy against him.

Fourth, Mr Barnes submitted that I was not precluded by the decision in *Jones* v *Sherwood Computer Services* from accepting that concept of jurisdiction in public law to be applicable to decisions of experts such as we have here. I do not agree. As I have shown, in the past analogies from jurisdictions over tribunals have been used in controlling experts and we have looked at *Dean* v *Prince* and *Burgess* v *Purchase & Sons*. Such methods of control have been expressly repudiated by the Court of Appeal in *Jones* v *Sherwood Computer Services*.

Finally, I turn to the question of what is the true meaning of the expression "vacant but fit for immediate occupation and use" as used in the second schedule to the lease. I have already referred to Mr Barnes' submission on behalf of the landlord. The expert is required to assume that the premises are fitted out ready for occupation by a hypothetical tenant. The building is not in a shell state but fitted out so that the tenant can just walk in and start business. It does not matter who paid for it. This has the consequence that there is to be no remission of rent to allow for fitting out to be done. But there is to be an uplift of rent because the hypothetical tenant does not have the financial burden of paying a capital sum to get in. The expert confined the expression requiring him to make no allowance for a fitting-out period. He was persuaded by arguments not to increase the rent on account of the state of the fitting out in readiness for immediate occupation, as I have explained.

Mr Morgan, for the tenant, pointed out the anomaly of charging the tenant £10 per sq ft for something which the landlord had not provided at the rent review date. He also points out that the expression was "fit for immediate use" not "fitted out". From this he deduced that the expression meant that the building was free from defects and ready for the tenant to go in, fit it out for his business and commence trading. A building is fit for occupation when it is ready to be occupied for fitting-out purposes. Ready for occupation does not mean ready for the tenant to go in and start trading immediately. If this were correct, it would mean that the tenant could negotiate for a reduction of rent over the fitting-out period, but Mr Morgan emphasised that he was not seeking to put forward any such claim in the present case. He accepted the expert's finding.

Of these views I prefer that proposed by the tenant for the following reasons. First, it is consistent with the purpose of the rent review clauses, whereas the landlord's construction is not. I refer to *Basingstoke and Deane Borough Council* v *Host Group Ltd* [1988] 1 WLR 348* at p 353D, where Nicholls LJ, giving the judgment of the Court of Appeal, said:

> The question raised on this appeal is one of construction of a rent review clause in a lease. In answering that question it is axiomatic that what the court is seeking to identify and declare is the intention of the parties to the lease expressed in that clause. Thus, like all points of construction, the meaning of this rent review clause depends on the particular language used interpreted having regard to the context provided by the whole document and the matrix of the material surrounding circumstances. We recognise, therefore, that the particular language will always be of paramount importance. None the less it is proper and only sensible, when construing a rent review clause, to have in mind what normally is the commercial purpose of such a clause.
>
> That purpose has been referred to in several recent cases, and is not in doubt. Sir Nicolas Browne-Wilkinson V-C expressed it in these terms in *British Gas Corporation* v *Universities Superannuation Scheme Ltd* [1986] 1 WLR 398, 401:
>
> "There is really no dispute that the general purpose of a provision for rent review is to enable the landlord to obtain from time to time the market rental which the premises would command if let on the same terms on the open market at the review dates. The purpose is to reflect the changes in the value of money and real increases in the value of the property during a long term."

The landlord's construction would give the landlord more than compensation for changes in the value of money.

Second, the tenant's construction fits in better with the remainder of the review clause. I note these points:

(a) The expert is required to assume that the premises are let with vacant possession for any use within the same class of the use classes order as a permitted user. This fits uneasily with the notion that he is required to assume that the premises have been fitted out for a specific office user.

(b) The expert, on the one hand, is required to assume that no work has been carried out by the tenant which has diminished the rental value and, on the other, to disregard the effect on rent, presumably upwards, for voluntary improvement. Strictly, these admonitions would not be required if one is required to assume that the premises were fitted out ready for the immediate profitable occupation of a hypothetical tenant, for that means that, in any event, one would ignore the fitting out done by the actual occupying tenant.

I was referred to a number of other cases on construction of similar, but not identical, review clauses. None of them precisely cover this case — cases on construction seldom do — and I do not propose to examine them further, save to say that the most recent, a decision of the Court of Appeal handed down on December 16 1991 but so far unreported, seems fatal to the landlord's approach here. The case is *Iceland Frozen Foods plc* v *Starlight Investments Ltd*†*. There there was to be an assumption that the premises are ready for immediate occupation and use. The landlords were arguing that the most profitable use was to subdivide the premises into seven separate units instead of the existing two and that is what the hypothetical tenant would bid for. As to that, Dillon LJ said at p 6 of the transcript§:

> As I see it, however, the first stage in this is that the assessment of the new rent is to be on the basis of the rental value of the landlord's premises in their actual condition for the time being. That is in accordance with a statement of the law by Fox J in *Trust House Forte Albany Hotels Ltd* v *Daejan Investments Ltd* (1980) 256 EG 915. At p 916 Fox J said:
>
> ". . . As to the protection of the landlord against inflation, I accept that such is the purpose of the clause, but what is it that is being protected against inflation? In the absence of clear language to the contrary, I would assume that it is the rental value of the landlord's premises in their actual condition for the time being.
>
> . . . You protect the landlord against inflation by reassessing every seven years the rent of the actual premises which he owns and not of other premises. I do not see why the landlord should get the benefit of assumed alterations which the landlord has not made."

That passage was approved by this court in *Orchid Lodge (UK) Ltd* v *Extel Computing Ltd* [1991] 32 EG 57. It was cited by Judge O'Donoghue in the court below in the present case, together with a statement by Hoffmann J in *Norwich Union Life Insurance Society* v *Trustee Savings Banks Central*

---

*Editor's note: Also reported at [1987] 2 EGLR 147.

†Editor's note: Also reported at [1992] 1 EGLR 126.

§Editor's note: Reported at p 126 *ante*.

A *Board* [1986] 1 EGLR 136 at p 137 where he referred to the "presumption that the hypothesis upon which the rent should be fixed upon a review should bear as close a resemblance to reality as possible".

Commenting on the proposed works of conversion, the Lord Justice said:

> . . . if the execution of these works will produce a higher rental value for the premises or an improved return from trading in the premises which will warrant paying a higher rent for the pleasure of converting the premises, the works of conversion will rank as tenant's improvements which are to be disregarded under the express terms of the rent review clause. Mr Dowding [counsel for the landlord] says: "But these are merely hypothetical alterations by a hypothetical tenant under a hypothetical lease. Under the rent review clause you are only required to disregard actual improvements carried out by
> B the real tenant." But it seems to me to make nonsense of the rent review clause as drawn and of the aim that the rent on review should be fixed so as to bear as close a resemblance to reality as possible, if it is not possible to charge rent for actual improvements but is possible to charge the actual tenant with rent for hypothetical improvements which have never been carried out.

I regard that as fatal to the submissions here of the landlord but, as I have also said, the precise meaning of the review clause is not before me. What is before me is whether the expert's decision is binding. As to that, I am inclined to make the declaration sought under para 2 of the notice of motion rather than strike out parts of the originating summons, but I will hear counsel.

C

Chancery Division

March 25 1992

(Before Mr Justice CHADWICK)

PRUDENTIAL ASSURANCE CO LTD v SALISBURYS HANDBAGS LTD

Estates Gazette June 13 1992

[1992] 23 EG 117

D **Landlord and tenant — Rent review — Whether hypothetical lease contains provisions for rent review — Term of hypothetical lease**

By an underlease dated June 24 1971 the defendant holds a term of 18 James Street, Harrogate, from the plaintiff until 2067 — The underlease, in providing for seven-year rent reviews from September 29 1982, did not define the term to be assumed at each review — The plaintiff contended that the hypothetical term at review must be the unexpired term of the actual lease and on the assumption that the hypothetical lease contains no provisions for rent review — The defendant contended that the hypothetical lease was for a seven-year
E **term, alternatively, if for any longer period, on terms which provide for seven-year rent reviews**

***Held*: The parties intended the hypothetical letting to be on terms which do not include the rent review provisions of the actual lease — But the term to be assumed for the hypothetical lease is the term the landlord might reasonably be expected to grant and the tenant might reasonably be expected to take in respect of the premises at a fixed rack-rent — The term to be assumed may vary at each review during the residue of the lease as circumstances may change — In the absence of agreement, the term must be determined by the arbitrator and,**
F **in so far as it falls within a range, the term to be taken is that which yields the best rent**

The following cases are referred to in this report.

Basingstoke and Deane Borough Council v *Host Group Ltd* [1988] 1 WLR 348; [1988] 1 All ER 824; (1987) 56 P&CR 31; [1987] 2 EGLR 147: 284 EG 1587, CA

British Gas Corporation v *Universities Superannuation Scheme Ltd* [1986] 1 WLR 398; [1986] 1 All ER 978; (1986) 52 P&CR 111; [1986] 1 EGLR 120; 277 EG 980

Equity & Law Life Assurance Society plc v *Bodfield Ltd* [1987] 1 EGLR 124; (1987) 281 EG 1448, CA

Lynnthorpe Enterprises Ltd v *Sidney Smith (Chelsea) Ltd* [1990] 2 EGLR 131; [1990] 40 EG 130

MFI Properties Ltd v *BICC Group Pension Trust Ltd* [1986] 1 All ER 974; [1986] 1 EGLR 115; (1986) 277 EG 862 G

National Westminster Bank plc v *Arthur Young McClelland Moores & Co* [1985] 1 EGLR 61; (1984) 273 EG 402

Philpots (Woking) Ltd v *Surrey Conveyancers Ltd* [1986] 1 EGLR 97; (1985) 277 EG 61

Pugh v *Smiths Industries Ltd* (1982) 264 EG 823

This was an originating summons issued by the plaintiff, Prudential Assurance Co Ltd, seeking declarations as to the proper construction of an underlease, held by the defendant, Salisburys Handbags Ltd, of a property at 18 James Street, Harrogate.

David Halpern (instructed by Berwin Leighton) appeared for the plaintiff; and Jonathan Brock (instructed by Theodore Goddard) H represented the defendant.

Giving judgment, **CHADWICK J** said: The plaintiff in these proceedings, the Prudential Assurance Co Ltd, is the person entitled by assignment to the reversion expectant on the determination of an underlease granted on June 21 1971 of property known as 18 James Street, Harrogate. The defendant, Salisburys Handbags Ltd, is the lessee under that underlease. The underlease was granted for a term of almost 97 years, commencing on December 25 1970 and continuing until November 4 2067, at an initial rent of £4,000 pa, subject to review.

The dispute between the parties is as to the true construction and effect of the provisions for rent review. The reddendum in clause 1 of J the underlease requires the lessee to pay to the landlord the yearly rent of £4,000, subject to variation, as provided in the second schedule, such rent to be paid without any deduction by equal quarterly payments in advance on the four usual quarter days in every year. Clause 2(1) contains a covenant to pay rent in the usual terms. The second schedule to the underlease is in these terms so far as material:

> The yearly rent payable under this underlease at the 29th day of September 1982 and thereafter on the 29th day of September in each succeeding seventh year of the said term shall be reviewed in manner hereinafter provided (each of the said dates being hereinafter referred to as "the date of review") namely:
>
> 1. From and after the date of review these presents shall be read and construed and shall take effect in all respects as if the yearly rent for time being payable hereunder had from such date of review been the fair rack rental K market value (as hereinafter determined and defined) of the demised premises immediately after such date of review but without prejudice to any of the other terms and conditions contained in this underlease and so that in no event shall the rent payable by the lessees to the landlords in respect of the demised premises be less than the rent payable immediately prior to the date of review.
>
> 2. For the purpose of this rent review the fair rack rental market value of the demised premises shall (subject as hereinafter provided) be the amount which shall be agreed between the landlords and the lessees (or in default of agreement as determined by arbitration as hereinafter provided) to be the best rent at which the demised premises as a whole (there being disregarded only permitted underletting of part) might reasonably be expected to be let in the open market by a willing lessor and subject to similar covenants and conditions (other than the amount of the rent and the provisions of this present clause for reviewing the rent) to those contained in this underlease . . . L

There then follow certain matters which might affect the property by reason of the lessee's existing occupancy under the underlease and which are to be disregarded in reviewing the rent; provisions for the determination of fair rack-rental market value by arbitration, if the landlords and lessees should fail to agree; and a provision relating to the cost of such determination. The provisions of para 2 of the second schedule are unusual in that they do not define the term which is to be assumed as at each review date.

I was told by counsel that the first review, as at September 29 1982, was effected under the provisions of the second schedule without the point which is now in dispute having been raised. It is not suggested that any form of estoppel arises from the determination then reached. The second review date was on September 29 1989. No M review as at that date has yet taken place. The dispute which has arisen is as to the hypotheses which have to be made, first, in respect of the term of the letting which is to be assumed for the purposes of para 2 of the second schedule and, second, the provisions for rent review to be included in the hypothetical lease.

The landlord's contentions are that the term of the hypothetical lease, upon which the demised premises must be assumed to be let as at the review date, is the unexpired portion of the term of the underlease itself — that is a term ending on November 4 2067 — and that the rent is to be fixed on review on the basis of a letting for that term under a hypothetical lease which itself contains no provisions for rent review.

A The tenant contends that the term of the hypothetical lease is seven years. Alternatively, it contends that it must be assumed that any letting for a longer period is on terms which provide for rent reviews at a period of seven years or at some similar period.

The approach which the court should adopt to the construction of rent review provisions in a lease was explained by Sir Nicolas Browne-Wilkinson V-C in *British Gas Corporation* v *Universities Superannuation Scheme Ltd* [1986] 1 WLR 398*. At p 403 between A and C he said:

> In my judgment the correct approach is as follows: (a) words in a rent exclusion provision which require *all* provisions as to rent to be disregarded produce a result so manifestly contrary to commercial common sense that they cannot be given literal effect; (b) other clear words which require the rent
B review provision (as opposed to all provisions as to rent) to be disregarded (such as those in the *Pugh* case (1982) 264 EG 823) must be given effect to, however wayward the result; (c) subject to (b), in the absence of special circumstances it is proper to give effect to the underlying commercial purpose of a rent review clause and to construe the words so as to give effect to that purpose by requiring future rent reviews to be taken into account in fixing the open market rental under the hypothetical letting.

That approach has subsequently been approved by the Court of Appeal in *Equity & Law Life Assurance Society plc* v *Bodfield Ltd* [1987] 1 EGLR 124 at p 125: see also *Basingstoke and Deane Borough Council* v *Host Group Ltd* [1988] 1 WLR 348† at pp 353-354 and *Lynnthorpe Enterprises Ltd* v *Sidney Smith (Chelsea) Ltd* [1990] 2 EGLR 131 at p 132H.

C Nevertheless, although it is proper, so far as the language of the rent review clause permits, to have regard to the underlying commercial purpose of a rent review provision, it must be kept in mind that the primary task for the court is to ascertain the intention of the parties to the lease by reference to the particular language they have used having regard to any context provided by the document as a whole and any relevant circumstances known to both at the time when the lease was executed: see, for example, *Philpots (Woking) Ltd* v *Surrey Conveyancers Ltd* [1986] 1 EGLR 97 at p 98H.

The hypothesis which is to be made for the purpose of determining the fair rack-rental market value of the demised premises at each review date is that it is to be the best rent at which the demised
D premises as a whole might reasonably be expected to be let in the open market by a willing lessor. But, as Walton J pointed out in an observation in *National Westminster Bank plc* v *Arthur Young McClelland Moores & Co* [1985] 1 EGLR 61 at p 62, an observation which has not, so far as I am aware, been disapproved:

> . . . there is no such thing as a fair market rent of any premises in the abstract. There is only a fair market rent upon a set of abstractions which may be actual, may be hypothetical and in most cases under rent reviews are a mixture of the one and the other.

So, in the present case, the open market letting is to be:

> . . . subject to similar covenants and conditions (other than the amount of the rent and the provisions of this present clause for reviewing the rent) to those contained in the underlease . . .

E At the date of the first rent review the amount of the rent then reserved by the underlease will be £4,000. At the date of each subsequent review the amount of the rent will be the amount of the fair rack-rental market value determined on the occasion of the previous review. This is the effect of para 1 of the second schedule. The first hypothesis required by the words in parenthesis in para 2 is that rent is not the rent at which the premises are to be let. This is a necessary hypothesis if the rent review provisions are to have any effect at all. It requires that the words ". . . covenants and conditions . . ." which appear in para 2 immediately before the parenthesis are not to be given a meaning so narrow that they exclude the reddendum in the underlease.

F The second hypothesis required by the words in parenthesis is that the premises are not let on covenants and conditions ". . . similar to the provisions of this present clause for reviewing the rent . . .". It was common ground that in that context "this present clause" refers not only to para 2 of the second schedule but to the whole of that schedule — save perhaps for the words "Second schedule before referred to" and "rent review" at the head of the first page. This must be so because para 2 does not itself, on a true construction, contain provisions for reviewing the rent. It contains nothing more than a definition of the fair rack-rental market value of the demised premises.

It seems to me, therefore, that the parties to the underlease must be G taken to have intended that the letting which is to be assumed for the purposes of the rent review will be on terms which do not include the rent review provisions contained in the second schedule or any similar provisions. In this respect the language of para 2 of the second schedule to the underlease which I have to consider is indistinguishable from that in subclause 5(4) in the subunderlease which was before Goulding J in *Pugh* v *Smiths Industries Ltd* (1982) 264 EG 823 at p 824. That decision was treated by Sir Nicolas Browne-Wilkinson V-C in the *British Gas* case as correct. He said, at p 402H:

> In the [*Pugh* case] the words used in the rent exclusion provision were so clear that imputed intention could not override them . . .

H In the *Pugh* case the hypothetical letting to be assumed at the review date was a letting for the residue of the term granted. The term granted was for 23 years expiring on June 24 2000. At the relevant review date, June 24 1980, the term had 20 years to run.

It was argued by Mr Brock, who appeared for the tenant in the related case of *FA Welch Holdings Ltd* v *Smiths Industries Ltd* heard by Goulding J at the same time and who appears for the tenant before me, that nevertheless the valuer concerned to establish the rent under the hypothetical lease might assume some sort of rent review clause to be included. Goulding J dealt with that argument in these terms at p 827:

> Mr Brock had an alternative argument that, although the valuer was directed to disregard the rent review clause in the lease he might none the less in J looking at his hypothetical 20 years' lease assume that some sort of rent review clause of a kind usual at the time in the market would be included. That seems to me something that the language of the subclause will not bear. The valuer has to find that rent which would be obtainable upon a specified day upon a letting with vacant possession for the residue of the term, in the present case 20 years. That language is not apt to allow the assumption of some rent review clause. Again, one asks is it sufficiently certain? What sort of intervals would you take? But I do not rest it on that last point of uncertainty, I think the language simply does not bear the suggested meaning.

I respectfully adopt that view. It seems to me that, in the present case, the necessary conclusion is that the parties intended that the letting, which is to be assumed for the purposes of rent review under this underlease, is a letting on terms which do not themselves include K provisions for rent review. In this respect, it seems to me, that the language which I have to construe falls within category (b) of the formulation in the *British Gas* case.

Nevertheless, there is an important distinction between the *Pugh* case and this present case. In the *Pugh* case the term of the hypothetical letting was prescribed; in the present case it is not. The question, therefore, is whether some term — and, if so, what term — is to be implied.

Prima facie, in a case where the term of the hypothetical lease is not prescribed by express words, the term to be implied is the unexpired residue of the actual lease. Hoffmann J, in *MFI Properties Ltd* v *BICC Group Pension Trust Ltd* [1986] 1 EGLR 115 in a passage subsequently described as giving recognition to the presumption of L reality, put it in these words, at p 116 between G and H:

> The terms upon which such hypothetical letting must be assumed to take place will naturally depend in each case upon the particular language of the rent review clause. In general, the purpose of such clauses would point to a *prima facie* assumption that the parties intended the hypothetical letting to be for the residue of the actual lease held by the tenant on the same terms and in circumstances which actually exist at the relevant date. Thereby, the tenant would be required to pay on the assumption that he was being granted the interest which he actually held on the rent review date. But the language of the clause may show clearly that a departure from reality was intended.

To make that *prima facie* assumption in the present case, as the landlord suggests, would be to attribute to the parties an intention (which they have not expressed) that on the first rent review in 1982 it M is necessary to ascertain the rent under a hypothetical lease having a term of some 85 years without provision for rent review. A similar exercise would have to be undertaken at seven-year intervals. So that, for some considerable period under the rent review provisions in this underlease, rent would be determined by reference to a lease for a lengthy term with no provisions for rent review. Such a lease was described by Hoffmann J in *MFI Properties* as being so far outside the experience and expertise of an ordinary valuer as to be properly regarded as extinct: see [1986] 1 EGLR 115 at p 117B. He said:

> A valuer asked to determine the rent at which premises could be let in the open market for a lengthy term without a rent review therefore has no comparables in the real world to which he can refer. All that he can do is to make some more

*Editor's note: Also reported at [1986] 1 EGLR 120.
†Editor's note: Also reported at [1987] 2 EGLR 147.

A or less arbitrary adjustment to the rent which would be commanded on a letting of the same premises with the conventional provisions for rent review. It seems to me improbable that the parties intended the rent to be determined in this way.

For the same reason it seems to me improbable that the parties to the present underlease could have intended that the rent provisions should operate on the basis of a hypothesis — namely that of a long lease without rent reviews — which would be outside the experience and expertise of an ordinarily experienced and skilful arbitrator to be nominated by the president of the Royal Institution of Chartered Surveyors under para 3 of the second schedule. In the absence of clear words, I decline to imply a term into para 2 of this schedule which attributes such an intention to these parties.

B What assumption, therefore, should be made as to the term of the hypothetical lease? In my view, guidance is to be obtained from the requirement that the rent to be ascertained is the best rent at which the premises might reasonably be expected to be let in the open market by a willing lessor. It cannot be right to assume a term for which the premises might not reasonably be expected to be let on a lease which contains no provision for rent review. It appears to me that the term to be assumed for the hypothetical lease is the term which the landlord might reasonably be expected to grant and the tenant might reasonably be expected to take in respect of a letting of these premises at a fixed rack-rent, proper regard being had to the tenant's desire for reasonable security and to the expectation that such security would C be obtained under the provisions of Part II of the Landlord and Tenant Act 1954.

The term to be assumed in respect of this hypothetical lease at each review date is one which may vary from time to time during the residue of the term of the underlease as circumstances change between now and the year 2067. Further, the term at which the premises might reasonably be expected to be let under the conditions which I have indicated may well fall within a range of terms, any one of which would be reasonable. The term to be chosen within that range is the one which yields the best rent.

The term to be assumed in respect of the hypothetical lease at each review date is a matter which, as it appears to me, the parties have decided to leave to be agreed between themselves at the time when the D review falls to be made or, in the absence of agreement, to be determined by the arbitration provisions which they have included in the second schedule. I have considered whether a bargain of this nature would be void for uncertainty. I have reminded myself of the concern expressed by Goulding J in relation to a slightly different point in the passage that I have read from his judgment in the *Pugh* case.

When this concern was raised in the course of argument I was assured by counsel for the landlord — who is, perhaps, as experienced a landlord as one could find in this field — that there was unlikely to be any difficulty in practice. A valuer could reasonably be expected to operate within the guidelines that I have indicated. Be E that as it may, if forced to choose between a bargain which fails for uncertainty and one of such unreality that it imputes to the parties an intention that there should be what Hoffmann J described as a more or less arbitrary determination of rent on review, I should hold that the bargain was uncertain rather than unreal.

The first question raised by the originating summons seeks determination whether on the true construction of para 2 of the second schedule to the lease the arbitrator is required to assume:

(a) A hypothetical lease which does not contain any of the provisions of the second schedule.
(b) A hypothetical lease which does contain all the provisions of the second schedule.
(c) A hypothetical lease containing some other and, if so, what provisions.

F It will be clear from my judgment that I answer that question in terms of (a), that is to say a hypothetical lease which does not contain the provisions of the second schedule.

Para 2 of the originating summons — which was added by amendment for which I gave leave — raises a similar question in relation to the term of the hypothetical lease. It asks whether the arbitrator is to assume:

(a) A hypothetical lease for a term equal to the unexpired residue of the actual lease at the review date; or
(b) A hypothetical lease for a term of seven years; or
(c) A hypothetical lease for some other and, if so, what term.

I answer that question in the terms of (c). The term to be assumed is that term which the landlord might reasonably be expected to grant and the tenant might reasonably be expected to take in respect of a G letting of these premises at a fixed rack-rent. In so far as that term falls within a range, the term to be taken is that which yields the best rent.

Declaration accordingly; no order as to costs.

Chancery Division

March 13 1992

(Before Mr Justice HOFFMANN) H

LAURA INVESTMENT CO LTD v HAVERING LONDON BOROUGH COUNCIL

Estates Gazette June 20 1992

[1992] 24 EG 136

Landlord and tenant — Rent review — Whether rent at review to reflect value of buildings erected after grant of lease

By a lease dated June 14 1972 a predecessor in title to the plaintiff demised to the defendants 5 acres of undeveloped land at Langley Close, Havering, for a term of 62 years — The J defendants covenanted to divide the land into plots to be underlet for the construction of factories and warehouses — At review the rent was to be the higher of the current rent and "the market rental value" — In respect of the first 20-year rent review, the plaintiff contended that, as buildings had now been erected on the demised premises, the "market rental value" is of the premises as they now exist and not on the assumption that they remained undeveloped

***Held*: The *prima facie* rule is that of a hypothetical letting of the actual premises as they exist at the review date, although this may be displaced by the particular language of a lease — There was nothing in the lease to displace the *prima facie* rule K and accordingly at review the buildings on the land are to be included in the valuation**

The following cases are referred to in this report.

Basingstoke and Deane Borough Council v *Host Group Ltd* [1988] 1 WLR 348; [1988] 1 All ER 824; (1987) 56 P&CR 31; [1987] 2 EGLR 147; 284 EG 1587, CA
British Gas Corporation v *Universities Superannuation Scheme Ltd* [1986] 1 WLR 398; [1986] 1 All ER 978; (1986) 52 P&CR 111; [1986] 1 EGLR 120; 277 EG 980
Goh Eng Wah v *Yap Phooi Yin* [1988] 2 EGLR 148; [1988] 32 EG 55; [1985] Malayan Law Journal 329, PC
Ipswich Town Football Club Co Ltd v *Ipswich Borough Council* [1988] 2 L EGLR 146; [1988] 32 EG 49
Ponsford v *HMS Aerosols Ltd* [1979] AC 63; [1978] 3 WLR 241; [1978] 2 All ER 837; (1978) 38 P&CR 270; [1978] EGD 137; 247 EG 1171, HL
Ravenseft Properties Ltd v *Park* [1988] 2 EGLR 164; [1988] 50 EG 52

This was an originating summons issued by the plaintiff, Laura Investment Co Ltd, seeking a declaration as to the proper construction of a lease held by the defendants, Havering London Borough Council.

Michael King (instructed by Binks Stern) appeared for the plaintiff; and Nicholas Patten QC and Timothy Fancourt (instructed by the solicitor to Havering London Borough Council) represented the defendants M

Giving judgment, **HOFFMANN J** said: This summons raises a question of construction on a rent review clause. The lease is dated June 14 1972 and was granted by the plaintiff's predecessor in title to Havering London Borough Council. It demised about 5 acres of undeveloped land for a term of 62 years. There was a short rent-free period and thereafter a rent of £16,000 pa subject to review. The council covenanted in clause 3(b) to divide the land into plots which were to be underlet for the construction of factories and warehouses.

Clause 5 provides for reviews of the rent as from the end of the 20th and 40th years of the term. The reviewed rent is to be the higher of the current rent and "the market rental value" at the rent review date. Since the grant of the lease the council have underlet to various

A tenants, who have erected buildings on their plots. The question in this summons is whether for the purposes of the rent review the "the market rental value" is that of the demised premises as they now exist or on the assumption that they have remained undeveloped.

The plaintiff says that the normal assumption is that one values the premises as they actually exist at the date of the rent review. The tenant says that this construction would be very unreasonable because the buildings were erected at the expense of its undertenants and their rent review clauses expressly say that the buildings are to be disregarded.

This rent review clause speaks of "the market rental value" without saying of what. It does not even say "of the demised premises". If it did, that would not have helped a great deal because,
B although the lease defines the demised premises to include any buildings on the land during the term, it qualifies that definition by applying it only "wherever consistent with the context".

The other provisions in the lease, such as the covenants, are not, in my judgment, of much assistance. As one would expect, they sometimes distinguish between the buildings and the land. Clause 3(f)(ii) says that, if any buildings are burnt down, the tenant is not to have a reduction in rent. Mr King, for the landlord, rightly says that this would be an odd provision if it were contemplated that no rent would ever be attributable to buildings. I think it would be unrealistic to put much weight on this because it may be simply a loose end derived from a precedent book.

C The answer must therefore depend upon the *prima facie* assumption one makes about the intentions of the parties to a rent review clause. If one assumes that in the absence of contrary language the parties intended the premises to be valued as they stood at the relevant date, then the plaintiff's construction is correct. If one assumes that in the absence of contrary language the tenant should not have to pay rent for improvements, which he or his undertenants have made at their own expense, then the tenant's construction is correct.

The courts have, over the years, formulated various *prima facie* assumptions as guides to the construction of rent review clauses. Although, of course, the construction of a particular document depends upon its own language, the *prima facie* assumptions laid
D down in the authorities are very important. They bring some consistency to the interpretation of leases and they enable parties buying and selling leases and reversions to obtain more reliable advice on what their rights and obligations are likely to be.

There are several cases which have dealt with the correct approach to the question which arises in this case. In *Ponsford* v *HMS Aerosols Ltd* [1979] AC 63*, the review clause required the tenant to pay "a reasonable rent for the demised premises". During the lease the tenant made improvements at its own expense and the question was whether they should be disregarded. The case appears to have been argued on the assumption that, if the clause had said "a market rent", there would have been no doubt that the improvements should
E be included in the rental valuation. The tenant tried unsuccessfully to persuade the House of Lords that "a reasonable rent" meant a rent which it was reasonable that he, in particular, should pay and that it would be unreasonable to make him pay rent for his own improvements. The speeches of all three law lords who formed the majority contained passages which support a *prima facie* assumption that one values the premises as they actually are. Viscount Dilhorne said at p 76B:

> Our task can indeed be simply stated. It is just to decide the meaning of the words "assessed as a reasonable rent for the demised premises" . . . Lessors and lessees are usually advised by lawyers on the terms of leases. If the parties to this lease had agreed that the effect of improvements was to be disregarded in assessing the rent, that could easily have been stated and if that had been
> F agreed, I expect it would have been . . . In the absence of any such express provision as Parliament thought it necessary to include in section 34, I do not think that one is entitled to conclude that by the use of the words "assessed as a reasonable rent for the demised premises" the parties were seeking to express their agreement that in assessing the rent the effect of improvements made by the lessees was to be disregarded.

Then at p 83A Lord Fraser of Tullybelton said of the reference to the demised premises:

> The premises would have included the improvements without express provision to that effect, on the principle that anything made part of the premises by the tenants enures to the landlord . . .

At p 85C Lord Keith of Kinkel said:

> At first impression the words "reasonable rent for the demised premises" G suggest that what has to be ascertained is simply the rent that is reasonable for the premises as such in their actual state, the situation being viewed entirely objectively. "The demised premises" must mean the demised premises as improved, by virtue both of the ordinary law and of the passage I have quoted from the licence agreement. So upon this view any contribution the improvements might have made to rental value would have to enter into the assessment.

Now, does it make any difference that the clause in *Ponsford* specifically said "for the demised premises" whereas this clause says only "market rental value"? The landlord in *Ponsford* had urged that the reference to "the demised premises" helped it. Lord Wilberforce, who dissented, said it added nothing. "What else," he said, "could the rent be for?" Lord Fraser, too, said that the words H did not add anything new, although they reinforced the view that the valuation was to be in accordance with objective rather than subjective criteria. That is not an issue here. It seems to me that it would detract from the value of a *prima facie* rule of construction if it were to be frittered away on such small distinctions.

A similar question came before the Privy Council in *Goh Eng Wah* v *Yap Phooi Yin* [1988] 2 EGLR 148. That was a demise of land with a cinema, which the tenants had built at their own expense. Lord Templeman said at p 149A:

> By well-established principles, that rent would issue out of the land including any buildings from time to time on the land and would *prima facie* be referable to the value of the land plus buildings.

J He then considered whether there was anything in the language of the lease which displaced that construction, but concluded that the parties intended that the rent to be fixed by the arbitrator should include them. He said at p 149H:

> . . . if the parties intended that the rent fixed by an arbitrator should ignore the buildings on the land, they should and would have given express instructions to the arbitrator for that purpose. In the absence of any such express instructions in the lease, then whatever the parties may in their heart of hearts have intended, the lease on its true construction does not authorise any deviation from the usual rule and it follows that the rent must be fixed by reference to the land and the buildings thereon.

In *Ipswich Town Football Club Co Ltd* v *Ipswich Borough Council* [1988] 2 EGLR 146 the Vice-Chancellor decided that the K terms of the rent review clause in that lease required the buildings to be disregarded. The lease to the football club distinguished in many places between what it called "the sports ground" and the buildings and erections upon it. The rent review clause required a determination of the rental value of the sports ground and the Vice-Chancellor concluded that this meant only the unimproved land. This was a case in which the *prima facie* rule of construction was rebutted by the particular language of the instrument.

In *Ravenseft Properties Ltd* v *Park* [1988] 2 EGLR 164, Judge Baker QC restated the *prima facie* rule. The clause in that case, as in this one, did not expressly say what the arbitrator had to value. It merely said he should decide "what increased rent, if any, should be payable (having regard to the level of rents then prevailing in the open L market)". Judge Baker said that *prima facie* this required him to value the premises as they stood at the rent review date. He referred to a passage from the speech of Viscount Dilhorne in the *Ponsford* case and said [at p 166D]:

> So when the arbitrator is required to assess what increased rent should be payable the implication there, in my judgment, is the rent of the demised premises, and it would require some express exclusion such as mentioned by Viscount Dilhorne in that passage for another view to be taken before one could give effect to the circumstance that it was the tenant who had paid for the relevant improvement to the premises rather than the landlord.

Up to this point all the cases are consistent and point to a *prima facie* rule which can, as in the *Ipswich* case, be displaced by the M particular language of the instrument, but which ordinarily requires one to make an assumption of reality. The hypothetical letting is of the premises as they actually stand on the rent review date and on the terms of the actual lease for the unexpired period of the term.

In respect of the terms of the lease, this assumption of reality has been firmly established by the cases of *British Gas Corporation* v *Universities Superannuation Scheme Ltd* [1986] 1 WLR 398* and *Basingstoke and Deane Borough Council* v *Host Group Ltd* [1988] 1 WLR 348.†

The only case which appears to point in a contrary direction is the

---

*Editor's note: Also reported at (1978) 247 EG 1171.

*Editor's note: Also reported at [1986] 1 EGLR 120.
†Editor's note: Also reported at [1987] 2 EGLR 147.

A decision of the Vice-Chancellor at first instance in *Sheerness Steel Co plc* v *Medway Ports Authority*. The case afterwards went to the Court of Appeal. The judgment in the Court of Appeal was given on November 19 1991 and I have been provided with a transcript*. In that case the rent review clause expressly provided for the disregard of certain improvements, that is buildings standing on the premises. It said nothing about any other kind of improvements. The question was whether the valuation should take into account extensive earth-moving operations by which the tenant had improved the property at his own expense. The Court of Appeal decided that the express exclusion of buildings and certain other provisions in the lease made it clear that only the buildings were to be disregarded.

B The Vice-Chancellor, who at first instance had come to a different conclusion, did so on the basis of what appears to be a new *prima facie* rule of construction. He said:

> . . . one does not expect under a rent review provision that the tenant will be called upon to pay an additional rent for improvements made at his own expense . . . If there are clear words showing an intention which conflicts with one's normal expectation, those words must be given effect according to their tenor. But if the words are not clear and if there is a doubt as to the actual intention as expressed in the words used, it is legitimate to construe the words containing such doubt or ambiguity so as to give effect to the assumed intention of the parties which I have mentioned.

In the Court of Appeal, Nourse LJ, who gave the only full judgment, cited that passage without comment and after drawing D attention to the provisions in the lease, which appeared to point in the opposite direction, said:

> To express it in terms of the approach stated by the Vice-Chancellor, I would say that here there are clear words showing an intention which overrides the normal commercial expectation that a tenant will not be called upon to pay an additional rent for improvements made at his own expense.

I do not regard that passage as an endorsement of the new *prima facie* rule which seems to me to run contrary to all the earlier authority which I have cited and to none of which the Court of Appeal referred. It may well be that the *prima facie* rule propounded by the Vice-Chancellor would be fairer than the *Ponsford* rule, which does not regard the fact that the tenant paid for the improvements as E sufficient to displace the normal assumption. It requires express provision (as in section 34 of the Landlord and Tenant Act 1954) or necessary implication (as in the *Ipswich* case). The *Ponsford* case is nevertheless the rule upon which landlords and tenants have been acting for many years. It has always been open to the parties to make the necessary express contrary provision and I think that it would defeat legitimate expectations in the market if the *Ponsford* case were now to be overthrown.

For those reasons I construe the rent review clause as requiring the buildings on the land to be included as in the valuation and I will make the declaration for the plaintiff.

Declaration as sought. F

C *Editor's note: Reported at [1992] 12 EG 138.

For further cases on this subject see p 283

LICENCE

A
Court of Appeal

April 17 1991

(Before Lord Justice BALCOMBE, Lord Justice WOOLF and Lord Justice STAUGHTON)

NORRIS v CHECKSFIELD

Estates Gazette January 11 1992

[1992] 01 EG 97

Licence — Service occupancy — Whether employee occupied
B **accommodation as a licensee — Whether notice to determine licence to occupy required — Whether licence a "periodic licence" within the meaning of section 5(1A) of the Protection from Eviction Act 1977 — Appeal from possession order granted by county court judge dismissed**

The defendant, who had previously worked for the plaintiff as a semi-skilled mechanic, was allowed by the plaintiff to occupy a bungalow on condition that he would be able to drive coaches for the latter's business and that he would apply for a public service vehicle licence for that purpose — Occupation of the bungalow would in the plaintiff's opinion make the defendant readily available for urgent transport work — The
C **defendant signed a document stating that occupation of the premises was a condition of his employment and that on termination of the employment the occupation would cease — £5 per week was to be deducted from his salary for his occupation — The defendant did not tell the plaintiff that he was disqualified from driving — When the plaintiff discovered this fact he dismissed the defendant summarily and took proceedings for possession — The county court judge granted the plaintiff a possession order and the defendant appealed**

***Held*: 1. It was now accepted law that an employee can be a**
D **service licensee although his occupation is not necessary for the purpose of his employment if he is genuinely required to occupy the premises for the better performance of his duties — Lord Reid, in *Glasgow Corporation* v *Johnstone*, went further and said that it was sufficient if the occupation was of "material assistance" to the employer — The judge was entitled to find that the defendant was a licensee — When the defendant went into occupation he was not in possession of a public service vehicle licence and there was no connection between his occupation of the bungalow and his work as a semi-skilled mechanic — Nevertheless, he had only been**
E **allowed into occupation on the basis that he would obtain the necessary qualification and work as a coach driver — There was sufficient factual nexus between the commencement of the occupation of the premises and the employment which would benefit from the occupation to satisfy the conditions of a service licence — It was proper to regard the defendant as going into occupation as a licensee in order the better to perform his duties when he became a coach driver**

2. Section 5(1A) of the Protection from Eviction Act 1977 (as amended by section 32 of the Housing Act 1988) requires a four weeks' notice in writing (containing prescribed information) to determine a periodic licence to occupy F
premises as a dwelling — The notice required by section 5(1A) did not apply to the present case — First, no notice was required to terminate the licence, as the employment could be ended summarily or on a week's notice and when the employment came to an end so did the licence — Second, the licence here was not in any case a "periodic licence"; it was a licence for a single period, the period of the employment — The subsection seemed to be aimed at licences which continued for a series of periods, analogous to weekly, monthly or quarterly tenancies — No notice was required to terminate the defendant's licence G

The following cases are referred to in this report.

A G Securities v *Vaughan; Antoniades* v *Villiers* [1990] 1 AC 417; [1988] 3 WLR 1205; [1988] 3 All ER 1058; (1989) 57 P&CR 17; [1988] 2 EGLR 78; [1988] 47 EG 193, HL

Australian Blue Metal Ltd v *Hughes* [1963] AC 74; [1962] 3 WLR 802; [1962] 3 All ER 335, PC

Glasgow Corporation v *Johnstone* [1965] AC 609; [1965] 2 WLR 657; [1965] 1 All ER 730; (1965) 63 LGR 171; [1965] RVR 111

Ivory v *Palmer* [1975] ICR 340, CA

Lace v *Chantler* [1944] KB 368

Smith v *Seghill Overseers* (1875) LR 10 QB 422

Street v *Mountford* [1985] AC 809; [1985] 2 WLR 877; [1985] 2 All ER 289; [1985] 1 EGLR 128; (1985) 274 EG 821, HL H

This was an appeal by Stephen Checksfield from the judgment of Judge Hammerton, at Hastings County Court, ordering him to give up possession to his former employer, Jesse William Norris (trading as J Davis & Son), of 2 Highfield Bungalow, Hawkhurst, Kent.

Martin Seaward (instructed by Holden & Co, of Hastings) appeared on behalf of the appellant; Martyn Zeidman (instructed by Menneer, Idle & Brackett, of St Leonards-on-Sea) represented the respondent.

Giving the first judgment at the invitation of Balcombe LJ, **WOOLF LJ** said: This is an appeal from a decision of His Honour Judge Hammerton at the Hastings County Court given on April 24 1990 J
when he granted the plaintiff an order for possession of 2 Highfield Bungalow, Highgate, Hawkhurst, Kent.

The appeal raises two issues as to the occupation of premises belonging to his employer by an employee. The first issue is whether an employee who has exclusive occupation of the premises and pays rent can be a licensee if his existing work does not, although his future anticipated work would, benefit as a result of his being in occupation. The second issue is as to the application of the requirements as to the giving of notice in section 5(1A) of the Protection from Eviction Act 1977 as amended by the Housing Act 1988 to a licence which is expressed to come to an end with the employee's employment.

The facts K

For the purposes of both issues I can summarise the facts quite briefly, basing myself on the note of the judgment of Judge Hammerton. In 1988 the defendant worked for the plaintiff as a semi-skilled mechanic. In 1989 he was invited to resume that employment and started in July 1989. At that time he was living in lodgings approximately one mile away. In August 1989 the defendant was seen by the plaintiff and asked if he would like to reside at the bungalow which was the subject of the proceedings, which had previously been occupied by another employee. The defendant wanted to occupy the premises and he was allowed into possession on two specific terms which the plaintiff made clear. The first was not of any relevance. The second was that the defendant would be able to drive coaches for the plaintiff's business and would apply for a PSV licence for this purpose. In the plaintiff's opinion, for this work it was clearly desirable that the defendant should be in the bungalow, since this would make him readily available in an emergency or if there was urgent work. The judge found that it was on "that condition" that occupancy of the bungalow was granted to the defendant.

Before the defendant moved into the premises he was asked to, and did, sign a document confirming the terms of his occupation. That document referred to the defendant's having a licence and so far as relevant provided:

> It is a condition of your employment that you shall occupy [the premises] or such other alternative property the employer may provide and that on termination of your employment your licence to occupy such property shall cease forthwith.

In addition it was arranged that the defendant would have the sum of £5 per week deducted from his salary in relation to his occupation.

At a later date, October 13 1989, the defendant was given a statement of the main particulars of his employment. These showed that he was to be employed as a semi-skilled mechanic, his normal hours of work were eight to five, Monday to Friday, Saturday if agreed. He was to give and receive one week's notice to terminate the employment but could be dismissed instantly for misconduct.

Possibly because he was unaware of this until October 1989, as he contended, the defendant did not tell the plaintiff that he was disqualified from driving. About the same time the plaintiff learned from the police that this was the situation and this, together with his no longer being a satisfactory employee, caused the plaintiff to dismiss the defendant summarily. About 10 days later, by letter delivered on December 11 1989, the plaintiff informed the defendant that he was required to vacate the bungalow. The proceedings for possession were instituted on December 18 1989.

First issue

In relation to the thorny issue as to when an employee is a licensee and not a tenant of premises belonging to his employer which he is allowed to occupy, Mr Seaward, who appears on behalf of the defendant, is prepared to accept Mr Zeidman's submission on behalf of the plaintiff. Mr Zeidman submitted that an employee can be a licensee, although his occupation of the premises is not *necessary* for the purposes of the employment, and if he is genuinely *required* to occupy the premises for the *better performance* of his duties.

In my judgment, this submission accurately reflects the law. We have been referred to a number of authorities which set out different tests. The most helpful decision is that of the House of Lords in *Glasgow Corporation* v *Johnstone* [1965] AC 609. In that case Lord Reid (with whose speech Lord Wilberforce agreed) at p 618A said:

> So, if necessity were the criterion, the appeal would succeed. But if it is sufficient for the respondents to show that their servant is bound to reside there, and that his residing there is of material assistance to them in the carrying out of their activities, then the appellants must fail on this point.

Lord Reid then went on to examine a number of English authorities and concluded this part of his speech by saying at p 619:

> In requiring that the occupation should be necessary I think that Mellor J's judgment is out of line with the other authorities, and the authorities on this topic appear to me to support the respondent's contention in the present case.

Lord Reid was therefore of the opinion that it would be sufficient if the employee's occupation was of "material assistance" to his employment. It need not be "*necessary*" for his employment. The same view was taken by Lord Evershed and Lord Hodson. Lord Guest stated the position which must exist for there to be a licence in the following terms, at p 629B:

> The residence must be ancillary to the duties which the servant has to perform (*Smith* v *Seghill Overseers*) or, put in another way, the requirement must be with a view to the more efficient performance of the servant's duties (*Fox* v *Dalby*).

As Mr Seaward correctly submitted, it would not suffice if the occupation was a "fringe benefit" or merely an inducement to encourage the employee to work better.

Unless the occupation fulfilled this test, the fact that the employee had exclusive possession and paid rent would almost inevitably establish a service tenancy: See generally *Street* v *Mountford* [1985] AC 809* and *AG Securities* v *Vaughan* [1990] 1 AC 417† at p 459 *per* Lord Templeman.

*Editor's note: Also reported at [1985] 1 EGLR 128.

†Editor's note: Also reported at [1988] 2 EGLR 78.

If in this case, as was contemplated, when the defendant went into occupation he had obtained a PSV licence and had changed the nature of his job so that he became a coach driver, the judge would undoubtedly have been entitled to regard the defendant as a licensee. He would then have entered into occupation under a document which described the relationship in terms of a licence and the occupation would be beneficial to the defendant's employment on the judge's findings. His occupation would enable him to assist his employer in cases of emergency or on short notice.

However, Mr Seaward submits that the employment situation which has to be considered is that which existed in fact at the time the licence was entered into, not the situation which might exist in the future. The situation which existed at the time the licence was entered into was that the defendant's occupation of the premises was irrelevant to the defendant's employment as a semi-skilled mechanic. In that employment he was not required to assist with emergencies and he could perform the work equally as well from the lodgings at which he was previously living or indeed from any other address which was within travelling distance to his place of work. Occupation was beneficial to the defendant but not beneficial to his employment.

There is no previous decision of the courts which directly conflicts with Mr Seaward's approach. However, I have no hesitation in coming to the conclusion that, notwithstanding this argument, the judge was entitled to come to the conclusion that the defendant was a licensee. Although the defendant was unable to obtain the necessary PSV licence to drive coaches, he was, on the judge's finding, allowed into occupation only on the basis that he would obtain the necessary qualifications and work as a coach driver. In my judgment, it would not be sensible, unless compelled to do so, to restrict an employer's ability to grant a licence to situations where the employment which would be benefited by the employee taking up occupation commenced simultaneously with or prior to the occupation of the premises. There may be many circumstances where it would be desirable for the employee to take up occupation before the relevant work commenced. What is required is that there should be a sufficient factual nexus between the commencement of the occupation of the premises and the employment which would benefit from that occupation. If for some reason it becomes apparent that the employee is not going to be able to fulfil the requirements of that employment within a reasonable time, then the position may be different. However, if the situation is one where it is contemplated, as was the position here, that the employee would, within a reasonable time, be able to take up the relevant employment, that will suffice. The fact that the employee during the interval may be performing some other duties which are not affected by the occupation of the premises does not prevent a licence coming into existence.

On the facts found by the judge in this case, this was, therefore, a case where it was proper to regard the defendant as going into occupation as a licensee in order better to perform his duties when he became a coach driver.

Second issue

Section 5 of the Protection from Eviction Act 1977, as amended, provides:

> **5.**—(1) Subject to subsection (1B) below, no notice by a landlord or a tenant to quit any premises let (whether before or after the commencement of this Act) as a dwelling shall be valid unless —
>
> (*a*) it is in writing and contains such information as may be prescribed, and
>
> (*b*) it is given not less than 4 weeks before the date on which it is to take effect.
>
> (1A) Subject to subsection (1B) below, no notice by a licensor or a licensee to determine a periodic licence to occupy premises as a dwelling (whether the licence was granted before or after the passing of this Act) shall be valid unless —
>
> (*a*) it is in writing and contains such information as may be prescribed, and
>
> (*b*) it is given not less than 4 weeks before the date on which it is to take effect.
>
> (1B) Nothing in subsection (1) or subsection (1A) above applies to —
>
> (*a*) premises let on an excluded tenancy which is entered into on or after the date on which the Housing Act 1988 came into force unless it is entered into pursuant to a contract made before that date, or
>
> (*b*) premises occupied under an excluded licence.
>
> (2) In this section "prescribed" means prescribed by regulations made by the Secretary of State by statutory instrument, and a statutory instrument containing any such regulations shall be subject to annulment in pursuance of a resolution of either House of Parliament.
>
> (3) Regulations under this section may make different provision in relation to different descriptions of lettings and different circumstances.

No notice was given to the defendant which complies with subsection (1A). However, that subsection is contravened only where a notice is given "to determine a periodic licence" which does not comply with the subsection. If no notice is required, the subsection has no application. It is contended on behalf of the plaintiff that the defendant's licence was not of a periodic licence and that it did not require a notice of termination but automatically came to an end as a result of the cessation of the defendant's employment. Although the document which the defendant signed before entering into possession was not a complete record of the terms on which he was allowed into occupation (for example, it did not mention the rent), it clearly states that the licence was to cease forthwith on termination of the defendant's employment. The draftsman's reference to a "periodic licence" in subsection (1A) creates so far as I am aware a new animal. "Periodic tenancies" referred to in subsection (1) are well known but I am not aware of any previous reference to "a periodic licence to occupy premises". Clearly the draftsman in subsection (1A) was creating a parallel situation to that in subsection (1) but wished to restrict it not only to licences which were required to be determined by notice but also to those licences which could be properly described as periodic.

There is no definition of a periodic licence in either the 1977 Act or the amending Housing Act of 1988. However, in section 45(1) of the Housing Act 1988 a "fixed term tenancy" is defined as meaning any tenancy other than a periodic tenancy. Mr Seaward draws attention to the contrast between a fixed term tenancy and a periodic tenancy which this interpretation provision identifies. He also referred the court to the case of *Lace* v *Chantler* [1944] KB 368 where this court held that a tenancy for the duration of the war does not create a leasehold interest because it is impossible to say at the outset how long the tenancy would last and therefore it was not of a sufficiently certain duration. Although Staughton LJ, with his intimate knowledge of the arcane areas of landlord and tenant law, was able to point out that *Lace* v *Chantler* has been overtaken by subsequent statutory intervention (Validation of War Time Leases Act 1944), Mr Seaward submits that the decision in that case should still be applied in this case because the date on which the defendant's employment would cease was equally incapable of identification at the commencment of the occupation in this case as it was in *Lace* v *Chantler*.

He submits that it is desirable in the case of a service occupancy that the statutory restriction on the period of notice required to determine the licence should apply, so as to protect the position of a former employee. He also submits that for the statutory requirement to apply in this sort of case would be beneficial because the prescribed particulars which are required to be given by section 5(2) of the 1977 Act would provide a valuable safeguard for the employee. (The prescribed information which has to be given includes a reference to the desirability of obtaining advice from a solicitor and the ability to obtain legal advice and assistance under the Legal Aid Scheme.) While I recognise the benefits which could follow for former employees if Mr Seaward is correct, I am unable to accept his argument and see real disadvantage in practice if he were correct.

Before the Act of 1977 was amended, a licence which was expressed to be terminable on an employee's employment coming to an end would come to an end without the requirement of any notice. That this is the position is supported by the decision of this court in *Ivory* v *Palmer* [1975] ICR 340. In that case it was held that a contractual licence which was ancillary to a contract of employment came to an end when the contract of employment was terminated even though the termination was wrongful. It was argued that that case can be distinguished from the situation here because no rent was payable for the occupation and the contractual licence was interwoven with the contract of employment. I cannot see why the fact that there is only one contract instead of two should make any difference to the situation as long as the contract dealing with the right to occupation makes it clear that it is to come to an end with the employment. The fact that rent is payable with reference to a period of time such as a week would be of significance only if there were no other express or implied event which terminated the licence. In that situation the period in relation to which the rent is payable may provide guidance as to the length of notice which would be reasonable before the licence could be terminated. If no notice is required, then the fact that the rent was payable periodically is of no relevance to when the licence terminates.

The question whether a requirement of reasonable notice is to be implied in a licence has to be answered in the light of the circumstances existing when the licence is created: see *Australian Blue Metal Ltd* v *Hughes* [1963] AC 74, at p 99, *per* Lord Devlin. In this case, therefore, in my judgment, no notice was required to terminate the licence. The employment could be ended summarily if appropriate or on a week's notice and when the employment came to an end so did the licence. That being the position, section 5(1A) of the 1977 Act can have no application. There is no notice which is needed which is relevant for the purposes of that subsection. Further, even if that approach is not correct, the question remains as to whether or not this was a "periodic licence". In my judgment it was not. This was a licence for a single period, namely, the period of employment. The subsection in referring to a periodic licence confines its operation to those licences which continue for a series of periods until terminated by a notice. The analogy is with weekly, monthly or quarterly tenancies. Having regard to the document which the defendant signed before he went into occupation, the fact that a weekly deduction was made from his wages by way of rent is not sufficient to turn this licence into a periodic licence.

The approach adopted above means that section 5 of the 1977 Act, as amended, will not apply to the great majority of, but not necessarily all, employment licences. Those licences will usually end with the employment. This appears to me to make good sense. If the premises are required to be occupied by an employee for the better performance of his employment, it is sensible that the premises should be required to be vacated as soon as the employment comes to an end so that they can be occupied by another employee. It is not difficult to envisage circumstances where great inconvenience could be caused even by a delay of four weeks after the end of a period of employment. A degree of support for this approach is provided by section 8(2) of the 1977 Act. That subsection provides:

> For the purposes of Part I of this Act a person who, under the terms of his employment, had exclusive possession of any premises other than as a tenant shall be deemed to have been a tenant and the expressions "let" and "tenancy" shall be construed accordingly.

This subsection does not apply to section 5 because that section is in Part II and not in Part I of the 1977 Act. As section 8 has been amended by the 1988 Act, if the arguments advanced on behalf of the defendant are correct and Parliament intended section 5, as amended, to apply to the situation here under consideration, then that object could have been achieved by amending subsection (2) so that it applied to Part II of the Act as well. Section 5 is the only section in Part II of the 1977 Act. In fact, Parliament did not amend subsection (2).

No notice being required to terminate the defendant's licence, I would dismiss this appeal.

Agreeing, **Staughton LJ** said: I was at one time troubled by a single sentence in the judgment of Lush J in the case of *Smith* v *Seghill Overseers* (1875) LR 10 QB 422 at p 431. In that case Smith was a mineworker and lived in a house which belonged to his employers. His occupation was to cease at the time when his employment ceased. Lush J said: "It is true that the holding is not for any fixed term". If one reads that with section 45(1) of the Housing Act 1988, one finds there that "fixed term tenancy" means any tenancy other than a periodic tenancy. It would follow that if the definition given by Lush J also applies, then a tenancy which is to expire when the employment expires is a periodic tenancy. However, it seems to me that in the Housing Act 1988 Parliament provided a different rule to that which had found favour with Lush J in 1875. Parliament decided that any tenancy which was not a periodic tenancy was a fixed term tenancy, although Lush J in 1875 thought that a tenancy for the duration of employment was not of that description. That has allayed my doubts on that aspect of the case and I agree that the appeal must be dismissed.

Balcombe LJ also agreed and did not add anything.

The appeal was dismissed with costs, not to be enforced without the leave of the court; application for costs to be paid by Legal Aid Fund adjourned to registrar; application for leave to appeal to the House of Lords was refused.

A Queen's Bench Division

December 21 1989

(Before Judge PETER CRAWFORD QC, sitting as a deputy High Court judge)

CORFIELD v D S BOSHER & CO

Estates Gazette February 1 1992

[1992] 04 EG 127

B Negligence — Conduct of rent review — Claim by plaintiff landlord against defendant solicitors — Allegation that defendants failed to advise time-limit for appealing arbitrator's award — Plaintiff's claim succeeded

Lease of commercial premises — Rent review on June 24 1985 — Determination of rent referred to arbitrator — Defendants were solicitors to the plaintiff landlord — Arbitrator held hearing and parties represented by counsel — Arbitrator wrote to the defendants on January 21 1986 informing them that his reasoned award could be taken up on payment of his fees — Defendants sent a copy of letter to the plaintiff and asked the plaintiff to send a cheque payable to C the arbitrator in respect of his fees — In fact the plaintiff sent cheque direct to the arbitrator — Arbitrator sent the award by a letter dated February 4 1986 direct to the plaintiff — Plaintiff considered award with his valuer — Defendants sent reminder to plaintiff on January 29 and February 10 not having realised that plaintiff had obtained award direct — Plaintiff wrote to defendants on February 11 1986 informing them that award had been taken up — Defendants receiving copy of award on February 11 1986

The award was unsatisfactory to plaintiff — Plaintiff's claim had been for annual rent of £31,500 and tenant's D submission had been for rent of £14,000 — Arbitrator determining rent at £19,547 — The plaintiff desired to take up his rights of appeal — Possible rights under section 23 of the Arbitration Act 1950 and section 1 of the Arbitration Act 1979 — Time-limits for both appeals being 21 days from publication of award — Award is made and published when the arbitrator has given notice to the parties that the award is ready — Date of publication January 22 1986 and 21 days expired on February 12 1986 — Defendants wrote to plaintiff on February 25 1986 stating that time for lodging of any appeal was running out — Legal executive of defendants E believing that time-limit for appeal was six weeks — Plaintiff's valuer criticised arbitrator in selection of comparables and failure of arbitrator to accept contention that premises should be valued on basis that a single tenant would occupy the whole; alternatively that the premises could be divided — The arbitrator had allowed 10% to reflect element of risk in respect of valuation based on division — Plaintiff's valuer asserting that allowance of 10% not reflected in evidence or arguments before arbitrator — Counsel advised that arguable point in relation to arbitrator's findings based on comparables and in relation to 10% deduction for risk element — Motion under Arbitration Act F 1950 and summons under Arbitration Act 1979 dismissed by Saville J on June 20 1986 as being out of time — Plaintiff claimed damages for negligence against the defendants' solicitors alleging failure to advise in connection with time-limits for appealing arbitrator's award

***Held*: By the second day of the arbitration hearing the defendants' legal executive was put on notice that an appeal was a distinct possibility if the reasoned award disclosed any grounds — Because an appeal was a distinct possibility the defendants, by their legal executive, were under a duty either to be aware of the appropriate time-limits which were G available if appellate rights were to be exercised or to take steps to inform themselves of them — Further the defendants were under a duty to inform the plaintiff if the time-limits were short or urgent action was required — The defendants, by their legal executive, were in breach of their duty to advise the plaintiff as to his rights of appeal and the time-limits for implementing any appeal, and failed to do so until the time-limit had expired — The defendants' submission that, irrespective of any breach on the part of the legal executive, no such breach was causative of the events which subsequently flowed was rejected — Although the plaintiff and his valuer were under no sense of urgency, the sense of urgency should H have been conveyed to them by the defendants — The failure to implement the appellate process in time was a direct consequence of the defendants' breach of duty**

In assessing damages, the court was called upon to make a two-stage valuation — The court had to determine not only what the prospects of success would have been in the appeal if it had been brought in time but what the results of further litigation would have been if the appeal had been successful — In relation to the first objection to the arbitrator's award, the plaintiff would have had to have persuaded the judge hearing an appeal whether the arbitrator's reasoning process involved J assessing evidence or creating evidence in relation to the effect of the locality — In respect of this arguable point the plaintiff's prospect of success would have been a one-third prospect — In relation to the 10% allowance for risk associated with a hypothetical conversion of the premises, the plaintiff had no prospect of success in showing misconduct for the purposes of section 23 of the 1950 Act — A court would not have been satisfied that there was a more or less strong *prima facie* case for the purposes of obtaining leave to appeal under section 1 of the 1979 Act — If the plaintiff had succeeded before the commercial court, that court would have set the K award aside or have remitted it to a fresh arbitrator

Before a fresh arbitrator the plaintiff would have had the opportunity of resubmitting and rearguing his contentions that the premises would have been let to a single tenant and that the zone A rents were a good deal closer to the plaintiff's valuer's estimate than to the tenant's valuer's estimate — The plaintiff would have had some success in persuading the arbitrator to increase the rent and, allowing for uncertainties, a figure of £2,000 per year over the seven years of the review was possible — In relation to the zone A rents there was a

possibility that further evidence might have assisted the plaintiff, and the plaintiff's chance was about £2 per sq ft amounting to an extra £1,340 pa — During the seven years of the review the plaintiff lost the chance of increasing the rent by £23,380 — The value of this chance was one-third, namely £7,790 — Allowing for the costs of a further arbitration the plaintiff was entitled to damages in the sum of £2,075

The following cases are referred to in this report.

Antaios Compania Naviera SA v *Salen Rederierna AB* [1985] AC 191; [1984] 3 WLR 592; [1984] 3 All ER 229; [1984] 2 Lloyd's Rep 235, HL
"Evimeria", The [1982] 1 Lloyd's Rep 55
Ipswich Borough Council v *Fisons plc* [1990] 2 WLR 108; [1990] 1 All ER 730; [1990] 1 EGLR 17; [1990] 04 EG 127, CA
Pioneer Shipping v *BTP Tioxide Ltd ("The Nema")* [1982] AC 724; [1981] 3 WLR 292; [1981] 2 All ER 1030; [1981] 2 Lloyd's Rep 239, HL
Top Shop Estates Ltd v *Danino* [1985] 1 EGLR 9; (1984) 273 EG 197
Zermalt Holdings SA v *Nu-Life Upholstery Repairs Ltd* [1985] 2 EGLR 14; (1985) 275 EG 1134

This was a claim by the plaintiff landlord, Mervyn Corfield, against the defendant solicitors, D S Bosher & Co, for damages for breach of contract and for breach of a duty of care owed in tort.

Mervyn Roberts (instructed by Stafford Young Jones) appeared for the plaintiff; Jonathan Brock (instructed by Blake Lapthorn, of Portsmouth) represented the defendants.

Giving judgment, **JUDGE CRAWFORD QC** said: In this action Mr Mervyn Ralph Corfield claims damages for negligence against D S Bosher & Co, a firm of solicitors who acted for him in connection with arbitration proceedings concerning rent review of premises which were owned by Mr Corfield and let to Savory & Moore Ltd at Hove, Sussex.

The evidence occupied two days and the submissions occupied the whole of the third day. In view of the fact that this is the last day of term, and the last day in which I shall be sitting in this jurisdiction for some time, I thought that it would be in the interests of the parties for me to give my judgment this morning, albeit in an extempore form, rather than to reserve it for the ampler consideration which some of the arguments perhaps merited; and I should make it plain that any infelicities in this judgment arise, in part at least, from that cause.

The premises in question are a commercial corner building at 105 Church Road, Hove, which, on April 30 1971, was let by the plaintiff to the predecessors in title of Savory & Moore, the tenants at the time of the relevant arbitration. The lease was in what I may describe as a common commercial form, providing for rent reviews at seven-year intervals. The relevant rent review was that for June 24 1985.

The plaintiff retained a Mr R Stapleton [FRICS], a valuer and local estate agent of considerable experience, as his professional adviser in that field. Negotiations between Mr Stapleton and the tenants, who were represented by another local valuer, Mr J Wareham [FRICS], were short and abortive and arbitration proceedings were begun. For the purpose of those proceedings the plaintiff instructed these defendants to act for him throughout the whole of the arbitration proceedings and in proceedings subsequent thereto in relation to a possible appeal. The defendants acted by their legal executive, Mr Cooper, of whom I need say no more at this stage other than he was a very experienced legal executive.

The arbitrator who was duly appointed, pursuant to the terms of the lease, was a Mr Michael G Clark, a Fellow of the Royal Institution of Chartered Surveyors, who practised not far from, but not in, Hove; and there was a preliminary hearing for directions before the arbitrator in November 1985. The arbitration was conducted by oral proceedings on December 3, adjourned to January 6 1986. Both parties were represented by counsel. Both Mr Stapleton and Mr Wareham attended and gave evidence; and both of those witnesses had submitted proofs in writing which had been exchanged prior to the oral hearings. In addition, the tenant's response to the arbitration also involved, at short notice and without having previously disclosed his proof, another expert witness, a building surveyor, which, to some extent, took the plaintiff and his advisers by surprise on the second day of the arbitration.

It evidently became apparent to the plaintiff and his advisers on the second day of the oral hearing, if not before, that the arbitration was not going well. The precise reasons for this none of the witnesses who gave evidence on this point was able to articulate with precision. This does not surprise me. It may have been comments from the arbitrator; it may just have been a feel for the way things were going. There was some discussion between the plaintiff and his advisers at the midday break on the second day of the hearing, as a result of which counsel, Mr Anthony Speight, who was appearing for the plaintiff, was instructed to, and did at the conclusion of the submissions, ask the arbitrator to make his findings in the form of a reasoned award.

The arbitrator did prepare his reasoned award with commendable dispatch and on January 21 1986 he wrote to the plaintiff landlord's solicitors — the defendants in this action — informing them that the award might be taken up upon payment of his fees and disbursements.

That letter was received, I find, the following day, January 22; and on that same day Mr Cooper of the defendants sent a copy of it to his client, the plaintiff, and asked the plaintiff to send to Mr Cooper a cheque payable to the arbitrator in the amount of the arbitrator's fees.

What the plaintiff did was to send the cheque himself direct to the arbitrator, in consequence of which the arbitrator, having received the cheque, sent the award direct to the plaintiff, thereby bypassing the defendants. The arbitrator did that by a letter dated February 4 1986, which I am prepared to infer, though there is no evidence on the file, reached the plaintiff a day or two later.

The plaintiff considered the award with Mr Stapleton. There was some measure of delay, caused apparently by Mr Stapleton's moving house at the material time and not being in his office so much as he might otherwise have been; and there was, from the point of view of the defendants, no reply to their letter of January 22. So, Mr Cooper sent a reminder on January 29 and a further reminder on February 10. In consequence of that the plaintiff wrote to Mr Cooper on February 11 informing him that the award had been taken up by the plaintiff personally. At about the same time, as a result of telephone conversations, a copy of the award was provided by Mr Stapleton or his firm to the defendants. It is plain from a letter of Mr Cooper of February 12 that he had received the award on the previous day, that is to say the 11th.

The award was unsatisfactory to the plaintiff in that the rent which the arbitrator had found was a great deal less than the rent for which the plaintiff and his expert witness had been contending. The plaintiff's claim had been for an annual rent of £31,500. The tenants' suggestion had been an annual rent of £14,000. The finding of the arbitrator was that the annual rent should be £19,547; substantially closer to the tenants' figure than the plaintiff's.

The plaintiff was aggrieved by this and desired to take up such rights as he had of appeal. These being arbitration proceedings, the only rights enjoyed by the plaintiff in the nature of appellate rights were the rights conferred by section 23 of the Arbitration Act 1950 to set aside the award on the ground of misconduct on the part of the arbitrator and the right conferred by section 1 of the Arbitration Act 1979 to appeal on a question of law.

The time-limit for the implementation of both those proceedings is 21 days from the publication of the award. That is by virtue of Order 73, rule 5 of the Rules of the Supreme Court, which provides that the application to the court must be made, and the summons or notice must be served, within 21 days after the award has been made and published to the parties. It is also clear law that an award is made and published to the parties when the arbitrator has given notice to the parties that it is ready. The arbitrator gave such notice by his letter of January 21, which was received by the defendants on January 22; and in this case, as I find, that is the date from which the period of 21 days began to run. It will have been observed that the period expired on February 12, which was the day on which Mr Cooper of the defendants wrote a letter saying that he had not received a copy of the award until the previous day and that there had been no comments from Mr Stapleton; that there appeared to be nothing further his firm could constructively do, and enclosing notice of charges. By that date, of course, no steps had been taken to pursue, or even to consider, an appeal.

After February 12, a period of nearly a fortnight elapsed in which nothing seems to have happened. On February 25 the defendants, via Mr Cooper, wrote to the plaintiff apologising for his delay in dealing with the plaintiff's letter of February 11, explaining in some measure what had happened, and saying that meanwhile the time for the lodging of any appeal that may be contemplated was running out.

Mr Cooper gave evidence before me, and in the course of his

A evidence he very frankly acknowledged that he had been under the impression throughout the whole of the material time that the period in which any appeal or quasi-appellate procedure should be implemented was six weeks. He thought that there was no particular urgency. In fact, there was very great urgency.

Mr Stapleton, having considered the award, but having done so without being aware of any pressures of time, criticised it on a number of grounds, of which only two remain relevant for the purposes of this action.

I turn now to the award pursuant to the request which had been made by counsel for the plaintiff on January 6. The award was in reasoned form. It set out the arbitrator's appointment; the course of the proceedings; various agreed facts; the issues which called for a
B decision between the parties. The arbitrator summarised the respective contentions.

It is unnecessary for me, in the course of this judgment, to repeat the evidence or the arguments in relation to the arbitration at any length. In substantial measure they follow a form which is common to such proceedings. Both the surveyors put forward valuations for the most commercially desirable part of the premises, that is to say, the usable part of the ground floor, which was referred to by both of them, as is the practice in these cases, as "zone A". Both of them sought to support their valuation for zone A by reference to comparable properties more or less nearby. It was agreed, apparently, that this was not an easy building to value. The premises
C themselves, although occupied on the ground floor as a retail shop and on the three upper floors partly as offices and partly as storage/accommodation by those same tenants, were not in an exclusively retail location. It was, apparently, in the location where much, if not most, of the surrounding properties were offices such as banks and building societies and so forth. It was a little way from what was acknowledged to be the main shopping street of Hove, which is George Street.

Mr Stapleton, for the plaintiff, asserted that the road in which the relevant premises lay, Church Road, was at the south end of George Street. There was another road at the north end of George Street, namely Blatchington Road, and Mr Stapleton said that George Street was the prime shopping location; that Blatchington Road, at its
D northern end, was almost as good; that Church Road, at its southern end, was almost as good again as Blatchington Road, although he conceded not quite; and Mr Stapleton put forward various comparable properties in Blatchington Road which he said should provide the arbitrator with a guide.

Mr Wareham took a contrary view and took the position that Blatchington Road was not in a relative sense comparable with Church Road. It was, he thought, too far away; and Mr Wareham relied on comparable properties at other locations in Church Road. These, in turn, Mr Stapleton criticised on various grounds. I am conscious that, in summarising the respective positions of the expert witnesses in the arbitration, I may not be doing full justice to every
E nuance; but I hope that I have not unfairly, albeit with a fairly broad brush, illustrated the differences between them on that issue.

The arbitrator was invited to, and did in fact, inspect not only the subject premises but also the comparables. He did so, as he says in his award, on January 7. In this part of his reasoned award he gives expression to matters which have formed the main ground of the complaint which was made against him by Mr Stapleton and in the abortive appellate proceedings, to which I will shortly refer. What he said was:

> I visited the area on 7th January 1986 and spent a full day inspecting the subject premises, a total of fifteen premises offered as comparables (including a limited inspection only of 119 Church Road) and assessing the level of business activity in the vicinity.
>
> F *The Zone A Rent*
>
> I observed a substantially higher level of activity in George Street and Blatchington Road as compared with Church Road. In my opinion no clear relationship has been established between the two shopping areas, nor do I believe the rental values of Hill House are relevant.

The arbitrator went on to discuss the evidence, and concluded that the value of zone A of the ground floor of the premises was £13.50 per sq ft. That is the first passage to which the plaintiff and his advisers in 1986 took exception. In a word, it is said that in the passage which I have read the arbitrator stepped outside his function of evaluating the evidence which had been called before him and embarked upon the separate and improper activity of taking evidence himself.

The second complaint which was made by Mr Stapleton in 1986 arose in this way. It was Mr Stapleton's contention on behalf of the G
plaintiff that the premises should be valued on the basis that a single tenant would occupy the whole, that is to say, the ground floor and the basement, and the first, second and third floors above. However, if a tenant could not be found to take the whole premises, it was possible, because of their construction, to divide them so that one occupant took the ground floor/basement and another occupant took the upper floors. To do that would require: first, planning permission; second, works of conversion of one kind or another; third, a fire staircase.

The surveyors were agreed that if the premises were to be divided in that way, such work would have to be carried out; and there was a measure of agreement as to how the valuation should be done. What H
there was disagreement about was the cost of conversion. The arbitrator heard evidence about this from Mr Stapleton on behalf of the landlord, the plaintiff, and from a Mr Christopher Legrande [FRICS], the building surveyor to whom I have referred, on behalf of the tenants.

The arbitrator dealt with this aspect of the dispute before him in this way: he said, under the heading "The rent of the upper floors":

> I have accepted the Claimant's argument that the upper floors, with independent access from Norton Road as existed before the 1980/81 works, could be let separately from the retail shop and basement. I also accept the strong probability that planning consent would be obtained for this.
>
> I have adopted the rental values agreed by the surveyors for refurbished offices, namely £6, £5 and £4 respectively, to include central heating and J
> carpeting.
>
> I have considered carefully the evidence given on the works to be carried out to comply with the Public Health Act and in particular Fire Precautions, and the cost of these works. In the absence of confirmation of the exact requirements of the Hove Fire Brigade, it has been necessary for me to form my own assessment of these from the evidence available to me. In particular I have made allowance for the provision of an independent fire escape but no allowance for a smoke detection system or emergency lighting.
>
> I assess the cost of these works, including a central heating system, carpeting and professional fees at £30,000

(and in parenthesis at this point I say that I note that that figure of £30,000 lies between the figures which were advanced for such works by the surveyors for the respective parties.)

The arbitrator continued: K

> Using the divisor of 6.86 as agreed by the surveyors, the amount to be deducted from the rental value at the agreed rates, namely £11,478 per annum following completion of the works, is £4,375 per annum. From the resultant figure of £7,103 per annum I have made an allowance of 10% to reflect the element of risk on the part of the Respondent, leaving a net value of £6,393 per annum,

and he found that that was the rent of that part of the premises.

In respect of that passage, it is said that the arbitrator's reference to an allowance of 10% to reflect the element of risk was not reflected in the evidence or arguments before him; that, in introducing that, he was introducing a matter on which the parties had had no opportunity to comment or make submissions; and that, in so doing, L
he misconducted himself.

I return now to the narrative. Following the defendants' letter of February 25, to which I have referred, a meeting eventually took place between the plaintiff, Mr Stapleton and Mr Cooper on March 5. I observe in passing that there had been arrangements to meet earlier, on March 3. They had proved abortive because, by reason of some confusion, Mr Cooper had not attended. There was a good deal of discussion about that in the evidence but, in the result, I think it forms no relevant part of what happened subsequently.

As a result of the meeting on March 5, Mr Stapleton was to provide Mr Cooper with his written comments about the award, which he did by a letter the same day; and Mr Cooper was to instruct counsel to advise. The papers were sent to counsel, Mr Speight, who had M
appeared in the arbitration. It does not appear precisely when, but it must have been within a day or two of March 5 because on March 10 there is an attendance note of Mr Cooper's which records:

> Counsel rang. Asked when the award had been known. Said we would check file and give him dates. Did so. Rang Counsel back and informed him of same. He was under the impression that the appeal time was three weeks and not six weeks, as was originally the case. However, an extension of time could be applied for. He would get into the matter and contact us directly he had settled the papers.

Counsel was right, as I have already indicated. The time-limit was three weeks and not six weeks, as it had been until the rules had been amended.

Counsel acted with commendable dispatch and prepared a written opinion, which is dated March 13, in which he refers to a decision to which he had been referred by Mr Stapleton and about which I shall have something to say in a minute, and advised: "That case clearly gives us an arguable point here" — that is to say, in relation to the arbitrator's findings, that he had assessed the level of business activity in the vicinity, and his observations thereon. Counsel also went on to say:

> I think that there is a seriously arguable complaint of the arbitrator's taking into account a 10% deduction for risk element.

That is the second point to which I have referred earlier.

Counsel dismissed various other criticisms made by Mr Stapleton and advised that proceedings should be by way of challenge to the award and misconduct, pursuant to section 23 of the Arbitration Act 1950. He dismissed the alternative approach to an appeal against an arbitration under section 1 of the 1979 Act, saying:

> I can find no indication of an error in the award in this case,

and he concluded by saying that:

> The fullest possible explanation should be given of why we are out of time in commencing this appeal.

That advice must have been received by the defendants within a few days of March 13, because on March 19 Mr Cooper swore an affidavit in support of the application, which was then taken out on March 20, to set aside the award under section 23 of the 1950 Act on both the grounds that I have mentioned. By that time, of course, the three-week period provided by Ord 73, rule 5, had long since expired; and so indeed had also the six-week period which Mr Cooper thought was operative in the circumstances.

Notwithstanding counsel's view that there was no point of law which justified an appeal under the 1979 Act, on May 8 the defendants issued a summons in the commercial court for an order to give leave to appeal from the decision of the arbitrator under that section. In an affidavit which he swore later, Mr Cooper said that the summons had been taken out at the suggestion of an official of the court. But he went on to make a claim in that affidavit that the applicant regarded the application as falling under section 23 of the 1950 Act and that the question of leave under the 1979 Act did not arise.

There seems to be some obscurity in the evidence before me as to precisely when, and by whom, the motion under the 1950 Act and the summons under the 1979 Act were heard. It seems likely that they were heard by Saville J in the commercial court on or about June 20 1986. I will go further than that: it is virtually certain that they were heard by Saville J on that date.

Saville J dismissed the application as being out of time. It follows that the plaintiff had exhausted such appellate rights as he had had without a full consideration of the merits. The plaintiff asserts that the circumstances in which he came to lose those rights amount to negligence and breach of duty on the part of the defendants in failing to advise him concerning the time-limits in relation to appeal.

The first issue with which I have to contend is whether the question of an appeal was expressly mentioned on the second day of the arbitration, on January 6 1986.

On that point I heard evidence from the plaintiff, Mr Corfield, and from the defendants. The plaintiff told me that at the midday adjournment on that day he was not happy with the conduct of the matter. It was not entirely clear from his evidence whether the plaintiff was not happy with the conduct of his own advisers, or with the conduct of the other side, or with the conduct of the arbitrator, or with the conduct of all of those. But certainly the plaintiff was not happy. He said that he alerted Mr Cooper to prepare himself for the possibility of an appeal, and he told me: "I asked the specific question: 'How long do we have for an appeal?', and Mr Cooper said: 'Six weeks after taking up the award'."

Mr Cooper did not accept that. He said that there was no such discussion and that there was no express mention of an appeal.

I am not prepared to find that Mr Corfield asked the specific question: "What is the time-limit for an appeal?" I notice that there were a number of persons present at that meeting. In particular, Mr Stapleton and Mr Speight were present, but neither of them gave any evidence about this particular point; and I have come to the conclusion that I am not able to make a finding about it. Or, put more strictly perhaps, I am not prepared to find — the burden of proof being on the plaintiff — that such an express reference was made. But I do find that enough was said at this meeting on that day to put Mr Cooper on notice that an appeal was a distinct possibility if the reasoned award, when it was published, disclosed any grounds. Indeed, Mr Cooper, in his cross-examination, agreed as much; and he told me that it was a possibility — I think he said it was a distinct possibility — that the plaintiff would wish to appeal if he had any right to appeal.

I find that, because of the admitted fact that at that time Mr Corfield certainly requested his advisers to obtain the reasoned award. Mr Corfield was plainly, I find, dissatisfied with the way in which the proceedings were going and was concerned by then at the possible outcome; and an appeal was something which was, in Mr Cooper's phrase, "a distinct possibility".

That being so, in my judgment Mr Cooper was under a duty either to be aware — to be rightly aware — of the appropriate time-limits which were available if such appellate rights were to be exercised; or, if he was not aware of them, to take steps to inform himself of them. Second, to inform the client if the limits were short or required urgent action.

How was that duty discharged by the events of January 26 to February 12? I have already outlined the material correspondence. It is quite plain from the correspondence and Mr Cooper's approach to the matter that he was under no sense of urgency. Why should he be? He was under the impression that the appropriate time-limit was six weeks. He thought it was six weeks. He was wrong. If it were six weeks, there was, initially at any rate, no particular urgency.

Having regard to the possibility of an appeal, and having regard to the actual length of time available for consideration and action, if thought appropriate, I find that the defendants, by Mr Cooper, were in breach of their duty to advise the plaintiff as to his rights of appeal and the time-limits for implementing any appeal. Mr Cooper did not do so; he did not raise the question at all until February 25, by which time it was much too late.

I should, at this point, interpose that I heard submissions on behalf of the defendants that, irrespective of any breach of duty on the part of Mr Cooper — none being admitted — no such breach was causative of the events which subsequently flowed. I reject that submission. It is true that the plaintiff and Mr Stapleton between them in fact caused a little delay by not communicating with the defendants when they — in particular Mr Stapleton — had the award. But they were not under any sense of urgency. Nor, in my judgment, was there any reason why they should have been. The sense of urgency should have been conveyed to them by Mr Cooper. I have no doubt whatever that if Mr Cooper had been aware that the time-limit was 21 days then he would have acted in a completely different way from January 22 onwards. If he had been aware of the true position, as in my judgment he should have been aware, then he would have conveyed the urgency to his client, and to his client's professional adviser, and the matter would have been dealt with entirely differently. I do not think, therefore, that it lies within the defendants' mouth in these proceedings to say that Mr Stapleton and the plaintiff added a week or two weeks to the delay when it was not for them to appreciate the severe constraints of time under which the appellate system operated.

I find, therefore, that the failure to implement the appellate process in time was a direct consequence of the defendants' breach of duty.

I now come to what appears to me to be by far the most difficult aspect of these proceedings, namely questions relating to damages. All such questions, in litigation of this kind, depend upon the evaluation of chances. Usually the court is called upon to make that evaluation once only. What is the value of the right which the plaintiff has lost? In this case, exceptionally in my experience, the court is called upon to make a two-stage valuation, because what the plaintiff lost was the opportunity of seeking to persuade an appellate court — in the context of this case the commercial court — to set aside the award and give the plaintiff a chance to argue his case before a fresh arbitrator. I have to determine not only what the prospects of success would have been in the appeal if it had been brought in time but what the results of further litigation would have been if the appeal had been successful.

I turn, first, to considering the prospects which the plaintiff had of persuading the commercial court that the arbitrator had misconducted himself, in the sense of section 23 of the 1950 Act. This falls under two heads: first, relating to the rent for zone A and the arbitrator's observations; the second relating to the rent for the

A upper floors and the arbitrator's introduction of the 10% element for risk. So far as the effect of the arbitrator's finding in paras 6 and 6.1.1 of the award is concerned, I start by observing that there is a distinction between the arbitrator assessing evidence and the arbitrator creating evidence. It is this distinction which was underlined by Leggatt J in his decision in *Top Shop Estates Ltd* v *Danino* [1985] 1 EGLR 9*. That, too, was a case arising out of a rent review arbitration. There the arbitrator had initiated what were described as a series of pedestrian counts in the course of considering his findings in relation to the two premises that were relevant in those proceedings. There were also comparables. The arbitrator said, in the course of his award:

> B I therefore propose to examine each comparable in the light of respective submissions and my own knowledge and observation. The day of my inspection was dry and reasonably sunny though cold. The centre was busy enough to confirm my personal assessment of relative positional qualities, these views being since confirmed by a series of pedestrian counts taken between . . .

various times, which the arbitrator set out. Later he said:

> I disagree with the landlord's surveyor's assessment of positional quality, however, and, from my own knowledge and observed pedestrian flow, the broad walk is quite considerably busier than the west side of the Broadway.

and so forth. Finally, in his findings, the arbitrator said that he took account of:

> My own knowledge, taken together with studies of pedestrian flow at busy
> C periods.

In relation to that evidence, there are two passages from the judgment of the learned judge to which I ought to refer. The first is this (at p 11):

> I think in the circumstances it would suffice if I say that the court will always take, unless there be cogent reasons to the contrary why it should not, a charitable view of what motivated an arbitrator in coming to particular conclusions, or conducting an arbitration in a particular fashion, even if, having reviewed the arbitration and the award following it, the court takes the view that the arbitrator has indeed (whether one describes it as "technically" or not) misconducted either himself or the proceedings. It might be said, as Mr Reynolds on behalf of the arbitrator submits, that in arranging upon his own initiative for a pedestrian count to be taken, the arbitrator was acting over-
> D zealously. It does not seem to me to be profitable to seek to select epithets appropriate to describe the conduct of the arbitrator in such circumstances. If he was acting over-zealously, then more relevantly, he was, in my judgment, also acting under a misapprehension of his function as an arbitrator, which is not to play the part of Perry Mason where he feels that the submissions or evidence of the parties might usefully be supplemented.

Later, Leggatt J continued (at p 12):

> It seems to me that that view of pedestrian counts conducted by or under the auspices of an arbitrator in aid of an inspection of premises, or as a check or confirmation of an impression which he otherwise gained or which he thought was established or suggested elsehow, is wholly illegitimate. There can be no warrant for it and, while, of course, I accept that the arbitrator was motivated by the most worthy intentions and did not think for a moment to do anything
> E which could possibly be regarded as unjust to either of the parties, none the less a moment's reflection will make plain that if either of the parties objected to the amount in which he assessed the rent, they were quite entitled to assume, and, indeed, bound to assume, in my judgment, having regard to the way in which the award and the conclusions of it were expressed, that the pedestrian count had played a part in the conclusion to which the arbitrator came. It is rightly said in this context by Mr Lightman that one is not merely dealing with a "Yes" or "No" answer by an arbitrator; one is dealing with the assessment of a rent, which may well, therefore, be based at any given point on the scale constituted by the difference between the amounts contended for by the respective surveyors. In those circumstances, even confirmation of a conclusion provisionally reached, or reached without recourse to pedestrian counts, might well be affected in some amount by pedestrian counts conducted in the way the arbitrator now says they were.

F Then the judge concluded that he had to hold, without there being implicit in his holding any imputation upon the personal conduct of the arbitrator, that he had misconducted the proceedings in such a way as made it necessary for the award to be set aside. He so ordered.

It is useful, in this context, if I also refer to the decision of Bingham J, as he then was, in *Zermalt Holdings SA* v *Nu-Life Upholstery Repairs Ltd* [1985] 2 EGLR 14, where he said, in the context of another rent review case (at p15K-L):

> I fully accept and understand the difficulties in which an expert finds himself when acting as an arbitrator. There is an unavoidable inclination to rely on one's own expertise and in respect of general matters that is not only not objectionable but is desirable and a very large part of the reason why an G arbitrator with expert qualifications is chosen. Nevertheless, the rules of natural justice do require, even in an arbitration conducted by an expert, that matters which are likely to form the subject of decision, in so far as they are specific matters, should be exposed for the comments and submissions of the parties. If an arbitrator is impressed by a point that has never been raised by either side then it is his duty to put it to them so that they have an opportunity to comment. If he feels that the proper approach is one that has not been explored or advanced in evidence or submission then again it is his duty to give the parties a chance to comment. If he is to any extent relying upon his own personal experience in a specific way then that again is something that he should mention so that it can be explored. It is not right that a decision should be based on specific matters which the parties have never had the chance to deal with, nor is it right that a party should first learn of adverse points in the decision against him. That is contrary both to the substance of justice and to H its appearance, and on the facts of this case I think the landlords' case is made out.

The plaintiff in this case contends that the conduct of the arbitrator in conducting his inspection of the comparables and his associated observations constituted the creation of evidence by the arbitrator in the sense analogous to those of the arbitrators in the two decisions to which I have referred.

On the part of the defendants, it is contended that the arbitrator did no such thing. All he did was to inspect the comparables and record what he had seen; and that an appellate court would have been bound to have come to that conclusion. It is plain that the arbitrator was expected to look at the comparable premises and the streets in J which they were situated. It was pointed out to me that on p4 of the written proof of Mr Wareham, who gave evidence on behalf of the defendants at the arbitration, he said:

> No doubt, the Arbitrator will wish to inspect the site and surroundings and he will see for himself the nature of Church Road and its relationship to the principal shopping areas of George Street and Blatchington Road.

It is said that in the passage of the award which is criticised on behalf of the plaintiff, that is no more than what the arbitrator was doing. The question that I have to decide is what were the prospects of the plaintiff's advisers in persuading the judge of the commercial court, if the matter had been brought there on time, that the circumstances here were the creation of evidence rather than merely the recording of evidence. K

It was submitted to me on behalf of the defendants that one matter to which I ought to pay consideration was that, if the matter had been brought before the commercial court in time, the arbitrator would have had the opportunity to submit an affidavit in which he could have explained what he had done and why. That is true; but, bearing in mind the limited function of such an affidavit, as pointed out by Bingham J in the *Zermalt Holdings* case, I do not think that that is a matter which I can effectively put into the scales.

Another matter about which I heard submissions was this: that Saville J, on June 20 1986 must have considered the merits and that therefore, in asking this court to consider the plaintiff's prospects, I ought to take the view that in fact the matter is, in a sense, already L decided because the judge of the commercial court considered the problem and rejected the plaintiff's case.

I find that a difficult submission to which to give effect in the circumstances. The only evidence before me that Saville J gave any consideration to the merits at all — apart from the inference to be drawn that in the ordinary course of commercial court practice the judge would have seen the papers, including the affidavits, in advance — was the evidence of Mr Cooper. I have not found it necessary in the course of this judgment to evaluate in detail Mr Cooper's evidence because I think it would be unhelpful to do so. But I think I must say, in relation to this issue, that I am not prepared to accept Mr Cooper's recollection of those events unsubstantiated by contemporary documents. I am bound to reflect that, by the time the M matter came before Saville J, not only had the three-week limit expired but also Mr Cooper's mistaken six-week period had expired. I am not prepared to find that the learned judge gave any consideration to the merits at all, the extent of the delay being so bad.

In a sense, the matter is one of impression. I bear in mind that the plaintiff had been advised by counsel that he had an arguable case, which was repeated by Mr Cooper of the defendants to the plaintiff in a letter of March 19 1986, where he says:

> I have also pursued with him,

that is to say counsel,

> the question of merit of the appeal and except to say that you have reasonably favourable prospects of success, he cannot gainsay the result.

---

*Editor's note: Also reported at (1984) 273 EG 197.

I bear in mind the submission that was made to me that on the following page of the award the arbitrator used the word "assessment" in plainly an unobjectionable way. It is urged on me that when he said: ". . . assessing the level of business activity in the vicinity . . .", he was doing no more than expressing a process which he applied to the evidence.

On behalf of the plaintiff, it was submitted to me that, when inferences are to be drawn in this case, I should not be quick to draw inferences against the plaintiff, having found that it was the defendants who were in breach of duty and who have, in a sense, put him in this position.

I also take into account, as I am bound to, the decisions since the 1979 Act that the policy of the courts should be for finality in arbitration.

At the end of the day, I have to come to a conclusion as to whether the plaintiff would have persuaded the commercial judge that this fell on the *"Top Shop Estates"* side of the line or on the "consideration of evidence" side of the line. It is plainly not so strong a case as *Top Shop Estates* or, for that matter, *Zermalt Holdings.* But I am not prepared to hold that the plaintiff had no chance in the commercial court that the judge would have set the award aside. On the other hand, I think that the odds were against the plaintiff; and I have come to the conclusion that the best evaluation I can make of the plaintiff's prospects of success in this issue is that he had a one-third prospect of success.

I turn now to the other matter in respect of which complaint was made, namely the 10%. The evidence was that no such conversion, in the context of which the 10% arose, would have been possible without the appropriate planning permission. No such planning permission had, at that date, been granted, it not being appropriate.

The arbitrator summarised in a sentence the evidence in respect of surveyors at para 5.2.8 of this award in this way:

> Mr Wareham disagreed with Mr Stapleton's assertion that there would be no difficulty in obtaining a Class II use for the whole building.

In his finding, at para 6.2.1 on this issue, he said:

> . . . I also accept the strong probability that planning consent would be obtained for this,

that is to say, the conversion of a portion. I note that he does not say "certainty". I think that the arbitrator was entitled to include a discount for the difference between a certainty and what he described as a strong probability; and that it could not be said that an arbitrator was wrong in law or misconducted himself by fixing such a discount at 10%. I think that there was an element of risk on the part of a hypothetical tenant and that the arbitrator could quantify that risk at 10%.

I agree that the arbitrator did not express himself in quite such explicit terms in para 6.2.5 of his award, but it seems to me that it would have been proper to do so. Indeed, I put it higher than that: an arbitrator ought to have made some allowance for the risk of planning permission being refused, having found, as he did find, that it was a strong probability only; and this is an unobjectionable way of putting that. I do not think that by referring to "risk" the arbitrator ought to be taken to be referring to contingencies in the cost of the conversion works, because these were allowed for elsewhere in his calculations. Nor do I think that the word "risks" is really appropriate to the voids in tenancies that arise where premises are sublet. But I do think it is appropriate in relation to the question of planning consent.

In my judgment, the plaintiff's prospects of success in persuading the commercial court that the arbitrator's award showed evidence of misconduct on this point was barely arguable; and I am not prepared to hold that he had any prospect of success under section 23 of the 1950 Act.

The point was argued in the alternative under section 1 of the 1979 Act, and I have been reminded of the decisions of the House of Lords in *The "Nema"* [1982] AC 784 and *The "Antaios"* [1985] AC 191 as recently applied to this class of litigation by the decision of the Court of Appeal in *Ipswich Borough Council* v *Fisons plc**, from which I think I need take only one sentence from the judgment of the Master of the Rolls, where he says that if he, that is to say, the judge, were to give leave under section 1(3)(*b*) of the, 1979 Act, he had to be satisfied that there was a more or less strong *prima facie* case that the arbitrator had erred on a question of law. I do not think that a judge would have been satisfied that there was a more or less strong *prima facie* case under this section.

Finally, before leaving this point, I should say that it was also submitted to me that this point is affected by section 1(4) of the 1979 Act, which provides that the High Court should not grant leave under subsection 3(*b*) unless it considers that, having regard to all the circumstances, the determination of the question of law concerned could substantially affect the rights of one or more of the parties to the arbitration agreement. On what was meant by "substantially affecting the rights of", I was referred to *Mustill & Boyd* on *Commercial Arbitration* at p 601, where the learned authors say that:

> In this particular context, we believe that . . . "— the word 'substantially' — " . . . it does not mean "more than de minimis", but denotes an effect on the rights of the parties which is of major importance, in the context of the dispute as a whole,

citing *The "Evimeria"* [1982] 1 Lloyd's Rep 55, to which I was also referred.

In my judgment, having regard to the size of this figure (which the arbitrator quantified at £710 pa) in relation to the context of the dispute and the other matters which the arbitrator decided, legally, factually and financially, this was not an issue substantially affecting the rights of the parties. Under that head I think that the plaintiff would not have succeeded.

If the plaintiff had succeeded, as to which I have held he had a one-third chance of success, in my judgment the order of the commercial court would have been to set the award aside. I cannot envisage that the commercial court, having decided in favour of the plaintiffs on the issue of the arbitrator's misconduct in conducting his own investigations, would have remitted the award to the same arbitrator for further consideration. If it had taken the step of finding misconduct, it would, in my judgment, have remitted the award to a fresh arbitrator and there would have been a fresh arbitration.

I now come to the second area in which I have to evaluate the plaintiff's prospects of success. What would have happened if there had been a subsequent arbitration before a fresh arbitrator? The plaintiff would have had the opportunity, with his advisers, of doing two things: first, of resubmitting and rearguing his contention that the premises would have been let to a single tenant and, therefore, none of the conversion works in relation to which the issue concerning the 10% which I have just discussed would have been required. Second, he would have had the opportunity to reargue before a fresh arbitrator his contention that the zone A rents were a good deal closer to Mr Stapleton's estimate of £18 per sq ft than to Mr Wareham's estimate of £11.50 per sq ft.

So far as Mr Clark's findings in the material arbitration are concerned, the arguments by Mr Stapleton that one tenant would have taken the entire premises was noted by the arbitrator but decided against the plaintiff without reasons. The arbitrator developed his award on the basis of the alternative submission that there would have to be a division of the premises. This puts this court in something of a difficulty because it is impossible to evaluate the arbitrator's reasons. The arbitrator might, I think I am bound to say, have awarded something for the possibility that a single tenant could be found, thus saving the full cost of the conversion. On the other hand, there is no basis upon which I can assess what went through the arbitrator's mind because in this part of his award he did not give reasons.

The plaintiff contends that the arbitrator in this case should have, and that a further arbitrator in a second arbitration would have, added back the full value of the upper floors, that is to say, a rent of £4,375 pa. It is conceded that some works of construction would have been necessary. So, although the evidence before me does not allow me to say precisely what those words would have been, something would have had to come off that. There would have to be some such calculation as was made by the valuers in relation to a divided tenancy, albeit in relation to smaller figures.

Doing the best I can, I think that justice would be done if I find that, if the plaintiff had had an opportunity to reargue this point, he would have had some prospect of persuading an arbitrator to increase the rent of the upper floors by some factor. I have come to the conclusion that, allowing for the uncertainties of quantum and the uncertainties of persuading an arbitrator to do that, the appropriate figure at which to evaluate that chance would be by taking a figure of £2,000 per year over the seven years of the review, that is to say, £14,000.

---

*Editor's note: Reported at [1990] 1 EGLR 17.

A The second matter on which the plaintiff and his advisers would have had the opportunity to reargue would be the rent for zone A. Again, I have to bear in mind that one experienced arbitrator had already found against Mr Stapleton's contentions. He might, on rearguing, have done better. Mr Stapleton told me, in the course of his evidence, that he would have done so because historically events had occurred which justified his appeal. I find it difficult to attach much weight to that assertion without having had the comments of another valuer and without Mr Stapleton considering the matter in detail. I also have to bear in mind that Mr Corfield, the plaintiff, might have done worse, in the sense that Mr Wareham, or somebody in the place of Mr Wareham, might have found other comparables which reinforced the position that had been adopted by him in the B arbitration in December 1985 and January 1986.

On the other hand, I have to take into account that the plaintiff has been deprived of his right of arguing this matter, and there is the possibility that further evidence might have assisted him. I think that the appropriate weight to be given to this is to say that the plaintiff had a prospect, which he has lost, of persuading an arbitrator to revalue the zone A rent up somewhat; and I assess that, in the sense of assessing the chance, at £2 per sq ft. I emphasise that I am not approaching this as a rent but I use it as a way of evaluating the plaintiff's chance. The arbitrator found that the superficial area to be treated, either which was zone A or which was to be treated as zone A, was 670 sq ft on p10 of his award. That is to say, taking zone A and in C smaller proportions zones B and C at £2 per sq ft, that would give the plaintiff £1,340 pa over the seven years remaining of the lease, which comes to £9,380. I am conscious that, in reaching figures of this kind, I have not strictly allowed for interest on payments that would have been made to date or discounts in relation to payments that would have been made in the future. I am also conscious of the fact that the arbitrator included various other minor allowances at the foot of p10 and the top of p11 of his award. I should say that I have taken those into account in reaching the figures which I have. It follows that I think that if the plaintiff had had the opportunity of rearguing the arbitration with fresh evidence and fresh submissions before a fresh arbitrator, I evaluate the chance which he has missed as £23,380. But, of course, he has not lost £23,380. What he has lost is, as I have found D in the earlier part of this judgment, a one-third chance of obtaining that, being the first stage of the chance. So that figure has to be divided by three, which is £7,790.

The plaintiff is also, in my judgment, entitled to recover the sum of £2,356, but divided by three, because if the appeal had been successful he would not have had to lay out that sum which he has paid to the respondents by way of costs of appeal, which was £785. So, in addition to the £7,790, the plaintiff would be entitled to £785, which is £8,575.

But, in order to recover that sum, the plaintiff would have had to undergo a further arbitration and to have paid the costs of it. It was agreed before me yesterday that the costs of a further arbitration E would have been £5,000 per side, plus arbitrator's costs of £3,000. The cost to the plaintiff of paying his own advisers and half the arbitrator's award would have been £6,500. I have to consider whether, if there had been a further arbitration, and if the result had been of the order which I have sought to express, what order the arbitrator would have made in relation to the costs of that arbitration.

I think that the arbitrator, if the result of the arbitration had been such as I have postulated, would have ordered that each party should have borne their own costs, and the costs of the arbitrator himself should be divided. I think that, in order to obtain the damages which represent the value of his loss, the plaintiff would have had to lay out a further £6,500, which falls to be deducted from that figure, which F comes to £2,075.

Judgment for the plaintiffs in the sum of £2,075.

Queen's Bench Division G

November 14 1990

(Before Mr Justice SCHIEMANN)

LAYZELL v SMITH MORTON & LONG

Estates Gazette April 4 1992

[1992] 13 EG 118

Negligence — Agricultural holdings — Negligent advice of solicitors regarding succession rights — Assessment of damages for loss of tenancy H

Prior to his death the plaintiff's father held an agricultural yearly tenancy of World's End Farm, Pebmarsh, Essex — The farm had a large Tudor farmhouse with some 164 acres — The plaintiff and his father managed the farm through a partnership — The farm was not very profitable because it required drainage — Not having sufficient capital himself, the plaintiff entered into an agreement with an uncle and a neighbouring farmer under which contractual services were provided for the farm in return for certain financial payments — On August 17 1986 the plaintiff's father died and on September 26 1986 the landlords gave notice to quit the holding — The defendants, the plaintiff's solicitors, failed to J **claim rights of succession for the plaintiff pursuant to section 39 of the Agricultural Holdings Act 1986 — The defendants acknowledge that this failure was negligent and constituted a breach of contract — There were two questions for consideration at the trial: (1) what prospects did the plaintiff have of succeeding his father as tenant of the farm; and (2) what compensation should he receive for the loss of those prospects — It was submitted on behalf of the defendants that the plaintiff did not satisfy condition (*a*) in section 36(3) of the 1986 Act, which provides for the eligible period during which an applicant must show that his only or principal source of** K **livelihood was derived from agricultural work on the holding — Further, that the plaintiff was not a suitable person**

***Held*: (1) Condition (*a*) in section 36(3) of the Act was satisfied, notwithstanding that the contractual services provided under the agreement with his uncle were at below market value, the plaintiff's only or principal source of livelihood came from the farm during the material time — (2) There was no reason for supposing that there was a significant chance that the agricultural land tribunal would have concluded that the plaintiff was not eligible to succeed — The evidence of unsuitability would have been regarded as trivial** L **by the agricultural land tribunal and there was no reason for supposing there would have been a significant chance that the tribunal would have concluded that the plaintiff was not suitable**

The sale-and-leaseback approach to the assessment of damages is the fairest way of compensating the plaintiff for what he has lost and to put him in a position where he can retrieve his losses — To achieve this the plaintiff would need initially £440,000 to acquire the freehold of a similar farm and he could recover £96,000 from selling the freehold and entering into a tenancy agreement similar to the one that he M **had lost — Damages of £344,000 were awarded**

The following cases are referred to in this report.

Groom v *Crocker* [1939] 1 KB 194; [1938] 2 All ER 394, CA

Taylor (CR) (Wholesale) Ltd v *Hepworths Ltd* [1977] 1 WLR 659; [1977] 2 All ER 784; [1977] EGD 637; (1976) 244 EG 631

In this action in the Chelmsford District Registry, the plaintiff, Roger Thomas Layzell, sought damages for breach of contract in respect of the defendants', Smith Morton & Long, a firm of solicitors, negligence in relation to the plaintiff's succession rights to World's End Farm, Pebmarsh, Essex.

Daniel Brennan QC (instructed by Gepp & Sons, of Chelmsford)

appeared for the plaintiff; Richard Seymour (instructed by Mills & Reeve, of Norwich) represented the defendants.

Giving judgment, **SCHIEMANN J** said: Roger Layzell grew up on World's End Farm at Pebmarsh, Essex. His father, Thomas Layzell, was the tenant of the farm and Roger hoped that in due course, on the death of his father, he, Roger, would be able to take over the farm. He thought he was entitled to do so because he thought he had a statutory right of succession under the Agricultural Holdings Act 1986 ("the Act") and its predecessors. In order to exercise any such right an application to the agricultural land tribunal had to be made within the period of three months beginning with the day after the death of Roger's father, namely August 18 1986. The application was not made in time. This was the fault of the defendant firm of solicitors. They acknowledged that they were negligent. I have to decide whether Roger Layzell is entitled to nominal damages or to some substantially greater amount and, if so, what amount. The defendants say that even if he had applied in time to the tribunal he would not have been granted a tenancy and that even if he had been granted a tenancy he would have been granted something which was worth little because the farm was completely unviable.

Background

The parties in their memoranda, and in court, referred to each other by their first names and I hope they will forgive me if I do the same. Roger was born on July 26 1944. His father, Thomas, held the farm under a tenancy agreement in writing dated May 28 1954. It was a yearly tenancy of a farm, with a large Tudor farmhouse with four huge bedrooms and four downstairs rooms, set in its own grounds with marvellous views and a paddock behind; in total by the time with which we are concerned the farm comprised some 164 acres. It was an arable farm growing cereals on some 155 of these acres.

The tenancy agreement contained an absolute covenant against assignment and an obligation on Thomas to reside constantly in the farmhouse. The residential obligation was, however, varied with the landlords' consent so as to permit Roger to live in the farmhouse. The father moved out in the mid-1970s and occupied himself as a bookmaker elsewhere. The accounts show the farming business as being carried on by a partnership between father and son. The son received each year what was described as a "partner's salary". The father received nothing under this head. This partner's salary was included among the fixed costs deducted from the gross margin so as to arrive at the net profit or loss for the year. This profit or loss was then split between the two partners in proportions which varied over the years.

The farm was not a very profitable enterprise, one of the problems being that it required draining. In 1979 the cost of such drainage was about £22,000, only half of which would be met by the Government. Roger did not have the other £11,000 in cash, but he was able to borrow it and his borrowings were supported by guarantee from his father, his uncle and a neighbouring farmer called Jonathan Nott. It is clear that Roger was well liked by his family and his neighbours and that he liked them in turn. To equip a 150-acre holding with appropriate machinery to conduct all farming operations would involve a considerable capital expenditure and the cost of servicing that capital would seriously deplete the farming profitability. In this context there are two courses open to a tenant:

(a) If he is to equip with sufficient machinery then he must consider becoming a contractor to other farmers in order to obtain the maximum use of this machinery and to earn a considerable income from such contracting services to set against his borrowing charges. (This was the option chosen by Jonathan.)

(b) To dispense with the capital expenditure and to employ a neighbouring farmer or contractor to provide him with the essential operations and machinery. (That was the option chosen by Roger.)

Roger was anxious not to spend money on machinery and so he entered into an agreement with Jonathan upon which something turns. Jonathan owned some 600 acres and had the relevant machinery and four men who worked on the farm. Roger owned 150 or so acres, but there was only one of him. Jonathan and Roger saw the arrangement as an effective pooling of their resources so that at those times of the year which are busiest for farmers they could work together and farm the 750 or so acres profitably. The agreement is contained in writing. The agreement was between Tom and Roger, described as "The Farmers", of the one part and Jonathan and various members of his family who were in partnership with him, described as "The Contractors", of the other part. The contractors agreed:

1. To provide all necessary plant and equipment to carry out agricultural operations on World's End Farm.

2. To provide all labour required, in addition to Roger's services, necessary to carry out the agricultural operations to which we have just referred. The farmers' obligations fall into two parts:

(i) Roger agreed to make himself available at all reasonable times for the purposes of the contractors' own business when he, Roger, was not engaged on his own farm.

(ii) The farmers agreed to make various financial payments to the contractors. Broadly speaking they agreed to pay:

(a) one-fifth of the contractors' total expenses incurred in providing fuels, labour, repairs and maintenance of the plant;

(b) a commission of 50% of the amount of the farmers' net profit for the relevant financial year. (The expression "net profit" is carefully defined in the agreement. For present purposes it suffices to note that in order to arrive at it one was entitled to deduct from the gross "a nominal salary payable to Mr R Layzell of £3,000 per annum".)

In 1984 that agreement was varied so as to provide broadly that Roger was to make himself available between July and October when not engaged on his own farm and to use his best endeavours to do the same for two weeks in spring; the nominal salary payable to him was increased to £4,500, then there was provision for counting as farm income a proportion of the amount that Roger might earn on non-farm activities.

Having entered into this agreement with Jonathan, Roger carried out extensive land drainage to improve the quality of his land. He told me, and I accept, that at that time he envisaged staying on the land for the rest of his life and handing it on to one of his daughters on his death should she wish to have it.

On August 17 1986 his father died. On September 26 1986 a notice to quit was served on behalf of the landlords on the defendant solicitors, who were acting on behalf of the plaintiff. The notice was expressed to expire on September 29 1987. Section 39(1) of the Act requires the tenant, if he wishes to claim rights of succession, to serve notice of application on the landlord within three months of the date of death. The defendants failed to serve such a notice on the plaintiff's behalf and acknowledge that this failure was negligent.

The plaintiff submits, and the defendants accept, that this negligence constituted a breach of contract by his solicitors. Before me, Mr Brennan QC, for the plaintiff, did not seek to found his case on tort because he accepts that such an approach is not open to a judge sitting at first instance. He reserved his right to advance arguments based on tort should this judgment be appealed. The result of the defendants' negligence is that Roger lost such prospects as he had of succeeding his father as tenant of the farm.

The questions which arise for my consideration are:

1. What prospects had Roger of succeeding his father as tenant of the farm?

2. What compensation should he receive for the loss of those prospects?

The prospects of success of an application for a direction

(a) *Eligibility*

In order to be entitled to make an application for a direction under section 39, Roger Layzell had to be an "eligible person" as defined in section 36(3) of the Act. That definition is:

> . . . any surviving close relative of the deceased in whose case the following conditions are satisfied: —
>
> (*a*) in the seven years ending with the date of death his only or principal source of livelihood throughout a continuous period of not less than five years, or two or more discontinuous periods together amounting to not less than five years, derived from his agricultural work on the holding or on an agricultural unit of which the holding forms part, and
>
> (*b*) he is not the occupier of a commercial unit of agricultural land.

It is common ground that Roger was a surviving close relative and was not the occupier of a commercial unit of agricultural land. The parties are at odds as to whether condition (*a*) above as to principal source of livelihood was satisfied in his case. The plaintiff maintains that it was but, in the alternative, draws my attention to section 41 of the Act, which provides that where the condition specified in condition (*a*) above, though not fully satisfied, is satisfied to a material extent the tribunal can determine that he is to be treated as an eligible person. This is to happen if it appears to the tribunal that in all the circumstances it would be fair and reasonable for the applicant to

A be able to apply for a direction entitling him to the tenancy of the holding. The defendants maintain that the plaintiff was not an eligible person, that condition (*a*) was not satisfied to a material extent, and that it would not have appeared to the tribunal that in all the circumstances it would be fair and reasonable for him to be treated as such. Alternatively, the defendants maintain that there was a risk that the tribunal would find that Roger did not fulfil the eligibility criterion.

The first issue here is whether in the seven years ending on August 17 1986 Roger's principal source of livelihood, throughout a continuous period of not less than five years, or two or more discontinuous periods amounting to not less than five years, derived from his agricultural work on the holding.

B It is clear from the accounts, and I so find, that, with one exception, the cash which Roger received during the period in question came from crops grown on World's End Farm. The exception to which I refer is Roger's earnings as a part-time lorry driver which, in the two years preceding his father's death, were £418 and £1,922 respectively. Roger's total cash income in the years 1981-86 was such that, on the face of the accounts, during 1985 and 1986 he received 88% and 83% of it from the sale of farm produce and in earlier years 100% from such sales.

The defendants submit, however, that Roger's principal source of livelihood did not derive from his agricultural work *on his farm.*

The substance of the submission is that the contractors provided C their services at substantially below market value, that their reason for so doing was Roger's willingness to work for the contractors off the farm without money payment, that if the contractors' services had been charged at market value then there would have been no income accruing to Roger from the farm and that in consequence, on a true analysis, Roger's principal source of livelihood was not "his agricultural work on the holding" but a combination of his work for the contractors off the holding and a perhaps unconscious act of generosity on the part of the contractors in undercharging for their services. Mr Seymour, who appeared for the defendants, accepted that if Roger had been his father's employee rather than his partner then Roger would have satisfied the test of eligibility. But, he submitted, the position was different by virtue of the fact that Roger D was a partner and not an employee.

From time to time in the submissions of Mr Seymour it was suggested that in truth it was the contractors who were in charge of all decisions on World's End Farm and that Roger was little more than an agricultural labourer. I find that this was not the case either under the agreement or in the way it was operated in fact. Roger remained in charge of the decisions as to what was to be done on his own farm. I accept the evidence of Roger and Jonathan on this point and reject the speculation of others. I find that the agreement was entered into at arm's length and that the contractors were not conscious of any act of generosity on their part. I find that the contractors were concerned to have an additional contribution to their fixed costs from another E party and were content as businessmen to enter into the profit-sharing part of the deal. I find, if it be relevant, that this was a perfectly sensible business decision for them to take and for Roger to take, that the value of the services which Roger rendered to the contractors off his own farm did not exceed £1,500 in any year and that Roger could have entered into an agreement with another contractor in the area which, in broad terms, would have secured the same services from that contractor in return for the same value given by the partnership and Roger in person.

In my judgment, one can be certain that the tribunal would have concluded, by use of section 41 if need be, that Roger was an eligible person. I say no more about this save to record that I find both Roger and Jonathan to be honest witnesses, who were doing their best to be F helpful to the court. And that I make similar findings in relation to Mr Anthony Smith [Mr Layzell's accountant] and Mr David Goldsmith [FRICS, an expert witness for Mr Layzell]. The chief motivation which the contractors had for entering into the agreement was a desire to find someone else who would help them to bear the continuing overheads of the expensive machinery which they had acquired. They were content to give the limited services of their four men on Roger's farm in return for a proportion of the profits which Roger hoped to make on his farm and a proportion of his time. It is wholly unrealistic to regard his principal source of livelihood as deriving from work that he did for the contractors on land other than his own farm.

I find that if Roger's notice of application had been served in time, G the landlords probably would have chosen to resist the application before a tribunal, although I record that there is no evidence of this fact from the landlords. All I have is that the landlords' agent would have recommended that the notice be challenged coupled with the fact that the landlords did not wish, once there had been failure to serve a notice of application, to grant a new tenancy to Roger. I find that, even if the application had been opposed by the landlords, the tribunal, which would have sat in the spring of 1987 before the disastrous harvest of that year, would have found that Roger satisfied the criterion of eligibility in the statute. The evidence before them would, in substance, have been the same as the material produced to me and I see no reason for supposing that there was a significant chance that they would have concluded that he was not H eligible.

(b) *Suitability*

Apart from determining whether an applicant is an eligible person, the tribunal must also under section 39 determine whether he is, in their opinion, a suitable person to become tenant of the holding. By section 39(8) in making such a determination the tribunal should have regard to all relevant matters including:

(a) the extent to which the applicant has been trained in, or has had practical experience of, agriculture;

(b) the age, physical health and financial standing of the applicant; and

(c) the views (if any) stated by the landlord on the suitability of the J applicant.

The defendants maintain that, if one has a look at all relevant matters including the financial history of the farm, one can see that Roger is not a suitable person to become tenant of the holding. Alternatively, the defendants maintain that there was a risk that the tribunal would find that Roger did not fulfil the suitability criterion.

As to suitability, I am satisfied from the evidence that I have heard that Roger had as much practical experience of agriculture as very many tenants who succeed to the holdings of their fathers, that his age and physical health leave nothing to be desired and that, while the farm is clearly not a substantial financial enterprise, he is a person of sufficient standing in the local community and among his relatives, K some of whom are better placed financially than he, for him to be regarded as someone of a financial standing which a tribunal would have found adequate. I record that I accept the evidence of his sister, Mrs McLeod, and his uncle, Mr G P Layzell, that in order to keep the farm in the family they would have provided such sums as were necessary in order to achieve this aim and that they had sufficient capital so to do. I also accept the evidence of his erstwhile bank manager that, with Roger's family and background, the bank would have stood behind Roger and continued his overdraft facility. I find that any evidence of unsuitability would have been regarded as trivial by the tribunal and that there is no reason for supposing that there is a significant chance that the tribunal would have concluded that he was not suitable.

L (c) *Notice to quit*

By section 44(1) it is provided that in a case such as the present in which a notice to quit had been served on behalf of the landlord, before giving a direction entitling an applicant to a tenancy the agricultural land tribunal should afford the landlord the opportunity of applying for their consent to the operation of a notice to quit.

By section 44(2) the provisions of section 27 were made applicable to an application for, and the granting of, such consent to the operation of the notice to quit. Section 27(1) provides that the tribunal shall consent to the operation of the notice to quit if, but only if, they are satisfied as to a number of matters. The only one of these matters which the defendants submit is relevant is set out in section M 27(3)(*b*), namely:

> that the carrying out of the purpose is desirable in the interests of sound management of the estate of which the land to which the notice relates forms part or which that land constitutes;

Section 27(2) provides that:

> Even if they are satisfied as mentioned in subsection (1) above, the Tribunal shall withhold consent under section 26 above to the operation of the notice to quit if in all the circumstances it appears to them that a fair and reasonable landlord would not insist on possession.

The defendants submit that World's End Farm is not economically viable and that it is likely that the landlords would have sought consent to the operation of the notice to quit and that, if sought, it is likely that consent would have been granted under section 27.

No significant evidence has been put before me to suggest that the tribunal could have been satisfied as required by section 27(1). In any event, I hold that there was not a significant chance, bearing in mind the provisions of section 27(2), that the tribunal would not have withheld consent to the operation of the notice to quit.

Damages

I make the following findings of fact:

1. It is all but impossible to obtain on the market a tenancy equivalent to the one lost by Roger through the defendants' negligence.

2. The 1987 freehold value of the farm and farmhouse was £425,000.

3. The 1987 value of the reversion was £225,000.

4. To rent a Tudor farmhouse such as the one at World's End Farm would at all material times cost about £2,500 pa.

5. The value in the years 1988 and 1989 of the gross income which would have been earned from the farm is £10,000 a year. The gross income which in fact was earned in those years by Roger's driving was £5,155 and £6,442 respectively.

6. In theory one could put Roger back in the situation in which he would have been had he not lost his tenancy by buying a freehold of a similar farm and then entering into a tenancy agreement with him similiar to the one lost. For Roger to achieve this situation he would need initially £440,000 in order to acquire the freehold and pay conveyancing and other charges, but then he could recover £96,000 from selling the freehold and taking the leasehold on these notional terms.

7. If £344,000 is paid to Roger as damages then, with financial help which is available to him, he would be in a financial position to buy the freehold of a farm similar to the one which he lost and that he would use that money to buy a farm.

Both parties accept the following passage from May J in *CR Taylor (Wholesale) Ltd* v *Hepworths Ltd* [1977] 1 WLR 659 at p 667D as correctly setting out the broad approach of the law:

> The various decided cases . . . show . . . merely the application in them of two basic principles of law to the facts of those various cases. These two basic principles are, first, that whenever damages are to be awarded against a tortfeasor or against a man who has broken a contract, then those damages shall be such as will, so far as money can, put the plaintiff in the same position as he would have been had the tort or breach of contract not occurred. But secondly, the damages to be awarded are to be reasonable, reasonable that is as between the plaintiff on the one hand and the defendant on the other.

Mr Brennan QC, who appears for the plaintiff, submits that the aim of compensation in the present case must be to put his client in a financial position to acquire a similar interest in a similar farm to that which has been lost through the defendants' negligence — essentially compensation for loss of an interest in land. Mr Seymour, however, submits that in essence what Roger has lost is a source of income and perhaps cheap housing and that the annual value of what Roger has lost is what should be compensated and that this can be done by employing a multiplier and multiplicand approach. He submits that the fact that this exercise would throw up a figure significantly less than the sale-and-leaseback approach shows that it is not reasonable as between plaintiff and defendant to adopt the sale-and-leaseback approach. A sale-and-leaseback exercise, in my judgment, provides the closest evaluation in money terms of what Roger has lost through his solicitors' negligence. It sidesteps, I recognise, the question whether the farm would have been profitable in the years 1987 onwards and whether a new farm will be profitable in the future. It also sidesteps the question of what is the value of the joy to be derived from a farmer's life. But it does give the market's assessment of the worth of his house and the surrounding farmland as an income-producing asset. To take that assessment seems to me to be the fairest way of compensating Roger for what he has lost and to put him in a position where he can retrieve his losses.

This is reasonable for Roger. Is it reasonable for the defendants in the circumstances? This is a matter of impression rather than the rigid application of rules. A defendant ought to be protected from a plaintiff acting eccentrically in a way in which he would never have acted if the defendant were not paying. On the other hand the court will not be astute to disadvantage an innocent plaintiff behaving reasonably in favour of lessening the burden on a guilty defendant. I find that Roger thought of World's End Farm as his family farm and, that having been lost, the court ought on the facts of this case to try to award him money which would enable him to be put in a similar position elsewhere. The position might well have been different if Roger had been much older or if this farm had not been the very centre of his existence, but he is in his early 40s, farming is and has been the centre of his life and it is not, in my judgment, reasonable to say that being a contract farmer for someone else without a farm of your own is essentially the same thing. I bear in mind that the farm was not a particularly profitable one. On the evidence which I heard there was no particular reason to suppose that in the future it would become more profitable. I bear in mind the line of cases, starting with *Groom* v *Crocker* [1939] 1 KB 194 (CA), which indicate that, when suing in contract, in general damages for mental anguish cannot be recovered. But, in my judgment, those cases have no great relevance to the essential questions, namely what damages are "such as will, so far as money can, put the plaintiff in the same position as he would have been had the . . . breach of contract not occurred"? and is it reasonable to award such a sum? I, therefore, award the sum of £344,000. Since that represents the figure presently required I am not asked to award interest on that sum.

Queen's Bench Division

November 28 1991

(Before Sir Michael OGDEN QC, sitting as a deputy judge of the division)

PK FINANS INTERNATIONAL (UK) LTD v ANDREW DOWNS & CO LTD

Estates Gazette June 20 1992

[1992] 24 EG 138

Negligence — Valuation — Whether surveyor should have verified planning assumption — Whether contributory negligence by plaintiffs in failing to send valuation to their solicitors to verify planning assumption — Whether failure to send valuation to their solicitors constituted a *novus actus*

At the request of Mr William Hancock, a director of Scotlane Ltd, Mr Appleton of the defendant company provided a valuation dated October 14 1985 of a large Victorian building on the outskirts of Hitchin — Scotlane Ltd bought the property and converted it into a nursing home — In November 1985 Mr Hancock sought a loan of £1m from the plaintiffs on the security of the property — Following agreement between Mr Hancock and the plaintiffs' marketing officer, Mr Appleton provided a valuation dated January 7 1986 in the same terms as that originally provided to Mr Hancock — It was apparant that before January 7 1986 Mr Appleton had received two documents from Mr Hancock, a planning consent for user of the property as a nursing home and another document entitled "Schedule" — The latter document recites the change of use to a residential nursing home and, in addition, conversion of a garage and stable-block to additional accommodation, community hall and 30 units of warden-assisted flats and bungalows — In his valuation Mr Appleton stated that he had made verbal planning inquiries but had not undertaken any official searches and for the purpose of his valuation he had assumed that consents were in existence for the additional nursing-home facilities in the stable-block and that consent had also been given for the sheltered housing — There was, in fact, no such planning consent for sheltered housing — The plaintiffs' contention was that Mr Appleton was negligent in not verifying the planning register or stating in his report that the existence and terms of the consents needed to be verified

***Held*: Judgment was given for the defendants — Mr Appleton was not negligent in failing to state in his report that there was a need for verification of the planning consents — The valuation, being provided for a financial institution, did not have to warn of the need for verification contrary to the Royal Institution of Chartered Surveyors Guidance Notes —**

A **The notes themselves make it plain that it is not necessary in all cases — The guidance notes are not to be regarded as a statute — Mere failure to comply with the guidance notes does not necessarily constitute negligence — The omission by Mr Appleton to state the source of his information regarding the 30 units of sheltered accommodation did not amount to negligence and had no causative effect — On the plaintiffs' concession that they were negligent in not sending the valuation to their solicitors, had the defendants been found negligent, contributory negligence by the plaintiffs would have been assessed at 80% — The defendants' contention that** B **the plaintiffs' failure to send the valuation to their solicitors constituted a *novus actus* was rejected**

The following cases are referred to in this report.

Knightley v *Johns* [1982] 1 WLR 349; [1982] 1 All ER 851; [1982] RTR 182, CA

McKew v *Holland & Hannen & Cubitts (Scotland) Ltd* [1969] 3 All ER 1261, HL

The plaintiffs, PK Finans International (UK) Ltd, claimed damages against the defendants, Andrew Downs & Co Ltd, of 7 Trebeck Street, London W1, in respect of a valuation dated January 7 1986 provided by Mr Andrew Appleton ARICS in respect of a property on the outskirts of Hitchin.

C Michael Nield (instructed by Herbert Smith) appeared for the plaintiffs; Martin Fodder (instructed by Reynolds Porter Chamberlain) represented the defendants.

Giving judgment, **Sir Michael Ogden QC** said: The defendants are a firm of surveyors and valuers. The plaintiffs allege negligence by Mr Appleton, of the defendants, in respect of a valuation of a freehold property upon which the plaintiffs advanced a loan of £1m. Unfortunately, Mr Appleton died a short while go.

The plaintiff company is part of a group of companies of which the parent company is one of the largest banks in Sweden. It has now ceased trading but it is important to note that at the material time it was a licensed deposit taker, and this was stated on the company's D writing paper.

The property in question lies on the outskirts of Hitchin. It is a large Victorian building, originally built as a private house but subsequently converted to other uses. Shortly before the events which give rise to this action, the property was bought by Scotlane Ltd, which proceeded to convert it into a nursing home called Foxholes Nursing Home and later Ambleside Nursing Home.

Scotlane was a company of which a Mr William Hancock was a director and one of the principal shareholders. Both parties to this action have asserted that Mr Hancock acted fraudulently in the events which I have to consider. He was not called as a witness and therefore I confine myself to saying of him that there are strong E indications that the assertion of fraud made against him is correct.

Mr Appleton had previously carried out several valuations for Mr Hancock when, towards the end of 1985, he was asked by Mr Hancock to make a valuation of the property in question, the conversion of which was within about 12 weeks of completion.

From a solicitor's attendance note of a meeting with Mr Appleton, in which the solicitors were taking notes for the purposes of this action, which note was admitted under the Civil Evidence Act 1968, it is plain that Hancock wanted a valuation done urgently, and this may be the reason why the original valuation was a fairly short one.

Mr Appleton inspected the property and his valuation is dated October 14 1985.

F On November 20 1985, Mr Hancock was introduced to Mr Swain, who was a marketing officer employed by the plaintiffs. Mr Hancock wanted to borrow £1m on the security of the property. Mr Swain duly recommended the loan to the plaintiffs' credit committee in a memorandum dated December 3.

Planning consent had been obtained for the use of the premises as a private nursing home. I accept Mr Swain's evidence that, in discussion with Mr Hancock, the latter either said specifically or plainly implied that there was planning consent for sheltered housing to be built on the property; there was, in fact, no such planning consent.

On December 27, Mr Swain wrote to Mr Hancock saying that the G matter had been considered by the credit committee, which required an independent valuation. Mr Swain in evidence recollected being given a copy of the October report, but said that he could not remember when this happened. However, in his statement, which constituted his evidence-in-chief, he said that he had seen or received it before he wrote his memorandum of December 3.

The credit committee required a further report, for two reasons: first, it wanted a report addressed to itself, since the October report contained the customary disclaimer for responsibility to a third party; and, second, the October report did not contain a forced-sale value.

The terms of a letter from Mr Hancock to Mr Swain, dated H January 2 1986, point to the probability that the two men had agreed that Mr Appleton could simply send the October valuation, redated, addressed to the bank but containing a forced-sale value. Certainly, when Mr Swain received it, he must have realised that he was looking at what was, for all practical purposes, the same valuation, save that the period of time to completion of the conversion had been altered from 12 weeks to four.

The valuation is dated January 7 but failed to contain a forced-sale value. Mr Swain telephoned Mr Appleton, following which the latter wrote a letter dated January 9, in rather enthusiastic terms, and returned the valuation amended to include a forced-sale value but still dated January 7.

J The attendance note (to which I have already referred) makes no reference to what, if anything, Mr Hancock told Mr Appleton about planning consents. From that part of it which deals with a telephone conversation with the local planning authority, it is plain that Mr Appleton was asking about planning permissions, in the plural, and because of a reference to "number of units", and to sheltered accommodation in particular. He would have asked those questions only if Mr Hancock had told him about it or provided documentary information about it.

I should interpolate at this point that it is obvious that a number of letters and other documents, which Mr Appleton must have had in his possession, have been lost or mislaid, which makes ascertainment of some dates very difficult. However, I am satisfied that before K January 7, Mr Appleton received two documents from Mr Hancock, one being a copy of the planning consent for user as a nursing home and the other (which was attached to the first) being a document entitled "Schedule". That document refers to the date of consent by the planning authority as being February 7 1985 and it recites the change of use to a residential nursing home and, in addition, conversion of garage and stable-block to additional accommodation, community hall and 30 units of warden-assisted flats and bungalows. The consent also covers a lodge building for use within the approved purpose.

Any experienced surveyor would appreciate at once, and I have no doubt that Mr Appleton, as an experienced surveyor, did so, that this L document is not a planning consent and must be a precis or summary prepared by someone, that person almost certainly being Mr Hancock.

At one stage, the plaintiffs suggested that its existence should have made Mr Appleton suspicious. I got the impression that that was not persisted in, certainly not in the final speech on behalf of the plaintiffs. In any event, I do not consider that this is so. A valuer does not work upon the assumption that his client is being fraudulent and, in any event, Mr Appleton knew that the existence and terms of the consent would have to be verified. I suspect that this document simply contained what Mr Hancock had already told Mr Appleton, just as he told Mr Swain.

In para 2 of the valuation of January 7, Mr Appleton said: M

> We have made verbal planning enquiries but have not undertaken any official searches and for the purpose of this valuation we have assumed that full planning consents were obtained for the development of the property as a nursing home, that consents are also in existence for additional nursing home facilities in a stable-block and that consent has been given for sheltered housing on plot number 426.

There is no reason to doubt that what he recounts is his understanding of what he was told on the telephone when speaking to someone in the local planning authority and there is no evidence that he was negligent in the questions he asked. The plaintiffs asserted that Mr Appleton should not have made oral inquiries at all, but this is unsustainable in view of the terms of para 7 of guidance note 6 of the

Royal Institution of Chartered Surveyors Guidance Notes* and bearing in mind that Mr Appleton must have assumed, as the expert witnesses agree he was entitled to assume, that verification would take place later.

The question of verification is at the heart of this case, since the plaintiffs now concede that Mr Appleton's valuation figures cannot be impugned as valuations, based upon the supposed existence of all the consents mentioned.

Both expert witnesses are surveyors of high standing and probity, but I consider that the plaintiffs' expert, Mr Colin Jennings [ARICS], tended to assert much higher duties on the part of a valuer than is justified. For example, at the commencement of his cross-examination, he said that the RICS guidelines were the valuer's "bible" and that if Mr Appleton conformed to the guidelines he was not to be criticised. However, later, when saying that Mr Appleton ought to have inspected the local planning authority register, he said: "Mr Appleton acted in accordance with the guide note, if looking to the letter of the guide note, but I consider that he was negligent in not doing more — namely inspecting the planning register". I reject the contention that he should have inspected the planning register. What he did, in my view, as I have already said, was in accordance with the guide notes — namely making an oral inquiry. I mention the point now, only because it constitutes one of several examples of pieces of evidence which caused me to prefer the evidence of the defendants' expert, Mr Desmond Hampton [FRICS], who, in any event, I regarded as a very impressive and reliable witness. In all respects in which their evidence conflicts, I accept that given by Mr Hampton.

I turn now to consider the terms of the report itself and in particular para 2. There is general agreement that, read fairly, it says that Mr Appleton made oral inquiries of the local planning authority and, based upon what he had been told during those inquiries, for the purpose of the valuation he assumed that there existed the consent specified, but he added the important proviso that no official searches had been undertaken. The plaintiffs' key contention is that Mr Appleton was negligent in not going on to say that the existence and terms of those consents needed to be verified. I reject this allegation.

I start by saying that, in my view, it was reasonable for Mr Appleton to assume the accuracy of what he had been told on the telephone, although, as both parties agree, some misunderstanding of some kind must have occurred, there being no evidence upon which I could begin to contemplate making a finding of negligence against Mr Appleton on that account.

In cross-examination, Mr Jennings said this, with which Mr Hampton agreed: "Mr Appleton, having made the assumption about planning, would expect his report to be passed to solicitors, who would verify the accuracy of the assumption". Unfortunately, the plaintiffs did not pass on the report to their solicitors. Mr Beak, who handled this matter at the plaintiffs' solicitors' firm, said that had he received the report he would, as a matter of course, have investigated to verify the consents, as indeed he did, in respect of the nursing home consent. I accept Mr Beak's evidence.

The RICS guidance note no 6, para 2.6, says as follows:

> Subject to these matters the valuer should (in most if not all cases) make it clear that information on which the valuation/report is based needs to be verified by the client's or other interested parties' legal advisers, before the valuation or the report is issued.

Mr Jennings' attitude was that in all cases the need for verification should be stated for fear that the client was a negligent client who did not know the basic principles and procedures which ought to be followed.

Mr Hampton, in giving evidence on this point, made the following observation:

> You are dealing with a professional member. It is like teaching your grandmother to suck eggs. Most of my clients would be rather insulted if I told them of the need for verification. It is a different case if you are dealing with a man from the street. Different standards are appropriate for different clients. Mr Appleton knew that the plaintiffs were a financial institution and would assume that they knew what they were about. One is assuming reasonably competent skills of a money-lending client. I would say what I have checked or not checked and what I have assumed. I would not say that it needed to be verified.

I accept Mr Hampton's evidence, which is, if I may say so, full of good sound common sense. This was not a valuation for a man in the street, it was for a financial institution which was a licensed deposit taker.

Mr Appleton was not told the size of the proposed loan, but obviously he would appreciate that it could be (as indeed it was) a loan of £1m. No sensible person who thought about it for a moment, still less any sensible financial institution, would contemplate making a large loan without verifying the existence and indeed the terms for planning consents, since they affected the valuation figure.

The guidance note makes it plain that it is not necessary in all cases, as Mr Jennings contends, to state the need for verification. In my view, therefore, there is no substance in this allegation against Mr Appleton whatsoever. He was not negligent in failing to add that since he had not conducted a search, this needed to be done. That was obviously the case and what happened here was a failure on the plaintiffs' part to do what it should have done, namely to pass the report to their solicitors.

During Mr Swain's evidence I got the clear impression that the plaintiffs' procedures left much to be desired, in that there appeared to be uncertainty about who should process matters and there also being an absence of a recognised procedure for doing so; such procedure should have stated precisely what should have been done, which would have included sending the valuation to the company's solicitors. As I have said, I reject this allegation completely.

A subsidiary complaint about the valuation was that reference to "30 units of sheltered accommodation" did not state the source of the information, namely Mr Hancock. This is a valid complaint in that while it did not matter when the October report was being made to Mr Hancock, when the plaintiffs became the clients, guidance note 6, para 6.1 requires that sources of information should be stated. However, Mr Nield, who appeared for the plaintiffs, was wholly unable to explain what difference it would have made if Mr Appleton had inserted in the report that the source of the information was Mr Hancock. Such evidence as there is (namely Mr Swain's) indicates that it would have made no difference whatsoever. Certainly there was no evidence that it would have made any difference.

I must add that, of course, these guidance notes are not to be regarded as a statute. I suspect that they are as much for the protection of surveyors as anything else, in that they set out various recommendations which, if followed, it is hoped will protect the surveyor from the unpleasantness of being sued. In any event, mere failure to comply with the guidance notes does not necessarily constitute negligence. I am not prepared to find Mr Appleton guilty of negligence because of this omission and in any event, as I have said, the omission had no causative effect.

By an amendment made at the commencement of the trial, three further allegations are made.

First, that the defendants failed to state clearly whether the valuation included the value of furnishings and fittings and to state the amount of such value. This allegation was torpedoed in evidence by Mr Swain. From what he said, it was quite plain that he understood the position perfectly clearly, as indeed I would have expected anyone in the plaintiffs' position to have done. As to the need to state the amount of the value, this was unsupported by evidence and in any event, in my view, in the context of a property valued at £1.5m, the value of fittings and furnishings is not a matter of consequence, and I reject the allegation.

Second, that the defendants failed to warn the plaintiffs that an ordinary mortgage would not give security over furnishings and all fittings. Again Mr Swain, in effect, said that he did not need to be warned; of course he knew the position. In my view, Mr Appleton was entitled to believe that any financial institution would know this absolutely elementary fact and I reject that allegation.

The third allegation related to the forced-sale valuation and was not pursued. Consequently, all the plaintiffs' allegations fail, but I consider that I should express views on other issues in case of an appeal.

It is conceded that the plaintiffs were negligent in not sending the valuation to their solicitors. Had I found the defendants negligent, I

---

*Editor's note: RICS Guidance Notes of the Valuation of Assets — 2nd ed, *The verification of information supplied to or adopted by a Valuer*. Para 7 is reproduced below:

Town Planning
The Valuer should state if he has made written or oral enquiries of planning authorities as to zoning and possible presence of any adverse planning proposals. The Valuer should state if he has inspected the statutory register and the information noted; he should also give details of planning consents made available to him and note any conditions which adversely affect value.

A would have assessed contributory negligence at 80%, leaving the percentage of damages to be paid by the defendants as 20%.

I reject the defendants' contention that there is no evidence for reliance by the plaintiffs on the fact that there was said to be consent in respect of sheltered housing. The very existence of that consent affected the value of the property. The plaintiffs would have lent only fractionally over 70% of the valuation figure and therefore the valuation figure was relied upon by the plaintiffs for that purpose.

I reject the defendants' contention that the plaintiffs' failure to send the valuation to their solicitors constituted a *novus actus*.

I was referred to two cases, *McKew* v *Holland & Hannen & Cubitts (Scotland) Ltd* [1969] 3 All ER 1621, at p 1623E, and *Knightley* v *Johns* [1982] 1 WLR 349, in particular the judgment of Stephenson
B LJ at p 366. On p 367 he said:

> The answer to this difficult question must be dictated by common sense rather than logic on the facts and circumstances of each case.

I respectfully agree. It is very much a matter of first impression, but, in my view, the circumstances of this case go nowhere near being something which constitutes a *novus actus* when the plaintiffs failed to pass on the report to their solicitors. One does not have to have, of course, the two acts which are effected to cause the damage occurring at the same point of time. This is an absolutely typical case, if I had found negligence on the part of the defendants, in which there was both negligence and contributory negligence. However, for the reasons given already, there must be judgment for the defendants.

C *Judgment for the defendants with costs.*

Queen's Bench Division

December 20 1991

(Before Judge BATES QC, sitting as a judge of the High Court)

HOOD AND ANOTHER v NATIONAL FARMERS UNION

D Estates Gazette June 27 1992

[1992] 25 EG 135

Negligence — Milk quota — Adverse decision of the Milk Quota Tribunal — Whether defendants negligent in failing to advise remedy — Whether tribunal in error in finding against plaintiffs

Until the completion of a contract for sale on May 1 1984, the plaintiffs farmed Archers Farm, Wiltshire, running a dairy herd of 100 cows on 112 acres — In June 1983 the plaintiffs exchanged contracts for the purchase of Greys Green Farm, Henley, Oxfordshire, a 420-acre farm that had
E not been used for dairy purposes for 10 years — The purchase was completed in September 1983 — Between August 1983 and April 1984 the plaintiffs constructed a milking parlour, silage plant and water supply on Greys Green Farm — They also provided 100 acres of grass leys — Although eight dry cows and 20 followers were moved in early 1984, the milking herd was not moved from Archers Farm to Greys Green Farm until April 4 1984 — The Dairy Produce Quotas Regulations 1984 gave effect to the EEC Council Regulations Nos 856/84 and 804/84, which imposed levies on milk production above prescribed quotas as from April 1 1984 —The plaintiffs
F received a provisional assessment of primary milk quota calculated on the basis of their milk production at Archers Farm in 1983 but none for Greys Green Farm

The Dairy Produce Quota Tribunal dismissed the reference to them of the plaintiffs' claim to make a development claim and an exceptional hardship claim for quota — The plaintiffs contended that the defendants failed to advise them that they had a right to issue judicial review proceedings to challenge the decision of the tribunal — The defendants submitted that the plaintiffs were not entitled to have made an exceptional hardship claim because: (1) the only relevant holding was Greys Green Farm, the plaintiffs were not farming that
holding on April 2 1984 (the relevant date) and were not supplying the Milk Marketing Board from that holding on that date; (2) a special case development claim had been made based on the same transaction or arrangement as that relied on in support of the hardship claim; and (3) para 17 of Schedule 2 to the 1984 regulations was not satisfied G

***Held*: Judgment was given to the plaintiffs — On the evidence it should have been clear to the defendants that the plaintiffs wanted to challenge the tribunal's decision — The defendants were negligent in failing to advise that a challenge could be brought by way of judicial review within three months — The plaintiffs were entitled to make an exceptional** H
hardship claim because: (1) on April 2 1984 Archers Farm was a production unit and the plaintiffs were producers; (2) article 4(1)(b) of Council Regulation 857/84 makes clear that development of milk production may lead to quota under both articles 3 and 4; it is no bar to an exceptional hardship claim that a development claim has also been made; (3) the requirements of para 17(b) of Schedule 2 to the 1984 regulations were satisfied — The plaintiffs had a 60% chance of establishing a claim

The following case is referred to in this report. J

Stubbs v *Hunt & Wrigley* [1992] 1 EGLR 17; [1992] 20 EG 107

This was a claim for damages by the plaintiffs, Alexander Robert Hood and Ann Marie Hood, against the defendants, the National Farmers Union, for negligence in relation to a decision of the Dairy Produce Quota Tribunal.

Alan Moses QC and Richard Gordon (instructed by Dawson & Co) appeared for the plaintiffs; and Stuart Isaacs QC and Richard McManus (instructed by Lloyd Cooper) represented the defendants.

Giving judgment, **Judge Bates QC** said: At all material times since 1970 the plaintiffs were dairy farming. They wanted to expand
from a milking herd of 100 cows on approximately 112 acres at K
Archers Farm, Wiltshire. The opportunity came in 1983 when Mr Hood's grandmother left him an inheritance. They looked for and found Greys Green Farm near Henley in Oxfordshire, which had approximately 420 acres and for which they paid £640,500 exclusive of the farmhouse.

Greys Green Farm has not been used for dairy farming for some 10 years and its milking parlour was gutted; but the plaintiffs planned to expand their existing herd initially to 200 milking cows and they constructed a milk parlour, silage plant and water supply for this purpose. This involved capital expenditure of nearly £50,000 and, of course, it took some time. In addition, in August 1983 they
cultivated, fertilised and sowed over 100 acres of grass leys. From L
December 1983 to January 1984 they inseminated cows to produce 40 heifer replacements, which was twice their usual number.

Contracts were exchanged for the purchase of Greys Green Farm in June 1983 and it was purchased on September 29 1983. Contracts on Archers Farm were exchanged in November 1983 and the sale was completed on May 1 1984.

The move of the milking herd at Archers Farm was delayed while the construction work at Greys Green Farm was completed, but just after Christmas of 1983 eight dry dairy cows and 20 dairy heifer calves were moved to Greys Green Farm. The move of the milking herd was made on April 4 1984.

The European Community by Council Regulation (EEC) No
856/84 of March 31 1984 amended Regulation (EEC) No 804/68 on M
the common organisation of the market in milk and milk products. The regulation entered into force on April 1 1984 and it recited, *inter alia*, as follows:

> Whereas the market in milk products in the Community is suffering from structural surpluses as a result of an imbalance betweeen supply and demand for the products covered by Regulation (EEC) No 804/68, as last amended by Regulation (EEC) No 1600/83.

A further recital:

> Whereas a careful examination of the different possible ways of re-establishing balance in the milk sector shows that, despite the administrative difficulties which its implementation may involve, the most effective method, and the one having the least drastic effect on the incomes of producers, is the

A introduction for an initial period of five years of an additional levy on quantities of milk delivered beyond a guarantee threshold.

In its operative part the regulation adopted in article 1 that the following article 5(c) in Regulation (EEC) No 804/68 be inserted and that provided that:

During five consecutive periods of 12 months beginning on the 1st April, 1984 an additional levy payable by producers or purchasers of cows' milk shall be introduced. The objective of the said levy shall be to curb the increase in milk production while at the same time permitting the structural developments and adjustments required, having regard to the diversity of the situations among individual Member States, regions and collection areas in the Community. However, the first period shall start on the 2nd April, 1984.

It provides that:

B The levy system shall be implemented in each region of the territory of the Member States in accordance with one of the following formulas:

The relevant one is Formula B for this case:

A levy shall be payable by every purchaser of milk or other milk products on the quantities of milk or milk equivalent which have been delivered to him by a producer and which, during the 12 months concerned, exceed a reference quantity to be determined.

The person to whom it is delivered, of course, in this case is the Milk Marketing Board.

It further provided that:

The purchaser liable to the levy shall pass on the burden in the price paid to those producers who have increased their deliveries, in proportion to their
C contribution to the purchaser's reference quantity being exceeded.

As we have seen, the reference quantities become subsequently in our legislation "quotas".

Para 3, the operative part, provided that:

Subject to paragraph 4, the sum of the reference quantities referred to in paragraph 1 may not exceed a guaranteed total quantity equal to the sum of quantities of milk delivered to undertakings treating or processing milk or other milk products in each Member State during the 1981 calendar year, plus one per cent.

Then they set out the guaranteed total quantity in thousands of tonnes, and for the United Kingdom that is 15,538.

Article 2 in this regulation provides that the regulation shall enter into force on the day of its publication in the official journal of the
D European Communities, which I believe was April 1 1984.

Council Regulation (EEC) No 857/84 is irrelevant. In the recitals that are provided, *inter alia*:

Whereas the Member States should be enabled to adapt the reference quantities to take into account the special situations of certain producers and to establish for this purpose, as necessary, a reserve within the above-mentioned guaranteed quantity . . .

— that, if I may interpose, is relevant and significant in this case —

the Member States should be enabled to adapt the reference quantities to take into account the special situations of certain producers.

The final recital in that regulation is:

Whereas the scheme must, as a matter of overwhelming public interest, enter
E into force on the 2nd April.

Article 2 provided that the reference quantity referred to in article 5(c) shall be equal to the quantity of milk or milk equivalent delivered by the producer during the 1981 calendar year (that was Formula A) or to the quantity of milk or milk equivalent purchased by a purchaser during the 1981 calendar year (that was Formula B) plus 1%.

Article 3 says: "For the determination of the reference quantities . . ." —and, as I say, "quotas" when it comes into our legislation, referred to in article 2 in connection with the applications of Formulas A and B — certain special situations shall be taken into account. This is particularly relevant in this case.

F (1) Producers who have adopted milk production development plans under Directive 72/159 (EEC) lodged before 1st March 1984 may obtain, according to the Member State's decision: if the plan is still being implemented, a special reference quantity taking account of the milk and milk product quantities provided for in the development plan, if the plan has been implemented after 1st January 1981, a special reference quantity, taking into account the milk and milk product quantities which they delivered in the year during which the plan was completed.

Investments carried out without a development plan can also be taken into account if the Member State has sufficient information.

Article 4 provides:

In order to complete the restructuring of milk production at national or regional level or at the level of the collecting areas the Member States may, in connection with the application of Formulas A and B:

— and then relevantly (b) — G

. . . grant an additional reference quantity to producers realising a milk production development plan approved after the entry into force of this Regulation under Directive 72/159 EEC, on condition that this plan meets the criteria referred to in Article 1(2) of Regulation (EEC) No 1946/81.

And (c):

Grant producers undertaking farming as their main occupation an additional reference quantity, whether their herd fulfils the conditions set out in paragraph (b) or not.

Then the terms are referred to and those to which we will subsequently come are defined in article 12. I just note now that under 12(c):

producer: a natural or legal person or group of natural or legal persons H
farming a holding located within the geographical territory of the Community: . . . supplying the purchaser;. . .

and, as I have already indicated, the purchaser in this case is the Milk Marketing Board. "Holding" is defined at (d):

all the production units operated by the producer and located within the geographical territory of the Community.

To give effect to the community milk quota and supplementary levy scheme — to which I have just been referring in these regulations — the Dairy Produce Quotas Regulations 1984 (SI 1984 No 1047) were made on July 24 1984.

As I have already noted, it was accepted in Council Regulation (EEC) No 856/84 that there would be administrative difficulties in J
re-establishing balance in the milk sector which was characterised by structural surplus. In his judgment in *Stubbs* v *Hunt & Wrigley** Phillips J referred to the regulations as being "of some complexity".

The plaintiffs say that when they heard the news on April 2 1984 of the introduction of the milk quotas they did not know what to do. They contacted everybody they could think of, but no one was able to give them any clear advice as to how they might be affected by their move on April 4 1984 to Greys Green Farm. The Milk Marketing Board's press statement on April 3 1984 apparently said: " . . . the scheme [was] implemented in haste [and] . . . many details of how the scheme will operate are still unknown".

In the defendants' closing submissions they say at para 2.1:

There is no dispute between the parties that chaos and confusion prevailed K
between about March 1984 and August 1984 due to the hasty introduction on 1st April 1984 of EEC Regulations 856/84 and 857/84 which introduced the Community milk quota and supplementary levy scheme and the later introduction in July 1984 of the United Kingdom Dairy Produce Quota Regulations, 1984 which implemented in the United Kingdom the Community milk quota and supplementary levy scheme.

In April 1984 the plaintiffs received a provisional assessment of primary milk quota calculated on the basis of their milk production at Archers Farm in 1983. They were uncertain about this because if the quota ran with the land, so to speak, it would belong to the purchasers of Archers Farm and so it would, as I understand it, if Mr Hood had made his application for it in time. He did not, so that L
capriciously, as it seems to me, the plaintiffs retained a primary milk quota. There was no question of obtaining any primary milk quota on Greys Green Farm, for there had been no milk production there for the last 10 years.

In his witness statement, which he confirmed in oral evidence, Mr Hood said that he and his wife put a great deal of thought into the effect of the milk quota system from their position as soon as details of the scheme were announced in the middle of April. That thought, if I may say so, was to good effect. For example, they formed the view, correctly, that in relation to an exceptional hardship claim there was no cut-off date as at March 31 1985. The plaintiffs made an exceptional hardship claim and, on the basis of special case treatment for those caught in the middle of an expansion programme when milk M
quotas were introduced, they made a development claim. A local panel of the Dairy Produce Quota Tribunal gave a decision on September 20 1984 on the latter and passed on the papers on the exceptional hardship claim to the tribunal themselves for consideration.

The plaintiffs considered the award of the local panel on the development claim was as good as they could expect, but they received information to the effect that if they accepted the local panel's allocation on the development claim, then the exceptional hardship claim could not go forward to the tribunal on its own. For

*Editor's note: Reported at [1992] 1 EGLR 17 *ante*.

A this reason, they took both matters to the tribunal and both matters were heard together by the tribunal, chaired by Lord Grantchester QC, on December 12 1984.

According to the plaintiffs the hearing before the tribunal lasted only about 10 minutes. Before they had gone in they were asked if they had any documents which needed copying for the tribunal. They handed over supporting documentation for this purpose but not the written submissions for the exceptional hardship claim because they had already taken three copies of them.

Mr Hood said in his witness statement that he expected to have the opportunity of taking the tribunal through the document and the various notes and documents referred to in it at the hearing. Mr Hood said that all Lord Grantchester wanted to know was how many cows
B they expected to be milking on March 31 1985. Mr Hood said the tribunal were well aware, or would have been if they had read their papers, that the plaintiffs had planned a breeding policy and that they had stock in the pipeline and had made investments in a farm which would have enabled them to be milking a 200-cow dairy herd within a short time.

In these circumstances, I do not consider that the plaintiffs were given a fair hearing at that tribunal.

A few days after the hearing Mrs Hood sent a letter to the Country Landowners' Association and on January 17 1985 the association replied:

I reply to your letter of 16th December 1984 to Mr William de Sellis. I
C apologise for the delay in replying to it. This is due to Christmas and New Year holidays and the exceptionally difficult travelling conditions I have experienced in the snow covered South East. The position is that there is no provision for appeal to the High Court under the Dairy Produce Quota Regulations. The only possible way of challenging the decision which I can think of would be to apply for an order from the High Court on the ground that the decision of the Tribunal was not within its statutory powers or that the requirements of the Regulations had not been complied with. The fact is that many dairy farmers are angry and frustrated about not being able to get the quota they expected to get, but it does not follow that the Tribunal has failed to act within its statutory powers or that the requirements of the Regulations have not been complied with.

A few days later, Mr John Davis, a group secretary with the National Farmers Union, called on the plaintiffs to see them about
D joining the NFU, having discovered that they had purchased Greys Green Farm. In his witness statement Mr Davis describes the NFU as a political-legal organisation representing farmers as a whole and one of the benefits of membership which he would have explained to the Hoods was the availability of expertise on most farming-related matters at NFU's headquarters in London, including access to experienced lawyers. He said that the plaintiffs spent some considerable time at that meeting telling him about their milk quota problem.

Mr Davis said that he was very sympathetic to the Hoods' position and he promised that, although he could not advise them, he would refer it to his county secretary, who would probably refer it to the
E NFU's headquarters for any advice which the experts there could give the Hoods as to what they could do to improve their position.

Mr Davis said that the Hoods were both forceful personalities and they left him in no doubt that they were furious about their treatment at the hands of the Dairy Produce Quota Tribunal and that they would do anything within their power to challenge the tribunal's decision.

I agree with Mr Davis' assessment of the Hoods as forceful personalities and, having seen and heard them in the witness box and read letters and documents prepared by them, I take the view, additionally, that they are both intelligent and honest witnesses and I accept their evidence. In particular, I accept that it was their position that they would do anything within their power to challenge the
F tribunal's decision.

Mr Davis said he asked the plaintiffs to let him have written details of their case for him to pass on to Mr Goldsworthy, who was the county secretary for Oxfordshire and Berkshire for the NFU and under whose control Mr Davis worked. From then on Mr Davis said he acted as little more than a post-box for transmitting the advice from the NFU's legal division to the plaintiffs when he received it from Mr Goldsworthy.

Mr Davis said that the impression he was left with from reading the opinions from the legal division and from his discussion with Mr Goldsworthy was that there was no immediate action which the plaintiffs could take to obtain redress but that the NFU was taking a number of cases through the courts which might set a precedent to enable the plaintiffs to make a successful claim at some time in the G
future. This, he said, would have been the message which he would have conveyed to the Hoods when he spoke to them, which he did at regular intervals.

Mr Davis said that occasionally Mr Hood would read an article in the press about other milk quota cases and mention it to him. This then led, he said, to further discussions between him and Mr Goldsworthy and letters to Mr Hood, such as his letter dated March 21 1986.

The first two paragraphs of that letter, which was addressed to Mr and Mrs Hood, are in the following terms headed "Quota Transfer":

Further to my letter of March 7th and subsequent telephone conversation with Mr Hood, I passed on your further observation to Mr W J Goldsworthy H
and with particular reference to your enquiry as to what had happened on the cases which had been referred for judicial review. He tells me now that having spoken to headquarters milk and legal departments again he understands that in the end there were very few cases where the circumstances merited judicial review proceedings being taken. We believe there were seven or eight. One case in particular which was referred back to the Tribunal which was the nearest in the circumstances to your own resulted in no change whatsoever. Two cases were considered and it is somewhat extraordinary that these cases are believed to have prospects of an amendment leading to an increased quota allocation.

Mr Davis said that at no time did he receive any information from Mr Goldsworthy or from headquarters which indicated that the plaintiffs' case had reached the end of the line. Certainly, in his letter J
of March 21 1986 that appeared to be the case. Right up until the time when Mr Hood consulted his own solicitors in June 1989, Mr Davis said that he continued to be under the impression that there were still cases going through the courts which could set a precedent for the plaintiffs to follow. That is a message he said he would have continued to convey to Mr Hood and, on their evidence, that was the impression and the message which the plaintiffs received.

Mr Goldsworthy made a witness statement dated November 29 1991 and he also gave oral evidence. He said in evidence that he had no doubt that the plaintiffs had a serious problem. He also said that in correspondence and discussions Mr Hood came across as someone exceptionally well-informed. It may be that because he took this view of Mr Hood Mr Goldsworthy said that he did not know how Mr K
Hood could not be aware of judicial review. Further, Mr Goldsworthy said he was under the clear impression that Mr Hood knew about the three months' time-limit in relation to judicial review proceedings. I am unable to accept this evidence of Mr Goldsworthy. Certainly, in a memorandum dated January 31 1985 from Miss Diane Smith of the legal division in NFU to Mr Goldsworthy reference is made to judicial review. The penultimate paragraph of the letter on p 9 indeed says (and I say letter, it may be a memorandum rather than a letter — anyway, it is from Miss Diane Smith to Mr Goldsworthy):

Obviously the implementation of the EEC provisions in the United Kingdom Regulations are vital and those Regulations provide that the quantity of dairy produce which the producer expects to sell by direct sale or deliver by wholesale delivery during the first quota year is to be taken into account. L
There is a problem here of the compatibility of the UK provisions with those of the EEC Regulation, but this would only be relevant with the amount of dairy produce which a producer expected to deliver and that is his expectations at the time of completing the Form MSL2 is greater than achieved by calculating the position as at 31st March, 1985. In such a case there may be a question to consider in respect of judicial review and in this regard I would refer you to Rule 23 of 85. In any event it is clear that no account is to be taken of the wider issue of post 1st April 1985 of the investments taken into account.

Miss Smith said in her evidence that it was intended that the advice she gave should be passed on by Mr Goldsworthy to Mr Davis and by him to the plaintiffs.

On the evidence before me I find that this intention was never M
implemented.

On February 6 1985 Mr Goldsworthy wrote to Mr Davis. On p 1 of that letter, the final paragraph says:

Obviously the implementation of the EEC provisions in the United Kingdom Regulations is vital and those Regulations provide that the quantity of dairy produce which the producer expects to sell by direct sale or deliver by wholesale delivery during the first quota year is to be taken into account. There is a problem here of compatibility of the United Kingdom provisions with those of the EEC, but this is only relevant where the amount of dairy produce which a producer expected to deliver (and that is its expectations at the time of completing the form MSL2) is greater than that achieved by calculating the position as at 31st March, 1985. In such cases there may (and I emphasise may) be a question to consider in respect of judicial review and in an attempt to clarify the situation NFU headquarters are taking one or two cases through the courts to try and produce some kind of clarification.

There was no clarification necessary, of course, in relation to a three-month time-limit, but nowhere that I have discovered in the evidence before me is this referred to in any context of conveying information advice from the NFU to the plaintiffs.

On February 22 1985 Mr Goldsworthy sent a memorandum to Miss Smith. The first two paragraphs are in the following terms:

> Further to your memo of 31st October and your subsequent telephone conversations, I have now obtained from the Hoods the decision taken by the Tribunal in relation to their applications both under the investment scheme and under exceptional hardship. You will see that under the investment case the Panel award was reduced and under exceptional hardship their case was rejected.
>
> I would like you to re-examine these papers in connection with the procedure to a judicial review so called. I can state that the Hoods quite definitely have decided that they will wish to proceed to such a review and any help and guidance which we can give them in that regard would be most appreciated. Equally, of course, if this case has the merits of the kind you are seeking to use to proceed through the various legal purposes, they would be happy to cooperate with that procedure.

I am unable to accept the sentence where he states: "I can state that the Hoods quite definitely have decided that they will wish to proceed to such a review", that is the judicial review as stated in this memorandum. Certainly the plaintiffs wanted to act. That had been their position from the beginning. Their problem was how to act. At that time, the time of this memorandum, Mr Hood said they did not know anything about judicial review and I accept his evidence.

The concluding paragraph in Mr Goldsworthy's memorandum to Mr Davis dated March 6 1985 is not consistent, in my judgment, with the knowledge of a determination on the part of the plaintiffs to proceed by way of judicial review. That paragraph is in these terms:

> I know that Mr Hood has indicated that he would wish to pursue such a case no matter what the opinion might be and would assure him that we will give every help we could to his solicitor should he decide to follow this course. The NFU is pursuing several cases through, probably ultimately to the European Court and maybe in doing so we may well be able to provide information which would be helpful to Mr Hood.

There is still no reference to the time-limit affecting judicial review proceedings.

On the evidence before me it is clear that the plaintiffs were indignant at the way they had been dealt with by the tribunal and what they wanted to do was challenge that decision. On the evidence the defendants were well aware of this and knew that the plaintiffs relied upon them for their advice.

Mr Hood said in evidence that he banged on endlessly to anyone who would listen; but I accept his evidence that he chose to leave the matter in the hands of the NFU because he did not think he could get better advice from any other source.

To challenge the tribunal's decision, the plaintiffs had to proceed by way of judicial review within three months of the date of the tribunal's decision.

I find as a fact that the NFU did not give the plaintiffs that advice and in that failure the defendants were, in my judgment, in breach of their duty to exercise reasonable care and skill in the advice which they gave; and the liability arises, in the circumstances of the case, both in contract and in tort.

The defendants submit that on the facts placed before the tribunal and the NFU the plaintiffs were not entitled to make an exceptional hardship claim successfully and, in such circumstances, the defendants cannot have been negligent. I now turn to this aspect of the case.

Notwithstanding the Dairy Produce Quotas Regulations 1984 that came into operation on July 24 1984, it would appear that the provisions in Schedule 2 to those regulations confirm that April 2 is the relevant date. For the purpose of this hearing this is accepted by the plaintiffs.

In order to make an exceptional hardship claim the plaintiffs must be producers within the meaning of article 4(1)(c). As we have already discovered, for that purpose "producer" is defined in article 12(c) of that regulation: "a natural or legal person or group of natural or legal persons farming a holding located within the geographical territory of the Community:. . . supplying the purchaser".

The defendants submit that the plaintiffs were not producers as so described because: (1) the only relevant holding was Greys Green Farm; (2) they were not farming Greys Green Farm as of April 2 1984; and (3) they were not supplying the Milk Marketing Board from Greys Green Farm on that day.

I do not consider that as at April 2 1984 the only relevant holding was Greys Green Farm. "Holding", as we have seen, is defined as "all the production units operated by the producer and located within the geographical territory of the Community". Clearly, on April 2 Archers Farm was a production unit and, because of this, apart from any other reasons, the plaintiffs were producers, in my judgment, in the relevant sense.

In the *Stubbs* case, to which I have already referred, Phillips J held that exceptional hardship quota could be awarded only if a producer were entitled to primary quota on April 2 1984 and the defendants submit that the plaintiffs were not so entitled and, accordingly, could not properly make an exceptional hardship claim. But notwithstanding the contingency relating to the claim by the purchaser of Archers Farm, it seems to me that, as a matter of fact, the plaintiffs were so entitled.

The defendants submit that no exceptional hardship claim could properly have been made in respect of Greys Green Farm because the plaintiffs have already been allocated secondary wholesale quota in respect of the special-case development claim based upon the same transaction or arrangement as that relied upon in support of their exceptional hardship claim. They stated that the purpose of the community milk quota and supplementary levy was to curb the increase in community milk production. Further, they submitted that any ambiguity in the 1984 regulations must be resolved in a manner consistent with the purpose of the community provisions and not in a manner inconsistent with them.

I agree, but the purpose of the regulations is not only to curb expansion but, as one might expect, also to permit structural development and adjustments of producers committed to expansion prior to April 2 1984 as was the case with the plaintiffs.

In my judgment, it is clear from the terms of article 4(1)(b) of Council Regulation (EEC) No 857/84 (to which I referred earlier in this judgment) that a development may lead to quota under both article 4 and article 3 (to which I also referred earlier). The question is whether the provisions therein are satisfied in respect of the claims made.

The defendants then submit that the plaintiffs' claim is still bound to fail because, in the view of the tribunal, none of the requirements of the 1984 regulations, Schedule 2, para 17(3) (b), (c), (d) or (e) thereof was satisfied and it cannot be argued that no reasonable tribunal, properly directing themselves, could have reached the decision reached by the tribunal in the present case.

I am unable to accept this submission. I do not accept their conclusion that none of the requirements of para 17(3) (b), (c), (d) or (e) was satisfied. It is beyond doubt, as it seems to me, that requirement (b) was satisfied; and, in my judgment, there were other errors of law in the decision, namely: (1) no reasons were given; (2) the tribunal erred in law in regarding the cut-off period as being March 31 1985; and (3) as I have already found, the plaintiffs were not, in my judgment, given a fair hearing.

Looking at all the circumstances and putting myself in the position of the divisional court, I have to ask whether the tribunal might not on a remitter have reached a different conclusion properly directing themselves as to the law. The answer is, that I think it would.

Further, I consider that, on a remitter, there is, notwithstanding the complexity of the regulation, a substantial chance (say 60%) of establishing a claim to exceptional hardship quota.

For further cases on this subject see p 289

RATING

A **Court of Appeal**

July 19 1991

(Before Lord Justice PURCHAS, Lord Justice GLIDEWELL and Sir John MEGAW)

FARMER (VO) v HAMBLETON DISTRICT COUNCIL AND ANOTHER

Estates Gazette March 21 1992

[1992] 11 EG 118

B **Rating — Poultry processing factory — Hereditaments used for processing feed and slaughtered birds — Ratepayer owning many poultry breeding and rearing farms — Whether subject hereditament "occupied together with" and "used solely in connection with" operations in an agricultural building — Whether buildings rateable — Appeal by respondent rating authority dismissed**

The second respondent, Buxted Poultry Ltd ("Buxted"), occupied a provender mill and poultry processing factory — Buxted made proposals to delete buildings from valuation list for rating purposes on ground that they were agricultural
C **buildings within the meaning of the Rating Act 1971 — Local valuation court deleted entries from the valuation list — Lands Tribunal dismissed appeals by valuation officer and rating authority — Rating authority appealed by way of case stated against decision of tribunal — Section 26 of the General Rate Act 1967 derates agricultural buildings — "Agricultural buildings" defined as buildings "occupied together with agricultural land . . . used solely in connection with agricultural operations carried on on agricultural land" — Effect of decision of House of Lords in *W&JB Eastwood Ltd* v *Herrod (VO)* abrogated by Rating Act 1971 — Rating Act 1971 defines agricultural building as including any building**
D **"which is occupied together with one or more buildings [used for the keeping or breeding of livestock] and is used in connection with the operations carried on in that building or those buildings" — Buxted submitted that the provender mill and chicken processing factory are buildings occupied together with buildings used for the keeping or breeding of livestock and used in connection with the operations carried on in those buildings — Buxted must satisfy qualification in section 2(3)(*a*) of the 1971 Act that a building is not an agricultural building unless "it is solely so used"**

Buxted owned and occupied 67 farms of which 48 were
E **rearing and the other 19 were breeding farms for poultry — The farms were situated between a quarter of a mile and 120 miles from the subject hereditaments — The provender mill produced pelleted feed for consumption by the poultry on the 67 farms — Some 6% to 8% of feed delivered to farms not owned or occupied by Buxted — Live poultry delivered to processing factory from the 48 poultry rearing farms for slaughter, plucking, cleaning and processing — Buxted submitted that the provender mill and/or the processing factory were occupied together with buildings used for the keeping or breeding of livestock and solely used in connection** F
with the operations carried on in those buildings

***Held*: There was no authority binding on the court in which the meaning of the phrase "occupied together with" in relation to either a building and land or a building and other buildings has definitively been considered — Two areas of land, or a building and land, can be said to be "occupied together with" if, even though they are a considerable distance apart, the operations on the land are carried out under the control of the same farmer with men and machinery sent out from the building or other land — But distance must be a relevant factor in deciding whether a building and other land, or two buildings in the present case, are occupied** G
together with each other — Tribunal falling into error in holding that the geographical test, that is to say the distance between the hereditaments the subject of the appeal and the buildings used for the breeding and rearing of livestock, was not a relevant consideration — A building can be said to be occupied together with another building used for the breeding or keeping of livestock, provided, first, that they have a single occupier; second, that the activities carried on in both are jointly controlled or managed; and third, that the physical communication between the two buildings is, by reason either of physical nearness or of some other factor, so close and H
convenient that they can properly be regarded as being occupied as parts of the same enterprise — Tribunal erred in law in adopting the wrong test

The provender mill was not "solely used" in connection with the operations carried on in the chicken breeding and rearing buildings — The 6% to 8% of processed feed going to farms not owned by Buxted was not *de minimis* — In any event there is no need or room to import the concept of *de minimis* — The poultry processing buildings were solely used in connection with the operations carried on at the 48 broiler rearing houses — It was a question of fact and degree for the J
tribunal to decide whether the sales office, canteen, shop and vehicle maintenance workshops were part of the processing factory — Appeal allowed in relation to the provender mill — The question of the relative geographical positions of the buildings is a question of fact and must be remitted to the tribunal for determination

The following cases are referred to in this report.

Eastwood (W & J B) Ltd v *Herrod (VO)* [1971] AC 160; [1970] 2 WLR 775; [1970] 1 All ER 774; [1970] RA 63, HL

Handley (VO) v *Bernard Matthews plc* [1989] 1 EGLR 215; [1989] 14 EG 67; [1988] RA 222

Hilleshog Sugar Beet Breeding Co Ltd v *Wilkes (VO)* (1971) 17 RRC 275; K
[1971] RA 275

In this appeal the Court of Appeal had to determine upon a case stated whether a decision of the Lands Tribunal (C R Mallett FRICS and T Hoyes FRICS) given on November 24 1989 on appeals to the tribunal against a decision of the North Yorkshire Local Valuation Court was correct. The second respondent, Buxted Poultry Ltd, had made proposals to delete a provender mill and poultry processing factory at Dalton, near Thirsk, North Yorkshire, from the valuation list on the ground that they were agricultural buildings.

Christopher Cochrane QC and Simon Bird (instructed by Sharpe Pritchard, for the solicitor to Hambleton District Council) appeared for the appellant first respondent; Guy Roots QC and Sebastian Head (instructed by Stephens & Scown) represented the second respondent, Buxted Poultry Ltd. The valuation officer did not appear and was not represented.

Giving the first judgment at the invitation of Purchas LJ, **Glidewell LJ** said: These are two appeals by way of case stated against a decision of the Lands Tribunal given on November 24 1989 on appeals to the tribunal against a decision of the North Yorkshire Local Valuation Court. Under section 3(4) of the Lands Tribunal Act 1949 an appeal to this court lies only for error of law. The second respondents, Buxted Poultry Ltd ("Buxted"), at all material times owned and occupied the buildings the subject of these two appeals, namely, a provender mill and a poultry processing factory at Dalton, near Thirsk, North Yorkshire. Buxted made proposals to delete both buildings from the valuation list for rating on the ground that they were agricultural buildings within the meaning of the Rating Act 1971. The valuation officer objected to both proposals and Hambleton District Council, the present appellants, objected to one of the proposals. The local valuation court deleted both entries from the list. The valuation officer and the district council as rating authority appealed to the tribunal, which dismissed the appeals. The district council now appeal to this court.

I start by saying something about the legislation and its background. The primary section which deals with the derating of agricultural land and buildings is now section 26 of the General Rate Act 1967. That is a successor to a provision which dates back originally to the Local Government Act 1929. Section 26(1) provides:

> No agricultural land or agricultural buildings shall be liable to be rated or be included in any valuation list or in any rate.

Subsection (2) is not relevant for present purposes.

Subsection (3) defines "agricultural land" in terms which include:

> In this section the expression "agricultural land" —
> (*a*) means any land . . . used for the purposes of poultry farming, . . .

Subsection (4) defines "agricultural buildings" as:

> (*a*) means buildings (other than dwellings) occupied together with agricultural land . . . used solely in connection with agricultural operations thereon; and
> (*b*) includes a building which is used solely in connection with agricultural operations carried on on agricultural land and which is occupied either —
> (i) by the occupiers of all that land; or
> (ii) by individuals who are appointed . . .

I need not go on with that. It is really dealing with a situation in which there is some co-operative operation on the land.

In *W & J B Eastwood Ltd* v *Herrod (VO)* [1971] AC 160 the House of Lords held that a number of buildings in which broiler chickens were hatched and reared, which were occupied together with some 1,150 acres of land, were nevertheless not agricultural buildings within the definitions to which I have just referred in section 26 of the 1967 Act.

The Rating Act 1971 abrogated this decision, apart from having other effects.

By section 1(1) of the 1971 Act:

> In section 26 of the General Rate Act 1967 (in this Part of this Act referred to as "the principal section") —
> (*a*) the expression "agricultural buildings" shall include any building which is an agricultural building by virtue of section 2, 3 or 4 of this Act; and
> (*b*) the expression "agricultural land" shall include land occupied with, and used solely in connection with the use of, one or more such buildings.

I shall have to come back to section 1 later. But for present purposes I go on to section 2.

Section 2(1) provides as follows:

> Subject to subsections (2) to (4) of this section, each of the following is an agricultural building by virtue of this section —
> (*a*) any building used for the keeping or breeding of livestock; . . .

Stopping there, that had the effect, to which I have already referred, of reversing the House of Lords decision in the *Eastwood* case. But the Act went further by the next subparagraph of subsection (1), which provides:

> (*b*) any building (other than a dwelling) which is occupied together with one or more buildings falling within paragraph (*a*) above and is used in connection with the operations carried on in that building or those buildings.

That subparagraph is the battleground on which these appeals have been fought. What is said by Buxted is that both the provender mill and the chicken processing factory are buildings occupied together with buildings used for the keeping or breeding of livestock and used in connection with the operations carried on in those buildings.

Subsection (1) refers to the following subsections of section 2 and it is necessary to go to subsection (3) which provides:

> A building occupied and used as mentioned in subsection (1)(*b*) of this section is not an agricultural building by virtue of this section unless either —
> (*a*) it is solely so used; or
> (*b*) it is occupied also together with agricultural land . . . and used . . . in connection with agricultural operations on that land, . . .

— (*b*) does not arise in the present case; (*a*) does.

There is another qualification in subsection (4):

> A building is not an agricultural building by virtue of this section unless it is surrounded by or contiguous to an area of agricultural land . . . which amounts to not less than two hectares.

It is conceded in this case and has always been that that condition is met.

In order to succeed, Buxted must satisfy both of the requirements in section 2(1)(*b*) as qualified by section 2(3)(*a*). That is to say, they must prove that the building in question, the provender mill or the processing factory, is "occupied together with" buildings used for the keeping or breeding of livestock; and "solely used in connection with" the operations carried on in those buildings in order to succeed.

The tribunal gave one decision which covered both the appeal in relation to the provender mill and that in relation to the poultry processing factory. The tribunal found, from an agreed statement of facts and from evidence put before it, that, as I have already said, the two hereditaments are situated in the rural village of Dalton, two miles south of Thirsk, and comprise a provender mill and a separate processing factory. It found that Buxted owned and occupied 67 farms of which 48 were rearing and the other 19 were breeding farms. They were situated between a quarter of a mile and 120 miles from Dalton, that is from the two relevant hereditaments. On those farms, in broiler houses, turkeys and chickens were bred and reared. The provender mill produced pelleted feed for consumption by the turkeys and chickens in the 67 farms. The quantity of feed produced was approximately 2,700 tons per week. The bulk of this was delivered to the 67 farms occupied by Buxted. Between 6% and 8%, which came up to 216 tons per week of the pellets, was not delivered to Buxted farms but was delivered to four other farms occupied by a group of independent turkey rearers. The tribunal found (and this is relevant to the next issue to which I have to come and so I cite it because it is one of the facts):

> The time spent in producing the feed is roughly pro-rata to the bulk, so that the 6% to 8% of feed which goes to the four independent farms takes up about 6% to 8% of the working time of the mill.
>
> It is admitted between the parties that the 67 poultry breeding and rearing farms are "buildings used for the keeping or breeding of livestock" within the meaning of section 2(1)(*a*) of the Rating Act 1971.

As I have already said, it is also admitted that both the subject hereditaments are surrounded by or contiguous to an area of agricultural land of not less than 2 ha.

So far as the poultry processing factory was concerned, the objects of the company were the breeding, rearing, slaughtering, processing, packing, selling and distribution of turkeys and chickens for human consumption. In 1985 a weekly average of 63,000 turkeys and 25,000 chickens were delivered live to the poultry factory from the 48 poultry rearing farms owned and occupied by Buxted. They were slaughtered, plucked, cleaned and processed. Some of them were chilled and then sold as fresh birds. A small number, particularly those which were somewhat misshapen, were sold after boning. The majority went through a process which finished with their being frozen and stored in a cold store and then sold in the frozen state. A proportion of them before they were sold were basted, that is to say the breast meat of each carcass was injected with liquid butter. The tribunal described the process and I do not think it is necessary to go into it. Buxted sold and delivered poultry which had been through this plant to all the 12 major retailers in the United Kingdom. The plant itself comprised a number of buildings connected to form a single factory, with integral walls which had been either demolished or altered in order to achieve the object of creating a continuous flow of the processing. The buildings on the hereditament included a building for offal-handling equipment, staff service facilities, including staff locker rooms, a canteen and a shop for use by the employees of the company; process-management offices and reception area; a small, three-room office occupied by sales staff; and vehicle maintenance facilities which serviced lorries and collection vehicles owned by the company and company cars used by employees in the course of their duties. Those are the essential facts.

I turn to consider whether on those facts Buxted satisfied both of the requirements in section 2(1)(*b*) of the 1971 Act as qualified by section 2(3)(*a*). I will deal with each phrase in turn.

Occupied together with

We were referred to a number of cases of which the first is the House of Lords decision, to which I have already referred, in the *Eastwood* case. The appellant company, Eastwood, owned and occupied 1,150 acres of land and a substantial number of broiler houses in some of which chickens were hatched and in the remainder of which, after their removal from the hatcheries, they were reared. Most of the birds never went outside the buildings, except for the short journey from one to the other. The cockerels, however, were from time to time put out to run on part of the 1,150 acres. The remainder of the land, the greater part of it, was arable and used for growing barley or other corn which was fed to the chickens as part of their feedstuff. The issue in that case was not the issue I am currently considering but the second issue: were the broiler houses used "solely in connection with" agricultural operations on the 1,150 acres? In that case it was conceded that the buildings were occupied together with the land. It was held both in this court and in the House of Lords that the broiler houses were not agricultural buildings because they were not used solely in connection with the agricultural operations on the land. However, in the course of his speech Viscount Dilhorne briefly considered the first test, that is to say: were the buildings occupied "together with" the land? At p 180 he said:

> In this case the respondent conceded, somewhat surprisingly, that all the buildings were occupied together with the 1,150 acres. But for this concession I do not think I should have found it easy to conclude that the packing station in Gainsborough nine miles or so away was occupied together with the agricultural land in the sense in which those words are used in the definition, and it may be that I would have had difficulty in coming to that conclusion in relation to the five layer houses at Norton Brisney some six miles away and some of the other buildings.
>
> In its context "occupied together with agricultural land" may connote more than common ownership. My impression on reading the definition of "agricultural buildings" is that it was an attempt by the draftsman to define a farm in statutory language and that it was intended to include buildings used and occupied together with the land for the purpose of farming the land, not buildings far distant and not used in connection with an operation on the land, even though owned by the same person.

That observation was clearly *obiter* and nothing similar is to be found in any of the other speeches in that case. However, the observation has been recognised as deserving of considerable respect.

There is, it seems, no authority binding on this court in which the meaning of the phrase "occupied together with" in relation to either a building and land or a building and other buildings has definitively been considered. There are, however, relevant decisions of the Lands Tribunal.

In *Hilleshog Sugar Beet Breeding Co Ltd* v *Wilkes (VO)* [1971] RA 275 heard before Sir Michael E Rowe QC, the then president of the tribunal, one issue was whether a plant research centre was occupied together with a number of widely scattered plots of land, some of which were up to 40 miles away from the centre. The plots of land were used for raising sugar beet plants. The president held that the research centre was occupied together with the land. His reasoning on that issue is set out at p 284 of the report. He quoted the 12th ed of *Ryde on Rating* where the learned editors say:

> The buildings need not be situated upon, or adjoin, the agricultural land; separation by distance of several miles has not been regarded as fatal to exemption. Nor is it necessary for land and buildings to be held by the occupier under the same title. "Occupied together with" seems to imply a functional rather than a geographical connection.

The president then quoted the passage from Lord Dilhorne's speech in the *Eastwood*'s case which I have read and, said, as I have, that it was *obiter* but should be treated with great respect. He then said at p 285:

> . . . I would readily accept that, even in these days of swift transport, distance might make it impracticable to "occupy together" buildings in one place and land in another because it would be quite uneconomic and inefficient for a farmer to try to control operations in both. But where, as in the present case, ratepayers can and do send out from the centre machines and men to carry out agricultural operations on the land and do it not merely occasionally but regularly year after year I think it is very difficult to deny the "togetherness" of the centre and all these plots and in so far as it is a question of fact, I find that the centre is occupied together with all and each of the plots.

He then gave an illustration from his own personal experience, which I think perhaps does not carry the matter much further.

Clearly that was a case very much on its own facts. The togetherness which the president found arose not merely from the fact that the centre and the pieces of land were all occupied by the appellant company, and thus had one occupier, but from the fact that the fields were all worked by men and machines sent out from the centre.

For my part, I agree with this approach. In my view, two areas of land, or a building and land, can be said to be "occupied together" if, even though they are a considerable distance apart, the operations on the land are carried out under the control of the same farmer with men and machinery sent out from the building or other land. So two farms at some distance from each may nevertheless be occupied together. But, in my view, distance must be a relevant factor, and how great a distance is relevant must depend upon the nature of the operation. It may well be practicable to send men and machinery 40 miles to plant out or fertilise some seed beds. It may be practicable for one farmer operating two farms several miles from each other to use the same tractors and the same combine harvesters driven by the same operatives at each. But if the two farms are 100 miles apart then it might well not be practicable to farm them in that way, in other words to use the same men and equipment to farm them both. In the latter case the president's test would not be satisfied. So, in my view, distance must be a relevant factor in deciding whether a building and other land, or two buildings in this case, are occupied together with each other.

We were referred to two decisions of the Scottish land valuation appeal court. I do not refer to them, not out of lack of respect but because the decisions relate to differently worded legislation, the difference between the Scottish and English statutes being explained as part of the reasons for the decisions.

So far, the cases to which I have referred were cases of a building or buildings on the one hand and land on the other. The concept of sending out men and equipment to carry out farming operations does not normally apply to the relationship between two buildings or one building and a series of other buildings. I therefore now come to the only decision which was cited to us in which two or more buildings were under consideration and the question considered was whether they were occupied together with each other. That was a decision of the Lands Tribunal in *Handley (VO)* v *Bernard Matthews plc* [1988] RA 222*. The member of the tribunal in that case was the president, Mr V G Wellings QC. The headnote reads as follows:

> A mill and premises, Gayton Road, Bawsey, King's Lynn, Norfolk, was held by the local valuation court to be exempt from rating under s 2 of the Rating Act 1971. The hereditament was used by the company for the production of pelleted food fed to turkeys on 29 farms owned and occupied by the company and situated between nine and 74 miles from the hereditament. The valuation officer appealed and said that the exemption did not apply because the requirement of s 2 that the hereditament must be "occupied together with" one or more livestock buildings was not satisfied. He said that regard must be had to a geographical test and the hereditament and the farms were too far apart sensibly to be regarded as occupied together.

The tribunal dismissed the appeal. It will be seen that the facts were very close to the facts of the present case in relation to the provender mill. The learned president, at the bottom of p 227 of his decision, said:

> Counsel for the valuation officer, in addressing me, conceded that the appeal hereditament was used in connection with the operations carried out in the livestock buildings on the 29 farms but he did not concede that it was solely so used nor that it was occupied together with buildings used for the keeping or breeding of livestock. He said that the words "occupied together with" required investigation not just of the functions of the various different buildings on the 29 farms but their geography as well. Thus even though the appeal hereditament and the livestock buildings were functionally connected with the business carried on by the company there came a point at which they

*Editor's note: Also reported at [1989] 1 EGLR 215.

were too far apart sensibly to be regarded as occupied together. He was prepared to accept that the appeal hereditament was occupied together with some of the livestock buildings but he did not identify them. His point was that as a matter of fact livestock buildings so far distant as 74 miles from the appeal hereditament were on the wrong side of the line. Accordingly, it could not be said that the appeal hereditament was occupied together with all the livestock buildings. On the assumption that that submission was correct, counsel for the valuation officer further contended that the appeal hereditament could not be said to be solely used in connection with operations in livestock buildings with which it was occupied together, for the reason that it was also used in connection with buildings with which it was not occupied (because they were too far away). On this subsidiary question, counsel for the company was in agreement with counsel for the valuation officer. Accordingly there is but one question for me to decide and that is whether the appeal hereditament, at the relevant date, was occupied together with the livestock buildings on all the 29 farms.

Counsel for the respondent company said that there was a danger, in attempting to devise tests, functional or geographical, to ascertain the meaning of the words "occupied together with", of putting a gloss on them. He did not wish to suggest that geography was irrelevant but the element of distance was subsumed in the functional test. He asked rhetorically: "What is the distance beyond which is too far?"

Counsel for the company in that case was Mr Guy Roots, leading counsel for the respondent company in the present case. During the course of argument in the present case I asked Mr Roots what he meant, if he was accurately reported, by saying that the "element of distance was subsumed in the functional test". I regret to say that I found his answer unclear.

The president, in his decision, concluded as follows [at p 229]:

> On the facts, I find that the appeal hereditament at all material times was and is occupied together with the livestock buildings on all the 29 farms. I am not impressed by any geographical test, at all events in the present case. If it is desired to ascribe a specific meaning to the words "occupied together with", it appears to me that the most likely meaning is "occupied at the same time as": see Stroud, *Judicial Dictionary* (5th edn). . . For these reasons the local valuation court came to the right decision, the appeal hereditament is exempt from rating and the appeal is dismissed.

In the present case it seems that the tribunal and counsel who appeared before them may well have considered (and I do not suggest incorrectly) that they were bound to follow the decision of the tribunal in the *Bernard Matthews* case.

The tribunal in the present case, at p 14 of their decision, said:

> . . . If we were to follow the narrow meaning of "occupied together with" it would appear that the buildings, if not contiguous with agricultural land, should be in the same neighbourhood, valley or hill side, so that labour, stock and machinery could readily be transferred from one to the other.
>
> We were not referred to any subsequent occasions where this narrow interpretation has been followed.

The tribunal then referred to the *Hilleshog Sugar Beet Breeding* case and to *Bernard Matthews*, quoted the conclusion of the president which I have just quoted, and then said:

> We agree with those views.
>
> In the instant cases, counsel for the valuation officer and for the rating authority accept that separate distances of 20, 40 or even 74 miles are not detrimental to the concept of "occupied together with" but, they say, 100 miles is too much.

I interpolate that Mr Cochrane, counsel for the valuation officer before us and before the tribunal, has explained that he made that concession in the sense that he was conceding that the tribunal in the present case were bound by the decision of the president in the *Bernard Matthews* case and thus before the tribunal he felt obliged to make the concession. But it was not one that he made in this court at all.

> In other words

— the valuation officer and the district council —

> do not seek to show that occupation within the same valley, hill side or neighbourhood, which we would term the narrow interpretation, has to be followed.

But Mr Cochrane has sought to show that in this case. Indeed, at one stage he propounded an even narrower test. The tribunal continued:

> On the evidence in these two cases there is no difference in the functional link between either of the hereditaments and the farms they serve, whether they are 20 or 120 miles apart. Distance does not appear to weaken this link nor deter the company from occupying those distant farms together with their buildings at Dalton.
>
> For these reasons we do not think that a geographical test has merit in these cases and we find that the provender mill and the factory are, as a matter of fact "occupied together with" the farms in the occupation of the company.

The president in the *Bernard Matthews* case and the tribunal in the present case were thus saying that the geographical test, that is to say the distance between the hereditaments the subject of the appeal and the buildings used for the breeding and rearing of livestock (the broiler houses), was not a relevant consideration. That, in my judgment, was the basis of both decisions. In so holding, in my view, both tribunals fell into an error which was an error of law. If the present tribunal's approach and that of the president were correct, all that is necessary to show that a building comes within section 2(1)(*b*) is to prove, first, that it and the building or buildings used for the breeding or keeping of livestock are both occupied by the same person or company and, second, that the second building is used in connection with the operations carried out in the livestock building.

This interpretation makes the phrase "together with" in the total phrase "occupied together with" otiose and in my view, therefore, cannot be correct. "Occupied together with" must be given some meaning. In my view, a building can be said to be occupied together with another building used for the breeding or keeping of livestock, provided, first, that they have a single occupier; second, that the activities carried on in both are jointly controlled or managed; and, third, that the physical communication between the two buildings is, by reason either of physical nearness or of some other factor, so close and convenient that they can properly be regarded as being occupied as parts of the same enterprise. I therefore conclude that the tribunal erred in law in adopting the wrong test to decide whether each of the buildings was "occupied together with" broiler houses.

It follows that I would allow both appeals. But, before considering what orders should follow from that, it is desirable to consider the other issue because different considerations here apply to the two different buildings. For the moment I content myself by saying that in the case stated the tribunal have posed two questions of which the second is:

> Was the Tribunal correct in holding on the facts and agreed and found by the Tribunal that the hereditament was occupied together with agricultural land?

Solely used

It follows that I would answer that question "No".

I turn to the second issue, whether each of these buildings in turn can be said to have been "*solely used*" in connection with the operations carried on in the chicken breeding and rearing buildings. I consider first the provender mill. The argument here turns upon the facts to which I have already referred, that 6% to 8% of the output of the provender mill was delivered to four farms not occupied by Buxted. It follows, therefore, that unless 6% to 8% of the output can be disregarded as being *de minimis*, on the face of it the mill was not used solely in connection with the operations at Buxted's poultry farms. However, by section 1(2) of the 1971 Act, in determining whether a building used in any way is solely so used, "no account shall be taken of any time during which it is used in any other way, if that time does not amount to a substantial part of the time during which the building is used".

The tribunal have found that producing the 6% to 8% of feed which goes to other farms "takes up about 6% to 8% of the working time of the mill" and that this is not a substantial part of the total production time of the mill. The flaw in this reasoning is that when section 1(2) refers to "any time during which [the mill] is used in any other way" it is clearly referring to a definable period of time. If a building is normally used for one purpose but, a let us say, on Friday afternoons or on the first Monday in each month is used for some other purpose, that is the sort of concept which, in my view, falls within section 1(2). But it is not suggested here that there was any definable period of time when the mill was being used to produce pellets for the other four farms. It was used to produce pellets the whole time and 6% to 8% of those pellets went to the other four farms. The "6% to 8% of the working time" was merely the consequence of the fact that the production process was continuous, and presumably at a constant rate, so that any proportion of the total production must occupy the same proportion of the working time of the mill. In my judgment, section 1(2) has no application at all to the question here under consideration.

That takes me back to section 2(3). As I have already said, since 6% to 8% of the produce of the mill was going to enterprises other than that of Buxted, the mill was not solely used in connection with the operations carried on in the 67 broiler houses unless the 6% to 8% can be disregarded as *de minimis*. There are two points here. First of all, does this concept apply at all to this particular part of the legislation?

A In my view it does not. This part of the legislation expressly includes the "substantial part of the time" provisions in section 1(2). There is no need or room to import the analogous concept of *de minimis*. But, even if that be wrong, I do not see how 6% to 8% of the production could be regarded as *de minimis*. Apart from my general view, I am supported by the fact that clearly Buxted do not and did not regard it as *de minimis*, because they would not have continued to supply the four other farms with about 200 tons per week of pellets if they had. I think that this is conclusive and, so far as the provender mill is concerned, on that ground also I would allow the appeal. In respect of the provender mill and to the first question asked by the tribunal, namely "was the Tribunal correct in holding on the facts agreed and found by the Tribunal that the hereditament was used solely in
B connection with agricultural operations?" I would answer also "No".

Different considerations apply to the factory. The question here was: "Was the factory solely used in connection with the operations carried on at the 48 broiler rearing houses?" I have already set out the description of the various buildings and activities at the factory. Before the tribunal counsel for the valuation officer and the rating authority contended that the use of the canteen and shop provided for the employees, the use of the sales office and Mr Cochrane before us added, "the use of the vehicle servicing facilities" were not uses in connection with agricultural operations on the 48 farms, though they conceded that those parts could be separately assessed if they did not
C qualify for exemption. Indeed, Mr Cochrane widened the argument also to comprehend the basting operation.

The tribunal found that the basting operation was "reasonably necessary to render the product marketable" and, so far as a number of the other buildings were concerned, they said:

> As to the offices occupied by the sales staff, we find that the size, type and degree of use are in keeping with the scale of operations at the factory and that they are ancillary thereto.
>
> In a similar manner we find that the staff facilities, such as staff canteen and staff shop, are ancillary to the factory and in keeping with the scale of the operations at the factory.
>
> For these reasons we find on the facts that the whole of the factory is an agricultural building or buildings used in connection with agricultural
> D operations on the 48 farms and that is the sole use.

Mr Cochrane before us contended that in applying the ancillary test to some of these buildings the tribunal were applying the wrong test and this was a concept not relevant here. On this aspect of the matter, for my part I do not agree with him. Given that the main part of the factory was solely used to slaughter and process birds from the 48 farms, the only issue was whether the sales office, the canteen, the shop and the vehicle maintenance workshop were so much part of the factory that it could properly be said that they were used for those purposes also. To decide whether those various areas were properly to be considered part of the factory, the ancillary test was, in my view, a proper one. It thus became a question of fact and degree for the tribunal with which, in my view, this court cannot properly interfere.

If this point were the sole issue, I would therefore find in favour of E
Buxted as regards the factory. But in the light of my view as regards the question of "occupied together with" and, so far as the provender mill is concerned, my conclusion on the "solely so used" issue, it follows, in my judgment, that in relation to the provender mill the proper course is to allow the appeal and reinstate the provender mill to the valuation list with the value which has already been agreed between the parties for that purpose. Differing considerations may apply as regards the factory because, although I think it right that the appeal should be allowed on the "occupied together with" test, it may be that Buxted may seek to argue that it would be right to send the factory appeal back to the Lands Tribunal for the finding of further facts. That, if my lords agree with my F
judgment, is a matter that maybe will have to be considered further.

For those reasons, I would allow both these appeals.

Sir John Megaw agreed and did not add anything.

Agreeing, **Purchas LJ** said: The appeal relating to the provender mill should be allowed on the grounds that the tribunal misdirected itself as to the effect of section 1(2) and section 2(3)(*a*) of the Rating Act 1971 in coming to the conclusion that the mill was "solely used" and I agree with Glidewell LJ that this hereditament should be restored to the valuation list. It is not necessary, in my view, for the purposes of this judgment, to consider the question whether the mill was occupied together with a building used for agricultural purposes. G
I further agree that the appeal relating to the poultry processing factory should be allowed on the ground that the tribunal misdirected itself in so far as it ignored the geographical relationship between the factory and the agricultural buildings. In coming to the conclusion that it was occupied "together with" those buildings I agree with Glidewell LJ that the relative geographical positions of the buildings falling for consideration under section 2(1)(*b*) of the 1971 Act is a factor which, together with other relevant matters, must all be considered in assessing as a question of fact whether the buildings are or are not within section 2(1)(*b*). This is not a matter which, in my view, and with respect to Glidewell LJ, can be determined by this court and the matter must, unless conceded by the respondents, be H
sent back for determination by the tribunal.

As to the second ground on which Mr Cochrane attacked the tribunal's finding, that the factory was "solely used" in the agricultural operation, I would not disturb the decision of the tribunal on this aspect. In my judgment, this was purely a question of fact for the tribunal and, therefore, not a matter upon which this court should interfere.

I would accordingly answer the questions in the same manner as Glidewell LJ has proposed and I also agree with the course that he has indicated in relation to the two hereditaments.

Appeal allowed, with costs of appeal and below: leave to appeal to House of Lords refused.

REAL PROPERTY AND CONVEYANCING

A **Chancery Division**

May 23 1991

(Before Mr Justice MUMMERY)

RE 1-4 WHITE ROW COTTAGES, BEWERLEY

Estates Gazette February 22 1992

[1992] 07 EG 119

Commons Registration Act 1965 — Common Land (Rectification of Registers) Act 1989, section 1(2) — Notice of
B **objection**

Four old, stone cottages, unoccupied for the past 20 years, derelict and unfit for human habitation, were now registered under the 1965 Act in the register of town and village greens — Notice of objection to their inclusion in the register were given in accordance with section 1 of the 1989 Act — An inquiry was held by a commons commissioner — The 1989 Act required as a condition of its application that there should be a "dwellinghouse" on the land — The commissioner decided that "dwellinghouse" in this context meant a building actually dwelt in or at least capable of being dwelt in — As the four old cottages were not within this category, the 1989 Act
C **did not apply to them, ie the objection to their conclusion in the register failed — The commissioner stated a case to the High Court**

***Held*: The commissioner had been wrong in referring to cases and some dictionary definitions and in concluding that the expression "dwellinghouse" contemplated actual use and enjoyment of the premises — He thought that this view was supported by the words "used and enjoyed" in section 1(3) in relation to gardens, private garages and outbuildings — As a matter of ordinary language "dwellinghouse" includes not only a house which is dwelt in but also a house which is**
D **constructed or adapted for dwelling in, although at the material time it might be vacant or even not fit and ready for occupation — The words "used and enjoyed" in section 1(3) in reference to a garden did not necessarily connote actual use and enjoyment; the purpose of the words was to define the link that must exist between the garden (private garage or outbuildings) and the dwellinghouse in order to qualify as ancillary land — The commissioner erred in law in deciding that the requirements of the 1989 Act were not satisfied in respect of the cottages — Declaration accordingly**

The following cases are referred to in this report.

E

Ford v *Barnes* (1885) 55 LJKB 24
Morleys (Birmingham) Ltd v *Slater* [1950] 1 KB 506

This was a case stated by a commons commissioner, Mr Martin Roth, following an inquiry and the commissioner's conclusion that derelict cottages at 1-4 White Row Cottages, Bewerley, N Yorks, were not "dwellinghouses" which came within the Common Land (Rectification of Registers) Act 1989. This conclusion, if correct, meant that objection could not be taken to their inclusion in a register of town and village greens under the Commons Registration Act 1965.

Martin Rodger (instructed by May May & Merrimans) appeared F
on behalf of the applicant, Richard Henry Piers, 17th Viscount Mountgarret.

Giving judgment, **MUMMERY J** said: On Greenow Village Green at Bewerley in North Yorkshire there are four stone-built cottages. They are well over 100 years old. They have never had any running water or indoor sanitation. There had been a communal stone-built privy about 25 yds away, but that collapsed some time ago and the stone has been removed. The cottages are in a derelict state. The stone structure and stone-tiled roofs are largely intact, but there are no windows or secure outer doors. The cottage at the north-east end of the row once had a two-storey extension at the rear; that has collapsed and some of the stone tiles have recently been removed. G

The cottages have not been occupied for the last 20 years or so. Four notices dated February 2 1970 were given under the Housing Act 1957 condemning the cottages as unfit for human habitation and incapable of being made fit at reasonable cost. Two of the cottages had already been vacated by then. The other two had ceased to be occupied by 1972.

In recent years numerous inquiries have been made by people interested in restoring the cottages for occupation. The obstacle to the sale and renovation of the buildings is the existence of the registration of them and of the land on which they stand as forming part of Greenow Village Green. Registration is in the land section of the register unit VG 104 in the Register of Town and Village Greens maintained by North Yorkshire County Council under the H
Commons Registration Act 1965 ("the 1965 Act").

In section 22(1) of the 1965 Act "town or village green" is defined as follows:

> . . . land which has been allotted by or under any Act for the exercise or recreation of the inhabitants of any locality or on which the inhabitants of any locality have a customary right to indulge in lawful sports and pastimes or on which the inhabitants of any locality have indulged in such sports and pastimes as of right for not less than twenty years.

The relevant period of 20 years referred to is the period immediately prior to the passing of the 1965 Act. That Act was passed on August 5 1965, so that the immediately preceding period of 20 years takes one back to August 5 1945. J

The buildings and land, along with the rest of the village green, were registered on the application of the parish council. In the ownership section of the register the parish council is registered as owner of the whole of the land comprised in the register unit. The registration was not disputed at the time. Registration became final on August 1 1972 and became conclusive evidence of the matters registered: see sections 7 and 10 of the 1965 Act. Under the 1965 Act there are no grounds available in this case either for amending the register under section 13 or for rectifying it under section 14 so as to exclude the four cottages or the land on which they stand from the register.

On July 21 1989 the Common Land (Rectification of Registers) Act 1989 ("the 1989 Act") came into force. It is a short Act. I need K
refer only to the first section of it. Section 1(1) provides:

> Within three years after the passing of this Act any person may, by notice in writing given to the registration authority maintaining a register of common land and of town and village greens under the Commons Registration Act 1965, object to the inclusion on either of the registers of the whole or part of any land in respect of which the requirements specified in subsection (2) below are satisfied.
>
> (2) Those requirements are —
>
> (*a*) that —
>
> (i) there is a dwellinghouse on the land and, if and so far as the land is not the site of that dwellinghouse, it is ancillary to that dwellinghouse; or

A (ii) the land is ancillary to a dwellinghouse which is not on the land; and (*b*) that the requirements of paragraph (*a*) above have been satisfied at all times since August 5 1945.

Subsection (3) provides:

> For the purposes of subsection (2) above land ancillary to a dwellinghouse means a garden, private garage or outbuildings used and enjoyed with the dwellinghouse; and in that subsection "dwellinghouse" includes a building consisting of two or more separate dwellings.

It is not necessary for me to refer to the remainder of that section or to the other two sections of the Act.

The procedure under the 1989 Act, as set out in the Act and as prescribed under regulations made under it, which came into force in B March 1990, was set in motion on July 5 1990, when notice of objection was given by Viscount Mountgarret to the inclusion of 1-4 White Row Cottages in the register. On September 7 1990 the objection was referred by North Yorkshire County Council to a commons commissioner, who held an inquiry at Harrogate on December 5 1990. At the inquiry he heard representations from counsel on behalf of Lord Mountgarret and from representatives of the parish council. There was also before the commons commissioner, Mr Martin Roth, a letter from the Open Spaces Society making representations against the notice of objection. The commons commissioner inspected the properties.

On December 14 1990 he gave his written decision in which he set out in lucid detail the background and procedural history of the C matter, his findings of fact, the relevant statutory provisions and a summary of the submissions which had been made to him. He stated his conclusion in these terms:

> I reach the conclusion that the word "dwellinghouse" in the 1989 Act means a building which is actually dwelt in or is at least capable of being dwelt in. I have no hesitation in deciding that the building which was occupied as four dwellings until some 20 years ago but which was then condemned as unfit for human habitation and has since remained unoccupied does not satisfy the requirements of the 1989 Act.

He went on to make observations about the shortcomings of the 1989 Act. He stated that, in his view, the result was unsatisfactory as it meant that the cottages would remain derelict. Both Lord Mountgarret and the parish council wanted to see the cottages D restored, but the properties had been rendered sterile by the existence of the registration. The council had even expressed the fear that the properties might be occupied by squatters.

It was therefore decided to take the matter further. On January 15 1991 a request was made on behalf of Lord Mountgarret to the commons commissioner to state a case on a point of law for the decision of the High Court. A case was stated on February 27 1991 in which the point of law was formulated by the commons commissioner in these terms:

> The question of law for the decision of this honourable court is whether on the facts found by me I erred in law in holding that the requirements specified in section 1(2) of the Act of 1989 were not satisfied in respect of any part of the E land to which this objection related.

I was informed by counsel that this is in fact the first occasion on which the 1989 Act has come before the High Court on the case-stated procedure.

The point of law stated may be rephrased as follows in respect of each of the four cottages by specific reference to the relevant statutory requirements in section 1(2): Has there been a dwellinghouse on the land in question at all times since August 5 1945? Are these cottages dwellinghouses when there is no one dwelling in any of them?

The commons commissioner considered two authorities in which the meaning of the word "dwellinghouse" had been discussed in different statutory contexts: *Morleys (Birmingham) Ltd* v *Slater* F [1950] 1 KB 506, a decision on the Rent Restrictions Acts, and *Ford* v *Barnes* (1885) 55 LJKB 24, a case on the Representation of the People Acts. He also referred to various dictionary definitions and concluded that the word dwellinghouse contemplated the actual use and enjoyment of the premises in question. On that basis the cottages were dwellinghouses during the respective periods of occupation but ceased to be dwellinghouses when they became empty. They have not been dwellinghouses "at all times" since August 5 1945 and therefore fall outside the scope of the 1989 Act.

He found support for this construction in the words "used and enjoyed" in section 1(3) in relation to gardens, private garages and outbuildings used and enjoyed with the dwellinghouse. In his view, those words meant actual use and enjoyment of the gardens, private garages and outbuildings in question. It was, therefore, "logical to G conclude that actual use and enjoyment of the dwellings is also contemplated".

In my judgment, the commons commissioner construed the provisions of section 1 of the 1989 Act too restrictively. In some statutory contexts and often in ordinary everyday language "dwellinghouse" is indeed used to describe a house in which people are actually living as a private residence. Actual residential occupancy is not, however, a necessary characteristic of a dwellinghouse.

As a matter of ordinary language dwellinghouse is capable of including not only a house which is dwelt in but also a house which is constructed or adapted for dwelling in although it may at the relevant H time be vacant or even not fit and ready for occupation. For example, a family may be forced out of their dwellinghouse by fire, flood or other natural disaster. The house may remain empty for a long period while building works are carried out on it. I do not think it would be a misuse of the English language to say of such a house that it was at all times, even when empty, a dwellinghouse.

Turning to the language of the 1989 Act I note three things:

(1) There is no statutory definition of "dwellinghouse"; section 1(3) says that a dwellinghouse includes a building consisting of two or more separate dwellings, but it does not define the word.

(2) There is no express statutory requirement in the 1989 Act that the dwellinghouse in question is resided in or occupied or inhabited or in actual use as a place of abode for the relevant period. This is to be J contrasted with legislation in which such requirements are sometimes imposed in express terms.

(3) In the absence of any statutory definition the word dwellinghouse should be construed in its ordinary meaning both in the context of the particular statutory provision and of the 1989 Act as a whole.

Adopting that approach, I conclude that the 1989 Act operates in the following fashion in this case. Nos 1-4 White Row Cottages were dwelling-houses at all times during the period from August 5 1945 to the respective dates on which they were vacated, that is 1968 in the case of two of the cottages and 1971 and 1972 respectively in the case of the other two. After they were vacated the cottages did not cease to exist; they were not demolished; they were not converted to other K uses, such as use as a shop or warehouse or office. The physical structure, character and nature of the cottages remained the same, save that they were empty, neglected and became tumbledown. I do not think that they ceased to be dwelling-houses simply because they were not dwelt in. They became dilapidated dwelling-houses but were still dwelling-houses, as distinct from other sorts of houses such as public houses or warehouses.

In my judgment, this result is more consistent with the object of the 1989 Act than the result which follows from the decision of the commons commissioner. The provisions of the 1989 Act are designed to provide a procedure for removing from the registers maintained under the 1965 Act certain land on which there is a dwelling-house or L land ancillary to a dwelling-house. Such a procedure for rectification of the register is not, as I have already observed, available under the 1965 Act, even in cases where it transpires that the land in question never was appropriate to be entered on the register but has been entered on the register in the absence of any objection made at the relevant time.

It is clear, having regard to the definition of "town or village green" in section 22 of the 1965 Act, that it never was appropriate to register 1-4 White Row Cottages as part of Greenow Village Green. This was so whether the cottages were at the relevant time dwelt in or were empty. I cannot discern any sensible purpose in Parliament seeking to draw a distinction between a house which is lived in for the whole of the relevant period and a house which is constructed for M living in but is not in fact lived in at all times during the relevant period and I do not think that Parliament did in fact draw any such distinction.

As to the reference in section 1(3) to a garden, private garage or outbuildings "used and enjoyed" with the dwellinghouse, I do not think that these words necessarily connote actual use and enjoyment any more than the word dwellinghouse in this context necessarily connotes actual residence. Again, as a matter of ordinary English, I do not think that "used" and "enjoyed" are synonymous with being used and enjoyed or actual use and enjoyment.

In the context of section 1(3) the purpose of the expression "used and enjoyed" is to define the link that must exist between the garden, the private garage or outbuildings and the relevant dwellinghouse in order to qualify as ancillary land. Thus, for example, a garden which is not used and enjoyed in relation to any dwellinghouse would not qualify as ancillary land within the meaning of section 1 of the Act.

In this case the commons commissioner found that there were small front gardens with no walls fronting on to a track which runs across the village green. They should be removed from the registration along with the cottages. I should add that, before me, the claim for removal of ground at the rear of the cottages was not pursued. This was rightly conceded, since the commons commissioner had found as a fact that there was no definable area of garden ground at the rear of the cottages.

For these reasons I decide that on the question of law stated for my decision the commons commissioner erred in law in holding that the requirements specified in section 1(2) of the Act of 1989 were not satisfied in respect of any part of the land to which this objection related. I hold that those requirements were satisfied in respect of 1-4 White Row Cottages and the front gardens of those cottages enclosed by the low walls that separate the gardens from the track across the village green.

Declaration accordingly, commissioner to notify the registration authority. No order for costs.

Chancery Division

June 25 1991

(Before Sir Nicolas BROWNE-WILKINSON V-C)

IDC GROUP LTD AND OTHERS v CLARK AND OTHERS

Estates Gazette February 29 1992

[1992] 08 EG 108

Easement — Constructive trust — Licence — Agreement for provision of fire-escape route through door in party wall into interior of flat — Whether agreement, which was made by deed, created merely a personal licence or whether it constituted an easement or a constructive trust enforceable against the sublessee of the flat in question, although she was not a party to the agreement

By an agreement made by deed between a number of interested parties a licence was granted allowing an opening to be made in a party wall between adjacent properties in St James's Square, London SW1 — The purpose was to provide a fire-escape route from the fourth floor of no 23 into the area in the fifth floor of no 22 now occupied by the sublessee of flat 35, the defendant, Mrs Clark — A doorway was duly constructed in the opening of the party wall and arrangements laid down for progress in the case of fire to the public street — At some stage the doorway was blocked, it is not known by whom — The plaintiffs, who were entitled to the freehold and leasehold interests in no 23, brought an action against Mrs Clark — They sought mandatory injunctions to open up the doorway and later their claim was, at least primarily, for damages as they had provided alternative escape routes at their own expense — Mrs Clark had taken third party proceedings against a firm of solicitors

Two preliminary issues were directed to be heard — The first was whether the deed constituting the agreement created any rights which were capable in law of binding the defendant, Mrs Clark — The second was whether, in the light of all the circumstances, any such rights were binding on her

The plaintiffs submitted that the deed operated to create a legal easement over the defendant's flat; alternatively, it was contended that she was subject to a constructive trust to give effect to the rights of the owners and occupiers of no 23

***Held*: The document did not create an easement — It was a professionally drawn document in which the word "licence" was deliberately chosen; it created only a personal licence, the defendant was not a party to or bound by the deed — Very special circumstances were needed to justify the inference that a constructive trust had been created — Equity will raise a constructive trust when it is held to be unconscionable for a person, having received property, not to give effect to the terms on which he has received it — Such an inference is not made lightly and in matters relating to title to land certainty is of prime importance — In the present case the factors relied on were insufficient to justify such an inference — Judgment was given to the defendant**

The following cases are referred to in this report.

Addiscombe Garden Estates Ltd v *Crabbe* [1958] 1 QB 513; [1957] 3 WLR 980; [1957] 3 All ER 563, CA
Ashburn Anstalt v *Arnold (No 2)* [1989] Ch 1; [1988] 2 WLR 706; [1988] 2 All ER 147; (1988) 55 P&CR 137; [1988] 1 EGLR 64; [1988] 23 EG 128, CA
Binions v *Evans* [1972] Ch 359; [1972] 2 WLR 729; [1972] 2 All ER 70; (1972) 23 P&CR 192; [1972] EGD 518; 222 EG 2147, CA
Halsall v *Brizell* [1957] Ch 169; [1957] 1 All ER 371; [1957] 2 WLR 123
Lyus v *Prowsa Developments Ltd* [1982] 1 WLR 1044; [1982] 2 All ER 953; (1981) 44 P&CR 213
Street v *Mountford* [1985] AC 809; [1985] 2 WLR 877; [1985] 2 All ER 289; [1985] 1 EGLR 128; (1985) 274 EG 821, HL

These were two preliminary issues in an action brought by the plaintiffs, who were the present freeholders and leaseholders of the property 23 St James's Square, London SW1, against Mrs Clark, the sublessee of a flat no 35 in the adjoining premises, 22 St James's Square, concerning the closing of the doorway between the fourth floor of no 23 and Mrs Clark's flat.

John Lindsay QC and John Rendall (instructed by Needham & James) appeared on behalf of the plaintiffs; John Boggis (instructed by Jaques & Lewis) represented the defendant, Mrs Clark; Peter Crampin (instructed by Reynolds Porter Chamberlain) represented a third party.

Giving judgment, **SIR NICOLAS BROWNE-WILKINSON V-C** said: These are two preliminary issues in an action brought by the plaintiffs, who were the leaseholders and freeholders of the property 23 St James's Square, SW1 ("no 23"). The action is brought against the defendant, Mrs Clark, who is the sublessee of a flat no 35 in the adjoining premises, 22 St James's Square. The action relates to an agreement made or contained in a deed dated March 28 1969, which provided for a fire-escape route from the fourth floor of no 23, through the party wall into the area on the fifth floor of no 22 now occupied by flat 35.

At the time of the deed dated March 28 1969, the ownership of the respective properties was as follows. As to no 23, the freehold was registered at the Land Registry in the names of the Friends' Provident Life Office. The headleaseholder of no 23 was a firm, Herring Son & Daw ("Herring"). On the other side of the party wall, that is to say no 22, the freehold, again registered with the Land Registry, was vested in a number of people who were the trustees of the Junior Carlton Club. The headleaseholders, under a lease of 150 years from December 25 1963, were Pall Mall Developments Ltd.

The deed was made between the trustees of the Junior Carlton Club of the first part, Pall Mall Developments, the headleaseholders, of the second part, Friends' Provident of the third part and the partners in Herring of the fourth part. It recites the title of the respective parties. Then the sixth recital reads as follows:

> Herring has requested Pall Mall to make an opening in the party wall between Nos 30 to 35 and No 23 at the level of the fifth floor of Nos 30 to 35, so as to provide No 23 with the fire escape routes hereinafter mentioned, and with the approval and consent of the trustees and the Friends' Provident, Pall Mall have carried out such works as the parties hereto acknowledge.

Then I must read the larger part of the operative part of the deed:

> (1) The trustees hereby grant their licence and consent to Pall Mall, having made the opening in the wall dividing Nos 30 to 35 and No 23 at the position indicated in red on the attached plan numbered 531/1 and to the erection of the doorway now erected in the said opening.
>
> (2) The trustees and Pall Mall do hereby each of them grant licence to Friends' Provident and to Herring and other of the occupiers for the time being of No 23, (a) to pass through the said opening and doorway in case of fire only and in such case to enter into Nos 30 to 35 by such means and to pass along its passages, corridors and stairways and through its entrance doorway to the public street and (b) in case of fire only to use the fire escape routes from the seventh floor parapet of No 23 to the roof and balconies of Nos 30 to 35 on the St James's Square frontage and a similar fire escape route on the same

level on the Pall Mall frontage, and to pass along its passages, corridors and stairways and through its entrance doorway to the public street, which said fire escape routes are more particularly shown on the attached plans numbered 1508/23, 1508/10 and 1508/11.

(3) Each of the parties hereto covenant with the other that they will use their best endeavours to ensure that no obstruction is made so as to prevent free access to the said fire escape routes.

(4) The expressions "the trustees", "Pall Mall", "Friends' Provident" and "Herring" shall include their respective successors in the title and assigns.

The opening referred to in clause 2(a) was the opening from no 23 into what is now the defendant's flat no 35. Someone — it is not yet established who — has at some stage blocked up the doorway. As a result, the route provided for by the 1969 deed as a fire escape has proved incapable of use.

The plaintiffs, who are those now entitled to the freehold and head-leasehold interests in no 23, therefore brought these proceedings against Mrs Clark, seeking mandatory injunctions to open up the doorway or, alternatively, damages in lieu. The plaintiffs now claim, at least primarily, damages, since during the currency of the action the plaintiffs have provided alternative fire-escape routes at substantial expense to themselves.

Before dealing with the issues that are directed to be heard by me, I must first deal with the subsequent devolution of the title to nos 22 and 23 since the 1969 deed.

I will deal first with no 22. The freehold, which is registered under title no NG1337311, was transferred first to a company, Corix Properties Ltd, which I will call Corix. It was then later transferred by Corix to Chattanooga Shipping & Investment Corporation SA. Chattanooga is the present registered proprietor. The 1969 deed is noted against the freehold title.

As to the headleasehold interest in no 22, Pall Mall Developments assigned their term to the Friends' Provident who in due course surrendered that term to Corix on October 27 1978. There was then a fresh headlease granted by Corix to a company, Blissfield Corporation NV. That new headlease is dated July 13 1979. It grants a headlease of the fourth to 10th floors only of no 22. It contains a covenant to comply with the provisions of the 1969 deed. The headlease is itself registered at the Land Registry under title number NGL 357395. Again, the deed of 1969 is noted against the head-leasehold title in the charges register.

Finally, I come to the underleasehold interest in no 22. On October 2 1981 Blissfield granted to a Mr Hughes a long underlease of flat no 35. That underlease expires in the year 2051. The underlease is expressed to be granted subject to:

the matters specified in Part II of the Fourth Schedule hereto.

Part II of the fourth schedule contains the following words:

The entries on the Property and Charges Register of title number NGL 357395 excluding any charges to secure money.

There is therefore in the underlease itself an express subjection of the demised premises to the matters referred to in the register of the headlease and the register of the headlease in turn refers to the 1969 deed.

In addition, the plan annexed to the underlease shows on it a door in the party wall in flat 35 against which are written the words "Existing fire escape to adjoining premises".

On March 21 1985 Mr Hughes entered into an agreement with the defendant, Mrs Clark, to assign to Mrs Clark the underlease of flat 35, 22 St James's Square. The agreement provided that:

the property is sold together with the rights, but subject to the rights exceptions, reservations, covenants, stipulations and conditions, whereof a copy has been already supplied to the purchaser or her solicitors contained or referred to in the said lease and the purchaser shall be deemed to purchase with full knowledge thereof and shall not raise any objection or requisition in respect thereof.

The contract also contains an admission by Mrs Clark that she had inspected the property. The property was transferred to Mrs Clark by a registered transfer dated March 29 1985.

The underleasehold title is itself registered with absolute title under title number NGL 409118. At no stage has the underleasehold register referred in any way to the 1969 deed.

There were directed to come before me three issues. It has been agreed that in principle I should be dealing with two only of them. The issues as directed read as follows:

(1) Whether the deed dated March 28 1969 referred to in paragraph 3 of the amended statement of claim created any rights capable in law of binding the defendant.

(2) Whether any rights created by the said deed are binding on the defendant as registered proprietor at HM Land Registry of the underlease dated October 2 1981 referred to in paragraph 4(2) of the amended statement of claim in view of the contents of the register relating thereto at HM Land Registry.

I will deal first with the issue as to whether the 1969 deed created any rights capable in law of binding Mrs Clark.

Broadly the case for the plaintiffs is put in two different ways. First, on the true construction of the deed it operated to create a legal easement over flat 35 in 22 St James's Square. Second, and alternatively, Mrs Clark in the circumstances is subject to a constructive trust to give effect to the rights of the owners and occupiers of no 23 conferred by the 1969 deed. I will deal first with the question of whether there is a legal easement. In my judgment, that is a pure question of construction of the deed set against the factual background, so far as it is known, in which it was executed.

Mr Boggis has appeared for Mrs Clark; in addition, Mr Crampin has appeared for a firm of solicitors against whom Mrs Clark has taken third-party proceedings. Both Mr Boggis and Mr Crampin accept that the nature of the rights granted is capable of constituting a legal easement, ie rights of entry in case of fire and escape in case of fire can exist as legal easements. The question is not whether they are capable of constituting an easement but whether on the true construction of the deed that is the effect of the words used by the parties.

Mr Boggis' and Mr Crampin's case is in essence the simplest possible. Clause 2 of the deed purports to do nothing more than to grant a licence to do certain acts. It is a professionally drawn document, drawn with some care. If a professional draftsman uses the word "licence", he usually means the word "licence". If he means to grant a right of way, being a right in the land as opposed to a personal liberty, a professional draftsman is well able to do that. Therefore, clause 2 takes effect according to its strict language and confers only personal rights. Since Mrs Clark was not a party to that deed, she cannot be liable in contract on it. Since it does not confer a property interest in the property or over the property, it is not enforceable against her.

Mr Lindsay, on the other hand, for the plaintiffs says that that is too simplistic a view. This is a deed executed for the benefit of a valuable central London property which needed a fire certificate. This fire-escape door was needed not only to ensure the safety of the individuals but also in order to get the necessary fire certificate to sanction the continued use of the premises for the business of the plaintiffs. A mere personal licence capable of being defeated by any assignment of flat 35 or any other part of no 22 would be quite inadequate to meet the full requirements of the plaintiffs. Mr Lindsay then points out that the transaction was carried through by way of deed not by way of mere consensual agreement. It is not called a licence, it is called a deed. The words used are "to grant a licence", the word "grant" being, as he submits, a word much more appropriate to the grant of property rights than to a mere agreement as to a personal non-property liberty. He relies on the fact that the rights, whatever they are, are granted not only to Friends' Provident and to Herring but also to the occupiers for the time being of no 23. He relies in particular on clause 4 whereby the trustees and Pall Mall are defined so as to include their respective successors in title and assigns. Mr Lindsay argues with great force that, if the document was not to create property rights, what purpose could there be in defining those on whom the burden of the deed lay as including their respective successors in title and assigns? Merely personal rights as opposed to property rights could not be enforced against such successors in title and assigns.

Finally, Mr Lindsay relies on the line of cases dealing with the difference between licences and tenancies; for example, *Addiscombe Garden Estates Ltd* v *Crabbe* [1958] 1 QB 513, *Street* v *Mountford* [1985] AC 809* and *Ashburn Anstalt* v *Arnold* [1989] Ch 1†. Those cases illustrate that, in drawing the distinction between a licence of premises on the one hand and a tenancy of those premises on the other, it is in no sense decisive if the parties label what they are doing as the grant of a licence. One has to look at the substance of the matter and see what has been granted: was it a tenancy or was it strictly a personal contractual licence?

Despite those submissions, I am not persuaded that this agreement

*Editor's note: Also reported at [1985] 1 EGLR 128.
†Editor's note: Also reported at [1988] 1 EGLR 64.

has to be construed otherwise than strictly in accordance with the words used. True it is that those interested in no 23 urgently needed as permanent as possible a fire-escape route in order to utilise no 23 properly. But by the same token the adjoining premises, the Junior Carlton Club, was already subject to a long development lease in favour of Pall Mall Development. Therefore, the question of redevelopment was very much on the agenda. To grant an indefinite freehold easement such as is alleged in this case would have been a very dangerous step for a developer to take, since it would have conferred rights on the owners of no 23 which it might have been very expensive to get rid of. The position is, to my mind, made even clearer on this aspect when one recollects that, although in this case I am concerned only with clause (2)(a) of the deed, clause (2)(b) confers other escape routes over the parapets of no 23 on to the balconies of the adjoining premises of no 22. Therefore, this document was envisaging very extensive rights not only over flat 35 but also over other parts of the building. The surrounding circumstances suggest, to my mind, that, so far as one can draw an inference, although the owners of no 23 wanted as much as they could get by way of fire escape, the owners of no 22 would be concerned not to tie up future development of their property. There is therefore nothing *a priori* pointing one way or the other as to what was the likely intention of both parties.

I do not find it surprising that the transaction was carried through by way of deed. It was a unilateral transaction. No consideration moved from the owners of no 23 to those granting the rights, whatever they were, over no 22. Therefore, a deed was necessary.

I do not find the use of the word "grant" very significant. In clause (1) the trustees as freeholders were giving their consent in writing to alterations carried out by the leaseholders without consent: "The trustees hereby grant their licence." That is not granting a legal interest in the land. Similarly, in clause (2) the use of the words "grant licence" is simply a phrase commonly used in connection with licences and carries the matter no further.

Again, I do not attach much importance to the fact that the licence is granted not only to Friends' Provident and to Herring but also to "other of the occupiers for the time being of No 23". I read clause (2) as including those words to show the people who were entitled to exercise the fire-escape rights, not those who were the grantees of those rights.

Much the strongest point, I think, is raised by clause (4), since there is great force in Mr Lindsay's submission that, if the rights conferred were purely personal rights, it is hard to see how the successors and assigns of the trustees of the Junior Carlton Club and Pall Mall could be affected by the matter in any event. That is undoubtedly a point pointing the other way, although there is some force in Mr Crampin's submission that the parties may have had in mind that clause (3) of the deed contains a negative covenant which might be enforceable against successors in title as being a restrictive covenant.

The indications being far from clear as to what the parties intended, I think that one has to take this as being a professionally drawn deed in which the draftsman has deliberately chosen to use the word "licence" and not the ordinary words associated with the grant of an easement. Therefore, since there is nothing sufficiently clearly pointing in any other direction, I reach the conclusion that the deed did create only a personal licence.

Mr Lindsay had an alternative argument whereby he sought to reach the same result by utilising section 60 of the Law of Property Act 1925. He said that this should be taken as being a conveyance of freehold land without words of limitation and, therefore, under section 60 it passed to the grantees the fee simple in the rights granted, ie a permanent easement. The basis of that submission is that under section 205(ix) "land" is defined to include:

> an easement, right, privilege or benefit in, over, or derived from land

He said that this escape route was a right, privilege or benefit in or over no 22. Therefore section 60 operated to transfer a freehold interest in it.

I am quite unable to accept that argument. It is dependent on saying that the rights created by the 1969 deed are "rights, privileges or benefits in or over land". The definitions in section 205 of the Law of Property Act 1925 apply only "unless the context otherwise requires". Section 60 is subject to the proviso that it applies only "unless a contrary intention appears in the conveyance".

In my judgment, if on the true construction of the document it creates personal interests only and not rights of the nature of property rights, then a contrary intention does appear. The Law of Property Act 1925, section 60, does not convert what would under the general law be a personal right into a freehold-property right. I therefore hold that the 1969 deed did not confer an easement on the owners and occupiers for the time being of no 23.

I turn, then, to the alternative way in which Mr Lindsay puts his case, constructive trust. As appears from what I have said in reciting the facts of the case, Mrs Clark, by the agreement to purchase, took the property subject to the rights, exceptions, reservations, covenants, stipulations and conditions contained or referred to in the underlease and purchased with full knowledge thereof. The provisions in the underlease themselves refer to the 1969 deed and its contents. Therefore, it is said that Mrs Clark has acquired the underlease subject to and with knowledge of the rights conferred by the 1969 deed and, even though they would not otherwise be enforceable against her, in the circumstances she holds subject to a constructive trust to give effect to the terms of the 1969 deed.

The law on constructive trusts in this area is not as yet entirely clear. It has, however, been clarified recently by a Court of Appeal decision. Before coming to that case, I must mention a decision of Dillon J, as he then was, in *Lyus* v *Prowsa Developments Ltd* [1982] 1 WLR 1044. In that case the plaintiffs had entered into a contract to buy a plot of registered land from a company. That company went into liquidation. Prior to the contract between the company and the plaintiffs, the company's bank had obtained a legal charge on the land. The bank as mortgagee sold the property to a purchaser, subject to and with the benefit of the contract between the plaintiffs and the company. That purchaser from the bank in turn sold on, taking a covenant to perform the contract in favour of the plaintiffs. If and to the extent that the plaintiffs' contract was enforceable against the company, it was said that the first purchaser was subject to a constructive trust to give effect to the original sale by the company in liquidation to the plaintiffs. Dillon J upheld that submission.

Having considered *Binions* v *Evans* [1972] Ch 359, Dillon J reached the conclusion that there were cases in which the purchase of property expressly subject to an obligation in favour of the third party could give rise to a constructive trust to give effect to that obligation to the third party. However, he distinguished expressly other types of case and said at p 1051H:

> By contrast, there are many cases in which land is expressly conveyed subject to possible incumbrances when there is no thought at all of conferring any fresh rights on third parties who may be entitled to the benefit of the incumbrances. The land is expressed to be sold subject to incumbrances to satisfy the vendor's duty to disclose all possible incumbrances known to him, and to protect the vendor against any possible claim by the purchaser if a third party establishes an overriding right to the benefit of the incumbrances against the purchaser. So, for instance, land may be contracted to be sold and may be expressed to be conveyed subject to the restrictive covenants contained in a conveyance some 60 or 90 years old. No one would suggest that by accepting such a form of contract or conveyance a purchaser is assuming a new liability in favour of third parties to observe the covenants if there was for any reason before the contract or conveyance no one who could make out a title as against the purchaser to the benefit of the covenants.

Dillon J then pointed out that the bank, whose charge was prior in time, was under no risk *vis-à-vis* the plaintiffs and could safely have sold free of the contract in favour of the plaintiffs. But the bank had elected to sell expressly subject to the plaintiffs' contract and with the benefit of it. He also referred to a letter to the bank's agents from the first purchasers giving an assurance that the first purchaser from the bank would take all reasonable steps in its power to make sure that the interests of contractual purchasers were dealt with quickly and to their satisfaction.

On the basis of that evidence, namely no reason for the bank to protect its position by disclosing an incumbrance which was not a fetter on its title and the express assurance that effect would be given to the contract in favour of the plaintiffs, Dillon J held that there was a constructive trust in that case.

That decision was approved by the Court of Appeal in *Ashburn Anstalt* v *Arnold* [1989] Ch 1. The Court of Appeal put what I hope is the *quietus* to the heresy that a mere licence creates an interest in land. They also put the *quietus* to the heresy that parties to a contractual licence necessarily become constructive trustees. They also held (at p 25 D-E) that the mere fact that property is sold subject to a contractual licence is not sufficient to create a constructive trust. They held (at p 26A) that the mere fact that somebody has purchased with notice of a claim does not give rise to a constructive trust.

However, the Court of Appeal plainly considered that *Lyus* v *Prowsa* was rightly decided.

The result, as it seems to me, is that in the normal case a conveyance of land subject to or with notice of prior incumbrances or prior interests will not operate so as to make enforceable under a constructive trust such prior incumbrances or interests which would otherwise be unenforceable. However, in certain circumstances equity raises a constructive trust because it is unconscionable for the person having received such property not to give effect to the terms on which he received it. As the Court of Appeal said, and with respect I would agree, at p 26E:

> In matters relating to the title to land certainty is of prime importance. We do not think it desirable that constructive trusts of land should be imposed in reliance on inferences from slender materials.

It is important always to bear in mind that it is of the greatest importance that the title to land should be capable of being ascertained in accordance with well-known procedures. To raise constructive trusts which do not fit into the conveyancing machinery currently operating, thereby giving rise to liabilities of which purchasers might otherwise not be aware, is a dangerous course to pursue.

In my judgment, the decision in *Ashburn Anstalt* does not warrant the creation of a constructive trust unless there are very special circumstances showing that the transferee of the property undertook a new liability to give effect to provisions for the benefit of third parties. It is the conscience of the transferee which has to be affected and it has to be affected in a way which gives rise to an obligation to meet the legitimate expectations of the third party.

In the present case the position stands strangely. The pleadings do not in terms allege a constructive trust. I have asked Mr Lindsay what are the factors upon which he would seek to rely if he were to have leave to amend the pleadings to raise a constructive trust against Mrs Clark. He says that he would wish to rely on the fact that Mrs Clark purchased under an agreement which made the flat subject to the rights and, in addition on the following matters: that Mrs Clark inspected the premises, saw the door in the premises, saw the plan attached to the lease, which showed the door there and the caption saying "Fire Escape", and that she knew that she was taking subject to the fire escape rights.

In my judgment, even if all those facts were alleged in the pleadings and proved, they would not be sufficient to raise a constructive trust in this case. There is no indication in those facts that there was any form of bargain between Mr Hughes and Mrs Clark that she was undertaking *de novo* to honour obligations relating to the fire escape. Short of that, it seems to me that the claim in constructive trust must fail.

Mr Lindsay puts his case finally by analogy with the decision of Upjohn J, as he then was, in *Halsall* v *Brizell* [1957] Ch 169. In that case inhabitants of a building estate laid out in the middle of the 19th century enjoyed under a deed rights of access over a road on the estate but by the same deed were required to pay their proportion of the cost of providing the road. Upjohn J expressed the view (although it was not his eventual decision for other reasons) that the defendants were not entitled to take advantage of the provisions concerning the user of the road and the other benefits created by the deed without undertaking the obligations which the deed itself imposed, ie payment for the services. He held that, although there was no privity of contract or estate, the defendants were bound by the positive obligation to pay for the services.

Because the matter went off on another basis, he never indicated how that obligation would be enforced. But my understanding is that his judgment was based on the old doctrine that a man cannot take a benefit under a deed without also performing the obligations it imposes. It was therefore of the essence of the decision in that case that the right to use the road and the obligation to pay for it arose under one and the same deed. Moreover, as I understand the principle, the obligation to pay could not be enforced by action as such. The only way in which it could be enforced would be by withholding from the defendants the right to use the road unless and until they paid the charges due under the deed.

In my judgment, that case has no application to the present case. Mr Lindsay seeks to say that Mrs Clark cannot enjoy her occupation of the flat without reliance on the underlease. The underlease refers to the 1969 deed. Therefore, Mrs Clark cannot be allowed to live in the flat without giving effect to the 1969 deed. In my judgment, that is fallacious. There is no single deed which confers on Mrs Clark the right to live in the flat and imposes on her the duty to permit the fire escape. The underlease confers the right to live in the flat: it does not impose the duty to permit the fire escape. That is not the principle in *Halsall* v *Brizell*. Moreover, I can see no way in which the plaintiffs in this case could seek to enforce the obligation. The only means of enforcement in a *Halsall* v *Brizell* type of case is to prevent the person from enjoying the rights which they seek to enjoy under the document, save upon the terms that they give effect to the positive obligations imposed on them. In the present case the owners of no 23 have no means of saying that Mrs Clark is not entitled to occupy her flat 35. I therefore also hold against the plaintiffs on that ground.

As I have held that the plaintiffs are not entitled to an easement and it is common ground that, if there were a constructive trust, which I have ruled against, it would in any event not be binding on the defendant, it is, in my judgment, unnecessary for me to decide the second issue referred to me because the defendants are entitled to judgment in the action in any event.

Judgment was given for the defendant with costs: plaintiffs to pay also the third party's costs. Leave to appeal was given if leave was necessary.

Court of Appeal

November 7 1991

(Before Lord Justice GLIDEWELL and Lord Justice BALCOMBE)

DANCE v TRIPLOW AND ANOTHER

Estates Gazette May 2 1992

[1992] 17 EG 103

Right of light — Prescription — Prescription Act 1832 — Burden of proof — Whether plaintiff had discharged burden of proof that he had not submitted to or acquiesced in interruption to right of light for one year — Appeal by defendants allowed

The plaintiff was the owner and occupier of "Hillcot", Crowborough Hill, Sussex — The defendants were the owners and occupiers of the adjoining house "Highdown" — In 1978, following the grant of planning permission, the defendants commenced building works to construct an extension; at that time the plaintiff knew it would be two storeys — There was a delay and construction of the walls did not commence until September 1988, but by November 1988 they were up and the roof was on causing an interruption to light enjoyed by the plaintiff — In early 1981 the plaintiff complained to the local planning authority about a proposed revision of the planning consent and in July 1981 the plaintiff consulted solicitors concerning the defendants' use of a weedkiller, which had drifted across the boundary — In February 1982 the plaintiff first consulted his solicitors about the interruption to the light caused by the extension; however, no communication of any kind was made to the defendants until the action was commenced in August 1984

In Brighton County Court Judge Coles QC, in giving judgment for the plaintiff, concluded on the evidence that the plaintiff had not acquiesced in the interruption to the right of light — The defendants appealed

***Held*: The appeal was allowed — The onus of proof was on the plaintiff to prove that, at the time when he commenced the action claiming interruption to the right of light the interruption had lasted for less than one year having regard to section 4 of the Prescription Act 1832: see *Presland* v *Bingham* — On the evidence, it was doubtful that it gave a solid basis for the judge's finding that the plaintiff objected to the defendants before consulting his solicitor in February 1982 — The plaintiff failed to discharge the burden of proof that he had not submitted to or acquiesced in the interruption — The failure to communicate with the defendants over a period of**

A **two and a half years could be construed only as amounting to submission to the interruption of the plaintiff's right of light — Discontent, not effectively communicated to the other party, cannot amount to non-submission or non-acquiescence**

The following cases are referred to in this report.

Davies v *Du Paver* [1953] 1 QB 184; [1952] 2 All ER 991; [1952] 2 TLR 890, CA
Glover v *Coleman* (1874) LR 10 CP 108
Presland v *Bingham* (1889) 41 Ch D 268

This was an appeal by the defendants, Mr and Mrs Dennis Triplow, the owners and occupiers of "Highdown", Crowborough
B Hill, East Sussex, from the judgment of Judge Coles QC given in Brighton County Court on May 8 1990, whereby she gave judgment for the plaintiff, Mr William Dance, the owner and occupier of the adjoining property "Hillcot", for damages for interruption to the plaintiff's right to light.

Derek Holwill (instructed by Cripps Harries Hall, of Heathfield) appeared on behalf of the appellants; John Reddish (instructed by Cooper & Burnett, of Tunbridge Wells) represented the respondent.

Giving judgment, **GLIDEWELL LJ** said: This is an appeal against a judgment of Judge Coles QC, given in Brighton County Court on May 8 1990, whereby she gave judgment for the plaintiff for £3,400 damages.

C The plaintiff and his late wife (who died in 1986) were the owners and occupiers of "Hillcot", Crowborough Hill, East Sussex. The defendants are the owners and occupiers of an adjoining house, "Highdown".

Hillcot is a bungalow with a spare bedroom, in which the window at the side of the house faces Highdown. In 1980 the defendants constructed an extension to their house on the side adjoining the plaintiff's bungalow, comprising a garage, a shower room and an additional bedroom over — that is to say, it was a two-storey extension.

The plaintiff's case is that this extension interfered with his right to light to the window of the spare bedroom, which he had gained by prescription.

D With that brief account of the action, I turn first to the relevant law, most of which is to be found in the Prescription Act 1832. Section 3, so far as material, provides:

> (i) Where the . . . use of light to and for any dwelling house . . . shall have been actually enjoyed therewith for a full period of twenty years without interruption, the right thereto shall be deemed absolute and indefeasible . . .

By section 4, the period of 20 years is the period (to use the language of the section) "next before some suit or action" — in other words, 20 years counting backwards from the beginning of an action — and also, by section 4:

> (iii) No act or other matter shall be deemed to be an interruption within the meaning of the statute, unless the same shall have been or shall be submitted to
> E or acquiesced in for one year after the party interrupted shall have had or shall have notice thereof . . .

The following matters are common ground between the parties. The plaintiff's bungalow, Hillcot, was constructed at some time in the 1930s; Highdown some time after the second world war. The defendants purchased Highdown in 1976 and the plaintiff purchased Hillcot a little later in that year.

The action was commenced by a county court summons issued on August 9 1984; therefore, the relevant period of 20 years under section 3 is the period of 20 years terminating on that date. It was found by the judge, and there is no dispute about this, that the plaintiff's light was obstructed and the easement of light which potentially had arisen was therefore, in fact, interrupted in and after
F November 1980, when the walls and roof of the extension to Highdown were completed.

It followed that the issue which had to be determined in the county court, and which arises before us, is: did the plaintiff submit to or acquiesce in this interruption of his potential light for at least one year so as to amount to interruption within the definition in section 4(iii) to which I have referred?

The judge found that the plaintiff had not submitted to or acquiesced in the interruption and therefore gave judgment in his favour. The defendants appeal on three grounds, which can be summarised as follows:

(1) The judge's findings of fact were not supported by the evidence.

(2) Alternatively, the facts as found by the judge led to the conclusion, as a matter of law, that the plaintiff had submitted to or G
acquiesced in the interruption.

(3) In particular, the finding of fact that there was no expression of objection to the obstruction of the light to the plaintiff's window between the time when he instructed solicitors in February 1982 and the date of the commencement of the action in August 1984 could amount only to submission or acquiescence.

I turn back to another matter of law, which concerns the onus of proof in proceedings such as this. It was established by this court in *Presland* v *Bingham* (1889) 41 ChD 268 that if the interruption of light flowing to a plaintiff's property is of a permanent character, then the onus is on the plaintiff to prove that, at the time when he commences the action, the interruption has lasted for less than one H
year. That derives from the judgment of Cotton LJ at p 272 and Lindley LJ at p 276 in that decision.

It follows that if the interruption has, in fact, lasted for more than a year, the onus is on the plaintiff to prove that he did not submit to or acquiesce in it. This is conceded by Mr Reddish, for the respondent plaintiff, in this court.

On the law relating to what constitutes submission or acquiescence, we have been referred to two authorities: first, *Glover* v *Coleman* (1874) LR 10 CP 108. In that case the plaintiff had, for over 20 years, enjoyed access of light to a window of his workshop. The defendant erected a building adjoining, which obstructed the light. The plaintiff, on several occasions, complained and protested to the defendant about his light being restricted. Fourteen months J
after the building had been erected the plaintiff commenced his action. It was held in this court that it was open to the jury to find that the plaintiff had not submitted to or acquiesced in the interruption for more than a year.

At p 119 Brett J said:

> Acquiescence, according to my view, would mean not an active agreement, but what may be called a tacit, a silent agreement, — a submission to a thing by one who is satisfied to submit. The question for the jury, therefore, upon a suggested acquiescence, would be, whether the plaintiff, although he has not specifically agreed that the thing should be done, has submitted to it, and has been satisfied to submit. Now, "opposition", on the other hand, I should say would mean dissent or dissatisfaction manifested by some act of opposition. "Submitted to", in this part of the section, seems to me to be something K
> intermediate between a submission by one satisfied to submit and a dissatisfaction manifested by some act of opposition. It seems to me that what it means is this, — submission without satisfaction and without any direct act of opposition, and although discontent be made apparent by some expression or some act. If that be so, the question for the jury would be this, — Although it be clear that the plaintiff did not submit to the interruption, being satisfied to submit, nevertheless, did he submit, though with discontent, and though it was made apparent? This would render necessary an inquiry into the state of mind of the person who is said to have submitted, and whether his state of mind was made apparent by either expressions or acts. Here, we have the fact of the interruption existing, and of the plaintiff not doing anything which amounts to active resistance. We have strong intimations by him, not merely to persons about him, but to the person who caused the interruption, that he was not satisfied to submit to it, — complaints made more than once, and L
> made in such a manner as to be described by witnesses as amounting to protests. We have the fact that the interruption was caused by the erection of a somewhat solid edifice, the fact that he did not knock it down the jury might fairly say was not evidence that the plaintiff acquiesced in or submitted to what the defendant was doing. Now, these protests, coupled with the fact of the substantiality of the building put up by the defendant, and taking into account that all took place within the space of a little more than a year, and giving consideration also to what was the condition in life of the plaintiff and to the fact that an action was commenced immediately after the year, seem to me all together to amount to facts upon which a jury might properly say that the plaintiff did not acquiesce in the interruption of his right, in the sense I have suggested; and that, although he did not more actively oppose it, and although in one sense he submitted, yet that he submitted without satisfaction, and with discontent in his mind which he made apparent to the M
> defendant by acts or sayings which were a reasonable mode of expressing his discontent, under the circumstances in which he was placed. I am of opinion that there was evidence upon which the jury might reasonably find that the plaintiff did not acquiesce, and that they might reasonably find that he did not submit . . .

At p 121 Grove J said:

> Now, what was that interruption? If the legislature had said that there must be actual enjoyment for twenty years, without regard to s 4, then the plaintiff undoubtedly had not that enjoyment, because for a portion of that period he had ceased actually to enjoy the access of light. But then comes the question, has he actually enjoyed it for the full period of twenty years without such interruption as is contemplated by the 4th section of the Act? The words of that section, as applicable to this question, are, that "no act or other matter

A shall be deemed to be an interruption within the meaning of this statute, unless the same shall have been or shall be submitted to or acquiesced in for one year after the party interrupted shall have had or shall have notice thereof and of the person making or authorising the same to be made". Therefore, unless this interruption has been submitted to or acquiesced in for one year, it is not such an interruption as within the meaning of the statute will prevent the enjoyment being "an actual enjoyment without interruption". Then we come to the words "submitted to or acquiesced in". Now, the statute does not prescribe any particular mode of resistance. The statute does not, indeed, use the word "resistance", although Mr Bosanquet in the course of his argument used it as opposed to submission or acquiescence. There being, then, nothing in the statute to guide us as to what shall be the nature of the submission or acquiescence, that must necessarily be a question of fact — has the party interrupted, by his words or conduct, submitted to or acquiesced in the
B interruption, or has he resisted or protested against it? There are, no doubt, various modes of resistance or of expressing dissatisfaction. I quite admit that mere grumbling, or complaining to a member of his family, or expressing dissatisfaction to a third person, would not be sufficient evidence of non-submission or non-acquiescence to leave to a jury. But where protests are made to the party causing the interruption — and we must, I think, take it that the protests or complaints here which were made to the son, were made to a person who represented the defendant, and were communicated by him to the defendant . . . the matter for our consideration is whether there is any reasonable evidence which ought to have been submitted to the jury, upon which they might have determined whether the interruption was or was not submitted to or acquiesced in by the plaintiff. I do not see how it can be other than a question for the jury.

C Following the views of Brett J, in my view the difference between acquiescence and submission is that a person acquiesces where he is eventually satisfied to submit to the interruption. Submission occurs when the plaintiff is not content to submit, but does not make his opposition to the interruption apparent directly. In other words, in order to disprove "submission", a plaintiff must prove both unwillingness on his part to accept the interruption and that his opposition is made clear to the person who is responsible for the interruption by some words or act.

The more recent authority is that of *Davies* v *Du Paver* [1953] 1 QB 184, a decision of this court. The plaintiff and the defendant were the owners and occupiers of adjoining sheep pastures. The plaintiff claimed, by long usage, a right to pasture his sheep on part of the
D defendant's land. The defendant erected a fence to prevent this. While the fence was being erected, the plaintiff protested vigorously and made it clear that he would bring legal proceedings if the fence were completed. However, after the fence was completed, 13 and a half months elapsed without any further protest being made to the defendant before action started. The issue was: had the plaintiff submitted to or acquiesced in the interruption of the right?

Singleton LJ took the view that the absence of open protest for more than a year was fatal to the plaintiff's claim. Birkett and Morris LJJ upheld the county court judge's decision that there had been no submission to or acquiescence in the interrruption. It is perhaps of no significance that the learned lord justices came from, respectively,
E Cumbria and North Wales.

Birkett LJ at p 203 said:

> On the facts of the case, it is clear that there was an interruption for one year after the plaintiff had notice of the interruption. The important question is: did he submit to it, or acquiesce in it, for one year after he had notice?
>
> The judge said that he had not found it easy to arrive at a conclusion, and I share his view. The first strong fact is that for a full year after the plaintiff had notice of the interruption nothing overt was done by him or his solicitors. The fencing seems to have been finished by August 9, 1950, and the first protest after that date was the issue of the summons dated September 27, 1951. The question, easy to state but difficult to answer, is: Is the lapse of that 13 months to be taken as conclusive evidence of "submission to, or acquiescence in," the admitted interruption? After all, "submission to or acquiescence in" is a state of mind evidenced by the conduct of the parties, and I am of opinion that it
> F was a question of fact for the judge to decide on all the facts of the case, and that it was not to be decided merely by saying that there was in fact a period of a year in which nothing was done, without considering all the surrounding circumstances. This is not a case under the Statute of Limitations. It is not a case where the words "a reasonable time in all the circumstances of the case" may be substituted for the words of the statute. It is a case where a judge must decide as a question of fact where there has been submission to, or acquiescence in, the interruption for one year after the party interrupted had notice of the interruption.
>
> On July 6, 1950, the solicitors for the plaintiff threatened proceedings against the defendant unless the interruption was removed; on July 28 they renewed their protest; on August 1 they stated that they had instructed the plaintiff to take such steps as he thought necessary to prevent further infringement; and then nothing outwardly was done until September 27,
> G 1951. I am of opinion that, following the decision in *Bennison* v *Cartwright**, it is a question of fact whether these circumstances show a submission to, or acquiescence in, the interruption for one year, despite the fact that one year had admittedly gone by during which no outward protest was made or any other action taken. The fact that there was a complete year in which the plaintiff did nothing is some evidence that he had submitted to, or acquiesced in, the interruption; but the fact also that he protested most strongly up to August 1950, and then actually took proceedings in September 1951, is some evidence that he did not. It was a question of fact for the judge.

Morris LJ said at the bottom of p 207:

> The fence was completed on August 9. Could it be said that the plaintiff had submitted to, or acquiesced in, the existence of the fence by that date? Having
> H regard to the events that had happened, and to the correspondence, I would have thought, had it been for me to decide this question of fact, that the answer would be in the negative. The parties were breathing fury on each side of a newly erected fence. Could it be said that the challenging protests of the plaintiff must, as the August days passed, be deemed to have signified nothing, and that his former claims and assertions should be regarded as supplanted by submission and acquiescence? As time went by, it might well be that silence and inaction could be interpreted as submission or acquiescence. But the date when submission or acquiescence begins must be determined as a question of fact, having regard to all the circumstances. Had there been a beginning by January 1, 1951, or by December 1, 1950, or by November 1, 1950? These are all questions of fact. Unless it is held that there was submission or acquiescence by September 27 1950, there would not be a period of one year. The judge referred to *Glover* v *Coleman* and stated: "On
> J the evidence I hold that neither of the plaintiffs submitted to or acquiesced in the interruption." This was a finding of fact which the judge was, in my view, entitled to make, and accordingly I consider that the first submission fails.

That last passage in Morris LJ's judgment makes it clear that he was saying in effect that if a protest had been made up to the beginning of August 1950 and was still effective so that it was clear to the defendant that the plaintiff was protesting until a date after September 27 1950, that was sufficient on the facts of that case. Thereafter, less than one year elapsed before the action started.

I turn to the evidence in the present case. It is right to start by saying that where the evidence of the plaintiff conflicted with that of the defendants, the judge preferred the evidence of the plaintiff. It was, therefore, her task to decide whether the plaintiff's evidence proved
K that he had not submitted for one year.

In December 1977 the plans for the extension of Hilldown were passed by the local authority. It is not absolutely clear whether that means that building regulation consent was given or planning permission — perhaps the former. In March 1978 the district planning officer sent a letter to the other neighbour of the defendants, giving notice of the application to construct the proposed extension. He asserted that a similar letter had been sent to the plaintiff, but the judge found as a fact that the plaintiff did not receive it.

In the autumn of 1978 the footings of the proposed extension were laid. There was a conversation between the plaintiff (or possibly his
L late wife) and the defendants at which the plaintiff (or his wife) were informed that "the extension would not concern us; it would only be a garage with a room over". However, the plaintiff said in cross-examination concerning this: "I knew that it would be two storeys and would affect the light."

After the footings were laid, no further building work was carried out for the rest of 1978 or 1979 — or, indeed, until the summer of 1980. At that time building work recommenced. By September the brickwork was well under way. In that month the plaintiff appreciated that the extension would have a door and a window at ground-floor level in the garage, facing towards his property. It transpires, although he did not know it at the time, that the approved
M plan showed the door at the rear and not facing towards the plaintiff's property. There was a conversation between the defendants and the plaintiff some time in September 1980, when the plaintiff protested about the lack of privacy which the door and window would cause him. The defendants did not agree to move the door; instead they submitted amended plans to the local authority to seek consent to retain the door at the side of the property. But they did agree to meet the plaintiff's complaint to this extent, that they agreed to put opaque glass in the window of the new garage (which they did).

By November 1980 the walls were up and the roof was on and so the interruption to the plaintiff's life commenced.

The plaintiff gave evidence that at about that time an estate agent

*Editor's note: Reported at (1864) 5 B&S 1.

A suggested to him that he might have a right to action in respect of loss of light.

When the plaintiff, in his evidence-in-chief, was dealing with the question of the door, he said:

> This is the time I went to the solicitor as well. I complained about the window to the planning officer and to Mr Triplow. I went on about the light.

In January 1981 there was a meeting of the planning authority to consider the revised plan showing the door at the side of the building. The plaintiff had objected to that and attended the meeting of the committee but was not allowed to speak (I think to his annoyance) but a councillor made the comment that there was no point in having a plan submitted in the first place unless the council were going to
B adhere to it. However, despite that, it would appear that the amended plan was agreed to by the planning committee.

In May 1981 the plaintiff applied for the rates on his property to be reduced on the ground that the extension had devalued Hillcot.

In June 1981 the plaintiff complained to the defendants about a compost heap which the defendants had made very close to the boundary with the plaintiff's land. Also in June 1981 the defendants used some strong weedkiller on their land close to the boundary with the plaintiff and, according to the plaintiff in cross-examination, the weedkiller incident broke down the relationship. It may well be that that is correct and it was this incident which finally soured the relationship between the parties. Certainly it seems to be the case that some of the weedkiller drifted across the plaintiff's boundary and
C covered some of his plants.

On July 13 1981 the plaintiff consulted solicitors concerning the defendants' use of the SBK weedkiller.

In his evidence-in-chief which, through no fault of counsel, was not given in the chronological order which I am seeking to follow, the plaintiff said: "I told a lot of people that the place was a monstrosity. I said this to Triplow. I said 'How (the hell) much higher is it going?' I had already been to the solicitors when I said this. This was after the meeting of the council because it was going up then over the door." That really gives two possible dates for his saying that to Mr Triplow: sometime after January 1981 (which was the meeting of the council at which the revised plan had been considered) or, alternatively, after July 1981 when he went to see his solicitor concerning the weedkiller
D incident — if the plaintiff's recollection in that respect was clear.

Later, under cross-examination, he said: "I first went to solicitor about weedkiller . . . The loss of light came afterwards when I went to the solicitors. That is when that came up." He is also recorded as saying: "No complaint re light to defendant because I went to lawyer", which I understand to mean that from the time he went to see his solicitors, or at any rate when he went to see them about the light, he made no more complaints himself to the defendants on that subject.

He said he had a conversation with the defendants, when the first defendant was on the roof and the plaintiff was in his garden, concerning the compost heap. He said he pointed at the compost
E heap and said: "And that can go too." Later he said that the conversation with his solicitors concerning light was at the beginning of February 1982 — indeed, it can be seen from the plaintiff solicitor's attendance note that that is correct.

After the plaintiff went to his solicitor to complain about loss of light, there was no communication, either verbal or written, between him and the defendants directly or by the plaintiff's solicitor and the defendants, as to the loss of light; there was no communication of any kind until action was commenced in August 1984. During that time it is clear that the plaintiff was applying for legal aid with a view to commencing proceedings, that various steps were being taken with a view to obtaining evidence on which to base the action and that eventually counsel was being instructed to settle the statement of
F claim.

During that period, on April 11 1984, the plaintiff's solicitor did write to the defendants about another, but totally different, matter. The defendants had decided that they wanted to replace a fence, which formed the boundary between the two properties, with a wall. However, the plaintiff took the view that the fence was on his land (as it probably was) and so his solicitor sent a letter to the defendants warning them not to move the fence between the two properties. That letter said nothing at all about interruption to rights of light.

I have already rehearsed the plaintiff's evidence as to when the question of right of light was mentioned. In his re-examination, the plaintiff said: "I was told that there was nothing I could do about the building — but I was going to do something about it — six years only four conversations. Up to the weedkiller (incident) there may have G
been other conversations. Prior to SBK just a few moans about the extension. Moaned about dark — what it looked (like) — a monstrosity. He said it's mine and stopping . . . I said I'd like the damn thing pulled down."

In her judgment, the judge outlined some of the passages in the evidence to which I have referred. She formed a view of the plaintiff which she expressed in this way:

> As an individual I consider that Mr Dance could be an obdurate and irascible person and certainly not the sort of person who would concede any right that he might have *vis-à-vis* his neighbours nor acquiesce in anything he does nor agree with.

She said, in relation to the meeting at the planning committee H
where the plan for the moving of the door was considered:

> Mr Dance attended hoping to be able to make representations but was not permitted so to do. I have no doubt at all that he was not acquiescing to anything but as far as he was concerned obviously by this time the building was a *fait accompli* albeit the tiles and the interior had not yet been constructed.

Later she said:

> He has said — and I am sure it is true — that he was very angry about this construction, that he called it and thought it was a monstrosity and affected his own property in many respects.

She said that the plaintiff sought a reduction in his rates because ". . . this extension had caused his second bedroom to become J
darker, to become damp and cold and to detract from his privacy".

She referred to the conversation about the compost heap in which she accepted the plaintiff said: "And you can remove that too."

She then continued:

> However, there is no doubt in my mind that Mr Dance, once he appreciated the size of the construction next to him at least as of November 1980 did nothing but complain about one thing or another. He felt, because he had been told (we have not been told by whom) that there was nothing he could do about the building itself that nevertheless he would be determined to do what he could about its effects and anything else that gave cause for action. The priorities it appears to me in his mind with reference to this building were: first, his lack of privacy due to the window and the door; second, what he considered was the damp caused by the channelling of the weather between the two buildings; third, the reduction of light into the second bedroom K
> which, according to him, precluded his wife from enjoying the afternoon sun and caused her to cease to use the room as a sewing room so that it was subsequently transferred into a store room; fourth, regarding other things: there was the element of the compost heap — that nuisance being abated; there was, fifth, the weed killer incident and the results thereof allegedly which he also pursued with vigour, that matter being dealt with by his solicitors on July 15 1981 . . . and finally the wall . . .

Then, coming to the end of the matters of February 1982, she said:

> Having put the matter in the hands of his solicitors Mr Dance said nothing to Mr Triplow taking the view that this was a matter for his lawyers to deal with.

Mr Reddish submits to the court that in various passages in the L
evidence, to which I have referred, albeit perhaps not as clearly as he might have done, the plaintiff was making it clear that he had complained to the defendants about the interruption of his light; first, as to the passage in which he said: "I went on about the light". I agree with Mr Reddish that the phrase "I went on" means "I complained on a number of occasions". Of course, the occasions are unspecified except that the plaintiff said: "This time I went to the solicitors as well." The judge found that to be incorrect; but if it has any relevance at all it places the time of the complaint somewhere around the time when the plaintiff went to the solicitors for the first time in June 1981. The same also applies to the phrase "How (the hell) much higher is it going?" because the plaintiff said that he had already been to the solicitors but that he said that to the defendant. M

As to the comment of the judge that the building was a *fait accompli*, for my part I do not regard that as the judge in effect saying that the plaintiff was "submitting"; on the contrary, I think what she meant was that the plaintiff had recognised that he was unlikely to be able to ensure that the extension was demolished, but he was going to take what steps he could to obtain compensation resulting from it.

Mr Reddish relies on the phrase "And you can remove that too" as meaning "You can remove that (the compost heap) too, in addition to the extension" — which is no doubt a rather sweeping assertion on the plaintiff's part if he really did not think there was any chance of getting the extension removed. But Mr Reddish relies on that as being the expression of objection to the extension generally.

Moreover, in the passage to which I have referred in which the judge summarised what she described as being the various matters within the plaintiff's mind and gave her view as to the order of priority of the various complaints made as to the defendants' conduct; if that is in fact a summary of what the plaintiff had in his mind she does not there find (although Mr Reddish submits it is clear she was inferentially finding) that the plaintiff had complained about all those matters, including, of course, the detraction from his light.

It is, however, clear that at least one of these matters was not the subject of complaint at all. I turn to the way in which, having summarised the evidence and then correctly directing herself, the judge went back, in effect, in her judgment, to her summary and conclusion and said:

> It is submitted by counsel on behalf of the defendants that there was positive acquiescence. I reject that. I accept that the evidence indicates that Mr Dance was aware by 1978 that a building was going to be constructed but he was unaware of the extent and there is no evidence that he was going to lose light in his second bedroom. To my way of thinking he never directed his mind to it.

I interpolate to say I propose to make a first comment on that. Then follows what I shall call subpara (2).

The judge continued:

> It is also submitted that he did not complain in October 1980 when the application for the amended planning permission went in regarding the extension itself. By this time Mr Dance was accepting that this was a *fait accompli*. He objected about the alteration of the door without planning permission; that objection seemed to have no effect because the plan went through retrospectively.
>
> (3) The fact that he accepted the offer of opaque glass in the window opposite his premises, in my view, in no way shows that he was acquiescing in the loss of light to the second bedroom.
>
> (4) It is submitted that the same applies with reference to his application to have the rates reduced. I do not concur. The lack of light was one of the factors according to Mr Dance that was taken into consideration when the local authority did indeed reduce the rates. That does not in any way indicate, in my view, that Mr Dance was acquiescing in this reduction of light; he wanted to make sure that this reduction in the value of his premises was reflected in the amount of rates that he was going to pay and I do not concur with the defendants' counsel's submission that there was no complaint prior to the issue of proceedings about the interruption of light.
>
> (5) He did complain I find as a fact albeit complaint coupled with other matters both to various authorities and obliquely and occasionally directly to Mr Triplow himself.
>
> (6) Having put the matter into the hands of his solicitor during the month of January or February 1982 he made no further complaint because as he says "it was in the hands of my solicitors".
>
> (7) The facts I have found do not, in my view, constitute acquiescence or submission and he was entitled having passed the matter to a lawyer to consider that the complaint would thereafter be diligently pursued on his behalf by him.

I have inserted the numbering in order to be able to comment on those passages.

The first passage — the finding that the plaintiff directed his mind to this — is contrary to the plaintiff's own evidence, to which I have already referred. As to the second passage, there is no finding at that point in the judgment that the plaintiff did complain or object to the defendant in October 1980 about the effect on his light. There was a finding that all he objected to was as to the door.

As to the third matter, the agreement to put in opaque glass "in no way shows that he was acquiescing in the loss of light to the second bedroom" and the fourth matter, namely application to have his rates reduced, not showing that he was acquiescing in the loss of light — of themselves, those propositions are correct, but to my mind they betray something which I regard as coming through in this part of the judgment, and to which I will return later, that the judge appears to be forgetting at that stage that she was dealing with an issue on which the onus was on the plaintiff and not on the defendant. It is not sufficient to say they do not prove that the plaintiff was acquiescing; the issue was, had he proved, on a balance of probabilities, that he was not?

As to the fifth matter, that is to say, "he did complain, I find as a fact, coupled with other matters", that is a finding of fact, but it does not specify the time to which it relates. As to the sixth matter, the judge is merely saying that the plaintiff's reason for not complaining after February 1982 was because the matter was in the hands of his solicitors.

In my judgment, accepting as I do that the judge preferred the plaintiff's evidence, I have grave doubts whether his evidence gave a solid basis for the finding that he did make objection to the defendants before he went to his solicitor in February 1982. The judge, with respect to her, appears not at all times to have borne in mind that the onus was on the plaintiff to prove this. In my judgment, the plaintiff failed to discharge this burden and therefore failed to prove that he did not submit or acquiesce.

I pass over the second ground of appeal for the moment and deal with the third, which concerns what happened after February 1982. It is common ground that there was no complaint or objection made by the plaintiff, or his solicitor, to the defendants between February 1982 and the date when the action was commenced; even though there was a communication about the fence in April 1984. I accept that if there had been evidence of strong complaint by the plaintiff periodically up to, and shortly before, February 1982, it would and should have been clear to the defendants at some time thereafter that the plaintiff was not submitting or acquiescing. That is following the judgments in *Davies* v *Du Paver*. But for how long after February 1982 should that have been clear? Of course, *Davies* v *Du Paver* does not answer that question, but the facts in that case were that it was said to be clear for six weeks. Can it reasonably be said that it would have been clear, and should have been clear, to the defendants on the material to which I have referred for at least 18 months after February 1982 to some date after August 9 1983 when there was less than a year to the date when the county court summons was issued?

The judge made no finding on that question. What she said was:

> The facts I have found do not in my view constitute acquiescence or submission and he was entitled having passed the matter to a lawyer to consider that the complaint would thereafter be diligently pursued on his behalf by them.

Well, it may be that he was so entitled, and there may be valid reasons why the plaintiff, having fallen out with the defendants, decided that the best course was for him to keep his mouth shut. But, if no representations or objections were going to be made on his behalf, what then is the position? The judge merely saying that "he had understood". Whatever the plaintiff may have intended — and I accept that he did intend to carry his objection to the stage of commencing proceedings — the approach of the judge to this wholly ignores the necessity for the plaintiff to prove (if he has shown no submission) that he, or his solicitor, was in some way making it clear that he was continuing to object.

In my judgment, the failure to communicate with the defendants over a period of two and a half years could be construed only as amounting to submission to the interference with the plaintiff's right to light. That is not a matter on which the judge reached a conclusion and it is therefore for this court to make a finding on it and I would so find.

For those reasons to which I have referred, therefore, on the evidence before us I can conclude only that the plaintiff has failed to prove that he did not submit or acquiesce. I would allow the appeal and set the judgment aside.

Agreeing, **Balcombe LJ** said: I add a few words on the point of general interest which arises in this case.

> "Submission to, or acquiescence in" is a state of mind evidenced by the conduct of the parties.

That is a quotation from the judgment of Birkett LJ in *Davies* v *Du Paver* [1953] 1 QB 184 at p 203.

To the like effect is the judgment of Brett J in *Glover* v *Colman* (1874) LR 10 CP 108 at pp 119 and 120; and Grove J in the same case at p 122, where he says (and this is a passage which Glidewell LJ has already quoted):

> I quite admit that mere grumbling, or complaining to a member of his family, or expressing dissatisfaction to a third person, would not be sufficient evidence of non-submission or non-acquiescence to leave to a jury. But where protests are made to the party causing the interruption . . . the matter for our consideration is whether there is any reasonable evidence which ought to have been submitted to the jury, upon which they might have determined whether the interruption was or was not submitted to or acquiesced in by the plaintiff.

The reason why the state of mind has to be evidenced by conduct is not merely to establish the existence of the state of mind — that could be established by the oral testimony of the person whose state of mind was in issue — but so as to establish that his state of mind was made known to the other party concerned. Discontent not effectively communicated to the other party cannot, in my judgment, amount to non-submission or non-acquiescence.

Judge Coles dealt with the silence which lasted from at latest February 1982 and possibly from as early as June 1981 by this passage in her judgment:

A Having put the matter into the hands of his solicitor during the month of January or February 1982 he [that is Mr Dance] made no further complaint because as he says, "it was in the hands of my solicitors".

The facts I have found do not, in my view, constitute acquiescence or submission and he was entitled having passed the matter to a lawyer to consider that the complaint would thereafter be diligently pursued on his behalf by him.

That passage goes only to Mr Dance's state of mind and completely ignores the necessity for communication to Mr Triplow.

Accepting that acquiescence or submission is a question of fact for the judge to decide on all the facts of the case, the period of at least two and a half years (and probably more) before action was brought on August 9 1984 (which is the relevant date for the purposes of the
B Prescription Act 1832) during which no complaint was made makes it clear, in my view, that Mr Dance is unable to establish that he did not submit to or acquiesce in the interruption of the access of light to his bungalow for at least one year before action was brought, and the judge's finding to the contrary is, in my judgment, plainly wrong. The effect of any complaints that may have been made before February 1982 must have worn off long before the action was brought. *Davies* v *Du Paver* is clearly distinguishable on its facts. In that case there had been strong complaints on July 6 and August 1 1950. The interruption was complete on August 9 1950. Action was brought on September 28 1951. The majority of this court there held that the judge was entitled to hold that there had been no submission or acquiescence.
C Birkett LJ said at p 204:

The fact that there was a complete year in which the plaintiff did nothing is some evidence that he had submitted to, or acquiesced in, the interruption; but the fact also that he protested most strongly up to August 1950, and then actually took proceedings in September 1951, is some evidence that he did not. It was a question of fact for the judge.

Where the facts show, as they do in this case, a period of at least two and a half years without complaint, there was simply no evidence to entitle the judge to hold that there was no acquiescence or submission on the part of Mr Dance during the relevant period.

Appeal allowed with costs.

D

Chancery Division

February 28 1992

(Before Mr Justice MORRITT)

HALIFAX BUILDING SOCIETY AND OTHERS v EDELL AND OTHERS

Estates Gazette May 9 1992

[1992] 18 EG 151

E **Building Societies Act 1986 — Ombudsman scheme — Whether Ombudsman has jurisdiction and duty to investigate and determine complaints in relation to valuations made by employees of a building society of properties to be charged to a society to secure further advances to an existing borrower**

The plaintiffs, the Halifax Building Society, the Woolwich Building Society, the Leeds Permanent Building Society and the Alliance & Leicester Building Society, are members of a scheme recognised by the Building Societies Commission — The first defendant is the Ombudsman appointed under the scheme and which he claimed gave him jurisdiction and a duty to investigate and determine complaints in relation to
F **valuations made by employees of a building society of properties to be charged to a society to secure further advances to an existing borrower from that society — The second to ninth defendants complained to the Ombudsman that valuations were carried out negligently by the plaintiffs — The plaintiffs denied that the Ombudsman had the jurisdiction and duty he claimed**

Section 83 of the Building Societies Act 1986 makes provision for the investigation of an individual's complaint against a society to be investigated under a scheme — The relevant matter of complaint was the grant or refusal to grant further advances to a borrowing member and referred to in para 3, Part II of Schedule 12 to the 1986 Act — Upon considering an application for a further advance, each of the plaintiff societies require a borrower to state whether he wishes a basic valuation, a house buyers' report or a structural survey G

***Held*: (1) In the case of the basic valuation, any failure to exercise due care and skill on the part of an employee of a building society can constitute maladministration within the meaning of Part III of Schedule 12 to the 1986 Act and of the scheme — Accordingly, the Ombudsman was entitled to investigate and determine complaints — (2) In the case of house or flat buyers' reports and valuations, the performance of an employed valuer was the performance of his contract of** H **employment and there was no scope for a contract by the valuer as principal with the borrower — If a society chooses an independent valuer, then there is a separate contract between the valuer and the borrower — If the society chooses to use its own employee, then there is no separate contract between the valuer and the borrower — In the latter case, but not the former, the society is responsible if the report is not made with due care and skill by its employee — It follows from this analysis that the provision of the report to the borrower is an action by the building society and that if it is prepared without appropriate skill and care the ground of the complaint,** J **namely breach of the society's contractual obligation, falls to be investigated by the Ombudsman — (3) In the case of the structural survey, there is no relevant distinction between the provision of a structural survey at the request of a borrower applying for a further advance and the provision of a house buyer's report in the same circumstances — An analysis of the contractual relationship produces the same result and the same consequences — Declarations granted accordingly**

The following case is referred to in this report.

R v *Local Commissioner for Administration for the North and East Area of England, ex parte Bradford Metropolitan City Council* [1979] 1 QB 287 K

By an originating summons the plaintiffs, the Halifax Building Society, the Woolwich Building Society, the Leeds Permanent Building Society and the Alliance & Leicester Building Society, sought declarations as to the jurisdiction of the first defendant, Mr Stephen Edell, the Building Societies Ombudsman, in relation to applications for further advances made by the second to ninth defendant borrowers.

Simon Berry QC and John McGhee (instructed by Church Adams Tatham & Co) appeared for the plaintiffs; and Timothy Lloyd QC and Katherine McQuail (instructed by Farrer & Co) represented the first defendants. The second to ninth defendants did not appear and L were not represented.

Giving judgment, MORRITT J said: Section 83(1) of the Building Societies Act 1986 confers on an individual the right as against a building society to have any complaint of his about action taken by the society in relation to a prescribed matter of complaint which affects him in prescribed respects investigated under a scheme recognised by the Building Societies Commission. There is one such scheme known as the Ombudsman Scheme, which was recognised by the Building Societies Commission on June 5 1987. Each of the plaintiffs is a member of the scheme and the first defendant is the Ombudsman appointed under it. The Ombudsman claims that the jurisdiction and duty to investigate conferred and imposed on him by M the scheme enables and requires him to investigate and determine complaints in relation to valuations made by employees of a building society of properties to be charged to the building society to secure further advances to an existing borrower from that society. The plaintiffs disagree and, by the originating summons now before me, seek the determination of certain questions of construction of the scheme so as to resolve the question of jurisdiction.

The brochures issued by each of the plaintiffs show that there are three types of valuation which may be made. There is the basic valuation required to enable the society to assess the adequacy of the security offered pursuant to section 13 of the Building Societies Act 1986. The second to seventh defendants have complained to the

Ombudsman that such valuations made in respect of further loans to them were carried out negligently by an employee of, in the case of the second and third defendants, the Halifax Building Society; in the case of the fourth and fifth defendants the Woolwich Building Society and in the case of the sixth and seventh defendants the Leeds Permanent Building Society. Such brochures also refer to two other types of valuation or survey. The second is a home or flat buyers' report and valuation. The eighth and ninth defendants have complained to the Ombudsman that such a report and valuation made by an employee of the Alliance & Leicester Building Society was carried out negligently. The third type of valuation and report is a full structural survey. It is accepted that such surveys are carried out only by employees of the building society in the case of the Woolwich Building Society, so that the question at issue can arise only in regard to the structural surveys provided by that building society.

None of the second to ninth defendants appeared at the hearing but all the arguments available to them were put by counsel for the Ombudsman. It was common ground that the scheme, though set up pursuant to a statutory obligation, was a private and not a public law scheme so that the procedure by way of originating summons rather than judicial review was appropriate.

Before considering the scheme, it is necessary to refer to various provisions in the Building Societies Act 1986. Section 83 provides, so far as material, as follows:

> (1) An individual shall, by virtue of and in accordance with schemes under this section, have the right, as against a building society, to have any complaint of his about action taken by the society in relation to a prescribed matter of complaint which affects him in prescribed respects investigated under the scheme.
>
> . . .
>
> (6) Schedule 12 to this Act has effect for the purposes of this section and, in that Schedule —
> (*a*) Part I prescribes the matters for which provision must be made by a scheme if it is to be a scheme which qualifies for recognition for the purposes of this section;
> (*b*) Part II prescribes the matters action in relation to any of which must be subject to investigation under a scheme if it is to qualify for recognition for the purpose of investigations in relation to that matter; and
> (*c*) Part III contains other requirements to which a scheme must conform if it is to be so recognised.

Subsection (7) enables the Building Societies Commission with the consent of the Treasury by order to vary Part II or Part III of Schedule 12.

Subsection (15) provides, so far as material:

> In this section, section 84, Schedule 12 and Schedule 13 . . .
> "action" includes any failure to act, and so as regards "exercise" in relation to any power; and "action", in relation to a society, includes action on its behalf by any body associated with it;
> "prescribed", in relation to matters of complaint, means prescribed for the time being in Part II of Schedule 12, and in relation to the respects in which a complainant is affected by any action, means prescribed for the time being in Part III of that Schedule as grounds for making action subject to investigation under the scheme;

Section 84 provides, so far as material:

> (2) Determinations of complaints under recognised schemes shall be made by reference to what is, in the adjudicator's opinion, fair in all the circumstances of the case and any direction given to a building society or associated body by an adjudicator may (if the complainant accepts the determination) require it or the complainant not to exercise or require the performance of any of the contractual or other obligations or rights subsisting between them.
>
> (3) Subject to subsections (4) and (5) below, a determination of the adjudicator under a recognised scheme which is, by virtue of the complainant's acceptance of it, binding on the building society or associated body shall not be questioned in any court of law.
>
> (4) Subsection (3) above does not apply where a society or associated body is authorised by the scheme to relieve itself of its obligation to take the steps it is directed to take or pay the compensation awarded by the society's undertaking an obligation to give the requisite publicity for the reasons for not doing so and the society undertakes that obligation.
>
> (5) Where a determination of the adjudicator under a recognised scheme is binding on the building society or associated body, the adjudicator shall, at the request of the society or associated body, state a case for the opinion of the High Court on any question of law and the High Court may direct the adjudicator to reconsider the complaint.

Schedule 12, Part II sets out the prescribed matters of complaint. The relevant matter is para 3 which states:

> The operation or termination of the account of a member borrowing on a class 1 or class 2 advance and the grant or refusal to grant a borrowing member of that description other or further class 1, or as the case may be, class 2 advances secured on the same or different land or other facilities normally available to borrowing members of his description.
>
> *Note:* The operation or termination of the account of a borrowing member includes any aspect of the relationship or the termination of the relationship between the society and a borrowing member as such, including in particular the exercise of the right of foreclosure or any other power over the land by virtue of the mortgage.
>
> *Note:* The grant of advances includes the terms on which they are granted.

Part III of Schedule 12 is headed "Minimum Requirements for Recognised Schemes: Other Provisions". Para 1 is headed "*Grounds of complaint*" and states:

> 1 The grounds for making action by a building society or associated body subject to investigation under the scheme must be that the action constitutes —
> (*a*) in the case of a building society, a breach of the society's obligations under this Act, the rules or any other contract, or
> (*b*) in the case of an associated body, a breach of the associated body's obligations under its rules (if any) or any contract, or
> (*c*) unfair treatment, or
> (*d*) maladministration,
> in relation to the complainant and has caused him pecuniary loss or expense or inconvenience.

Para 2 sets out the permissible exclusions from investigation, and paras 3 to 5 set out the functions of the adjudicator. Para 3(*a*), (*c*) and (*d*) provide as follows:

> A scheme must, as regards the duties and powers attached to the investigation of complaints under the scheme —
> (*a*) impose on the adjudicator a duty, subject to the provision made in pursuance of sub-paragraph (*c*) below, to investigate and determine any complaint duly made;
> . . .
> (*c*) confer power on the adjudicator to advise, mediate or act as conciliator before proceeding further with an investigation;
> (*d*) confer power on the adjudicator to extend the scope of his investigation to other matters related to the action complained of . . .

Para 4 provides, so far as material:

> (3) Subject to sub-paragraph (4) below, a scheme may preclude the adjudicator, in his determination, from questioning the merits of any decision taken by the society or associated body with reference to —
> . . .
> (*b*) the creditworthiness, for the purposes of any advance or other service or facility, of the complainant;
> but not otherwise.
>
> (4) Where investigation of a complaint on the ground of maladministration involves consideration by the adjudicator of any decision taken with reference to the creditworthiness of the complainant the scheme must confer power on the adjudicator to direct the society or associated body to take its decision again and reach it by proper procedures.

Pursuant to section 85 and Schedule 14, if both the complainant and the society agree, the adjudicator may act in relation to a complaint within section 83 as an arbitrator with all the legal consequences that that entails.

The scheme conforms with the provisions of the Act. It provides all the minimum requirements and no more. It provides for the appointment of the Ombudsman to be called the Building Societies Ombudsman as the adjudicator required by the Act. Clauses 14, 17 and 18 provide as follows:

> 14. **Conditions for Investigation**
> The Ombudsman shall (in accordance with and subject to the following Clauses of this Scheme) investigate any complaint received by him from an individual if: —
> (a) the complaint relates to action taken in the United Kingdom by a building society or a body associated with it;
> (b) at the time that the complaint is received by the Ombudsman, the building society or (as the case may be) associated body is a Participant in the scheme;
> (c) the action was taken in relation to one of the activities specified in Clause 17;
> (d) the grounds of the complaint are included in the grounds specified in Clause 18; and
> (e) the complainant alleges that the action has caused him pecuniary loss, expense or inconvenience.
>
> . . .
>
> 17. **Activities Covered**
> The activities which may be subject to investigation under this Scheme are those specified in Part II of Schedule 12 of the 1986 Act, which are described there under the following headings: —
> — Share Accounts
> — Deposit Accounts

A — Borrowing Members: class 1 or class 2 advances
— Borrowers: loans by appropriate mortgage companies
— Borrowers: mobile home loans
— Borrowers: other loans
— Banking services
— Trusteeship
— Executorship.

18. **Grounds of Complaint**
The grounds of any complaint must be that the action complained of constitutes, in relation to the complainant: —
(a) in the case of a Participating Society, a breach of its obligations under the 1986 Act, its rules or any other contract; or
(b) in the case of a Participating Associate, a breach of its obligations under its rules (if any) or any contract; or
B (c) unfair treatment; or
(d) maladministration.

It is unnecessary to refer to any other provisions of the scheme because I have already quoted the corresponding provisions in the Act. The provisions of clauses 14, 17 and 18 relevant to the issues before me have been helpfully set out in para 7 of the affidavit of Mr Thomas McKeown [solicitor for the plaintiffs] in which I have interpolated three numbers to indicate the particular passages which were dealt with in argument. With that interpolation, the relevant provisions read as follows:

> The Ombudsman shall . . . investigate any complaint received by him from an individual if [1] . . . the complaint relates to action taken in the United Kingdom by a Building Society . . . [2] in relation to . . . the grant or refusal to C grant a borrowing member . . . other or further . . . advances secured on the same or different land provided that the grounds of complaint [are] that [3] the action complained of constitutes in relation to the complainant: —
> (a) in the case of a participating Society, a breach of its obligations under the Act, its rules or any other contract or . . .
> (b) unfair treatment or
> (c) maladministration.

Before referring to the arguments, it is necessary to refer to other provisions of the Act. Section 13(1) provides as follows:

> It shall be the duty of every director of a building society to satisfy himself that the arrangements made for assessing the adequacy of the security for any advance to be fully secured on land which is to be made by the society are such as may reasonably be expected to ensure that —
> D (*a*) an assessment will be made on the occasion of each advance whether or not any previous assessment was made with a view to further advances or re-advances;
> (*b*) each assessment will be made by a person holding office in or employed by the society who is competent to make the assessment and is not disqualified under this section from making it;
> (*c*) each person making the assessment will have furnished to him a written report on the value of the land and any factors likely materially to affect its value made by a person who is competent to value, and is not disqualified under this section from making a report on, the land in question;
> but the arrangements need not require each report to be made with a view to a particular assessment so long as it is adequate for the purpose of making the assessment.

E Section 45 sets out the criteria for the prudent management of a building society. By subsection (3), para 4, and subsection (10) such criteria include maintenance of the requirements of section 13.

By section 34, a building society was authorised to provide the services described in Schedule 8. As originally enacted, paras 1, 2, 3, 8 and 14 were:

> 1. Money transmission services.
> 2. Foreign exchange services.
> 3. Making or receiving of payments as agents.
> . . .
> 8. Arranging for the provision of credit, whether on behalf of the borrower or the person providing credit, and providing services in connection with current loan agreements to the party providing credit.
> F . . .
> 14. Surveys and valuations of land.

Paras 1, 2, 3 and 8 corresponded to the headings and description of the matters of complaint originally contained in Schedule 12, Part II, paras 6, 7, 8 and 9. Schedule 8 was varied by an order made in 1988. Paras 1, 4, 5 and 6 now read as follows:

> 1. Banking services.
> . . .
> 4. Trusteeship.
> 5. Executorship.
> 6. Land services.

"Land services" is defined as follows:

> "Land services" means services relating to the acquisition, management, development or disposal of land.

When Schedule 12, Part II was amended by order in 1990, matters of complaint in paras 6, 7 and 8 corresponded to paras 1, 4 and 5 of Schedule 8 as amended. Thus, both before and after amendment the matters of complaint in Schedule 12 embraced some only of the services a society was authorised to provide and omitted both surveys and valuations of land and land services. G

It is common ground between the plaintiffs and the Ombudsman that the only relevant matter of complaint is that part of Schedule 12, Part II, para 3, which refers to the grant or refusal to grant further advances to a borrowing member. Thus, it is also common ground that the complainant must have been an existing borrower at the time of the grant or refusal in question. Likewise, because the complaint must be about action taken by the society, the Ombudsman has no jurisdiction to investigate complaints relating to valuations made by a surveyor who is not an employee of the society in question. Thus it is clear that the Ombudsman could, if his arguments are right, investigate only complaints arising from negligent valuations in the case of some lenders and some borrowers. H

I propose to consider the questions raised in respect of basic valuations first. In the case of each plaintiff, the application form for a further advance requires the borrower to state whether he wishes a basic valuation, a house buyers' report or a structural survey, the brochure having pointed out the limited scope of the first two. If he opts for a basic valuation, for which he pays a lower fee than for either of the other two, he will normally receive a copy. Such a valuation is, in any event, required by the building society to ensure compliance with the criteria of prudent management set out in section 45 and by the directors of the building society to ensure compliance by them of the duties imposed by section 13. J

For the plaintiffs it is submitted that the action of the employee valuer in exercising his professional skill and judgment in providing a basic valuation for consideration by the assessor is not action taken by the society but a service rendered to the society by a professional man as such, even though he is also an employee.

Second, it is submitted for the plaintiffs that such action is not in relation to the grant or refusal of a further advance because the making of the valuation is not part of the administration or managerial process leading to the grant or refusal but is only a condition precedent to the actual process of making a grant. K

Third, it is submitted that maladministration is the only relevant ground for complaint in respect of basic valuations and that that does not comprehend the exercise of professional skill and judgment by an employed surveyor. The plaintiffs rely on the ordinary meaning of that word as connoting faulty administration or inefficient or improper management of affairs as approved by Eveleigh LJ in *R* v *Local Commissioner for Administration for the North and East Area of England, ex parte Bradford Metropolitan City Council* [1979] 1 QB 287, at p 314, in the context of the Local Government Act 1974.

I was referred to a number of other Acts in which the word "maladministration" is used in that sense, namely Parliamentary Commissioner Act 1974, Local Government Act 1974, National Health Service Act 1977 and Social Security Act 1990. L

The plaintiffs submit that if maladministration includes the provision by an employee of a negligent valuation, then the result would be anomalous because, as is common ground, the provision of such a report by an independent surveyor could not be maladministration.

I will consider each of these points in turn, though it must always be borne in mind that the provisions of the scheme must be construed as a whole so that each phrase provides part of the context for the others.

As a body corporate, a building society can act only through the human agency of its officers and employees. The performance of his contract of employment by an employee involves the provision of services to his employer. But if such performance involves an action within the scope of his employment affecting persons outside the building society, then such action is by the building society. I see no justification for concluding that if the service provided involves the use of professional skill and judgment, the provision of such a service to the society cannot also constitute action by the society. In my judgment, it is plain that the provision of a basic valuation by an employee to a building society to enable the assessment of the adequacy of the security offered by a borrowing member for a further advance involves action taken by the society about which the member may complain if the other prescribed conditions are satisfied. M

A The next question is whether such action is taken in relation to the grant or refusal to grant a further loan. It was common ground that the words "in relation to" are general words linking the action to the prescribed matter of complaint. It is apparent from section 13 that a basic valuation is necessary to the consideration of an application for a further advance. It is quite true, as the plaintiffs submit, that provided the person making the valuation is competent to value, it matters not who instructed him to produce the valuation or whether he is an employee of the building society or not. Thus, there is no obligation on a building society itself to provide the valuation. But this consideration seems to me to be immaterial. The existence of such a valuation is a *sine qua non* to the grant of a further advance. B The provision of such a valuation is in relation to the grant of a further advance because that is its *raison d'être*. The fact that it may have been provided by the member, not the society, and produced by an independent surveyor, not an employee of the society, merely shows, as conceded, that it is not in those cases action taken by the society. But this is irrelevant to the question whether it is in relation to the grant of a further advance. In my judgment, it plainly is.

So the crucial question is whether, if the basic valuation is produced by an employee of a building society, any failure to exercise due care and skill can constitute maladministration. It was common ground that this word connotes faulty administration, inefficient or improper management of affairs. Likewise, it was common ground that faulty administration contemplated the procedures by which C decisions are made as opposed to the merits of the decisions. This distinction is borne out by Schedule 12, Part III, para 4(4).

To my mind, the answer to the question depends on ascertaining in the context of the Act as a whole what is the administration referred to. A building society is not concerned with public or local government or the provision of a public service. Thus, the administration referred to is not of a governmental nature and the other legislation to which I was referred is of no assistance. It is clear from Schedule 12, Part II, that the relevant maladministration, and hence administration, is not the conduct of the building societies' business as a whole but only in those spheres of their activity falling within the prescribed matters of complaint. Those spheres are financial and commercial. It seems to me that the administration in D question must be the procedure by which the building societies' business is conducted within those spheres.

In the case of the grant of a further advance, those procedures include all the steps necessary to assess the adequacy of the security offered for an advance. This requires ensuring that there is a basic valuation and considering its contents. If the borrowing member does not produce a basic valuation, then the building society must procure one itself. Thus, in my judgment, the administration of the society's business in the field of making advances includes, if necessary, obtaining a basic valuation from, if it so chooses, one of its employees. If it chooses to get one from one of its employees, then it seems to me that the preparation of that report by one of its E employees is as much a part of the administration of the societies' business in the field of making advances as is consideration of the report by the assessor required by section 13(1), para (*b*). The fact that it is not an essential part of the process of granting an advance in all cases that a basic valuation should be supplied by an employee of the society does not mean that where the basic valuation is supplied by an employee its provision is not part of the process of administration. It follows from this that the failure of an employee to exercise reasonable skill and care in making his basic valuation is faulty and, therefore, maladministration.

It is true that this conclusion may appear anomalous in that the right conferred on a borrowing member by section 83(1) is confined to cases where the basic valuation is produced by an employee of the F society. But anomalies are bound to arise with any investigative scheme of limited scope. Moreover, the anomaly arises not from the word "maladministration" but from the limitation that the complaint must be about action taken by the society. In my judgment, the meaning of the words used is too clear to permit the exclusion of negligent basic valuations by an employee of the society by any process of construction. The fact that this conclusion enables the Ombudsman to investigate complaints that a professional man has failed to use due care and skill otherwise that in accordance with the common law is no reason for reaching a contrary conclusion either. It is inherent in the business of lending money that a variety of professional skills will be deployed. I see no ground for excluding the valuer if his acts or omissions are those of the society.

Likewise, the fact that surveys and valuations of land or land G services are permissible activities for a building society but not, as such, matters of complaint within Schedule 12, Part II, does not mean that the provision of a basic valuation by an employed valuer in relation to the grant of a further advance must be excluded from the administration of granting or refusing advances. The terms are not mutually exclusive.

In my judgment, the Ombudsman is right. There is nothing relating to the facts of any of the specific complaints of the second to seventh defendants which requires separate consideration. Accordingly, I refuse to make the declaration sought by subparas (a), (b) and (c) of para 1 of the originating summons and I will answer the questions raised in paras (a) to (e) of para 2 in the affirmative.* H

The next set of issues relates to house or flat buyers' reports and valuations. The brochures issued by each of the plaintiffs encourages a borrowing member to have such a report for his or her own protection. A higher fee is charged for such a report and where the surveyor who carries it out is an employee of the society, the fee is retained by the society. Where an independent surveyor is used, he is paid by the society but the society retains part of the amount paid by the borrowing member to cover the costs of its administration. The surveyor conducts one inspection and provides a fuller report than is required to ensure compliance with section 13. The report is sent to the borrowing member. Either that report or a summary of it or a basic valuation prepared from the same materials is used by the building society for the purposes of section 13. J

As with the basic valuation, the issues arise only where the house buyers' report is made by an employee of the society, for it is accepted that in other cases there is no relevant action by the society.

In addition to the three points on the basic valuation, namely action by the society in relation to the grant or refusal of an advance and maladministration, one further point arises in the case of a house or flat buyers' report. That is whether a report prepared by an employee valuer without due care and skill is a breach of contract by the society so as to constitute a ground of complaint within Schedule 12, Part III, para 1(*a*).

---

*Editor's note: Para 1(a) to (c) and para 2(a) to (e) of the originating summons are reproduced below: K

1. A declaration that upon the true construction of the Scheme the *First* Defendant is not entitled to investigate or determine:
   (a) the complaint against the First Plaintiff received by him from Michael Robert Allen and Christine Allen *the Second and Third Defendants respectively* relating to the allegation that the Report and Valuation for Mortgage Assessment carried out on the 9th November 1988 for the First Plaintiff in respect of the property known as 16 St Mary's Grove, Newton Regis, near Tamworth was prepared negligently in that it failed to deal properly or at all with the major structural problems at the premises which had caused the roof to sag and the ceiling in the master bedroom to crack;
   (b) the complaint against the Second Plaintiff received by him from Jeffrey Leonard Brommage and Heather Maureen Brommage *the Fourth and Fifth Defendants respectively* relating to the allegation that the Report and Valuation for Mortgage dated the 13th September 1988 prepared for the Second Plaintiff in respect of the property known as 3 Woodland Close, Failand, Bristol was prepared negligently in that it failed to deal properly or at all with a major structural fault in the premises namely the floor and wall deflections at first floor level caused by overstressing of the floor beams due to insufficient support under the solid block walls; L
   (c) the complaint against the Third Plaintiff received by him from Lawrence Frederick West and Christa West *the Sixth and Seventh Defendants respectively* relating to the allegation that the Report and Mortgage Valuation dated the 25th January 1988 prepared for the Third Plaintiff in respect of the property known as The Winnats, Audrey Street, Station Road, Ossett was prepared negligently in that it failed to deal properly or at all with the extensive woodworm, dangerous wiring, rising damp and rotten window frames at the said premises;
2. The determination of the Court on the following question namely, upon the true construction of the Scheme, whether and if so in what circumstances the *First* Defendant is entitled to investigate and determine a complaint relating to an allegation of failure to exercise the requisite degree of professional skill and care on the part of a valuer or surveyor employed by the building society against which the complaint is made in relation to a report by him on the condition or value of any property where the report in question consists of: M
   (a) a written report prepared pursuant to section 13 of the Building Societies Act 1986 for a building society on the value of land which is proposed as security for an advance to be made by the society and on any factors likely materially to affect its value made by a person who is competent to value and is not disqualified under section 13 aforesaid from making a report on the land in question;
   (b) a written report known as a "Valuers Report and Valuation for Mortgage" prepared for the First Plaintiff and more fully described at pages 74-78 of the bundle referred to in the affidavit herein of Thomas William McKeown;
   (c) a written report known as a "Report and Valuation for Mortgage" prepared for the Second Plaintiff and more fully described at pages 79-83 of the said bundle;
   (d) a written report known as a "Report and Mortgage Valuation" prepared for the Third Plaintiff and more fully described at pages 84-91 of the said bundle;
   (e) a written report known as a "Valuation for Mortgage Advance" prepared for the Fourth Plaintiff and more fully described at pages 92-99 of the said bundle; . . .

A There are variations in the literature used by each of the four plaintiffs and it is necessary to refer to the various brochures, application forms, conditions of employment and report forms.

In the case of the Halifax Building Society, the brochure states in relation to valuations generally:

> We have an extensive panel of valuers and surveyors who are well equipped to complete these reports speedily and effectively. Please bear in mind, however, that we cannot accept responsibility for the contents of the reports as the valuers act independently from the Society — except in certain parts of the country where we employ in-house valuers.

With regard to the house buyers' report, the borrower is informed:

> The valuer advises whether or not the price agreed for the house is realistic, B taking into account its repair and condition and the property market in general. A summary of the report will be sent to the Society for mortgage assessment.

The conditions of engagement of the valuer are set out indicating the nature of the service he will perform and, so far as relevant, state:

> 1. The Surveyor will advise the Client as to his opinion of the state of repair and condition of and the value of the property specified by the Client on the standard form of House Buyers Report and Valuation published on behalf of The Royal Institution of Chartered Surveyors.
>
> 8 The Report is provided for the sole use of the named Client and, where the Surveyor is so notified, his Mortgagee, and is confidential to the Client and his professional advisers. The Surveyor accepts responsibility to the Client alone for the stated purposes that the report will be prepared with the skill, care and C diligence reasonably to be expected of a competent Chartered Surveyor, but accepts no responsibility whatsoever to any person other than the Client himself. Any such person relies upon the Report at his own risk.

In the form of application for a mortgage the applicant is invited to state which type of inspection he requires. The form of declaration to be signed by the applicant includes the following:

> I/we agree to a valuer being instructed to prepare the report and valuation and I/we accept the terms and conditions of engagement.

A six-page report is provided to the borrower signed by the surveyor. Where he is employed by the society, that will be indicated below the signature. A copy was formerly, but is no longer, supplied to the society. The society always received and still receives from the same surveyor a basic valuation containing a summary of some but D not all the matters contained in the report to the borrower. The borrower does not receive a copy of the basic valuation.

In the case of Woolwich Building Society the brochure states:

> Who prepares the report?
>
> One of the Woolwich's surveyors, all of whom are members of the Royal Institution of Chartered Surveyors, or The Incorporated Society of Valuers and Auctioneers, or an approved surveyor nominated by the Society.

Under the heading "What does the report cover?", it is stated:

> The full terms and conditions upon which the Report is prepared are set out opposite with the scale of fees charged shown separately. The Report will be used by the Woolwich instead of the mortgage valuation report as it includes all the information which usually appears in that report.

E The relevant conditions of engagement are:

> 1. The Surveyor will advise the applicant and Woolwich Building Society (the Society) as to his opinion of the state of repair and condition of, and the value of, the property specified by the applicant on the standard form of Home Purchase Report published by the Society in conjunction with the Royal Institution of Chartered Surveyors.
>
> . . .
>
> 7. The Report is provided for the sole use of the named applicant and the Society and is confidential to the applicant and his professional advisers. The Surveyor accepts responsibility to the applicant and the Society only for the stated purposes that the Report will be prepared with the skill, care and diligence reasonably to be expected of a competent Chartered Surveyor, but accepts no responsibility whatsoever to any person other than the applicant F and the Society. Any such person relies upon the Report at his own risk.

The form of application states:

> I/we request the society to arrange for a Woolwich Home Purchase Report to be prepared and confirm I/we have read, understand and accept the conditions of engagement. The fee of [blank] pounds is enclosed.

As indicated in the brochure, a copy of the whole report is supplied to the building society. The last page contains a summary of the information required for the purposes of section 13, though some consideration is also required of the first page. The report is signed by the surveyor, who will indicate under his signature whether he is an employee of the building society or an outside valuer. Where the property in question is a flat, the society receives a separate section 13 basic valuation as well.

In the case of the Leeds Permanent Building Society, the brochure G now in use states:

> Two alternative valuation and survey services are available to you, both of which are primarily for your use — House Buyers Report or Structural Survey. If you wish to obtain either of these more detailed reports we can request one from the Surveyor or Valuer we would normally use to obtain our Report and Mortgage Valuation.

In respect of the house buyers' report and valuation it is stated:

> The House Buyers Report and Valuation, without being a Structural Survey, is a report on the state of repair and condition of the property. It contains a summary for mortgage purposes and statistics and can be used by the Society instead of a Report and Mortgage Valuation.

The relevant conditions of engagement state: H

> 1. The surveyor, who will be a Chartered Surveyor or an Incorporated Valuer, will advise the applicant(s) and Leeds Permanent Building Society as to his opinion of the state of repair and condition of, and the value of, the property specified by the applicant(s) on the standard form of Leeds Permanent/RICS House Buyers Report and Valuation.
>
> . . .
>
> 8. The Report is provided for the sole use of the named Applicant(s) and Leeds Permanent Building Society and is confidential to the Applicant(s) and his professional advisers. The Surveyor accepts responsibility to the Applicant(s) and the Leeds Permanent Building Society for the stated purposes and the report will be prepared with the skill, care and diligence reasonably to be expected of a competent Chartered Surveyor or Incorporated Valuer, but accepts no responsibility whatsoever to any person other than the Applicant(s) and the Leeds Permanent Building Society. Any J such person relies upon the Report at his own risk.

The report consists of seven pages, the first six, described as the inspection report, signed at the foot of p 6 by the surveyor with an indication whether or not he is employed. The seventh page is described as a "Summary for mortgage purposes and Statistics" and is also signed by the surveyor with the appropriate section 13 certificate. Both the society and the borrower receive all seven pages.

In the case of the Alliance & Leicester Building Society, the brochure states:

> The Society will be pleased to help you get a report suitable for your needs; details of the House or Flat Buyers Reports (Option 2) and Building Survey (Option 3) are set out overleaf. The fees for a fuller report are, of course, higher than for the Society's valuation report, because the surveyor will spend K more time both inspecting the property and preparing a fuller report suitable for your needs.

In relation to option 2 it is stated:

> With these reports the Society will instruct, on your behalf, a suitably qualified surveyor who will write to you direct with the details of the Terms and Conditions of Engagement. The Surveyor will be responsible to you for the stated purpose of the report.

The relevant conditions of engagement may be either those of the Royal Institution of Chartered Surveyors or the ISVA. The former are those I have quoted in the case of the other building societies. The relevant condition of the latter states:

> Responsibility is accepted in the preparation of this report for the skill and L diligence reasonably to be expected of a competent surveyor and valuer but the information it contains is for the confidential information only of the clients for whom it is prepared and of any building society, bank or other lender to whom written application for a mortgage advance has been made or will be made within 28 days after the date of this report.

In the form of application in relation to the house buyers' report it is stated:

> In choosing this option, you authorise the society to request an appropriate surveyor to undertake a report and valuation as described in the guidance notes for mortgage applicants and you are satisfied and understand the conditions of engagement.

The report when made takes the same form as that in the case of the Leeds Permanent Building Society. Both the society and the M borrower receive a copy of all seven pages.

For the plaintiffs it was submitted that in each case either the only relevant contract was between the borrower and the valuer or, if there were a contract to which the building society was a party, its obligation was limited to instructing a competent valuer. In neither case, it was submitted, was the society contractually bound to provide a report prepared with due skill and care. In each case the valuer, though employed by the society, contracted as principal and accepted liability to the borrower. In consequence, it was submitted, the relevant action was taken not by the society, the report was not required by the society for its own purposes and was thus not in relation to the grant or refusal of an advance, and a report negligently prepared could not constitute maladministration or a breach of the society's obligations under any contract.

As conceded by the Ombudsman, these submissions are well founded in the case of a valuer who is not employed by the society. The forms of document in use by each of the plaintiffs are directed primarily to cases where independent valuers are instructed but they must also be considered against the background of a contract of employment between the valuer and the society. In these cases the society not only fixes and receives the fee payable for the report but retains the fee for itself. The valuer gets his usual salary and no more. It is entirely a matter for the society whether it selects one of its own employees or instructs an independent firm. In the ordinary case, the borrower plays no part in the choice and has no right to do so. As recognised by the conditions of engagement, the report is made to the borrower and the valuer accepts liability to him. But I do not think that these circumstances demonstrate a contract between the valuer as principal and the borrower. The performance of his task by the employed valuer is the performance of his contract of employment. There is little scope and seemingly no consideration for a contract by the valuer as principal. The conditions of engagement show that the valuer accepts liability in tort to the borrower and the society but this has no additional consequence where the valuer is liable to the society under his contract of employment and the society would be vicariously liable to the borrower for the valuer's negligence if it were not already liable to the borrower under a contract. In the case of Halifax Building Society the brochure states that the society accepts responsibility for its employed valuers.

I accept the submission for the Ombudsman. In accordance with that analysis, there is a contract between the society and the borrower that the society will procure a house buyers' report either through its own employee or, at the society's option, through an independent valuer. If it chooses an independent valuer, then there is a separate contract between the valuer and the borrower, the society's obligation being confined to using care in the selection of the valuer. If the society chooses to use its own employee, then there is no separate contract between the valuer and the borrower. The valuer does the work pursuant to his contract of employment and the society provides the report to the borrower pursuant to the contract between them. In the latter case, but not the former, the society is responsible if the report is not made with due care and skill by its employee.

It follows from this analysis that the provision of the report to the borrower is an action by the building society and that if it is prepared without appropriate skill and care the ground of the complaint, namely breach of the society's contractual obligation, falls within Schedule 12, Part III, para 1(*a*). Thus, it is unnecessary to consider the alternative ground of maladministration.

But it does not necessarily follow that the action of the society is in relation to the grant or refusal of an advance. As all the brochures make plain, the house buyers' report goes beyond what is required by the society for the purposes of section 13. The want of due skill and care may arise in respect of a matter which finds no reflection in the summary or separate basic valuation. Thus, it is true, as the plaintiffs submitted, that the provision of a house buyers' report is not in all its details a necessary part of the process of assessing the adequacy of the security.

But the phrase "in relation to the grant or refusal of an advance" does not require the action complained of to be a necessary part of the process. As a matter of ordinary English it includes actions, which, while not strictly necessary, relate to the grant or refusal of an advance. In the case of each of the plaintiffs, the application form indicates that the borrower must ask for one of the three valuations referred to. If the borrower chooses the house buyers' report, the work undertaken by the valuer results in both a report to the borrower and the necessary report to the society. A single fee is paid to the society. In reality it is all one process not susceptible of division, which is in relation to the grant or refusal of an advance because the borrower selected that form of report in his application for an advance and, in consequence of the society's decision to use its own employee, it is contractually liable for any want of due skill or care.

Accordingly, I refuse to make the declarations sought in para 1(d) of the originating summons and I will answer the questions raised in para 2(f) to (j), but limited to the documentation currently used by each of the four plaintiffs, in the affirmative.

This leaves the structural survey. The question arises only in the case of the Woolwich Building Society because none of the other plaintiffs uses an employee to produce it. In their brochure the Woolwich Building Society state:

> The Woolwich can now offer homebuyers a structural survey report. This will provide a much more detailed examination of the structure and condition of the property being purchased and will be carried out by one of the Woolwich's own structural surveyors.

The form of application states:

> I/we request the society to arrange for a structural survey and, if appropriate, a mortgage valuation report to be prepared and confirm that I/we have read and that I/we understand and accept the conditions of engagement. The fee of [blank] pounds is enclosed.

The relevant conditions of engagement state:

> 1. The Surveyor will advise the client and Woolwich Equitable Building Society (the Society) as to his opinion of the state of repair and condition of, and the value of, the property specified by the client on the report form published by the Society.
>
> . . .
>
> 15. The report is provided for the sole use of the named client and the Society and is confidential to the client and his professional advisers. The Surveyor accepts responsibility to the client and the Society only for the stated purposes that the report will be prepared with skill, care and diligence reasonably to be expected of a competent Chartered Surveyor, but accepts no responsibility whatsoever to any person other than the client and the Society. Any such person relies upon the report at his own risk.

The report is provided on a form headed "Woolwich Building Society: Structural Survey". A copy is supplied to the society. But, as the brochure makes plain, a structural survey may be provided as part of the provision of surveying and valuation or land services wholly unconnected with any application for the grant of an advance. In those cases the report is not in relation to the grant or refusal of an advance, so that even if negligently prepared it is not a matter of complaint within Schedule 12, Part II, para 3. But where there is an application for the grant of an advance a structural survey is one of the three forms of valuation from which the applicant is required to choose. The same valuer will also produce a basic valuation for the purposes of section 13. In such a case, as the brochure states:

> The fee for the structural survey normally includes the cost of a separate mortgage valuation report, which all mortgage applicants need to have carried out. This report is prepared to enable the Woolwich to determine how much to lend. If you are intending to take out a Woolwich mortgage a copy of this report will be provided to you and will accompany our formal mortgage offer. If, however, you are not intending to obtain a Woolwich mortgage then the fee will reflect the fact that there will not be a need for a mortgage valuation report.

I do not see any relevant distinction between the provision of a structural survey at the request of a borrower applying for a further advance and the provision of a house buyers' report in the same circumstances. An analysis of the contractual relationship produces the same result and the same consequences. Accordingly, in relation to para 2(k) of the originating summons I will answer the question in the affirmative, subject to the question being limited to the existing documentation in use by the Woolwich and subject to the request for the structural survey being part of an application for a further advance.

For further cases on this subject see p 303

TOWN AND COUNTRY PLANNING

A
Court of Appeal

February 5 1992

(Before Lord Justice GLIDEWELL, Lord Justice BELDAM and Lord Justice NOLAN)

CAMBRIDGE CITY COUNCIL v SECRETARY OF STATE FOR THE ENVIRONMENT AND ANOTHER

Estates Gazette May 30 1992

[1992] 21 EG 108

B **Town and country planning — Meaning of development — Meaning of building operations — Meaning of "other operations" — Whether demolition development**

On October 3 1989 the respondents, Cambridge City Council, issued two enforcement notices alleging the demolition of part of two semi-detached houses involving the removal of all slates or roof tiles, battens and other materials from the main roofs of the premises — The second appellant, Milton Park Investments Ltd, appealed, contending that the matters alleged as a breach of planning control did not constitute or involve development — The inspector allowed the appeal and quashed both enforcement notices, finding
C **that the projected total demolition of the two houses was not to form part of preparatory works for the erection of a building and would not have been of such scale or complexity as to constitute an engineering operation; he concluded that the operations carried out did not constitute development — Mr David Widdicombe QC (sitting as a deputy judge of the Queen's Bench Division) allowed the appeal on the ground that the inspector erred in law because he failed to consider whether the demolition of the houses was a building operation, holding that although the demolition was not development merely because it was part of larger proposals**
D **for the development of the site but because the demolition of the houses was "development", namely "a building operation" being an "other operation normally undertaken by a person carrying on business as a builder" — The first appellant, the Secretary of State for the Environment, and the second appellant appealed**

***Held*: The appeals were allowed — Certain propositions derived from the authorities — It is a question of fact for the inspector as to whether the works carried out to the premises were operations normally undertaken by a person carrying on business as a builder — That was a matter upon which the inspector had made no finding of fact and about which there**
E **was no evidence before him or the judge — Accordingly, there was no material upon which the judge could find that the works carried out were building operations — In relation to the issue on which the judge found it unnecessary to reach a conclusion, namely whether in this case the works of demolition were "other operations" within the meaning of section 22 of the Town and Country Planning Act 1971, demolition of itself is not an "other operation" — The fact that when the demolition is complete it will or may be followed by works or a change of use which will constitute development within the statutory definition would not alter such a** F
conclusion

The following cases are referred to in this report.

Coleshill & District Investment Co Ltd v *Minister of Housing and Local Government* [1969] 1 WLR 746; [1969] 2 All ER 525; (1969) 68 LGR 334; 20 P&CR 679; [1969] EGD 736; 211 EG 727, HL

Iddenden v *Secretary of State for the Environment* [1972] 1 WLR 1433; [1972] 3 All ER 883; (1972) 26 P&CR 553, CA

Kirkness v *John Hudson & Co Ltd* [1955] AC 696; [1955] 2 WLR 1135; [1955] 2 All ER 345, HL

Marks & Spencer Ltd v *London County Council* [1953] AC 535; [1953] 2 WLR 932; [1953] 1 All ER 1095; (1953) 3 P&CR 409; 51 LGR 329; 161 EG 436, HL G

Parkes v *Secretary of State for the Environment* [1978] 1 WLR 1308; [1979] 1 All ER 211; (1978) 77 LGR 39; 36 P&CR 387; [1978] EGD 973; 248 EG 595; [1979] JPL 33, CA

Prest v *Secretary of State for Wales* (1982) 81 LGR 193, CA

This was an appeal by the Secretary of State for the Environment and Milton Park Investments Ltd from a decision of Mr David Widdicombe QC (sitting as a deputy judge of the Queen's Bench Division), reported at [1991] 1 EGLR 177; [1991] 09 EG 119, who had allowed an appeal by the respondents, Cambridge City Council, under section 246 of the Town and Country Planning Act 1971 arising out of two enforcement notices issued by the respondent to the second appellant. H

Gerald Moriarty QC and Michael Kent (instructed by the Treasury Solicitor) appeared for the first appellant; Lord Silsoe QC and Robert McCracken (instructed by Ginn & Co, of Cambridge) appeared for the second appellant; Robert Carnwath QC and Christopher Lewsley (instructed by Nabarro Nathanson) represented the respondents.

Giving judgment, **GLIDEWELL LJ** said: This appeal raises two questions, the first of which is of general application. They are:

(i) Do works for the demolition of a building invariably constitute "development" within the meaning of the Town and Country Planning Act 1990 and its predecessors? J

(ii) If not, do the particular works of demolition in this case constitute such development?

On October 3 1989 Cambridge City Council ("the council") served two enforcement notices on Milton Park Investments Ltd ("Milton Park") relating to a pair of semi-detached houses, 21 and 23 Milton Road, Cambridge. Milton Park appealed to the Secretary of State for the Environment against the enforcement notices. The Secretary of State delegated the decision on that appeal to one of his inspectors, Mr Peter Norman MA MRTPI. The appeal was conducted by written representations, together with a site visit by the inspector.

By letter dated May 15 1990 the inspector allowed Milton Park's appeals and quashed both enforcement notices.

On June 12 1990 the council appealed to the High Court against the inspector's decision on a point of law under section 246 of the Town and Country Planning Act 1971 (now section 289 of the Town and Country Planning Act 1990). On January 11 1991 Mr David Widdicombe QC, sitting as a deputy judge of the Queen's Bench Division , gave judgment allowing the appeal by the council*. He remitted the matter to the Secretary of State with his opinion. K

The Secretary of State and Milton Park now appeal to this court against Mr Widdicombe's decision.

*Editor's note: Reported at [1991] 1 EGLR 177; [1991] 09 EG 119.

History

The factual background to this matter is significant. Milton Park is a developer. Early in 1989 it was engaged in the development of offices known as the Westbrook Centre, Milton Road, Cambridge. Milton Park purchased the two semi-detached houses, 21 and 23 Milton Road, with the intention of demolishing them and on the cleared sites thereafter "providing additional car parking, enhancing the visual aspect of the development by high-quality landscaping and improving" visibility for drivers of vehicles leaving the development and joining Milton Road. While the construction of the offices was in progress, Milton Park used the two houses as site offices, with temporary planning permission for that use. When that temporary permission came to an end, Milton Park intended to demolish the houses over the Easter weekend, March 24-27 1989. It informed the chief planning officer of the council of its intention by telephone on March 23 1989. He reminded the company that it had not complied with its obligation to give six weeks' notice of intention to demolish under the Building Act 1984. Milton Park therefore instructed its demolition contractor to carry out only a limited amount of demolition over the weekend, including the removal of damaged slates from the roofs and the removal of certain equipment from the interior so as to make the houses unattractive to squatters.

Shortly afterwards, the council sought and were granted an interim injunction prohibiting Milton Park from carrying out any further work of demolition to the two houses. However, this order was set aside on the *inter partes* application.

The council then, early in May 1989, served a building preservation notice on Milton Park in respect of the houses. Milton Park sought leave to challenge this in the High Court by way of judicial review, but the council then agreed to withdraw the preservation notice.

The enforcement notices were served on October 3 1989. It is noticeable that they were not accompanied by stop notices. Nevertheless, we were told that no further work of demolition has taken place, and the houses are in the same condition as they were at the end of March 1989.

The enforcement notices

The notices were in identical terms. The breach of planning control alleged in each was "the demolition of part of the premises involving the removal of all slates or roof tiles, battens, and other materials from the main roofs of the premises . . . and the making of holes in the ceilings of some of the first floor rooms in the premises without having previously obtained the benefit of planning permission". The notices required Milton Park to cease the alleged breach of planning control, to replace the slates or roof tiles on the roofs of the premises so as to match those existing and to make good and repair the ceilings. The period for compliance with the notices was six months.

Milton Park appealed to the Secretary of State on the grounds set out in section 88(2) (*a*) (*b*) and (*c*) of the 1971 Act (section 173 of the 1990 Act). The argument on the appeal, and at all stages up to this court, has centred on ground (*b*), namely that the matters alleged as a breach of planning control do not constitute or involve development.

The relevant statutory provision

Since the planning legislation of 1990 is now in force, I shall give the section numbers in this legislation as well as those in the 1971 Act, under which provisions this appeal has been, and is being, conducted.

Section 22(1) of the 1971 Act contains the classic definition of development, which until now has endured since the coming into force of the 1947 Act. So far as it is material to this case, this is:

"development" . . . means the carrying out of building, engineering, mining or other operations in, on, over or under land . . . [1990 Act (section 55(1)]

By section 290(1) of the 1971 Act (section 336(1) of the 1990 Act) various words and phrases used in that definition of development are further defined. The relevant definitions are:

"building operations" includes rebuilding operations, structural alterations of or additions to buildings, and other operations normally undertaken by a person carrying on business as a builder.

"building" . . . includes any structure or erection, and any part of a building, as so defined . . .

"engineering operations" includes the formation or laying out of means of access to highways.

It will be seen that in none of these definitions do the words "demolition" or "demolish" appear. The only reference in the 1971 Act to demolition was in relation to the demolition of buildings of architectural or historic importance (section 75). To this was added by amendment in 1977 the demolition of buildings in conservation areas (section 277A). These provisions are now to be found in sections 55 and 74 to 76 of the Planning (Listed Buildings and Conservation Areas) Act 1990.

However, Parliament has now legislated in the Planning and Compensation Act 1991 to amend the definition of "building operations" where that expression is used in the definition of development. The 1991 Act adds, by section 13, a new subsection to section 55 of the 1990 Act which provides:

(1A) For the purposes of this Act "building operations" includes —
(*a*) demolition of buildings;
(*b*) rebuilding;
(*c*) structural alterations of or additions to buildings; and
(*d*) other operations normally undertaken by a person carrying on business as a builder.

This is subject to a power given to the Secretary of State by the same section to exclude any description of building from the scope of the section. The 1991 Act has received the royal assent, but section 13 is not yet in force.

Representations by the council

The argument advanced in written representations by the council to the inspector was that in its decision in *Coleshill & District Investment Co Ltd* v *Minister of Housing and Local Government* [1969] 1 WLR 746 the House of Lords had decided that the partial demolition of a permanent building was, or at least could be, "development" within the meaning of the Acts. Thus the partial demolition of these two houses was development for which planning permission was required.

The inspector's decision

In his decision letter, the inspector considered whether he should accept the council's argument that the demolition at the two houses was only partial or the argument for Milton Park that they intended to demolish the houses completely and had ceased the demolition works only because they had been required to do so by the various steps taken by the council. In para 7 of his decision letter, the inspector said of a letter dated May 15 1989 written by Milton Park's parent company to the Secretary of State that:

[It] clearly indicates that the appeal properties were acquired by your clients with a view to providing additional car parking for the adjoining office development, enhancing its visual aspect by landscaping, and improving highway safety for vehicles using the access road. The total demolition of the properties is implicit in those proposals, and I am satisfied that that was your clients' intention.

This was a clear finding of fact rejecting the argument for the council. Since no other issue had been canvassed before the inspector, he might have allowed the appeal and quashed the enforcement notices on this finding alone. But he did not do so. He continued by considering whether, as a matter of law, the total demolition of the two houses could or did constitute development.

In the following paragraphs of his decision letter the inspector said:

8. In the light of the *Coleshill* case I accept that demolition — whether total or partial — is capable of being development. The Courts had previously decided that total demolition and clearance of a site may constitute works for the erection of a building, and therefore development (*London County Council* v *Marks & Spencer Ltd* [1953] 3 P&CR 409, HL); however your clients propose to use the sites of the appeal properties for car parking, landscaping and highway improvements rather than for the erection of any building, and it therefore seems to me that that judgment does not lend support to the Council's case. The facts in *Coleshill*, on which the authority seek to rely, were highly unusual in that what was to be demolished, namely blast walls and embankments erected as external parts of an explosives store, were not buildings in the ordinary sense of the word; moreover, as the Council emphasise, that case was concerned with partial demolition only.

9. In Milton Road the operations undertaken so far have been the first stages in the projected total demolition of the houses, not their partial demolition with a view to alteration, and in my view the facts are comparable to those in *Iddenden* rather than in *Coleshill*. The projected total demolition was not to form part of preparatory works for the erection of a building, and nor has it been nor would it be of such scale or complexity as to constitute an engineering operation in its own right. In the light of the foregoing, and having taken into account all the points made in the representations, I conclude that the operations carried out do not constitute development requiring planning permission. The appeals will therefore succeed on ground (b) and no other aspect of the matter falls to be considered.

He therefore quashed both enforcement notices.

A I shall comment later on paras 8 and 9 of this letter. I note, however, at present that in para 9 is another finding of fact, that the "projected total demolition . . . (would not) constitute an engineering operation in its own right". Since the council were entitled to appeal against the inspector's decision only on a question of law, they could thus not challenge the findings that the demolition was intended to be total and that it did not constitute an engineering operation.

The appeal to the High Court

The notice of motion pleaded that the inspector erred in law in two respects:

1. In failing to consider whether the demolition of the houses B may have constituted works for, or have been coupled with, the laying out and formation of the car-parking spaces and the landscaping and highway improvements intended, or the change of use of the land which these would entail, and thus constitute development within section 22 of the 1971 Act.

2. In saying in para 9 of his decision that "the facts are comparable to those in *Iddenden* rather than in *Coleshill*".

The judge's decision

Mr Widdicombe QC most helpfully summarised his conclusions. I shall have to consider certain parts of the reasoning in his judgment in more detail later, but at present it suffices to set out his summary, C which was as follows:

(1) The inspector erred in law in his understanding of the *Marks & Spencer* case, but the error is not material because the *Marks & Spencer* case is of little relevance in any event.

(2) The inspector erred in law in his understanding of the *Iddenden* case, and the error is material because he failed to consider whether the demolition of the houses was a building operation.

(3) The demolition of these houses is not "development" merely because it is part of larger proposals for the development of the site; and

(4) The demolition of these houses is "development", namely "a building operation", being an "other operation normally undertaken by a person carrying on business as a builder".

It is for this reason that the judge allowed the appeal against the inspector's decision and remitted the matter to the Secretary of State. D Having reached this decision, he did not find it necessary to express a view on the question whether the demolition of the houses was an "other operation in, on, over or under land".

The relevant authorities

Before I come to consider the judge's judgment in more detail and whether the inspector was correct in his conclusions, it is necessary to refer to the three previous decisions which are relevant to our consideration of the issues on this appeal.

The first in point of time was *Marks & Spencer Ltd* v *London County Council* [1952] Ch 549, (CA); [1953] AC 535, (HL). Shortly before 1939, permission was granted for the redevelopment of a site E in Marylebone under the Town and Country Planning Act 1932. The old buildings on the site were demolished in 1939, but the erection of the new buildings was prevented by the outbreak of war. After the end of the war, when the 1947 Act was in force, the question arose whether a new planning permission for the erection of the proposed buildings was necessary. This depended on section 78(1) of the 1947 Act, under which planning permission was deemed to be granted where "any works for the erection or alteration of a building have been begun but not completed before the appointed day". The question at issue was whether the works for the demolition of the building were within the phrase "works for the erection of a building". The Court of Appeal by a majority held that they were and the House of Lords upheld this decision unanimously. We are not F concerned with the phrase "works for the erection of a building". In his judgment in the Court of Appeal, with which all the members of the House of Lords expressly agreed, Jenkins LJ referred to the phrase "works for the erection or alteration of a building" as including "in terms, operations which are not in themselves building operations" (at p 563). The phrase under consideration in the *Marks & Spencer* case was thus clearly wider than the phrase with which we are concerned in the present case, namely "building operations".

The decision of the House of Lords in *Coleshill & District Investment Co Ltd* v *Minister of Housing and Local Government* [1969] 1 WLR 746 concerned a group of buildings constructed during the 1939 to 1945 war as an ammunition depot. It consisted of four magazines and two explosive stores, around each of which were erected about 4 ft from the walls of the buildings themselves blast G walls about 9 ft high. On the outside of each of the blast walls was a sloping embankment extending from the top of the wall to 8 ft or 10 ft from its base, and covered with grass. In 1962 the minister decided that use of the buildings for commercial storage would not constitute development and they were so used. In 1966 the owners decided to remove the embankments and blast walls. The embankments were removed by using a mechanical excavator. Before the blast walls were demolished, the local planning authority served an enforcement notice alleging that the removal of the embankments constituted development without planning permission. The owners appealed to the minister and also asked for a determination whether the removal of the blast walls would constitute development requiring H permission. After an inquiry, the minister concluded that the blast walls and the embankments were an integral part of each of the buildings, that the removal of the embankments was an engineering operation within the definition of "development" in section 12(1) of the 1962 Act (which was identical with the definition in the 1990 Act). He also held that the removal of the blast walls would constitute an alteration of the buildings, materially affecting their external appearance, and thus also constitute development. On appeal against these decisions the Divisional Court allowed the appeal, but the Court of Appeal restored the minister's decisions. The decision of the Court of Appeal was upheld in the House of Lords. Put shortly, the conclusion was that it was open to the minister as a matter of law to conclude as he did and that his decisions that the blast walls were an J integral part of each of the buildings and that the removal of the embankments was an engineering operation was in each case a question of fact for him.

Lord Morris of Borth-y-Gest said in his speech at p 755F:

The definition of "development" is set out in section 12(1) and section 221 of the Act. Subject to certain exceptions, and leaving aside the making of any material change in the use of any building or other land, "development" means the carrying out of building operations (which include rebuilding operations, structural alterations of or additions to buildings, and other operations normally undertaken by a person carrying on business as a builder) or of engineering operations (which include the formation or laying out of means of access to highways) or of mining operations or of other operations, in, on, over or under land . . . Wide though the definition is, it does not K include any and every operation on land. In section 43 is set out the procedure to be followed when it is desired to have a determination whether planning permission is required. Any person who proposes to carry out "any" operations on land may apply for a ruling. Had development meant "any" operations in, on, over or under land there would not have been included in section 12(1) the words "building, engineering, mining or other". I think that the word "other" must denote operations which could be spoken of in the context of or in association with or as being in the nature of or as having relation to building operations or engineering operations or mining operations.

Lord Upjohn said at p 761F:

The main submission of the appellants pushed to its ultimate length was that no act of demolition "*per se*" or "*simpliciter*" (whatever those L qualifications mean — I think they were intended to mean without a view to redevelopment), whether of a complete building or of a part of a building, could in law constitute development for the purposes of section 12. I cannot understand this argument pushed so far. Counsel for the appellants rightly pressed your Lordships with the view that in the Act the word "demolition" was only used in relation to cases where corrective action was required after service of enforcement notices or in relation to "specially listed" houses of historic interest which may not be demolished without a special order. But there is nothing in section 12 or elsewhere which makes it plain that demolition *per se* or *simpliciter* is necessarily excluded from the very wide words of section 12(1) if otherwise the relevant operation fits within those words as a plain matter of the use of the English language. In fact there is nothing to exclude demolition. In this case the Minister has decided that this demolition is an engineering operation and so within the section. I do not understand M how, if this demolition be properly so described, the fact that it is only demolition *per se* takes it out of that section.

Lord Wilberforce said at pp 764H-765H:

Finally there is the appellants' suggestion that the relevant operations must at least be of a constructive character, leading to an identifiable and positive result. I think that this is near the heart of the matter, and that there is an important element of truth in the argument. I would accept, and think it important to emphasise, that the planning legislation should be approached with a disposition not to bring within its ambit, unless specific words so require, operations in relation to land which do not produce results of this kind, that is to say, results (I deal only with operations, not with use) of a positive, constructive, identifiable character. In my opinion, the appellants succeed in showing that neither the development of the legislation, nor the

A successive descriptions of "development", nor the policy of control and, while it lasted, of charge on development, nor common sense or common expectation, require or suggest that the mere removal of a structure, or a building, or of a part of a building should be subject to the code. And I think that they derive important support for this argument from the Minister's circular No 67 of February 15, 1949, in which he stated that he was

"advised that the demolition of a building does not of itself involve development, although, of course, it may form part of a building operation, or lead to the making of a material change in the use of the land upon which it stood".

The advice referred to may not have been quite correct (I return to this point) but in giving this information to planning authorities, the minister was undoubtedly reflecting a common-sense and accepted opinion as to the general nature of development. I accept of course that, as an interpretation of
B the Act of 1947, under which it was issued, the circular had no legal status: but it acquired vitality and strength when, through the years, it passed, as it certainly did, into planning practice and textbooks, was acted on, as it certainly was, in planning decisions and when the Act of 1962 (and, I may add, the Town and Country Planning Act of 1968) maintained the same definition of "development" under which it was issued.

So far, for the appellants, so good: they establish, in my opinion, this general approach to the construction of development and I think that it assists them in the limitation of the words "other operations". But, as I said above, this can only be so unless specific words so require. The governing statutory words remain those of sections 12 and 221, and where these fairly apply, they must prevail. They cannot be prevented from applying to a particular operation, which comes within them, by the mere fact that, in addition, the
C descriptive label "demolition" would fit that operation.

The Minister's circular to be fully accurate, should then have said not that it (the demolition of a building) may form part of a building operation, but that, what might be described as demolition may fall within one of the specific types of operation described in section 12(1) and rank as development accordingly.

The Minister has held here that the removal of the embankments was an engineering operation, and that removal of the walls would be a building operation. Neither of these conclusions appears to me to have been inevitable: he might have held that both or either would be demolition and that neither the one fell under "engineering" nor the other under "building". But both were marginal decisions given in relation to a very special case, and I think were open to him to make as he did.

Lord Pearson said at p 770A:

Thirdly, there is the removal of the embankments. I would have been inclined
D to regard this also as a structural alteration of the magazine treated as a single building, structure or erection, and on that basis there would be a building operation within the definition and it would be development requiring planning permission. There is also the possibility that the removal of the embankments might come within "other operations normally undertaken by a person carrying on business as a builder", but without expert knowledge or expert evidence I do not feel able to form a view on this possibility. The Minister, however, did not decide on either of these grounds. He decided on the ground that the removal of the embankments was an "engineering operation" . . . In my opinion, there was evidence on which the Minister could reasonably accept the view of the local authority and the inspector that the removal of the embankments constituted an engineering operation and, therefore, was development requiring planning permission.

The argument for the appellants was mainly based on the proposition that
E demolition, or demolition in or by itself, or demolition *per se*, does not constitute development. I think there is in this proposition some truth but only a limited amount of truth. On the one hand, there is in section 12 and in the relevant definitions in section 221 no mention of or reference to demolition or removal or any such operation. Therefore, an operation does not qualify as development by virtue of being a demolition or removal operation. It is not right to say "This is a demolition or removal operation: therefore, it is development." On the other hand, there are not in section 12 or in the relevant definitions in section 221 any words excluding operations from being development if they are demolition or removal operations. An operation is not disqualified for being development because it is a demolition or removal operation. It is not right to say "This is a demolition or removal operation: therefore, it cannot be development." Notwithstanding that an operation is a demolition or removal operation, one still has to see whether it comes within
F the scope of development as defined in section 12 assisted by section 221. It may be within the definition of "building operations," eg because it constitutes a structural alteration of a building or because it is such an operation as to be normally undertaken by a person carrying on business as a builder. It may be an engineering operation. Whether it is or not any of those things depends on the facts of the particular case.

In *Iddenden* v *Secretary of State for the Environment* [1972] 1 WLR 1433, a decision of the Court of Appeal, the material facts were that the applicants had bought land on which there were some Nissen huts and a lean-to structure in which they manufactured small metal components. They applied for permission to demolish existing buildings and to erect new ones. Permission was refused but, despite that, the owners demolished the Nissen huts and the lean-to structure and erected three prefabricated buildings, in which they continued to G
manufacture the metal components. The local planning authority served an enforcement notice requiring the applicants to demolish the unauthorised buildings and discontinue the unauthorised use of them.

The owners appealed to the Secretary of State, who upheld the enforcement notice. The owners then further appealed to the divisional court, contending that the enforcement notice was defective in not requiring them after the demolition of the new buildings to replace the original Nissen huts and lean-to structure. That court and the Court of Appeal dismissed the appeal.

The argument for the appellants was that the breach of planning control consisted in both the demolition of the original buildings and the erection of the new ones, as well as their use. The issue was H
whether the demolition was part of the breach of planning control.

The basis of the judgments dismissing the appeal in this court was that section 15(5) of the Town and Country Planning Act 1968 (now section 173 of the 1990 Act) required that the enforcement notice:

shall specify —

(a) the matters alleged to constitute a breach of planning control;
(b) the steps required by the authority to be taken in order to remedy the breach, that is to say steps for the purpose of restoring the land to its condition before the development took place . . .

It was argued that this wording obliged the local planning authority to insert in the enforcement notice a requirement not only that the new buildings must be pulled down but also that the old J
buildings should be restored.

In his judgment, Lord Denning MR, with whom Buckley and Orr LJJ agreed, said at p 1439B:

The words of the section are "to remedy the breach". That means the breach of planning control. What was the breach in this case? If it was the pulling down of the old Nissen huts or the old workshop, then it may be that in order to remedy the breach they would have to be restored. But I do not think their demolition was a breach of planning control. Whilst some demolition operations may be development — see *Coleshill & District Investment Co Ltd* v *Minister of Housing and Local Government* [1968] 1 WLR 600, [1969] 1 WLR 746 — the demolition of buildings such as these were not development. Mr Iddenden did not need planning permission to pull them down. No doubt the pulling down of the old and the erection of the new was all one combined operation by the workmen. But in planning law they are different operations. K
Mr Iddenden only required planning permission for his new buildings and their user. The only breach by him of planning control was the unauthorised erection of the new buildings, the concrete base, and the user of the factory. The enforcement notice required him to remedy that breach. It told him the steps he had to take. It was perfectly good.

Submissions

Mr Carnwath QC, for the council, submits that the inspector was required to consider whether the works of demolition constituted a building, engineering or other operation. His finding of fact that the works of demolition to the houses which had been carried out when the enforcement notices were served were part of a total demolition of the houses, and were thus not a partial demolition forming part of L
a structural alteration to the buildings, and that the works were not an engineering operation, is not open to challenge and is accepted. However, the inspector did not consider, as he should have done, the questions whether the works of demolition which had been carried out constituted either:

(i) an other operation "normally undertaken by a person carrying on business as a builder", and thus within the definition of "building operations" within section 336; or
(ii) an "other operation . . . on . . . land" within the meaning of the definition of development in section 55(1).

As to the first contention, the learned judge was correct and was entitled to come to the conclusion to which he came. Alternatively, if M
he was not entitled to conclude that the work of demolition was an "operation normally undertaken by a person carrying on business as a builder" without evidence to that effect, the matter should be remitted to the Secretary of State to be further considered. As to the second point, any operation on land which has a permanent physical effect on the land is development. In order to determine whether works of demolition are within the meaning of "development" as being "other operations", it is necessary to ascertain the purpose for which the demolition is being carried out. If it is carried out in preparation for works which will have a permanent physical effect on the land, then the demolition itself constitutes development. In this case, since the demolition of the houses was intended to be followed

A by the laying out of the site for car-parking and improvement of access, together with some landscaping, the demolition constituted development.

It suffices to say that Mr Moriarty QC for the Secretary of State and Lord Silsoe QC for Milton Park controvert these submissions.

In my view, the decisions in the *Marks & Spencer, Coleshill* and *Iddenden* cases establish the following propositions:

(a) Works for the demolition of a building may, but do not necessarily or inevitably, constitute "development" within the meaning of section 55 of the 1990 Act.

(b) Such works constitute "development" if, but only if, they are properly to be regarded as either:

B (i) "building operations" as defined in section 336(1) of the Act; or

(ii) "engineering operations"; or

(iii) "other operations . . . on . . . land".

(c) Demolition works may be building operations if they are a part of "structural alterations of . . . buildings", as in *Coleshill*. In such a case the demolition will inevitably be partial only, since if it were total there would be no building left to be altered.

(d) Demolition works of a particular type or scale may be "operations normally undertaken by a person carrying on business as a builder".

(e) Demolition works of particular structures, eg the embankments in *Coleshill*, may be "engineering operations".

C (f) Whether works of demolition are within any of these categories of development is a question of fact for the decision-maker — the Secretary of State or, as in the present case, the inspector to whom he has delegated the decision.

(g) The definition of development does not comprehend every operation on land. This is made clear by the words in section 64(1):

> If any person who proposes to carry out any operations on land . . .
> (*a*) wishes to have it determined whether the carrying out of those operations . . . would constitute or involve development of land . . .

Thus "other operations . . . on . . . land" in the definition in section 55(1) does not mean *all* other operations (*per* Lord Morris in *Coleshill* at p 655H).

D (h) "Other operations" in that definition are operations which, while not of one genus comprising also building and engineering operations, nevertheless "must at least be of a constructive character, leading to an identifiable and positive result", or be "similar to building operations or to engineering operations" (*per* Lord Wilberforce at p 764H and Lord Pearson at p 771E in *Coleshill*).

(i) Whether particular works of demolition constitute development within the statutory definition must be decided in relation to those works, and not to other projected works to which the demolition is a preliminary. "In planning law they are different operations": *per* Lord Denning MR in *Iddenden* at p 1439. The *Marks & Spencer* decision does not controvert this proposition. In

E that case the works of demolition were held to be "works for the erection or alteration of a building" which, as Jenkins LJ made clear, is a phrase with a wider meaning than "building operations" and apt to include preliminary works of demolition.

Conclusion

The judge described the question whether the demolition of the houses was development within the definition in the Act as "a question which like a ghost has haunted planning law for many years". He said: "The time has now come when the ghost must be laid to rest."

If that is an accurate description of the task in which we are engaged, it is ironic that we should be charged with it at the very

F moment when Parliament, in the 1991 Act, has breathed new life into the phantom. However, we must construe the law as it is at present, for which purpose I turn to apply the principles I have tried to formulate to the present appeal.

It is accepted by Mr Carnwath, for the council, that the inspector's findings of fact lead to the conclusion that the works of demolition carried out at 21 and 23 Milton Road were neither "structural alterations of buildings" nor "engineering operations".

Was the judge right to find that they were "operations normally undertaken by a person carrying on business as a builder"? I have already said that this was a question of fact for the inspector. The judge said:

> I take the view that the demolition of these houses is a "building operation", that is an "other operation normally undertaken by a person carrying on G
> business as a builder" . . . the pulling down of ordinary dwellinghouses like these seems to me unquestionably within the scope of a builder's business, whether he goes on to develop the site afterwards or not.

Mr Moriarty and Lord Silsoe argue that the judge was not entitled to make this finding of fact. It was a matter upon which the inspector had made no finding of fact and about which there was no evidence before him or the judge. There was thus no material upon which the judge could find as he did.

I agree with these submissions. It is clear that this issue was not canvassed before the inspector — there is no word of it in the written representations. If he had been asked to consider it, he might have said that he required evidence of the normal scope of a builder's

business, or he might have been able to make a decision from his own H expert knowledge. In the absence, however, of such a finding by the inspector and of evidence, there was no material upon which the judge could reach his conclusion. Despite his undoubted expertise, he was not sitting as a tribunal finding facts.

Mr Carnwath, however, argues that if we reach this conclusion, we should also hold that this a matter which the inspector was obliged to consider, that he was thus in error in failing to do so and that (subject to his argument on other issues) we should remit the matter to the Secretary of State for him to make the necessary findings. He draws our attention to a passage in the judgment of Watkins LJ in *Prest* v *Secretary of State for Wales* (1982) 81 LGR 193 on the duty of the Secretary of State or his inspector. J

I accept that when determining an appeal under the planning legislation an inspector has a wider task than that of a judge determining a civil action. The inspector must consider what appear to him to be "material considerations", whether they have been canvassed by the parties or not. But there are limits to this duty. The inspector is under no obligation to cast about, trying to think of every issue which might be relevant. The issues raised by the parties in their representations are a good indication of the matters they consider relevant. Thus I reject the submission that the inspector here should, of his own volition, have considered and made a finding about the normal scope of a builder's business.

It follows that, in my view, the judge was wrong to allow the appeal

to him on the basis he adopted. It is therefore necessary for us to K consider the issue on which he found it unnecessary to reach a conclusion, namely whether in this case the works of demolition were an "other operation . . . on . . . land". Applying the principles I have sought to set out at (g), (h) and (i) above, I conclude that demolition of itself is not an "other operation" within the meaning of section 55(1). The fact that when the demolition is complete it will or may be followed by works or a change of use which will constitute development within the statutory definition does not alter this conclusion. If I understand his judgment correctly, the judge thought that the principle in the *Marks & Spencer* case, that demolition of a building intended to be followed by the construction of a new building amounted to "works for the erection of a building", applied

equally to demolition intended to be followed by development such L as the laying out of a car park. If the judge did mean this, I respectfully agree with him. As Jenkins LJ made clear in his judgment in the *Marks & Spencer* case in the passage to which I have referred earlier, the phrase "works for the erection of a building" has a wider meaning than the phrase "building operations". The wider phrase, as indeed Mr Carnwath argues, has no relevance in the present case. I therefore conclude that the demolition works at 21 and 23 Milton Road were not an "other operation" within section 55(1).

For completeness, I should say that when at para 9 of his decision letter the inspector said that "the facts are comparable to those in *Iddenden* rather than in *Coleshill*", he was, in my view, doing no

more than constrasting the relevance of total demolition with that of M partial demolition. I see no error there.

For these reasons I would allow this appeal and restore the order made by the inspector, quashing both enforcement notices.

Beldam LJ agreed and did not add anything.

Nolan LJ said: I also agree with the judgment of Glidewell LJ and wish to add some words of my own solely upon the question of statutory construction.

An enforcement notice may be served "where it appears to the local planning authority that there has been a breach of planning control"; and a breach of planning control occurs if, *inter alia*, "development has been carried out . . . without the grant of

A planning permission": see section 87(1) and (2) of the Town and Country Planning Act 1971, now replaced by section 172 of the 1990 Act.

The enforcement notices served upon Milton Park in the present case alleged that in the case of each of the two houses concerned there had been a breach of planning control consisting of "the demolition of part of the premises involving the removal of all slates or roof tiles, battens and other materials from the main roofs of the premises, the removal of all slates or roof tiles and roof timbers from the roof of the single storey rear extension to the premises, and the making of holes in the ceilings of some of the first floor rooms in the premises without having previously obtained the benefit of planning permission". The
B sole question is whether the works of demolition thus described amounted in the circumstances to development as defined by the Act. Since an enforcement notice is penal in nature, the statutory definition must be viewed strictly.

In *Coleshill & District Investment Co Ltd* v *Minister of Housing and Local Government* [1969] 1 WLR 746 Lord Morris paraphrased and summarised the statutory definition in these terms, at p 755:

> Subject to certain exceptions, and leaving aside the making of any material change in the use of any buildings or other land, "development" means the carrying out of building operations (which include rebuilding operations, structural alterations of or additions to buildings, and other operations normally undertaken by a person carrying on business as a builder) or of engineering operations (which include the formation or laying out of means
> C of access to highways) or of mining operations or of other operations, in, on, over or under land.

Thus the definition does not expressly refer to demolition but, as *Coleshill* shows, it will include demolition if, in the circumstances of the particular case, the process of demolition involves or forms part of building, engineering, mining, or other operations in, on, over or under land. Mr Carnwath QC, representing the council, accepts that on the basis of the inspector's findings of fact the demolition works in the present case did not involve or form part of a structural alteration to the buildings or an engineering operation. He submitted, however, that the learned deputy judge was entitled to regard them as building operations in the generic sense of "operations normally undertaken by a person carrying on business as a builder".

D Surprisingly, to my mind, there appears to be no previous case in which this point has arisen. I should have thought that there must be many occasions, of which the present case may be one, where the works of demolition do not require the employment of a demolition expert but would normally be undertaken by builders. Whether, even so, works of simple demolition which are not linked to any of the specified types of operation in the definition can fall as a matter of law within the scope of operations amounting to "development" is another matter and one which I consider below. But at this point it does not arise because there is no evidence that the works described in the enforcement notices in fact were operations normally undertaken by builders. In the absence of such evidence, I agree with Glidewell LJ
E that it was not open to the judge to reach his own conclusion upon the matter.

Mr Carnwath submitted alternatively that the works of demolition fell within the definition as "other operations . . . on . . . land". He referred to the judgment of Lord Denning MR in *Parkes* v *Secretary of State for the Environment* [1978] 1 WLR 1308 at p 1311 where Lord Denning spoke of the word "operations" as comprising "activities which result in some physical alteration to the land, which has some degree of permanence to the land itself". If that remark were of general application, then it would follow that almost any process of demolition, whether total or partial, would constitute development within the meaning of the Act. The Court of Appeal in *Parkes*, however, were concerned only to decide whether the storing
F and sorting of scrap materials amounted to an operation on land or a use of land (they decided that it was a use of land) and Lord Denning was clearly not concerned to expand the meaning of the statutory phrase "other operations" on land in any general sense, still less to consider whether it included demolition. It is, again, surprising to my mind that this appears to be the first case in which the courts have had G to decide whether the statutory definition of "development" includes purely destructive works, such as those in the present case, which neither involve nor form part of building, engineering or mining operations. In the absence of authority, I would construe the definition as not including them, for the simple reason that demolition must be a very common form of operation on land and, if Parliament had intended to make it subject to planning control, as a general rule I would have expected the 1971 Act to refer to it expressly, as it did in the case of works of demolition on listed buildings. In saying this I do not overlook the fact that the members of the House of Lords in *Coleshill* derived no assistance from the use of the word "demolition" in the listed building provisions, but the explanation for that, as it seems to me, is simply that they were H considering a different question on different facts. There is, none the less, some support in the speech of Lord Wilberforce for the proposition that works such as those in the present case would not fall within the scope of the phrase "other operations". I refer to the passage already quoted by Glidewell LJ at [1969] 1 WLR 764H where Lord Wilberforce said:

> Finally there is the appellants' suggestion that the relevant operations must at least be of a constructive character, leading to an identifiable and positive result. I think that this is near the heart of the matter, and that there is an important element of truth in the argument. I would accept, and think it important to emphasise, that the planning legislation should be approached with a disposition not to bring within its ambit, unless specific words so J require, operations in relation to land which do not produce results of this kind, that is to say, results (I deal only with operations, not with use) of a positive, constructive, identifiable character. In my opinion, the appellants succeed in showing that neither the development of the legislation, nor the successive descriptions of "development", nor the policy of control and, while it lasted, of charge on development, nor common sense or common expectation, require or suggest that the mere removal of a structure, or a building, or of a part of a building should be subject to the code.

At p 765F Lord Wilberforce, expressing himself with his usual precision, added by way of comment upon a circular published by the minister:

> The Minister's circular to be fully accurate, should then have said not that it (the demolition of a building) may form part of a building operation, but that, what might be described as demolition may fall within one of the specific types K of operation described in section 12(1) and rank as development accordingly.

Since the works of demolition in the present case have no constructive element and would not fall within any of the specific types of operation described in the definition, they cannot, in my judgment, constitute development.

Lord Silsoe argued additionally that the inclusion of demolition in the new definition of development which has been introduced by section 13 of the Planning and Compensation Act 1991 is a further indication that it did not form part of the previous definition. To my mind, however, this approach falls foul of the rule (jealously guarded in the tax cases as a barrier against disguised retrospective legislation) that, save where express provision is made to that effect, the L provisions of a later Act cannot be used as an aid to the construction of an earlier Act, even though the Acts are to be read as one, save for the purpose of resolving an ambiguity, that is to say a word or phrase fairly and equally open to different meanings: see *Kirkness* v *John Hudson & Co Ltd* [1955] AC 696. An example sometimes given of such an ambiguity is "twelve o'clock", which might equally mean midday or midnight. The pre-1990 meaning of "development" may have been difficult to fathom, but it was not ambiguous.

I would not, therefore, accept the argument based on the 1991 Act, but for the reasons already given, as well as those given by Glidewell LJ, I would allow the appeal.

Appeal allowed; order for costs below set aside; second appellants M *to have their costs against the first appellants below and of the appeal; no order for the first appellants' costs below, but costs of the appeal against the respondent. Application for leave to appeal to the House of Lords refused.*

VENDOR AND PURCHASER

A
House of Lords

January 23 1992

(Before Lord KEITH OF KINKEL, Lord ACKNER, Lord GOFF OF CHIEVELEY, Lord JAUNCEY OF TULLICHETTLE and Lord BROWNE-WILKINSON)

WALFORD AND OTHERS v MILES AND ANOTHER

Estates Gazette March 21 1992
[1992] 11 EG 115

Sales of land — Oral agreement — Collateral "lock-out"
B **agreement — Whether implied term to negotiate in good faith — Whether agreement enforceable**

At the material time the respondents, Mr and Mrs Miles, owned a company PNM Laboratories Ltd together with premises in Blackfriars Road, London, let to the company — In 1986 the respondents decided to sell the company and its business premises — Martin Walford and his brother Charles Walford, the first and second appellants, were interested in purchasing the company and its premises — On March 16 1987, following negotiations, the first appellant faxed a letter headed "subject to contract" to the respondents' solicitor
C **recording that the first respondent had given his assurance that, provided he received a clear indication of the intention to proceed with the purchase not later than March 25, he would not treat with any third party or consider any other alternative offers — In a telephone conversation on the following day with the first appellant, the first respondent agreed that if the appellants provided a comfort letter from their bankers confirming that they were prepared to provide the finance to effect the purchase by Friday of that week, he would terminate negotiations with any third party with a view to concluding agreements with the appellants — The first respondent further**
D **agreed that even if he received a satisfactory proposal from a third party before the following Friday he would not deal with that third party nor would he give further consideration to any alternative — On March 30 1987 the respondents informed the appellants that they had decided to sell the company and premises to a third party**

The appellants brought proceedings relying upon an oral agreement, collateral to the negotiations — The consideration for this oral agreement was the respondents' agreeing to continue the negotiations and not to withdraw and, further, the comfort letter from the bankers — The Court of Appeal allowed an appeal by the respondents on the grounds that the
E **agreement alleged was no more than an agreement to negotiate and was therefore unenforceable — The appellants appealed**

***Held*: The appeal was dismissed — The reason why an agreement to negotiate, like an agreement to agree, is unenforceable, is simply because it lacks the necessary certainty — The concept of a duty to carry on negotiations in good faith is inherently repugnant to the adversarial position of the parties when involved in negotiations — Each party to the negotiations is entitled to pursue his own interest so long as he avoids making misrepresentations — To advance that** F
interest he must be entitled, if he thinks it appropriate, to threaten to withdraw from further negotiations or to withdraw in fact, in the hope that the opposite party may seek to reopen the negotiations by offering him improved terms — A duty to negotiate in good faith is as unworkable in practice as it is inherently inconsistent with the position of a negotiating party — Accordingly, a bare agreement to negotiate has no legal content

However, there is clearly no reason in English contract law why A, for good consideration, should not achieve an enforceable agreement whereby B agrees, for a specified G
period of time, not to negotiate with anyone except A in relation to the sale of his property — This is a negative agreement giving A an exclusive opportunity, for a fixed period, to try to come to terms with B, an opportunity for which he has, unless he makes his agreement under seal, to give good consideration — The agreement alleged in the appellants' statement of claim contained the essential characteristics of a basic valid lock-out agreement save that it did not specify for how long it was to last — Lacking the necessary certainty it was therefore unenforceable

The following cases are referred to in this report. H

Albion Sugar Co Ltd v *William Tankers Ltd* [1977] 2 Lloyd's Rep 457
Channel Home Centers, Division of Grace Retail Corporation v *Frank Grossman* (1986) 795 F 2d 291
Courtney & Fairbairn Ltd v *Tolaini Brothers (Hotels) Ltd* [1975] 1 WLR 297; [1975] 1 All ER 716, CA
Hillas & Co Ltd v *Arcos Ltd* (1932) 147 LT 503; [1932] All ER Rep 494; 38 Com Cas 23; 43 Lloyd's Rep 359, HL
Mallozzi v *Carapelli SpA* [1976] 1 Lloyd's Rep 407, CA; [1975] 1 Lloyd's Rep 229
Nile Co for the Export of Agricultural Crops v *H&JM Bennett (Commodities) Ltd* [1986] 1 Lloyd's Rep 555
Scandinavian Trading Tanker Co AB v *Flota Petrolera Ecuatoriana* [1981] 2 Lloyd's Rep 425
Star Steamship Society v *Beogradska Plovidba* [1988] 2 Lloyd's Rep 583 J
Trees Ltd v *Cripps* (1983) 267 EG 596
Voest Alpine Intertrading GmbH v *Chevron International Oil Co Ltd* [1987] 2 Lloyd's Rep 547

This was an appeal by Martin Walford and Charles Walford from a decision of the Court of Appeal ([1991] 2 EGLR 185; [1991] 27 EG 114 and [1991] 28 EG 81) allowing an appeal from Judge Bates QC ([1990] 1 EGLR 212; [1990] 12 EG 107), who gave judgment for the appellants against the respondents, Mr and Mrs Miles, in relation to the sale of a company and premises in Blackfriars Road, London SE1, by the respondents.

Philip Naughton QC and Angus Moon (instructed by Wedlake Bell) appeared for the appellants; Stanley Brodie QC and Edward Cohen (instructed by Tarlo Lyons) represented the respondents. K

In delivering the leading speech, LORD ACKNER said: Mr Martin Walford, the first-named appellant, is a solicitor in private practice. He is the brother of the second-named appellant, Mr Charles Walford, who is a chartered accountant. They own the third-named plaintiff, a company which plays no part in this dispute. The respondents, Mr and Mrs Miles, husband and wife, at the material time owned a company, PNM Laboratories Ltd, together with premises in Blackfriars Road, London SE1, which were let to the

A company, where it carried on a photographic processing business. The company's auditors at the material time were Mr Patel and Mr Khanderia, who carried on their profession under the name of Patel, Khanderia & Co.

In 1985, because of illness, Mr Miles decided to sell the company and its business premises. Negotiations took place with Mr Patel and Mr Khanderia via the medium of a company, Status Guard Ltd, in which Mr Patel and Mr Khanderia had a 30% interest. These negotiations proved unsuccessful.

Towards the end of 1986 Mr and Mrs Miles decided once more to try to sell the company and its business premises. Mr Patel put forward an offer of £1.9m. Meanwhile, the Walfords had heard that B the business was up for sale. There was a meeting between Mr Martin Walford and Mr Miles on April 23 1987. Although the Walfords knew nothing about the photographic processing business, they thought that they had found a bargain. The Miles' were prepared to warrant that at the date of completion the cash resources in the company's bank account would not be less than £1m and that the trading profits for the 12 months following completion would not be less than £300,000 before tax. The Walfords considered that the business and its premises were, in the words of Mr Naughton QC in opening this appeal, "dramatically undervalued". They were accordingly enthusiastic to purchase it at this price.

Following a meeting on March 12 1987 at Mr Martin Walford's offices, the main terms of the purchase were agreed in principle and C on March 16 Mr Martin Walford faxed a letter expressly headed "subject to contract" to Mr Randall of Tarlo Lyons Randall Rose, the solicitor for Mr and Mrs Miles. The purchasers were not to be the Walfords but a company controlled by them (the third-named plaintiff). In his letter of March 16 Mr Martin Walford recorded that Mr Miles had given his assurance that, provided he received a clear indication of the intention to proceed with the purchase not later than the close of business on Wednesday March 25, he would not treat with any third party or consider any other alternative offers. Mr Randall replied on March 17 that he did not have any instructions to proceed with the sale and that Mr Miles had not given the assurance alleged. This letter was, however, overtaken by oral exchanges which occurred on the same day between Mr Martin Walford and Mr Miles. D On March 18 Mr Martin Walford wrote to Mr Randall confirming what had been agreed during that telephone conversation. It is common ground that the penultimate paragraph of that letter correctly sets out the agreement which had been reached as to negotiations between third parties. It reads as follows:

> The last matter discussed between Mr Miles and me related to our ability to make payment for the shares in the business and the property. He asked me to provide a comfort letter from our bankers confirming that they are, subject to contract, prepared to provide the finance of £2,000,000 to enable Acquisition Corpn to effect the purchase. Mr Miles agreed that if such a letter were in your hands by close of business on Friday of this week he would terminate negotiations with any third party or consideration of any alternative with a view to concluding agreements with me and my brother and Acquisition E Corpn and he further agreed that even if he received a satisfactory proposal from any third party before close of business on Friday night he would not deal with that third party and nor would he give further consideration to any alternative.

The letter from Lloyds Bank which Mr Walford enclosed, dated March 18, was addressed to Mr Randall and expressed to be given without responsibility. It confirmed that the bank had offered the brothers loan facilities to enable them and the company controlled by them to make the purchase for £2m.

On March 25 Mr Randall wrote to Mr Martin Walford acknowledging the receipt of this letter and enclosure and confirmed that, subject to contract, his client agreed to the sale of the property and the shares at a total price of £2m. On the same day, Mr Randall F wrote to Mr Patel's solicitors informing them that his client had concluded terms for the sale of the property and the shares in the company to another party and that he was waiting to receive a draft contract. He pointed out that everything was still subject to contract and that the transaction might not go through, but that if it did not his client would be interested to pursue discussions. On March 26 Mr Walford's solicitors sent preliminary inquiries and a draft share purchase agreement to Mr Randall, who sent them on to Mr Miles the following day. Mr Martin Walford was anxious to meet Mr Miles upon that day, but Mr Miles was not available.

On March 30 Mr Randall wrote to Mr Martin Walford informing him that, after careful consideration, his clients had decided to sell to a company associated with the auditors, adding that he hoped that the Walfords would accept his clients' decision without question. G This they refused to do. They treated Mr Randall's letter as a repudiation of what they alleged to be a contract and then issued these proceedings. Meanwhile, the shares in the company and the property in Blackfriars Road had been sold for £2m to Statusguard Ltd, the corporate vehicle through which Mr Patel and Mr Khanderia had tried to make the purchase in 1985.

Mr Miles did not give evidence before the trial judge, Judge Bates QC sitting as a judge of the High Court. It was Mrs Miles who informed the court that she and her husband spent the afternoon of Friday March 27 at the company's premises and during that time decided not to sell to the Walfords or their company. Her explanation was that they were concerned whether they and their staff would get on well with Mr Martin Walford. If they failed to do so, they might H lose staff and then fail to produce the £300,000 profit which was the subject of the warranty. Moreover, they were both concerned that her husband's health might suffer during the year that he would be continuing to work in the business, providing the Walfords with the expertise which they lacked. They therefore decided either to continue in business themselves or to ask Mr Patel if he was still interested. There was a telephone conversation on the evening of Friday March 27 with Mr Patel, who confirmed that he was still interested and agreed readily to increase his offer by £100,000, thus matching the price which the Walfords had offered. Mr Patel, in his evidence, said that after he had received Mr Randall's letter of March 25 he had no contact with Mr Miles until the telephone conversation J on March 27 to which I have just referred. The trial judge did not believe him. He concluded that Mr Miles and Mr Patel had continued to keep in touch, notwithstanding the oral agreement of March 17 recorded in the letter of March 18.

The pleaded case

The Walfords relied upon an oral agreement, collateral to the negotiations which were proceeding to purchase the company and the land that it occupied "subject to contract". The consideration for this oral agreement was twofold — first, the Walfords' agreeing to continue the negotiations and not to withdraw and, second, their providing the comfort letter from their bankers in the terms requested. K

For this consideration it was alleged in para 5 of the statement of claim as follows:

> . . . the First Defendant on behalf of himself and the Second Defendant would terminate negotiations with any Third Party or consideration of any alternative with a view to concluding an agreement with the Plaintiffs and further that even if he received a satisfactory proposal from any Third Party prior to the close of business on 25 March 1987, he would not deal with that Third Party or give further consideration to any alternative.

As thus pleaded, the agreement purported to be what is known as a "lock-out" agreement, providing the plaintiffs with an exclusive opportunity to try to come to terms with the defendants, but without expressly providing any duration for such an opportunity. L

For reasons which will become apparent hereafter, it was decided to amend this paragraph by the following addition:

> It was a term of the said collateral agreement necessarily to be implied to give business efficacy thereto that, so long as they continued to desire to sell the said property and shares, the First Defendant on behalf of himself and the Second Defendant would continue to negotiate in good faith with the plaintiff.

Thus the statement of claim alleged that the defendants not only were "locked-out" for some unspecified time from dealing with any third party but were "locked-in" to dealing with the plaintiffs, also for an unspecified period.

In the statement of claim it was further alleged that by reason of the wrongful repudiation by the Miles', the Walfords lost the M opportunity of completing the sale and purchase of the shares and property and that the true market value of the shares and the property was of the order of £3m. Accordingly, the Walfords claimed that they lost the difference between the price which they had agreed to pay of £2m and the true market value. In addition to the above, there was a claim for damages for misrepresentation by the Miles' in continuing to deal with third parties. This consisted of the expenses incurred in the negotiations and in the preparation of contract documents.

The decision of first instance and the Court of Appeal

After the close of pleadings, directions were given, *inter alia*, that the assessment of any damages to which the Walfords might be

A entitled for the alleged loss of opportunity of completing the sale and purchase of the shares and property should await the determination of the issue of liability*. At the trial it was contended on behalf of the Miles' that the agreement alleged in para 5 of the amended statement of claim was no more than an agreement to negotiate and as such was unenforceable. The judge did not deal with this contention. He held that there was a collateral agreement whereby the Miles' undertook to terminate negotiations with any third party or consideration of any alternative and that even if Mr Miles received a satisfactory proposal from any third party before the close of business on the Friday night he would not deal with that party or give further consideration to any alternative, and that this agreement had been repudiated by the Miles'. He therefore ordered that the damages for
B the alleged loss of opportunity be assessed. He further held that the promises of Mr Miles under the collateral agreement were misrepresentations and awarded the Walfords £700 on account of special damages, being the agreed wasted expenditure.

In the Court of Appeal, by a majority (Dillon and Stocker LJJ), the appeal was allowed (save to the extent of the award of the damages for misrepresentation) on the grounds that the agreement alleged was no more than an agreement to negotiate and was therefore unenforceable. Bingham LJ, who dissented, would have held that the agreement was enforceable on the ground that it could be construed as an agreement by the Miles' not to deal with any party other than the Walfords and not to entertain any alternative proposal. He would
C have set aside the award of damages for misrepresentation on the grounds that it was not justified by the evidence or the trial judge's findings. Before your lordships the respondents were not contesting the £700 award.

The validity of the agreement alleged in para 5 of the statement of claim as amended

The justification for the implied term in para 5 of the amended statement of claim was that in order to give the collateral agreement "business efficacy", Mr Miles was obliged to "continue to negotiate in good faith". It was submitted to the Court of Appeal and initially to your lordships that this collateral agreement could not be made to work unless there was a positive duty imposed upon Mr Miles to
D negotiate. It was, of course, conceded that the agreement made no specific provision for the period it was to last. It was, however, contended, albeit not pleaded, that the obligation to negotiate would endure for a reasonable time and that such time was the time which was reasonably necessary to reach a binding agreement. It was, however, accepted that such period of time would not end when negotiations had ceased, because all such negotiations were conducted expressly under the umbrella of "subject to contract". The agreement alleged would thus be valueless if the alleged obligation to negotiate ended when negotiations as to the terms of the "subject to contract" agreement had ended, since at that stage the Miles' would have been entitled at their whim to refuse to sign any contract.

E Apart from the absence of any term as to the duration of the collateral agreement, it contained no provision for the Miles' to determine the negotiations, albeit that such a provision was essential. It was contended by Mr Naughton that a term was to be implied giving the Miles' a right to determine the negotiations, but only if they had "a proper reason". However, in order to determine whether a given reason was a proper one, he accepted that the test was not an objective one — would a hypothetical reasonable person consider the reason a reasonable one? The test was a subjective one — did the Miles' honestly believe in the reason which they gave for the termination of the negotiations? Thus they could be quite irrational, so long as they behaved honestly.

F Mr Naughton accepted that as the law now stands and had stood for approaching 20 years, an agreement to negotiate is not recognised as an enforceable contract. This was first decided in terms in *Courtney & Fairbairn Ltd* v *Tolaini Brothers (Hotels) Ltd* [1975] 1 WLR 297, where Lord Denning MR said at pp 301-302:

> If the law does not recognise a contract to enter into a contract (when there is a fundamental term yet to be agreed) it seems to me it cannot recognise a contract to negotiate. The reason is because it is too uncertain to have any binding force. . . It seems to me that a contract to negotiate, like a contract to enter into a contract, is not a contract known to the law . . . I think we must apply the general principle that where there is a fundamental matter left undecided and to be the subject of negotiation, there is no contract.

*Editor's note: First instance decision reported at [1990] 1 EGLR 212; Court of Appeal decision reported at [1991] 2 EGLR 185.

In that case at p 302B Lord Denning rejected as not well founded G (and Lord Diplock expressly concurred with this rejection) the dictum of Lord Wright in *Hillas & Co Ltd* v *Arcos Ltd* (1932) 147 LT 503 at p 515 that:

> There is then no bargain except to negotiate, and negotiations may be fruitless and end without any contract ensuing; yet even then, in strict theory, there is a contract (if there is good consideration) to negotiate, though in the event of repudiation by one party the damages may be nominal, unless a jury think that the opportunity to negotiate was of some appreciable value to the injured party.

The decision in the *Courtney* case was followed by the Court of Appeal in *Mallozzi* v *Carapelli SpA* [1976] 1 Lloyd's Rep 407. In that case Kerr J [1975] 1 Lloyd's Rep 229 had applied the dictum of Lord H
Wright in *Hillas* v *Arcos* before the *Courtney* case had been decided and held that there was an obligation on the parties at least to negotiate bona fide with a view to trying to reach an agreement. In that case a contract for the sale of grain contained a clause which provided:

> Cif FREE OUT ONE SAFE PORT WEST COAST ITALY — excluding Genoa. First or second port to be agreed between Sellers and Buyers on the ship passing the Straits of Gibraltar.

The Court of Appeal, however, held that it was impossible to say that the provision in the contract was legally enforceable or that there was any legally binding obligation to negotiate.

The decision that an agreement to negotiate cannot constitute a legally enforceable contract has been followed at first instance in a J
number of relatively recent cases — *Albion Sugar Co Ltd* v *William Tankers Ltd* [1977] 2 Lloyd's Rep 457; *Scandinavian Trading Tanker Co AB* v *Flota Petrolera Ecuatoriana (The Scaptrade)* [1981] 2 Lloyd's Rep 425; *Trees Ltd* v *Cripps* (1983) 267 EG 596; *Nile Co for the Export of Agricultural Crops* v *H & J M Bennett (Commodities) Ltd* [1986] 1 Lloyd's Rep 555; *Voest Alpine Intertrading GmbH* v *Chevron International Oil Co Ltd* [1987] 2 Lloyd's Rep 547; *Star Steamship Society* v *Beogradska Plovidba* [1988] 2 Lloyd's Rep 583.

In the Court of Appeal and before your lordships, Mr Naughton submitted that the *Courtney* and the *Mallozzi* cases were distinguishable from the present case, because that which was referred to negotiation with a view to agreement in those cases was an K
existing difference between the parties. In the present case, so it was contended, by the end of the telephone conversation on March 17 there was no existing difference. Every point that had been raised for discussion had been agreed. However, this submission overlooked that what had been "agreed" on the telephone on March 17 was "subject to contract". Therefore, the parties were still in negotiation even in relation to those matters. Further, there were many other matters which had still to be considered and agreed.

Before your lordships it was sought to argue that the decision in *Courtney's* case was wrong. Although the cases in the United States did not speak with one voice your lordships' attention was drawn to the decision of the United States' Court of Appeal, Third Circuit, in *Channel Home Centers, Division of Grace Retail Corporation* v L
Frank Grossman (1986) 795 F 2d 291 as being "the clearest example" of the American cases in the appellant's favour. That case raised the issue whether an agreement to negotiate in good faith, if supported by consideration, is an enforceable contract. I do not find the decision of any assistance. While accepting that an agreement to agree is not an enforceable contract, the Court of Appeal appears to have proceeded on the basis that an agreement to negotiate in good faith is synonymous with an agreement to use best endeavours and as the latter is enforceable, so is the former. This appears to me, with respect, to be an unsustainable proposition. The reason why an agreement to negotiate, like an agreement to agree, is unenforceable is simply because it lacks the necessary certainty. The same does not M
apply to an agreement to use best endeavours. This uncertainty is demonstrated in the instant case by the provision which it is said has to be implied in the agreement for the determination of the negotiations. How can a court be expected to decide whether, *subjectively*, a proper reason existed for the termination of negotiations? The answer suggested depends upon whether the negotiations have been determined "in good faith". However, the concept of a duty to carry on negotiations in good faith is inherently repugnant to the adversarial position of the parties when involved in negotiations. Each party to the negotiations is entitled to pursue his (or her) own interest, so long as he avoids making misrepresentations. To advance that interest he must be entitled, if he thinks it appropriate, to threaten to withdraw from further

A negotiations or to withdraw, in fact, in the hope that the opposite party may seek to reopen the negotiations by offering him improved terms. Mr Naughton, of course, accepts that the agreement upon which he relies does not contain a duty to complete the negotiations. But that still leaves the vital question — how is a vendor ever to know that he is entitled to withdraw from further negotiations? How is the court to police such an "agreement"? A duty to negotiate in good faith is as unworkable in practice as it is inherently inconsistent with the position of a negotiating party. It is here that the uncertainty lies. In my judgment, while negotiations are in existence either party is entitled to withdraw from those negotiations at any time and for any reason. There can be thus no obligation to continue to negotiate until there is a "proper reason" to withdraw. Accordingly, a bare B agreement to negotiate has no legal content.

The validity of the agreement as originally pleaded in the statement of claim

Para 5 of the statement of claim, as unamended, followed the terms of the oral agreement as recorded in the penultimate paragraph of the letter of March 18. It alleged that for good consideration (and this certainly covered the provision by the plaintiffs of the "comfort letter") Mr Miles on behalf of himself and his wife agreed that they:

> would terminate negotiations with any Third Party or consideration of any alternative with a view to concluding an agreement with the Plaintiffs and, further, that even if he received the satisfactory proposal from any Third Party prior to the close of business on 20 March 1987 he would not deal with C that Third Party or give further consideration to any alternative.

Despite the insistence by Mr Naughton upon the implied term pleaded in the amendment involving the obligation to negotiate, Bingham LJ, in his dissenting judgment, considered that that obligation could be severed from the agreement. He concluded that the agreement, as originally pleaded, was a valid and enforceable agreement and entitled the Walfords to recover whatever damages they could establish resulted in law from its repudiation.

Before considering the basis of Bingham LJ's judgment, I believe it helpful to make these observations about a so-called "lock-out" agreement. There is clearly no reason in the English contract law why A, for good consideration, should not achieve an enforceable D agreement whereby B agrees, for a specified period of time, not to negotiate with anyone except A in relation to the sale of his property. There are often good commercial reasons why A should desire to obtain such an agreement from B. B's property, which A contemplates purchasing, may be such as to require the expenditure of not inconsiderable time and money before A is in a position to assess what he is prepared to offer for its purchase or whether he wishes to make any offer at all. A may well consider that he is not prepared to run the risk of expending such time and money unless there is a worthwhile prospect, should he desire to make an offer to purchase, of B not only then still owning the property but being prepared to consider his offer. A may wish to guard against the risk that, while he is investigating the wisdom of offering to buy B's E property, B may have already disposed of it or, alternatively, may be so advanced in negotiations with a third party as to be unwilling, or for all practical purposes unable, to negotiate with A. But I stress that this is a negative agreement — B, by agreeing not to negotiate for this fixed period with a third party, locks himself out of such negotiations. He has in no legal sense locked himself *into* negotiations with A. What A has achieved is an exclusive opportunity, for a fixed period, to try to come to terms with B, an opportunity for which he has, unless he makes his agreement under seal, to give good consideration. I therefore cannot accept Mr Naughton's proposition, which was the essential reason for his amending para 5 of the statement of claim by the addition of the implied term, that without a positive obligation on B to negotiate F with A the lock-out agreement would be futile.

The agreement alleged in para 5 of the unamended statement of claim contains the essential characteristics of a basic valid lock-out agreement, save one. It does not specify for how long it is to last. Bingham LJ sought to cure this deficiency by holding that the obligation upon Mr Miles and his wife not to deal with other parties should continue to bind them "for such time as is reasonable in all the circumstances". He said:

> the time would end once the parties acting in good faith had found themselves unable to come to mutually acceptable terms . . . the defendants could not . . . bring the reasonable time to an end by procuring a bogus impasse, since that would involve a breach of the duty of reasonable good faith which parties G such as these must, I think, be taken to owe to each other.

However, as Bingham LJ recognised, such a duty, if it existed, would indirectly impose upon the Miles' a duty to negotiate in good faith. Such a duty, for the reasons which I have given above, cannot be imposed. That it should have been thought necessary to assert such a duty helps to explain the reason behind the amendment to para 5 and the insistence of Mr Naughton that without the implied term the agreement, as originally pleaded, was unworkable — unworkable because there was no way of determining for how long the Miles' were locked out from negotiating with any third party.

Thus even if, despite the way in which the Walfords' case was pleaded and argued, the severance favoured by Bingham LJ were H permissible, the resultant agreement suffered from the same defect (although for different reasons) as the agreement contended for in the amended statement of claim, namely that it, too, lacked the necessary certainty and was thus unenforceable.

I would accordingly dismiss this appeal with costs.

Lords Keith of Kinkel, Goff of Chieveley, Jauncey of Tullichettle and **Browne-Wilkinson** agreed with the speech of Lord Ackner and the reasons given in it and did not add any observations of their own.

The appeal was dismissed.

LANDS TRIBUNAL

CAPITAL TRANSFER TAX

LADY FOX'S EXORS v COMMISSIONERS OF INLAND REVENUE

(DET/1/1990)

November 11 1991

(V G WELLINGS QC (President) and T HOYES FRICS)

Estates Gazette May 16 1992 and May 23 1992

[1992] 19 EG 173 and [1992] 20 EG 121

Valuation of agricultural land — Deceased freeholder farming land in partnership — Land subject to tenancies — Method of valuation — Whether freehold interest and partnership share to be treated as single unit of property — Finance Act 1975, s38

Freehold interest in Cambridgeshire agricultural estate of over 3,000 acres owned by deceased, Lady Fox — Prior to her death in March 1981 she farmed the land with two partners and it was subject to certain tenancies which she had granted to herself and her partners — Notice of determination of value for capital transfer tax purposes issued by Commissioners of Inland Revenue purported to aggregate or lot together her freehold interest and her share of the partnership business as a single unit of property for the purposes of s38 of Finance Act 1975 — The amount asserted to be the value of the alleged single unit was £5,650,000, calculated by deducting £560,625 for risk and profit and the shares in the partnership of Lady Fox's partners from the vacant possession value of £6,125,000 and rounding the result — After apportionment according to a mathematical formula the value attributed to the freehold was £4,280,768 — The executors claimed that the freehold value was the price the land might reasonably have been expected to fetch if sold in the open market, namely £2,751,000 — Further, they challenged the single-unit basis of valuation as contrary to s38

***Decision*: The appeal was allowed and the value of the freehold interest determined at £2,751,000, its investment value**

(1) The deceased's share in the partnership was not itself an interest in land — The tenancies were assets of the partnership but it did not follow that a share in the partnership was anything other than personalty or a chose in action — Accordingly, the freehold interest and the share did not constitute a single unit of property for the purposes of s38 and the commissioners had no power to lot the two together

(2) The freehold interest and the share did not constitute a natural unit of property within the meaning given in *Duke of Buccleuch* v *Commissioners of Inland Revenue* — The commissioners' purpose in lotting the two items together was to try to obtain a higher value than the sum of values of the separate parts, but that aim did not convert the two into a natural unit of property

(3) If it were permissible to lot the freehold interest and the partnership share together as a single unit of property then s38 required that it be valued as a single unit and apportionment was neither admissible nor appropriate

(4) Accordingly, the notice of determination was based on false premises which vitiated the district valuer's valuations for the commissioners

(5) If the tribunal's decision were wrong, the district valuer's valuation approach paid realistic regard to the circumstances actually prevailing and likely to have been within the knowledge of the hypothetical purchaser — While the general basis of value to be adopted was prescribed by statute as open market value, there was no prescription of the valuation method to be applied — If adjustment of the agreed value with vacant possession by deductions and/or apportionment were adopted, it fell to be considered on the criterion of whether it produced a realistic answer judged against the prevailing factual circumstances — With no argument or evidence as to adjustment available, no adjustment would be made

The following cases are referred to in this report.

Buccleuch (Duke of) v *Inland Revenue Commissioners* [1967] 1 AC 506; [1967] 2 WLR 207; [1967] 1 All ER 129, HL

Cooper v *Critchley* [1955] Ch 431

Inland Revenue Commissioners v *Crossman* [1937] AC 26

Richard Bramwell QC and Alun James (instructed by Farrer & Co) appeared for the appellant executors; Mary Arden QC and Nicholas Warren (instructed by the solicitor of Inland Revenue) for the respondent commissioners.

Giving their decision, THE TRIBUNAL said: This is an appeal against a notice of determination by the Commissioners of Inland Revenue under section 19 of the Finance Act 1975 and para 6 of Schedule 4 to that Act. The purpose of the notice was to determine the value to be attributed to the freehold interest in land situate and known as the Croxton Park Estate, Cambridgeshire, immediately before the death on March 27 1981 of the owner, the Hon Myra Alice Lady Fox (whom we will call "Lady Fox"). A copy of that notice of determination is contained in Appendix A to this decision [see end]. The appeal is by Lady Fox's executors. A copy of the notice of appeal is contained in Appendix B to this decision [see end]. Both the notice of determination and the notice of appeal are revised versions of earlier similar notices, the revisions having been agreed between the parties under para 6(5) of Schedule 4 to the Act of 1975.

At all material times prior to her death, Lady Fox farmed the land to which the notice of determination related in conjunction with two partners and that land was subject to certain tenancies which Lady Fox, as freeholder, had granted to herself and her partners.

The notice of determination, for the purpose of attributing a value to Lady Fox's freehold interest in the land, purports to aggregate or lot together her freehold interest in the land and her share of the partnership business "as a single unit of property for the purposes of section 38 Finance Act 1975". The notice then proceeds to assert "the relevant amount" (£5,565,000), which is not defined in the notice but which, we understand, is alleged to be the value of the alleged single unit of property, calculated by deducting £560,625 (for risk and profit and the shares in the partnership of Lady Fox's partners) from the vacant possession value of £6,125,000 and rounded in the result. The relevant amount is then apportioned according to a mathematical formula set out in the notice and, finally, the value to be attributed to the freehold interest is stated to be £4,280,768.

The notice of appeal states that, for the purposes of section 38 of the Finance Act 1975, the value of the freehold interest in the land specified in the notice of determination is the price which it might reasonably have been expected to fetch if sold in the open market at

A the date of the deceased's death and that that value is £2,751,000. Further and in the alternative, the notice of appeal challenges the basis of valuation described in the notice of determination as being contrary to the provisions of section 38 in that, it is said, the freehold interest and the share in the partnership do not fall to be aggregated together as a single unit of property and that apportionment is neither permissible nor relevant.

The notice of determination and the notice of appeal are concerned with the value of the freehold interest for the purposes of capital transfer tax or, as it is now called, Inheritance Tax.

Mr Richard Bramwell QC, for the appellants, called Mr Donald Hugh Reeves MA FRICS, a partner in Strutt & Parker, a firm of, *inter alia*, chartered surveyors, of London and elsewhere. Miss Mary
B Arden QC, for the Crown, called Mr Michael David Clegg FRICS, district valuer of Cambridge.

The facts

From an agreed statement of facts and from the evidence we find the following facts:

(1) The Croxton Park Estate, Cambridgeshire, is an agricultural estate situated approximately 14 miles to the west of Cambridge on the A45 Cambridge to St Neots Road. The estate comprises mansion house, parkland, woodland and agricultural land and buildings. The mansion house is a Grade II* listed building. Its gross external area is approximately 2,000 m². The land immediately surrounding the
C mansion house comprises parkland including lake, woodland, avenues, drives and monuments and various buildings and features of historic and architectural interest.

(2) At the valuation date, the mansion house and many buildings on the estate were in rather a poor state of repair.

(3) The arable land is mainly Grade II, MAFF agricultural land classification, with little variation in quality.

(4) The whole estate comprises an area of approximately 3,405.66 acres as shown on one of the district valuer's plans. The area to which the notice of determination referred is 3,163.36 acres or thereabouts as shown on another plan. That area is the area covered by four tenancies as follows:

Tenancy 1:	2,705.942 acres
D Tenancy 2:	28.16 acres
Tenancy 3:	166.618 acres
Tenancy 4:	191.789 acres

The total of those areas is 3,092.509 acres or 70.851 acres less than 3,163.36 acres. The difference is attributable to changes in Ordnance Survey maps since the 1901 edition thereof (referred to in the tenancy agreements).

(5) The 3,163.36 acres comprise:

arable land	— 2,780 acres or thereabouts;
parkland and pasture	— 292 acres or thereabouts;
spinney	— 91 acres or thereabouts.

(6) At all material times Lady Fox owned the freehold of the
E Croxton Park Estate.

(7) The lands included in the notice of determination were subject to four agricultural tenancies as follows:

Tenancy 1

Agreement dated September 1 1972 between Lady Fox (landlord) and Lady Fox and Major Norman Cochrane Fraser trading in partnership as Croxton Park Farms (tenants). This related to the Croxton Park Home Farm of 2,705.942 acres and the rent from September 1 1972 was £250 pa. Obligations on the part of the tenants included:

(i) an obligation not to assign underlet or part with possession of the farm or any part thereof;

F (ii) an obligation to renew as necessary and leave the farmhouse, cottages and all buildings in good repair and condition.

Tenancy 2

Agreement dated December 7 1973 between Lady Fox (landlord) and Lady Fox and Major Norman Cochrane Fraser, trading as Croxton Park Farms (tenants), increases the holding referred to in Tenancy 1 by 28.16 acres and increases the rent from October 11 1973 to £260 pa.

Tenancy 3

Agreement dated August 23 1977 between Lady Fox (landlord) and Lady Fox, Major Norman Cochrane Fraser and Edward John Crees (tenants) increases the holding in Tenancies 1 and 2 by 166.618 acres (known as West Farm) and the rent to £270 pa from November 3 G
1976.

Tenancy 4

Agreement dated January 16 1978 between Lady Fox (landlord) and Lady Fox, Major Norman Cochrane Fraser and Edward John Crees trading in partnership as Croxton Park Farms (tenants). This relates to Weald Farm, Croxton Park, Barns Farm, Meadow Farm and Hill Farm comprising 191.789 acres and the rent from October 11 1977 was £4,500 pa. The tenants' obligations included an obligation:

(i) not to assign let or part with possession of the holding or any part thereof and at all times to reside in the Croxton Park farmhouse provided that the requirement of personal residence should be satisfied in the case where there is more H
than one tenant by the personal residence of one of them;

(ii) to repair the buildings comprised in the tenancy but all the buildings at Barns Farm, Meadow Farm and Hill Farm were declared to be redundant.

(8) By the relevant date, Major Fraser had retired from the army and was employed by Beacham Animal Health Ltd. He entered into partnership with Lady Fox on September 1 1972. He retired from his employment in the late 1970s but he was still a partner at the relevant date. At that date he was 61 years of age, having been born on February 24 1920.

(9) Edward John Crees joined the partnership on November 21 1974. He was a partner in Strutt & Parker. He was their farm
management partner and was responsible for giving consultancy J
advice to Lady Fox. He was not a chartered surveyor but farmed on his account at a farm owned by him at which he lived. At the relevant date he was 57 years old, having been born on September 7 1923.

(10) Strutt & Parker acted as consultants to Lady Fox and to the partnership from 1969 onwards and from 1970 Mr Reeves was deeply involved in that consultancy. As from the date of Lady Fox's death, he was involved on their behalf in their capacity as managing agents to the appellants and played a major part in actual disposals of parts of the estate which occurred after the relevant date.

(11) The original partnership deed was dated September 1 1972 and a further deed was dated November 21 1974. The parties, Lady
Fox, Major Norman Cochrane Fraser and Edward John Crees, K
agreed to farm in partnership. Originally the profits and losses were to be shared in the following proportions: Lady Fox 98%, Major Fraser 1% and Mr Crees 1%, but with effect from April 1 1975 these proportions were altered to: Lady Fox 92.5%, Major Fraser 2.5% and Mr Crees 5%. For ease of reference we were provided with a consolidation of the partnership deed, which, *inter alia*, contained the following provisions:

(i) Lady Fox agreed to grant to the partnership a tenancy from year to year upon the terms and conditions set out in a draft attached to the original partnership deed.

(ii) the capital of the partnership should be £5,600, which should be subscribed in cash as to £5,000 by Lady Fox and as to £300
each by Major Fraser and Mr Crees. L

(iii) all additional capital required by the partnership from time to time should be provided by Lady Fox alone and Major Fraser and Mr Crees should not be obliged to subscribe for the same.

(iv) any difference arising as to any matter connected with the partnership business should be decided by Lady Fox provided that, save with the consent of both Major Fraser and Mr Crees, no change should be made in the nature of the partnership business and no other partner should be admitted to the partnership.

(v) Lady Fox should be responsible for the day-to-day control and running of the business and should devote to the partnership
business such time and attention as she should in her absolute M
discretion think fit and should be entitled to engage in any other business not competing with the partnership business.

(vi) Both Major Fraser and Mr Crees should devote to the partnership business such time and attention as they should in their absolute discretion think fit and should be entitled to engage in any other business not competing with the partnership business.

(vii) Any losses to be borne by either Major Fraser or Mr Crees should be confined to "the limit" (that is to say £300 in respect of each accounting year) and any losses in excess of the limit should be borne by Lady Fox alone.

(viii) A partner might dissolve the partnership with regard to himself

A by giving six months' notice of dissolution in writing to the other partners.

(12) The tenancy agreements were assets of the partnership.

(13) The number of persons employed by the partnership (not including Lady Fox, Major Fraser and Mr Crees) was 17.

(14) In the years ended March 31 1980 and March 31 1981 the partnership business made net profits before tax of £140,197 and £127,440 respectively.

(15) No planning permission with respect to any of the land referred to in the notice of determination existed at the valuation date which would have affected the value of the land. The land was situated in a rural area of high landscape value and planning permission for any material development was unlikely.

B (16) There were no outgoings, easements or restrictions etc which would have affected the value of the land.

(17) An application was made for heritage relief on October 23 1981 but was abandoned on February 24 1983 following a decision by the main beneficiary under Lady Fox's will not to make Croxton Park his home.

(18) The aggregate rent of £4,770 reserved under the terms of the tenancies could have been increased had notices under section 8 of the Agricultural Holdings Act 1948 been served. Such notices were given on a yearly basis. Shortly after the valuation date, Mr Reeves advised the executors that an increase of approximately £100,000 pa could have taken effect from the following September 29 or October 11 (as C appropriate). The district valuer's estimate was that the revised rents could have been in the region of £120,000 pa.

(19) After the valuation date the tenancies were surrendered as follows:

(i) Weald Farm (comprised in tenancy no 4 and part of no 1) was surrendered on September 29 1982 in consideration of a rent reduction of £1,000 pa on the rest of the estate.

(ii) On October 7 1983 land comprised in part of tenancy no 3 was surrendered in return for an abatement of rent for one year in the sum of £40,000.

(iii) The remaining land subject to the tenancies (on 2,364.869 acres) was surrendered on March 9 1984 for £827,000.

D This sum, less professional fees of £16,963, was divided as follows:

To the personal representatives	
of Lady Fox	£749,284 (92½%)
Major Fraser	£20,251 (2½%)
Mr Crees	£40,502 (5%)

(20) After the valuation date, certain disposals took place and they included the sale of Croxton Park and certain other land for £6,700,000; Weald Farm was sold on September 29 1982 for £999,000 and houses at Croxton on June 5 1984 for £460,000.

E

Appendix A

(Revised by agreement)
Capital Transfer Tax
Notice of Determination under Finance Act 1975, Schedule 4, paragraph 6

To: Robin Gray Esq
c/o Messrs Farrer & Co
66 Lincoln's Inn Fields
LONDON WC2A 3LH

Date: 2 October 1990
Ref: F300444/81D

A. In relation to —

1. the Deed of Partnership made first day of September 1972 between The Honourable Myra Alice Lady Fox ("the deceased") of the one part and Major Norman Cochrane Fraser ("Major Fraser") of the other part, as amended and extended by the Deed of Accession made the twenty-first day F of November 1974 between the deceased of the first part, Major Fraser of the second part and Edward John Crees ("Mr Crees") of the third part, relating to the "Croxton Park Farms" partnership;

2. the property comprised in an Agreement of Tenancy made the first day of September 1972 between the deceased of the one part and Major Fraser of the other part, as amended by a Deed of Variation made the twenty-first day of November 1974 between the deceased of the first part, the deceased and Major Fraser of the second part, the Agricultural Mortgage Corporation Ltd of the third part, and the deceased, Major Fraser and Mr Crees of the fourth part, relating to the Croxton Park Home Farms;

3. the property comprised in a Memorandum of Agreement made the seventh day of December 1973 between the deceased of the one part, and the deceased and Major Fraser of the second part, relating to the "Golden Acres";

4. the property comprised in an Agreement made the twenty-third day of August 1977 between the deceased of the one part, and the deceased, Major Fraser and Mr Crees of the other part, relating to the West Farm Eltisley; G

5. the property comprised in an Agreement made the sixteenth day of January 1978 between the deceased of the one part, and the deceased, Major Fraser and Mr Crees of the other part, relating to the Weald Farm;

6. the transfer of value on the death of the deceased on 27 March 1981;

B. the Commissioners of Inland Revenue have determined that —

(1) in ascertaining the value (for the purposes of the charge to Capital Transfer Tax on the death of the deceased) of the property to which the deceased was beneficially entitled at her death there shall be attributed to her freehold interest in the land specified in paragraphs A2 to 5 above (on which the farming business under the Deed of Partnership mentioned in paragraph A1 above was carried on) an appropriate proportion of the value of H

i. her freehold interest in the land referred to in paragraphs A2 to 5 above on which the said farming business was carried on and

ii. her interest in the said farming business (including any interest of that business in the said land)

such freehold interest and interest in the said farming business being aggregated together as a single unit of property for the purposes of section 38 Finance Act 1975

(2) the amount of that value in so far as it is attributable to the said land or any interest therein is referred to hereafter as "the Relevant Amount"

(3) the appropriate proportion of the value referred to in B(1) above is the following fraction of the Relevant Amount

$$\frac{(VP - T)}{VP - (7.5\% \times T)}$$

where VP equals the market value of the freehold interest with vacant J possession of the said land on the basis set out in the said section 38 and T equals the value which should properly be attributable to the interest of the said partnership in the said land in a case where the whole interest in the said land was being sold by the persons between them entitled to the interests making up that whole for a price equal to VP

(4) for the purposes of valuing the estate of the deceased (whether or not for the purposes of paragraph B(1) above)
VP = £6,125,000
T = £1,500,000 and
The Relevant Amount is £5,565,000.

(5) the value to be attributed to the said freehold interest is £4,280,768.

(Signed) P R Twiddy for the Commissioners of Inland Revenue

K

Right of appeal

If you wish to appeal against this determination (or any part of it), you should within 30 days after service of this notice give notice of appeal in writing, specifying the grounds of appeal, to the Capital Taxes Office, Minford House, Rockley Road, London W14 0DF.

If you do not appeal the determination will be conclusive in accordance with paragraph 6(5) of Schedule 4 to the Finance Act 1975.

Appendix B

L

Revised/notice of appeal
The Honourable Myra Alice Lady Fox deceased
Ref: F300444/81D

This notice of appeal relates to the Notice of Determination sent to Robin Gray ("the Appellant") c/o this firm dated October 2nd 1990 as amended by agreement between the parties.

An appeal is hereby lodged on behalf of the Appellant against the said determination The Grounds Of The Appeal Being:

1. (a) For the purposes of section 38 of the Finance Act 1975, the value of the deceased's freehold interest in the land specified in paragraph A of the said Notice of Determination is the price which it might reasonably have been expected to fetch if sold in the open market at the date of the deceased's death M

(b) On this basis the value of the deceased's freehold interest in the said land is £2,751,000

2. *Further and in the alternative*, the basis of valuation described in paragraph B of the said Notice of Determination is contrary to the provisions of the said section 38 in two respects:

(a) for the purposes of that section the deceased's freehold interest in the said land and her interest in the farming business do not fall to be aggregated together as a single unit of property;

(b) where under the said section aggregation is appropriate (as, for example, where two or more chattels form a set) there is for the purposes of Capital Transfer Tax no statutory authority for apportioning the value of the aggregate between the constituent parts and that, accordingly, the only method of valuing the property defined in para 1(a) of this Notice of Appeal is the method described in that sub-paragraph.

A 3. *Further and in the alternative:*

Neither "VP" nor "T" is relevant for the purpose of valuing the deceased's estate.

Dated this 22nd day of October 1990
(Signed) Farrer & Co
Solicitors to the Appellant

Valuations

Mr Clegg's valuations are set out in Appendix C to this decision [see p 215]. No valuation as such was provided by Mr Reeves.

Agreed values

Certain agreed values are set forth in the agreed statement of facts. B These include:

the vacant possession value of the land referred to in the notice of determination as being:	£6,125,000;
the investment value of the freehold reversion in isolation:	£2,751,000.

The agreed statement of facts shows how these agreed values devalue per acre.

The issues

The issues are agreed to be as follows:

(1) Should the freehold reversion and the partnership share be treated as one unit of property for the purposes of section 38 of the Finance Act 1975?

(2) If the freehold reversion and the partnership share are to be so C treated, should the value to be attributed to the freehold reversion be ascertained in the manner and in the amount contended for by the Crown or in some other and if so what manner and/or amount?

(3) In any event, is the value of (or to be attributed to) the freehold reversion for the purposes of section 38 of the Act of 1975 the sum of £4,280,768, or £2,751,000 or some other and if so what sum?

It is agreed that these issues are without prejudice to the need for any alternative determination under rule 54(3) of the Lands Tribunal Rules 1975.

Relevant provisions of Finance Act 1975

Section 22 provides, *inter alia*, that:

> On the death of any person after the passing of [the] Act tax shall be charged as D if, immediately before his death, he had made a transfer of value and the value transferred by it had been equal to the value of his estate immediately before his death . . .

Section 38(1) provides that:

> Except as otherwise provided by this Part of this Act, the value at any time of any property shall for the purposes of capital transfer tax be the price which the property might reasonably be expected to fetch if sold in the open market at that time; but that price shall not be assumed to be reduced on the ground that the whole property is to be placed on the market at one and the same time.

The requirements as to notices of determination are set out in para 6 of Schedule 4 to the Act. Para 7 of that Schedule is as follows:

> (1) A person on whom a notice under paragraph 6 above has been served may, within 30 days of the service, appeal against any determination specified E in it by notice in writing given to the Board and specifying the grounds of appeal.
>
> (2) Subject to the following provisions of this paragraph the appeal shall be to the Special Commissioners.
>
> (3) Where it is so agreed between the Appellant and the Board, or the High Court, on an application made by the Appellant, is satisfied that the matters to be decided on the appeal are likely to be substantially confined to questions of law and gives leave for that purpose, the appeal may be to the High Court.
>
> (4) Neither the Special Commissioners nor the High Court shall determine any question as to the value of land in the United Kingdom on any appeal under this paragraph, but on any such question the appeal shall be to the Lands Tribunal or, as the case may be, the Lands Tribunal for Scotland or for Northern Ireland.

F The provisions of the Act of 1975 as to capital transfer tax are superseded by the Capital Transfer Act 1984 as from January 1 1985. By section 100 of the Finance Act 1986, the latter Act is now known as the Inheritance Tax Act 1984. Because Lady Fox died before January 1 1985, the Finance Act 1975 is still relevant for present purposes.

Jurisdiction

Under para 7 of Schedule 4 to the Act of 1975, it is plain that on an appeal under that paragraph any question as to the value of land in England or Wales is solely within the jurisdiction of the Lands Tribunal. It is equally plain that the Lands Tribunal has no jurisdiction in respect of any question other than one relating to the value of land. It seems that originally the appeal came before Mr R H Widdows, Commissioner for the Special Purposes of the Income Tax Acts. He gave some written observations on a preliminary issue as to G jurisdiction, dated November 3 1989. In his observations he said, *inter alia*:

> Assuming as I do that the act contemplates the possibility of divided jurisdiction, is the question of aggregation raised by the notices of determination one of law or of valuation? As at present advised I should suppose it to be primarily a valuation matter because both notices refer to section 38 of the Finance Act 1975 and because the purpose of aggregation is presumably to obtain a higher value than the sum of the two separate values.

He added:

> But if the question of aggregating the deceased's interest in the farming business with her interest in the land on which the farming is conducted is a valuation matter, is it wholly or partly a question of the value of land?

He said that he saw no clear answer to that question at that stage. H

We, too, have had doubts as to whether the Lands Tribunal has jurisdiction in this appeal, because, on its face, the notice of determination appears to lot together land and personalty. However, we have accepted the assurances of counsel that the tribunal has jurisdiction and have done so because the purpose of the notice of determination, in the end, is to determine a value for the land and nothing but land.

First issue

We were referred to all the authorities including the decisions of the House of Lords in *IRC* v *Crossman* [1937] AC 26 and *Duke of Buccleuch* v *Inland Revenue Commissioners* [1967] 1 AC 506.

Mr Bramwell's first submission was that, as a matter of law, for the J purposes of section 38 of the Finance Act 1975, land cannot be lotted together with property other than land. Although that section was silent on the matter, the point was implied because the jurisdictional rules in para 7 of Schedule 4 were framed on that assumption. Accordingly, the Commissioners of Inland Revenue had no power to make a determination lotting together Lady Fox's freehold interest with her share in the partnership. Miss Arden disagreed. She referred to section 23(1) of the Act of 1975, which provides that a person's "estate" is the aggregate of all the property to which he is beneficially entitled, and submitted that the tribunal must accordingly proceed on the basis that all the property of Lady Fox was on the market at one and the same time. She contended further that Lady Fox's share in K the partnership was in itself an interest in land and referred to the decision of the Court of Appeal in *Cooper* v *Critchley* [1955] Ch 431. She said, in the alternative, that any question as to the value of Lady Fox's freehold reversion and of her interest (as a 92.5% partner) in the tenancies constituted a question "as to" the value of land within the jurisdiction of the Lands Tribunal under para 7(4) of Schedule 4 to the Act of 1975.

We do not accept that Lady Fox's share in the partnership was itself an interest in land. The tenancies were assets of the partnership but it does not follow that a share in the partnership was anything other than personalty or a chose in action. Accordingly, we agree with Mr Bramwell that the freehold interest and the share did not constitute a single unit of property for the purposes of section 38 of L the Act of 1975 and that as a matter of law the Commissioners of Inland Revenue had no power to lot the two together in the notice of determination.

We agree further with Mr Bramwell's second submission, namely that the freehold interest and the share did not constitute a natural unit of property within the meaning of the speeches in the House of Lords in the *Buccleuch* case. The purpose of the Commissioners of Inland Revenue in lotting the two items together was, as Mr Clegg said in evidence and as Mr Widdows assumed in his observation, to endeavour to obtain a higher value than the sum of the values of the two separate parts. However, that aim does not convert the two parts into a natural unit of property. Mr Reeves said that there was no reason to suppose that the purchaser would pay more than the agreed M investment value of the freehold reversion. Indeed, he said, there was reason to believe that he would pay less, since a lot consisting of the freehold reversion and the partnership share would not appeal to any known sector of the market. He was not aware of any such lotting ever having been made and he could think of no analogous case. Nor did he know of any incidence of a share in a farming partnership having been sold in the open market.

Third, we agree with Mr Bramwell that if it were permissible to lot the freehold interest and the share in the partnership together as being a single unit of property then section 38 requires that that single unit be valued as a single unit and that apportionment is neither admissible nor appropriate.

A For these reasons, in our judgment, the notice of determination was based on false premises which vitiate Mr Clegg's valuations. Accordingly, we allow the appeal and determine that the value of Lady Fox's freehold interest is to be ascertained by reference to its investment value, that is to say, the agreed value of £2,751,000. We so determine.

Contrary to the way in which the third issue is expressed, Mr Bramwell said that we are not at liberty to adjust that sum. We have not heard any argument or evidence as to what adjustment (if any) might be made to that sum and, accordingly, our determination remains as we have stated it, although, had we been free to do so, we might have made an adjustment to the sum of £2,751,000.

B

Second issue

On the assumption that we are wrong in our decision on the first issue and that the freehold reversion and the partnership share are to be treated as one unit of property for the purposes of section 38 of the Act of 1975, the question is whether the value to be attributed to the freehold reversion is to be ascertained in the manner and in the amount contended for by the Crown or in some other and if so what manner and/or amount.

In order to put the evidence of the two valuation witnesses into context, it is appropriate to refer to the background to valuation arising from the statutory provisions with respect to particular aspects of this appeal. The parties agree that:

C (1) All the assets, real and personal property, owned by the deceased at date of death on March 27 1981 fall to have a value ascribed to them.

(2) The basis of that value is the best price obtainable on an assumed sale in the open market immediately prior to death.

(3) The assumed sale is to be by a common vendor and by virtue of (2) above is to take place at one and the same time.

(4) The hypothetical purchaser is to assume all the privileges and obligations which resided in the deceased to the extent of "stepping into her shoes". Thus as owner of the freehold reversion there would be power to increase the passing rents for the several agricultural tenancies and as owner of a 92.5% share in the farming partnership the authority to exercise the power of

D resignation or dissolution contained in the deeds of partnership.

(5) The circumstances surrounding any assumed sale are to be those actually existing in March 1981 without allowance for any interval during which to adjust or vary by negotiation and agreement the terms of any existing relationship, be it with the landlord or farming partner.

(6) The information available to a hypothetical purchaser relating to the assets is to be that capable of being ascertained by a reasonably informed and properly advised purchaser but no more.

As we have stated, two values are agreed in respect of the freehold interest in the land: first, the value with vacant possession if it had

E subsisted on March 27 1981 is agreed in the sum of £6,125,000; second, the value subject to the agricultural tenancies subsisting on March 27 1981 was £2,751,000. This value is expressed to be in isolation in the sense that the hypothetical purchaser of the 92.5% share in the farming partnership would have no particular desire to purchase the freehold reversion to the extent of outbidding the generality of investors for let agricultural land.

Mr Reeves spoke to the valuation of £2,751,000 above, as he considered this was the sum which a prudent purchaser would pay. He accepted that this was considerably lower than the vacant possession value but this was a long-established feature of the market for freehold let agricultural land. Purchasers were always mindful of the possibility of securing vacant possession at some future date and

F this prospect was thereby built into the prices paid. The securing of vacant possession by an agreed surrender or the death of a tenant was subject to high risk and only where it was a virtual certainty, by, for example, prior binding agreement, would a purchaser pay markedly in excess of general investment value. He did not consider the opportunity to "step into the shoes of Lady Fox" by also purchasing her partnership share afforded such virtual certainty, because it yielded less than the full benefits of freehold ownership with vacant possession. The privileges to be derived from continuing the combination of the two assets, the reversion and the share, left the freehold encumbered by the tenancies and inhibited any subsequent sale with vacant possession. In reality this reversion was no different from that to any other let farm and purchasers in the market would G bid accordingly.

In his opinion, the suggested sale of the two assets together was artificial and rested upon the assumption that the desire to own an agricultural estate as an investment and to farm in partnership a large area of land commonly resided in the same individual. He saw the two markets as different. Mr Reeves had not, therefore, regarded it as necesssary to the valuation process or otherwise to consider the value of the agricultural tenancies as assets of the farming partnership.

Mr Clegg produced a primary valuation (reproduced at Appendix C) of the freehold reversion and the partnership share (so far as relating to the value of the agricultural tenancies) in the sum of £5,565,000 of which he apportioned mathematically £4,280,768 to H the freehold reversion: the value of the tenancies relative to the 92.5% partnership share was thus £1,384,232. He produced an alternative apportionment (also in Appendix C) as a check in the sum of £4,316,250 for the freehold reversion, with £1,248,750 ascribed to the tenancy element of the 92.5% share. In a secondary valuation of the freehold reversion only (see also Appendix C) Mr Clegg arrived at a sum of £4,165,000.

APPENDIX C

Valuations by Michael David Clegg FRICS

Valuation 1

Vacant possession value as agreed		£6,125,000	J
Deduct			
(1) *Value of Crees (5%) Fraser (2½%) interest in the partnership which is attributable to the tenancy*			
Value of interest as a whole (see note*)	£1,500,000		
Crees and Fraser 7½% share of £1,500,000	£112,500		
Allow 10% for share	− 11,250		
Value of Crees & Fraser's interest	101,250		
(2) *Allowance for risks and profit*			
7½% of vacant possession value	459,375	560,625	
		5,564,375	
Value of freehold reversion and deceased's interest in partnership	say	£5,565,000	K

*Note: *Valuation of interest*	
Vacant possession value agreed	£6,125,000
Investment value agreed	£2,751,000
Vacant possession premium	£3,374,000
45% of premium as value of tenancy	£1,518,300
Say	£1,500,000

Apportionment 1

The value of £5,565,000 is required to be apportioned as between the deceased's interest in the freehold reversion and the partnership share in so far L as it is attributable to the land.

On the assumption that the property is to be sold with vacant possession, the price of £6,125,000 needs to be divided between landlord and tenant. The landlord's proportion is the vacant possession (VP) value less the tenant's interest (T). These figures are represented by £6,125,000 (VP) — £1,500,000 (T) = £4,625,000 for the landlord.

However, the tenant's interest of £1,500,000 is held by the partnership in the proportions of 92½% to Lady Fox and 7½% to Crees and Fraser = £1,387,500 Lady Fox and £112,500 Crees and Fraser.

The apportionment to arrive at landlord's share of the £5,565,000 is therefore:

$$\pounds 5{,}565{,}000 \times \frac{\text{Landlord's interest}}{\text{Vacant possession value } \textit{less} \text{ Crees and Fraser share}}$$ M

Represented by

$$\pounds 5{,}565{,}000 \times \frac{VP - T}{VP - (7\tfrac{1}{2}\% \times T)}$$

$$\pounds 5{,}565{,}000 \times \frac{\pounds 6{,}125{,}000 - \pounds 1{,}500{,}000}{\pounds 6{,}125{,}000 - \pounds 112{,}500}$$

$$\pounds 5{,}565{,}000 \times \frac{\pounds 4{,}625{,}000}{\pounds 6{,}012{,}500}$$

Apportionment attributable to the freehold reversion £4,280,768

A **Apportionment 2**

A simple method of apportionment would be to take the value of £5,565,000, and deduct Lady Fox's partnership share, in so far as it is attributable to the land, as follows:

Value of deceased's interests		£5,565,000
Less partnership share		
(Value of interest as a whole £1,500,000)		
Lady Fox 92½%	£1,387,500	
Less 10% for share	£138,750	£1,248,750
Apportionment attributable to the freehold reversion		£4,316,250

Valuation 2

B Valuation of deceased's freehold reversion only, as at March 27 1981.

Value of freehold, with vacant possession		£6,125,000
Less Value of tenancy		£1,500,000
		£4,625,000
Allow for risks and profit 7½% of vacant possession value		£459,375
		£4,165,625
	Value say	£4,165,000

These valuations and the apportionments had as their starting point the agreed value of £6,125,000 for the freehold interest with vacant possession. This was because, in Mr Clegg's opinion, in the
C factual circumstances relating to the land there was an "overwhelming likelihood" that the tenants would agree to surrender their tenancies to the reversioner for a realistic sum; broadly, £1,500,000 of which 92.5% would accrue to the purchaser of the deceased's share in the farming partnership. To be willing and able to pay the value ascribed to the reversion by Mr Clegg it was essential that the hypothetical purchaser secure vacant possession without undue delay and crucial to achieving this was the ability also to buy at the same time the 92.5% share in the farming partnership. In effect, within the provisions of the partnership deeds were mechanisms for securing vacant possession and thus the partnership share was the key to the release of latent or marriage value. Sooner or later the dissolution of the farming partnership could be secured to
D ensure sole control for continued farming or, if desired, the ability to give vacant possession to any purchaser, thereby removing the constraints which led to the caution of Mr Reeves. In any event, in the interim period the hypothetical purchaser of the combined asset could enjoy 92.5% of "the fruits of occupation" and all the advantages of ownership save for the ability to sell with vacant possession; for the immediate absence of these advantages Mr Clegg accorded a discount of 7.5% (some £460,000), which he accepted was a professional judgment.

The approach adopted by Mr Clegg was largely conditioned by his opinion that at the valuation date, taking the reversion and the 92.5% share together, the only barrier to realising the vacant possession
E value of the freehold interest was the minority 7.5% shareholding of two individuals. Neither of these shareholders was resident on the estate nor involved in day-to-day farming of the land. These tenants and shareholders had other business interests and activities, they were not particularly young men and, in Mr Clegg's opinion, they could be expected to depart upon payment to them by the freeholder and major shareholder of a sum of about £100,000.

In support of this approach to securing vacant possession, Mr Clegg referred, as being comparable, to a transaction in a farm on the Cambridgeshire-Essex border where an agricultural tenant of some 437 acres surrendered to a new purchaser of the reversion for a consideration of 24.6% of the vacant possession value; in the instant F
case Mr Clegg attributes £1,500,000, 24.5% of the agreed vacant possession value, to the tenancies as assets of the farming partnership.

As we have stated, Mr Reeves produced no formal valuation related to the issue under discussion. This was because he regarded the lotting together of the reversion and the partnership share as artificial and directed his attention to investment value (£2,751,000) only. As the tenancies were non-assignable it was accepted by both experts that any value they might have had derived solely from the ability to surrender to a reversioner interested in securing vacant possession. This was how the market operated and Mr Clegg supported his figure of value by reference to his comparable G
transaction. We accept this evidence as to the aggregate value of the tenancies (£1,500,000) and the unchallenged amount (£1,248,750, say £1,250,000) ascribed to the 92.5% partnership share of the deceased, so far as referable to the tenancies as partnership assets.

On the assumption that in law the respondent may lot or sell together the freehold reversion and the 92.5% share in the farming partnership as a single unit of property, we are persuaded that the approach to valuation adopted by Mr Clegg pays a realistic regard to the circumstances actually prevailing and likely to have been within the knowledge of the hypothetical purchaser at the valuation date. While the general basis of value to be adopted is prescribed by statute as open market value there is no prescription of the method of valuation to be applied. If adjustment of the agreed value with vacant H
possession by deductions and/or apportionment is adopted, it falls to be considered on the criterion of whether it produces a realistic answer when judged against the prevailing factual circumstances. On the evidence before us the most likely prospect in March 1981 was that the minority shareholders would negotiate a surrender of the tenancies; as indeed they in fact did during the years 1982-84.

The comparable cited by Mr Clegg supports the level of consideration ascribed by him to the tenancies as assets of the farming partnership and thus the gross sum in his secondary valuation for the reversion. However, it is apparent from the documents in evidence relating to this comparable that a guaranteed certainty of vacant possession was available to the purchaser by virtue of the prior agreement of the terms of surrender. On the J
evidence in the instant case the route to vacant possession was indeed novel, indirect and possibly uncertain in outcome. There was no guarantee that the minority shareholders would accept the allocated sum of *circa* £100,000 or that any acceptance would be immediate upon purchase of the reversion for a considerable sum, Moreover, it was unlikely that the return on capital from rents and/or farming in the short term would cover the cost of finance for purchase, with the result that delay would prove expensive for the hypothetical purchasers. For these reasons, if Mr Clegg's valuation were adopted, we would wish to investigate further the question whether the 7.5% discount for risk and profit made by Mr Clegg is sufficient. Mr Bramwell said that we could not adjust either value (£2,751,000 or K
£4,280,768): it was an all-or-nothing case. Miss Arden disagreed. Certainly we had no argument or evidence as to adjustment and accordingly we refrain from making any adjustment.

Third issue

For reasons already given, it does not appear that this issue arises at the present stage in the proceedings.

It is ordered that the respondent pay to the appellants their costs in the appeal, such costs if not agreed to be taxed by the Registrar of the Lands Tribunal on the High Court Scale on the standard basis.

LANDS TRIBUNAL

COMPENSATION

A **BATCHELOR v KENT COUNTY COUNCIL**

(REF/117/1986)
October 28 1991
(T HOYES FRICS)

Estates Gazette January 25 1992 and February 1 1992
[1992] 03 EG 127 and [1992] 04 EG 138

Reference remitted to tribunal — Acquisition of land for highway purposes — Land required to provide access to development land — Court of Appeal deciding that the principle in *Pointe Gourde* and rule (3) of section 5 of the Land B **Compensation Act 1961 did not apply to limit compensation — Whether order land had ransom value — Whether alternative accesses might have been available — Whether *Stokes* principle applies to valuation of access land**

By its decision on February 19 1988 the tribunal had determined the value of the order land, and adjoining land (the subject of a counternotice under section 53 of the Land Compensation Act 1973), as £500,000 and £150,000 respectively — Following an appeal by way of a case stated, the Court of Appeal decided that the tribunal could not be faulted for failing to observe the *Pointe Gourde* principle and that rule (3) in section 5 did not apply to limit the C **compensation — However, the Court of Appeal remitted the tribunal's determination of the value of the respective lands for failing to give reasons**

Following the confirmation of a compulsory purchase order and the service of a notice to treat, the acquiring authority took possession of the order land on August 1 1984 — The claimant's counternotice in respect of the adjoining land was accepted and possession taken on December 1 1986 — To the south and east of the two parcels of land was a development area identified as such on the development plan — Planning permission was granted in outline for the D **development of 1,750 dwellings on part of this land — The acquiring authority, pursuant to the terms of an agreement with the developers, constructed a roundabout on the order land to provide suitable access to the development area — The claimant sought £21.09m for the order land on the basis that it held the "key" to the development of something less than 200 acres of adjacent land — The acquiring authority contended that the order land was not the only possible location for the necessary improved access and their valuer spoke to a valuation of £75,000**

***Decision*: The scheme of junction improvement carried out on the order land was undoubtedly the best solution to the** E **traffic problem in 1984 — It was the most acceptable scheme in town planning terms — Those concerned with the residential development would not have indefinitely held up that development for want of highway improvements — None the less there were identified alternatives to the order land for highway improvement — The "no-scheme" world, prior to the confirmation of the compulsory purchase order in September 1983, comprised a number of developers assembling land in order to implement long-established development proposals — The order land was the preferred, but not the only, location for junction improvements — It was** F **endowed with increased value — The suggested distinction, for valuation purposes, to be drawn between the owner of back land buying "access" land and the owner of "access" land buying back land was rejected in favour of the "reverse of the coin" inherent in the *Stokes* principle — On the evidence, 15% should be attributed to the order land under the *Stokes* principle — The development value of the development land determined at £157,000 per acre — Value of order land determined at £2,150,000 — Value of adjoining land determined at £10,000**

The following cases are referred to in this report. G

Adams v *Birmingham City Council* [1977] RVR 42, LT
Birmingham Corporation v *West Midland Baptist (Trust) Association (Inc)* [1970] AC 874; [1969] 3 WLR 389; [1969] 3 All ER 172; (1969) 67 LGR 571, HL
Dicconson Holdings Ltd v *St Helens Metropolitan Borough* (1977) 249 EG 1075 & 1178
Horn v *Sunderland Corporation* [1941] 2 KB 26; [1941] 1 All ER 480, CA
Margate Corporation v *Devotwill Investments Ltd* [1970] 3 All ER 864; (1970) 22 P&CR 328; 69 LGR 271; [1971] EGD 313; 218 EG 559, HL
Ozanne v *Hertfordshire County Council* [1988] RVR 133; [1988] 2 EGLR 213; [1988] 36 EG 195 & 37 EG 123, LT
Pointe Gourde Quarrying & Transport Co Ltd v *Sub-Intendent of Crown Lands* [1947] AC 565, PC
Provincial Properties (London) Ltd v *Caterham and Warlingham Urban District Council* [1972] 1 QB 453; [1972] 2 WLR 44; [1972] 1 All ER 60; (1972) 70 LGR 151; 23 P&CR 8, CA H
Rathgar Property Co v *Haringey London Borough Council* (1977) 248 EG 693, LT
SJC Construction Co Ltd v *Sutton London Borough Council* (1974) 28 P&CR 200; [1974] EGD 621; 230 EG 906
Stokes v *Cambridge Corporation* (1961) 13 P&CR 77; [1961] EGD 207; 180 EG 839, LT

The Court of Appeal remitted this reference by Hubert Dorrington Batchelor, the claimant, to the tribunal [1990] 1 EGLR 32; [1990] 14 EG 129; the parties agreeing that the issue to be determined was the appropriate values of two parcels of land situated on the eastern side of the junction of New Cut Road and Bearsted Road, Maidstone, J Kent.

Robin Purchas QC and Meyric ap Michael Lewis (instructed by Hallett & Co, of Ashford, Kent) appeared for the claimant; Malcolm Spence QC and Adrian Trevelyan-Thomas (instructed by Sharpe Pritchard for the solicitor to Kent County Council) represented the acquiring authority.

Giving his decision, MR HOYES said: This reference was initially heard by this tribunal (W H Rees Esq FRICS) in October 1987 and a decision was published on February 19 1988. That decision was subject to appeal by the acquiring authority by a case stated to the Court of Appeal, which decided that the principle in *Pointe Gourde* and rule (3) of section 5 of the Land Compensation Act 1961 were not K to apply so as to limit compensation.

By its decision of July 26 1989* the Court of Appeal remitted to this tribunal the assessment of compensation for the "adjoining land" and the determination of the value of the "order land" which on February 19 1988 had been determined at £150,000 and £500,000 respectively. The parties have agreed that "the issue to be determined is what the appropriate values are of the Order land and the Adjoining land".

The reference was made by the claimant to determine the amount

*Editor's note: Reported at [1990] 1 EGLR 32.

A of compensation to be paid on the compulsory acquisition of two adjoining parcels of freehold land ("the two parcels"), a little under 2 acres in all, between the M20 and Maidstone, Kent. At the further hearing the claimant sought £21.09m and the valuer for the acquiring authority spoke to a valuation of £75,000. This very large difference comes about because the claimant maintains that one of the two parcels of land holds the "key" (improved highway access) to something less than 200 acres of adjacent land allocated for residential purposes. The acquiring authority, while accepting the need for improved access, contend that the relevant part of the reference land was not the only possible location and accordingly this, together with other matters, significantly affects the amount of compensation payable.

B Counsel for the claimant called Mr D S Shaw MA CEng, MICE, who spoke with regard to highway matters, Dr P G Biddle MA FArbor Assoc in relation to trees and landscaping issues, Mr J A Little FRIBA MRTPI, whose evidence concerned town planning matters, Mr R H Bulteel FRICS, in respect of a land transaction elsewhere, and Mr R W R Pocock FRICS, whose valuations are referred to below. Counsel for the acquiring authority called Mr G M Heard CEng MICE, whose evidence related to highway matters, Mrs V A Randall BSc (Econ) MRTPI and Mr J Atkins MRTPI in relation to town planning matters, Mr J C Parker BSc (Hort) ALI in respect of trees and landscaping issues, Mr H J M Thomas FRICS and Mr A H Cherry MBE FRICS FSVA concerning the land transaction elsewhere, Mr J V Walker, a director C of Ward Holdings plc, and Mr N A Hicks BSc FRICS, whose valuations are also referred to below.

From an agreed statement of facts the land, which is subject to Kent County Council (New Cut/Bearsted Road Improvement) Compulsory Purchase Order 1983 ("the order land"), is 0.86 of an acre. The order was made on January 25 1983 and was confirmed on September 12 1983 after a public inquiry. Notice to treat dated November 28 1983 was served on the claimant and possession was taken on August 1 1984, the agreed valuation date for this parcel.

A counternotice dated January 10 1984 was served on the acquiring authority under section 53 of the Land Compensation Act 1973 in respect of 0.97 of an acre of adjoining land ("the adjoining land") and required them to purchase the claimant's interest. This D counternotice was accepted on March 7 1984 and possession was taken on December 1 1986, the agreed valuation date for this parcel.

The two parcels are situated upon the eastern side of the junction of New Cut Road and Bearsted Road, Maidstone, and are located within approximately three-quarters of a mile of junction 7 of the M20. On the south and east sides of the reference land is the Weavering Street development area (now known as Grove Green), which lies within the "Informal Action Area Plan B-Vinters Park and Land to the East". On the west side of New Cut Road is Vinters Park Crematorium. Trees on the west side of New Cut Road are protected by a tree preservation order. To the north of Bearsted Road is an area of great landscape value and to the north of the M20 lies the E North Downs area of outstanding natural beauty. The two parcels were farmed together (grade 2 agricultural land) and were last cropped for peas in 1984.

On the Maidstone and Vicinity Town Map 1970, approved by the Secretary of State for the Environment in September 1977, the two parcels are shown as part of a larger area allocated primarily for residential use. In 1973 Kent County Council approved an informal action area plan for the purpose of development control in the area to the east of New Cut Road, allocated for residential development, including the two reference parcels identified within phase three of that development. Under this plan some 155 acres of land were proposed for development by about 1,750 houses and additional areas were identified for local distributor roads, public open space, F schools and other communal facilities. The plan also indicated a basic road pattern for the development area involving two new access points to an improved New Cut Road, with improved junctions for that road to the south with Ashford Road and to the north with Bearsted Road in the approximate location of the order land.

On March 30 1982 planning permission was granted in outline for the development of approximately 1,750 dwellings on land within the action area plan to the south of the two parcels. Conditions upon that planning permission were, *inter alia*, that:

(a) The development should be carried out in accordance with the approved Informal Action Area Plan;

(b) Public open space in accordance with the allocations and provisions of the Informal Action Area Plan should be reserved, laid out and landscaped;

(c) No dwellings within Phase 2 should be occupied until the off-site road G works (subject to a section 60 agreement in respect of that phase) have been completed.

The reason for the latter condition was to secure a satisfactory, comprehensive and phased scheme of development. On March 30 1982 the owners of the land within that planning permission entered into an agreement with Maidstone Borough Council as local planning authority, under section 52 of the Town and Country Planning Act 1971 (the planning agreement), which also provided that the owners would enter into an agreement under section 278 of the Highways Act 1980 (the highway agreement) with the acquiring authority as highway authority. This highway agreement was to relate to the improvement of New Cut Road and also its junction with H Bearsted Road (principally upon the order land) and to require the owners not to occupy or permit or cause to be occupied any dwelling in certain housing areas until the improvement works to the junction of New Cut Road and Bearsted Road had been completed.

The acquiring authority, as highway authority, on February 1 1983 accordingly entered into an agreement between the owners and developers of the land to the south for the improvement of the junction, to include the order land, to enable that development to proceed. On the first valuation date 205 houses had been completed at Grove Green and detailed planning permission had been given for the remainder of phase one of the development.

The scheme J

In about 1984 a roundabout was constructed on the order land replacing the former T-junction of New Cut Road with Bearsted Road, and there is no dispute that this is the scheme underlying the acquisition of the order land. It is now common ground that in the "no-scheme" world the market for the reference land comprised the owners and/or developers of the Grove Green area seeking to implement the 1982 planning permission. The owner of the largest area was referred to as "Wards": others are: "Abbey Homesteads", "McLean Homes/Barratts", "Steetly" and "Hillreed Homes". In order to develop all of their holdings, at least Abbey Homesteads, if not the others as well, would probably have to, and on the evidence did, come to terms with Wards for access to new estate roads.

The following were referred to: K

Stokes v *Cambridge Corporation* (1961) 13 P&CR 77; 180 EG 839 (LT)
Margate Corporation v *Devotwill Investments Ltd* [1970] 3 All ER 864; (1970) 218 EG 559 (HL)
Horn v *Sunderland Corporation* [1941] 2 KB 26
Birmingham Corporation v *West Midland Baptist (Trust) Association (Inc)* [1969] 3 All ER 172 (HL)
Adams v *Birmingham City Council* [1977] RVR 42 (LT)
SJC Construction Co Ltd v *Sutton London Borough Council* (1974) 28 P&CR 200; 230 EG 906 (LT)
Rathgar Property Co Ltd v *Haringey London Borough Council* (1977) 248 EG 693 (LT)
Provincial Properties (London) Ltd v *Caterham and Warlingham Urban District Council* [1972] 1 QB 453 (CA) L
Dicconson Holdings Ltd v *St Helens Metropolitan Borough* (1977) 249 EG 1075 & 1178 (LT)
Ozanne v *Hertfordshire County Council* [1988] RVR 133; [1988] 2 EGLR 213; [1988] 36 EG 195 & 37 EG 123 (LT)

The highway evidence

From the evidence of Mr Shaw and Mr Heard I conclude that the scheme of junction improvement carried out upon the order land was undoubtedly the best solution to the emerging traffic problem in 1984. Nevertheless, in highway engineering terms the order land was not the only possible location for the required works, albeit that the M four alternatives spoken to embraced a variety of different problems relating to extra construction costs, compliance with accepted design standards, land assembly from a number of owners, a more adverse impact by the completed works upon the locality, and for these reasons possible delay in their becoming available by 1984. In my judgment, and in that of the highway experts since at least 1980, when two of the alternatives were considered and rejected, the preferred solution was upon the order land and any alternative was very much second best in terms of engineering, acceptability and implementation.

Improvements to the junction would have become essential and would have been carried out by the highway authority by 2000 or

A possibly before, even if there had been no residential development at Grove Green.

In the absence of the scheme upon the order land, the prospect of direct highway access to facilitate residential development, upon it and the adjoining land, is less clear cut. It was accepted that there was no prospect of access from New Cut Road as it existed in 1984 and that some relaxation of the then standards may well have been necessary to sanction an access to Bearsted Road, whether in the centre or at the eastern end of the combined site. With the junction of New Cut Road and Bearsted Road approaching capacity, with or without the development at Grove Green, and already possessing features detrimental to the safety of pedestrians and motorists, it is unlikely that permission for a totally new access for housing would B have been forthcoming. Furthermore, if it were appropriate in town planning terms to approve residential development of the 1.83 acres, the combined site, safe access could physically have been made by about 1987 into the Grove Green area to the south. This position would further militate against the approval, in both 1984 and 1986, of any new direct access on to Bearsted Road since a more acceptable alternative would shortly have been available.

Dr Biddle and Mr Parker gave evidence as to the species, quality and amenity value of trees in the locality. In particular they identified those which would be lost in the implementation of alternative schemes of junction improvement and the future amenity screening problems arising therefrom. The witnesses were at one that the C individual trees and woodland groups were a significant feature of the local landscape, fully justifying the protection accorded by the tree preservation order and other planning policies. Any loss of trees to roadworks was to be regretted and especially those which provide an amenity screen to the crematorium. Mr Parker was more optimistic than Dr Biddle that any resulting loss of privacy at the crematorium could be made good by new planting and landscaping. He considered that an effective replacement tree screen could be achieved in 10 to 15 years.

The town planning evidence

Mr Little explained the relevant planning policies and proposals as they affect the two parcels and the land nearby; he described the D physical characteristics of the latter and highlighted the classifications accorded to the landscape north of Bearsted Road.

In his opinion, had no roundabout been provided, no building beyond the 205 houses mentioned above would have taken place at Grove Green. He also gave his opinion concerning the four alternative schemes of junction improvement, each involving the construction of a roundabout, the plans for which had been prepared by the acquiring authority. The first showed the roundabout to be wholly on the land of the crematorium, the second on land north of Bearsted Road, the third and fourth on land mainly fronting the south side of Bearsted Road, some distance east of the two parcels. The last two, variations upon a common theme, would involve the E construction of a long length of road in place of the northern section of New Cut Road, across land within the Grove Green area.

Mr Little's view was that planning permission would not have been granted for a roundabout at either the first or second alternative positions owing to the loss of mature trees and other environmental factors. He was almost as pessimistic about alternatives three and four owing to the expected effect upon established dwellings nearby, on the landscape in the locality and upon the open-space provision within the Grove Green area. Mr Little rated the prospect of planning permission being granted for any of the alternative improvement schemes as negligible; he considered the order land essential for the junction improvements and its purchase, in the "no scheme" world, by the developer of Grove Green as imperative. The planning and F highway authorities have always seen the order land as the only location for the improvements, and have for many years sought to ensure its availability by the threat of using compulsory purchase powers. To grant planning permission for any of the alternatives could be done only in an attempt to save money for the developer, which is not part of the function of the town planning system. From his evidence, I find that the most acceptable scheme, in town planning terms, is that which has been formed on the order land.

However, any delay in making the junction improvements could not only affect financially the developer of Grove Green by virtue of the embargo but would have a serious town planning dimension in that the required five years' supply of residential land in Maidstone would cease to be available, this reserve being largely represented by the Grove Green area. If the order land were assumed not to be G available, the options open to the planning authority would be either to secure improvement of the junction upon one of the alternative sites or to identify other building land elsewhere in the Maidstone area. The reality is that all the four alternative junction sites would be contentious, for different reasons, and Maidstone is surrounded by classified landscape and high-quality agricultural land which the planning authority have hitherto sought to protect against residential development. For these reasons Mr Little regarded the order land as essential to securing the further progress of housing development at Grove Green.

His evidence in respect of the town planning circumstances of the order land, in the absence of its being required for junction H improvements, together with the adjoining land, was to the effect that permission could reasonably have been expected for residential development with highway access directly from Bearsted Road or, in the alternative, with access into Grove Green over land allocated for public open space. This opinion appeared to be founded principally upon the allocation in the 1970 town map.

The evidence of Mrs Randall related to the planning policy background of the Grove Green area, in particular the decision taken in the late 1960s to concentrate the supply of new residential land for Maidstone mainly in this area. Following this decision the allocation was included within the 1967 town map revision. The informal action area plan was published in 1973 to direct and guide implementation of the proposals. In practical terms the town map and the action area J plan fall to be read together. It was the latter which identified the improvement of New Cut Road, including the junction with Bearsted Road upon the order land, and the basic road layout to be adopted for the Grove Green area. Another important objective of the action area plan was to seek the visual separation of new housing at Grove Green from Bearsted Road and the area of high landscape quality to the north. This was to be achieved by a linear area 500ft deep, including the residue of the reference sites not required for junction improvements, allocated for public open space and landscaping.

In town planning terms, therefore, the order land was identified for junction improvements and the adjoining land for open space. For these reasons it cannot be assumed under section 16(2)(*b*) of the Land Compensation Act 1961 that "planning permission might K reasonably have been expected" for any other use, such as residential development upon the combined site. The previously agreed assumption of residential development put before this tribunal in 1987 was erroneous.

Mrs Randall was of the opinion that if the order land had not been physically available in the "no-scheme" world she would have advised, albeit reluctantly, that planning permission be granted for one of the alternative schemes of junction improvement. The maintenance of a continuing supply of residential building land in Maidstone was a major planning objective which had to succeed despite the accepted environmental damage arising from the alternatives; efforts would have been made at the detailed design L stage to minimise the impact by additional landscaping works. If pursuing alternative junction improvement proposals would have involved interruption to the progress of new housing at Grove Green, Mrs Randall, like Mr Heard, would have supported, for a period, a relaxation of the embargo upon further housing contained in the planning and highway agreements.

The evidence of Mr Atkins was that, in terms of his responsibility for planning policy at Maidstone Borough Council, he supported the opinions of Mrs Randall. The provision of new housing at Grove Green was the prime objective and any adverse environmental impact which might arise from alternative junction improvements was regrettable, but capable of some amelioration.

Bearing in mind the building land supply problem which would M emerge for the local planning authority, I am unable to accept that all concerned would have held up the development of Grove Green indefinitely for want of the off-site highway improvements. There would be difficulty in obtaining planning permission for any of the alternatives, particularly because trees would have to be felled, extra expense would be involved and delay may have occurred. However, in the circumstances prevailing in 1984 the order land would not have been the only possible location for the necessary junction improvements. Either the first or the second alternative put forward by the acquiring authority would probably have been adopted had the order land not been physically available.

In the situation of my finding that there were identified

alternatives to the order land, there would be some competition between the respective vendors, including the claimant. If the first alternative were selected, the sole landowner would be the borough council which, as planning authority, would have an emerging residential land shortage problem. However, in the "no-scheme" world the situation of the borough council would be subject to conflict in terms of securing the best price for land sold, following their own planning policies with respect to protecting the character and amenity of the area and the proper stewardship, for posterity, of public assets in the form of Vinters Park and the crematorium. All the alternatives, save for the order land, involve damage to established interests the protection of which largely resides with the borough council either as planning authority or in some other capacity.

In relation to the order land, if not used for junction improvements, together with the adjoining land I prefer the evidence of Mrs Randall, which was based upon the real situation of development control in the locality in 1984 and 1986. Her evidence had regard to the expectations issuing from all the development plan documents rather than to the historic town map allocation which, in terms of section 16 of the Land Compensation Act 1961, is only the starting point for informing valuation. The 1973 action area plan puts flesh on the skeletal allocation of the 1970 town map; that flesh for the reference land is an identification as active-use open space for the benefit, in particular, of the inhabitants of Grove Green. A direct link was envisaged via the outline planning permission of March 30 1982 (see the recital of the proposal and condition (1), cited as (a) above), between the reference land and the development of the Grove Green area. For reasons that are unclear, the reference land was excluded from the section 52 agreement, which specifies the requirement for public open space at clause (12) and relates its provision at area (D) of the sketch map, which includes both reference parcels, to the progress of housebuilding in phase three of the overall scheme. If the effect of the exclusion of the reference land is to sever the intended linkage with the 1973 action area plan proposals, the issue then arises whether, having regard to all other material considerations in 1984 and 1986, planning permission for residential development could reasonably have been expected. On the evidence, I find that permission would not have been granted for a small isolated pocket of dwellings, regardless of density, in such a sensitive location. Despite the town map allocation the appropriate use of the reference land in isolation, for valuation purposes, is therefore agriculture or accommodation land: see *Provincial Properties (London) Ltd* v *Caterham and Warlingham Urban District Council* [1972] 1 QB 453 at p 461.

The valuation evidence

Mr Pocock produced four valuations (reproduced at annexure 1) in respect of the order land together with some supporting calculations relevant to the possible availability of alternative locations for the junction improvements. His primary approach, contained in valuation A, was to follow the principle in *Stokes* v *Cambridge Corporation* (1961) 13 P&CR 77* by adopting 50% of the development value of the land comprising phases two (less 40 houses), three and four of the Grove Green proposals; gross area 182.52 acres. This valuation was in the sum of about £20.7m.

Valuation B was also upon the *Stokes* principle, but had regard primarily to the financial advantage likely to accrue to the developer of Grove Green by not having to wait 10 years to 1994 for the junction improvements to be carried out as a statutory function by the highway authority. This valuation was in the sum of £13.2m, reduced by one-third to £8.8m if there is found to be competition from alternative locations for the junction improvements.

Valuation C was also rooted in the *Stokes* principle but sought to have specific regard to the possibility that the developer seeking to implement the required junction improvements may identify, and get planning permission for, an alternative site. The risk for the claimant inherent in such a situation resulted in a valuation of £13.8m, the £20.7m of valuation A less one-third.

Valuation D was promulgated upon an entirely different basis, namely the value of the order land for residential development, with all elements of ransom or premium value disregarded; this figure was £211,042. The value of the adjoining land was approached upon this basis throughout to produce a sum of £387,081.

Mr Pocock sought to support the above valuations by reference to transactions in the Maidstone and Gravesend areas and the adoption of 50% as the basic ransom or premium element by reference to *Stokes* and other decisions of this tribunal together with three transactions in East Anglia, one spoken to in some detail by Messrs Bulteel, Thomas and Cherry.

In cross-examination it was put to Mr Pocock that there was a difference in principle between the disposal of land required for access to the owner of the backland, the "access site", and the disposal of the backland to the frontage or access owner at a discount. By practice and custom the latter follows the *Stokes* principle, whereas the former should not; a multiple of market value in isolation and/or an *ad hoc* sum as an inducement is the correct

*Editor's note: Also reported at (1961) 180 EG 839.

ANNEXURE 1

R W R Pocock FRICS

Valuation A

The order land		£
Development value		229,855 per acre
Existing use value		3,000 per acre
Net increased value		£226,855 per acre
Calculation		
½ of 182.52 acres @ £226,855 per acre		20,702,787
The adjoining land		
0.97 acre @ £399,053 per acre		387,081
	Total	£21,089,868

Valuation B

The order land			£
Development value			229,855 per acre
Existing use value			3,000 per acre
Net increased value			£226,855 per acre
Calculation	£		£
The immediate net increased value of 182.52 acres @ £226,855 per acre	41,405,574		
PV of £41,405,574 deferred 10 years @ 10% (0.386)	15,982,552		
The difference	£25,423,022		
One half of this is			12,711,511
Plus			
The advantage to a developer of securing certain access, say			500,000
			£13,211,511
As to the *adjoining land*, the valuation is as in valuation A			£387,081
		Total	£13,598,592

Valuation C

The order land		
Calculation		£
Two-thirds of ½ of 182.52 acres @ £226,855 per acre		13,801,927
As to the *adjoining land*, the site value for residential purposes as at December 1 1986.		
Calculation		
0.97 acre @ £399,053 per acre		387,081
	Total	14,189,008

Valuation D

i) As to the order land, the site value for residential purposes as at August 1984.		
Calculation		£
0.86 acre @ £245,398 per acre		211,042
ii) As to the adjoining land, the site value for residential purposes as at December 1 1986.		
Calculation		
0.97 acre @ £399,053 per acre		387,081
	Total	£598,123

A approach to price. This was not accepted by Mr Pocock since he regarded the one, relative to the other, as "the reverse of the coin", nor did he consider that the *Stokes* principle was any less relevant where the area of backland was very large, as in the instant case. Mr Pocock was challenged, with somewhat inconclusive results, upon the comparables used and the land values for residential development deduced therefrom, namely £229,855 per acre as at August 1 1984 and £399,053 per acre as at December 1 1986.

Mr Hicks produced two valuations (reproduced at annexure 2) based upon alternative assumptions. His primary valuation, KCC1, in the sum of £50,000 (revised by KCCIB to £75,000 during the hearing), for the order land and the adjoining land together, was

B

ANNEXURE 2
N A Hicks BSc FRICS
Valuation KCC 1 B

		£
1.83 acres @ £5,000 per acre, say		10,000
multiplier of 3 as inducement		3
		£30,000
	£	
Most expensive option	+ 128,800	
Wards share of highway costs	× 65%	
	£83,720	
Benefit of cheaper option to be shared C equally between the parties	× 50%	
Additional payment for claimant's land		41,860
		71,860
But, to reflect the imprecise nature of this type of negotiation, say	Total	£75,000

(This sum would be apportioned *pro rata* based on areas as between the order land and the adjoining land)

Valuation KCC2

	£	£
Order land — 0.86 acre @ £150,000 per acre	129,000	
Less abnormal development costs of £70,100 shared D equally with adjoining land (pumped sewage system and link road), say	35,000	
	94,000	
Deferred 3 years as road network on Wards land would not be in place until 1987 PV £1 in 3 years @ 10%	0.75	
	70,500	
Less 50% to reflect cost of negotiating with Wards rights of access and to connect with sewage system	35,250	
Value at 1.8.84		35,250
Adjoining land — 0.97 acre @ £300,000 per acre	291,000	
Less abnormal development costs of £77,200 shared E equally with order land	38,600	
	252,400	
PV £1 in 1 year @ 10%	0.91	
	229,684	
Less 50% for access and sewage rights	114,842	
Value at 1.12.86		114,842
Total value		150,092
	say	£150,000

based upon a multiplier of three times agricultural value plus, as a further inducement, 50% of the saving in construction costs for the junction improvements which could be made by using the order land F in preference to the most expensive of the four alternatives which he considered to be available. His secondary valuation, KCC2, was a detailed calculation based upon the assumption of residential development of the 1.83 acres with road access and foul drainage facilities being provided via the Grove Green development. This valuation was £150,092 and had as its starting point £150,000 per acre (in 1984) for the order land and £300,000 per acre (in 1986) for the adjoining land. In its basic assumption it was identical to valuation D by Mr Pocock, but there the similarity ceased.

In support of his values per acre Mr Hicks referred to sale transactions by the acquiring authority in or close to Maidstone and his own analysis, involving several adjustments, of Mr Pocock's comparables. Mr Hicks cited a purchase by the county council in Swale District to support his multiplier and/or inducement approach G to the value of the land required for access to backland.

Although denying that the order land was endowed with a ransom or premium value due to its ability to facilitate highway improvements for the benefit of the Grove Green area, Mr Hicks submitted supplementary valuations (not reproduced) to match Mr Pocock's valuation A, in the sum of £2.335m and valuation B at about £1.183m. These figures result from a different approach to Mr Pocock upon several aspects of the calculations.

In the cross-examination of Mr Hicks upon the Swale District transaction, it became apparent that the vendor may have desired to dispose of an acknowledged unsatisfactory residential investment, and this influenced the price accepted. Although criticised in respect H of details of his analysis of the residential land transactions, Mr Hicks was not prepared to move from his considered opinion of £150,000 and £300,000 per net acre at the respective dates.

Mr Bulteel and Mr Thomas both spoke to the facts of a transaction in "access land" during the period 1988-90; Mr Bulteel as adviser to owners of backland and Mr Thomas as adviser and negotiator for the vendor of the "access land". Although both negotiated with a third-party prospective developer, Mr Cherry, and not with each other, their evidence was essentially complementary.

In essence, access was required to some 400 acres gross, 240 acres net developable, and the backland owners were prepared to pay in mid-1988 by way of deduction from full residential development value, about 25%. The vendor was able to provide the sole access and J also the land required for widening the existing public highway to which a connection was to be made. Putting aside costly and tortuous alternatives which were considered and rejected because of possible delay, the owner of the access land was in a virtual monopoly situation. It was accepted by both witnesses that the *Stokes* principle applied to the disposal of "access land" but not necessarily the percentage of development value hitherto adopted in respect of smaller areas of backland. It was clear from this evidence that the percentage was applied to the value of the net developable area, 240 acres, after allowance for the cost of main infrastructure and deferment to have regard to the anticipated development period for the specific site.

K Negotiations had continued over several years before contracts were signed and by which time the value of residential building land was falling. However, I am satisfied that the willing purchaser-willing vendor concept is best represented by the price for the "access land" and the backland value agreed in 1988, since after this time the third-party developer became contracted to purchase the backland without having exchanged contracts to purchase the "access land". Once committed to the backland the developer had little flexibility in relation to the "access land", even though the agreed 1988 price, on exchange of contracts in 1990, represented a higher percentage of the then market value of the backland.

Mr Cherry negotiated the purchase of the backland and the "access land", starting in about mid-1986. His evidence confirmed L that of Mr Thomas and Mr Bulteel. He regarded the alternative access proposals instigated by Mr Bulteel as not practical, and considered them to be no more than a "negotiating point". While the precise capital value of the backland with the benefit of access was not central to his evidence, it was apparent in answer to my questions that Mr Cherry had regard to the net developable area, the cost of major on- and off-site infrastructure and deferment to arrive at his opinion of market value. It was a percentage of this figure which he agreed with Mr Thomas in 1988, and that sum (or a figure very close to it) was paid in 1990 despite the fall in land values for residential development by that date.

The evidence of Mr Walker related to the purchase of the site of the M Grove Green development in the early 1980s from several owners, its subsequent development and the sale of completed houses. He said that it was an attractive site which had proved very popular with house-purchasers, but he was unable to give annual sales figures. Surprisingly, having regard to the successful sale of houses in phase one, over about 18 months from 1982, and the documents in evidence referring to the advancement of phase two, Mr Walker said that his company was under no urgency in 1984 to proceed with further phases. During the progress of the development scheme he had negotiated payments for rights of access over company land with other owner-developers, and the consideration paid had been based upon the "ransom or *Stokes* principle".

His view throughout concerning the junction improvements had

A been that they were for the acquiring authority to deliver, and Wards had agreed to pay 65% of the cost. In consequence his company had conducted only cursory discussions with the claimant to acquire his land, in which it perceived no ransom value, and no consideration had been given to the alternative locations now put forward by the acquiring authority. His reason for the absence of activity and the perception of lack of ransom in the order land was founded upon the view that Wards could have secured access for residential development directly from Bearsted Road via frontage land in their ownership. Mr Walker did not produce any sketch scheme to demonstrate feasibility wholly within company-owned land nor did he indicate that any such scheme had been discussed with the highway B and planning authorities. I therefore attach neither relevance nor weight to this suggestion.

In answer to my questions, he said that it was customary when acquiring any site for housing development to secure all access points at an early stage. He also stated that the market for new houses in Maidstone was more consistent than in Gravesend and that house prices were higher in Maidstone. I have inspected the reference land including its locality and the majority of the comparables cited and relied upon in evidence.

Decision

The "no-scheme" world

It is accepted that the order land falls to be valued disregarding the C scheme underlying the acquisition, identified above, and the fact that the acquisition is taking place under compulsion. It is not disputed that the market for the order land effectively comprised prospective developers of the Grove Green area of which Wards were the most prominent. It is also accepted that if the order land is found to be endowed with some measure of premium value, the amount is entirely in issue, that sum is to be treated as having accrued independent of the scheme. In essence, what falls to be ascertained is the bargain which would have been made between the claimant and a prospective developer-purchaser had the acquiring authority not intervened.

In considering this matter, Mr Purchas, for the claimant, regarded D the following as central to a proper conclusion:

1. The facts affecting the order land and the parties comprising the market, including the claimant, were those actually existing at August 1 1984.

2. The order land had been available at all material times; a difference of opinion regarding price does not affect availability.

3. Section 15(1) of the Land Compensation Act 1961 accords to the order land at all times the assumption of planning permission for the junction improvements.

4. The acquiring authority, as highway authority, were in existence to perform their statutory functions but they would not use compulsory purchase powers to acquire land for the junction improvements at other locations.

E Against this background Mr Purchas urged that I find the order land to be the only possible site for the junction improvement and that in August 1984 the need in Wards for their provision was urgent. Those were the facts relevant to negotiations in the market at the agreed valuation date, and *Birmingham Corporation* v *West Midland Baptist (Trust) Association (Inc)* [1969] 3 All ER 172 requires that they be applied. There was no suggestion that the interest of the claimant in the order at the date of notice to treat, November 28 1983, was in any way different to that at the date of valuation eight months later.

Mr Spence, for the acquiring authority, contended that the above was only a partial disregard of the scheme and urged a broader F approach. The first step was to decide whether, on the evidence, an alternative location for the junction improvements was impossible, *per* Lord Morris in *Margate Corporation* v *Devotwill Investments Ltd* [1970] 3 All ER 864. If, as I have found, it was not technically and physically impossible, then it was necessary to disregard not only the actual compulsory purchase order but also all the procedural steps taken, including the committee decision on November 21 1980 to use compulsory purchase powers to assist Wards in securing the junction improvements.

Having thus erased from mind all traces of the scheme, Mr Spence then submitted that the relevant time for considering the facts was from November 21 1980 onwards. He said disregarding the scheme required that all prospective developers of Grove Green had "a clean sheet to adopt any feasible means of junction improvement they chose". Underlying the approach of Mr Spence is the notion that the G availability of the order land is affected in the "no-scheme" world by the price sought by the claimant. If that is a high price the prospective purchaser will look to the alternatives. If that process had begun in November 1980 with concurrent negotiations in respect of all sites, the pre-eminence in 1984 of the order land would have been markedly reduced or largely eliminated and no urgency would reside in Wards. Furthermore, he submitted that, on the authority of the *Margate* case, the use of compulsory purchase powers to aid the acquisition of land in the alternative locations is assumed to be available. These things taken together should lead to the conclusion that the order land is no more than the marginally convenient location meriting only modest compensation.

Although the statutory provisions and authorities cause the parties H to notionally reside in a world conditioned by assumption and a measure of unreality, they do not, in my judgment, require the abandonment of the actual facts as subsisting at the valuation date. To cast aside the actual facts and substitute a different factual matrix founded upon speculation and supposition is contrary to the *West Midland* case and the principle stated by Lord Morris in the *Margate* case when he said, at p 868:

> The decisions to be made by the Lands Tribunal were decisions in the field of fact. Some of the witnesses appear to have felt that they were asked to apply their minds to situations having an unreality only to be expected in fairyland. I do not think that their approach should have been so hesitant. Their evidence had to be directed to the realities of the conditions that existed and to the J actual facts of the period so that the Lands Tribunal should be assisted to reach decision on the basis of the assumptions that had to be made.

I therefore decline to follow the route identified by Mr Spence and the joint invitation to consider an alternative award will be responded to later.

The order land

As entry on to the order land on August 1 1984 preceded the agreed date of valuation for the adjoining land on December 1 1986, the compensation payable for the latter will be influenced by the assumptions made and the basis adopted in respect of the order land.

From the supplementary statement of agreed facts and issues put in evidence following the decision of the Court of Appeal it was agreed that, "in valuing the order land under rule (2) of section 5 of K the Land Compensation Act 1961 account can be taken of any relevant increase in the market value, if in fact there was any such increase, having regard to the use of the land for the purpose of facilitating development of land to the south".

Chronologically, the proposal for, and the subsequent progression of, the development of the Grove Green area began with the allocation in the 1970 town map, amplified by the informal action area plan in 1973, followed by the grant of outline planning permission in March 1982 together with the associated planning and highway agreements. The "no-scheme" world, prior to the confirmation of the compulsory purchase order in September 1983, in concept and reality, comprised a number of developers assembling L land in order to implement the long-established development proposals for the area.

On the town planning and highway evidence the junction improvements required further to implement the 1982 planning permission and to satisfy the obligations under the planning agreement were becoming a matter of some urgency in mid-1984; the orderly progression of the residential development was shortly to be interrupted. Phases two (save for 40 houses) three and four could not be built, as traffic conditions would become progressively unacceptable at best and intolerable at worst. Indeed, had there been no development proposals for the Grove Green area, junction improvements would have been required and constructed but possibly not before the year 2000. M

The order land, as found above, was the preferred but not the only location for the junction improvements; the identified alternatives were second best, but in the real world the required works would not have been held up indefinitely for want of the order land. Since the order land had been earmarked by the planning and highway authorities for many years prior to 1984 as the most appropriate location for the works, and residential developers were anxious to proceed with implementing the approved proposals, it is, in my judgment, beyond doubt that the order land was endowed with increased value because of its potential for "facilitating development of land to the south". In the market of the "no-scheme" world its use and value is almost entirely bound up with this large area of land.

A Material to quantifying this potential value under rule (2) of section 5 of the Land Compensation Act 1961 is a clear and full understanding of the respective negotiating strengths of the claimant as vendor and Wards as the most likely purchaser at August 1 1984. It is the apparent omission or inability of the valuation witnesses in this reference to arrive at a common understanding upon this which results in such a wide divergence of opinion as to value.

The following matters appear relevant:

(1) the intended road pattern as detailed upon the "planning agreement" plan for the Grove Green proposals and the significance of the order land thereto;

(2) the available alternatives capable of providing competition and their relative costs;

B (3) the manner in which developer-purchasers assemble large housing sites in the market and how "strategic" parcels of land are regarded;

(4) the principles underlying valuation and the prices offered for "strategic" parcels;

(5) the extent and value of the land to be benefited by "strategic" parcels;

(6) the stage reached in the development programme at August 1 1984.

The evidence of Mr Hicks fails adequately to address the extent of potential value residing in the order land. The "warts and all" approach of Mr Pocock, which I take to mean a very broad-brush C approach without regard to such matters as the cost of major on- and off-site infrastructure and the extent of land sterilised against building, is too simplistic in the absence of specific evidence relating to transactions in comparable large sites.

1 Road pattern and order land

Mr Pocock rightly excluded phase one of the development, since its progress was not constrained by any lack of improvements on the order land. Moreover, this phase had independent access to New Cut Road and Weavering Street and required the separate junction improvements referred to earlier to New Cut Road and Ashford Road to the south.

The other phases require the formation of three new junctions to D Weavering Street, one to each of Bearsted Road and New Cut Road, a connection to Grovewood Drive (the phase one spine road) together with some general widening of New Cut Road as a continuation of the junction improvements upon the order land; a number of "strategic" points are therefore involved. In terms of timing and the amount of land upon which development is constrained, 182.5 acres gross or 123 acres net, the works upon the order land are the most significant, but the other junctions and works become equally essential over the years to reducing areas as development progresses. For these reasons there was no absolute monopoly of improved road access, essential for the development of the 182.5 acres, residing in the order land.

E *2 Alternatives and costs*

Four roundabout schemes were put in evidence as possible alternatives to the works upon the order land. Ignoring the cost of land, the market value of which with informed vendors is likely to be similar, these alternatives involved extra construction costs, compared with the scheme as built, ranging from £7,000 to £128,800. As already found, these alternatives constitute competition to the order land and their existence would act to moderate the price which would be paid for the order land. Mr Pocock in valuation C puts this as a deduction of one-third in the expected payment to the claimant, which I accept as realistic.

3 Site assembly and "strategic" parcels

F The evidence of Mr Bulteel and Mr Thomas, confirmed by Mr Walker, demonstrated that a developer assembling and purchasing a large site usually identifies at the outset any "strategic" parcels of land, such as those required for access. Like Mr Cherry, he then estimates their likely premium value or cost and proceeds to negotiate a purchase or secure options to purchase. The developer is then able to formulate offers and negotiate for the "backland" in the knowledge of the approximate cost of the "strategic" parcels, thereby avoiding overpayment. This practice appears not to have been followed to conclusion with respect to the order land owing to the assistance afforded with acquisition by the acquiring authority.

4 Valuation principles

I reject the suggested distinction, for valuation purposes, sought to be drawn between the owner of backland buying "access" land and G the owner of "access" land buying backland; I prefer "the reverse of the coin" approach of Mr Pocock. Moreover, I do not consider the inducement approach of Mr Hicks relevant save in circumstances where there are a number of alternatives for access of about equal merit, and thus real competition, as in *Rathgar Property Co Ltd* v *Haringey London Borough Council* (1977) 248 EG 693 or the Swale District comparable.

The evidence of Mr Pocock is accepted that, since the *Stokes* case in 1961, it has become common practice in the land market and in some references before this tribunal to attribute a percentage of the development value of backland to "access" land as a premium value; he puts the proportion at 50%. This principle is also adopted by Mr H Hicks in his secondary valuation (KCC2) but not in his primary valuation (KCC1B) which proceeds on the "inducement principle". There is evidence that for small "single access" sites, market practice is to attribute up to 50% as a premium value to the "access" land. The transaction, where the parties were professionally advised by Messrs Bulteel, Thomas and Cherry, indicates 25% for a very large "single access" situation; in negotiation Messrs Thomas and Cherry rightly attached no weight to the alternatives canvassed. In the circumstances of a "multi access" site, such as Grove Green, it is unrealistic to attribute 50%, as Mr Pocock does, to only one of the "strategic" land parcels. Logic and practice (see *Ozanne* v *Hertfordshire County Council* [1988] RVR 133) dictate a sharing of the premium value but not necessarily equally. On the evidence, for a J large site the starting figure is 25% and, in my judgment, 15% should be attributed to the order land as a premium value to reflect its relative importance among the "strategic" parcels to the incomplete part of the Grove Green development in 1984.

5 Extent and value of benefited land

The gross area of 182.52 acres (123 acres net, 67.4% of the total) adopted by Mr Pocock is agreed. He supported his valuation of this 182.52 acres by reference to two comparables with areas of 9.68 acres gross (8.68 acres net, 89.67% of the total) and 5.77 acres net. In the event he relied mainly on the former, which he analysed to show £229,855 per acre, and adopted that figure as the foundation of his K valuation of the benefited land (see valuations A, B and C). Mr Pocock adopted the larger site since he regarded it as a "bulk site", namely of interest almost exclusively to national housebuilders. This may be a fact of the market but it does not overcome the problem of the large difference in physical area compared with the 182.5 acres. This difference in area is of economic importance in terms of financing ownership of the site pending and during a lengthy period of development. Mr Pocock said that on the 9.6-acre site 79 houses had been built and he estimated the development period at between 18 and 24 months. On the 182.5 acres there remained to be built in 1984 some 1,500 houses, which he estimated as requiring about five years to complete; a perhaps optimistic estimate when judged by performance up to 1984. Mr Pocock did not take account of these L differences of time-scale in his valuations of the 182.52 acres by adopting any deferment, the conventional approach supported by the evidence of Messrs Bulteel, Thomas and Cherry.

The size of site and development period apart, Mr Pocock has not compared like with like owing to the widely differing ratios at the two locations of gross site area to net developable land, namely that land which can be devoted to the sites of dwellings, their fronting access roads and small, incidental areas of open space, but excluding distributor roads, main areas of public open space and sites for non-commercial communal facilities. As the ratio of "unremunerative" land is greater at Grove Green (32.6%) than for the comparable adopted (10.33%), logic dictates that an adjustment should be made M to the analysed per-acre value.

Mr Hicks analysed this comparable on a net developable area basis at £256,336. To express this figure in terms of per gross acre at Grove Green, it is necessary to discount by 32.6%, which produces £172,770. Mr Hicks made further adjustments in his analyses of eight transactions to allow for comparative location and the date of transaction relative to August 1 1984 to produce a value per net acre, ranging from £115,000 to £166,000. From this analysis he selected £150,000 per net acre as appropriate to Grove Green, which when discounted by 32.6% becomes £101,100 per gross acre.

With the benefit of inspection, I incline towards Mr Pocock's 9.6 acres gross comparable as the best contemporary transaction and I

A reject any discount for the inferior location of Grove Green as contended by Mr Hicks but not supported by Mr Walker.

To create the areas of developable land at Grove Green with road access and main services at the boundary so as to have similar facilities to the comparable would require considerable capital expenditure upon off-site works such as the junction improvements on the order land, other widening of New Cut Road, and upon site the construction of some 2 miles of principal spine road (including junctions) and the laying out of extensive areas of open space in satisfaction of the planning agreement. No figures were put in evidence relating to these costs other than for the works on the order land at *circa* £210,000 at 1983 prices, but in the context of comparative valuation a deduction from the £172,770 per acre above B needs to be made. On the very slender evidence available to me, I determine the per gross acre value after allowance for expenditure at £160,000, which after deducting the agricultural value at £3,000 per acre (an undisputed figure) gives a development value of £157,000 per acre.

6 Urgency

On the evidence, I conclude that in August 1984 Wards were keen for the junction improvements on the order land to begin so as to avoid delay to their Grove Green housing development. A start would also reduce their financing costs relating to land already purchased and infrastructure works carried out. I am not convinced that it was C apparent in August 1984 that Wards could benefit financially due to rising land and house prices by slowing down the rate of housebuilding as suggested by Mr Walker; I consider this to be a matter of hindsight.

Furthermore, even if negotiations in respect of land in the alternative locations had been pursued concurrently with those for the order land, a failure to reach agreement with the claimant may have resulted in delay while they were further investigated, the works designed and taken through town planning procedures. The path of any applications could be expected to be more contentious than proposals upon the order land. This is a factor favouring the order land and material to the strength of the competition from the alternatives, but, even so, I am not persuaded that it merits the *ad hoc* D addition of £500,000 as proposed by Mr Pocock in valuation B.

Bringing together the factors at 1-6 above, I arrive at the following calculation:

		£
182.52 gross acres @ £157,000		28,655,640
PV £1 in 3 years @ 10% (half estimated development period)		0.751
		21,520,386
Ransom value @ 15%		0.15
		3,228,058
Discount for risk one third		0.666
		2,149,886
Value of order land	Say,	£2,150,000

E

The adjoining land

Upon the basis that as at August 1 1984 the order land had been purchased by a developer for the junction improvements, the adjoining land would thus cease to be influenced by whether an alternative site may be selected. In effect it would be an isolated small parcel devoid of the possibility of highway access through the order land to the original Bearsted Road but with frontage to Bearsted and New Cut Roads as realigned. The original agricultural access which served it at the intersection of New Cut and Bearsted Roads would also be lost and a new access would be required for continued agricultural use. In the circumstances of the counternotice being accepted there was no actual need to provide this access. In the "no- F scheme" world, as in the scheme world, the claimant would with justification require any purchaser of the order land also to take this parcel.

Against this background the sole issue is the expected price; this, to a considerable extent, is influenced by the anticipation or otherwise of planning permission for a different use. Mr Pocock's valuation D proceeds on the basis of residential use with direct access from Bearsted Road. The secondary valuation (KCC2) by Mr Hicks is also in respect of residential development, albeit with access over land to the south owned by Wards. Having regard to the finding on the town planning evidence, neither of these valuations is appropriate to the circumstances prevailing at December 1986.

There remains the first part of Mr Hicks' primary valuation (KCC1B), disregarding the later part involving the junction G improvements, where he adopts £5,000 per acre as paddock value. As an alternative, Mr Pocock stated that for open space land within the ambit of the Grove Green area the developer could be expected to pay the same price per gross acre as for any other land within the planning permission area (in 1984 £229,855 adjusted above to £160,000 per acre). Since the developer obtained an outline planning permission in 1982 which specifically excluded this parcel, I regard the true interest in purchasing in 1986 as fairly slight, confirmed in evidence by Mr Walker, and I reject the approach of Mr Pocock. I also reject his alternative suggestion that if it is found that there is no immediate residential development potential in the land it nevertheless exists in the longer term, and in 1986 a developer would pay 50% of the then residential value, which he considered to be £193,540 for 0.97 of an H acre; £399,053 per acre full value.

In the real or "no-scheme" world it is unreasonable to expect a farmer willingly to sell land at or close to agricultural value where it is identified in planning policy documents for non-agricultural use, namely public open space or other amenity purpose associated with nearby residential development. In the absence of evidence as to transactions in such land I take a lead from, but do not follow further, Mr Hicks' primary valuation and multiply the agreed paddock value of £5,000 by two and determine the compensation payable for the adjoining land at £10,000 (10 thousand pounds).

The award in respect of the two parcels comprising the reference is therefore: J

		£
1. The order land		2,150,000
2. The adjoining land		10,000
	Total	£2,160,000

To this sum will be added a surveyor's fee in accordance with Ryde's Scale (1984).

Matter of alternative award

The key to the validity of the argument by Mr Spence is whether the order land was available at all material times. It has not been shown to be unsuitable for the junction improvements; indeed the evidence K was wholly the other way. If in law availability ceases at a certain level of price, this sum has not been identified; the claimant throughout has sought market value. Mr Pocock endeavoured to quantify the financial advantage to the developer-purchaser by proceeding in August 1984 rather than waiting a decade or possibly more by which time the public need for the improvements would be greater owing to increased traffic. This is a measure of market value which has regard not only to time but also to the potential of the order land to facilitate development at Grove Green.

If, in certain circumstances, it has to be assumed that the order land were not available, then on the evidence the alternatives would have come to the fore so as to satisfy the need in the planning authority and the developer to progress the Grove Green housing development. L There might have been competition between the three main alternatives, but each owner-vendor must be assumed to be well informed and would accordingly approach the matter of price in like manner to the claimant and not upon the basis advocated by the acquiring authority; the *Stokes* rather than the "inducement" principle would be paramount whether acquisition were by agreement or under compulsion.

If the order land is to be assumed to have been available throughout, as contended by Mr Purchas, I am not persuaded that the competition from the alternatives would be any greater, if viewed from 1980 onwards, than it was on the facts in August 1984; the *ad hoc* addition suggested by Mr Pocock for certainty and urgency has M been rejected as being without justification.

For these reasons, and upon the evidence, if I were to adopt the concept of the "no-scheme" world as urged by Mr Spence, there would be no variation in the above award for the order land.

The costs of transfer and interest upon the compensation money are the subject of specific statutory provisions.

Having opened the amended sealed offer dated September 12 1990 lodged by the acquiring authority, I find the amount thereof to be less than the award. The acquiring authority will pay the claimant's costs of the reference, including those relating to the initial hearing in 1987. Such costs unless agreed to be taxed by the Registrar of the Lands Tribunal on the High Court Scale.

ROCHESTER POSTER SERVICES LTD v DARTFORD BOROUGH COUNCIL

(REF/16/1989)
July 25 1991
(V G WELLINGS QC, President)

Estates Gazette February 8 1992
[1992] 05 EG 172

Land at Dartford, Kent, used as advertising site — Agreement granting right to erect and maintain poster panels on outer perimeter — Whether agreement created compensatable interest — Whether, alternatively, claimant in lawful possession — Tenancy or licence — Concession in agreed statement of facts affecting decisions — Decisions in claimants' favour

Premises used by claimants for stationing of advertising posters on hoardings erected by them under agreement made in March 1980 — Agreement to remain in force for 12 years and thereafter from year to year with option to renew for another nine years from expiry of original period — Parties referred to throughout agreement as "Grantor" and "Licensee" respectively — No reference to "demised" premises — Provision for three-year rent reviews in schedule to which there was no reference in agreement — Statement of agreed facts placed before tribunal included statement in para 10 that "claimant had exclusive possession of the site pursuant to the terms of the agreement" — Two preliminary questions of law: (1) whether the agreement created an interest in land for which the claimant was entitled to compensation under s1 of the Land Compensation Act 1961 or s6 of the Compulsory Purchase Act 1965; (2) alternatively, whether the claimant had been in lawful possession of the land and was entitled to a disturbance payment under s37 of the Land Compensation Act 1973 — In submitting that the agreement created a tenancy, the claimants relied strongly on the para 10 concession

***Decision*: (1) On the first issue, in view of the concession in para 10 of the agreed statement of facts, which appeared to beg the question, it was difficult to avoid the inference that the agreement from the date when the panels were erected intended the claimant company to have exclusive possession of the perimeter of the land to which they were affixed — Further, it was difficult to see how anyone other than the company could have exclusive possession after the panels were erected and maintained in situ — The outer perimeter was let to the claimants, in the tribunal's judgment (though not without some feeling of doubt), for, at the least, a tenancy from year to year and, under the doctrine in *Walsh* v *Lonsdale*, for a term of 12 years and thereafter from year to year with a further option of another nine years — Accordingly, the claimants had an interest for which they were entitled to compensation under s1 of the Land Compensation Act 1961**

(2) On the second, alternative, issue, the para 10 concession made it impossible to argue that the claimants were not in exclusive possession, since it stated the contrary — Accordingly, if, contrary to the decision on the first issue, the claimants had no interest in the outer perimeter, they were nevertheless in lawful possession of that land

The following cases are referred to in this report.

Addiscombe Garden Estates Ltd v *Crabbe* [1958] 1 QB 513; [1957] 3 WLR 980; [1957] 3 All ER 563, CA
Facchini v *Bryson* [1952] 1 TLR 1386, CA
Provincial Bill Posting Co v *Low Moor Iron Co* [1909] 2 KB 344
Street v *Mountford* [1985] AC 809; [1985] 2 WLR 877; [1985] 2 All ER 289; [1985] 1 EGLR 128; (1985) 274 EG 821, HL
Taylor, Garnett, Evans & Co v *Overseers of the Poor of Pendleton* (1887) 19 QBD 288
Walsh v *Lonsdale* (1882) 21 ChD 9
Wilson v *Tavener* [1901] 1 Ch 578

Jonathan Brock (instructed by Rowe & Maw) appeared for the claimant company; Miss Kim Franklin (instructed by the solicitor for Dartford Borough Council) for the acquiring authority.

Giving his decision, **THE PRESIDENT** said: These are preliminary issues of law arising under a reference claiming compensation for compulsory acquisition of land, that is to say, the outer perimeter of the land situate at and known as 37 Bullace Lane, being on the southern side of Bullace Lane and on the western side of Suffolk Road, Dartford, Kent.

By an agreement in writing (which I will call "the agreement") dated March 28 1980 and made between John Henry Woolls and Violet Louise Reynolds, trading as Forrest Amusements (in the agreement called "the Grantor"), and Rochester Poster Services Ltd (called "the Licensee"), the company was granted the exclusive right to erect and maintain poster panels on the outer perimeter of the premises situate and known as 37 Bullace Lane, Dartford.

The preliminary issues are:

(1) whether the agreement creates an interest in land for which the claimant is entitled to compensation under section 1 of the Land Compensation Act 1961 or section 6 of the Compulsory Purchase Act 1965;
(2) in the alternative, whether the claimant was in lawful possession of the land the subject of the agreement, from which it was displaced by the compulsory purchase order, and is entitled to a disturbance payment under section 37 of the Land Compensation Act 1973 therefor.

In stating the first preliminary issue I have substituted statutory references agreed at the hearing for those in the issue as originally ordered.

There is an agreed statement of facts and a copy of it is contained in Appendix A to this decision. A copy of the agreement is contained in Appendix B.

First issue

It is agreed between counsel that the company is not entitled to compensation for compulsory purchase unless it had an interest, legal or equitable, in the land to which the agreement related. The important question, therefore, is: did the agreement create a tenancy or a mere licence?

The following features of the agreement should be noted:

(1) The references to "the Grantor" and "the Licensee".
(2) The permitted posters are described and their approximate measurements given.
(3) The licensee agrees to pay for the exclusive right granted by the agreement the sum of £2,800 pa.
(4) The agreement is to remain in force for 12 years from the date on which the panels are erected and is to continue from year to year thereafter until determined by either party giving to the other six months' previous notice in writing of desire in that behalf.
(5) There is to be an option for a further period of nine years from the expiration of the original period.
(6) The grantor is to allow the licensee uninterrupted access to the advertisement site.
(7) The expressions "the Grantor" and "the Licensee" are, where the context admits, to include their respective successors in title.
(8) At the end of the agreement there is a schedule which is not referred to in the body of the agreement but which contains a provision as to "a rent review" every three years throughout the duration of the agreement.

The agreed statement of facts contains details of the compulsory purchase and of the use made by the claimant of the premises. In para 8 it is stated that at all material times the claimant used the premises for the stationing of advertising posters on hoardings erected by the claimant on the premises. Para 9 gives details of the measurements of the hoardings and the structure to which they were attached. Para 10 contains an important statement: it says:

> The claimant gained access to the premises at all times from the public highway and at all material times until compulsory acquisition the claimant had exclusive possession of the site pursuant to the terms of the agreement. . .

A Planning permission for the erection of the poster hoardings and associated featherboard fencing was given by the Dartford Borough Council on September 1 1980.

I was referred to the decision of the House of Lords in *Street* v *Mountford* [1985] AC 809*, in which decision it was held that, where residential accommodation had been granted for a term at a rent with exclusive possession, the grantor providing neither attendance nor services, the legal consequence was the creation of a tenancy. Lord Templeman's speech contains a number of important observations. He said, at p 818:

> there can be no tenancy unless the occupier enjoys exclusive possession; but an occupier who enjoys exclusive possession is not necessarily a tenant. He may be owner in fee simple, a trespasser, a mortgagee in possession, an object of
> B charity or a service occupier. To constitute a tenancy the occupier must be granted exclusive possession for a fixed or periodic term certain in consideration of a premium or periodical payments. The grant may be expressed, or may be inferred where the owner accepts weekly or other periodical payments from the occupier.

Lord Templeman, at p 822, approved the decision of the Court of Appeal in *Addiscombe Garden Estates Ltd* v *Crabbe* [1958] 1 QB 513, where Jenkins LJ said, at p 522:

> the whole of the document must be looked at; and if, after it has been examined, the right conclusion appears to be that, whatever label may have been attached to it, it in fact conferred and imposed on the grantee in substance the rights and obligations of a tenant, and on the grantor in substance the rights and obligations of a landlord, then it must be given the
> C appropriate effect, that is to say, it must be treated as a tenancy agreement as distinct from a mere licence.

Having referred to the decision of the Court of Appeal in *Facchini* v *Bryson* [1952] 1 TLR 1386 at p 1389, Lord Templeman, in his speech, continued:

> Exclusive possession is of first importance in considering whether an occupier is a tenant; exclusive possession is not decisive because an occupier who enjoys exclusive possession is not necessarily a tenant. The occupier may be a lodger or service occupier or fall within the other exceptional categories mentioned by Denning LJ in *Errington* v *Errington and Woods* [1952] 1 KB 290.

At p 826, Lord Templeman continued:

> My Lords, the only intention which is relevant is the intention demonstrated
> D by the agreement to grant exclusive possession for a term at a rent. Sometimes it may be difficult to discover whether, on the true construction of an agreement, exclusive possession is conferred. Sometimes it may appear from the surrounding circumstances that there was no intention to create legal relationships. Sometimes it may appear from the surrounding circumstances that the right to exclusive possession is referable to a legal relationship other than a tenancy. Legal relationships to which the grant of exclusive possession might be referable and which would or might negative the grant of an estate or interest in the land include occupancy under a contract for the sale of the land, occupancy pursuant to a contract of employment or occupancy referable to the holding of an office. But where as in the present case the only circumstances are that residential accommodation is offered and accepted with exclusive possession for a term at a rent, the result is a tenancy.

E Mr Brock [for the claimant] also referred, *inter alia*, to the decision of the Queen's Bench Divisional Court in *Taylor, Garnett, Evans & Co* v *Overseers of the Poor of Pendleton* (1887) 19 QBD 288. That case had some similarities to the present case except that there was a purported letting. By an agreement between an owner of land and an advertising agent the owner agreed to let and the agent to take an advertising station at a yearly rent, the tenancy to commence from completion of erection and continue for seven years. The dimensions of the hoardings to be erected were specified and advertising hoardings supported on posts fixed into the ground were erected. It was held that the agreement created a tenancy. Wills J said, at p 293:

> . . . I think the case is very clear. I have no doubt that the document gives an exclusive occupation. The agreement provides for letting and taking the
> F advertising station. Those are words appropriate to a demise. No doubt this is not conclusive, and it is necessary to look at the whole document, but there is nothing in the context to require us to reject the ordinary meaning of the words, and an advertising station is clearly capable of being the subject of a demise. Then the letting is to be at a yearly rent, which is an expression applicable primarily to a tenancy. It is further provided that the tenancy is to remain in force for a period of seven years. It is true that no tenancy for seven years can be created by a document which is not under seal; still such a document may operate as a demise, and after payment of rent a tenancy from year to year would come into existence.

Of the other cases mentioned by Mr Brock, I need refer only to *Wilson* v *Tavener* [1901] 1 Ch 578 (licence held to exist) and *Provincial Bill Posting Co* v *Low Moor Iron Co* [1909] 2 KB 344 G
(advertising hoardings held to be fixtures and opinions expressed that a tenancy existed).

Finally in his argument, Mr Brock relied strongly upon the concession in para 10 of the agreed statement of facts that "the claimant had exclusive possession of the site pursuant to the terms of the agreement". Accordingly, he submitted that the agreement created a tenancy. In support, he pointed to the provision for a yearly payment, the grant of a 12-year term from the date on which the panels were erected and yearly continuation thereafter, the option, the statement in clause 8 of the agreement as to the meaning of the expressions "the Grantor" and "the Licensee" and to the rent review clause.

If a tenancy were created by the agreement then, like the tenancy in H
the case of *Taylor*, the tenancy would be void under section 52 of the Law of Property Act 1925 because, being for a term in excess of three years, it was not made by deed. However, the tenancy would operate at common law as a tenancy from year to year from the date when the panels were erected and rent was paid. Furthermore, in equity it would operate as a contract to grant a tenancy for a term of 12 years (and tenancy from year to year thereafter) from the date of the erection of the panels; and, because that contract would no doubt be enforced by a court of equity by decree of specific performance, under the doctrine of *Walsh* v *Lonsdale* (1882) 21 ChD 9, the lease for such a term would be deemed to have been granted.

Miss Franklin [for the acquiring authority] helpfully provided a J
skeleton of her argument, which was that the agreement does not create an interest in land but a mere licence entitling the claimant to do something which otherwise would have been a trespass. She pointed out that the parties were described as "grantor" and "licensee" and not "landlord" and "tenant". There was no reference to "the demised premises" and the sums payable were not described as "rent" save in the schedule to which there was no reference in the body of the agreement. She submitted that the agreement granted the right to use the outer perimeter for the erection of hoardings but that it was not *let* to the company. The concession in para 10 of the agreed statement of facts amounted to an admission that the company had actual possession attributable to its ownership of the panels which were affixed to the land. She said that the K
concession did not determine the intention of the agreement because that was a question of the construction of it.

In my opinion, the first issue is difficult. It is made particularly difficult by the concession in para 10 of the agreed statement of facts. That concession appears to beg the question. As a consequence it is difficult to avoid the inference, as Mr Brock suggested, that the agreement from the date when the panels were erected, intended the company to have exclusive possession of the perimeter of the land to which they were to be affixed. As I have observed, the agreement defines the panels and their size and it is difficult to see how anyone other than the company could have exclusive possession after they were erected and maintained *in situ*. Of course, since planning L
permission for their erection was not given until September 1 1980, presumably they were not erected until after that date, so that there was an interval in which apparently the company had no interest in land but in respect of which the annual sum was nevertheless payable. Thereafter, in my judgment (though not without some feeling of doubt), the outer perimeter was let to the company for, at the least, a tenancy from year to year and, under the doctrine of *Walsh* v *Lonsdale, supra*, for a term of 12 years and thereafter from year to year with a further option of another nine years from the expiration of the original period. The result is that I answer the first preliminary issue by holding that the claimant company had an interest in land which was taken by the acquiring authority by compulsory purchase and in respect of which the company is entitled to compensation M
under section 1 of the Land Compensation Act 1961.

Second issue

Miss Franklin endeavoured to submit that the company did not have a right to possession of the land. She said that the company was no more in possession of the outer perimeter than a firm of estate agents is in possession of the patch on which it plants its advertising board or the owner of a dustbin is in possession of the area where he keeps it. However, in my opinion, the ground was cut from underneath Miss Franklin's argument by the concession in para 10 of the agreed statement of facts. That concession makes it impossible to argue that the company was not in exclusive possession of the outer

*Editor's note: Also reported at [1985] 1 EGLR 128.

A perimeter, because it states the contrary. Moreover, it states that that possession was pursuant to the terms of the agreement. That possession was not criminal or tortious. It was plainly lawful. Accordingly, if it is necessary in the present case to refer to section 37 of the Land Compensation Act 1973 and if (contrary to my decision on the first issue) the claimant had no interest in the outer perimeter, being the land from which it was displaced, it was nevertheless in lawful possession of that land and I answer the second issue accordingly.

APPENDIX A

Agreed statement of facts

1. By an Agreement dated 28th March 1980 ("the Agreement") between B John Henry Woolls and Violet Louise Reynolds trading as Forrest Amusements of the one part and Rochester Poster Services Limited of the other part, Rochester Poster Services Limited were granted the right to erect and maintain poster panels on the outer perimeter of premises ("the premises") known as 37 Bullace Lane, Dartford.
   The premises are identified in Plan No RSE 45C dated 14th June 1985, in the plans submitted with Planning Application D80/18/ADV (revised) dated 1st September 1980 and in the copy photographs annexed hereto.
2. On 26th February 1985 the Dartford Borough Council ("the Council") made a Compulsory Purchase Order over land including the premises.
3. On 6th November 1986 that Compulsory Purchase Order was confirmed by the Secretary of State for the Environment.
4. On 2nd April 1987 by general vesting declaration the land and premises, the subject matter of the Agreement, were vested in the Council.
5. C Pursuant to notice to treat and notice of entry, both dated 14th May 1987, the Council took possession of the land the subject matter of the Agreement.
6. By letter dated 26th August 1987, Rochester Poster Services Limited by its agents, Messrs Healey & Baker, claimed compensation for loss of rights under the agreement pursuant to the Compulsory Purchase Acts.
7. The parties are in dispute as to the entitlement of Rochester Poster Services Limited as a matter of principle to compensation under the said Acts and have requested the Tribunal to determine two preliminary issues as set out in the Order of the Registrar of the Lands Tribunal dated 11th April 1991.
8. At all material times the claimant used the premises for the stationing of advertising posters on hoardings erected by the claimant on the premises.
9. The hoardings consisted of a series of 7 ft irons (average) attached to large D up-right 6-10 ft wooden posts making four hoardings each 20 ft long by 10 ft high. The irons and posts were implanted into the ground of the property and irons concreted into the earth at the boundary of the land, substituting, where it existed, for a normal fence. The uprights were reinforced by back braces 4" × 2" which had irons inserted into the concrete. Crossrails 3" × 2" from post to post laterally provided the support for the fence of each advertising display board consisting of exterior plywood framed by 4" × 1" wood surrounds. An interlocking featherboard wooden fence (with featherboard aprons below the panels if and where appropriate) linked the advertising hoardings together, fencing the land off from public access for the length of the property's boundary. The plans and photographs annexed hereto identify the fence and the hoardings concerned which composed the advertising station.
10. The claimant gained access to the premises at all times from the public E highway and at all material times until compulsory acquisition the claimant had exclusive possession of the site pursuant to the terms of the agreement dated 28th March 1980 referred to in para 1 herein.

APPENDIX B

The agreement

THIS AGREEMENT IS BETWEEN Forrest Amusements (hereinafter called "the Grantor") of 12 Spring Vale, Dartford, and Rochester Poster Services Ltd, 84 High Street, Chatham, Kent (hereinafter called "The Licensee").

WHEREAS the Grantor grants to the Licensee the exclusive right to erect and maintain poster panels on the outer perimiter (*sic*) of the premises situate and known as 37 Bullace Lane, corner of Home Gardens, Dartford.

AND WHEREAS The Licensee proposes to erect four such panel(s) each F measuring approximately 10 ft high × 20 ft wide and to be responsible for any damage caused in erection, removal or maintenance thereof: such materials to remain the property of the Licensee and to be removed by him at the expiration of the Licence.

1. The Licensee agrees to pay for the exclusive right aforesaid the sum of Two Thousand Eight Hundred Pounds per annum to be paid quarterly in arrears.
2. The Licensee agrees to pay any rates levied by the Local Rating Authorities on the advertising site and all outgoings and expenses whatsoever caused by or in connection with the use of the property for advertising purposes.
3. This agreement shall remain in force for twelve years from the date the panels are erected and thereafter from year to year until determined by either party giving to the other six months' previous notice in writing of desire in that behalf.
4. At the request of the Licensee the Grantor will grant to the Licensee the G option of a renewal of this agreement for a further period of nine years from the expiration of the original period.
5. The Grantor shall allow the Licensee uninterrupted access to the advertisement site.
6. In the event of war or state of emergency or if the advertisement site becomes obscured or for any other reason or should any competent authority prevent the continued display of advertisement(s) then the Licensee shall be allowed to terminate this Agreement by giving to the Grantor 28 days' notice in writing.
7. The Licensee confirms that he has effective Insurance for any third party risks arising out of the erection maintenance or removal of the said advertisement site.
8. In the Agreement the expressions "the Grantor" and "the Licensee" shall where the context admits include their respective successors in title. H

I hereby acknowledge copy of this Agreement.
(Sgd) J Woolls/V Reynolds
Witness M Forrest
Address 12 Spring Vale, Dartford, Kent
Date 28/3/80

CAMERON v NATURE CONSERVANCY COUNCIL

LANDS TRIBUNAL FOR SCOTLAND J
(Ref/LTS/COMP/1990/9)
June 5 1991

(Lord ELLIOTT and A R MACLEARY FRICS with R MACDONALD BSc Agr MSc, a member of the Scottish Land Court, sitting as agricultural assessor)

Estates Gazette February 22 1992, February 29 1992 and March 7 1992
[1992] 07 EG 128; [1992] 08 EG 120 and [1992] 09 EG 147

Diminution in value of Perthshire agricultural and forestry estate consequent on notification of SSSIs — Reference under s50(3) of Wildlife and Countryside Act 1981 — First such K reference — Offer of compensation in draft management agreement — Basis of valuation — Difference between restricted and unrestricted values — Parties' rival schemes — Consideration of various practical valuation matters, including basis of capital value, treatment of risk, value of forestry land and agricultural improvements costs

Two SSSIs, together forming 38% of estate, notified by NCC in 1984-85 — Claimant purchased estate in August 1986 for £800,000, which included £571,020 for land — SSSIs renotified — Entry taken and application for forestry grant made in October 1986 — NCC advised in November 1986 of L claimant's intention to proceed with proposed integrated forestry and land improvement scheme — Forestry grant application refused following NCC objections — In January 1990 NCC offered claimant a management agreement with proposal for him to manage SSSIs in accordance with stipulated management policy and subject to certain prohibitions and restrictive conditions — Claimant refused offer and opted under s50(3) of 1981 Act for determination by tribunal, as arbiter, of amount of payment due to him, such payment to be by way of lump sum — Agreed valuation date October 21 1987 — Claim for £872,700 — DV's figure £213,250 M

Two schemes considered in detail: (1) the claimant's scheme (the "Cameron" scheme) formulated on the basis of no SSSIs and (2) a scheme put forward for the NCC (the "Elliot/Scruton" scheme) with SSSI restrictions in place — Tribunal noted that payments under 1981 Act are to be determined "in accordance with guidance given by the Ministers" (s50(2)), such guidance being contained in an HM Stationery Office publication, *Financial Guidelines for Management Agreements* — Tribunal further noted that claimant's purchase price showed some "peculiarities" which did not appear to reflect a true open market situation: (1) that

A **the seller had wanted to sell lock, stock and barrel without a count of livestock and with the purchaser taking over existing labourforce and liability for any redundancies, so that less than full open market value appeared to have been paid: (2) that the existence of the SSSIs was known to the purchaser**

***Decision*: 1. In the absence of the SSSIs a forestry grant would probably have been available for claimant's entire scheme**

2. The prohibitions in the NCCs proposed management agreement made it impossible for the claimant to carry out his own scheme as originally conceived and there was no legal
B **duty on him to put forward an alternative scheme avoiding the SSSIs**

3. The claimant, when originally buying the estate, would have modified his offer to a figure somewhere between its unrestricted and restricted values and thus his actual purchase price could not be regarded as providing any reliable guide to either value

4. Two separate values of the claimant's interest were required: (i) open market value if SSSIs had not been notified and (ii) open market value resulting from notification of SSSIs and restrictions flowing therefrom, each valuation being made as if on a compulsory purchase under s12 of the Land
C **Compensation (Scotland) Act 1963 subject to any necessary modifications**

5. There was no difficulty in accepting the Cameron scheme as a whole as a type of possible development plan, although not every particular detail might be acceptable — A hypothetical purchaser considering making a bid on an unrestricted basis with professional advice would envisage a programme along its lines

6. The Elliot/Scruton scheme could not be accepted in its entirety for reasons concerning viability of certain forestry blocks and implications for sheep management — The
D **resultant amended scheme, however, retained integrity — Faced with SSI restrictions a hypothetical purchaser would contemplate a type of limited development of this kind and would wish to test it financially after taking professional advice**

7. Accordingly, the basis of the unrestricted value was the Cameron scheme, determined at £1,787,532, and that of the restricted value was the Elliot/Scruton scheme as amended, determined at £1,232,513

8. The lump sum payment due to the claimant was £550,000

The following cases are referred to in this report.

E *Inland Revenue Commissioners* v *Clay* [1914] 3 KB 466; (1914) 83 LJKB 1425; 111 LT 484; 30 TLR 573

McEwing & Sons Ltd v *Renfrewshire County Council* (1959) 11 P&CR 306; 1960 SC 53; 1960 SLT 140

Raja Vyricherla Narayana Gajapatiraju v *Revenue Divisional Officer, Vizagapatam* [1939] AC 302

Brian Gill QC and Patrick S Hodge (instructed by J R Stevenson & Marshall, of Dunfermline) appeared for the claimant: A M Philip QC and H W Currie (instructed by Archibald Campbell & Harley WS, of Edinburgh) appeared for the respondents.

Giving their decision, THE TRIBUNAL said: This is a claim by Mr John Bell Cameron, Balbuthie, Fife, for payment of a sum of £872,700 as representing the diminution in value of his owner's
F interest in Glenlochay Estate, Perthshire, in consequence of the notification by the Nature Conservancy Council ("NCC") of two sites of special scientific interest ("SSSI"). Section 50(3) of the Wildlife and Countryside Act 1981 ("the Act") provided for disputed questions of payment due, following a management agreement, to be referred to an arbiter for determination. But parties have agreed that the present dispute should be referred to the Lands Tribunal, as arbiter, under section 1(5) of the Lands Tribunal Act 1949. They have also requested that the tribunal sit with a member of the Scottish Land Court as agricultural assessor. We have therefore done so with Mr R Macdonald BSc Agr MSc.

As this is the first such case to be so referred, it is appropriate to mention not only the basic facts as proved in evidence but also the procedure governing SSSIs under the 1981 Act, including the G
financial guidelines governing payments to compensate for the effect of the restrictions on use imposed.

The case was heard in Edinburgh on January 7 1991 and following days, when the claimant was represented by Mr Brian Gill QC and Mr Patrick S Hodge, advocate, who led in evidence the claimant himself; his shepherd, Mr George Bickerton; Mr Leonard Yuill BSc, general manager for Scotland of the Economic Forestry Group plc; Mr Archibald Macleod OBE NDA, former adviser with the West of Scotland College of Agriculture and a member of the Red Deer Commission; Mr John Macnaughton, who farms at Inverlochlairig and is a member of the Wool Marketing Board and member of the Forestry Commission Regional Advisory Committee; Mr James H
Sharp, sheep farmer; Professor John Cunningham, formerly Professor of Agriculture at Glasgow University and former director of the Hill Farming Research Organisation; Dr Timothy Fletcher, deer farmer; Mr John Jones-Perrott FRICS, of West Highland Estates; and, finally, Mr Michael Read-Thomas FRICS, of Davidson & Robertson, chartered surveyors.

The NCC were represented by Mr A M Philip QC and Mr H W Currie, advocate, who led in evidence on their behalf Mr Walter Elliot OBE, sheep farmer and agricultural consultant who is also a member of the Wool Marketing Board; Mr Richard Scruton, independent forestry consultant; and Mr Archibald Maclaren FRICS, district valuer, Dumbarton. The tribunal also inspected the relevant parts of Glenlochay Estate both under snow conditions and, later, J
clear of snow.

The factual background to the case as established in evidence and the statutory procedures followed were as follows:

Mr Cameron, the claimant, purchased Glenlochay Estate, extending to about 8,200 ha (20,276 acres), in August 1986 at a total price of £800,000 including livestock and deadstock. The price paid for the land was £571,020. Entry thereto was taken shortly thereafter.

As at the date of purchase, two SSSIs had already been notified by the NCC to the previous owner in terms of section 28(1) of the Act specifying: "(a) the flora, fauna or geological and physio-graphical features by which the land is of special interest; and (b) any K
operations appearing to the Council to be likely to damage that flora or fauna or those features".

The first SSSI, at Meall na Samhna, is situated on the south-east corner of the estate and was notified in 1984. The second was at Ben Heasgarnich and was notified in 1985. These two sites together extend to about 3,162 ha and cover approximately 38% of the whole estate. They were said to be designated by the NCC not only because of the presence there of rare or restricted plants and plant communities but also because the selection of these upland areas was regarded by the NCC as representing "the maximum ecological diversity characteristic of the district running from the valley floor up to high altitudes".

These SSSIs were later renotified by the NCC to the claimant L
following his subsequent purchase of the estate. He thus became prohibited from carrying out thereon any operations specified in the notification unless he gave written notice of a proposal to carry out any operation specifying its nature and the land on which it was proposed to carry it out and had subsequently obtained the NCC's written consent thereto — or the operation was to be carried out in accordance with a management agreement: see sections 28(5) and (6).

The claimant, who is a prominent farmer and former president of the Scottish National Farmers Union, was aware of the two SSSI notifications when he purchased the estate. These were mentioned on p3 of the particulars of sale. In para 8 of his formal offer to purchase dated August 26 1986 it was stated: "It is understood that there are M
two sites of special scientific interest on the estate but that no management agreement has been entered into with the Nature Conservancy Authority."

As Mr Cameron explained in evidence, he was then considering the purchase of Glenlochay Estate with a view to an integrated farm and forestry development for which he considered the glen to be well suited and, indeed, which had been neglected in this regard. The particulars of sale also mentioned "forestry potential" on p1. His proposed scheme, as spoken to by himself and supporting agricultural witnesses, involved the planting of several new or enlarged forestry shelter blocks and the construction of related hill parks. This was both to improve the sheepstock and also the deer cull

A on what is not only a well-known hill sheep farm but also a Highland sporting estate.

Mr Cameron took entry on October 1 1986 and applied, as from that date, for grant under the Forestry Grant Scheme for planting seven forestry blocks. Of these blocks nos 1, 5, 6 and a small part of block 3 lay within one or other of the two SSSIs. The proposed planting was to be carried out through the Economic Forestry Group plc ("EFG"), who indicated that they could readily find an eager purchaser for the proposed forestry blocks owing to the considerable tax advantages which still prevailed.

By letter dated November 11 1986 the claimant then gave the NCC notice, in terms of section 28(5)(*a*) of the 1981 Act, that he intended to go ahead with his proposed integrated forestry and land
B improvement scheme. They thereupon wrote to the Forestry Commission by letter dated December 18 1986 intimating objections to the claimant's blocks 1, 5, 6 and part of block 3 (ie those within the SSSIs) and also inviting refusal of blocks 2, 4 and 7 being accepted under the Forestry Grant Scheme. As a result, the claimant's application for forestry grant was refused by letter sent by the Forestry Commission to EFG, then acting on the claimant's behalf, stating as follows:

> In reaching their decision, the Commission took into account your client's view that the removal of the proposed planting areas falling within the SSSIs would render the scheme unviable. While they were of the opinion that the landscaping objections made to your client's proposals were not
> C insurmountable, the Commission has considered, however, that the proposals were incompatible with the objectives of the SSSI designations and that, as a consequence, the application as a whole could not be admitted.

By subsequent letter to the claimant's agent dated October 30 1987 EFG confirmed that their estates company would have been prepared to purchase the proposed woodland part of Glenlochay Estate for a sum of £430,000. This letter also stated that the purchaser was to have sole responsibility for the construction and maintenance of all deerproof perimeter fences around the various new blocks.

In the absence of the SSSIs, the tribunal consider that forestry grant would probably have been available to the claimant for his entire scheme. The parties are also agreed that Mr Cameron's integrated forestry and agricultural scheme was a well-thought-out
D and viable one.

By letter dated January 19 1990, with draft agreement annexed, NCC in terms of section 50(1)(*b*) then served an offer through their solicitors to enter into a management agreement with the claimant. The said draft agreement proposed that the claimant would manage the two areas of land contained in the SSSIs in accordance with a proposed management policy as stipulated therein. There was also appended a list of "operations likely to damage the features of special interest", which included, *inter alia,* "cultivation and reseeding; grazing; stock feeding; application of fertilisers, lime or other substances; the release into the site of any wild, feral or domestic animal; afforestation; drainage; the construction of fences
E and the introduction of game and water fowl management and hunting." It was also stated that in consequence of these restrictions the NCC would make a payment to the claimant of a lump sum of £210,000 and also reimburse him for certain expenses incurred in erecting, replacing or maintaining certain fencing.

The tribunal agree with the claimant that the said prohibitions made it impossible for him to carry out his scheme as originally conceived: for this involved not only the fencing off of the new forestry blocks but also the creation and control of several related hill parks (partly using the forestry fencing) which were crucial to increase sheep numbers and improve both the sheep stock and the deer.

Certain restrictive conditions contained in management
F agreements can presumably be changed by the NCC by a new management agreement if parties reach mutual agreement as, for instance, on the control of grazings or the like. If the NCC, however, had any scope for modifying their present position they did not seek to do so. An SSSI boundary designation itself would seem to require formal redesignation and notification if needing to be altered. There was certainly no legal duty on the claimant to put forward an alternative integrated farming and forestry scheme avoiding the SSSIs.

The question of "minimising loss" is, however, mentioned in para 11 of the Financial Guidelines. But this is only in the special circumstance of an owner or occupier increasing the sum eventually payable to him by entering into any contract or commitment with third parties after the date of notification. The claimant does not G appear to have done so in the present case; nor was it contended on behalf of NCC that he had.

This tribunal, however, are only strictly concerned with the valuation of the resultant compensation due to the claimant. Section 50(3) provides:

> If the offeree so requires within one month of receiving the offer, the determination of those amounts shall be referred to an arbitrator (or, in Scotland, an arbiter) to be appointed, in default of agreement, by the Secretary of State.

Mr Cameron, as offeree, has now so required and, by letter dated January 26 1990, has refused to accept the offer contained in the management agreement. Instead he has opted under section 50(3) of H the 1981 Act for the determination by an arbiter (now the tribunal) of the amount of payment consequently due to him. He has also opted that such payment should be by way of a lump as opposed to an annual sum. Under the same subsection, where the amount determined by the arbiter exceeds that determined by the offeror the latter:

> shall (a) amend the offer so as to give effect to (the arbiter's) determination; or (b) except in the case of an offer made to a person whose application for a farm capital grant has been refused in consequence of an objection by the offeror, withdraw the offer.

The latter would normally appear to have the effect of allowing the owner's proposals to proceed. But, in the present case, not only has J Forestry Grant (not mentioned in the statute) been refused owing to the NCC but the tax advantages of forestry planting have also been removed. Section 50(2) of the 1981 Act provides that payments "shall be of such amounts as may be determined by the offeror in accordance with guidance given by the Ministers". This ministerial guidance is contained in *Financial Guidelines for Management Agreements* issued by Her Majesty's Stationery Office, which, however, is not a statutory instrument. Para 1 of the *Financial Guidelines* provides, under reference to section 50 of the 1981 Act:

> where management agreements are offered, in specified circumstances, by the Nature Conservancy Council . . . for the purposes of conservation or amenity, payments under such agreements shall be determined by the body concerned (the offeror) in accordance with guidance given by Ministers. K

Para 2 states:

> these provisions have statutory application in the following circumstances: — Sites of Special Scientific Interest (SSSI) — if the Nature Conservancy Council (NCC) has been notified by an owner or occupier of a proposal to undertake an operation specified as likely to damage an SSSI and, in consequence offers to enter into a management agreement with him.

Para 14 deals with lump sum payments to owners or owner-occupiers in the following terms:

> Payment, at the commencement of the agreement, of a single lump sum for a Management Agreement over a 20 year period (or such other period possibly operating in perpetuity, as may be agreed between the parties). *The amount should be equal to the difference between the restricted and unrestricted value of the owner or owner-occupier's interest,* calculated having regard to the L rules for assessment in respect of the compulsory acquisition of an interest in land, as set out in Section 12 of the Land Compensation (Scotland) Act 1963 so far as applicable and subject to any necessary modifications. . . (tribunal's emphasis).

Section 38 of the *Financial Guidelines* regulates the effective date of the management agreement. Para 39 states that where, as here, a lump sum payment has been chosen, the date of valuation shall be the same as the effective date. This is now agreed by the parties to be October 21 1987.

In the result, the tribunal have now to determine two separate values for the claimant's interest in Glenlochay Estate as at October 21 1987, being (1) its open market value if the SSSIs have not been M notified and (2) its open market value resulting from the notification of the SSSIs and the restrictions that flowed therefrom. Each separate valuation (whether restricted or unrestricted) requires to be made as if on a compulsory purchase in terms of section 12 of the Land Compensation (Scotland) Act 1963, subject, however, to any necessary modifications. The only applicable provision here appears to be that contained in section 12(2) ("rule (2)"), which provides:

> The value of land shall, subject as hereinafter provided, be taken to be the amount which the land if sold in the open market by a willing seller might be expected to realise.

There is no scope for the application of section 12(6) or any other subsection.

The case law interpreting section 12(2) therefore applies. It posits a willing seller and, by implication, also a willing purchaser. In the compulsory purchase case known as "the *Raja* case" [1939] AC 302*, Romer LJ said at p 312:

> The compensation must be determined, therefore, by reference to the price which a willing vendor might reasonably expect to obtain from a willing purchaser. The disinclination of the vendor to part with his land and the urgent necessity of the purchaser to buy must alike be disregarded. Neither must be considered as acting under compulsion . . . but this does not mean that the fact that some particular purchaser might desire the land more than others is to be disregarded. The wish of a particular purchaser, though not his compulsion, may always be taken into consideration for what it is worth.

In *Inland Revenue Commissioners* v *Clay* [1914] 3 KB 466, which, although a revenue case, also involved the open market criterion, Swinfen-Eady LJ said (at p 475):

> The market is to be an open market as distinguished from an offer to a limited class only such as the members of the family, the market is not an auction sale. The section means such amount as the land might be expected to realise if offered under conditions enabling every person desirous of purchasing to come in and offer and if proper steps were taken to advertise the property and let all likely purchasers know that the land is in the market for sale.

In such an open market situation, purchasers will also be likely to raise their offers for the land if there are good prospects for development and the land has not yet been improved to its full potential: see *McEwing & Sons Ltd* v *Renfrewshire County Council* 1960 SC 53 *per* Lord President Clyde at p 60.

Before assessing the merits of the two rival schemes, it is first, however, necessary to consider if any assistance can be derived by updating Mr Cameron's own purchase price paid for the estate about a year before the valuation date. This was for a total sum of £800,000 of which the land element was said to be £571,020. However, there are peculiarities to this price which do not appear to reflect a true open market situation. First, the seller wanted to sell the estate to Mr Cameron lock, stock and barrel without a count of the livestock and with his taking over the existing labour force with liability for any redundancies. The price was paid by cheque on the spot prior to the advertised closing date for offers. The purchaser, therefore, appears to have paid less than full open market value.

But there is a more fundamental objection to use of this purchase price for indicating the unrestricted value of the land. This is because the previous owner had been notified of the SSSIs, which were also mentioned in the particulars of sale and were therefore known to the purchaser. Mr Cameron himself was not closely examined on this point; but we must assume that he was in general aware of the appropriate location and extent of the two SSSIs — although he would not know the extent of the restrictions until he later made his own development scheme known to NCC, who then responded with their offer of a management agreement.

The tribunal consider that when originally buying the estate in August 1986, knowing of the two SSSIs, Mr Cameron would have modified his offer to a figure somewhere between its unrestricted and restricted values. In these circumstances we cannot regard his actual purchase price as providing any reliable guide as to either full unrestricted or restricted value. Nor, in the end of the day, did we understand either party to place much reliance upon the price paid for the estate in August 1986.

In addition to the claimant's own detailed scheme, which was supported in evidence by several witnesses, we have heard evidence of an alternative scheme put forward on behalf of the NCC and which was referred to throughout as the "Elliot/Scruton" scheme after the names of its co-authors. But the tribunal do not consider that they are bound to decide in favour of either the "Cameron" or the "Elliot/Scruton" scheme. While both are certainly relevant to our two valuations, our basic task is to determine, as the amount now payable to the claimant, the difference between the two separate open market values of the land as at October 21 1987. Hence we are not strictly obliged to follow one detailed scheme or the other in its entirety. The real question is what a hypothetical purchaser would have been likely to offer for Glenlochay Estate on the open market in October 1987 without or, alternatively, with the relevant restrictions.

The tribunal therefore proceed to examine (1) the Cameron scheme formulated on the basis of the absence of the SSSIs and (2) the Elliot/Scruton scheme with the SSSIs in place to determine how these plans, in whole or in part, might indicate what a hypothetical purchaser would be prepared to pay in offering for the estate on the open market in October 1987.

We are proceeding on the basis that such a hypothetical purchaser would take professional advice sufficient to reveal the latent potential for development suggested in each of the schemes and the parties are content that this is a reasonable assumption to make in these circumstances.

The Cameron scheme

We deal first with the Cameron scheme, described in the evidence as an integrated land-use plan comprising agriculture, forestry and sport. The agricultural development was central to the whole programme involving the creation of parks associated with seven forestry blocks providing shelter and improved grazing, increasing the sheep numbers from 4,500 to 5,250 over five years, and lambing percentages from 75% to 90%. Cattle numbers would not increase but better-quality calves would be produced. The annual cull of stags and hinds would increase.

The forestry proposals, extending to 710 ha, comprised six new blocks of forestry and an existing plantation extending to 46 ha but enlarged on to bare land on both sides. These blocks were to be sold to EFG for the sum of £430,000.

Summarised, the evidence for the claimant was that to achieve the target increase in sheep numbers and in lambing percentage proposed on an extensive hill farm of this nature, management practices required the control of nutrition of the stock, particularly at pre-mating to increase ovulation; in mid-pregnancy and pre-lambing to ensure foetal growth and adequate lactation; and improved pasture to encourage growth rates before and after weaning. It was thus necessary to create parks and improve the quality of pastures within these parks by applying lime and inorganic fertilisers with some reseeding in a mosaic pattern. A better utilisation and control of hill pasture on the lower ground was also proposed. In short, capital was to be concentrated on improving certain areas to make maximum impact on sheep and cattle performance in order to achieve the targets set out. The parks would also enable animals requiring attention such as lean ewes and ewes with twins to be given special treatment.

The hoggs, which had previously run with the breeding ewes on their hirsels, were now housed in a building at the main steading at Kenknock. Thus one important aspect of the development plan was already in place at the valuation date.

The tribunal are of the opinion that this plan was well devised and the targets set were achievable. Apart from the scale of the operation, the management programme itself was unexceptional and is practised on many hill farms throughout Scotland along somewhat similar lines.

Glenlochay Estate has certain physical advantages to encourage a development programme of this nature. It is a glaciated valley with a high percentage of brown forest soils on the valley floor and on the slopes. Acid bent/Festuca grasslands predominate and morainic hummocks provide good natural shelter. These grasslands are improvable by surface treatment. The restriction of forestry planting to the 500m contour would create obvious competition between sheep grazing and forestry; but the plan was devised to minimise the loss of grazing through increased productivity in the parks between the forestry blocks. Substantial capital investment in land improvement, fencing and drainage was proposed, offset by the sale of the forestry blocks to commercial interests.

The claimant's plan envisaged a five-year programme of build-up of stock. This was to be achieved by retaining an additional 200 ewe lambs to those required for stock replacement over a four-year period. By the fifth year the required number would be available with surplus ewe lambs from the additional ewes in the flock to allow for a better selection. The land improvement works would also be undertaken within this period, but the plan was also to rely on shelter being available from the forestry blocks during the five-year build-up programme. Evidence was led by the claimant that during the early period the forestry fences and the herbage inside the blocks would provide sufficient shelter.

The tribunal are not persuaded on this last point and prefer the evidence of Mr Elliot in this regard. We consider that nutrition is the key to the targets sought, but the availability of shelter is also an important factor. Thus, while proposed stock numbers and land improvement are technically achievable during a five-year period,

*Editor's note: *Raja Vyricherla Narayana Gajapatiraju* v *Revenue Divisional Officer, Vizagapatam*.

A the tribunal consider that a longer build-up programme over at least 10 years would be necessary for the benefits from the shelter component of the plan to be attained. These would begin to build up only from around year five. The longer time-scale would have the added advantage of providing a better selection of ewe hoggs. Additional feed would also be required, particularly in the early years, to achieve higher fecundity and to maintain the increasing numbers on a lesser acreage. The tribunal are aware from evidence and from our inspection that the sheepstock on Glenlochay already has access *ad lib* to treacle from special feeders. The additional costs of concentrates, and hay in severe weather, are a matter of some conjecture, but the tribunal are of the opinion that the total extra cost of livestock feed over 10 years could be contained within a global
B figure of £21,000.

The management of the spring calving Luing herd would continue as before, but better-quality calves would be produced. They are managed in two herds of 50 cows at each end of the glen, grazing in summer on Meall na Samhna and Ben Heasgarnich and fed in winter on cobs, and silage made on the farm. Winter feeding areas are conveniently located for transporting silage wagons to the sites along existing roads. Normally, towards the end of the year they are brought down to their winter quarters.

It was also anticipated by the claimant that, owing to the availability of better-quality herbage in the parks from which deer would not be excluded, the effect would be an increase in the number
C of stags for stalking and hinds, the latter mostly for sale to deer farms for breeding.

The tribunal are thus of the opinion that the Cameron scheme carried no undue risk apart from the normal hazards associated with hill farming and that it could be implemented over the amended time-scale.

It is, then, necessary for the tribunal to calculate what a hypothetical purchaser would have been likely to offer as a capital sum for Glenlochay Estate on the open market in October 1987 contemplating a scheme of the Cameron type. Such a purchaser would be aware that the pre-existing sheepstock was well known for quality and breeding, and stocking density would confirm productive grazings. He would be aware that it had a well-known
D deer forest. He would also be made aware that a major part of the estate was already roaded by the North of Scotland Hydro-Electric Board, for which he would not be responsible for maintenance but over which he would have access for estate management. The estate is conveniently located within easy distance of livestock markets in Stirling and Perth.

Parties were agreed that in valuing the worth of the estate in 1987 the preferred approach for sheep and cattle was to assess a rate per ewe and per cow using a multiplier of 25 years' purchase. Evidence of comparable rates, derived from either straight farm rentals or capital values, were put to the tribunal by both parties.

For the claimant, his expert witness, Mr Jones-Perrott, quoted
E Corrylorn Farm, Kilniver, Oban, where discussions had taken place with the tenant on the basis of a rental of £3.33 per ewe at Martinmas 1984. These discussions were aborted when the farm tenant purchased the farm. Craig Farm, Dalmally, was reviewed in 1983 by arbitration and a rent of £3.00 per ewe and £20.00 per cow fixed. Callachly Farm, Mull, was agreed between landlord and tenant in 1987 at a rental of £2.96 per ewe. Torran Farm, Ford, Argyll, was let by open tender by the landlords, the Forestry Commission, from March 1984 at £9.23 per ewe. Two probate values were also proffered, namely Barguillean Farm, Taynuilt, where rates were agreed with the district valuer at £7.00 per ewe and £30.00 per cow, and Glenstrae, Dalmally, where a rent of £4.50 per ewe was agreed.

Mr Reid-Thomas carried out his calculations on a capital as
F opposed to a revenue basis. He has adopted a figure of £175 per ewe based on an analysis of sales of comparable estates. This equates with a rate of £7 per ewe.

Mr Elliot, for the NCC, relying on his experience, expressed the view that a fair rent for the sheep element of Glenlochay at the date of valuation would have been in the region of £3.80 per ewe. He found it difficult to accept that rents in the range of £7 to £9 as spoken to by Mr Reid-Thomas could eventually be achieved at Glenlochay. Referring to his own farm, which had also been improved, where shelter was established and which was in a "kinder" part of the country, he had agreed a rental of £6.20 per ewe in May 1986. His calculations of rental figures over time (in Production DV 7) produced a maximum of £6.34 per ewe for Glenlochay following 12 years' maturation of the plan. The tribunal consider these figures are based on the view G
that a tenant farmer would take when considering offering a rent for a property.

The district valuer based his capital values per ewe in accord with the analysis of sales of comparable estates as specified in Production DV 6. As a check, his capital values of £100 per ewe at the valuation date; £160 per ewe post the Cameron scheme and £135 per ewe post the Elliot/Scruton scheme, were used for both valuations and produced rates per ewe of £4.00, £6.40 and £5.40 respectively. He stated that these rates, derived from capital values, bore reasonable comparison with Mr Elliot's estimate of rental for the same stages of £4.25, £6.34 and £5.35 per ewe respectively (Production DV 7). The one farm rent the district valuer quoted was of £3.00 per ewe fixed by the Land Court for the farm of Inverlair, Roy Bridge, on a joint H
application by the landlord and tenant in terms of section 7 of the Agricultural Holdings (Scotland) Act 1949 as amended by section 2 of the Agricultural Holdings (Amendment) (Scotland) Act 1983.

The tribunal found that these comparables provided useful guidelines but with the comment that in relation to farm rentals fixed or agreed under the Agricultural Holdings Acts different criteria, such as goodwill, may obtain to those involved in deducing a rate per ewe or per cow from estates sales prices.

We have given consideration to both the sources and the nature of the evidence and are of the opinion that a hypothetical purchaser would base his calculations on an eventual capital value of £175 per ewe. This represents a rate of £7.00 per ewe capitalised at 25 years' J
purchase. An eventual capital value of £1,000 per cow based on a rate of £40.00 also capitalised at 25 years' purchase is appropriate.

A hypothetical purchaser would also have regard to the sporting aspect of the estate. The claimant considered that eight extra stags could be shot following the improvements proposed and that the shelter would improve the fecundity of the hinds and thus increase the overall number of deer on the estate. This is a plausible contention, supported to some extent by the Deer Commission, although there was a curious ambivalence in the advice given by the commission, at one point stating that withdrawal of some 1,200 hoggs off the grazings and maintained inside would reduce the grazing pressure by the theoretical equivalent of 500 deer. If this was indeed the situation, it tends to cast doubt on the other conclusions K
reached, that it would be impracticable to expect a sustained increase in deer numbers as well as a significant addition to the sheep stock. Dr Fletcher, expert witness for the claimant, strongly supported the claimant's view, in particular that he would have no difficulty in trapping 100 hinds each year and that 60-100 hinds was a reasonably sustainable crop. He was less certain about the increase in stag numbers, as he quite reasonably stated that his expertise was in deer farming and the physiology of deer rather than wild deer management.

The tribunal are of the opinion that deer, and hinds in particular, would benefit from the shelter and better-quality grazing and more calves would be born. The tribunal are less convinced that this would necessarily result in a higher number of stags shot in the stalking L
season. During the stalking season in late summer and autumn the stags are out on the high ground and there is no guarantee that additional stags produced as a result of the better pastures and higher fecundity of the hinds would be within the Glenlochay marches during the stalking season. The tribunal do concede, however, that the possibility exists and that an increase could be justified of four stags without detriment to the deer management area of which Glenlochay forms part. The tribunal are of the opinion, however, that the situation is different in relation to hinds which winter on Glenlochay. We consider that a hypothetical purchaser at the valuation date would estimate that, with trapping facilities in place, 35 hinds could be trapped. M

In relation to the capital value of stags, a number of comparable sales of deer forests were referred to in evidence, ranging from £4,000 per stag to £10,250. The capital value of the stags on Glenlochay was stated by the district valuer as £2,500, derived from an analysis of the purchase price of the estate. It would seem extraordinary that Glenlochay Deer Forest, with its reputation and accessibility in west central Perthshire, should show such a discrepancy viewed against the comparables. The tribunal are of the opinion that, from the evidence of the comparables, £7,000 fairly represents the capital value per stag at the effective date which a hypothetical purchaser would be prepared to offer.

Having considered the Cameron scheme, the tribunal accept that it

A is a viable proposal. The merits of the proposed development have been recognised by the NCC's advisers. In the conclusion to their report, Messrs Elliot and Scruton say: "We recognised at the outset that the scheme put forward by Mr Cameron would have been an ideal development for this glen and acknowledge its superiority" (Production DV 5 p 8). We have no difficulty, therefore, in accepting the scheme as a whole as a type of possible development plan, although we may not agree with every particular detail. However, we can have some confidence that a hypothetical purchaser considering making a bid for Glenlochay Estate on an unrestricted basis, and taking professional advice, would envisage a programme of development along these lines.

B *Elliot/Scruton scheme*

We then turn to the question of what a hypothetical purchaser would be likely to offer for Glenlochay Estate with the restrictions imposed by the SSSIs.

The tribunal have given detailed consideration to the productions and the evidence adduced by witnesses for the NCC and have inspected their alternative proposals in detail on the ground. Briefly the Elliot/Scruton scheme envisaged five forestry blocks including the existing plantation enlarged by some 15 ha. Some 70 ha of land were to be improved by surface treatment on the low ground although the areas were not specified, and no parks were to be created. Cattle numbers were to remain the same. Sheep numbers C were to increase from 4,500 to 4,900 over a period of 12 years, the increase to be outwith the SSSIs, with lambing percentages there increasing to 90% but remaining constant within the SSSI areas. This would be in accord with the proposed management agreement, which also required the cattle to be removed from the two SSSIs from the end of October until the beginning of May. It was thus necessary to find an alternative winter feeding area for the cattle presently fed within the Meall na Samhna SSSI. The evidence was that an alternative feeding area could be developed at Innischoarach.

The tribunal are of the opinion that the Elliot/Scruton scheme is one which a hypothetical purchaser of the estate at the valuation date, if required to comply with the management agreement, could have considered. But a detailed inspection of the estate has now D satisfied the tribunal that the plan could not be implemented in its entirety. The main criticism of the plan centres around the size and location of the forestry blocks 4, 5A and 7, and the assumption that sheep numbers and lambing percentages could be increased as proposed without the need for grazing control through the provision of holding parks.

We deal first with the forestry blocks. Blocks 2 and 3 in the plan are acceptable, and indeed are similar to the claimant's plan, the latter being a standing plantation of timber with an additional 15 ha plantable on the south side. Block 5A extends to 27 ha, which the NCC considered would afford a degree of shelter to the adjacent hirsel to the west. There was strong opposition to this proposal by the E claimant on the grounds that the land comprised excellent sheep-grazing and that it would encourage sheep from hirsel C to draw towards a dangerous ravine. Neither was it regarded by his forestry expert as a commercial proposition.

Following inspection, the tribunal conclude that there is substance in these objections both on sheep husbandry and commercial grounds and because the unit costs of establishing such a small block would be unacceptably high. We thus exclude it from the scheme.

In relation to block 4, the claimant objected on the grounds that the fence lying on the west boundary created a trap for sheep moving between the higher ground and the low ground. He also alleged that the block as proposed split the existing hirsel in two, sterilising the land on the south end of the hirsel and reducing the stock to 300 ewes. F Our inspection confirmed this to be the case and while it might have been possible marginally to adjust fence lines, there did not seem to be any possibility of doing so while retaining the existing crossing point for ewes across the dangerous burn running through the block. The tribunal are thus of the opinion that the extended block is incompatible with the maintenance of existing sheep numbers on this hirsel. Some 21 ha are also unplantable. The original block 4 in the Cameron scheme could, however, remain in an amended scheme.

Block 7 is an enlargement to 152 ha of the original block 7 in the Cameron plan extending north-eastwards in the valley bottom to include the area of "Old Sheilings", and further up the sides of the valley except at the south-west end. The fencing on the south-east side of the block would enclose the existing sheep runs along this side of the valley and would also fence off the natural downfall at the mouth G of the glacial "Hanging Valley", necessitating a further detour to the end of the fence and down a very steep gradient to the valley floor. The terrain above this proposed fence line slopes steeply, with very frequent water courses or runnels. Following our inspection we agree with the evidence of expert witnesses for the claimant that the movement of sheep, particularly with lambs, above such a fence would present unacceptable shepherding difficulties for the sheep hefted west of Beinn nan Imirean.

The sheep on hirsel F2 within the glacial "Hanging Valley" known as the "Doctor's Glen" would also have their natural downfall to the main valley blocked off by the proposed fence, necessitating a detour north-eastwards above the fence along broken ground interspersed H with water courses to the end of the fence at the 500m contour, before descending down the steeply sloping gradient to the valley floor.

On the north-west side of the proposed block, the evidence was that the fence followed the same contours as in the Cameron plan. This may have been the intention, but our inspection did not bear this out, and sheep hefted west of Stob a Bhiora on hirsel E would meet this fence and require to go uphill for a considerable distance to the end of the fence before descending to the valley floor. A substantial area of the sheltered green ground around the "Old Sheilings" would be incorporated in the block, thereby increasing grazing pressure on the remainder. The sheep fanks are also located here so that there are frequent gatherings of sheep from these hirsels to this point.

We find Mr Bickerton's evidence on the effects of planting this J block entirely reliable following our inspection and we are of the opinion that the proposed block 7 would effectively sterilise the land carrying hirsels E and F2, reducing numbers to uneconomic levels and with implications for the employment of shepherds.

The tribunal then considered whether the original block 7 could be reintroduced into the Elliot/Scruton scheme as we have seen fit to do with block 4. We have rejected this on the evidence of the forestry expert for the claimant, who held that such a block was only marginally viable in the context of the claimant's more extensive forestry plan. It would be financially unviable in an alternative scheme. The tribunal thus exclude this whole block from the scheme as unacceptable both on farm management and on financial grounds and hence not one which a hypothetical purchaser in 1987 would have K considered.

The net effect is that the area of land for afforestation in the Elliot/Scruton scheme is reduced to 269 ha, including 46 ha of solum of the existing plantation block 3. Nevertheless, the tribunal are of the opinion that at the valuation date these three blocks of land together would be regarded as sufficiently extensive and so located as to be commercially viable and accept Mr Yuill's evidence to this effect.

But the core of the development in this, as in the claimant's scheme, is the agricultural component. The effect of the exclusion of blocks 5A, 7 and part of 4 would be to restrict the increase in existing sheep numbers to 150, ie from 4,500 to 4,650. Because of the lack of L shelter on the hirsels at the west end of the estate, no increase in lambing percentage would be possible there. It should, however, be possible to improve areas west of block 4 and within block 2 and higher lambing percentages could be expected on hirsels A and F1, achieving 90% by year 10.

The cattle would remain at 100 cows as in the claimant's scheme. But because they must be off the SSSIs by October 31 until May 1, additional provision for feed would be required to compensate for the loss of some six weeks' grazing in a normal year. It is acknowledged that in some years heavy snow would require earlier feeding of silage. Provision would thus require to be made for setting aside an additional 4 to 5 ha of arable land for silage. It would also be necessary to find an alternative feeding area for the 50 cows on Meall M na Samhna grazing. The tribunal inspected the alternative of Innischoarach and found it unacceptable. It lies in the flood plain of the River Lochay and is subject to flooding. The higher ground immediately south of Innischoarach would be eminently suitable, but this area is included in the SSSI and is thus not available. In the absence of any other alternative site with a good access road, and ground that did not readily poach, the tribunal find it necessary to make provision in the valuation of the cattle for a shelter at an appropriate place.

It is then necessary to determine capital values for the agricultural element as before. Having considered again the comparable evidence spoken to by the expert witnesses, the tribunal are of the opinion that

A a hypothetical purchaser at the valuation date with the possibility of an amended type of Elliot/Scruton scheme, would base his calculations on an eventual capital value per ewe of £118.75, representing a rate of £4.75 per ewe. The tribunal are also of the opinion that, because of a loss of grazing, the need for additional feeding, and the requirement for an alternative feeding area, a capital value per cow for the Ben Heasgarnich herd of £850 based on a rate of £34 and a capital value of £650 per cow based on a rate of £26 for the 50 cows on Meall na Samhna would be appropriate.

Some additional sheep feed and land improvement would also be required on the hirsels where increased lambing percentages and numbers were still possible. This would be based on a 10-year development plan for improving pastures and providing shelter. The
B cost of the additional feed could be fairly stated in the global sum of £5,000.

In relation to deer, a similar capital value to that in the claimant's plan of £7,000 is appropriate. The improved areas and the limited shelter would still attract deer into the glen in winter, but not, in the tribunal's view, in the same numbers as in the Cameron scheme. There would thus be fewer hinds to trap, and calving rates would be less than in the claimant's scheme. In these circumstances, there is less likelihood of any increase in the numbers of stags shot in the stalking season, but, because of the uncertainty, the tribunal consider that an additional two stags could be justified. In relation to hinds, the tribunal are of the opinion that given the limited land
C improvement and shelter in this plan, as against the claimant's plan, an additional seven hinds could be trapped.

In conclusion, we find that we are unable to accept the Elliot/Scruton scheme in its entirety for the reasons that we have given. Our reservations are on the viability of forestry blocks 4, 5A and 7 and the implications for sheep management. The removal of these blocks reduces the extent of the forestry component of the scheme and has consequences for the potential of sheep performance at the west end of the estate. In our view, however, the amended scheme which results retains integrity. The reduced tree planting would still be viable financially and would bring additional shelter and financial benefits. While the optimum benefits which may be achieved for the whole enterprise may now be diminished, no real
D disadvantages accrue as a consequence. Faced with the restrictions of the SSSIs, we believe that a hypothetical purchaser of the estate would contemplate a type of limited development of this kind and would wish to test it financially having taken professional advice when considering making his open market bid.

Valuation method

We now turn to the question of how the unrestricted and restricted open market values of the Glenlochay Estate, with the potential of the Cameron and amended Elliot/Scruton schemes respectively, are to be valued. The valuation supporting the claim by Mr Cameron is attached as Appendix A. The valuations prepared in connection with
E the assessment of payment due on behalf of the NCC are attached as Appendix B. The tribunal's valuations and calculation of the payment due are attached as Appendix C.

The parties agree that potential can be taken into account in estimating the open market value of the estate in both its restricted and unrestricted conditions. The method by which this is to be done is in dispute. The district valuer believes that common and unchangeable items would be identified and excluded. The difference between unrestricted and restricted values is then analysed with reference only to the elements of potential value in each of the alternative schemes. He would then make an end deduction of 55% in respect of additional risk for unrealised potential. The claimant submits that this approach is fatally flawed because it obscures the
F actual open market value in each set of conditions and also because it attributes additional risk to normal agricultural operations where these risks have already been discounted in the valuation. Consequently there is double discounting in the district valuer's valuation.

There is also dispute as to detailed factors within the valuation such as the propriety of deducting interest lost on lamb sales forgone and the method of handling costs of agricultural improvements.

The submission of the claimant is that since the amount to be identified is the difference between the restricted and unrestricted values of the same asset, then the method should be to value the whole asset on an unrestricted basis and on a restricted basis and then deduct one value from the other. This would meet the requirements

APPENDIX A G

Valuation of claim by John B Cameron

Value of Glenlochay Estate with scheme completed

A	Land & timber			£430,000
B	1. 62 stags @ £8,000		£496,000	
	2. Additional 8 stags @ £8,000	£64,000		
	3. Additional 50 hinds available for live sale	50,000		
		114,000		
	4. Deferred 5 yrs @ 4%	0.822	93,700	589,700
C	5,500 ewes @ £175	£962,500		
C1	100 cattle @ £1,000	£100,000		
		1,062,500		H
D	*Less* deduction for deferment of agricultural improvements £522,500 @ 4% for 5 years	93,000		969,500
				1,989,200
E	*Less* cost of agricultural improvements		35,000	
F	Interest on extra capital		15,000	50,000
				1,939,200

Value of estate with NCC restrictions, but some improvements

G	Timber		£57,500	
H	62 stags @ £8,000		496,000	
I	4,750 ewes @ £100/ewe	£475,000		J
	50 cows @ £750/cow	37,500		
	50 cows @ £550/cow	27,500	540,000	
			1,093,500	
	Less cost of agricultural improvements	20,000		
	Interest on extra capital	7,000	27,000	1,066,500
	Amount of claim			£872,700

APPENDIX B

Valuation of payment due calculated by district valuer K

A.	*Afforestation*			
	Area details			
	Cameron scheme		710 ha	
	Less			
	1. Already planted	46 ha		
	2. Unplantable	21 ha	67 ha	643 ha
	Alternative scheme			
	Less			
	1. Already planted	46 ha		
	2. Unplantable	33 ha	79 ha	418 ha
A.	Difference in net plantable area between schemes			225 ha

Comparison of extra cost on roads L

1.	Cameron scheme (per EFG letter of intent).	
a.	Cost of providing new roads (as estimated by D Scruton). 1,250m @ £12/m	£15,000
b.	Cost of repairs to existing road — as set out in EFG letter. 10,300m @ £3/m	£30,900
	Total cost on roads	£45,900
	Which on plantable area of 643 ha represents £71.38/ha	
2.	Elliot/Scruton scheme	
	Same cost on roads	£45,900
	Which on plantable area of only 418 ha represents £109.8/ha	
Extra costs — Elliot/Scruton scheme		£109.8/ha
Extra costs — Cameron scheme		£71.38/ha M
Difference in costs on roads between both schemes		£38.42/ha

Accordingly, purchaser for lesser plantable area of 418 ha in Elliot/Scruton scheme would only be prepared to pay £576.2 *less* £38.42 = £537.78 per plantable ha.

Therefore, total figure received by J Cameron under EFG letter of intent

46 ha plantation — timber value only		£33,000	
46 ha as above solum value 643 ha plantable	ie 689 ha solum for planting or already planted @ £576.2/ha	397,000	
21 ha unplantable		Nil	
710 ha total price as per letter			£430,000

A *Less*			
Total figure receivable under Elliot/Scruton scheme			
46 ha plantation — timber value only		33,000	
46 ha as above solum value / 418 ha plantable { ie 464 ha solum for planted or already planted @ £537.78/ha		249,530	
33 ha unplantable		Nil	£282,530
Shortfall between schemes			£147,470
Offered already	£155,000		
B. *Agriculture*			
1. No scheme at all, ie as at date of valuation 4,500 ewes @ £100/ewe capital		£450,000	
B Cameron scheme — 5,250 ewes @ £160/ewe capital		840,000	
Producing a total expected quality/quantity difference		390,000	
Deferred 5 years @ 4% (average time for potential to be realised)		0.822	
		£320,600	
Less			
a. Present value of cost of providing additional feeding for cattle and sheep from years 1-12 till shelter from timber plantations fully available	£29,843		
C b. Prime cost of provision of fencing necessary to implement scheme plus element to reflect maintenance thereafter (net of grant)	21,000	£50,843	
		£269,757	
Proportion of potential reflected in hypothetical land price — 45%			£121,391
2. No scheme at all (ie as at present) 4,500 ewes @ £100/ewe capital		£450,000	
Alternative scheme		£661,500	
D 4,900 ewes @ £135/ewe capital		£211,500	
Deferred 5 years @ 4% (average time for potential to be realised)		0.822	
		£173,850	
Less			
a. Present value of cost of providing additional feeding for cattle and sheep from years 1-12 till shelter from timber plantations fully available.	£16,099		
E b. Prime cost of provision of fencing necessary to implement alternative scheme plus element to reflect maintenance thereafter (net of grant).	£6,300	£22,399	
		151,451	
Proportion of potential reflected in hypothetical land price — 45%		68,153	
Summary (agriculture)			
Potential under Cameron scheme	£121,391		
Potential under alternative scheme	£68,153		
Difference	£53,238		
Cattle enterprise — no difference	NIL		
	£53,238		
C. *Sporting*		Nil	
D. *General*			
F Loss in capital value of owner's interest at Glenlochay as a result of constraints imposed by management agreement but not specified or allowed for in potential as assessed above.			£5,000

Summary

A. Afforestation		£155,000	
B Agriculture		53,238	
C. Sporting		Nil	
D. General		5,000	
Total	£213,238		
Say	£213,250		

The above figure excludes surveyor's fees.

NOTE: The fencing referred to at B. Agriculture 2b. above excludes sections between SSSI and non-SSSI land now included as an extra in the management agreement.

of the *Financial Guidelines for Management Agreements*. This is the G approach adopted by Mr Reid-Thomas in his valuation.

The NCC submit that the guidelines require the difference to be established between the restricted and unrestricted values of the interest in land. The NCC say that this is what the district valuer's valuation does. In their submission, the district valuer's presentation is helpful because it eliminates common factors and therefore concentrates on those areas where differences lie between the unrestricted and restricted schemes such as the extent of forestry and the numbers of sheep.

On this point the tribunal accept the submission of the claimant. Although the approach of the district valuer is designed to be helpful in isolating the differences in the two schemes, its correctness in H quantifying the difference in value between them does depend on the common variables having the same open market value in each case. Where the unrestricted and restricted values depend on different integrated land-use schemes, as here, there may be differences in the values of the variables which compose both the common and the differentiated elements. Again, certain features of the glen may have latent value which will vary between different schemes. For example, the glen is exceptionally well roaded and it is agreed between the parties that this will be highly attractive to a potential purchaser. As an element common to both schemes the roads will nevertheless have greater value the more extensively the glen is developed for both agriculture and forestry. There is, therefore, uncertainty in this approach, and an alternative which addresses separately the question J of open market value for each of the two conditions is to be preferred, since it will avoid that uncertainty while meeting the requirements of the relevant legislation. In any case, we do not believe that it is an approach that would be adopted by a hypothetical purchaser.

Basis of capital value

The unrestricted and restricted values of the interest in land are capital sums. In order to determine these capital sums parties have had recourse to the comparative method of valuation. The only exception to that approach is the adoption by both parties of the offer proposed by the EFG for the 643 ha of plantable land and existing timber in the sum of £430,000. The parties are agreed that the K appropriate capitalisation rate for both sheep and cattle is 4%, which gives a years' purchase of 25.

Treatment of risk

There is one very important difference between the parties, however, in respect of the treatment of risk with respect to the agricultural aspects of the estate as viewed by a hypothetical purchaser considering either the unrestricted or restricted open market values.

Although the same yield of 4% is accepted as the rate of return on the agricultural element and on the deferment of the receipt of future returns from that element, the district valuer makes an additional L discount of 55% from the present value of the potential of that element, in each scheme, for the reason that he believes that the hypothetical purchaser would not pay for the whole of the unrealised potential. The justification for this argument is that these prospective benefits could not be realised without the application of the purchaser's own efforts and enterprise between the point of purchase and the full maturing of the future benefits to which risk and uncertainty is attached. The claimant submits that it is normal for purchasers of agricultural land to contemplate improvement of agricultural land and such risk and uncertainty as may exist is therefore allowed for in the rate of return expected and that comparable sales will also include such allowances. No further adjustment of any kind is therefore necessary. He also submitted that M there is no evidence at all that such arbitrary end allowances are made in the market place.

The rate of return expected on agricultural land is low in comparison with the return to be expected on other investment assets. There are a number of reasons for this, including the enduring nature of land, special tax advantages and fiscal support for agriculture. In any agricultural land purchase the return on capital anticipated takes into account those macro-and micro-economic variables which will affect expected returns, such as changes in product prices and interest rates. Any future benefits will have their present value discounted to allow for their deferment. Both parties accept this, but the district valuer perceives additional factors which

A have to be taken into account beyond those that will affect otherwise similar agricultural enterprises subject to the same risks.

The district valuer has said in his evidence that the schemes proposed on both the unrestricted and restricted bases represent development of enterprises different in scale and innovation from those which might be anticipated in the normal run of sheep farm purchases. They also carry particular risks. If the forestry is not planted immediately, then the agricultural development programmes upon which they fundamentally depend will be prejudiced, and the hypothetical purchaser may have no control over forestry operations when the forestry land has been disposed of out of the estate.

As we have explained earlier, the tribunal do not accept that the B schemes represent innovate (and therefore risky) agricultural practices.

Neither did there at the valuation date appear to be any unusual risk attached to the implementation of the forestry operation. Expert witnesses for both parties spoke of the eagerness of prospective investors at the date of valuation to commence operations before any expected change in the favourable tax regime. Delay would, in any case, be against their own interests. The rate of return on land for forestry operations as calculated on an equated yield basis, anticipating planting in the first period, is lower than that of agriculture, at a maximum of 3.8%, but that lower yield would of itself have built into it the possible risk of failure to commence C immediately. Since the forestry investor has not seen fit to discount further risk for failure to plant immediately, it is not clear why the agricultural investor should feel obliged to do so. In any case, the terms of sale of the forestry land, as evidenced by the offer from EFG, would require a contractual obligation to commence planting operations immediately.

The district valuer has made a deduction of 55% to allow for these future risks because he does not believe that deferment of the future benefits (over five years at 4%) is itself sufficient to cater for what he considers to be abnormal risks. The tribunal are of the opinion that the correct factor to adjust for risk is the yield percentage on the investment. Mr Reid-Thomas agrees with a deferment factor of 0.822. However, he would estimate that the receipt of future benefits D would be over a shorter period of time and he would have used a higher rate of return. By the mathematical principle of equivalence, a shorter period of deferment will result in a higher yield rate. For example, if the average time for the potential to be realised was three years and four months this would give a rate of return of 6%.

The proportion of potential which the district valuer has deducted would give a rate of return of 17.76% on the Cameron scheme for a deferment period of five years, with a deferment factor of 0.44165, if the amount were to be deducted there rather than at the end of his valuation. This gives a years' purchase of 5.62 as against 25 for otherwise similar agricultural investments. For the Elliot/Scruton scheme the equivalent figures are 18.49%, 0.42814 and 5.41. This E means that the risk premium above 4% to be attached to the future benefits of the schemes would be 14.76% and 14.49% respectively. Effectively, this means that the hypothetical purchaser would be treating the future benefits of the scheme as several times more risky than all of the risks attached to the on-going agricultural operations.

The tribunal accept that the future benefits of the schemes should be discounted at a higher rate than 4%, but believe 17.76% and 18.49% to be excessive without much greater justification. The district valuer was unable to support this scale of adjustment by reference to market transactions and Mr Reid-Thomas had not heard of adjustments of this scale being made in the valuation of similar assets. The tribunal are of the opinion that deferring future benefits with a risk premium of 1% (25% above the all-risks yield of 4% on F similar agricultural enterprises) will be sufficient to deal with any additional risk concerning the probability of successfully completing the investment programme. We therefore adopt a rate of return of 5% for the calculation of the present value of future agricultural benefits under the schemes.

Period of deferment

For the reasons given earlier in our decision, the tribunal are of the opinion that the period for deferment should be 10 years, in recognition of the fact that the benefits will be greater at the beginning of the period and less at the end, with an average deferment period of 4 years, which at 5% return gives a deferment factor of 0.823.

We now turn to the prices to be established for the remainder of the G elements comprised in the valuation of the unrestricted and restricted open market values of the estate at the date of valuation.

Value of forestry land

Regarding the valuation of the forestry land, as has already been indicated the Cameron scheme can claim the value of £430,000 offered by the Economic Forestry Group, and that is the amount which the tribunal credit to the unrestricted open market value of Glenlochay Estate.

So far as the value of the forestry component of the amended Elliot/Scruton scheme is concerned then, that is now determined on the basis that the total area of plantable forestry has been diminished H as a consequence of the tribunal's evaluation of the agricultural and financial merits of that proposal.

The district valuer has made a deduction to the capital amount per ha which a potential purchaser would offer for the forestry land in the Elliot/Scruton scheme for the reason that, on a reduced area, the Elliot/Scruton scheme would be burdened with extra initial costs in respect of the provision of necessary infrastructure, since the same roads would be required to service each scheme. This is a logical adjustment to make. With a further reduced area these initial costs per ha of land would be even higher, but this is off-set by the fact that the remaining forestry land in that scheme is now predominantly in the eastern and lower part of the glen. It is more accessible and can be planted, managed and cropped with less extensive roads, particularly J that running up to block 7. On the other hand, the quantum of land is no longer so attractive. However, the evidence of Mr Yuill was that, although he considered the whole of the forestry land in the Elliot/Scruton scheme to be unattractive to an invester, he nevertheless considered that the plantable land at blocks 2, 3 and 4 would be attractive and that he would offer £640 per ha for it. We see no reason, therefore, for considering any other figure for the value of that land.

In both schemes the same rate per ha of £1,250 is accepted for the 46 ha of land and standing timber in block 3. The timber itself is valued at £33,000 and this implies a rate of £532.61 per ha of plantable land reflecting the additional establishment costs. K

Costs of agricultural improvements

The tribunal have considered the cost of the agricultural improvements involved if either scheme is to be realised. Information has been given on the costs incurred by Mr Cameron between the acquisition of the estate and the effective date and including the construction of the hogg wintering shed. These costs are not relevant to the restricted and unrestricted open market value at the date of valuation and are ignored. The expenditure does have a bearing on the value of the agricultural element of the estate as at that date and this is reflected in the value of the ewes at that time, which both sides value at a capital sum of £100 each. The tribunal accept this figure.

So far as costs may be incurred after the effective date, the tribunal L recognise that these will include fencing, drainage, lime and phosphate application and extra feeding for the sheep and cattle.

The district valuer has estimated that 5,000 linear metres of stock fencing will be required to implement the Cameron scheme. The claimant has suggested a total of 4,760 linear metres. Since the total length is uncertain, the tribunal accept the claimant's figure, since this will be based on more particular and detailed local knowledge of the area and the plan.

Although the NCC will fence certain boundaries of the SSSIs, thus separating hirsels, additional fencing will also be needed to implement the Elliot/Scruton scheme. Here the claimant has suggested that 4,640 linear metres will be required for the M Elliot/Scruton scheme. However, the Elliot/Scruton scheme does not have the same fenced parks as the Cameron scheme and we choose to accept the NCC's own estimate of 1,500 linear metres as being appropriate. The claimant suggests a figure of £2.80 per metre as an appropriate cost for stock fencing and the district valuer uses a figure of £3.00. In the circumstances it seems fair to compromise at £2.90 per linear metre as the cost of fencing in the first year of the scheme. Grant at 30% is allowed.

The district valuer has deducted the, net of grant, cost of fencing in year one, and he has allowed for the deduction of an identical amount of capital to provide for renewal of these fences in perpetuity. This is done by allowing for a sum to be set aside as a sinking fund which will

A accumulate to the cost of refencing after a period of 14 years at a rate of return of 7%.

The claimant submits that a deduction for the cost of on-going maintenance is inappropriate, since it is built into every open market calculation when an offer is made for a farm. The prospective bidder will allow for the burdens of maintenance which the possession of the agricultural enterprise will require.

The tribunal agree that the on-going maintenance of equipment such as fencing will be allowed for in market bids. The position here, where a capital expenditure is anticipated which is designed to bring a return in enhanced profitability due to the exploitation of development potential, is quite different from that which obtains in B the compulsory purchase of land, where the claimant is expected to be relieved of any additional present or future burden falling upon him as a result of the acquisition.

The future maintenance costs will be affordable out of the future profits from the enhanced enterprise and sinking fund provision is therefore unnecessary.

The NCC do not make an allowance for the costs of drainage, although Mr Elliot conceded that some may be necessary. We accept the claimant's estimate that 9,000 chains will be involved in the realisation of the Cameron scheme and we suggest 3,000 chains in the amended Elliot/Scruton scheme. We also accept a cost of £1 per chain and a deduction for agricultural grant of 30%.

The cost of liming and slagging required by the Cameron scheme C has been estimated by the claimant at £8,260 based on the treatment of 200 acres at a cost of £59 per acre with a reduction of 30% for grant received. He has estimated that the NCC may have partially to regenerate about 100 acres at a cost of £4,130.

The NCC do not make a deduction in their valuations for liming and slagging the land in either scheme. The advice of Mr Elliot was that, on comparison of cost, extra feeding was to be preferred. Extra feeding would also have more immediate effect. They envisaged 120 ha being treated in the Cameron scheme and 70 ha in the Elliot/Scruton scheme at initial costs of £185/ha, and later at £74/ha. These produced respective totals of £23,310 and £39,960, although they would have allowed grant to be deducted from that.

For the reasons given earlier in our opinion, the tribunal consider D that both feeding of the stock and some regeneration of the better grazing land will be desirable and would be anticipated by a potential purchaser. Since regeneration is a firm intention of the claimant's plan then, we accept his estimate of 200 acres being treated with lime and inorganic fertiliser and with some reseeding. Since the amended Elliot/Scruton scheme will produce only 150 additional ewes to the Cameron scheme's 750 additional ewes, we adjust the acreage required to be regenerated in the former scheme to 40 acres.

We accept the claimant's estimate of £59/acre as the cost of carrying out this work, but in addition we believe that a potental purchaser would allow for the cost of grass seed an amount of £2,000 on the Cameron scheme and £400 on the Elliot/Scruton scheme. It is E again assumed that, at the date of valuation, grant would be available, at 30%, to reduce the cost of this activity.

Interest on capital

The claimant has added to the cost of the agricultural improvements proposed in his scheme an amount in respect of interest on capital expended. The interest is in respect of, first, the cost of agricultural improvements such as fencing, drainage etc and, second, the retention of lambs in order to increase flock numbers. These would normally have been sold. Therefore, he has effectively tied up capital during the early part of the development period which will be burdened with interest.

The district valuer, in his evidence, does not agree with the F deduction of interest on extra capital. However, the tribunal recognise that deductions for interest on unproductive capital are normal in analogous valuations such as, say, the determination of the residual valuation of land for some proposed form of development. The tribunal are of the opinion that it has been properly included by the claimant. We have based our calculations on the assumption that the agricultural improvement costs will be unproductive for a period of two years and that lamb sales will have to be forgone over a period of three years to build up the number of ewes required in each scheme.

Determination of payment

Having now finally considered those practical matters related to the method of the valuation to be used in determining the open G market values of the unrestricted and restricted estate, we proceed to set out our calculations in Appendix C.

The basis for our valuation of the unrestricted value is the Cameron scheme and of the restricted value it is our amended version of the Elliot/Scruton scheme. These two schemes have provided the conceptual and physical characteristics upon which our valuations are based, since they represent the kind of development proposals which a hypothetical purchaser could reasonably be expected to have in mind when contemplating the potential of the estate. We recognise, nevertheless, that such a purchaser would not necessarily have available all of the details of these schemes and their implementation as has been available to the tribunal as arbiter. But H with professional advice we accept that schemes of this type would be considered and calculations of this nature would be made by a potential purchaser approaching Glenlochay afresh.

The result of our valuations is that we determine an unrestricted open market value of Glenlochay Estate in October 1987 to be £1,787,532 and the restricted open market value of the estate at the same date to be £1,232,513.

The determination of the amount of the lump sum payment due to Mr John Cameron as owner of Glenlochay Estate in terms of the Act is the difference between the unrestricted and restricted open market values, say £555,000.

We also certify the case as appropriate for the employment of senior and junior counsel and further certify as expert witnesses all J those witnesses we were requested to certify including Mr John Jones-Perrott, who was objected to but who, we consider, did give expert evidence.

APPENDIX C

Tribunal's calculations

Unrestricted value

A	Forestry land and timber						
	Already planted	46	1,250/ha		£57,500		
	Unplantable	21	—				
	Blocks 1-6	545	640/ha	(say)	348,000		K
	Block 7	98	250/ha		24,500		
		710				£430,000	
B	Deer						
	62 stags @ £7,000				434,000		
	4 stags @ £7,000		£28,000				
	35 hinds @ £125 × 8 YP		35,000				
			63,000				
	Def'd 4 years @ 5%		0.823		51,849	485,849	
C	Cattle and sheep						L
	Hirsel A	700					
	B	500		5,250			
	C	500		ewes @			
	D	700		£175/ewe			
	E	500	918,750				
	F1	1,100					
	F2						
	G	1,250					
	100 cattle @ £1,000/cow		100,000		1,018,750		
	Deduction for present value						
	Value at effective date (see C3)		515,000				
	Value of future stock		503,750				M
	Less def'd 4 yrs @ 5% (0.823)		414,586		89,164	929,586	
						1,845,435	
D	Agricultural improvement costs						
	Fencing: 4,760m @ £2.90/m		13,804				
	Drainage: 9,000 @ £1/chain		9,000				
	Fertiliser: 200 acre @ £59/acre		11,800				
	Reseeding: 50 acres @ £40/acre		2,000				
			36,604				
	Less grant @ 30%		10,981		25,623		

Interest on capital Agricultural improvements		25,623		
Interest for 2 yrs @ 12%		6,150		
750 lambs @ £25/lamb	£18,750			
Interest @ 10%	1,875			
Year 1	1,875			
Year 2 def'd @ 10%	1,705			
Year 3 def'd @ 10%	1,550	5,130	11,280	
E Cost of feed:			21,000	57,903
			Total	£1,787,532

Restricted value

A Forestry land and timber				
Already planted	46	1,250/ha	£57,500	
Unplantable	18	—		
Blocks 2-4	205	640/ha	131,200	
	269			£188,700
B Deer				
62 stags @ £7,000			434,000	
2 stags @ £7,000		£14,000		
7 hinds @ £125 × 8 YP		7,000		
		21,000		
Def'd 4 years @ 5%		0.823	17,283	451,283
C Cattle and sheep Hirsel A 700, B 500, C 500, D 500, E 500, F1 550, F2 400, G 1,000	4,650 ewes @ £118.75/ewe	552,188		
50 cattle @ £850/cow		42,500		
50 cattle @ £650/cow		32,500		627,188
Deduction for present value				
Value at effective date	(see C3)	515,000		
Value of future stock		112,188		
Less def'd 4 yrs @ 5%	(0.823)	92,331	19,857	607,331
				1,247,314
D Agricultural improvement costs				
Fencing: 1,500m @ £2.90/m		4,350		
Drainage: 3,000ch @ £1/chain		3,000		
Fertiliser: 40 acres @ £59/acre		2,360		
Re-seeding: 10 acres @ £40/acre		400		
		10,110		
Less grant @ 30%		3,033	7,077	
Interest on capital Agricultural improvements		7,077		
Interest for 2 yrs @ 12%		1,698		
150 lambs @ £25/lamb	£3,750			
Interest @ 10%	375			
Year 1	375			
Year 2 def'd @ 10%	341			
Year 3 def'd @ 10%	310	1,026	2,724	
E Cost of feed			5,000	14,801
			Total	£1,232,513

Agriculture element at date of valuation

C Cattle and sheep Hirsel A 500, B 500, C 500, D 500, E 500, F1 1,000, F2, G 1,000	4,500 ewes @ £100/ewe	£450,000	
50 cattle @ £750/cow		37,500	
50 cattle @ £550/cow		27,500	£515,000

Calculation of payment due		
Open market value without restriction		£1,787,532
Open market value with restriction		£1,232,513
Difference between restricted and unrestricted values		£555,019
	Say	£555,000

We reserve all questions of expenses in respect of which parties may submit written submissions, with a right, however, to either party to move for a further hearing on expenses.

C & J SEYMOUR (INVESTMENTS) LTD v LEWES DISTRICT COUNCIL

(REF/76/1990)
July 22 1991
(T HOYES FRICS)

Estates Gazette March 21 1992 and March 28 1992
[1992] 11 EG 127 and [1992] 12 EG 152

Blight notice — No entry taken by date of hearing — Land values falling after deemed notice to treat — Whether valuation at date of deemed notice to treat — Whether later date would provide financial equivalence — Whether principles of *Stokes* v *Cambridge Corporation* in respect of access to be applied

Subject land a freehold site in the centre of Seaford, Sussex — Blight notice served following publication of draft local plan identifying site as an extension to an adjacent car park, owned by the acquiring authority, to the south — Right of way over service yard to the north, in claimants' control, not included in notice — Deemed notice to treat came into existence on November 14 1988 — Certificate of appropriate alternative development later issued for residential development — Entry not made by last day of hearing, May 10 1991 — Claimants submitted that valuation date should be November 18 1988, the date of formal acceptance of the notice, and not the date of the tribunal's determination as adopted since *Birmingham Corporation* v *West Midland Baptist (Trust) Association* — Land values, it was argued, were significantly lower in mid-1991 than in November 1988 — The claimants could not be regarded as willing sellers in a depressed market and the principle of financial equivalence would not be fulfilled if the acquiring authority could buy on values prevailing at the bottom of the market — Claim for £190,000 — Acquiring authority's amended figure £26,000 — It was not disputed that vehicular access was necessary to facilitate development

***Decision*: Total compensation of £139,000 at May 10 1991 was awarded**

(1) A claimant sells and buys in the same market and he is disadvantaged only if he is deemed to sell, as in *West Midland*,

in a lower-priced market but replace or reinvest in a later, higher-priced market — In the present case the claimants were losing a small building site at the date of valuation and, in principle, compensation could be used to purchase another site — Following the Court of Appeal decision in *W&S (Long Eaton) Ltd* v *Derbyshire CC*, the date of valuation was the last day of the present hearing, May 10 1991

(2) The development to be assumed for the purpose of s17 of the Land Compensation Act 1961 was four town houses with vehicular access over the yard to the north — The aggregate value for residential development therefore resided, but not necessarily equally, in two parcels of land, the subject site and that part of the yard over which a right of way was needed

(3) In the circumstances of no right of way passing to the acquiring authority, compensation fell to be determined by reference to the value of land taken under rule (2), s5, Land Compensation Act 1961 and injurious affection to retained land under s7, Compulsory Purchase Act 1965

(4) It was necessary to apply the principles of *Stokes* v *Cambridge Corporation* to the land taken and then assess the loss of access potential as injurious affection to retained land with an allowance to reflect the fact that the yard remained unencumbered by rights of way for the benefit of the land taken

(5) After allowing for a *Stokes* payment of £67,000, compensation for the land taken was £77,000

(6) For injurious affection to retained land, compensation was £62,000, being £67,000 for loss of potential to provide access to land taken *per Stokes* less, adopting a pragmatic approach, £5,000 for loss of encumbrance to claimants' yard

The following cases are referred to in this report.

ADP&E Farmers v *Department of Transport* [1988] 1 EGLR 209; [1988] 18 EG 80, [1988] 19 EG 147 & [1988] 20 EG 104
Birmingham Corporation v *West Midland Baptist (Trust) Association (Inc)* [1970] AC 874; [1969] 3 WLR 389; [1969] 3 All ER 172; (1969) 67 LGR 571, HL
Clinker & Ash Ltd v *Southern Gas Board* 18 P&CR 372; [1967] EGD 532; (1967) 203 EG 735; [1967] RVR 477; [1967] JPL 473, LT
Fitzwilliam's (Earl) Wentworth Estates Co Ltd v *British Railways Board* (1967) 19 P&CR 588; 204 EG 612 & 728; [1968] JPL 287, LT
Horn v *Sunderland Corporation* [1941] 2 KB 26; [1941] 1 All ER 480, CA
Stokes v *Cambridge Corporation* (1961) 13 P&CR 77; [1961] EGD 207; 180 EG 839, LT
W&S (Long Eaton) Ltd v *Derbyshire County Council* (1975) 31 P&CR 99

D Randall Sinnatt, solicitor, of Thompson & Thompson, of Bexhill on Sea, appeared for the claimant company; J Woosnam, district solicitor to the Lewes District Council, for the compensating authority.

Giving his decision, **T Hoyes** said: This is a reference to determine the compensation payable following the service of a blight notice relating to land in the centre of Seaford, East Sussex.

The claimant company seeks compensation in the sum of £190,000 for the freehold interest with vacant possession. The acquiring authority contend that compensation should be £15,530, amended during the hearing to about £26,000.

The subject land is situated upon the eastern side of Saxon Lane, a short distance south of its junction with High Street. Adjacent to the northern boundary is other land in the control of the claimant, namely an open service yard and shops, 34-40 High Street, over which there are also flats. Directly to the south is a surface car park owned and operated by the district council. The frontage to Saxon Lane embraces a mixture of residential, commercial and other uses.

In June 1987 the district council published a draft local plan which identified the subject site as an extension to the adjacent car park. On September 14 1988 the claimants' predecessor in title, Clear View Rental & Relay Ltd (an associated retail company), served a blight notice under section 192 of the Town and Country Planning Act 1971 to which the district council did not object. Under section 194 a deemed notice to treat therefore came into existence on November 14 1988. On May 31 1989 application was made for a certificate of appropriate alternative development under section 17 of the Land Compensation Act 1961 specifying development for four terraced dwellings by reference to attached plans. In September 1989 the district council issued a certificate under section 17 for residential development. The acquiring authority have not yet entered upon the subject land and the valuation date to be adopted is a matter in contention between the parties.

Mr D Randall Sinnatt, solicitor for the claimant company, called Mr C H Seymour, managing director, and also managing director of the associated company above, and Mr B K Perrin CBE FRICS. Mr J Woosnam, solicitor to the district council, called Mr D F Jarman, a principal planning officer, Mr D A Gruber, a landscape officer, and Mr E J D Brown BSc FSVA, the estates surveyor and valuer to the council.

From the evidence and an agreed statement the following facts are found:

1 The local plan was adopted by the district council in April 1990.
2 Outline planning permission was refused on December 8 1987 for development of the subject land by four town houses with vehicular access directly from Saxon Lane because (in summary form):
   (a) the site was required for public car parking;
   (b) the proposal involved overdevelopment of the site;
   (c) the removal of part of the boundary wall to Saxon Lane would be detrimental to the Seaford Conservation Area.
3 Outline planning permission for a broadly similar development of the site, save for vehicular access being over the service yard to the north, was refused on March 8 1988 because:
   (a) the site was required for public car parking;
   (b) of a loss of private car parking in the yard.
4 On January 20 1988 the council adopted a non-statutory document, *Residential Design Guidelines*, for development control.
5 The location, species, size and condition of trees upon the subject site are not disputed.
6 The freehold land subject to the blight notice is 711.26m²; a right of way over the yard to the north was not included in the notice.
7 The site is surrounded by 3.4m-high retaining walls in local flint and its surface is about 2m above the carriageway of Saxon Lane. It has pedestrian access to Saxon Lane and to the yard to the north; it is a neglected garden, the ownership of which was severed some time ago from another house to the east (Stone House in Crouch Lane).
8 The carriageway of Saxon Lane is 4.7m wide, there is no footway along the frontage of the subject site and vehicular traffic operates one way from the south, entering via Steyne Road.
9 It is common ground between the expert witnesses that the valuation of the land taken should proceed upon the assumption of some scheme of residential development.
10 The district council accept that Mr Seymour controls both Clear View Rental & Relay Ltd and C & J Seymour (Investments) Ltd. They raised no issue about the separated ownership of the subject site and the adjacent yard after January 15 1990 in relation to the assessment of compensation, if any, for injurious affection.

The issues to be determined in this reference are:
(1) The date of valuation to be adopted.
(2) The circumstances to be assumed in respect of vehicular access to the subject land.
(3) The form of development to be assumed for valuation purposes consequent upon the section 17 certificate.
(4) The value of the land taken and the compensation, if any, payable for injurious affection to land accepted as "held with".

The following cases were referred to:

Birmingham Corporation v *West Midland Baptist (Trust) Association (Inc)* [1969] 3 All ER 172, HL
Clinker & Ash Ltd v *Southern Gas Board* (1967) 18 P&CR 372; 203 EG 735, LT
Fitzwilliam's (Earl) Wentworth Estates Co Ltd v *British Railways Board* (1967) 19 P&CR 588; 204 EG 612 & 728, LT
Horn v *Sunderland Corporation* [1941] 2 KB 26
Stokes v *Cambridge Corporation* (1961) 13 P&CR 77; 180 EG 839, LT

Date of valuation

Mr Randall Sinnatt submitted that the practice followed since *West Midland* and the subsequent explanatory Circular 35/70 with respect to the date of valuation, namely the date of determination by this tribunal, since the acquiring authority have not entered on the subject land, should not be adopted in this reference. His reason is that land values are acknowledged to be significantly lower in mid-1991 than at November 18 1988, the date which he suggested should be adopted. Therefore, to follow established practice will deprive the claimant of some compensation to which it is statutorily entitled.

He selected November 18 1988 because that is the date on which the acquiring authority confirmed in writing that they would accept the blight notice dated September 6 1988. From that date the claimant, in terms of compensation, was constrained from creating new interests in the land or improving it and could deal with it only subject to the deemed notice to treat, effectively restricting the rights and privileges of any purchaser to the receipt of the statutory compensation. In the no-scheme world prior to November 18 1988 it had been open to the claimant to secure planning permission, on appeal if necessary, and then dispose of the site in the buoyant market of 1988. Only if that opportunity is restored, in financial terms, can the claimant be properly regarded as a willing seller under the definition of open market value for land taken in rule (2) in section 5 of the Land Compensation Act 1961. No willing seller would dispose of land in a depressed market and the principle of equivalence which lies at the root of statutory compensation, *per* Scott LJ in *Horn*, is not fulfilled if the acquiring authority are allowed to buy on values prevailing at the bottom of the market.

Mr Randall Sinnatt urged that it was for this very reason that the previous practice of valuing invariably as at the date of notice to treat was overturned by the House of Lords in *West Midland*. The evil and the injustice to the claimant in that case arose from delay by the acquiring authority in a rising or inflationary market. Delay in entering upon land by an acquiring authority in a falling market causes equal injustice to a claimant where the date of entry or, in this reference, the date of determination is the prescribed valuation date.

The response by Mr Woosnam can be put as follows. First, the acquiring authority had not procrastinated in the acquisition process, albeit not initiated by the council, save for a short initial delay arising from an agreement on September 6 1988 with the claimant to withdraw a blight notice dated July 14 1988 and substitute the operative notice. Second, since late 1988 the acquiring authority had sought to agree the sum of compensation payable, but differences of expert opinion as to value have prevented this. Third, if the claimant were concerned about the passage of time in a falling market there was always available, after November 14 1988, the procedure of making a reference to this tribunal; the claimant delayed the reference until May 17 1990. Fourth, the alleged impediment to agreement of compensation, a lack of available funds by the district council, was unfounded as the money had been available for some two years; in any event, delay for this reason would have been an unmeritorious act by the acquiring authority. Fifth, the practice arising from *West Midland* and detailed in Circular 35/70 should be followed; the parties could not have foreseen the downturn in the market and when it emerged the claimant took no action by way of early reference. The claimant in this reference, as in acquisitions initiated by an acquiring authority, must take its chance with respect to market conditions; the date of valuation is a matter of well-established practice and not for choice by a claimant. Willingness to sell in rule (2) in section 5 of the Land Compensation Act 1961 falls to be judged at the date determined by matters other than the state of the market and is an issue going to quantum at the predetermined date by, for example, excluding from evidence of value forced or other sales of land at an undervalue at that date.

While Circular 35/70 sets out in convenient form the recommended practice for valuers to follow, having regard to the principles of *West Midland*, it is not an authority binding on this tribunal. No doubt mindful of this, Mr Randall Sinnatt referred me at some length to the judgments in the House of Lords.

Putting aside the particular facts in *West Midland* and concentrating on the fundamental principles of their lordships' reasons for departing from the established practice as to the date of valuation, I am not persuaded by the argument of Mr Randall Sinnatt that the claimant will suffer any injustice by the adoption of the date of determination as the date of valuation; following the decision of the Court of Appeal in *W & S (Long Eaton) Ltd* v *Derbyshire County Council* (1975) 31 P&CR 99 I take this to be the last day of the hearing, namely May 10 1991.

In *West Midland* Lord Donovan said at p 188:

> . . . in any developing community there must be power to take land from private owners for public purposes; and in a society where private ownership of land is permitted, justice requires that compensation should be paid for such taking. When this compensation takes the form of a monetary payment, the date which at once suggests itself as the date when values should be ascertained and compensation computed is the date of the taking. It is then that the private owner sustains his loss; and it is at that date, one would have thought, that the amount of his loss should be calculated at prices then prevailing.

Following the same line of reasoning, but additionally adopting the notion of a claimant replacing the land taken by using the compensation moneys, Lord Morris said at p 184:

> but if someone is having his land compulsorily taken from him, it is likely that any alternative arrangements that he makes will be made in relation to the time when his land is actually taken or in relation to the time when compensation is actually payable. Someone who receives rule (2) compensation may wish to buy other land, or he may wish to invest the sum he receives. He will buy or he will invest when he receives the price of what he has been forced to sell.

The short answer to the issue is that a claimant sells and buys in the same market and he is disadvantaged only if he is deemed to sell, as in *West Midland*, in a lower-priced market but replace or reinvest in a later, higher-priced market. In this case the claimant is losing a small building site in May 1991 and, in principle at least, the compensation can be used to purchase another site. That is the practical implication of the principle of equivalence expressed in money terms referred to by Scott LJ in *Horn* and cited by Lord Reid at p 177 of his judgment in *West Midland*.

Mr Randall Sinnatt is in reality not seeking equivalence at a particular date but compensation for an opportunity lost in a higher value market because the claimant was, in practical terms, constrained from dealing in that market owing to the deemed notice to treat. He rightly anticipates that from November 18 1988 onwards no buyer would be forthcoming at 1988 values because title would confer only the right to receive compensation at best at 1988 prices and in a falling market at a lower figure.

The answer to this issue is not only to be found in *West Midland* as indicated above but was also fully considered in *W & S (Long Eaton) Ltd*, a case entirely in point (but not cited in argument) which arose from a purchase notice and concerned the date of valuation in the falling market for residential building land during the period 1973-74.

Vehicular access to subject site

It is agreed that the value of the subject land should be assessed by reference to its use for some form of residential development. It is not disputed that to facilitate development vehicular access is necessary. From the town planning evidence detailed later, the most practical and acceptable vehicular access would be over other land, the yard to the north, controlled by the claimant. Therefore, the aggregate value for residential development resides, but not necessarily equally, in two parcels of land, the subject site and that part of the yard over which a right of way is needed.

The land the subject of the deemed notice to treat is defined in the blight notice by reference to a red verge upon a plan without any mention of a right of way over adjacent land. The acquiring authority do not require any such right to implement the scheme of car parking, as the subject site can be, and is intended to be accessed, from the existing public car park to the south.

Mr Woosnam contended that the subject site fell to be valued without the benefit of vehicular access, as that was the factual and legal situation; the correct valuation approach was that adopted by Mr Brown and he applied the *Stokes* principles. Mr Randall Sinnatt sought to distinguish the circumstances in *Stokes* from the present references because here the claimant effectively owns and controls both the subject land and the land needed for access. He referred to *Earl Fitzwilliam's Wentworth Estates*, where this tribunal said at p 599:

> It seems clear to us that in the *Stokes* case the tribunal did have regard to the pattern of surrounding ownerships. But in that case the vendor did not own the key and therefore the site which he had to sell was land-locked. In that situation the tribunal thought it right to make the deduction mentioned in the report.
>
> We do not think that any principle emerges from the *Stokes* case which requires us to assume that the claimants, as willing sellers of the subject land, would have sold without access, or to blind ourselves to the undoubted fact that they owned the land over which access could be provided.

Adopting this approach, Mr Perrin makes no deduction in his valuation for the fact that the notice to treat excludes any right of

A access; he assumes the automatic passing of such right as urged by Mr Randall Sinnatt. In addition to compensation for land taken, Mr Perrin also assesses compensation for injurious affection to other land and premises, some flats, to the north. However, in computing this sum he makes no deduction or allowance for the fact that no right of access will pass to the acquiring authority over any part of the yard, so that in reality it will remain unencumbered by rights of access vested in others, namely the eventual owners of dwellings assumed to be built upon the subject land.

On the facts of the present case, including the acceptance by the acquiring authority that the claimant owned and controlled both parcels of land to the effect that they were "held with", the approach B of both valuers is incomplete. The statutory provisions accord to the claimant compensation for the greater of existing use or potential value with respect to both land taken and the depreciation in value of land retained. The taking of the subject land for the acquiring authority's scheme denies the claimant the site of dwellings and at the same time destroys the potential residing in part of the yard to access those dwellings and thereby enhance the value in the market of the land taken. This is a compensatable loss to be categorised as injurious affection to retained land, and in the computation of that loss regard needs to be had to the fact that after the taking of the subject land the yard of the claimant remains unencumbered by rights of way to dwellings on adjacent land. Mr Brown, following *Stokes*, disregarded the loss completely and Mr Perrin, following C *Fitzwilliam*, made no allowance for the continuing advantage of the absence of any encumbrance consequent upon the acquiring authority's neither needing nor taking the right of way. In the circumstances of no right of way passing to the acquiring authority, the matter falls to be dealt with, on the statutory provisions, by reference to the value of land taken (rule (2), section 5, Land Compensation Act 1961) and compensation for injurious affection to retained land (section 7, Compulsory Purchase Act 1965): see *A D P & E Farmers* v *Department of Transport* [1988] 18 EG 80; [1988] 1 EGLR 209.

The evidence of Mr Seymour related to his personal purchase of the subject site with other land in 1968, his sale of the Stone House and the transfer of the residue of the purchase in December 1970 to D his trading company, Clear View Rental & Relay Ltd, the intermittent use of the subject site and his longer-term intention to use it for development, possibly by two houses, one of which would be for his own occupation. In his view, the site enjoyed a most convenient residential location and such use was considered acceptable in town planning terms by the district council until 1987. He considered car parking an inappropriate use of land in a conservation area and accordingly sought planning permission for development by four town houses. He had taken advice on the viability of such a scheme, for which he was satisfied there was a ready demand, and he produced a skeleton calculation to demonstrate viability upon the basis of a site value of £160,000.

E The claimant (an investment company) when purchasing the subject site from its associated company in January 1990 secured an agreement to grant vehicular access to facilitate the residential development of the site. To make good any car parking which might be lost in the yard due to the provision of vehicular access thereover for the subject site, Mr Seymour had terminated, in early 1988, the use and occupation of part of the yard and a garage building by a motor vehicle repairer. In effect, this reduced the number of vehicles resorting to the yard and increased the area available for access and/or car parking. He had also effected some minor improvements to the access point on to Saxon Lane. The scheme of residential development prepared for the subject site would retain all existing worthwhile and prominent trees and would be less destructive of trees F than the car-parking proposal of the acquiring authority.

Town planning evidence

The approach of Mr Perrin was that in the no-scheme world, ignoring the public car-parking proposal for the subject site, the reasons for refusal of the second planning application for four town houses become without foundation and therefore this is the form of development to be assumed for valuation purposes.

The evidence of Mr Jarman was less clear cut. Although he accepted residential development of the site he considered it should be in the form of two detached, three-storey, five-person, three-bedroom houses or, in the alternative, four three-person, two-bedroom flats. He regarded four three-storey town houses as overdevelopment of the site despite the fact that he had not so advised G his council on the second planning application. He also considered that two houses or four flats would create less vehicle movement through the existing access and yard, but this must be entirely speculative as the houses were to have double garages and the number of flats was identical to the number of town houses proposed by the claimant. Mr Jarman also proposed that any loss of car parking in the yard should be made good by providing two or three vehicle spaces within the subject site. This suggestion disregarded the departure of the motor vehicle repairer after the second refusal of planning permission and also the subsequent improvements made to the access.

Mr Jarman stated that the development of the site by four town houses would not meet the requirements of the *Residential Design* H *Guidelines* adopted by the council in January 1988 in terms of car-parking provision and the amount of private open space; more relaxed standards apply to flats. He also considered that the existing trees on the site would be more adversely affected by the development proposals of the claimant.

Mr Gruber referred to his survey of the trees and an agreed schedule with supporting plan. His evidence was that in a conservation area close to Seaford Town Centre, not well endowed with trees, every effort should be made to retain sound and significant trees. Only the better trees on the Saxon Lane frontage were of sufficient importance to justify preservation. In cross-examination he accepted that, provided care was taken not to J damage the trees during building works or to disturb main roots during excavation, by retaining an adequate clear working distance, their presence need not inhibit the scheme of development proposed by the claimant.

I did not find the evidence of Mr Jarman convincing in the circumstances of this reference. Had the claimant pursued, and not withdrawn, the appeal against the second refusal of planning permission it would almost certainly have been allowed, if it could be assumed in the context of a planning appeal, that the public car-parking proposal had to be disregarded. In such a situation the reasons given for refusal would be groundless and the matters raised by Mr Jarman outside the stated reasons would therefore be inadmissible. The evidence of Mr Perrin is therefore accepted and the K evelopment to be assumed for the purpose of the section 17 certificate, and therefore valuation, is by four town houses with vehicular access over the yard to the north.

Valuation evidence

Mr Perrin valued the subject land as at November 18 1988 on the basis of its development by four town houses or two detached dwellings in the sum of £180,000, to which he added £10,000, an admitted estimate, for injurious affection by virtue of noise, other disturbance and reduced security to the remaining land controlled by the claimant owing to the use of the land taken for public car parking. In support of his valuation, he relied upon offers to purchase solicited from developers prior to service of the blight notice and L upon a 1986 transaction, adjusted by him for the passage of time, in a small site fronting nearby Steyne Road. As evidence of a buoyant market in 1988, but not in direct support of his valuation, he referred to other transactions close by in somewhat larger areas of land for high-density residential development, mainly by flats, in proximity to the seafront. In his opinion, a small site conveniently situated for the town centre and the seafront would always be in demand, especially for retirement homes.

If the date of valuation has to be in May 1991, Mr Perrin accepted in cross-examination that his primary figure should be reduced to £144,000. He considered that the low point in the market, at least for willing sales of small sites, had passed and that a slowly improving M market lay ahead. Although comparable transactions were not available in November 1988 or in May 1991, owing to the rather unique characteristics of the subject site, he regarded the comparative approach, informed by experience and judgment, as preferable to a residual valuation.

Mr Brown spoke to two valuations and he adopted May 1991 as the relevant valuation date; a residual valuation for two houses in the sum of £15,530 (revised during the hearing to £25,939, say £26,000) and a comparative-based valuation for four flats, incorporating a number of adjustments, at £12,810 (accepted during the hearing as requiring revision possibly to *circa* £24,000). In both valuations Mr Brown applied the *Stokes* principles and deducted 50% of the value

A in excess of existing use value, which he put at £7,000, to have regard to the exclusion of access rights from the deemed notice to treat. No sum was included for injurious affection to the retained land.

In cross-examination Mr Brown accepted that the subject site had locational advantages over the comparable he used at Sutton Drove [Seaford], and in terms of two-bedroom flats he quantified this difference at about £10,000 per unit.

Decision

I have inspected the subject site, its environs and also, from the highway, the other comparables cited in evidence. I did not find the Sutton Drove site referred to by Mr Brown to be comparable in many important respects and I regard it only as a datum below which the
B value of the subject site should not be determined. Moreover, from the evidence it was by no means certain that the £65,000 (about £11,000 per flat unit) realised at auction in March 1991 was by a willing seller as required by rule (2) in section 5 of the Land Compensation Act 1961; the possibility was raised that it may have been a forced or distress sale. As both the valuations by Mr Brown were in any event, even after adjustment during the hearing, significantly less than this sum I find them of no assistance.

It is therefore not necessary for me to deal with the appropriateness or otherwise of a residual valuation, or to consider the relevance of the observations of this tribunal relating to that valuation method in *Clinker & Ash* as urged by Mr Randall Sinnatt.

C There remains the evidence and valuation by Mr Perrin, whose direct approach using nearby comparables I found more convincing subject to (a) adjustment of his valuation date to May 10 1991 and (b) what is said later about injurious affection. As already indicated, Mr Perrin is entitled to adopt the development of the site by four town houses as the basis of his valuation and I also accept his opinion that in this location there would be no measurable difference in value if the site were developed by two larger, detached dwellings.

Making the best judgment that I can upon the site values of the comparables cited in evidence and mindful of the stated prices paid for houses and flats in the locality, I am persuaded that the subject site with vehicular access over the yard to the north had a value of £144,000 at May 10 1991.

D I reject the £10,000 suggested by Mr Perrin for injurious affection to the retained land as being without foundation in terms of the car-parking extension proposals of the acquiring authority, since the northern boundary wall will remain undisturbed and access into the extension will be from the existing car park to the south.

In order to have proper regard to the statutory provisions relating to compensation, it is necessary to apply the principles of *Stokes* to the land taken and then assess the loss of access potential as injurious affection to retained land with an allowance to reflect the fact that the yard remains unencumbered by rights of way for the benefit of the land taken. In the absence of direct evidence on the latter point, I can adopt only a pragmatic approach; this I do by fixing the deduction at £5,000. In evidence, two values were placed upon the existing
E (garden) value of the land taken; Mr Brown £7,000 and Mr Perrin about £13,000. I determine the figure to be adopted at £10,000.

The calculation of the compensation is therefore:

1. *Land taken*	
Value of subject land with vehicular access	£144,000
Value for existing use	£10,000
Development value	134,000
Deduct 50% *per Stokes*	67,000
Value of land taken	£77,000
2. *Injurious affection to retained land*	
Loss of potential to provide access to land taken per *Stokes* above	£67,000
Deduct for lack of encumbrance upon yard	5,000
Compensation for injurious affection	£62,000
Summary	
(1) Land taken	£77,000
(2) Injurious affection	62,000
Total	£139,000

F

Accordingly, the award is the sum of £139,000 to which will be added a surveyor's fee based upon Ryde's Scale (1984).

I heard submissions as to costs. In the absence of a sealed offer by the acquiring authority to the claimant it is ordered that the acquiring G authority pay the costs of the claimant in the reference, such costs if not agreed to be taxed by the Registrar of the Lands Tribunal upon the High Court Scale.

W WATERS & SONS LTD v WELWYN HATFIELD DISTRICT COUNCIL; SAME v DEPARTMENTS OF THE ENVIRONMENT AND TRANSPORT

(REF/27-31/1988)

December 17 1991 H

(C R MALLETT FRICS and T HOYES FRICS)

Estates Gazette April 11 1992 and April 18 1992

[1992] 14 EG 117 and [1992] 15 EG 90

Petrol filling station, garage and showroom — Basis of valuation — Another petrol filling station opposite agreed "held with" for injurious affection purposes — Both premises somewhat outmoded — Whether total or partial redevelopment to be assumed — Identity of scheme — Risk factor for possible by-pass — Risk of increased competition and creation of "milking stations" — Estimated annual J consumption for bid purposes — Price per gallon of estimated throughput — Modernisation costs — Amount of injurious affection compensation — Whether amount agreed for disturbance properly payable

Subject premises (Stonehouse), almost entirely freehold, situated on east side of A1 trunk road at Hatfield, Herts — Acquisition to improve road and facilitate construction of Hatfield Tunnel — In early 1980s the A1(M) motorway stopped immediately to the north and south of Hatfield, the two ends being joined by the A1 trunk road through Hatfield centre — Linking the two sections could have been achieved in K several ways, including by-passing, tunnelling under or bridging over the town — Immediately opposite subject premises was another petrol filling station, garage and workshop (Aero) in claimants' ownership and occupation — Both sets of premises used for integrated motor-related business — Agreed, for this reason, that Aero "held with" Stonehouse — Valuation date agreed as February 14 1984 — Opening was 24 hours a day for 364 days a year and petrol was discounted at about 3p a gallon off the local prevailing price — Combined petrol sales volume in early 1980s was some 5m gallons annually, about 2.75m being at Stonehouse — L Following opening of Hatfield Tunnel in December 1986, annual throughput at Aero during 1987, after modernisation, was 60% of previous volume — During 1970s some uncertainty surrounded both premises owing to various proposed road-improvement schemes which doubtless influenced decisions concerning new investment and modernisation — It was common ground that commercial prudence indicated the need for some modernisation at Stonehouse in February 1984 but it was not agreed whether total redevelopment as a petrol filling station only, with consequent loss of showroom and workshop facilities, or reconstruction of existing forecourt, but retaining other M facilities, would have been more advantageous

Parties agreed cost of partial modernisation would be £370,000 for forecourt with 10 pumps capable of throughput of at least 3m gallons pa — Acquiring authorities added £50,000 for new storage tanks and £36,000 for roof works to some residual buildings — Original total claim for £4.293m — Authorities' figure £1.8m — Both figures amended during hearing — For injurious affection, claimants' figure was £790,000 and authorities' £375,000, both on before and after basis — The authorities sought to avoid payment on items of disturbance which related to a search for an integrated motor

A **trade site or for a site for a petrol filling station**

Decision: **The total awarded was £3.541m plus surveyor's fee**

(1) There is no statutory definition of "scheme" and the word could be used suitably to describe either the initial overall aim to link the motorway routes or any of the various detailed plans — The initial overall aim did not involve and could not have involved compulsory purchase until the preferred route had been firmly established — Therefore, the underlying scheme was the detailed scheme involving the tunnel route

B **(2) There would have been a risk that the alternative by-pass route might have been adopted and that the effect would have been to reduce the value of the Stonehouse premises in the market by 10%**

(3) In absence of compulsory acquisition the most profitable and acceptable arrangement for the claimants would be to dispose of the forecourt, slightly extended, to a major oil company for modernisation and to retain the other buildings, slightly reduced, for their own occupation for motor-trade-related activities — An oil company would no doubt have preferred to take the whole site for a petrol filling station only, but the potential throughput was sufficiently
C **attractive for them to have accepted less than the ideal — The unit to be valued was the single unit and the evidence established that the value of that unit was enhanced by the fact that there was market demand for two distinct uses**

(4) No adjustment for special risk of increased competition from existing petrol filling stations or from creation of "milking stations" was justified

(5) The appropriate method of arriving at the capital value of the modernised forecourt was to estimate the likely annual throughput and apply that figure to a price per gallon derived from analysis of comparable transactions, to produce a
D **capital value for either the bare site or the site with plant and buildings**

(6) The appropriate figure of annual throughput for the partially modernised Stonehouse site was 2,837,560 gallons, which represented not the maximum achievable in the future but the figure that would be likely to form the basis of a bid from an oil company at the valuation date

(7) The claimants' valuer's per-gallon figure of £1 was to be preferred to district valuer's 45p but required adjustment because it was derived from evidence upon petrol filling stations that had continued trading for a number of year after
E **acquisition by an oil company without any major modernisation works being carried out — £1 per gallon therefore represented value where modernisation could be deferred — No dispute that Stonehouse would have had to be modernised immediately and this factor had to be taken into account**

(8) An addition to modernisation costs for roof works was not justified but £50,000 should be added for new storage tanks to arrive at total of £420,000 — The proper figure of costs to be deducted was the difference between £420,000 and that figure deferred 7 years at 8%, namely £175,140, as
F **representing relative development cost of Stonehouse site compared with petrol stations capable of continued operation**

(9) The claimants' valuation for injurious affection to the Aero site was to be preferred — A throughput of 1.8m gallons pre-acquisition should be valued at 85p per gallon and the reduced output of 1.276 gallons post-acquisition at 75p, to give loss of value of £573,000

(10) The claimants were entitled to compensation for costs of removal and all costs properly incurred in seeking and finding alternative premises to accommodate the whole of their displaced business and accordingly the agreed sum of £140,561.45 was payable as disturbance compensation

The following cases are referred to in this report. G

ADP & E Farmers v *Department of Transport* [1988] 1 EGLR 209; [1988] 18 EG 80; [1988] 19 EG 147; [1988] 20 EG 104

Horn v *Sunderland Corporation* [1941] 2 KB 26; [1941] 1 All ER 480, CA

Margate Corporation v *Devotwill Investments Ltd* [1970] 3 All ER 864; (1970) 22 P&CR 328; 69 LGR 271; [1971] EGD 313; 218 EG 559, HL

Mayplace Garage (Bexleyheath) v *Bexley London Borough Council* (1989) 29 RVR 158, LT

Payne v *Kent County Council* [1986] 2 EGLR 218; [1986] 280 EG 645; 31 RVR 199, LT

Pointe Gourde Quarrying & Transport Co Ltd v *Sub-Intendent of Crown Lands* [1947] AC 565, PC

Trocette Property Co Ltd v *Greater London Council* (1974) 72 LGR 701; 28 P&CR 408; [1974] EGD 547; 231 EG 1031, CA

H

Matthew Horton QC (instructed by Longmores, of Hertford) appeared for the claimants; Duncan Ouseley (instructed by Welwyn Hatfield District Council and the Treasury Solicitor) for the acquiring authorities.

Giving their decision, THE TRIBUNAL said: These five references are to determine the compensation payable on the compulsory acquisition of petrol filling station, garage and showroom premises situated upon the east side of the A1 at Hatfield, Hertfordshire.

It is agreed between the parties that the five references be heard together (in effect to be treated as a single reference) leading to the award of one sum of compensation. It is further agreed that another petrol filling station and premises on the west side of the A1, opposite J
the subject premises, is land "held with" for the purposes of the assessment of compensation, if any, for injurious affection. The issues in the references relate mainly to matters of fact and value. The claimant company sought at the outset compensation in the aggregate sum of £4,293,000 for land taken, injurious affection and disturbance and the respondent acquiring authorities contended that compensation should be £1,800,000. These figures were amended during the hearing.

Counsel for the claimant called Mr F Wagstaff, a director and former general manager of the company, Mr R D Cook FCA, financial controller of the company, Mr M Dyke, Mr P W Squire FRICS and Mr P R Jones BSc FRICS. Counsel for the acquiring authorities called Mr D A Bradbury CEng MICE, Mr J A Ward DipArch ARIBA, K
Mr C J Conway ARICS, Mr A C Stay, of the St Albans district valuer's office, and evidence by affidavit by Mr A M Tyler ARICS.

On December 30 1980 the district council made the Welwyn and Hatfield District Council (Hatfield Corridor) Compulsory Purchase Order No 2 1980 under the Town and Country Planning Act 1971 in respect of Lands Tribunal references 27 and 28 and this order was confirmed by the Secretary of State for the Environment on November 9 1982. On January 25 1983 the A1(M) Motorway (Roestock to Stanborough Section and Connecting Roads) Compulsory Purchase Order (No CE1) 1983 was made under the Highways Act 1980 by the Secretary of State for Transport in respect of Lands Tribunal references 29, 30 and 31. Notices to treat were L
served by the Secretary of State for Transport on April 25 1983 (reference 31), May 9 1983 (references 29 and 30) and by the district council on May 13 1983 (references 27 and 28). Following service of notices of entry, possession was taken of the subject land in stages at various times between October 3 1983 and March 12 1984, but for the purpose of the combined reference to this tribunal it is agreed that February 14 1984 should be taken as the date of entry and valuation.

The subject premises together with land in other ownerships have been acquired to enable the Department of Transport to improve the A1 trunk road to current motorway standards and in particular to facilitate the construction of what is now known as the Hatfield Tunnel, which has been built in the substrata. The involvement of the district council relates to the redevelopment, currently in progress, of M
the surface land over and adjacent to the tunnel for retail and other commercial uses.

With the exception of reference 29, relating to the leasehold interest to AD 2011 in a strip of land on the southern boundary used for access purposes, the whole of the land taken, 3,260m² (0.806 acre) was freehold in the ownership and occupation of the claimant company.

The premises, known as Stonehouse Garage, had a frontage to A1, at this location a six-lane dual carriageway, of 190 ft with access and exit points thereto for south-bound traffic only. For many years they had been occupied by the company for its activities of retailing petrol, servicing and repairing cars, selling new and used cars and administering a sizeable motorcar-related business conducted both here and at a number of other premises in South Hertfordshire. Immediately opposite and accessible only to north-bound traffic were other premises known as the Aero Garage in the ownership and occupation of the company. Aero Garage was used mainly for the retailing of petrol, the sale of new cars by different manufacturers to those sold at Stonehouse and the wholesaling of spare parts. Aero was dependent upon Stonehouse, *inter alia,* for workshop facilities to repair and service cars and for general administration of the activities conducted from the Aero site. In effect the two premises together with two smaller satellite premises nearby were used for an integrated motor-related business and for this reason it is agreed that Aero is "held with" Stonehouse within the context of the assessment of compensation upon compulsory acquisition under the Land Compensation Acts 1961 and 1973 and the Compulsory Purchase Act 1965.

The Stonehouse buildings were developed and altered in a somewhat piecemeal fashion over the years 1932 to 1958. The last major rearrangement of the petrol forecourt was in 1970, although the pumps were replaced by stages from 1977 to 1979. Up to the date of entry the method of operating the forecourt was for customers to be served by attendants rather than serving themselves as is now common, if not universal, practice. The stations opened 24 hours per day for 364 days of the year and petrol was discounted, namely sold at about 3p per gallon lower than the general price prevailing in the locality. The business strategy of the claimants appears to have been that convenience of location on a very heavily trafficked trunk road, the certainty of being open and a low price would attract a high volume of petrol sales even in competition with modern self-service stations providing extended facilities, such as a retail shop or kiosk selling a wide range of non-motor-related goods. An identical approach was also adopted by the company at the Aero site. The combined petrol sales volume at the two sites in the early 1980s amounted to some 5m gallons per year, about 2.75m at Stonehouse and 2.25m at Aero. The Aero site is still owned and operated by the company, but following the opening of the Hatfield Tunnel to traffic on December 10 1986 the annual throughput achieved during 1987, following modernisation, was 1.276m gallons, about 60% of the previous volume.

The Stonehouse and Aero premises were of some age in terms of their buildings and equipment, with the result that the layout and appearance of the forecourts were somewhat outmoded for high-volume petrol stations on a principal trunk road such as the A1. During the 1970s a measure of uncertainty surrounded both premises owing to the various road-improvement schemes which were being proposed and no doubt this influenced decisions concerning new investment and modernisation by the company. It was common ground that commercial prudence indicated the need for some modernisation at Stonehouse in February 1984, but it was not agreed whether total redevelopment as a petrol filling station only, with the consequent loss of the showroom, workshop etc facilities, or a reconstruction of the existing forecourt, but retaining the other facilities, would have been more advantageous. This situation affects the approach to valuation adopted by the parties and the ingredients of the total compensation sum contended for. Pending a determination upon this issue of principle the parties have agreed figures for certain elements of the overall claim, namely:

1. The value of the freehold interest in the existing land and buildings at Stonehouse (excluding the forecourt area) upon the assumption that the buildings would have remained on site in their existing form for the foreseeable future, £535,000.
2. Substantially the same buildings as above but with parts demolished to facilitate the construction of a rearranged forecourt, £465,000, revised during the hearing to £450,000.
3. Items of disturbance subject to their not being inconsistent in principle with the basis of compensation ultimately determined for land taken:

	£
(a) Removal expenses	14,949.91
(b) Adaptation of new premises	2,046.50
(c) Double overheads	27,590.22
(d) Fees and costs re new premises	2,809.75
(e) Fees and costs incurred re abortive sites and premises	45,761.07
(f) Management time re subject premises, new premises and abortive sites and premises	47,404.00
Total	£140,561.45

The issues in dispute material to the amount of compensation payable for the Stonehouse premises can be summarised as follows:

1. The market value of the freehold interest in the site of 3,260m^2 at February 14 1984. Relevant to this are:
   (a) An assumption of total redevelopment as a petrol filling station only.
   (b) An alternative assumption that the existing buildings largely remain for the continued operation of an integrated motor business but that some reconstruction and modernisation is carried out to the forecourt; at the hearing accorded the somewhat misleading term "partial redevelopment".
   (c) The future sustainable petrol throughput of the premises: (1) in their exact 1984 state; (2) upon assumption (a) above and (3) upon assumption (b) above.
   (d) The effect, if any, upon the estimated post-1984 throughput of an assumption that an alternative by-pass scheme might have been adopted at that date, the "risk factor". Reference was made to *Pointe Gourde* and *Margate Corporation* v *Devotwill Investments Ltd.**
   (e) The conditions likely to have been imposed upon any planning permission required for (a) and (b) above with reference to highway improvement works.
   (f) The estimated cost of the works in (a) and (b) above.
   (g) The capital value or price which the potential to sell the estimated annual gallonage as determined at (c) above would command in the market.

(2) The amount of injurious affection suffered by the claimants consequent upon the accepted reduction in the petrol throughput of the Aero site as a result of the opening of the Hatfield Tunnel.

(3) Whether the whole or a part only of the sum agreed for disturbance items is properly payable having regard to the basis of valuation selected for the land taken. Reference was made to *Horn* v *Sunderland Corporation.*

Evidence for claimant

The evidence of Mr Wagstaff, who has been involved with the company for some 44 years, was directed towards the history, management and detailed operation of the business at Stonehouse, Aero and other premises in the locality. He emphasised that it had always been the aim of the company to maximise the amount of petrol sold and to remain independent in the sense of not being tied for supplies to any particular oil company. On the two sites the company had massive, 70,000 gallons, storage capacity which ensured a supply for customers 24 hours per day even during occasional periods of industrial disputes which affected deliveries by oil companies. The new vehicle franchises held by the company were Peugeot at Stonehouse and Alfa Romeo and Honda at Aero; a very large quantity of spare parts was stored at the Aero site for wholesale and retail to customers throughout a wide area of the northern home counties. He had been aware of possible improvement works to the A1 for some 30 years but until 1974 he expected the major impact to be at the Aero site rather than Stonehouse where previous capital expenditure had been concentrated.

Since 1983, he had spent considerable time searching for and investigating alternative premises to replace the mixed motor-trade facilities being lost at Stonehouse. The search for an existing high-volume petrol filling station or a site upon which to construct a station had proved abortive but the majority of the other facilities had been secured elsewhere, including upon the Aero site by effecting some reorganisation of that site. A considerable part of Mr Wagstaff's evidence related to the extent, nature, relevance and reasonableness of his investigation, with professional assistance, of alternative premises and sites which he stressed had to be conducted against the six months' time-period specified in the notice of entry served upon the company. As an agreement was reached by the parties during the hearing as to the costs attributable to the search for alternative premises, if as a matter of law that item is compensatable in whole or part, it is not necessary to record further that aspect of the evidence.

Mr Wagstaff agreed that in 1984 self-service was becoming widespread in petrol retailing but the company had not changed from attendant operation as it wished to be certain that self-service had become a permanent feature of the trade before incurring the capital expenditure to make the change. He accepted that during the years 1979-82 there was some decline in petrol throughput at Stonehouse,

*Editor's note: Reported at (1970) 218 EG 559.

A which he attributed in part to not being a self-service station. Although all the buildings at Stonehouse were in use in 1984, some modest demolition, minor internal adaptation and reallocation of space would have been possible to enable the forecourt to be reconstructed for self-service operation. He also stressed that the business of the company was an integrated motor-trade operation and that in 1984 it wished to continue trading in that manner at the Stonehouse and Aero sites. In his opinion, the Stonehouse premises were of adequate size to run a self-contained integrated motor-trade operation divorced from the Aero and other nearby sites occupied by the company. The relocation of present central administrative and other functions would make sufficient space available for storage on B site of spare parts etc. The capital required for adapting the premises and reconstructing the forecourt could have readily been obtained from an oil company keen to secure a tie in respect of the supply of such a large volume of petrol by means of a sale-and-leaseback of the whole site or of the forecourt area only.

Mr R D Cook, who had been employed by the company for some 18 years, first as chief accountant and since 1983 as company secretary, gave evidence relating to the financial aspects of the operation of the petrol sales at Stonehouse. It was his opinion that if the forecourt had been reconstructed and changed to self-service operation, but retaining the car showroom and the repair workshop in accordance with a scheme prepared by Mr Dyke, an annual throughput of 3m gallons was both achievable and sustainable.

C He considered the price charged up to 1984, 3p below the prevailing price, was needlessly low and that it could have been increased without reducing the volume of sales. Furthermore, night sales, between 11.30 pm and 7.30 am, accounted for only 9% of the total volume and if the station had been closed during these hours it would have had little impact upon net profit. He produced in evidence calculations showing how, following reconstruction and adoption of self-service operation, net profits could be increased substantially by a price rise of 1p per gallon and a significant contribution from retail sales at a kiosk or shop. These calculations were based upon an achieved throughput in 1983 of 2.25m gallons, but if the calculations had been based upon the preceding 5 years' average throughput of 2.837m gallons then the net profit would have D been increased further by about 40%.

Although in cross-examination these calculations of increased net profit were shown to be marginally over-optimistic, Mr Cook nevertheless held to the view that they clearly demonstrated the financial benefits of reconstructing the forecourt for self-service operation. The achievable net profit was very considerable and undoubtedly would attract a potential occupier, including an oil company. Like Mr Wagstaff, Mr Cook was also confident that if the Stonehouse site had to operate as an independent integrated motor business there would be adequate space within the buildings following minor internal reorganisation to accommodate the needs of both a reconstructed forecourt and the other uses required at the E site.

Mr M Dyke, who had been involved in the design of petrol filling stations for some 10 years for oil companies and other operators, produced in evidence plans of the Stonehouse forecourt in 1984, a scheme for reconstruction to self-service operation (revised somewhat by an additional plan during the hearing) and a scheme for the total redevelopment of the site as a petrol filling station only. The latter plan was prepared mainly by way of response to a total redevelopment scheme produced on behalf of the acquiring authorities.

Briefly summarised, the evidence of Mr Dyke was that the Stonehouse site was capable of achieving and maintaining a throughput of 3m gallons per annum. This would be very attractive F to an oil company in 1984, as interest increased when the throughput was 1m gallons per annum or more. To attain 3m gallons per annum it would be necessary to reconstruct the then existing forecourt area to permit self-service operation and provide a kiosk for control and retail sales. His scheme provided for the replacement of the 11 semi-obsolete pumps on two islands by nine multi-nozzle pumps on separate islands which would permit a maximum of 18 cars to fill at any one time. The existing underground storage tanks of 43,000 gallons capacity would remain undisturbed, a new canopy would be constructed and also a sales kiosk of about 270 sq ft. These works would cost £165,000 (altered during the hearing to £262,000 following an agreement reached between quantity surveyors) and take about eight to 10 weeks to construct.

Mr Dyke criticised the total redevelopment scheme prepared for G the acquiring authorities, with an estimated cost of £572,000, mainly on details of layout and the over-extensive site area used, in particular some back land which he considered could be used for a further motor-related function such as a tyre and exhaust centre. His own alternative scheme was not significantly different in concept but had an estimated cost of £501,500. In cross-examination, Mr Dyke's reconstruction scheme for the existing forecourt was challenged as to practicability and acceptability to an oil company occupier on the basis of the size of the kiosk, the adequacy of the nine pumps to dispense 3m gallons per annum, the availability in February 1984 of the multi-nozzle pumps proposed, possible problems for cars in manoeuvring around and between the pumps without obstructing the required through-flow of vehicles and that the estimated costs did H not allow for the replacement of the ageing storage tanks. Some of these criticisms were accepted but in the context that, from experience, they could be either overcome or ameliorated at the stage of detail design to the extent that the resulting forecourt would be acceptable to an oil company operator.

The evidence of Mr P W Squire, who for the past 10 years has been wholly involved in the sale and valuation of motor trade and petrol filling station properties, can be summarised as follows:

(1) The number of petrol filling stations in the United Kingdom fell by one-third between the years 1975 and 1988 (31,246 to 20,016), the percentage owned by oil companies rose from 27% to 33.5% and the average annual gallonage sold per station J increased from 143,000 to 320,000.

(2) Oil-company-owned petrol filling stations generally have throughputs in excess of the average for all stations as they own and continue to purchase the higher volume stations.

(3) The prime objective of all oil companies, major and minor, is to increase market share so as to spread production and distribution overheads over a greater volume; effectively oil companies comprise the most important sector of the market for higher volume stations.

(4) Oil companies increase their market share by purchasing freehold or long leasehold premises and, failing this, as second best, enter into exclusive supply agreements or "ties" for three to five years with independent owners and operators. It is not K uncommon following the purchase of a freehold or leasehold to license the station back to the former owner, who will be obliged to retail the products of the oil company.

(5) He had no experience of offering for sale a petrol filling station with a past recorded throughput as high as 3m gallons, which volume is comparable to that achieved at two motorway service areas within his knowledge.

(6) Past throughput and the prospect of its being maintained or increased is an important factor to an oil company purchase; an established station will command a premium price over a new greenfield location.

(7) Oil companies would prefer to purchase a single high-volume L station of 3m gallons rather than three stations with a throughput of 1m gallons each.

(8) It is customary to analyse sale transactions on the basis of the capital sum paid per existing gallon of throughput. In the market this sum appears not to be reduced or varied for a station operating 24 hours per day or where petrol is being sold at a discount.

(9) An average figure for the cost of redeveloping a petrol filling station is £500,000, roughly 50% of which relates to building works and 50% to plant and machinery; the latter attracts a capital allowance of 25% for four years and is available as a deduction against trading profits. The cost of redevelopment is of less significance in the market for high-volume stations and is M frequently a general bargaining factor rather than an actual deduction from purchase price.

(10) The Stonehouse forecourt when reconstructed in accordance with the plans prepared by Mr Dyke would be capable of a throughput of 3m gallons per annum and would have been so regarded by oil companies in 1984.

(11) The "starter gate" layouts for petrol pumps and the size of sales kiosk incorporated into the redevelopment scheme prepared for the acquiring authorities and by Mr Dyke were post-1984 innovations.

(12) While accepting that in 1984 the Stonehouse forecourt was of outmoded layout and appearance, on the basis of his experience

and the evidence he produced of three comparables, he was of the opinion that the freehold forecourt would have commanded a price in 1984 of not less than £1 per gallon and perhaps up to £1.25. Logically, a higher price would be paid where the throughput was 3m gallons per annum rather than 1m gallons, but there was no market evidence of freehold or long leasehold transactions in stations with this higher volume.

(13) His valuation of the Stonehouse forecourt, as existing in 1984 prior to any reconstruction, was at least 2,837,560 gallons (the average sales for 1978 to 1982) at £1, ie £2,837,560.

In cross-examination, Mr Squire's evidence was challenged on the basis that the three comparables upon which he relied related to stations of more modern layout and appearance than Stonehouse and in any event they did not support the £1-per-gallon figure to which he had spoken when analysed in detail having regard to surplus land and other factors. Mr Squire's response was threefold. First, as a matter of practice he divided the sale price by the actual gallonage being achieved at the date of sale, since achieved and maintainable throughput was predominant in the mind of oil company bidders. Second, the cost of redevelopment or reconstruction was a minor factor in the price offered for a 1m-gallon-per-annum-plus station and would be especially so with a very high volume station such as Stonehouse. Third, his comparables related to stations of about 1m gallons annual throughput whereas the subject premises historically had achieved in excess of 3m gallons.

It was also put to Mr Squire that the three comparable transactions upon which he relied resulted from the sale of a group or package of filling stations to a single purchaser for an aggregate consideration or sum. His response was that he personally negotiated the sales to the single purchaser and agreed an individual price as quoted for each property. How the purchaser subsequently viewed the transaction was outside his knowledge and should not impugn the validity of the transactions as comparables. In reply to a question by the tribunal, Mr Squire said that the £1 per gallon minimum capital value which he had adopted assumed that the past average throughput would be maintained and was not exposed to risk from abnormal competition or other adverse factors.

The evidence of Mr P R Jones, who has some 20 years' professional experience, including advising oil and petrochemical companies, in summary form was as follows:

(1) His firm advised the claimant company prior to the confirmation of the compulsory purchase orders and his personal involvement began following confirmation of the orders. He saw the Stonehouse premises prior to demolition.

(2) The Stonehouse and Aero sites were operated as a single entity and petrol sales for the combined site averaged 5.1m gallons per annum for the period 1976 to 1983 — 5.4m was achieved in 1979 and 5.37m in 1980; the respective volumes for Stonehouse alone were 2.84m and 3m gallons.

(3) The only petrol filling stations known to him with a sales volume of this order are in Dover and some motorway service areas, but evidence is generally scarce owing to the desire of the oil companies to preserve confidentiality by not disclosing commercially sensitive trading information.

(4) There would have been keen competition among oil companies in February 1984 to purchase the Stonehouse site owing to the very high volume of past sales.

(5) The reconstructed forecourt as detailed in the proposals by Mr Dyke would function satisfactorily and, if necessary, the sales kiosk could be enlarged by including a small part of the existing workshop.

(6) Petrol filling stations with annual throughputs of 0.5m to 1.25m gallons act as a benchmark value but do not assist directly in the valuation of premises selling *circa* 3m gallons per annum. The concern of oil companies is to increase market share, and therefore high-volume petrol filling stations free of tie are particularly attractive.

(7) Oil companies would be interested in acquiring the whole premises freehold but a more likely way of effecting disposal would be a sale of the freehold, a leaseback of the workshop and showroom and a licence back to operate the petrol filling station.

(8) When analysing sale transactions to obtain a site price or value it is necessary to deduct the value of the on-site building works and machinery etc used in the operation. If these are obsolete the transaction price effectively represents the site value.

(9) He based his initial valuation upon £1 per gallon of the average past achieved throughput (2.837,560m gallons) and deducted from that sum a cost of £165,000 as initially estimated by Mr Dyke for reconstructing the existing forecourt.

(10) This £1 per gallon was based upon his own experience and also upon three transactions put in evidence by Mr Squire, which he now understood were interpreted by Mr Squire as supporting a site value of £1 per gallon and thus was not a value for a site with buildings and plant.

(11) During the hearing Mr Jones therefore increased his valuation by the sum of £165,000, which he had previously deducted for the cost of reconstruction works.

(12) The bulk of the available evidence related to petrol filling stations with annual throughputs of *circa* 1m gallons, some transactions being before and some after the agreed valuation date.

(13) Detailed analysis of these transactions demonstrates a rising market over the years and a higher price per gallon as the aggregate sales volume rises; in his view, 3m gallons per annum must be worth more per gallon than 1m gallons per annum (he supported this opinion by reference to schedules and graphs).

(14) All the activities of the claimant company on the Stonehouse site had been disturbed by the taking of the premises, but post-February 1984 they had all been relocated elsewhere save for the petrol forecourt.

(15) His valuation of the total premises, the forecourt, all buildings and other land was based upon their continued use in substantially the same form as in February 1984 save for reconstruction of the forecourt per the scheme prepared by Mr Dyke.

(16) This was an "existing use" basis of value for the freehold interest and in addition the claimant was entitled to the costs involved consequent upon being disturbed from occupation and for effecting removal elsewhere.

(17) This approach was not contested in broad principle by the acquiring authorities but it had not been possible to agree this as the basis of valuation because they considered that a higher value, in excess of existing use value plus disturbance, resulted from an assumed total redevelopment of the site, to facilitate which it would have been necessary for the claimant to vacate the whole site of its own accord.

(18) His valuation of the whole site on this alternative total redevelopment basis was £2.8m; some £600,000 lower than his primary valuation.

(19) The acquiring authorities regarded the reconstruction of the forecourt as constituting "partial redevelopment" and therefore involving a degree of inconsistency with a comprehensive claim for compensation for disturbance.

(20) The items relevant to the comprehensive claim for disturbance have now been agreed (see above).

(21) In respect of compensation for injurious affection to the Aero site, the land "held with", there is no dispute in principle with the acquiring authorities; the agreed approach is to proceed on a before and after basis. Differences, however, exist in relation to the sales volumes to be adopted and the price per gallon to be applied thereto.

(22) He made no specific allowance in his valuations for the risk of potential trade being attracted away from the Stonehouse premises by new competition elsewhere or by the possible diversion of traffic on to a by-pass road in the "no scheme" world.

In cross-examination Mr Jones was challenged, *inter alia*, in respect of the following matters:

(a) The principle and reality of reconstructing the forecourt area. He did not accept that total redevelopment as at Dover was necessary.

(b) It was suggested that the £1 per gallon capital value adopted was "out of line" and excessive by reference to his own comparables, those of Mr Squire and others introduced by the acquiring authorities. He did not accept that this was so when regard was had to the relative transaction dates and the higher volume of throughput at Stonehouse. Although the Dover site had an estimated 3m gallons per annum throughput he was not sufficiently informed about the transaction and in particular whether there was any competition to buy among purchasers, bearing in mind that at the date of sale it was tied to an oil company (the ultimate purchaser) for some 18 months.

(c) Failure to make any allowance for the risk issue. In his view, the extent of any allowance in February 1984 would depend entirely upon the state of the market at that time and the attraction of the opportunity to acquire a very high volume throughput station would result in its being minimal.

(d) Discounting the sale price of petrol and 24-hour working. In his view, it was not common oil company practice to make an allowance for these factors which tended to enhance throughput. It was, however, common practice in rating valuations to make specific allowances and his firm had agreed the assessment upon the Aero site after making deductions for these factors. In his view, the size of the throughput notionally being offered in the market was such as to render these considerations insignificant.

(e) He considered that a comprehensive disturbance claim was appropriate if the "existing use" primary valuation, based upon a reconstructed forecourt, was adopted for the purpose of compensation for land taken. However, no items relating to the reorganisation of existing space at Stonehouse to facilitate the reconstruction were, or should be, included in the agreed figures of claim.

(f) In the injurious affection claims different values per gallon were used "before" (90p) and "after" (65p). This was to have regard to the relative annual volume of sales at the selected dates of 1983 (before) on an assumed self-service operation of 2m gallons and 1987 (after) at 1.276m gallons; the achieved volume by self-service operation.

Evidence for acquiring authority

The evidence of Mr Bradbury related to the traffic conditions upon the A1 trunk road in the 1970s and early 1980s and the reasons for selecting the tunnel option which gave rise to the acquisition of the reference premises. He also commented upon the traffic implications of the forecourt modernisation scheme prepared by Mr Dyke and in particular the need, in his opinion, for a deceleration lane to reduce congestion on the A1 caused by traffic queueing to enter a modernised petrol station.

Mr Conway, responsible for development control at the district council, agreed that Mr Dyke's proposal was acceptable in town planning terms. Like Mr Bradbury, his main concern would have been to secure an improved access to the site and he agreed that a deceleration lane was necessary in respect of which he would have sought a section 52 agreement had an application for planning permission been made for the modernisation.

Mr Ward, who has considerable experience in the design and layout of petrol filling stations for oil companies, said that self-service operation became common in the late 1970s. His detailed evidence was directed to the layout he had prepared for the total redevelopment of the reference premises, solely as a petrol filling station. He spoke about the detailed design criteria for modern stations to meet the requirements of oil companies, including such matters as the number and type of pump, vehicle manoeuvring space, required underground tankage and size of retail shop. He also referred to the requirements of the petroleum officer with respect to new and improved petrol stations.

Mr Ward commented upon the proposals (as revised) by Mr Dyke, and while he felt they were not ideal he accepted in cross-examination that a 10-pump layout as proposed could dispense 3m gallons per annum and that this volume could increase to 3.4m gallons on an ideal layout such as would be achieved by total redevelopment by, for example, his own scheme.

Mr Stay was of the view that the best price that could be obtained for the reference premises was to sell the whole site to an oil company, who might be expected to demolish everything on the site and build a modern self-service petrol filling station without any of the attendant workshops, offices, and buildings used for other motor trade purposes. His primary valuation, therefore, was what he called a total redevelopment basis, but he also produced a valuation in response to the claimant's valuation on what he called a partial redevelopment basis.

To establish the market value he made an extensive search for evidence of transactions relating to sites and developed petrol filling stations with a throughput or estimated throughput of over half a million gallons per annum.

In several cases purchasing companies had declined to authorise disclosure of the details of the transaction or the throughput, actual or estimated, and therefore he was able to refer to only 10 petrol filling sites and nine developed petrol filling stations where transactions had taken place between September 1980 and April 1987.

He analysed these transactions in order to arrive at the capital value of a bare site in terms of pounds per gallon estimated annual capacity (EAC). To achieve this he first deducted from the purchase price his estimate of the value of the plant and buildings. In the case of rented premises he had capitalised the rent assuming an 8% yield. He then adjusted the actual throughput where necessary to allow for 24-hour opening and price discounting, normally deducting 7½% for each. In his analysis he had adjusted the achieved throughput where, in his view, it would be likely to change on being acquired by an oil company and/or on the plant and buildings being improved.

He valued the Stonehouse premises on the assumption that the whole site would be available to accommodate a modernised petrol filling station. He assumed an EAC of 2,900,000 gallons and a rate of 50p per gallon less 10% for the by-pass risk factor (see Appendix IIa).

He prepared a tentative check valuation on the partial redevelopment basis adopted by the claimant at £1,153,350 (see Appendix IIb at p 247), based upon an EAC of 2,125,000 gallons at 60p, less 10% for the by-pass risk factor, net 55p.

APPENDIX IIa

Mr A C Stay's valuation

Stonehouse premises on the basis of sale for redevelopment of whole site as self-service petrol filling station.

Value of land taken

(i)	Estimated initial throughput of motor fuel on basis 16-hour opening and normal prices	Say	2,900,000 gals pa	
	Having regard to risk factor — possible by-pass (50p − 10%)		at 45p	£1,305,000
(ii)	*Less* costs of works to highway, to form deceleration splay and alteration of statutory undertakers' services		Say	30,000
				1,275,000
(iii)	Spare land at rear for car wash and car parking etc Say, one-third acre at £250,000			83,333
	(Demolition costs absorbed)			£1,358,333

Mr Stay had also looked at the trading accounts of the claimant company and concluded that there was evidence to support his view that the intensively manned attendant service, with discounted prices and 24-hour opening, yielded very little profit and was therefore uneconomic. He concluded that a better return could have been obtained by selling the whole site for redevelopment by an oil company, who would have paid a price in the region of his valuation figure of £1,358,333.

He sought support for his approach to valuation from guidelines which formed the basis of agreement of the rating assessments of petrol filling stations under the 1973 valuation list (which remained in operation until 1990) and similar guidelines issued for the 1990 revaluation. He also referred to the rating assessment of the Aero premises which had been negotiated and agreed with members of Mr Jones' firm.

He produced his own analysis of the transactions in comparable petrol filling stations referred to for the claimant and he rejected as reliable evidence those transactions where the petrol filling stations had been sold as part of a "package" involving a number of other petrol filling stations.

He made reference to a report of the Monopolies and Mergers Commission on refining capacity which demonstrated that the highest percentage of refinery utilisation was achieved in the year 1984. In his view, this demonstrated that the demand for petrol filling stations in 1984 was likely to be low, as oil companies did not have unused refining capacity. This conclusion was contrary to Mr Squire's experience that there was an active market in petrol filling stations in that year.

Decision

We have inspected the locality of the subject premises (now

APPENDIX IIb

Mr A C Stay's tentative check valuation

Stonehouse premises as existing but with a forecourt rearrangement

With a 10% risk factor for possible by-pass

1. *Land and buildings*			
Agreed total value			£535,000
Less: (a) 2-storey building (Refs 6, 8 & 9)	£70,000 (agreed)		
(b) Part of main showroom (Ref 4 & forecourt display (Ref 11))	10,000 (agreed)		
(c) Part of main workshop (Ref 3) for shop etc and new ramp from N side right of way Say	31,786		
		£111,786	
Add back			
(i) value of first-floor offices	£36,823		
(ii) additional value of used car display (formerly workshop)	7,824		
		44,647	
Net reduction		67,139	
			Say £67,100
			£467,900
2. *Petrol forecourt*			
Average throughput 1978-82	2,837,560 gals pa		
Making only a 15% adjustment for poor layout and congestion	425,634		
	2,411,926 but say 2,500,000 gals		
Less 15% for 24-hour opening and cut price		375,000 gals	
		2,125,000 gals	
2,125,000 gals pa at 55p			£1,168,750
(as developed) (reflecting by-pass risk factor)			£1,636,650
3. *Cost of modernisation works to be deducted*			
As partially agreed between quantity surveyors			483,300
		Value	£1,153,350

demolished), the Aero site and a number of the comparables cited in evidence.

We were referred to the following authorities and Lands Tribunal decisions:

Pointe Gourde Quarrying & Transport Co Ltd v *Sub-Intendent of Crown Lands* [1947] AC 565.

Margate Corporation v *Devotwill Investments Ltd* (1970) 218 EG 559, HL.

Horn v *Sunderland Corporation* [1941] 2 KB 26, CA.

Trocette Property Co Ltd v *Greater London Council* (1974) 231 EG 1031, CA.

Payne v *Kent County Council* [1986] 2 EGLR 218; (1986) 31 RVR 199, LT.

Mayplace Garage (Bexleyheath) Ltd v *Bexley London Borough Council* (1989) 29 RVR 158, LT.

ADP & E Farmers v *Department of Transport* [1988] 1 EGLR 209; [1988] 18 EG 80, LT.

There are a number of issues in the reference and we shall deal with them in turn.

The scheme

In the early 1980s the A1(M) motorway stopped immediately to the north and to the south of Hatfield and the two ends were joined by the A1 trunk road through the centre of the town. It must have been evident at that time that this situation could not continue indefinitely and that there would eventually be a scheme linking the two sections of the motorway so as to improve traffic conditions in and around Hatfield.

This could have been achieved in a number of ways: by widening and improving the existing A1 route through the town centre, by by-passing the town altogether, or by tunnelling under or bridging over the town. If the by-pass scheme had been adopted it is easy to see that traffic past the Stonehouse premises would have been severely reduced; but the likelihood of the by-pass scheme being preferred was not great. Mr Stay makes a 10% end deduction in his valuation to allow for this risk.

The amount of the deduction is not disputed but the claimant contends that as a matter of law there should be no such deduction because in the "no-scheme world" there would be no risk as the by-pass scheme falls to be disregarded by reason of section 6 of the Land Compensation Act 1961 and/or what is generally referred to as the *Pointe Gourde* principle. It is essential to the success of that argument that the relevant scheme is the original and general intention to link the two ends of the motorway by whatever means.

The acquiring authority contend that the plan to build the tunnel route is the relevant scheme and nothing further.

It is acknowledged by the authorities that the identity of the relevant scheme is a matter of fact. There is no statutory definition of "scheme" and we think that the word can be used suitably to describe either the initial overall aim to link the motorway routes or any of the various detailed plans designed to achieve this.

The need to identify the scheme arises so as to give effect to the principle that compensation should not include any increase or decrease in the value of the land which is solely attributable to the scheme giving rise to the compulsory acquisition (see *Pointe Gourde* and section 6 of the Land Compensation Act 1961).

In the present case the initial overall aim to link the motorways did not and could not have involved the compulsory purchase of the subject premises until the preferred route had been firmly established. Therefore it seems to us that the scheme underlying the compulsory purchase of the subject premises is the detailed scheme involving the tunnel route. That is the scheme that gave rise to the compulsory acquisition and that is the scheme that has to be disregarded for the purposes of assessing compensation.

On this basis we find that there would have been a risk that the alternative by-pass route might have been adopted and that the effect would have been to reduce the value of the subject premises in the market by 10%.

The Stonehouse premises — unit to be valued

There is no dispute that the most likely purchaser of the Stonehouse premises would be one of the major oil companies seeking to exploit the potential of the petrol filling station but with no particular desire to acquire the workshops and other buildings used by the claimant in connection with other aspects of the motor trade. It is accepted that the petrol filling station would have had to be completely modernised in order to maintain or improve the existing throughput. What is disputed is the degree of modernisation required to achieve this objective.

The claimant, after several attempts, finally adopted (for valuation purposes) a scheme of partial modernisation whereby the whole forecourt was rearranged and increased in size with nine multi-nozzle pumps which would provide for a throughput of 3m gallons per annum together with a new and larger shop and cash desk. This could be achieved only by the demolition of a small part of the showroom building at the front and resiting the offices used in connection with other motor-trade activities. This partial modernisation would have resulted in something less than the ideal petrol filling station layout normally sought by a major oil company.

It was suggested by the acquiring authorities that this scheme would therefore be unacceptable to prospective purchasers and they submitted a plan for demolishing all the buildings and plant followed by redevelopment of the entire site as a petrol filling station with associated facilities together with ample space for further extension.

During the hearing these two plans were referred to as partial and total redevelopment. As neither plan would involve a change of existing use we think it preferable to refer to the proposals as partial modernisation and total modernisation.

Counsel for the acquiring authorities also suggested that as the whole site was the subject of compulsory acquisition it must be valued as one unit and not split into two or more units, each with a separate value. He submitted that the Compensation Acts required a willing vendor and this assumed that all the land was available for sale as a single unit at the valuation date. Therefore, there was no opportunity for suggesting, for valuation purposes, that the claimants or a similar operator might, after a freehold sale of the whole to an oil company, continue to use the residual land and buildings for motor-trade-related activities and possibly manage and operate any modernised petrol filling station on the subject site. In support of this contention he referred to the Court of Appeal decision in *Trocette*.

We accept the evidence that, in the absence of compulsory

A acquisition, the most profitable and acceptable arrangement for the claimants would be to dispose of the forecourt, slightly extended, to a major oil company for modernisation and to retain the other buildings, slightly reduced, for their own occupation. This could have been achieved in a number of ways including the sale freehold of the whole site to an oil company subject to separate leasebacks or licence of the forecourt and the other buildings; there is evidence that this is not an unusual arrangement. An oil company would no doubt have preferred to take the whole site for use as a petrol filling station only, but we are satisfied that the potential throughput was sufficiently attractive for them to have accepted less than the ideal.

In our view, *Trocette* is not wholly in point. However, it is not B contrary to the principles in that case to have regard to evidence of events and disposal procedures current in the market at the relevant date. The true position is that the compulsory purchase orders define the land to be taken and the Land Compensation Acts prescribe the basis of compensation but not the method of valuation to be adopted in assessing compensation.

In the present case the unit to be valued is the single unit. The evidence establishes that the value of that single unit is enhanced by the fact that there is a market demand for two distinct uses. It is proper to take account of that demand in valuing the single unit.

Risk of competition or milking

The established throughput of any particular petrol filling station C may be at risk from increased competition from other existing petrol filling stations and the possibility of new stations being created. These are known as "milking stations". The evidence is that milking is usually more effective from upstream competition than downstream; in the instant reference, to the north of the Stonehouse premises.

Mr Stay was of the view that there was a substantial risk of a milking station being established upstream at Oldings Corner on the northern boundary of Hatfield and an even greater risk from the existence of two small downstream petrol filling stations within sight of the subject premises, which had potential for redevelopment. He also referred to two modern self-service stations established on the A414 and A405. Mr Stay considered the milking risk at the subject D premises to be above normal and allowed for it by reducing the estimated annual consumption (EAC). In cross-examination he expressed the view that the allowance should be in the order of 15-20%.

There is no evidence of comparable adjustments for milking risks being considered in the analysis of the transactions relating to comparable premises. Moreover, there is no evidence to support the view that the existing petrol filling stations constituted a risk to the subject premises. Indeed, the agreed compensation on the acquisition of these two small stations immediately to the south did not include any value for redevelopment potential. It seems to us likely that town planning and highway policies would have deterred E the creation of a petrol filling station on the motorway immediately to the north. On the evidence it would appear that the subject premises were, if anything, in a position of less than normal risk.

Therefore, in our view, no adjustment for special risk is justified at the subject premises. We are, however, left in some doubt as to what adjustment, if any, should have been made in the analysis of some of the comparable transactions, where there are indications that the petrol filling station was subject to a more than normal risk of milking.

Basis of valuation

Partial modernisation

On the evidence, we are of the view that it would be reasonable to F suppose that any oil company acquiring the extended forecourt on St Valentine's Day 1984 would have modernised the petrol filling station in the manner shown on Mr Dyke's final revised plan (reference no 106D amended). This layout would have provided a self-service petrol filling station under a canopy with nine pumps together with a shop and cash desk. The layout would have been less than the ideal hoped for by a major oil company but it would, in the circumstances of the unusually large throughput, have been acceptable. The existing storage tanks were sufficient for the throughput of some 3.32m gallons per annum which had been achieved in 1976. They were 20 years old and still serviceable, although we accept Mr Ward's view that in practice when making a major change in the layout of the station it would have been prudent to renew the underground tanks. G

It is accepted that the appropriate method of arriving at the capital value of the modernised forecourt is to estimate the likely annual throughput and to apply to that figure a price per gallon derived from an analysis of comparable transactions; this will produce a capital value for either the bare site or the site with plant and buildings.

The relevant valuations of Mr Jones and Mr Stay are set out respectively in Appendices I and IIb [see p 247 for Appendix IIb] attached hereto.

As the parties are at issue on practically every stage of the valuation we will examine each in turn.

APPENDIX I

Mr P R Jones' valuation H

Stonehouse premises on an existing use basis assuming a forecourt rearrangement

Adopting a 10% risk factor for possible by-pass

Land and buildings		£535,000	
Deduct			
(a) Two-storey buildings (refs 6, 8 and 9)	£70,000		
(b) Part of main showroom (ref 4) and car display	10,000		
(c) Part of main workshop (ref 3)	28,000	108,000	£427,000
Add back (per DV 36A)			
(i) Value of first-floor offices	36,823		
(ii) Used car display (formerly workshop)	7,824		
	44,647	Say	45,000
			472,000
Petrol forecourt			
Average annual sales 1978-82	2,837,560 gals		
Less 10% risk factor	283,756		
	2,553,804 at £1.00		
		Say	2,555,000
		Total	£3,027,000
		Say	£3,030,000

J K

Notes: 1. Based upon Mr Dyke's drawing 106D.
2. No deduction for modernisation costs following Mr Squire's evidence.

Estimated annual consumption (EAC)

Mr Squire is experienced in the sale of petrol filling stations to oil companies. In his view, oil companies would usually base their bid upon an EAC directly related to the actual throughput achieved over the last few years before the transaction. Oil companies would normally expect a modest increase in the throughput on taking over from a private operator and they might anticipate a further increase if L the station was ripe for modernisation. Nevertheless, their estimate of throughput for the purpose of purchase would normally be based on the historic throughput in the hope that it would generally be exceeded.

He was aware that the average annual throughput at Stonehouse for the period 1978 to 1982 was some 2.84m and he had never experienced a throughput of this magnitude other than at motorway service stations. He did not attempt to estimate the future throughput at Stonehouse because in analysing his comparable transactions he adopted the historic throughput figures without any uplift for the factors mentioned above.

Mr Jones accepted Mr Squire's evidence as to throughput and his price per gallon. He based his valuation of the subject premises on the M average throughput of 2,837,560 gallons per annum.

Mr Stay in his analysis of his comparable transactions started with the actual throughput achieved immediately prior to the transaction but adjusted for the effect of price discounting, self-service operation, 24-hour opening, oil company operation and modernisation. He estimated that the subject premises, as modernised in accordance with Mr Dyke's final plan (drawing 106D amended) and operated by an oil company, with no discounting and on a 16-hour working day, would reach 2.125m gallons per annum.

Mr Ward, who is well experienced in designing petrol filling stations to meet the requirements of oil companies, agreed in cross-examination that the 106D (amended) layout could have achieved 3m gallons per annum and may have reached 3.4m gallons. His own design of a petrol filling station occupying the entire site could have achieved a similar throughput. These were not positive estimates of throughput but a designer's interpretation, after considering details of traffic flow, turn-off rate and average fill, of what might be expected by an oil company for that site contemplating normal trading over a 16-hour day. The figures are compatible with Mr Squire's estimates but cannot be reconciled with Mr Stay's.

Mr Squire was firmly of the view that an oil company would not adjust the historic throughput figures to allow for discounting, 24-hour sales and any expected improvement in throughput following its own occupation.

Mr Cook was of the view that the decline in throughput from 3m gallons per annum had been largely due to the lack of modernisation as a self-service station. In his view, the level of sales could still have been maintained if the discount had been reduced. Furthermore, sales outside the 16-hour period accounted for about 9% of the total throughput and most of this could have been recovered by a small extension of the trading hours.

Having considered the evidence, we have come to the view that the appropriate figure of throughput for the partially modernised Stonehouse site is 2,837,560 gallons. That figure represents not the maximum throughput which could be expected to be achieved in the future but the figure that would be likely to form the basis of a bid from an oil company at the valuation date.

Site value — price per gallon

Mr Squire's approach to the problem was to look at the evidence of comparable transactions and divide the sale price by the known annual throughput; he thereby arrived at his opinion of £1.00 per gallon. The reference premises had a much higher throughput than any of his comparables and it was his view that that would be particularly attractive to oil companies. Nevertheless, he did not seek to increase his figure of £1.00 per gallon.

He said that in his experience oil companies would pay the same price for a petrol filling station whether the buildings and plant had to be replaced immediately or not. He explained that the reason for this was that oil companies were in the habit of modernising petrol filling stations at frequent intervals. Buildings were altered or replaced long before the expiry of their normal life-span and new plant was in any event written off against corporation tax within a few years.

We can see that these factors tend to operate to reduce the value of buildings and plant but not, we think, to render them valueless. It seems to us that there is a real difference between the reference premises and most of Mr Squire's comparables. In the main the comparables were established petrol filling stations where the purchasing oil company did not need to modernise the plant and buildings in order to maintain or increase the throughput. As a matter of fact no major modernisations have been carried out even after a lapse of several years since those stations were purchased.

On the other hand, it is not disputed that modernisation of the reference premises would have been essential and could not be deferred.

Mr Jones accepted Mr Squire's figure but initially applied it to a valuation of the unmodernised subject petrol filling station. After hearing Mr Squire's evidence to the effect that £1.00 per gallon represented the site value only, he amended his valuation by adding back the £165,000 which he had deducted for the cost of modernisation. Mr Jones applied this figure of £1.00 per gallon to the average past achieved annual throughput, namely 2,837,560 gallons, reduced by 10% for the by-pass risk factor.

Mr Stay's figure of 45p per gallon is derived from his analysis of a number of transactions in petrol filling stations, the large majority having a substantially smaller throughput. His figure represents the site value, and therefore when analysing the transactions in comparable petrol filling stations he deducted his estimate of the value of the buildings and plant on the site. He then divided the resultant value of the buildings and plant on the site. He then divided the resultant value of the site not by the actual historic throughput but by his estimated figure throughput. In an extreme case this resulted in the actual throughput being increased by some 30%. Furthermore, in estimating the value of the buildings and plant Mr Stay appears to have been guided more by cost than value and has, in our view, failed to appreciate fully that buildings and plant at petrol filling stations can reach a state of obsolescence within a few years.

The simple truth is that, in analysing factual evidence, the greater the reliance on estimate and opinion the less reliable will be the end result. Reliance on Mr Stay's price per gallon is further undermined by the fact that there is no evidence to support his adjustments of achieved throughput or his estimates of the value of the buildings and plant. On the other hand, we think that Mr Squire's approach is too robust to distinguish between petrol filling stations that require immediate improvement and those that can continue trading at the same throughput for some years before modernisation becomes necessary.

Mr Squire was also criticised for including in his comparables certain petrol filling stations which were not sold individually but formed part of what was termed "a package deal" involving a number of petrol filling stations. For this reason Mr Stay rejected this part of Mr Squire's evidence.

Mr Squire's evidence, not successfully challenged, was that he negotiated the prices for all the petrol filling stations which formed the package deal. His approach was to seek offers for each petrol filling station individually before considering the effects of the package deal and it is these contemporary figures upon which he relies to support his opinion. This evidence is not the same as evidence of an individual transaction and should be treated with caution but, in our view, it is useful evidence and should not be rejected.

Faced with these alternative approaches, and with no evidence to support any other approach, we have come to the conclusion that Mr Squire's figure of £1.00 per gallon is to be preferred but requires adjustment because it is derived from evidence upon petrol filling stations that have as a matter of fact continued trading for a number of years after acquisition by an oil company without any major modernisation works being carried out. The £1.00 per gallon must, therefore, represent the value of a petrol filling station where modernisation can be deferred for a number of years. As there is no dispute that the subject site would have had to be modernised immediately, in our view this factor must be taken into account in the valuation.

Deduction for modernisation costs

On the basis of partial modernisation the parties are agreed that the cost would be £370,000 for a forecourt with 10 pumps capable of a throughput of at least 3m gallons per annum.

The acquiring authorities add to this figure £50,000 for the cost of new storage tanks and £36,000 for the cost of roof works to the residual buildings on the site.

On the evidence, we are not satisfied that the addition for the roof works is justified. The existing tanks were 20 years old and the envisaged development of the forecourt was substantial. The cost of plant would have been written off against corporation tax over a few years. In these circumstances it would be reasonable to suppose that an oil company would be likely to install new tanks immediately. Therefore we add £50,000 to the cost and arrive at a total of £420,000.

It seems to us that the real difference between the Stonehouse site and the comparable petrol filling stations quoted by Mr Squire is that the development cost had to be incurred immediately rather than be deferred for, say, seven years.

Therefore, in our view, the proper figure to be deducted is the difference between £420,000 and that figure deferred 7 years. No valuation evidence was presented on this factor and therefore we calculate the deferment at 8% over 7 years giving a figure of £244,860. Deducting this from £420,000 we arrive at £175,140 and that figure should be deducted as representing the relative development cost of the Stonehouse site when compared with petrol stations capable of continued operation.

Value of Stonehouse premises

Initially there was a difference between the parties as to the extent and value of the residual land and buildings at Stonehouse after the modernisation of the forecourt. This value has now been agreed at £450,000.

We set out in Appendix III [see p 250] our calculation of the value of the Stonehouse premises at £2,827,500. It is unnecessary for us to look in detail at the alternative valuations, which stem from the assumption that the Stonehouse site might have been modernised wholly as a petrol filling station because, on any version of the valuation, the figure would be less than our calculation on the partial modernisation basis.

Injurious affection to the Aero site

There is no dispute that on the compulsory acquisition of the Stonehouse premises the Aero site suffered injurious affection. Mr Jones calculates the loss at £790,000 and Mr Stay at £375,000.

A

Appendix III
Lands Tribunal award

Stonehouse premises on an existing use basis assuming a forecourt rearrangement

Land and buildings (as agreed)		£450,000
Petrol forecourt —		
Average sales per annum 1978-82 incl	2,837,560 gals	
Less for risk of by-pass 10%	283,756	
	2,553,804 at £1 Say	2,553,000
		3,003,000
B *Deduct*		
Cost of forecourt improvements and new tanks as found	£420,000	
Less present value of cost deferred 7 years at 8% 0.583	244,860	175,140
		2,827,860
	Say	£2,827,500

Mr Jones' approach to the valuation is to value the premises before and after the acquisition of the Stonehouse site. Mr Stay originally chose a more complex route but later acknowledged the correctness of Mr Jones' approach and produced his own valuation of £375,000 C after adopting that method. The two valuations are at Appendices IV and V respectively.

We prefer Mr Jones' valuation with reservations. We have already decided that the appropriate figure at Stonehouse for a developed petrol forecourt was £1.00 per gallon upon a sustainable annual throughput of some 2.5m gallons. We have little evidence to guide us but it is our view that by comparison a throughput of 1.8m gallons at the Aero site should be valued at 85p per gallon and the reduced output of 1,276,100 gallons should be valued at 75p.

D

Appendix IV
Mr P R Jones' valuation of Aero forecourt
To ascertain extent of injurious affection

Adopting a 10% risk factor for possible by-pass

1. *Valuation prior to acquisition of Stonehouse site*			
Petrol sales 1978 to 1982 = annual average of	2,250,780 gals		
Following conversion to self-service	say 2,000,000 gals		
Less 10% risk factor	200,000		
	1,800,000 at 90p		£1,620,000
Less cost of self-service equipment			30,000
E			£1,590,000
Deduct			
2. *Valuation post acquisition*			
Petrol sales 1987	1,276,100	gals at 65p £829,465	
Deduct cost of self-service equipment		30,000	
		799,465	
		Say	800,000
		Loss of value	£790,000

Making these adjustments we arrive at the figure of £573,000 as F shown in Appendix VI.

Appendix V
Mr Stay's valuation of Aero premises
To ascertain extent of injurious affection

Before			
Petrol sales: Average of 5 years 1978-82:	2,251,541	gals	
Less 15% for 24-hour opening and cut prices:	337,731		
	1,913,810		
Say	1,914,000 gals		
Allowing for by-pass risk 5p — 10%	at 45p per gal		£861,300
After			
Petrol sales 1987:	1,276,156 gals		
Less 15% for 24-hour opening and cut prices:	191,423		
	1,084,733		
Say	1,085,000 gals		
Risk of by-pass eliminated 50p			
Now a less attractive site – 5p — 45p	at 45p per gal		488,250
			373,050
		Say	£375,000

G

Appendix VI
Lands Tribunal award

Aero premises — injurious affection

H

1. *Valuation prior to acquisition of Stonehouse site*		
Average sales per annum 1978-82 incl	2,250,780 gals	
Following conversion to self-service say	2,000,000 gals	
Less for risk of by-pass 10%	200,000	
	1,800,000	
	at 85p	
	£1,530,000	
Less cost of self-service equipment	30,000	£1,500,000
Deduct		
2. *Valuation post acquisition*		
Petrol sales 1987	1,276,100 gals	
	at 75p	
	£957,075	
Less cost of self-service equipment	30,000	
	927,075 Say	927,000
	Loss of value	£573,000

J

Land taken		£2,827,500.00
Injurious affection		573,000.00
Disturbance		140,561.45
	Total	£3,541,061.45
	Say	£3,541,000

Disturbance compensation

K In this reference there are two categories of disturbance. The first covers the cost of removal, adaptation of new premises, double overheads, fees and costs in relation to new premises. The second category relates to fees and costs incurred in relation to the search for alternative premises and to management time. The amount is agreed. The issue is whether the whole amount is payable in law.

The acquiring authorities seek to avoid payment on those items which relate to the search for an integrated motor trade site or for a site for a petrol filling station. As we understand their argument, because the compensation for acquisition includes an additional value in respect of the potential for improvement and alteration (modernisation), then any expenses that are a consequence of realising that potential must be disregarded, following *Horn*. L

They argue that where premises are sold voluntarily on the open market the purchaser does not expect to pay the vendor's removal expenses. However, where premises are compulsorily acquired the acquiring authority has a statutory obligation to pay the claimant's costs of removal to other premises under the provisions of rule (6) of section 5 of the Land Compensation Act 1961, which states that "the provisions of rule (2) shall not affect the assessment of compensation for disturbance or any other matter not directly based on the value of land".

As we understand it, the principle underlying the decision in *Horn*'s case is that if a claimant seeks compensation which includes a potential value, over and above the existing use value, which could be realised only by discontinuing the existing use, then the costs of M relocating the existing use elsewhere are to be disregarded.

In the present case the award for land taken is based upon the existing use, namely the continued use of a workshop etc buildings for motor-trade activities and the modernisation of the forecourt for petrol sales. The latter involves consequential adaptation to the existing buildings and some temporary disruption to trading activities, whereas the taking of the premises requires that they be vacated, totally and permanently.

The claimant is not entitled to compensation for temporary disruption to business or for the cost of adapting buildings to facilitate modernisation and none is claimed. However, the claimant is entitled to be paid compensation in respect of the costs of removal

A from the premises and for all costs properly incurred in seeking and finding alternative premises to accommodate the whole of its displaced business. The agreed sums relate solely to these matters and accordingly the agreed total of £140,561.45 is payable to the claimant as compensation for disturbance.

Our award is therefore the sum of £3,541,000 to which will be added a surveyor's fee in accordance with Ryde's Scale (1984). We had opened a sealed offer which was below our award.

We heard submissions as to costs. The acquiring authorities are to pay the costs of the claimants in the references. Such costs if not agreed to be taxed by the Registrar of the Lands Tribunal on the High Court Scale.

B

DEANE v BROMLEY LONDON BOROUGH COUNCIL

(REF/158/1990)

November 4 1991

(C R MALLETT FRICS and M St J HOPPER FRICS)

Estates Gazette May 30 1992

[1992] 21 EG 120

Tree preservation order — Application to reduce height of C trees — Consent given subject to conditions — Work to be carried out by contractor approved by Arboricultural Association — Claim for compensation under section 203 of the Town and Country Planning Act 1990

In 1985 a tree preservation order was made by Bromley London Borough in respect of 40 trees of various species in the spacious garden of a house in Beckenham — The order included 26 horse chestnuts listed as a group which formed a row along the boundary with allotment gardens owned by the borough — The trees were closely planted and had been pollarded — In 1989 a request by the claimant to prune the D trees or arrange for the work to be carried out was registered by the council as an application to reduce the height of the trees — Consent was given in February 1990 subject to conditions including one that the work be carried out by a contractor approved by the Arboricultural Association — The claimant obtained estimates from three firms, one of whom carried out the work for £1,250 plus VAT — He maintained that but for the condition requiring him to employ an approved contractor he would have been able to do the work himself, with help from fellow residents and friends, at no cost — Alternatively, he would have been able to employ a competent but unapproved contractor at considerably less E cost — He submitted that the main beneficiaries were the allotment holders on the adjoining council land, which had been severely screened by the trees from sunlight

***Decision*: (1) Taking the total cost of the contracted work to be £1,437.50 including VAT and deducting an estimated cost of, say, £400 plus VAT for hiring additional plant and wear and tear of equipment, the award was £977.50 plus costs of £106**

(2) The tribunal had no evidence to support the claimant's view that the work could have been done more economically by employing an unapproved contractor

(3) The claimant had on the premises most of the equipment F to enable him to do the work — With fellow residents he had successfully taken down dangerous and overhanging branches

(4) The claimant was firmly of the view that he could do the work properly and the tribunal had no evidence to suggest otherwise

(5) Unaccountable delay in the council's giving consent to the application had exacerbated the claimant's problem by adding another summer's growth to the trees

(6) The pollarding had benefited the claimant in ensuring safety and providing a supply of fuel but had also benefited the allotment holders

No cases are referred to in this report. G

The claimant appeared in person; M Mullins (instructed by the solicitor for Bromley London Borough Council) for the compensating authority.

Giving their decision, THE TRIBUNAL said: This is a reference under section 205 of the Town and Country Planning Act 1990 to determine the amount of compensation, if any, payable on the granting of an application, subject to conditions, to reduce in height trees which are the subject of the London Borough of Bromley Tree Preservation Order No 246.

Mr E G Deane, the claimant, appeared in person and gave evidence. Mr Mullins of counsel appeared on behalf of the London H Borough of Bromley and called Mr R Nicholls, an arboriculturist responsible for the protection of trees in the council's ownership.

The parties have helpfully agreed a statement of facts from which we extract the following:

No 107 South Eden Park Road, Beckenham, is a large, detached Victorian house set in a spacious garden. To the south and west there are houses fronting a cul-de-sac laid out in the 1930s and to the north are allotment gardens owned by the London Borough of Bromley. Mr Deane has owned the property since 1979 and is joint owner with his wife. The house is now divided into flats and Mr and Mrs Deane occupy one of the flats.

Tree Preservation Order No 246 was made in January 1985 in respect of trees in the garden of 107 South Eden Park Road. The J order covers two cedars, three yews, two oaks, a redwood, a birch, a chile pine, a cypress, a fir, a pine and 27 horse chestnuts. One of these chestnuts is specified individually and 26 are listed as a group; it is these 26 which are the subject of this reference.

Mr Deane objected to the making of the order and on May 19 1985, after considering his objection, the London Borough of Bromley confirmed the order with amendments.

The 26 chestnut trees with which we are concerned form a row of trees along the northern boundary of the garden and continuing to the west beyond the boundary of the subject premises. The trees have been closely planted and never thinned but have in the past been pollarded. K

Mr and Mrs Deane maintain the garden themselves and from time to time their fellow residents have assisted, particularly with the lopping of overhanging and dangerous branches, which were subsequently used for fuel.

In 1989 a major limb from one of the chestnut trees fell on to an adjoining allotment. On September 11 1989 Mr Deane asked the council either to prune the trees or to arrange for the work to be carried out. This was registered by the council on September 18 1989 as an application to reduce the height of the horse chestnut trees.

There were four letters written by objectors to the proposed reduction in height and two letters written in support, one from the allotment holders and the other from a person or persons supporting safety work. Finally, on February 9 1990 consent was given to reduce L in height 26 horse chestnut trees subject to the following conditions:

01 The development to which this permission relates must be begun not later than the expiration of five years beginning with the date of this permission.
98 The work shall result in the reduction of not more than 50 per cent in the overall height of each tree.
99 The work should be carried out by a contractor approved by the Arboricultural Association.

The reasons given for these conditions were:

01 Section 41 the Town and Country Planning Act 1971.
98 In the interest of safety and the visual amenities of the area.
99 In the interest of good Arboricultural Practice.

Mr Deane subsequently obtained three estimates for the work as M follows:

£2,160 plus VAT, from Trees
£1,735 in total, from ISC Tree Specialist
£1,250 plus VAT, from Dolwin & Gray.

In all cases the brushwood was to be chipped and left on site and the remaining timber cut into lengths and stacked on site. Of the three companies, both Trees and Dolwin & Gray were approved by the Arboricultural Association; ISC Tree Specialist is a local company.

Mr Deane requested that the council pay for the work, but on September 20 1990 he was informed that the council were unable to offer any financial assistance for carrying out the work.

The work was carried out by Dolwin & Gray at the end of

A September 1990. All 26 trees were pollarded, the brushwood was chipped and left on site and the remaining timber cut into lengths and stacked on site.

The basis of Mr Deane's claim is that condition 99 attached to the consent required him to employ an approved contractor and this cost him £1,250 plus VAT. He maintains that but for that condition he would have been able to do the work himself at no cost. Although Mr Deane has passed the age when most people would have retired, he and his wife are very active in maintaining their garden and they have the equipment necessary to do the work. They could have expected help from fellow residents and friends. In the alternative, he maintains that if he had been able to employ a competent but B unapproved contractor the cost would have been considerably less. Therefore he felt that the council should pay the extra cost of having the work done by an approved contractor.

Mr Deane does not deny that some work was necessary to some branches of the horse chestnuts in the interest of safety. As the trees had previously been pollarded, it was a sensible solution to pollard again. He maintains that the main beneficiaries were the allotment holders on the adjoining council land immediately to the north which was severely screened by the chestnut trees.

Mr Mullins, for the council, called Mr Nicholls, an experienced arboriculturist, who had visited the site before and after the work was done. In his view, the cost of £1,250 plus VAT was not unreasonable and he would have been prepared to recommend a figure of £100 or C £150 in excess of that figure, as the contractor would have had to have been one of considerable experience in the type of work involved.

Mr Mullins submitted that there was no evidence to support the contention that a non-approved firm could have done the work for less and there was no evidence to support the contention that it would have been possible for Mr Deane to do the work himself, together with voluntary help, and at a nil cost. He submitted that Mr Deane had an obligation to ensure that the trees were safe and that he was the main beneficiary from the work that had been done. There was no evidence that Mr Deane had suffered any loss and the claim should be dismissed.

We have viewed the subject premises. We have seen the chestnut trees in their pollarded state when they had not yet started into new D growth. To the west we have seen the extension of the line of chestnuts on to other land which have been pollarded in the past and are now showing good growth.

The gardens of the house, in which there are some fine specimen trees, appeared to be well maintained. The 26 horse chestnuts along the northern boundary have been planted much too closely and this fact, combined with severe pollarding, has led to distorted growth. Any visual attraction when the trees are in leaf must be marred by the break in continuity of treatment of the chestnuts to the west. The 26 chestnuts do not screen any unsightly view. There is no doubt that the allotment gardens to the north of the trees were severely screened from the sunlight.

E It seems to us that the pollarding of trees has benefited Mr Deane in that safety is ensured together with a supply of fuel for him and his tenants; but there is little doubt that it has also benefited the allotment holders.

One would have hoped that, in a spirit of neighbourliness, the parties could have agreed to apportion between them the cost of pollarding. This has not happened and we are concerned only with the narrow issue of compensation as provided by section 203 of the Town and Country Planning Act 1990, which reads as follows:

> A tree preservation order may make provision for the payment by the local planning authority, subject to such exceptions and conditions as may be specified in the order, of compensation in respect of loss or damage caused or incurred in consequence —
>
> F (a) of the refusal of any consent required under the order, or
> (b) of the grant of any such consent subject to conditions.

Tree Preservation Order No 246 contained such a provision.

Condition 99 of the consent provided that the work be carried out by an approved contractor and we have no evidence to support Mr Deane's view that the work could have been done more economically by employing an unapproved contractor.

We are left with Mr Deane's contention that he could, with voluntary assistance, have carried out the work himself at no cost. Mr Nicholls is of the view that considerable experience was required in order to do the work properly, but he took no steps to inquire whether Mr Deane had such experience. Mr Deane has on the premises most of the equipment to enable him to do the work. He and his fellow residents have from time to time successfully taken down dangerous G and overhanging branches.

Mr Deane is firmly of the view that he could do the work and do it properly over a period of time and we have no evidence to suggest otherwise. His application to reduce the height of the trees was made in the early autumn of 1989, but for some unaccountable reason most of the winter passed before consent was given. This delay in effect added another summer's growth to the trees before pollarding could take place. The delay added to Mr Deane's problem.

We doubt that the considerable task of pollarding 26 trees could have been achieved without some cost, in particular the cost of hiring machinery or transport to deal with the brushwood and some wear and tear to equipment.

We have no details of costs, but doing the best we can with the H evidence we have taken the total cost of the contracted work to be £1,437.50 including VAT and have deducted from this our estimated cost of hiring additional plant and wear and tear of equipment, say £400 plus VAT, which results in a figure of £977.50 and that is the amount of our award.

We have received written submissions as to costs. The claimant's costs in this reference are to be paid by the compensating authority in the sum of £106.

HODDOM & KINMOUNT ESTATES v SECRETARY OF STATE FOR SCOTLAND J

LANDS TRIBUNAL FOR SCOTLAND
(LTS/COMP/1990/5)
October 28 1991
(Sheriff A C HORSFALL QC, A R MACLEARY FRICS and J DEVINE FRICS)

Estates Gazette June 6 1992
[1992] 22 EG 118

Compulsory acquisition of part of farm — Basic claim for compensation agreed — Additional claim for non-allocation K of milk quota refused — Admissibility of claim under s 61 of Lands Clauses Consolidation (Scotland) Act 1845 — Date of actual loss prior to date of acquisition — Loss of milk quota held too remote a consequence to be compensated

The subject property, part of a farm at Annan, near Dumfries, Scotland, was compulsorily acquired in 1986 for construction of the Annan bypass — The proposal to construct the bypass through the farm was intimated to the claimants in 1981 — At that time the farm was operated as a dairy farm by a tenant of some 20 years' standing — After notice by the tenant, the tenancy terminated in November L 1983 — Between then and April 1984, when milk quotas were introduced, the farm was vacant — The claimants considered it not practicable to let as a dairy farm because of uncertainty regarding the road works — Milk quotas were introduced with effect from April 2 1984 but no allocation was made to the subject farm, since it was not registered with the Milk Marketing Board nor was it operating as a dairy unit — Subsequent attempts to obtain an allocation failed — The claimants averred that, as a result of the refusal, they were unable (1) to farm as a dairy unit in conjunction with their neighbouring farm and (2) to let as a dairy farm although it was still equipped as such — Therefore, the value of the farm M was significantly reduced — The Secretary of State submitted that the claim was irrelevant because (a) loss of milk quota was too remote a consequence, (b) the hypothesis on which it was advanced (ie the farm would have been tenanted at vesting date) was inconsistent with the rest of the claim (ie the farm was in vacant possession) — The date of vesting and valuation was August 29 1986 — At that date the claimants had vacant possession — The claim was in respect of loss in value of the farm on grounds of severance and injurious affection under s 61 of the Lands Clauses Consolidation (Scotland) Act 1845

***Decision*: (1) It was reasonably foreseeable that threat of**

A **impending road works would result in the farm's ceasing to be used as a dairy farm**

(2) Whether it was also foreseeable that thereafter the claimants would leave the farm unused as a dairy farm, that was sufficiently explained in the claimants' averments, read as a whole, to enable it to be looked at in the same light as the tenant's giving up his tenancy — If the latter was a foreseeable consequence of the impending roadworks and compulsory acquisition, then so was the former

(3) The position was different, however, in considering whether it was foreseeable that the cause would co-exist with B **the subsequent introduction for a milk quota system and that the cumulative effect of those causes would be that the claimants would fail to obtain the added value which the land would then have acquired had it still been in use as a dairy farm**

(4) The alleged loss would not have been incurred but for the happening of an extraneous event completely unconnected with compulsory acquisition, namely the introduction of the milk quota system

(5) The claimants averred specifically that because no notice was given of the imposition of milk quota there was no possibility open to them of taking action to avoid the non-allocation of a quota during the time when they themselves C **had vacant possession before milk quota was imposed — It could hardly be said, therefore, that the acquiring authority could have foreseen that the claimants might suffer loss in such a way**

(6) Accordingly, the loss of milk quota was too remote a consequence of the acquisition to be compensated

(7) On the question of the alleged inconsistency between the claims, the claim for loss of milk quota following the departure of the tenant was not inconsistent with a claim for compensation of the land held with vacant possession

The following cases are referred to in this report.

D

Aberdeen City District Council v *Sim* 1983 SLT 250; (1982) 264 EG 621

Harvey v *Crawley Development Corporation* [1957] 1 QB 485; [1957] 2 WLR 332; [1957] 1 All ER 504; (1957) 55 LGR 104; 8 P&CR 141, CA

Prasad v *Wolverhampton Borough Council* [1983] Ch 333; [1983] 2 WLR 946; [1983] 2 All ER 140; (1983) 82 LGR 265; 47 P&CR 252; [1983] EGD 627; 265 EG 1073, CA

Smith v *Strathclyde Regional Council* 1982 SLT (Lands Tr) 2; [1981] EGD 491; (1980) 257 EG 391

Venables v *Department of Agriculture for Scotland* 1932 SC 573

Anthony MacIver (instructed by Bonar Mackenzie WS, of Edinburgh) appeared for the claimants; Gerard Moynihan (instructed by the Solicitor to the Secretary of State for Scotland) for E the acquiring authority.

Giving their decision, **The Tribunal** said: This reference is to determine a claim for compensation arising out of the compulsory acquisition of land on Violetbank Farm, Annan, for the construction of the Annan bypass. The basic claim for compensation for the land compulsorily acquired has been agreed. This claim is solely in respect of the additional loss which the claimants, as proprietors, allege that they have suffered because of their failure to secure the allocation of a milk quota for the farm. They contend that that failure resulted from the making of the compulsory purchase order. The district valuer has refused to accept any such claim.

The proposal to construct the bypass on a route through F Violetbank Farm was intimated to the claimants' factor in the summer of 1981. At that time the farm was tenanted and operated as a dairy farm by Mr Mathew Jamieson, who has been the tenant for more than 20 years. Although this is not stated specifically in the claimants' averments, it can be inferred that Mr Jamieson was aware of the proposed road works at about the same time as the claimants. On September 16 1982 he gave notice of termination of the tenancy, which was terminated on November 25 1983, although Mr Jamieson remained in occupation of the farmhouse.

The claimants' averment of the reason for Mr Jamieson's decision to give up the farm is in these terms:

> The tenant of the dairy farm had become concerned that his dairy operation would be disrupted by the construction of the road which seemed imminent. The positioning of the route of the road meant that approximately two-thirds G of the area of the farm would be cut off or isolated from the remaining one-third which contained the steading buildings, the net result being that the tenant farmer would have to drive the herd across the construction works four times a day. This would have entailed taking the cattle down a cutting which was some 20 feet deep and was not a practical possibility. The alternative was a detour of some 5 miles by road and this also was not a practical alternative. The tenant took the view that this would present an insurmountable problem. The tenant decided therefore to retire and to disperse the dairy herd. . . . The stock was sold in the summer of 1983 after they had calved and the farm implements were sold in November/December 1983.

The claimants then aver:

> Between 28th November 1983 and 1st April 1984 the farm was vacant and was not occupied. It was considered to be not practicable to let as a dairy farm at H the time because of the uncertainty regarding the road works. The said farm has been let as grass parks from 1st April to 31st October in each year since 1984.

With effect from April 2 1984 milk quotas were introduced for dairy farms. On that date Violetbank Farm was not registered with the Milk Marketing Board nor was it operating as a dairy unit. It is averred that "consequently the farm was not allocated a milk quota by the Department of Agriculture and Fisheries for Scotland". The claimants aver that since then they have done all that they can to obtain the allocation of a milk quota to Violetbank Farm but that such an allocation has been refused. They aver that since April 2 1984 their intention to take the farm in hand and to farm it as a dairy unit in conjunction with their neighbouring farm of Hollybush was defeated J by the refusal of the allocation of a milk quota of Violetbank and that for the same reason they have been unable to let Violetbank as a dairy farm, although it is still equipped as such. They claim that as a consequence the value of the farm has been significantly reduced.

They submit that, in assessing the loss, the non-allocation of a milk quota should be taken into account, being a loss directly attributable to the subsequent compulsory purchase. They aver that had the roadworks not been proposed and constructed then Violetbank Farm would have continued in operation as a dairy unit and would have received a milk quota in respect of the milk production there. They further aver that, since no notice was given by the Government of the imposition of a milk quota, there was no possibility open to K them of taking action to avoid the non-allocation of a quota.

The compulsory purchase order was made on March 13 1986 and the acquiring authority took possession of the land on August 29 1986, which it is agreed was the date of vesting and so of valuation. The claimants admit that at that date they had vacant possession of Violetbank Farm.

Their claim for compensation in respect of the non-allocation of a milk quota is stated to be in respect of loss in value of the farm on the grounds of severance and injurious affection, under section 61 of the Lands Clauses Consolidation (Scotland) Act 1845. That section provides that, in estimating the purchase money or compensation to be paid by the promoters of an undertaking, regard shall be had not only to the value of the land to be purchased or taken by the L promoters of the undertaking but also to the damage, if any, to be sustained by the owner of the lands by reason of the severing of the lands taken from the other lands of such owner, or otherwise injuriously affecting such land by the exercise of the powers of that or the special Act or any other Act incorporated therewith.

It was submitted on behalf of the Secretary of State that the claimants' averments were irrelevant for two reasons: first, because the loss of a milk quota was too remote a consequence of the compulsory purchase order and, second, because the hypothesis on which this part of the total claim was advanced was inconsistent with the rest of the claim. The claim in respect of the milk quota proceeded on the hypothesis that Violetbank Farm would have been tenanted at M the vesting date had it not been for the compulsory purchase order, whereas the rest of the claim proceeded on the basis of the farm being then in vacant possession.

It is not in dispute that in deciding for what a dispossessed landowner is entitled to be compensated the general rule is as stated by the Lord Justice-Clerk (Alness) in *Venables* v *Department of Agriculture for Scotland* 1932 SC 573, at p 581, namely that:

> the sound principle would seem to be that the person dispossessed should get compensation for all loss occasioned to him by reason of his dispossession.

Nor is it in dispute that the tests to be applied are as stated by Romer LJ in *Harvey* v *Crawley Development Corporation* [1957] 1 QB 485, at p 494, where he said:

A any loss sustained by a dispossessed owner (at all events one who occupies his house) which flows from a compulsory acquisition may properly be regarded as the subject of compensation for disturbance, provided, first, that it is not too remote and, secondly, that it is the natural and reasonable consequence of the dispossession of the owner.

It was accepted that that applied equally to a claim for severance and other injurious affection, that in that context the word "consequence" had a causal rather than a temporal meaning and that there must therefore be a causal connection between the dispossession and the loss: see *Prasad* v *Wolverhampton Borough Council* [1983] Ch 333. The loss might accordingly be one incurred before the date of dispossession, so long as it was the result of something done in reasonable anticipation of the dispossession: see B *Smith* v *Strathclyde Regional Council* 1982 SLT (Lands Tr) 2; (1980) 257 EG 391 and *Aberdeen City District Council* v *Sim* 1983 SLT 250; (1982) 264 EG 621.

It was submitted, however, that in the present case the only connection which had been averred between the compulsory purchase and the claimants' loss was not causal but was purely temporal. It was not the compulsory purchase alone which had caused the loss. The loss, it was submitted, was due to a number of concurring factors of which the compulsory purchase was only one. The other factors were the surprise introduction of a milk quota system and the fact that the farm was unoccupied and was not in dairy production on April 2 1984 when quotas were introduced. Those, it was submitted, were factors which could not have been foreseen.

C It was contended that in principle the test of remoteness of loss should be the same here as in the case of a loss of which a breach of contract was only one of a number of causes. In *Gloag on Contract*, 2nd ed, at p 690, the learned author states:

> It may be suggested as a rule which will cover many of the cases that if the loss of which the pursuer complains is due to several causes, of which the defender's breach of contract is one, the defender will not be liable for it unless the co-existence of the other causes, and their cumulative effect, should have been within the contemplation of a reasonable man who was considering whether it would be to his advantage to break his contract or not.

Here, it was submitted, the co-existence of the other causes of the claimants' loss, other than the compulsory purchase, and the D cumulative effect of those causes, could not have been within the contemplation of a reasonable man who was considering whether to acquire this land under compulsory powers. The loss, it was said, must therefore be regarded as too remote a consequence of the compulsory purchase for there to be any right to compensation.

On behalf of the claimants it was submitted that the loss due to the non-allocation of a milk quota was not too remote. Their averments of the history of the whole matter demonstrated that the loss of a milk quota was a direct consequence of the proposed roadworks and the compulsory purchase order. On the claimants' pleadings, it could not be said that the true cause of the loss was the concurrence of a number of separate factors. As it was averred, everything that happened after August 1981 was the result of the road proposals. The fact that the E farm did not have a milk quota had been caused initially by the tenant's giving up his tenancy, and it was averred that he had done that because he considered it impractical to carry on in view of the threat of the roadworks and of the compulsory purchase. The claimants had then averred that, after the tenant had given up, the farm could not be let in 1983 as a dairy unit because of the impending roadworks and that it would have been impractical. It was submitted that that giving up of the tenancy and cessation of dairy farming at Violetbank Farm were things which, on the claimants' averments, the acquiring authority could have foreseen as likely consequences of the roadworks.

One thing is left unclear. Even if the initial act which led in the end F to the failure to obtain a milk quota was the giving up of the tenancy by Mr Jamieson, and even if, because of the impending roadworks, it would then (in 1983) have been impractical to relet the farm as a dairy unit, there is no specific explanation of why the claimants themselves did not take the farm in hand and farm it as a dairy unit between November 1983 and April 1984.

It is clear that the claimants appreciate the importance of the fact that they did not do so, since they aver that:

> the claim in respect of the non-allocation of milk quota does not assume that the tenancy would have continued but for the compulsory purchase order. The question of a tenancy does not affect the allocation of milk quota. Had the Applicants [the claimants] transferred their own cows on to Violetbank Farm on 29th November 1983, at the termination of the tenancy, then the Applicants could have obtained an allocation of milk quota in respect of the G period from 29th November 1983 to 2nd April 1984, and then had this allocation apportioned over a whole twelve-month period. In such circumstances, Violetbank Farm would still have been valued on the basis of vacant possession but would in addition have had the benefit of milk quota.

So far as the period after April 2 1984 is concerned, the claimants' averments are clear and specific. They aver that after that date the farm could not be let as a dairy unit as, in addition to the roadworks, it had no milk quota allocated to it. They further aver that at that stage their intention to take it in hand and to farm it as a dairy unit in conjunction with their neighbouring farm of Hollybush was defeated by the refusal of an allocation of milk quota to Violetbank. That, however, can refer only to the period after April 2 1984 and does not explain why they did not take the farm in hand as a dairy unit before H that date. Had they done so, then on their own averments they could have obtained an allocation of milk quota.

The question, then, is not only whether it was foreseeable that because of the impending roadworks the tenancy would be given up but whether it was also foreseeable that thereafter the claimants would leave the farm unused as a dairy farm, so that it did not qualify for the allocation of milk quota when the quota system was subsequently introduced. The further question which remains to be answered is whether the introduction of the milk quota system was itself foreseeable.

With regard to the action of the claimants in not taking the farm in hand but leaving it unused between November 1983 and April 1984, J we are prepared to hold that that is sufficiently explained in the claimants' averments, read as a whole, to enable it to be looked at in the same light as Mr Jamieson's giving up the tenancy. If the latter was a foreseeable consequence of the impending roadworks and compulsory acquisition, then so was the former. In addition to the more detailed and specific averments relating to the reasons for Mr Jamieson's decision, the claimants have averred separately and more generally that there was insufficient land for the grazing of the dairy herd on the same side of the proposed roadworks as the steading and that driving the herd across the construction works was not a practical possibility. It can reasonably be inferred that that was the reason for their failure to work the farm themselves as a dairy unit after Mr Jamieson gave up.

K In our opinion, it could be said to have been reasonably foreseeable that the threat of the impending roadworks would result in Violetbank ceasing to be used as a dairy farm, whether because the tenant gave up for the reasons averred or because for the same or similar reasons the claimants did not themselves continue that use. The position, however, is different when one comes to consider whether it was foreseeable that that cause would co-exist with the subsequent introduction of a milk quota system and that the cumulative effect of those causes would be that the claimants would fail to obtain the added value which the land would then have acquired had it still been in use for dairy farming.

The situation here differs from that in the cases of *Smith* and L *Aberdeen District Council* in respect that in each of those cases the loss consisted of expenditure reasonably incurred in anticipation of the dispossession which was likely to come about and, in the case of *Smith*, with a view to mitigating the resulting loss. That the person about to be dispossessed would incur that expenditure was something which the acquiring authority should have had in contemplation. Here, however, the alleged loss would not have been incurred but for the happening of an extraneous event completely unconnected with compulsory acquisition, that being the introduction of the milk quota system. The loss consists of the value of the quota of which the claimants aver that they would have had the benefit but for a sequence of events arising from the anticipation of the possible consequences of the impending dispossession. They do not aver, M however, that the introduction of such a quota system was foreseeable. Indeed, it is implicit in their averments that it could not have been foreseen. They aver specifically that because no notice was given of its imposition there was no possibility open to them of taking action to avoid the non-allocation of a quota during the time when they themselves had vacant possession of the farm in the months before the milk quota system was imposed. If that was so, then it can hardly be said that the acquiring authority could have foreseen that the claimants might suffer loss in such a way. In our opinion, therefore, this particular loss was too remote a consequence of the acquisition for the claimants to be entitled to be compensated for it by the acquiring authority.

With regard to the alleged inconsistency between the principal

A claim and the claim in respect of the milk quota, it was accepted that, since a claim for compensation, although calculated under several heads, is one global sum, a claim for severance and other injurious affection must be consistent with the remainder of the claim: see the *Stair Encyclopaedia of the Laws of Scotland*, vol 5, p 83. Originally, the alleged inconsistency was said to be that, whereas the principal claim had been made on the basis that at the vesting date the farm was in vacant possession, the claim in respect of the milk quota was on the basis that had it not been for the compulsory purchase order the farm would have been tenanted at that date. As the argument developed, however, it appeared that what was alleged to be inconsistent was really that the claim in respect of the milk quota was based on the hypothesis that, had it not been for the compulsory purchase order,
B the farm would have been tenanted on April 2 1984, which was the crucial date for the allocation of a milk quota.

On behalf of the claimants it was submitted that there was no such inconsistency in their pleadings. They had averred that the claim in respect of the non-allocation of milk quota did not assume that the tenancy would have continued but for the compulsory purchase order and that the question of a tenancy did not affect the allocation of milk quota. They had averred that, if they had transferred their own cows on to Violetbank Farm at the termination of Mr Jamieson's tenancy, they could have obtained an allocation of milk quota, so that the farm would still have been valued on the basis of vacant possession but would also have had the benefit of a quota.
C That, it was submitted, sufficiently averred that the giving up of the tenancy by Mr Jamieson in November 1983 had nothing to do with the failure to obtain a milk quota when quotas were introduced in the following April. Even if it were as a result of Mr Jamieson's giving up that the farm was then in hand, it was submitted that it was by then the responsibility of the claimants whether it was or was not allocated a quota. They had averred that because of the impending road works it was impractical for them to relet the farm as a dairy unit. That was why they had failed to obtain a milk quota, but it was nevertheless a result of the impending road works and so of the subsequent compulsory acquisition.

In those averments there is no necessary inconsistency. The claimants' case is that Mr Jamieson's giving up the tenancy because
D of the proposed road works was only the first in a chain of events which resulted in the farm not being occupied and not being in production as a dairy farm on April 2 1984, with the result that it was not eligible for the allocation of a milk quota. That case is, on the face of it, entirely consistent with their having themselves vacant possession of the farm both then and on the vesting date.

On the assumption that the claimants' case was based on the hypothesis that but for the compulsory purchase order the farm would still have been tenanted on April 2 1984, it was further submitted on behalf of the Secretary of State that their averments were also irrelevant because it was not they who would have benefited from the allocation of a milk quota. In assessing the value of their interest it was to be assumed that but for the compulsory acquisition Mr Jamieson's tenancy would have continued. That resulted from E
the provisions of section 44 of the Land Compensation (Scotland) Act 1973, which provided that the security of tenure which the tenant would enjoy were the land not required for use by the acquiring authority was to be taken into account. The quota would then have been allocated to Mr Jamieson as the producer and not, as the claimants had averred, to the farm. Admittedly, if Mr Jamieson had subsequently given up the tenancy, the quota would have remained with the farm, but Mr Jamieson would have been entitled to payment for it from the claimants. They would therefore have gained nothing from the allocation of a quota, and so long as the farm continued to be tenanted they themselves would have lost nothing as a result of a quota not being allocated. F

It is clear, however, from the terms of the Dairy Produce Quotas Regulations 1984 and 1988 that, while it may be that a quota must be registered in the name of the producer, who in the case of a let dairy farm will be the tenant, the quota itself attaches to the land and cannot be taken with him by the tenant at his waygoing. Although the departing tenant now has a limited right to compensation for that part of the quota which has become attached to the tenancy as a result of his efforts or of his expenditure, even that limited right was introduced only after the date of vesting of the claimants' land on August 29 1986. Before that date the full value of the quota was retained by the landlord at the tenant's waygoing: see *Gill, Agricultural Holdings in Scotland*, 2nd ed, para 512. Even if the G
allocation of a milk quota had been lost solely as a result of Mr Jamieson's giving up his tenancy when he did, then, so far as the quota regulations are concerned, it would still be the claimants who had suffered that loss. Their averments could not, therefore, be held to be irrelevant on that ground.

In a more general way, the claimants' averments were also criticised because, it was alleged, they suggested that the farm had been permanently sterilised, at least as a dairy farm, by the two concurring factors of the introduction of a milk quota and the farm's being unoccupied on the critical date. Yet it was not suggested that the farm was not still a perfectly viable farm for other purposes. However, even if it were true that the claimants' averments came to H
this, that the farm had been permanently sterilised as a dairy farm by those factors, that would not mean that those averments were irrelevant. Whether the farm might be used equally profitably for other purposes would be a matter for proof when it came to assessing the amount of any loss resulting from the non-allocation of a milk quota. In any event, counsel for the claimants denied any intention of claiming that the farm had been permanently sterilised for that purpose. It was now permissible to buy and sell milk quotas and a quota could therefore be bought for Violetbank. That, however, would cost money and the position would still be that the claimants would have lost for ever the value of the quota which, at no cost to them, would in all probability have been allocated to the farm as at April 2 1984.

SUPPLEMENT
CASES NOT REPORTED IN "ESTATES GAZETTE"

AGRICULTURE

A **Court of Appeal**

October 24 1991

(Before Lord Justice GLIDEWELL, Lord Justice BALCOMBE and Lord Justice WOOLF)

LUTTENBERGER v NORTH THORESBY FARMS LTD AND OTHERS

Agricultural Holdings Act 1986 — Notice to pay rent — Notice to quit — Summons for summary judgment under Ord
B **14 of the Rules of the Supreme Court seeking possession — Whether notice to quit contained a false statement — Whether arguable defence**

The first defendant, North Thoresby Farms Ltd, is the tenant of Eastfield House Farm, North Thoresby, Lincolnshire; the farm is a holding of 630 acres and occupied and farmed by the second and third defendants, Mr and Mrs Popoff — The plaintiff, Mrs Luttenberger, is the owner of the freehold as personal representative of Miss Eva Robinson and Helen Batt (both deceased) — The first defendant failed to pay the half-year's rent of £15,264 on October 10 1989 — On November 17 1989 the plaintiff's solicitor served the first
C **defendant with a notice to pay rent pursuant to the provisions of the Agricultural Holdings Act 1986, Schedule 3, Part I, Case D — On November 24 1989 a statutory demand for the payment of the rent was served on the first defendant — That was not complied with and a petition to have the defendant company wound up was presented — On January 19 1990 the plaintiff's solicitor served a notice to quit on the first defendant alleging a failure to comply with the notice to pay rent and requiring possession of the holding to be delivered up on April 6 1991 — On December 22 1989 the defendant's then solicitors sent a cheque dated October 10 1989 to the plaintiff's**
D **solicitors for the sum of £14,805.60 being a half-year's rent less the drainage rate payable by the landlord — The cheque was not received until January 19 1990 — Contrary to the bank's mandate from the first defendant company, the cheque did not have two signatures — The plaintiff alleged that the bank stated the cheque would not be honoured — The defendant's evidence was that no such statement was made and the cheque was honoured later the same day — On February 7 1990 the defendant's then solicitors served a notice requiring that the issue whether the rent had been paid be put to arbitration, but neither they nor the defendant took any further steps in time or at all to agree to or request the**
E **appointment of an arbitrator — The defendant did not give up possession on April 6 1991 and the plaintiff issued proceedings claiming possession — On the hearing of a summons for summary judgment under Ord 14 of the Rules of the Supreme Court, the master granted the plaintiff an order for possession — Mummery J allowed an appeal against that order, set it aside and gave the defendants unconditional leave to defend — The plaintiff appealed from that order**

***Held*: The appeal was dismissed — Proof of posting the rent on December 22 1989 is *prima facie* proof of payment of the rent subject to the requirement that the cheque should be** F
honoured — There was a series of disputes of fact as to whether the cheque had been dishonoured or not — There was a series of issues of fact which ought to be tried as to whether the plaintiff's solicitor knew or ought to have known that the rent had been paid by the cheque and was only waiting at the Post Office — Those issues went to the allegation that the statement in the notice to quit that the rent was unpaid had been made recklessly and without honest belief in its truth — There was a triable issue

The following cases are referred to in this report. G

Beevers v *Mason* (1978) 37 P&CR 452; 248 EG 781, CA
Rous v *Mitchell* [1991] 1 WLR 469; [1991] 1 All ER 676; *sub nom Stradbroke (Earl of)* v *Mitchell* [1991] 1 EGLR 1; [1991] 03 EG 128 & 04 EG 132.

This was an appeal by the plaintiff, Alexandra Helena Luttenberger, from a decision of Mummery J given on July 1 1991 whereby he allowed an appeal from the decision of Master Gowers, who had made an order for possession in respect of Eastfield House Farm, North Thoresby, Lincolnshire, against the defendants, North Thoresby Farms Ltd, Michael Dimitri Popoff and Mrs Lesley Popoff, on a summons under Ord 14 of the Rules of the Supreme Court. H

Christopher Wilson-Smith QC and Olav Ernstzen (instructed by Prince & Co, of Malton) appeared for the appellant; Robert Reid QC and John McGhee (instructed by H V Beckett & Co, of Market Rasen) represented the respondents.

Giving judgment, **GLIDEWELL LJ** said: This is an appeal against a decision of Mummery J given on July 1 1991. The basic facts are summarised at the start of the judge's judgment with admirable clarity. I cannot do better than adopt what he said and I therefore intend to incorporate his summary of the facts into this judgment.

Eastfield House Farm (to which I shall refer as "the farm") is a holding of about 630 acres in North Thoresby, Lincolnshire. It has been occupied and farmed by the second defendant, Mr Popoff, and his wife, the third J
defendant, Mrs Popoff, for the last 30 years. The freehold was owned by two sisters who were great aunts of Mr Popoff. There was Miss Eva Robinson, who died on January 7 1986, and Mrs Helen Butt, who died on February 28 1987.

The plaintiff, Mrs Luttenberger, is the personal representative of both sisters, and she brings these proceedings in that capacity. She is also the sister of Mr Popoff. It would be an understatement to say that Mr Popoff and Mrs Luttenberger are not on good terms and it is sad to see that the relationship between their respective solicitors is little better.

This acrimonious dispute is about possession of the farm. On Mrs Luttenberger's application for summary judgment, Master Gowers made an order on June 14 1991 for possession of the farm. He refused a stay of execution. This is an appeal from that order.

Mr Popoff was originally the tenant of the farm. On September 28 1979, the K
first defendant company, North Thoresby Farms Ltd (to which I shall refer as "the company") was formed. The company is owned and controlled by Mr and Mrs Popoff. On June 30 1982 the company entered into a tenancy agreement of the farm with Mr Popoff's two great aunts. The term was from April 6 1982 to April 6 1983 and thereafter continued from year to year. It was determinable at the end of any year of the tenancy by not less than 12 months' notice in writing. The rent was payable by equal half-yearly payments in arrears on April 6 and October 10. The rent was originally fixed in the tenancy agreement at £17,000 pa. That was increased to £24,500 pa as from April 6 1985 and to £30,528 pa as from April 6 1988.

On October 10 1989, the half-year's rent due from the company to Mrs Luttenberger was £15,264, subject to an adjustment I shall later mention. It

A was not paid at that time. On October 25 1989, Miss Prince, who is Mrs Luttenberger's solicitor, sent a reminder to the company about the rent. The rent was still not paid. On November 17 1989 Miss Prince served on the company a notice to tenant to pay rent due. That notice required the rent to be paid in two months. The notice was given pursuant to the provisions of the Agricultural Holdings Act 1986, Schedule 3, part 1, Case D. The notice was in the form prescribed by regulations made in 1987. On November 24 1989, a statutory demand for payment of the amount of rent due was served on the company. That was not complied with. On December 19 1989 a petition was presented by Mrs Luttenberger to have the company wound-up on the grounds, *inter alia,* of deemed insolvency. The hearing of the petition was fixed for February 7 1990. It has not been finally disposed of. Indeed, I understand that it has been adjourned generally.

At about 8.30 am on January 19 1990 — that is very soon after the B expiration of the two-month period referred to in the notice to pay dated November 17 1989 — Miss Prince, on behalf of Mrs Luttenberger, served a notice to quit on the company. The notice alleged a failure by the company to comply with the notice of November 17 requiring payment of rent due. The notice required possession of the farm to be delivered on April 6 1991 that is, at the expiration of 12 months from the end of the year of the tenancy then current.

Possession was not delivered up on April 6 1991 and the plaintiff therefore alleges that the defendants have since remained in possession of the farm as trespassers. A writ claiming possession was issued on April 8 1991 and the statement of claim on April 9. A summons for summary judgment under Ord 14 of the Rules of the Supreme Court was issued on April 25. On that application Master C Gowers granted the plaintiff an order for possession. Mummery J allowed an appeal against that order, set it aside and gave the defendants unconditional leave to defend. It is against that decision that the plaintiff now appeals.

Before saying any more about the facts, I propose to turn to the scheme of the Agricultural Holdings Act 1986 as it is relevant to the situation in this case.

In *Rous* v *Mitchell,* a decision of this court reported at [1991] 1 WLR 469*, I attempted to deal with this matter in my judgment beginning at p 475B. I said and I repeat:

> Lettings of agricultural holdings have for many years been controlled by statute. The Act in force in relation to these proceedings is the Agricultural Holdings Act 1986. Section 2 of that Act provides, in general, that a letting of D an agricultural holding may not be for a lesser interest than a tenancy from year to year.
>
> By section 25(1) of the Act, the normal rule that a tenancy from year to year may be terminated by at least six months' notice is amended so that an agricultural tenancy requires at least 12 months' notice determining at the end of a year of the tenancy.
>
> Section 26 of the 1986 Act places restrictions on the operation of notices to quit, which in effect divide such notices into two classes. The difference between the two classes is that notices to quit in the first class need not state or be based upon the reason why the landlord seeks to recover possession of the holding, whereas a notice to quit in the second class must be based upon such a reason and must state the E reason in the notice itself.
>
> This case is not concerned with notices in the first class. The second class is governed by section 26 of the Act, which provides in effect that the consent of the agricultural land tribunal is not necessary "in any of the Cases set out in Part 1 of Schedule 3 to this Act":
>
> By Case D of Schedule 3, one of the cases where the consent of the tribunal to the operation of the notice to quit is not required is where:
>
> At the date of the giving of the notice to quit the tenant had failed to comply with a notice in writing served on him by the landlord, being either . . . —
>
> (*a*) a notice requiring him within two months from the service of the notice to pay any rent due in respect of the agricultural holding to which the notice to quit relates, or (*b*) [is irrelevant] and it is stated in the notice to quit that it is F given by reason of the said matter.

Again quoting from *Rous* v *Mitchell* at p 476D:

> The combined effect of section 29 of and paragraph 1 of Schedule 4 to the Act of 1986 is that the Lord Chancellor may make an order "requiring any question arising under the provisions of section 26(2) of, and Schedule 3 to, this Act to be determined by arbitration under this Act".
>
> In exercise of that power the Lord Chancellor made the Agricultural Holdings (Arbitration on Notices) Order 1987. Article 9 of that Order provides:
>
> "Where it is stated in a notice to quit an agricultural holding or part thereof that the notice is given for one or more of the reasons specified in Case A, B, D or E and the tenant wishes to contest any question arising under the provisions of section 26(2) of, and Schedule 3 to, the 1986 Act relating to any of the G reasons so stated, he shall within one month after the service of the notice serve on the landlord notice in writing requiring the question to be determined by arbitration under the 1986 Act".
>
> Article 10 further provides:
>
> A notice under article 9 requiring arbitration under the 1986 Act shall cease to be effective three months after the date of the service of that notice unless before the expiry of those three months —
>
> (a) an arbitrator has been appointed by agreement between the parties, or
>
> (b) (in default of such agreement) an application has been made by the tenant or the landlord under paragraph 1 of Schedule 11 to that Act for the appointment of an arbitrator,
>
> for the purposes of that arbitration.

In this case the defendants' former solicitors duly served, on H February 7 1990, a notice that they required the issue whether the rent had been paid within the two-month period to be referred to arbitration. But thereafter neither they nor the defendants themselves took any steps in time, or at all, to agree to or request the appointment of an arbitrator.

Mummery J approached his task on bases which were and are common ground between the parties. First, it is for the defendants in Ord 14 proceedings to satisfy the court that there is an issue or question in dispute which ought to be tried. Second, despite the failure of the defendants or their former solicitors to apply for the appointment of an arbitrator within three months of the notice requiring arbitration, the defendants in these proceedings are entitled, if they can, to prove that the notice to quit contained false J statements fraudulently made, in which case the notice will be invalid and of no effect. That is derived again from the decision in *Rous* v *Mitchell* [1991] 1 WLR 469, in which, at p 487G, I summarised the authorities referred to in that decision in this paragraph:

> In my view these authorities establish the proposition that a landlord's counter-notice under section 26(6) of the Act of 1954 is invalid and of no effect if the statement contained in it of the landlord's intention is fraudulent, not honestly made. The notice with which we are here concerned is a notice of the same sort as a notice under section 26(6). It is not a mere notice to quit. In my view, a notice to quit which states that it is given for one of the reasons set out in the various cases in Schedule 3 to the Act of 1986 is invalid and of no effect if it contains a statement which is false and made fraudulently by the giver of the notice, ie knowing the statement to be untrue, or reckless whether it is true K or false.

The third basis upon which the learned judge approached this matter was that the defendants accepted that, in order to establish that the notice to quit was invalid, they must prove both of two separate facts: first, that the statement in the notice to quit — that the defendants had not complied with the notice to pay rent, dated November 7 1989 — was, when the notice to quit was served on January 19 1990, false and, second, that that false statement was made by Miss Prince, the solicitor for the plaintiff, on behalf of the plaintiff recklessly, that is without any honest belief in its truth.

Mummery J held that the evidence before him disclosed questions in dispute which ought to be tried in relation to both issues. In this L appeal, Mr Wilson-Smith for the plaintiff argues that the judge was wrong to do so and that, on the evidence on both issues, no court could properly reach a conclusion in the defendants' favour. However, Mr Wilson-Smith needs only to persuade us that there is no question which ought to be tried on either one of the issues in order to succeed in this appeal. I therefore deal in turn with the evidence before the court on each issue.

First, on the question "Was the statement in the notice to quit that the rent due had not been paid by January 19 1990 false at that date?", the defendants' case was that the rent had in fact been paid on December 22 1989. The defendants' evidence, which in this respect is not challenged, shows that the defendants' solicitors posted on December 22 1989 by recorded delivery to the plaintiff's solicitors M a cheque dated October 10 1989, drawn by the defendant company on its bank for £14,805.60, being a half-year's rent less the drainage rate, which was payable by the landlord. In the past, rent had commonly been paid and accepted by cheque in favour of the plaintiff's solicitors sent by post.

The decision of this court in *Beevers* v *Mason* (1978) 37 P&CR 452* is authority for two propositions: first, that where, by reason of a previous course of conduct, a landlord has accepted payment of rent by cheque sent by post, that method of payment may be held to be a method agreed between the parties as acceptable; and, second, when

*Editor's note: Also reported *sub nom Stradbroke (Earl of)* v *Mitchell* [1991] 1 EGLR 1.

*Editor's note: Also reported at (1978) 248 EG 781.

A the cheque is posted, subject only to its being honoured, the rent is paid. So proof of posting on December 22 is *prima facie* proof of payment of the rent but subject to the requirement that it should be honoured, that is to say the cheque should be honoured. Here the plaintiff's case is that the evidence shows that the cheque was not honoured. The cheque was not received at Miss Prince's office until the afternoon of January 19 1990, which was a Friday. The reasons for the delay are relevant not so much to this issue as to the second issue and I will return to consider them.

When Miss Prince received the cheque, she delivered it that same afternoon to the Bridlington branch of the Yorkshire Bank (the bank with which her practice banked) for special presentation. As I understand it, that process has no effect on the procedure by which B payment is made on the cheque until the cheque itself arrives at the bank on which it is drawn. In this case the evidence shows that the cheque arrived at the Louth branch of the Midland Bank (with whom the defendant company banked) on the morning of Wednesday January 24 1991. The cheque then received bore only one signature, that of Mrs Popoff. The Midland Bank's mandate from the defendant company required two signatures on cheques drawn by the company.

What happened thereafter is in dispute, the resolution of which, if we had to resolve it, which we do not, is made more difficult by the fact that the deponents to affidavits on either side dealing with this matter are not the persons who could give first-hand evidence. C Shortly, the plaintiff's case is that on making inquiry that morning a Yorkshire Bank employee was told by a Midland Bank employee that the cheque was not being honoured. The defendants' case in sum is a denial that that was said.

According to Mr Miller (of the Midland Bank) in an affidavit sworn on the defendants' behalf, later on January 24 1990, that is to say after the first telephone conversation between the employees of the two banks, Mr Popoff authorised payment of the cheque despite the fact that it bore only one signature. Moreover, Mr Miller deposes that despite having done so, Mr Popoff called at the branch of the bank that same afternoon and signed the cheque. So the Midland Bank on that same day honoured the cheque. A banker's payment was sent in the post that night which was received by the Yorkshire D Bank and credited to the Miller & Co account.

Mr Wilson-Smith makes two points in relation to this issue. First, he submits that a document bearing only one signature when the mandate between the bank and its customer calls for two is not a cheque. Alternatively, what purports to be a cheque with one signatory when the mandate requires two is not the same document as the document which results from a second signature being added. Mr Wilson-Smith refers us to sections 3(1) and 73 of the Bills of Exchange Act 1882. I intend no discourtesy to him in not quoting those sections because, in my view, they do not assist his argument nor do they help us to decide this issue. It is my view that a document with one signature which in other respects appears to be a cheque is a E cheque. But the result of a mandate requiring two signatures may well be that the bank on which it is drawn is not under any obligation to meet it.

Mr Wilson-Smith's second submission is that, assuming the document was a cheque, it was dishonoured during the course of the telephone conversation on the morning of January 24 between the employees of the respective banks. As to this, the judge said that there were relevant disputes of fact which he could not resolve on this application for summary judgment. Mr Wilson-Smith took us carefully through the relevant affidavits. He sought to show that in certain respects the affidavit on behalf of the defendants did not traverse directly certain assertions made in the affidavit sworn on behalf of the plaintiff. Despite that and despite the criticism which I F have already made, that both those affidavits were sworn by a gentleman who had no first-hand knowledge of the telephone conversation or indeed the other conduct on January 24 1990, I for my part am satisfied that the two affidavits do give a sufficiently clear account to make it apparent that there is a series of disputes of fact such as were referred to by the learned judge. On this issue I therefore entirely agree with Mummery J's decision.

The second issue is: is there a triable issue on the question whether the assertion in the notice to quit that the rent was unpaid was made by Miss Prince recklessly without honest belief in its truth? The evidence discloses a state of affairs to which I think one can at least refer as extraordinary.

For some time before December 1989 Miss Prince had been in dispute with the Post Office about recorded delivery mail. She G maintained that the Post Office was required to deliver recorded delivery mail to her office without requiring a signature by way of receipt. For some time, months I think, the Post Office refused to deliver such mail without the signature of a receipt.

On a date late in December 1989, I believe on the 28th, the envelope containing the cheque was sought to be delivered by the Post Office to the offices of Prince & Co. Either Miss Prince herself or somebody else in that office on her instructions declined to sign the receipt. The Post Office declined thereupon to deliver. A card was left, making it clear that the envelope would be available for collection at a named Post Office upon somebody being willing to give a receipt; and there the matter and the cheque both rested. H

On January 12 1990, Miss Prince was told by the plaintiff herself that she had learned that Mr Popoff had asserted that the rent had been paid by recorded delivery letter. That date was still within the two-month period for payment of rent. On January 15 1990, Miss Prince wrote two letters to the defendant company and to Epton & Co, who were then the solicitors for the defendants, referring to this conversation. She caused those letters to be delivered by hand to their addresses on the same day. In a letter to the defendant company she said:

> Mrs Luttenberger has been informed by Mr John Hewson of North Thoresby that, because Mr Hewson saw the said Winding-up Petition advertised last week, Mr Hewson asked Mr M D Popoff what Mr Popoff was doing about it and that Mr Popoff replied to Mr Hewson that the rent on Eastfield House J Farm has recently been paid up to date "by Recorded Delivery". This letter serves to confirm that no payment of such rent or such amount, in any form, has been received by our client or by us, whether by Recorded Delivery or by any other means. No part of the £15,264.00 mentioned in the said Statutory Demand has ever been paid by North Thoresby Farms Limited or by anyone else.

It is perhaps noticeable that the letter makes no mention of the fact that the reason why the recorded delivery letter had not been received was because of the dispute between Miss Prince and the Post Office. But it is, of course, right also to say that Miss Prince cannot have known for certain that the recorded delivery letter which she knew existed did indeed contain the cheque.

Despite having received those letters on the late afternoon or K evening of January 15, neither Mr Popoff nor anyone else on behalf of the defendant company, and certainly not Epton & Co, made any response or did anything about the matter contained in the letter. Since, as I have said, January 15 was still within the two-month period for payment of rent, prompt action on behalf of the defendants would have prevented these proceedings starting at all — none of us would have been here. One can only assume, though we have not heard any explanation of this, that there was some incompetence on the part either of the defendants or of their solicitors.

According to Miss Prince, on January 18, by which time the two months had elapsed, she arranged for Post Office officials to visit her office to discuss the recorded delivery dispute on the following day. L They arranged to come on the afternoon of January 19. Before they did so, on the morning of January 19 the notice to quit asserting that the notice to pay rent had not been complied with was served. On January 19, according to Miss Prince, there was a discussion at her firm's offices, not with her but with one of her employees or associates. The envelope containing the cheque was handed over by the Post Office officials and it was opened and it was found to contain the defendant company's cheque. Arrangements were then made between Miss Prince's firm and the Post Office for the future delivery of recorded delivery mail.

As I have said, that cheque was then presented and I have rehearsed the history, or at least the evidence as to the history, of the M presentment of the cheque thereafter.

The defendants' case in short is that Miss Prince, knowing from about December 28 1989 that a recorded delivery letter was waiting to be delivered to her firm, knowing from January 12 1990 that Mr Popoff was asserting that the rent had been paid by recorded delivery and knowing on January 18 1990 that the Post Office intended to bring a recorded delivery envelope (the one she knew they had) to the meeting the following day, deliberately decided to serve the notice to quit alleging non-payment of rent before ascertaining what the recorded delivery envelope contained.

Mr Wilson-Smith in reply makes the following points. He submits that no court could properly find that Miss Prince wilfully shut her

A eyes to the possibility that the recorded delivery envelope contained the rent cheque for the following reasons: first, the defendant company has a history of late payment; second, payment by recorded delivery, as opposed to ordinary post without delivery being recorded, was a departure from the ordinary procedure; third, the parties to this action were well aware that there had been problems, which Mr Wilson-Smith described as irregularities, in relation to certain previous rent cheques; fourth, there had been no acknowledgement of the demands for payment in October and November 1989; next, the arrangement made on January 18 for the Post Office to attend for a discussion was accelerated at Miss Prince's specific request; next, at all times she was demanding her post and being refused it only because its delivery was made contingent on a
B signature; and last, because the letter of January 15, of the two letters, put the defendants and their solicitors on notice. They having been put on notice and no response having been received, Miss Prince was, it is submitted, entitled to assume that the rent had not been paid.

In my view, the matters to which Mr Wilson-Smith refers, if in the event they are proved, are all factors to be taken into account by a court deciding the issue. It may be that such a court will decide that on balance those factors should result in a decision that Miss Prince was not recklessly fraudulent. But Mummery J concluded, referring to C
the points made on behalf of the defendants to which I have referred earlier, that there is a series of issues of fact which ought to be tried. He expressed the fatality of that issue in these words at p 17C of the transcript of his judgment:

> That issue is whether Miss Prince did deliberately abstain from inquiring into the contents of that recorded delivery letter, and whether she sought to shut her eyes to a fact she did not want to know and therefore acted recklessly.

The learned judge concluded that there were issues that ought to be tried on both the questions as to whether the notice to quit contained a false statement and, if so, as to whether that false statement was made recklessly by Miss Prince on behalf of her principal. I agree on this issue also. Indeed I go further.

Recognising that the master thought it right to enter judgment D
under Ord 14, nevertheless, in my view, this is an action to which it should have been obvious that the provisions of Ord 14 had no real application. It is an action which bristles with issues which should be determined as soon as possible.

I would dismiss the appeal.

Balcombe and **Woolf LJJ** agreed and did not add anything.

Appeal dismissed with costs.

For further cases on this subject see p 1

COMPULSORY PURCHASE AND COMPENSATION

Court of Appeal

November 14 1991

(Before Lord Justice MUSTILL, Lord Justice MANN and Sir Christopher SLADE)

PROCTER & GAMBLE LTD v SECRETARY OF STATE FOR THE ENVIRONMENT AND OTHERS

Compulsory purchase order — Whether compulsory purchase order confirmed for a purpose different from that for which it was made — Whether confirming authorities had had regard to an irrelevant consideration — Whether there was evidence that the subject land was required for the purpose for which the order was made

On June 26 1989 the Tyne & Wear Development Corporation ("TWDC") made a compulsory purchase order ("the order") under powers conferred by section 142 of the Local Government, Planning and Land Act 1980 — In the non-statutory statement of reasons for making the order the TWDC stated that it was vital to secure the regeneration of an area called East Quayside by the comprehensive redevelopment of the whole area in accordance with a planning permission dated March 3 1989 — The appellants, Procter & Gamble Ltd, are the owners of New Sandgate House, a plot within the order land having a frontage to City Road — On submission of the order to the Secretaries of State for the Environment, Transport and Energy the appointed inspector held a public local inquiry at which evidence was given of a need to widen City Road, a need perceived by Newcastle City Council in their capacity as local highway authority — At the inquiry the TWDC stated that New Sandgate House was required for road widening and that this requirement precluded its retention as sought by the appellants — Following the recommendations of the inspector, the Secretaries of State confirmed the order on July 3 1990 — On appeal from the decision of Auld J it was submitted on behalf of the appellants that the Secretaries of State had confirmed the order for a purpose which was different from, or was additional to, the purpose for which it was made and that such a confirmation was unlawful — The different or additional purpose was the achievement of a scheme of highway improvements which was required to facilitate the development of sites in the urban development area of which East Quayside was but one — In so far as the Secretaries of State had regard to such a scheme of highway improvements, they had had regard to an irrelevant consideration

***Held*: The appeal was dismissed — Despite the absence of authority, a compulsory purchase order made for one purpose cannot lawfully be confirmed for another purpose or for a purpose additional to that for which it was made — The purpose for which the order was made was the regeneration of the East Quayside part of the urban development area — Importance must be given to the expressed purpose selected by the maker of an order — The inspector's conclusion that the road improvements were reasonably necessary to secure the regeneration of East Quayside makes it impossible to say that the purpose for which the order was confirmed in regard to New Sandgate House was not within the terms of the purpose for which it was made — Although the means of achieving the expressed purpose had plainly changed between making and confirming from an office development to a road improvement, the expressed and actual purpose remained constant — Accordingly, the Secretaries of State had not had regard to an irrelevant consideration in taking into account a scheme of highway improvement — There was ample evidence to enable the conclusion of the inspector that without the agreed improvements the highway authority would object to the present regeneration proposals**

The following cases are referred to in this report.

Carington (Lord) v *Wycombe Railway Co* (1868) 3 Ch App 377
Galloway v *Mayor and Commonalty of London* (1866) LR 1 HL 34
Grice v *Dudley Corporation* [1958] Ch 329; [1957] 3 WLR 314; [1957] 2 All ER 673; (1957) 55 LGR 493; 9 P&CR 58; [1957] JPL 723
Meravale Builders Ltd v *Secretary of State for the Environment* (1978) 36 P&CR 87; 77 LGR 365; [1978] JPL 699
Simpsons Motor Sales (London) Ltd v *Hendon Corporation* [1964] AC 1088; [1963] 2 WLR 1187; [1963] 2 All ER 484; (1963) 62 LGR 1; 14 P&CR 386; [1963] EGD 207; 187 EG 581; [1963] RVR 522, HL

This was an appeal from a decision of Auld J who, on June 4 1991, had dismissed an application by the appellants, Procter & Gamble Ltd, under section 23(1) of the Acquisition of Land Act 1981, seeking to have the Tyne & Wear Development Corporation (East Quayside, Newcastle upon Tyne) Compulsory Purchase Order 1989, made by the second respondents, the Tyne & Wear Development Corporation, quashed following its confirmation by the first respondents, the Secretaries of State for the Environment, Transport and Energy.

Malcolm Spence QC and Nicholas Nardecchia (instructed by the solicitor to Procter & Gamble Ltd, of Newcastle upon Tyne) appeared for the appellants; Michael Barnes QC and Alison Foster (instructed by the Treasury Solicitor) appeared for the first respondents, the Secretaries of State for Environment, Transport and Energy; David Mole QC and Nicholas Huskinson (instructed by Nabarro Nathanson) represented the second respondents, the Tyne & Wear Development Corporation.

Giving judgment, **MANN LJ** said: On June 4 1991 Auld J dismissed an application which had been made by Procter & Gamble Ltd under section 23(1) of the Acquisition of Land Act 1981. The company had by their application sought to have the Tyne & Wear Development Corporation (East Quayside, Newcastle upon Tyne) Compulsory Purchase Order 1989 ("the order") quashed in so far as it affects their office building and premises known as New Sandgate House. They now appeal against the dismissal of that application.

The grounds for an application to quash a compulsory purchase order are prescribed in section 23(1) and (2) of the 1981 Act and the relevant one of them is the ground that the authorisation of the compulsory purchase was not empowered to be granted. It is the appellants' case that the authorisation of the purchase of New Sandgate House was not empowered to be granted in that: (1) the order was confirmed for a purpose different from that for which it was made; (2) the confirming authorities had had regard to an irrelevant consideration; and (3) there was no evidence that New Sandgate House was required for the purpose for which the order was

A made. Arguments (1) and (2) are interrelated but argument (3) is independent of them.

The order was made by the Tyne & Wear Development Corporation (the "TWDC") on June 26 1989 in the exercise of the power of compulsory acquisition conferred by section 142 of the Local Government, Planning and Land Act 1980. I shall later come to that power but it is a power exercisable in accord with the procedure specified in Part II of the 1981 Act (1980 Act, section 142(2A)). Under that procedure a compulsory purchase order made by the TWDC had to be submitted to and confirmed by the confirming authority. In this case and for reasons which are immaterial, the Secretaries of State for the Environment, Energy and B Transport were jointly the confirming authority. The order was submitted for confirmation and there were objections to confirmation by, among others, the appellants. A public local inquiry was held by Mr J P MacBryde ARIBA MRTPI MCIT FRSA. The inquiry lasted from November 14 to December 1 1989 and on February 16 1990 the inspector reported with his recommendation that the order be confirmed subject to the agreed exclusion of a plot owned by a water authority. On July 3 1990 the Secretaries of State issued a decision letter in which they agreed with the inspector's conclusions, accepted his recommendation and confirmed the order accordingly.

The background to the making of the order by the TWDC is fully set out by Auld J in his judgment and I gratefully accept his C unchallenged account as the basis for my own shorter narrative. East of the centre of Newcastle upon Tyne are extensive urban areas, which are either run down or derelict. On April 2 1987 the Secretary of State for the Environment designated 2,375 ha of those areas as an urban development area under section 134 of the Act of 1980. On May 15 1987 and in the exercise of his power under section 135(1) of the Act, he established the TWDC for the purpose of regenerating the designated area. I shall come later to the object and the general power to acquire land of a development corporation. On October 1 1987 and by virtue of an order made under section 149(1) of the Act, TWDC became the local planning authority for the designated area.

Within the designated area there is an area of about 12 ha known as D East Quayside. It is a strip of land on the north bank of the River Tyne and it is the subject of the order. The northern boundary of the land is the south side of City Road (A186), which is a highway running from Milk Market in the west and approximately parallel to the river. New Sandgate House has a frontage to City Road near the junction with Gibson Street from the north and is designated as plot 11 on the map referred to in the order. Other highways in the East Quayside area are Quayside, which runs along the river bank, and a road called successively Sandgate, St Mary's Street and St Ann's Street, which divides laterally the eastern half of the area before turning north to form a junction with City Road opposite Crawhall Road from the north.

The TWDC saw East Quayside as having a significant potential for E regeneration and they sought proposals for its redevelopment. They held a competition and the winning proposal was one submitted by Newcastle Quayside Developments plc ("NQD"). Their scheme was for a comprehensive redevelopment for a variety of uses and it involved the demolition of the buildings on plot 11 and their replacement by new office buildings. NQD sought an outline planning permission for their scheme and this was granted by the TWDC on March 3 1989. The siting of the new buildings was not a reserved matter with the result that the scheme's location of the new buildings was permitted.

The order was made on June 26 1989. Article 1 of the order provided so far as is material:

F the . . . Corporation are, under section 142(1)(*a*) of the Local Government, Planning and Land Act 1980, hereby authorised to purchase compulsorily for the purpose of securing the regeneration of part of the Tyne and Wear Urban Development Area the land which is described in the schedule hereto and is delineated . . . on the [sealed] map.

A non-statutory statement of reasons for making the order was issued by TWDC. The conclusion of that statement was:

The Corporation considers that it is vital to secure the regeneration of East Quayside within the shortest possible timescale but the only way to guarantee that is by comprehensive redevelopment of the whole area under the control of the Development Corporation in association with a developer and that it is in the public interest that this Order be confirmed. The provision of development on what is a flagship site in Newcastle will instil confidence in both the public and private sectors. A large number of jobs would be created within the proposed development together with attractive residential areas on G a prime site. The leisure elements and the commercial trade generated by both the office developments and the hotel are vital economic components. The proposed development would act as a magnet for tourism and inward investment and contribute towards regeneration over a wider area than its immediate locality. The Corporation wishes this proposal to be brought forward without delay and the Secretary of State will be asked to confirm the Order as a matter of urgency.

The non-statutory statement indicated at para 9.1 that an application would be made for an order under what was then section 209 of the Town and Country Planning Act 1971 [section 247 of the Town and Country Planning Act 1990] extinguishing existing public rights of way in order to enable development to be carried out in accordance with the permission of March 3 1989. Such an application H was made and related among other rights of way to Quayside, Sandgate, St Mary's Street and St Ann's Street. It was also the subject of the inquiry held by Mr MacBryde and he recommended that an order be made. His recommendation was accepted by the Secretary of State for Transport, who on July 3 1990 made the Stopping up of Highways (County of Tyne and Wear) (No 15) Order 1990.

On October 10 1989, in discharge of the obligation imposed on them by rule 4(4) of the Compulsory Purchase by Public Authorities (Inquiries Procedure) Rules 1976, the TWDC served on the objectors to the order a statement of their reasons for making the order. The conclusions of the TWDC were set out in para 6.3 of the statement and among them were these:

The Corporation's comprehensive regeneration strategy for this site is now J fully set out in the development framework. The application of the principles underlying that framework enabled the Corporation to grant outline planning permission for a comprehensive redevelopment scheme to Newcastle Quayside Developments plc.

The Corporation is confident that its objective of securing comprehensive regeneration of the site can be achieved by the implementation of a proposal such as that for which outline planning permission has already been granted. That proposal satisfies the general requirements of the Corporation in terms of the substance and quality of its vision for East Quayside. If the compulsory purchase order is confirmed and the stopping up Order made the Corporation would have sufficient authority to proceed to regenerate the site as quickly as possible.

Nowhere in either the non-statutory statement of reasons or the K rule 4 statement is there any mention of a need to widen City Road or to take part of East Quayside for such an improvement. The scheme approved on March 3 1989 did not accommodate such an improvement. A need to widen City Road was, however, to become a factor of critical importance. It was a need perceived by Newcastle City Council (the "NCC") in their capacity as local highway authority and I must give some account of what occurred prior to the inquiry as it can be derived from the evidence produced at the inquiry.

On October 25 1989 the regional capital and urban development subcommittee of NCC received a report from the acting city engineer (Mr Peter McNamara) and the director of administration upon the subject "East Quayside Development: Highways Agreement and Access Agreement". The report recited that a formal objection to the L stopping-up order had been made by the NCC and described the process of establishing the NCC highways requirements to the point where the terms of a highway agreement with the TWDC could be settled and objections to the stopping-up order could be withdrawn. The officers wrote (para 2.3 of EQ43):

Our full and detailed requirements for highway improvements have now been finalised and are shown on Drawing No XH11/01/01, which is displayed, and which forms part of the Highways Agreement. It must be stressed that the improvements are required as a consequence of the development proposals, which received planning approval on the 10 March 1989 and take account of two factors:

(i) the loss of highway capacity resulting from the closure of two existing highways (ie the quayside and St Mary's Street). M
(ii) the traffic that will be generated by the new development and the need to provide adequate access to and from the site.

The improvements shown upon drawing XH11/01/01 included a widening of City Road so as to encroach on plot 11 and thereby to preclude both the retention of the existing office buildings and the erection of office buildings in the situation approved on March 3 1989. The subcommittee accepted the officers' recommendation that the highways agreement be entered into and heads of agreement with the TWDC were concluded on the same day. Those heads recited that their purpose "is to secure the improvements in the area which are necessary as a consequence of the stopping up of the highways and the development". Head (1) was a provision that the TWDC "is to

A acquire and dedicate as highway the additional land . . . which is needed to carry out the improvements". The NCC withdrew their objection to the making of a stopping-up order on November 13 1989.

The case put at the inquiry by the TWDC was that plot 11 was required for road widening and that this requirement precluded the retention of the existing office building as was sought by the appellants. Other and independent grounds were also advanced against the case for retention, but I need not discuss them. In support of their highway case the TWDC called Mr David Stewart, who is a senior engineer in Mott MacDonald, who were retained as consulting engineers by the TWDC. He said in his statement (UDC 6 para 2.5):

> B Part of the land covered by the Compulsory Purchase Order is required to enable this highway improvement scheme to be achieved. This scheme has been designed to enable the increase in traffic levels created by the comprehensive redevelopment of this site along with other developments within the urban development area and the closure of the internal road network to be accommodated on City Road.

Mr McNamara was called to provide information for the inspector and, as will appear, he said that the widened City Road would accommodate traffic generated by redevelopment elsewhere in the urban development area.

The appellants had notice that a highway case was to be made in support of the order and there is no suggestion that the deployment of that case involved any element of unfairness. Indeed the appellants
C called as a witness Mr John Tomalin, who is a consulting traffic engineer and who propounded alternative traffic schemes.Those schemes were in the event rejected by the inspector as being unsatisfactory and no complaint is or could be made about their rejection.

It is unnecessary to refer to all of the submissions which the appellants are reported as having made to the inspector but I must quote the record of one argument (para 149). It is that the TWDC:

> . . . are seeking to obtain land for a collateral or indirect purpose and they are seeking to dispose of the land [ie part of plot 11] for such a purpose to a third party, namely the local highway authority. The highway improvement is not merely intended for the purpose of accommodating the traffic diverted as a result of the concurrent stopping-up order or the traffic generated by the new
> D development within the East Quayside area. It is also quite clearly seen to be required by the totality of traffic originating from other riverside projects sponsored by the TWDC as they have indeed argued in justification of the precise scale and nature of the improvements.

The reply of the TWDC to this argument is recorded (para 151) as being:

> The question of the improvement of City Road is extremely relevant to the larger issue of East Quayside regeneration because it is fully accepted that such works are necessary to accommodate traffic displaced by the new development, for excellent environmental reasons, and traffic generated by the land in its new (and much more intensive) use. If such traffic cannot be carried on an improved City Road, the regenerative scheme certainly cannot, or rather will not, proceed to completion.

E The inspector found as a fact (para 156 (18)) that without the agreed improvements "the local highway authority would object to the present regeneration proposals, the subject of outline planning consent granted on 3 March 1989". This finding is now challenged by the appellants and I shall return to it. Those of the inspector's conclusions which are relevant are contained in para 157(1) and (13). They are:

> (1) The submissions made on behalf of Procter and Gamble relate to the use of section 142 powers by the acquiring authority. My conclusions are that these powers appear to have been properly used in relation to the acquisition of the land which is needed for incidental highway improvements. I fully accept the arguments put forward by the acquiring authority, namely that the powers which are conferred by statute are wide and bear little comparison
> F with those the subject of private acts. In addition, the point is made that the essential legal test is the end to which the land so acquired is to be put. In this case I am left in no doubt that the road improvements required by the local highway authority are reasonably necessary to secure the regeneration of East Quayside in that they are the subject of detailed negotiations and agreement between the TWDC and Newcastle City Council.
>
> . . .
>
> (13) The decisive objection to retaining New Sandgate House is probably that of the pattern of traffic management which it would impose upon the area which surrounds East Quayside. Any solution of the problem of absorbing more traffic along the restricted width of City Road would seem to require area traffic management involving some degree of one-way working within residential areas. Such proposals are seemingly not acceptable to the City Council and are in marked contrast to those comprehensively agreed between the latter and the development corporation. The need to amend details of siting of the component blocks of the NQD layout to fit in with the highway G
> improvement line cannot truly be compared with the constraints imposed by an existing and substantial office building: the latter is fixed and the former is inherently fluid.

The Secretaries of State agreed with those conclusions. A criticism was made of the last sentence of conclusion 157(1) as being one where the conclusion ("are reasonably . . . East Quayside") did not follow from the attached reason ("the subject . . . Newcastle City Council"). I acknowledge the formal defect but the conclusion itself is clear and its clarity is reinforced by the preceding reference to "incidental highway improvements", which must mean improvements following on the redevelopment of East Quayside. In any event the formal defect is I think no more than an infelicity in H
language, for, as will appear later in this judgment, if the inspector had used the phrase "as is shown by the" instead of "in that they are the subject of" he would not have exposed himself to criticism.

Mr Malcolm Spence QC, who appeared for the appellants, made it his first submission that the Secretaries of State had confirmed the order for a purpose which was different from, or was additional to, the purpose for which it was made and that such a confirmation was unlawful. The different or additional purpose he said was the achievement of a scheme of highway improvement which was required to facilitate the development of sites in the urban development area of which East Quayside was but one. Closely related to that submission was Mr Spence's second submission, which was that, in so far as the Secretaries of State had regard to such J
a scheme of highway improvement, they had had regard to an irrelevant consideration.

We were not referred to any authority which addressed the question of whether a compulsory purchase order made for one purpose can lawfully be confirmed for a different purpose or for that purpose and for an additional purpose. We were, however, referred to some of the authorities for the proposition that the authorisation of a compulsory purchase by statute cannot be used so as to acquire for a purpose other than that for which the authorisation was given: see, for example, *Galloway* v *Mayor and Commonalty of London* (1866) LR 1 HL 34 at p 43 *per* Lord Cranworth; *Lord Carington* v *Wycombe Railway Co* (1868) 3 Ch App 377 at pp 380-381 *per* Lord K
Cairns LJ. The same rule should sensibly apply where the authorisation is conferred by a compulsory purchase order and there is authority for the proposition that it does: see *Grice* v *Dudley Corporation* [1958] Ch 329 at p 344 and also *Simpsons Motor Sales (London) Ltd* v *Hendon Corporation* [1964] AC 1088, at p 1117*.

Despite the absence of authority, I have no hesitation in thinking that a compulsory purchase order made for one purpose cannot lawfully be confirmed for another purpose or for a purpose additional to that for which it was made. Confirmation is the ratification by the confirming authority of what the acquiring authority have resolved to do. By hypothesis the acquiring authority will not have resolved to acquire for any purpose other than that for which the order was made. I think it appropriate that a rule analogous L
to that established for regulating the exercise of an authorisation should apply to the confirmation of an authorisation. There may be room for argument as to whether the rule would preclude confirmation for a purpose which although not the same as that for which the order was made is none the less within that purpose. Common sense may suggest that there could be confirmation to that extent, but I find it unnecessary in this appeal to express an opinion on the point.

It was conceded before Auld J by leading counsel for the Secretaries of State (who did not appear for them in this court) and by Mr David Mole QC, who appeared for TWDC, that "a compulsory purchase order cannot be confirmed for a different object from that for which it was made". In this court Mr Michael Barnes QC, who M
appeared for the Secretaries of State, said that the concession although not literally incorrect, was capable of being misunderstood. The capacity of the concession to give rise to misunderstanding arose, said Mr Barnes, from its failure to make explicit that the only purpose for which a compulsory purchase order can lawfully be made and therefore confirmed is the *statutory* purpose for which the power of acquisition is given. If this proposition is pursued in the present case it will be found that the power to acquire land within an urban development area given by section 142(1)(*a*) of the Act of 1980 is given to a development corporation without an expression of the

*Editor's note: Also reported at (1963) 187 EG 581.

A purpose for which it is exercisable. However, I think it plain that the power is given for the purpose of achieving the single statutory object of a development corporation, which is "to secure the regeneration of its area" (section 136(1)). That it is plain is because Parliament, in granting to a development corporation their general powers, has empowered them to acquire land "for the purpose of achieving the object". On this analysis, said Mr Barnes, the only purpose of the order could be and was to secure the regeneration of the Tyne and Wear Urban Development Area, and the qualification "part of" in the order was, in his word, "surplusage".

If Mr Barnes' proposition and subsequent analysis is correct then that would be decisive of the appeal, for it was inherent in Mr B Spence's submission that the purpose of the acquisition for the improvement of City Road was that of securing the regeneration of the Tyne and Wear Urban Development Area when viewed as a whole, that is, as including all development sites. Confirmation would have been for the same statutory purpose as that for which the order was made. The purpose would have been and remained the purpose for which acquisition was empowered. The case would in that respect be unlike *Meravale Builders Ltd* v *Secretary of State for the Environment* (1978) 36 P&CR 87 where Willis J quashed an order made in the exercise of powers conferred for the purpose of providing housing accommodation and expressed to be made and confirmed for that purpose, on the ground that the actual purpose also included the provision of a road which was to serve a function independent of C servicing the housing accommodation which was to be provided. The decision shows that the courts can look behind the expressed purpose of both making and confirming in order to identify on the evidence the actual purpose which axiomatically must be one within the purpose for which the power to make and confirm is granted. It is to be observed that Willis J would have reached a different conclusion had the road been "fairly and reasonably incidental to the provision of housing" (36 P&CR 87 at p 95). *Meravale* was cited to the inspector (report para 150) and it seems probable that his own use of the word "incidental" in para 157(1) of his report was taken from the judgment, thereby reinforcing my view as to the clarity of the conclusion in that paragraph.

I doubt whether Mr Barnes' proposition is correct although I do D not doubt his subsequent analysis. His proposal is that as a matter of law the purpose must be not merely within but must also be the same as that for which the power is conferred regardless of the purpose promoted by the maker of the order. This is a proposition which is supported neither by authority nor by perceptible justification. My doubt is such that I am well content to assume in favour of the appellants that the proposition is wrong. However, I do not decide that it is, for I find that a decision on the proposition is unnecessary for the disposal of this appeal.

On the basis of my assumption the question which next arises is as to the identification of the purpose of the order. Mr Spence proposed it was either "the regeneration of part of the Tyne and Wear Urban E Development Area" (that is to say East Quayside) as expressed in the order or the implementation of the scheme for which planning permission was granted on March 3 1989, that is to say the NQD scheme. If the latter was the purpose then the appellants would succeed, because the widening of City Road is incompatible with the approved layout. Auld J rejected the second proposal and I think he was right to do so. Indeed Mr Spence did not strenuously suggest to the contrary. Undoubtedly the TWDC had the NQD scheme in mind when they resolved to make the order, but nowhere do they confine their purpose to the implementation of that scheme. I note in particular the reference in para 6.3, 4 of the rule 4 statement, which I have already quoted, to "the implementation of a proposal *such as that* for which outline planning permission has already been F granted" (emphasis supplied). This is an approach which I would have expected, because the means of achieving the regeneration or redevelopment of an area will almost inevitably alter with the lapse of time if only in order to accommodate requirements which hitherto were unforeseen. The accommodation of the requirements of the highway authority in the present case is a good example of the process of change. In my judgment, Mr Spence was correct in his primary contention that the purpose for which the order was made was the regeneration of part of the urban development area, that is to say of East Quayside. That was the expressed purpose of the maker. On my assumption that Mr Barnes' wider proposition is wrong, I attach decisive importance to the expressed purpose selected by the maker of an order. That is the natural focus of attention for both layman and lawyer. That purpose must of course be either identical to or one G selected from within the purpose for which the making of an order is empowered.

The inspector concluded (report para 157(1)) that "the road improvements . . . are reasonably necessary to secure the regeneration *of East Quayside*" (emphasis supplied). The Secretaries of State accepted that conclusion. It is a conclusion which, in my judgment, makes it impossible to say that the purpose for which the order was confirmed in regard to plot 11 was not within the terms of the purpose for which it was made. The means of achieving the expressed purpose had plainly changed between making and confirmation from an office development to a road improvement but, in my judgment, the expressed and actual purpose remained H constant. Mr Spence's first argument accordingly fails. It also follows that his argument to the effect that the Secretaries of State had regard to an irrelevant consideration in taking into account a scheme of highway improvement necessitated by developments of which East Quayside was but one must also fail, for they did not take into account such a scheme.

Mr Spence's third submission was that there was no evidence to support either the inspector's accepted and critical conclusion in para 157(1) of his report or his accepted finding of fact that without the agreed improvements the highway authority "would object to the present regeneration proposals" (para 156(18)). I am unable to accept this submission.

There was, in my judgment, ample evidence to enable the J conclusion to be made. I have already quoted from the report by the NCC's officers to the meeting on October 25 1989 and from the heads of agreement entered into on the same day. The quotations show a direct linkage between the regeneration of East Quayside and the road improvements. This linkage is also found in two statements to the inquiry made by Mr McNamara (NCC/2/WR and /4/WR). In the first statement (made on November 23) the acting city engineer said, after referring to the October 25 report:

> The . . . proposals [on drawing XH11/01/01], and the associated Highways Agreement, have evolved as a direct consequence of the NQD development proposals which were given planning approval by the Urban Development Corporation on the 10th March 1989.

The second statement (made on November 30) was a response to K six written questions of which the first was from the inspector, who requested an estimate of the effect of retaining New Sandgate House upon the effective capacity of City Road. Mr McNamara's answer deserves quotation in full. It was:

> If we look at the traffic flows [on a specified diagram] we can see that the eastbound traffic flows on the new link from the Milk Market to City Road are such that *if the various roads within the East Quayside Development are stopped up*, then two lanes will be required on City Road for eastbound traffic, east of Gibson Street. This would be the case even if there was no contribution to that eastbound movement from the East Quayside Development via the Milk Market junction, which is a contribution of 89 vehicles. This then would mean that if New Sandgate House were to be retained, only two lanes would be available for the westbound movements on L City Road between Crawhall Road and Gibson Street. The reassigned existing traffic westbound on this section of City Road is shown in Fig Q2 as 1,475 vehicles. *The addition of the development traffic (overall) increases this by nearly 900 vehicles*, increasing the right turn to Gibson Street by 200 vehicles, the straight ahead movement to Swan House by 450 vehicles and the left slip to Milk Market by 250 vehicles. *I am informed that 700 of these vehicles are generated by the East Quayside Development* (very nearly 50% of the total existing traffic). It is a matter of agreement that the total westbound traffic on City Road requires 3 lanes, ie one for each movement. It would be difficult for me to postulate at what level of traffic City Road could operate westbound with only two lanes, but on the basis of [certain figures], 1,400-1,700 vehicles/hour would appear to be the limit and this has been confirmed by Mott MacDonald's computer tests. *The implications of this is that all, or the greater part, of the 700 vehicles generated by East Quayside Development in this area would overload the system.* M It is probable that a certain proportion of this could transfer to the Milk Market Junction. Again it is difficult for me to postulate what level this could be but if between a quarter and a third were considered as an acceptable "switch" within the development, say 200 vehicles/hour, this would leave up to 500 vh/hr which would have to find other routes or which would have to disappear completely, ie by reducing the development content. 500 vehicles is over 40% of the total existing traffic generated by the East Quayside Development. *Some proportion of the development content up to that figure could therefore be at risk.*

The emphases are mine. They show an asserted connection between the stopping-up of highways and the generation of traffic within East Quayside on the one hand and the agreed road improvements on the other. True it is that the loading from

A developments elsewhere in the urban development area would also be accommodated as seven ninths of the additional west-bound flow was identified as generated within East Quayside and the east-bound problem was said to arise by reason of the proposed road closures in East Quayside. Whether the asserted connection should be accepted as correct was a question for the inspector and the Secretaries of State, but I think that ample evidence was available to enable them to conclude as they did. Our attention was not drawn to any other evidence which might have been capable of displacing the interdependence asserted in the report, the heads of agreement and the estimates by Mr McNamara. The evidence to which I have referred is what led me to think in an earlier part of this judgment that it is permissible to regard the formal defect in the reasoning in para B 157(1) of the inspector's report as being no more than an infelicity of language.

In my judgment, that evidence also entitled the inspector to find as a fact, as he did, that without the agreed improvements the NCC would have objected to the regeneration proposals. I have already stated that the council did object to the making of the stopping-up order and did not withdraw their objection until after the heads of agreement had been concluded. Mr McNamara remarked in his second statement that there was no realistic alternative to the agreed improvements which did not involve diverting traffic through a residential area upon which would be inflicted an unacceptable level of traffic noise. He observed that in the event of the land required for C the highway improvement not being forthcoming "the City Council would need to reconsider its position and they may indeed wish to oppose any alternative form of development". Here in total is adequate material for enabling the inspector to find as he did.

I would dismiss this appeal.

Agreeing, **Sir Christopher Slade** said: I gratefully adopt Mann LJ's statement of the facts together with his definitions. As he has explained, it is in essence the appellant's case that the authorisation of the purchase of New Sandgate House was not empowered to be granted because:

(1) the order was confirmed for a purpose different from that for which it was made,

D (2) the confirming authorities had had regard to an irrelevant consideration, and

(3) there was no evidence that NSH was required for the purpose for which the order was made.

I wish to add a few observations principally in relation to the first of these three points.

As was eventually common ground in the court below, I agree that a compulsory purchase order made for one purpose cannot lawfully be confirmed for another purpose. The starting point must therefore be a consideration of the purpose for which the TWDC made the order in the present case.

As the premise of the first of his alternative arguments in response E to the appeal, on behalf of the Secretaries of State, Mr Michael Barnes QC submitted that the order was made and confirmed for the purpose of securing the regeneration of the *whole* of the Tyne and Wear Urban Development Area. This, he submitted, was the only purpose for which the order could lawfully have been made and confirmed, having regard to section 136(1) of the Local Government, Planning and Land Act 1980, which provides that "the object of an urban development corporation shall be to secure the regeneration of its area". This being so, it was argued, it is irrelevant that, because of the requirements of the local highway authority, the method by which, at the date of the order, it was contemplated this broad purpose would be achieved changed between that date and the date of the confirmation: compare *Simpsons Motor Sales (London) Ltd* v F *Hendon Corporation* [1964] AC 1088. The making of the order and its confirmation must have been equally lawful.

I, for my part, cannot accept this particular argument because I cannot accept the premise. As Mann LJ has pointed out, while the power to acquire land within an urban development area given by section 142(1)(*a*) of the 1980 Act is given to a development corporation without any expression of the purpose for which it is exercisable, it must be deemed to have been given for the purpose of achieving the statutory object of an urban development corporation, set out in section 136(1), ie "to secure the regeneration of its area". There may well be cases, however, where on the facts a particular acquisition, while undoubtedly appropriate to secure the regeneration of *part* of the relevant area, will have no effect on the remainder of the area. In the absence of authority precluding a G different construction, I would without much hesitation hold that the 1980 Act empowers an urban development corporation to acquire land in their urban development area for the purpose of securing the regeneration of *part* of such area; the greater includes the less. I cannot believe that Parliament would have intended otherwise.

I turn to consider the purpose for which the order in the present case was in fact made. As to this I feel no doubt. By its very terms, the order was expressed to be made "for the purpose of securing the regeneration of *part of* [emphasis supplied] the Tyne and Wear Urban Development Area". In my judgment, it is plain that the part this referred to was East Quayside, the subject of the order. The order thus purported to be made, and was in truth made, for the purpose of H securing the regeneration of East Quayside. I can see no sufficient reason for attributing to the TWDC a purpose in making the order other than that therein stated. This was indeed the principal submission of Mr Spence QC for the appellants. In his skeleton argument, there had featured prominently a different submission, namely that the purpose was the more restricted purpose of regenerating East Quayside by a particular method, that is to say, by carrying out the permitted NQD scheme. However, without formally abandoning it, Mr Spence did not press this different submission on this court and, for the reasons given by Mann LJ, I think he was right not to do so.

I now turn to consider the purpose for which the Secretaries of State confirmed the order. In para 6 of their letter of July 3 1990 they J said:

> After careful consideration of all the material before him the Secretary of State wholly agrees with the Inspector's conclusions and accepts his recommendation. He is satisfied that the Order must be confirmed so that the regeneration of the Order land may be secured. The Secretary of State has accordingly decided to confirm the Order with the modifications shown in it and this letter conveys his decision to that effect.

According to their intentions as thus expressed, the Secretaries of State were confirming the order for the same purpose as that for which it had been made, namely the regeneration of East Quayside.

For the appellants, however, it is said that the matter is by no means as simple as that. Between the dates of the order and the confirmation circumstances had changed. At the date of the order K the TWDC had contemplated that the buildings on Plot 11 would be demolished and replaced by new office buildings. The attitude of the local highway authority (the "NCC") had subsequently necessitated a substantial revision of their plans. As at the date of confirmation, though the buildings on Plot 11 were still to be demolished, the site was intended to be used not for the purpose of erecting new office buildings, but for the purpose of widening City Road. In para 157(13) of his report, which Mann LJ has already quoted, the inspector referred to the pattern of traffic management which the retention of New Sandgate House would impose upon the area surrounding East Quayside and referred to this as "the decisive objection to retaining New Sandgate House". Mr Spence skilfully deployed this and other L points to found an argument that the order was confirmed not for the purpose of securing the regeneration of East Quayside but for the purpose of carrying out highway works as required by the NCC.

I cannot agree. The road improvements required by the NCC were themselves "reasonably necessary to secure the regeneration of East Quayside". The inspector made an express finding of fact in these terms in para 157(1) of his report (quoted by Mann LJ*) and further found in para 156(18) that "without such agreed improvements, the local highway authority would object to the present regeneration proposals, the subject of outline planning consent granted on 3rd March 1989". I agree with Mann LJ that there was ample evidence to support both these findings. Though Mr Spence has criticised the last two lines of para 157(1) beginning with the words "in that" as M involving a *non sequitur*, and I agree that they could have been more accurately expressed, they do not, in my view, involve a *non sequitur* if they are read in conjunction with para 156(18) and the other earlier passages in the report dealing with the highways position and the attitude of the highway authority; they are simply to be read as a shorthand reference to that background.

Mr Spence suggested that the reference in para 157(13) to the pattern of traffic management which the retention of New Sandgate House would impose upon "the area which surrounds East Quayside" involved the introduction of an immaterial

*Editor's note: See p 268 *ante*.

A consideration, namely the effect of the order and scheme for highway improvement on developments other than East Quayside. I do not think this is so. In its context, the reference was, in my opinion, included because it was relevant to the attitude of the NCC, as local highway authority, and that authority's attitude was highly material to any scheme for the regeneration of East Quayside itself.

For these and the further reasons given by Mann LJ, I do not think that any of the three grounds of appeal is well-founded. I, too, would B accordingly dismiss this appeal.

Mustill LJ agreed and did not add anything.

Appeal dismissed with costs. Application for leave to appeal to the House of Lords refused.

For further cases on this subject see p 35

LANDLORD AND TENANT

GENERAL

A

Court of Appeal

October 7 1991

(Before Lord Justice STOCKER, Lord Justice BUTLER-SLOSS and Sir George WALLER)

WHITE v WAREING

Claim for possession against residential tenant — Arrears of rent — Appeal against order of county court that possession should be given — Whether valid notice to quit served — Whether tenancy a statutory tenant — Whether county court considered reasonableness of making order

B **The appellant is the monthly tenant of residential premises — The rent had been registered since 1989 at £175 a month — There were arrears of rent in 1990 — The respondent landlord served a notice to quit on February 1 1991 — The summons for possession indicated two months' arrears amounting to £350 but the particulars of claim related to £150 — The summons was adjourned on June 24 1991 for the respondent landlord to prove service of the notice to quit — The court made an order for possession within 28 days suspended upon the appellant's paying the arrears on terms — The appellant appealed that decision taking the following points — The service of the** C **notice to quit had not been proved — The notice to quit was in any event invalid — The respondent landlord was not entitled to any sum other than that claimed in the particulars of claim — The county court had not considered whether it was reasonable to make an order for possession under section 98(1) of the Rent Act 1977**

***Held*: The county court judge was entitled to find that the notice to quit had been served relying upon the documents on the court file and an affidavit of service — Had the tenancy been contractual, the notice to quit was admittedly defective — There was little doubt that it was a statutory tenancy — It was pleaded as a statutory tenancy and it was reasonable to** D **assume in the case of a tenancy lasting for a long time that it had been converted to a statutory tenancy — Accordingly, the form of the document called a notice to quit was of no importance save that it was important that the tenant should be given notice that he was expected to leave and the lack of a letter before action or similar warning may affect the issue of reasonableness under section 98 of the 1977 Act — There was an obvious technical error in the particulars of claim which was of no consequence — The county court had exercised its discretion under section 98 as was made clear from the evidence — The appeal was dismissed**

E

The following case is referred to in this report.

Morrison v *Jacobs* [1945] KB 577

This was an appeal by the defendant, Alan Wareing, from the decision of Judge Hitching, sitting in Bow County Court in an action by the plaintiff, Nicholas York White, for arrears of rent and possession of 41 Gordon Road, Wanstead, London E18.

Renee Calder (instructed by Ouvry Goodman & Co, of Sutton) appeared on behalf of the appellant; Graham Read (instructed by Edell & Jones) represented the respondent.

Giving the first judgment at the invitation of Stocker LJ, **BUTLER-SLOSS LJ** said: The appellant is the tenant of 41 Gordon Road, F Wanstead. He has been the tenant for many years. There have been a number of applications between him and his landlord with respect to rent assessments before the rent assessment panels and boards and indeed he is the subject of a decision of this court in 1985, referring to an assessment in March 1983*. Again there was an assessment of July 1989. He has held this property on a monthly tenancy, the 26th day of the month being a rent day, at a rent since 1989 of £175 a month. It seems clear to this court that he is and has for many years been a statutory tenant. He was not actually so described in the evidence, or in the judgment of Judge Hitching, although it was pleaded in the particulars of claim that he held as a statutory tenant. G

There were arrears of rent in the latter part of 1990. No payment was made after January 1 1991 and the landlord issued a notice to quit on February 1 1991, together with a covering letter. There was a subsequent letter from the landlord to the tenant, undated, subsequent to February 1 but prior to the issue of the summons for possession on May 1. That letter indicated the intention of the landlord to take proceedings and that the tenant was expected since the notice to quit to vacate.

The particulars of claim, together with the summons for possession, indicated two months' arrears before the issue of proceedings. Those two months were shown in the summons as coming to £350, but shown in the particulars of claim as £150, together with a claim for mesne profits. It seems clear that the £150 in H the particulars of claim was a typing error, but it was not corrected before the judge.

The tenant filled in a form for replying to a summons for possession of land in which he ticked the "No" box in answer to the question "Do you admit the plaintiff is entitled to possession of the premises?" and gave as his reason "No notice to quit served". Also he did not admit the money claim because he said cheques had been sent by post. He did not choose to file a defence to the particulars of claim. Consequently the only defence before the court was the form for replying to a summons and the only matter raised before the court was the lack of service of notice to quit, together with the dispute as to the amount of money owing. J

On June 24 the summons for possession came before Judge Cox. He made an order that the matter be adjourned to July 1 for the plaintiff to prove service of the notice to quit. On July 1 it came before Judge Hitching and his decision is the subject of the appeal to this court.

The tenant acted for himself, although today he is represented by counsel. He asked for an adjournment and indicated his intention at the adjournment to raise certain legal matters relating to the form of the notice to quit. The judge said that the notice to quit was on the file, having been received in copy form by Judge Cox on the previous occasion and that it was in proper form. The tenant was informed he could have an adjournment only on terms that he paid the costs thrown away. He chose not to have an adjournment on those terms. K The landlord gave evidence and the tenant gave evidence.

The judge held that possession should be given on this property and it was to be a 28-day order but that "the judgment for possession be suspended upon the Defendant paying £570.00 in 14 days and [the] balance of £305 within [a] further 14 days thereafter". The defendant appeals this decision.

Counsel on behalf of the tenant before us in this court today has exhaustively covered every point available. They can be summarised as five points, which include leave that we have granted at a very late

---

*Editor's note: *Wareing* v *White* [1985] 1 EGLR 125; (1985) 274 EG 488.

A stage to allow a supplemental notice of appeal to include whether the judge erred in law in not considering whether it was reasonable to make an order for possession in all the circumstances as he is required to do by section 98 of the Rent Act 1977.

The first point raised was the service of the notice to quit. This was the only point raised in the form for replying to the summons and the point upon which the matter was adjourned from Judge Cox to Judge Hitching. At the first hearing on June 24 there was handed in to Judge Cox both a copy of the notice to quit and a copy of the covering letter to the defendant. Before Judge Hitching there was also an affidavit that had been sworn on June 24 by the clerk receptionist in the employ of the solicitors, who swore that she did on February 1 serve the defendant with a notice to quit by posting the same, together with the B covering letter, and delivered the letter to the subpost office in East Ham. But the letter or envelope has not been returned through the dead letter service. Criticism was made of this affidavit, including the fact that it does not specifically exhibit either the copy notice to quit or the covering letter, which had of course been handed to the judge in the previous week. Indeed it had already been handed to the judge, no doubt, the very same day this affidavit was sworn and was consequently not available to the deponent.

It is said that the judge on July 1 was not entitled to make the assumption that the copy notice to quit signed by the solicitors, the copy letter coming from the possession of the solicitors and the affidavit from somebody in the employ of the solicitors could not be C put together to make together sufficient evidence to show, on the balance of probabilities, that the solicitors served this notice to quit by posting it.

In my judgment, a judge, particularly in the county court, where matters have to come as best they can, is perfectly entitled to take such pieces of paper together and find from them sufficient evidence from which to infer that this notice to quit was properly served by post. To come to any other conclusion would be flying in the face of common sense. There is nothing whatever to show that what the landlord said, which was that he told his solicitors to get on and do it, and what the solicitors indicate through their clerk receptionist and by producing copies of the relevant documents did not, in fact, take place. I, for one, am satisfied that this notice to quit was properly D served in the ordinary way by posting it and did not come back to the solicitors.

It is not wholly irrelevant — though only a minor point, as the judge pointed out — that a subsequent letter from the landlord reached its destination without the slightest difficulty. This was a question of fact on which the judge was entitled to make a finding, and this he did on the standard proof which is required in the county court in civil matters as in any other civil court, ie the balance of probabilities. There was just the hint of a suggestion that some other standard was required and I, for one, do not agree with that.

The next point is the validity of the notice to quit. This depends upon whether this tenancy was a contractual tenancy or a statutory E tenancy. These are the second and third points which are raised. It is conceded by counsel on behalf of the respondent landlord that if the notice to quit is required to be effective as a notice to quit and not as a piece of paper indicating a requirement for the tenant to leave, as a notice to quit it is defective and it is not necessary therefore to go into in what way it is defective. But it is said that it is not necessary, if it is a statutory tenancy, for there to be a notice to quit in proper form. It has not been suggested by counsel on behalf of the tenant that that is an incorrect proposition of law and it is undoubtedly supported by a decision of this court in *Morrison* v *Jacobs* [1945] KB 577, where Scott LJ, who gave the leading judgment, said that in respect of the Rent and Mortgage Interest Restrictions (Amendment) Act 1933 it was not necessary for a landlord before suing to serve any notice to F quit on the tenant, since the landlord of course has to prove, under what is now section 98 and the appropriate schedules to the Rent Act 1977, the grounds upon which possession can be obtained in a Rent Act case.

So the point comes to be considered, is this a statutory tenancy and was there evidence of any sort, or inferences which the judge was entitled to make, that it was a statutory tenancy?

There is little doubt that it is now clear that it is a statutory tenancy. The degree of enthusiasm for and reliance upon the notice to quit, particularly before Judge Cox, may have been due to some degree of confusion on the part of the advocate agent representing the landlord as to how important the notice to quit in itself was. But by the time the case came before Judge Hitching it is quite clear, to me at least, that he was considering this as a statutory tenancy. First of all it was G pleaded as a statutory tenancy and there was no defence to say it was not. It was never suggested to him that it was a contractual tenancy. It is reasonable to assume, in the case of a tenancy lasting for a long time, as this had lasted, where there had been a number of references to the rent tribunal, among other things, that this was a tenancy that had converted to a statutory tenancy. I am satisfied, particularly bearing in mind that it was pleaded as a statutory tenancy, that this is the way it was prosecuted before Judge Hitching and it is clear that when he called Mr Wareing back at the end of the evidence to ask him about his means and ability to pay, bearing in mind that the claim here was in respect of failure to pay the rent, he was considering the reasonableness of whether or not he should make a possession order, which is, of course, one requirement under section 98 of the Rent Act H 1977 — that he will not make an order unless the court considers it reasonable to make such an order.

The majority of cases in the county court these days are statutory tenancies. It is, I would have thought, rare to have a contractual tenancy. When it is pleaded as a statutory tenancy it is perfectly reasonable for the judge so to assume it unless it is raised that it is not. Indeed, in this particular case, any other conclusion than a statutory tenancy would have been an absurd one. In those circumstances, the form of the document called a notice to quit was of no importance. What was important was that the tenant should be given notice that he was expected to leave, and those are the wise words in the note to the Rent Act 1977, section 3, where it says, referring to *Morrison* v J *Jacobs*, that it is not required to give a statutory tenant any notice to quit. However, the lack of a letter before action or similar warning may affect the issue of reasonableness under section 98(1) and costs. Looking, therefore, at the notice to quit as being purely and simply an important indication to the tenant that he has to leave because an action for possession is to take place, that document fulfilled its purpose. The judge found as a fact that it had been served and the judge found that this tenant owed money.

The point is made by counsel on behalf of the tenant that since the particulars of claim indicated only £150, that is all the landlord is entitled to other than the mesne profits. But I do not agree with her: (1) because the summons gave the exact correct amount; and (2) because the evidence before the judge dealt with the correct amount K and the monthly payments that were unpaid and this was purely and simply a technical error, first in a typing error and second in the advocate for the landlord not asking the judge technically to amend the particulars of claim. Justice should not be seen not to be done for what is such an obvious technical error and that point also on behalf of the tenant falls, in my judgment, by the wayside.

The last point which is raised is that section 98(1) and the test of reasonableness had not in fact been an exercise undergone by the judge, who had not exercised his discretion to consider, in all the circumstances, whether or not he should grant a possession order despite the fact that the facts of non-payment of rent were proved. It is clear, however, from the notes of evidence that the defendant was L recalled and specific questions were asked of him of his means; and means are, of course, in a non-payment of rent case, for which the application for possession is made, quite the most relevant of the questions. It was also shown that he was single, living in the property, that he had no other assets than £494 in the bank, but that he was in a position to borrow some money from friends. The judge said at the end of his judgment that he would make an order for possession. I have no doubt that, as an experienced judge, in asking those particular questions he was clearly putting his mind to the issue of reasonableness.

In dealing with the large number of claims that come through the county court, a circuit judge cannot be expected on every single occasion to say: "I am specifically putting my mind to section 98(1), M and I consider in this particular case that it will be reasonable to grant a possession order". In an ideal world perhaps he would do it, but one has so very often in these extempore judgments, as this clearly was, to take from it what the judge was thinking, and from that and from the evidence it is clear that he put his mind to section 98(1) and the requirement of reasonableness, as one would expect any experienced judge trying these possession actions to do. This was a perfectly straightforward case of a statutory tenancy application for possession based on the non-payment of rent, and a large number of red herrings have strayed across the path; none of them is relevant in the event.

I would dismiss this appeal.

A **Sir George Waller** said: I entirely agree with the judgment of Butler-Sloss LJ. I would only put in my own words the two main points which were argued on behalf of the appellant.

First, the question of whether or not this is a statutory tenancy and if the judge took notice of it. It seems to me that the fact that the appellant was pleaded as being a statutory tenant meant that the case was being dealt with on the basis that he was a statutory tenant; the facts that the rent was registered and had recently been fixed at £175 per month all go to support the view that it was a statutory tenancy, and the allegation in the particulars of claim was not challenged.

Second, was it reasonable to make the order for possession? In my judgment, as Butler-Sloss LJ has already said, the fact that the learned judge went through the position of the tenant specially after B making the decision about the general principles showed that he was considering whether or not it was reasonable and I entirely agree with Butler-Sloss in her judgment.

Stocker LJ said: I agree with both judgments. There is ample evidence on which the judge could conclude that he was dealing with a statutory tenancy which pre-existed the service of the notice to quit. It is also clear, for the reasons given by Butler-Sloss LJ, that the judge considered the question of reasonableness under section 98. Indeed, for my part, I would regard it as difficult to accept that a judge of his experience in this field could possibly have ignored that requirement before making an order for possession.

C *Appeal dismissed; application for costs against the Legal Aid Board adjourned for 10 weeks; legal aid taxation for the appellant; application for leave to appeal to the House of Lords refused.*

Official Referees' Business

November 12 1991

(Before Judge Esyr LEWIS QC)

PRIMERIDGE LTD v JEAN MUIR LTD

D

Landlord and tenant — Service charges — Whether landlords entitled to costs on indemnification basis following payment into court

The defendants hold an underlease of premises at 22 Bruton Street, London W1, from the plaintiffs — By the terms of the lease the defendants covenanted to pay service charges — A dispute having arisen between the parties, the plaintiffs commenced two actions seeking the recovery of arrears of rent and service charges — Shortly before the trial of both actions the plaintiffs accepted a sum of £8,600 paid into court by the defendants — In accordance with Ord 62, r5, of the Rules of E the Supreme Court, the plaintiffs became entitled to their costs "up to the time of giving notice of acceptance" — In the application before the court the plaintiffs contended that it was appropriate for the court to order that the plaintiffs' costs be taxed on an indemnity basis by virtue of a contractual provision of the underlease "To pay the Landlord all proper costs charges and expenses . . . incurred by the Landlord . . . In connection with any breach of a covenant by or the recovery of arrears of rent due from the Tenant . . ."

On behalf of the defendants it was submitted that: (1) the plaintiffs were now seeking to enforce a contractual claim F which had not been pleaded; (2) the wording of the contractual provision was not apt to entitle the plaintiffs to costs on an indemnity basis; and (3) because the application was delayed, the court should exercise its discretion against the plaintiffs

***Held*: The application was dismissed — A court, in the exercise of its discretion as to costs, may take into account a relevant contractual term notwithstanding a failure to plead it — The discretion cannot be fettered by a contractual term — To recover costs on an indemnity basis, it is necessary for the plaintiffs to show that the contractual term plainly and unambiguously entitles them to be indemnified for their costs — "All proper costs" means no more than such costs as may be recovered under an order of the court; accordingly, there G was no such contractual term — The application would not have been rejected on the grounds of delay**

The following cases are referred to in this report.

Adelphi Hotel (Brighton) Ltd, Re, District Bank Ltd v Adelphi Hotel (Brighton) Ltd [1953] 1 WLR 955; [1953] 2 All ER 498

Bank of Baroda v Panessar [1987] Ch 335; [1987] 2 WLR 208; [1986] 3 All ER 751

This was an application by the plaintiff lessors to have their costs taxed on an indemnity basis following a payment into court by the defendant lessees in an action by the plaintiffs claiming arrears of H rent and service charges for premises at 22 Bruton Street, London W1.

Andrew Jordan (instructed by H Omar & Co) appeared for the plaintiffs; Stephen Jourdan (instructed by Beachcroft Stanleys) represented the defendants.

Giving judgment, **Judge Esyr Lewis QC** said: The defendants hold an underlease of the first floor and part of the basement of business premises at 22 Bruton Street, London W1, from the plaintiffs for a term commencing from December 25 1987. Under clause 5.8.2 of the underlease the defendants were obliged to pay to the plaintiffs 17.79%.

> of the costs expenses and outgoings incurred by the Landlord in the repair J maintenance and renewal of and the provision of services to the Building . . .

A dispute arose between the plaintiffs and defendants and other sublessees of parts of the premises as to claims made by the plaintiffs for payment of maintenance and service charges. The plaintiffs began two actions against the defendants for recovery of maintenance and service charges said to be due to them under the provisions of clause 5.8.2. In the first action the plaintiffs claimed rent, which had been withheld, together with service and maintenance charges alleged to have fallen due between January and June 1988 and interest. This action was commenced by a writ dated September 9 1988. In a second action, commenced by a writ dated December 17 1990, the plaintiffs made similar claims for sums which were said to have become due to them in the period after June 1988. K The defendants served defences and counterclaims in each action. The counterclaims were disposed of before the plaintiffs accepted on April 3 1991 the sum of £8,600 paid into court by the defendants in respect of both actions on March 15 1991. The acceptance of the sum in court brought the two actions to an end shortly before the date of trial. At the time of the acceptance the plaintiffs were claiming a sum of £14,000 plus interest in their first action and a sum of £7,700 plus interest in the second action.

Ord 62, r 5, provides, *inter alia*, that a plaintiff who accepts a payment into court in respect of all his causes of action "shall be entitled to his costs of the action incurred up to the time of giving notice of acceptance": see Ord 62, r 5(1) and (4). L

Ord 62, r 3, makes the following relevant provision:

> (4) The amount of his costs which any party shall be entitled to recover is the amount allowed after taxation on the standard basis where —
> (*a*) an order is made that the costs of one party to proceedings be paid by another party to those proceedings, or
> (*b*) an order is made for the payment of costs out of any fund (including the legal aid fund), or
> (*c*) no order is required,
>
> unless it appears to the Court to be appropriate to order costs to be taxed on the indemnity basis.

In this application the plaintiffs contend that in the circumstances of this case it is appropriate that the court should order that the plaintiffs' costs should be taxed on the indemnity basis. The basis on M which the plaintiffs' costs should be taxed is important to both parties in the light of the provisions of Ord 62, r 12. The relevant parts of this rule are as follows:

> 12.—(1) On a taxation of costs on the standard basis there shall be allowed a reasonable amount in respect of all costs reasonably incurred and any doubts which the taxing officer may have as to whether the costs were reasonably incurred or were reasonable in amount shall be resolved in favour of the paying party; and in these rules the term "the standard basis" in relation to the taxation of costs shall be construed accordingly.
>
> (2) On a taxation on the indemnity basis all costs shall be allowed except insofar as they are of an unreasonable amount or have been unreasonably incurred and any doubts which the taxing officer may have as to whether the costs were reasonably incurred or were reasonable in amount shall be resolved

A in favour of the receiving party; and in these rules the term "the indemnity basis" in relation to the taxation of costs shall be construed accordingly.

The effect of para (2) is to enlarge the scope of the costs which may be recoverable on taxation by the party having the benefit of a costs' order on the indemnity basis. *Prima facie,* this party is to be allowed "all costs" subject to the exception and any doubt as to whether his costs were reasonably incurred or were reasonable in amount is to be resolved in his favour.

The plaintiffs submit that they are entitled to an order for the taxation of their costs on an indemnity basis by virtue of a contractual provision contained in clauses 5.8.3 and 5.8.3.4 of the underlease under which the defendants covenanted, in so far as is relevant:

B To pay to the Landlord all proper costs charges and expenses (including professional advisers' costs and fees) incurred by the Landlord . . . (5.8.3)

In connection with any breach of covenant by or the recovery of arrears of rent due from the Tenant hereunder (5.8.3.4).

I should say that, by virtue of the provisions of clause 3.2 of the underlease, service and maintenance charges payable by the tenant are "reserved as rent" and thus fall within the provisions of clause 5.8.3.4.

The substance of the submission made by Mr Jordan, on behalf of the plaintiffs, is that clauses 5.8.3 and 5.8.3.4 of the underlease give them a contractual right to recover their costs on the indemnity basis and is essentially to the same effect as the provisions of Ord 62, r 12(2). It is further submitted that, as the plaintiffs have a C contractual right to be indemnified for the costs incurred in enforcing their right to recover arrears of rent, including service and maintenance charges, the court should make an order for taxation on the indemnity basis rather than requiring the plaintiffs to pursue their claim under the contract in fresh proceedings should a taxation of their costs on the standard basis leave them short of what they are entitled to recover under the contractual term of the underlease.

In support of his submissions, Mr Jordan, has referred me to the decision of Walton J in *Bank of Baroda* v *Panessar* [1987] 2 WLR 208. In that case the defendants entered into a guarantee to pay the bank in certain circumstances sums owed by two companies to the bank and secured by debentures and "all costs charges and expenses which you (ie the bank) may incur in enforcing or obtaining payment D of the sums due to you from the principal . . ." The guarantors unsuccessfully defended the claim against them and the bank successfully sought an order for their costs on an indemnity basis. Having said that the claim for costs on an indemnity basis under the terms of the guarantee raised a novel point, Walton J said at p 224C:

Well, treating this as an entirely novel point, it seems to me that it is not possible for a person in the position of the bank to exclude the discretion of the court, but one nevertheless starts from the position that the contractual position between the parties is that the costs will be paid on an indemnity basis. I cannot think that the words "all costs" mean anything other than that. So one starts from the position that this is the contractual position but not, as I venture to think, binding on the court. The court might very well take the view that, in the circumstances of any particular case, that was a E contractual provision which it ought to overlook and it ought not to give effect to. However, when one comes to consider whether this is such a case, I think the circumstances are all one way. Mr Higham has pointed out to me that, in this case, the defences of the defendants raise the points that the guarantees on which the bank sued were not signed properly by all parties, that no demand for payment was ever made and also that alternative finance was available; and, really, those were persisted in to an extent which was totally and utterly unreasonable.

After describing in more detail the basis on which the guarantors' defence was run, Walton J said at p 225B:

Therefore, so far as the bank is concerned, I have no hesitation whatsoever at the end of the day in saying that I think, basing myself first on exact terms of the guarantee, but also, if it were necessary, on the way in which the action has F been conducted on behalf of the defendants — and I must make it perfectly clear that I am not making the slightest criticism of counsel — that it is just and proper that in this case the award of costs should be on an indemnity basis.

Walton J thus founded his decision that the plaintiffs should be awarded their costs on an indemnity basis both on the contractual term and on the manner of their defence. It is to be noticed that Walton J's opinion was that the contractual term could not exclude the exercise of the court's discretion in making an award of costs and was not binding on the court. He dealt with the matter, it seems to me, on the basis that the contractual term as to costs was a matter which could be taken into account in the exercise of his discretion.

Mr Jourdan, for the defendants, has submitted that the court should not make an order for costs on the indemnity basis on three main grounds. First, he submits that the plaintiffs did not plead in G either of the two actions that they were entitled to costs on an indemnity basis and are now seeking to enforce a contractual claim of which the defendants had no warning before this application was launched. Second, he submits that the wording of clauses 5.8.3 and 5.8.3.4 of the underlease is not apt to entitle the plaintiffs to costs on the indemnity basis and contrasts the phrase "*all costs charges and expenses which you may incur* in enforcing the sums due to you . . ." used in the contract considered by Walton J in the *Bank of Baroda* case with the phrase "To pay to the Landlord *all proper costs charges and expenses*" . . . used in the underlease in this case. Third, he submits that this application is made very late in the day and that the delay in making it is a ground on which the court should in any event H not exercise its discretion to award costs on the indemnity basis. He points to the fact that the present summons, issued on August 5 1991, was not served until October 3 1991 although under Ord 62, r 29(1)(b) proceedings for taxation should have been begun within three months of the plaintiffs' acceptance of the payment into court. He has also referred to Ord 62, r 28(4), which provides that where a party fails without good reason to commence or conduct proceedings for the taxation of his costs in accordance with Ord 62 the taxing officer may allow the party entitled to costs less than he would otherwise have allowed or may wholly disallow the costs.

Conclusions

I can find no indication in the report of the *Bank of Baroda* case J that a claim for costs on an indemnity basis based on the contract of guarantee was pleaded. As I have said, it seems to me that the existence of the relevant contractual term was a matter taken into account by Walton J in exercising his discretion as to what was the appropriate order in the circumstances of that action. Since the question of costs arises in normal circumstances only at the conclusion of an action, whether after a hearing or on the determination of the action in some other way, as here, I cannot see any logical reason why circumstances relevant to the making of an appropriate order should have to be pleaded. If indeed the plaintiffs have a contractual right to recover their costs on an indemnity basis, I consider that it would be absurd for the court wholly to ignore it in considering what order is appropriate. I accordingly reject Mr K Jourdan's first submission.

I consider, as Walton J indicated in his judgment in the *Bank of Baroda* case, that the court has a discretion in the matter of costs which cannot be fettered by a contractual term agreed by the parties.

Usually, the appropriate order for costs, where the order has to be made by the court, is made in the light of the result of the proceedings where they have been contested and the manner in which they have been conducted and must take into account matters such as a payment into court. In the present case, since the proceedings ended with the acceptance of the payment into court, there is, in my judgment, no material which I am able to take into account in deciding whether the plaintiffs are entitled to costs on the indemnity L basis rather than on the standard basis, as provided for by Ord 62, r 3 (4)(c), except the provision of the underlease relied on by the plaintiffs. It is not, in my view, a relevant circumstance that the plaintiffs, by accepting the payment in, received very substantially less than the amount they claimed. This is, I consider, a neutral matter whether viewed from their standpoint or the defendants. I am, however, left in no doubt by the judgment of Walton J, which I respectfully follow, that I am entitled to take into account the terms of the relevant provision of the underlease in deciding the issue before me.

Mr Jourdan has referred me to a passage in *Fisher and Lightwood on The Law of Mortgage* (10th ed), which has given me assistance. It reads at p 660: M

Party and party costs. The basic rule is that costs in a mortgage suit should be recovered by the mortgagee on the standard basis . . ., unless there is a contract between the parties plainly and unambiguously providing for taxation on some other basis (for example, the indemnity basis) . . . but even such a contract cannot exclude the exercise of the court's discretion . . .

One of the cases referred to in a footnote to that passage is a decision of Vaisey J in *Re Adelphi Hotel (Brighton) Ltd, District Bank Ltd* v *Adelphi Hotel (Brighton) Ltd* [1953] 2 All ER 498. There, where mortgagees sought to recover costs on an indemnity basis under a contractual provision contained in a mortgage, Vaisey J said this at p 502, after construing the contract term relied on by the mortgagees:

A It suffices for me to say that I put my judgment on this part of the case, involving as it does a point of construction, on the ground that, as every taxation in which more than one party (in addition to the solicitor) is interested is *prima facie* a taxation as between party and party, any other basis of taxation is only justified when the party asking for it can show that he is entitled to it, either on some well-recognised principle, or under some contract plainly and unambiguously expressed. I can find no such principle nor any such contract here. The relevant part of the order of Apr 23, 1953, is framed with sufficient aptness. There is no need to use the words "as between party and party" or any other words to express what is, in my judgment, implied. I should myself have put the direction slightly differently, thus: Let the costs of the plaintiff of this action together with any other costs, charges and expenses properly incurred by it as mortgagee be taxed. But the variation is not of substance; it is merely of form, and the order may properly stand as it is
B without variation. The motion must be dismissed with costs.

I consider that the principle stated in *Fisher and Lightwood* and exemplified by Vaisey J's decision applies with equal force in this case. In my judgment, in order to recover their costs on an indemnity basis, it is necessary for the plaintiffs to show that the contractual provision upon which they rely plainly and unambiguously entitles them to be indemnified for their costs of their action and thus to obtain an order for costs on the indemnity basis rather than on the standard basis as provided in Ord 62, r 12(2). It seems to me that "all proper costs" means, in relation to legal proceedings to enforce the plaintiffs' claim, nothing more than such costs as the plaintiffs are entitled to recover under an order of the court or, as here, under a
C provision in the rules. I am accordingly unable to find that the contractual provision upon which the plaintiffs rely does entitle them to be indemnified for their costs of the proceedings and thus to an order for costs on the indemnity basis. The language used in the relevant contractual term in the *Bank of Baroda* case was different, namely "all costs charges and expenses which you may incur . . .". The vital difference is, of course, the presence of the adjective "proper" in the provision on which the plaintiffs rely.

The conclusion I have reached on Mr Jourdan's second submission is enough to dispose of the application. In deference to his argument, however, I will briefly express my view on his third submission, which relates to the delay by the plaintiffs in bringing this application. I would not have rejected this application on the grounds of delay. In
D my judgment, the matter of delay should be left for consideration by the taxing master, who is given a wide discretion to deal with it under the provisions of Ord 62, r 28(4).

For the reasons I have given, this application is dismissed.

Application dismissed.

Court of Appeal

November 21 1991

(Before Sir Donald NICHOLLS V-C, Lord Justice RUSSELL and
E Lord Justice LEGGATT)

TARGETT v TORFAEN BOROUGH COUNCIL

Landlord and tenant — Whether landlord liable for personal injuries to tenant arising out of accident on access steps to demised property — Held landlord liable

In 1982 the defendant local authority granted the plaintiff a weekly tenancy of a dwelling-house at 8 Butlers Court, Garndiffaith, Pontypool — The premises were designed by the borough architect two years previously — Access to the front door of the premises was by way of two flights of steps of
F stone manufacture — There were hand rails provided for the first flight of steps, but for the lower flight of four steps there was no hand rail — No artificial light was provided in the immediate vicinity of the steps although there was a street lamp about 25 yards away — On February 4 1984, during the hours of darkness, the plaintiff returned to his house, missed his footing on the lower flight of steps and fell suffering injury to his leg — In Newport (Gwent) County Court Mr Recorder Crispin Masterman found primary liability of the defendants established on the basis of their common law negligence in failing to construct or provide a safe and proper means of access, without a hand rail on either side and without any adequate lighting of the steps — He assessed contributory G
negligence by the plaintiff at 25% — The defendant local authority appealed

***Held*: The appeal was dismissed — The facts could not be distinguished from those in *Rimmer* v *Liverpool City Council* — The *Rimmer* case was binding on the court and had not been overruled by the House of Lords in *Murphy* v *Brentwood District Council* — A weekly tenant is in an entirely different position from an owner of a defectively constructed house who discovers the defect thereby rendering it no longer a latent defect — The duty of care owed by the defendants was not necessarily negatived by knowledge of the defect by the** H
plaintiff if the plaintiff has no practical alternative to using the steps

The following cases are referred to in this report.

Anns v *Merton London Borough Council* [1978] AC 728; [1977] 2 WLR 1024; [1977] 2 All ER 492; (1977) 75 LGR 555; [1977] EGD 604; 243 EG 523 & 591; [1977] JPL 514, HL
Billings (AC) & Sons Ltd v *Riden* [1958] AC 240; [1957] 3 WLR 496; [1957] 3 All ER 1, HL
Cavalier v *Pope* [1906] AC 428
Denny v *Supplies & Transport Co Ltd* [1950] 2 KB 374
Department of the Environment v *Thomas Bates & Son Ltd* [1991] 1 AC 499; [1990] 3 WLR 457; [1990] 2 All ER 943; [1990] 2 EGLR 154; [1990] 46 EG 115, HL J
Donoghue v *Stevenson* [1932] AC 562, HL
Dutton v *Bognor Regis Urban District Council* [1972] 1 QB 373; [1972] 2 WLR 299; [1972] 1 All ER 462; (1971) 70 LGR 57; [1972] 1 Lloyd's Rep 227, CA
Grant v *Australian Knitting Mills Ltd* [1936] AC 85
Murphy v *Brentwood District Council* [1991] 1 AC 398; [1990] 3 WLR 414; [1990] 2 All ER 908; [1990] 2 Lloyd's Rep 467; (1990) 89 LGR 24, HL
Rimmer v *Liverpool City Council* [1985] QB 1; [1984] 2 WLR 426; [1984] 1 All ER 930; (1983) 82 LGR 424; [1984] EGD 827; 269 EG 319, CA
Slater v *Clay Cross Co Ltd* [1956] 2 QB 264; [1956] 3 WLR 232; [1956] 2 All ER 625, CA

This was an appeal by Torfaen Borough Council, the owners of a house at 8 Butlers Court, Garndiffaith, Pontypool, which had been let to the plaintiff, Raymond Targett, on a weekly tenancy, from a K
decision of Mr Recorder Crispin Masterman sitting at Newport (Gwent) County Court. The learned recorder had allowed the plaintiff's claim for damages for personal injuries and found contributory negligence on the part of the plaintiff to the extent of 25%.

Andrew Green (instructed by Dolmans, of Cardiff) appeared for the appellants; Lloyd Williams (instructed by Watkins & Gunn, of Pontypool) represented the respondent.

Giving judgment, **RUSSELL LJ** said: This is an appeal from a judgment of Mr Recorder Crispin Masterman sitting at Newport (Gwent) County Court who, on November 20 1990, found in favour of the plaintiff in an action claiming damages for personal injuries. L
The learned recorder found contributory negligence on the part of the plaintiff and reduced his damages by 25%.

The defendant local authority appeal on the ground that no liability against the authority was established. Alternatively, it is contended that the learned recorder should have found contributory negligence to a greater extent than he did.

At all material times the defendants were the owners of a dwelling-house, 8 Butlers Court, Garndiffaith, Pontypool, which premises were let to the plaintiff on a weekly tenancy in late 1982. They had been designed by the borough architect in or about the year 1980. The plaintiff and his family were the first occupants of the house.

Access to the front door of the premises was over sloping land and there had been designed and constructed two flights of steps of stone M
manufacture. Towards the top of the sloping land there were hand rails provided for the first flight of steps, but for the lower flight of four steps giving access to the ground floor of the house there was no hand rail. No artificial light was provided in the immediate vicinity of the steps. There was a street lamp about 25 yds away but such illumination as was provided by this light was restricted by the shadows of trees and effectively there was no illumination of the steps save for the top of them.

Before the plaintiff moved into the house he had complained of the lack of lighting to the clerk of works but nothing had been done.

On the evening of February 4 1984, during the hours of darkness, the plaintiff returned to his house and was attempting to descend the lower flight of steps when he missed his footing and fell, thereby injuring his leg. The plaintiff gave evidence to the effect that the weather conditions were bad, that it was dark and raining and that if there had been a hand rail he would have used it.

In the statement of claim allegations of breach of statutory duty under section 4 of the Defective Premises Act 1972 were made on behalf of the plaintiff as well as an allegation that the defendants were in breach of their implied covenant under section 32 of the Housing Act 1961. However, these allegations were abandoned at trial and no more need be said about them. The learned recorder found primary liability established on the basis of the common law negligence of the defendants and, in his judgment, he expressly adopted three allegations pleaded in the statement of claim as follows:

(i) Failing to construct or to provide or to maintain a safe and proper means of access to the said dwellinghouse

(ii) Causing or permitting the said flight of steps to be constructed without a handrail on either side thereof

(iii) Failing to provide or to arrange for the provision of any, or any adequate lighting of the steps when the same were constructed or thereafter.

Mr Andrew Green, on behalf of the appellant council, first submitted that the facts did not demonstrate that the steps were potentially dangerous. I am quite unable to accept that submission. Indeed I find it difficult to reconcile the submission with the concession made during argument by Mr Green that the steps without adequate lighting or a hand rail presented a foreseeable risk of causing injury to anyone using them during the hours of darkness. This was not the view of an engineer called by the defendants, but the recorder was entitled to and did reject the engineer's opinion. In my judgment, there are no grounds for disturbing the recorder's findings of fact to the effect that the steps as constructed without a hand rail and designed without lighting were dangerous.

The court was referred to two authorities and I must now turn to them. In *Rimmer* v *Liverpool City Council* [1985] QB 1* the Court of Appeal were concerned with a dwelling-house that the local authority had designed and built. The plaintiff became the tenant of the house. He suffered personal injuries when he fell against a panel of glass within the property which, as the county court judge found, was too thin and potentially dangerous. The plaintiff was aware of the thinness of the glass prior to his accident and had complained to the local authority about it. At first instance the judge found in the plaintiff's favour and the local authority appealed. Reliance was placed upon *Cavalier* v *Pope* [1906] AC 428 wherein the House of Lords had decided that, in the absence of a contractual duty to maintain and repair, a landlord was not liable for the dangerous state of an unfurnished house let to a tenant injured by the dangerous conditions. In *Rimmer's* case the judgment of the court was delivered by Stephenson LJ, who, after citing from the speech of Lord Salmon in *Anns* v *Merton London Borough Council* [1978] AC 728†, said at p 13:

The landowner, who designs or builds a house or flat, is no more immune from personal responsibility for faults of construction than a building contractor, or from personal responsibility for faults of design than an architect, simply because he has disposed of his house or flat by selling or letting it. The council through their architects' department designed, and through their direct works department built, the plaintiff's flat with its dangerous glass panel. They owed him, not as tenant but, like his wife or his child, as a person who might reasonably be expected to be affected by the provision of the glass panel in the flat, a duty to take such care as was reasonable in all the circumstances to see that he was reasonably safe from personal injury caused by the glass panel. They knew the thickness of the glass, and on the judge's finding they ought to have known that, placed where it was, it was dangerous to the occupants of the flat. The plaintiff himself considered it dangerous; but even if he had the right (which we doubt), he certainly was under no duty to protect or remove it, and he had been told it was standard and nothing could be done about it.

In those circumstances his knowledge that the glass was dangerously thin cannot exonerate the council.

"Knowledge or opportunity for inspection, per se and without regard to any consequences they may have in the circumstances, cannot be conclusive against the plaintiff,"

said Sir Raymond Evershed MR, with the agreement of Jenkins LJ, in *Denny* v *Supplies & Transport Co Ltd* [1950] 2 KB 374, 382.

Later, after citing further authorities, Stephenson LJ continued at p 14:

From these authorities we take the law to be that an opportunity for inspection of a dangerous defect, even if successfully taken by A who is injured by it, will not destroy his proximity to B who created the danger, or exonerate B from liability to A, unless A was free to remove or avoid the danger in the sense that it was reasonable to expect him to do so, and unreasonable for him to run the risk of being injured by the danger. It was not reasonable or practical for the plaintiff to leave the flat or to alter the glass panel. He remained in law the council's neighbour, although he had complained that the glass was too thin. We reach our decision without treating *Cavalier* v *Pope* [1906] AC 428 as overruled, for Pope did not design or construct the floor through which Mrs Cavalier fell. He was not a builder owner, but what may be called a bare landlord, or a landowner as such.

Mr Green sought to distinguish the facts of *Rimmer* from the instant case. He submitted that this court was not concerned with a case of defective manufacture at all in that the construction of the steps as such complied with building regulations and with the acknowledged practice of competent architects and builders at the material time. The absence of the hand rail did not render the steps of faulty manufacture. Likewise the absence of proper illumination did not amount to a defect of manufacture or construction. The steps remained capable of use as a means of communication and nothing was inherently defective about them. For my part I am quite unable to accept these submissions. During the course of them it was put to Mr Green by way of illustration that an electrical cable without insulation remained a means of communicating electricity; the absence of the insulation did not eliminate the ability of the cable to pass electricity but its effect was to create danger. It was a manufacturing defect just as in the instant case the absence of the hand rail was a manufacturing defect. The absence of appropriate lighting in the vicinity of the steps was a design fault. I cannot distinguish the facts of this case from those in *Rimmer* and, unless that decision was *per incuriam* (which plainly it was not) or unless it has since been overruled by the House of Lords, it is binding upon this court and we must follow it.

Mr Green's alternative submission was to the effect that *Rimmer* can no longer be regarded as good law but on the contrary must be taken as having been overruled by the House of Lords in *Murphy* v *Brentwood District Council* [1991] 1 AC 398. The facts of that case can be taken from the headnote:

The appellant council, pursuant to its duty under section 64 of the Public Health Act 1936, referred the plans for the building of a house to consulting engineers acting as independent contractors. In reliance upon the consulting engineers' advice, which had not taken account of calculation errors in the design of the foundation, the council passed the plans and, as a result, the house was built with a defective foundation. While the plaintiff was occupying the house, the foundation cracked and, consequently, there was extensive damage to the walls and pipes of the house. Instead of repairing it at the estimated cost of £45,000, the plaintiff sold the house for £35,000 less than its market value in undamaged condition and claimed damages for negligence against the council. The purchaser was living in the house without having carried out any repairs. The judge held that, in passing the plans, the council had relied on negligent advice and, therefore, had itself been negligent; and that the defects in the house, having become an imminent danger to the safety and health of the plaintiff while the plaintiff had been occupying the house, gave the plaintiff a cause of action against the council; accordingly, he awarded the plaintiff £38,777 in respect of the diminution in the value of the house and the expenses incurred by him as a result of the damage to it.

The Court of Appeal dismissed an appeal by the council, but the House of Lords allowed the appeal.

It is important to observe that *Murphy's* case was not a personal injury case nor was it a case where there was any relationship of landlord and tenant between the plaintiff and the defendants. The claim related solely to economic loss. None the less, Mr Green relied upon passages from the speeches of Lord Keith of Kinkel and Lord Bridge of Harwich. Mr Lloyd Williams, on behalf of the respondent, cited extensively from the speech of Lord Oliver of Aylmerton.

Mr Green drew our attention to a passage appearing at p 464F of the report in *Murphy*. Lord Keith of Kinkel said:

But there can be no doubt that, whatever the rationale, a person who is injured through consuming or using a product of the defective nature of which he is well aware has no remedy against the manufacturer. In the case of a building, it is right to accept that a careless builder is liable, on the principle of *Donoghue* v *Stevenson*, where a latent defect results in physical injury to anyone, whether owner, occupier, visitor or passer-by, or to the property of any such person. But that principle is not apt to bring home liability towards an occupier who knows the full extent of the defect yet continues to occupy the building.

Reliance was also placed by Mr Green upon the words of Lord Bridge appearing at p 475:

*Editor's note: Also reported at (1983) 269 EG 319.

†Editor's note: Also reported at (1977) 243 EG 523 & 591.

A If a manufacturer negligently puts into circulation a chattel containing a latent defect which renders it dangerous to persons or property, the manufacturer, on the well known principles established by *Donoghue* v *Stevenson* [1932] AC 562, will be liable in tort for injury to persons or damage to property which the chattel causes. But if a manufacturer produces and sells a chattel which is merely defective in quality, even to the extent that it is valueless for the purpose for which it is intended, the manufacturer's liability at common law arises only under and by reference to the terms of any contract to which he is a party in relation to the chattel; the common law does not impose on him any liability in tort to persons to whom he owes no duty in contract but who, having acquired the chattel, suffer economic loss because the chattel is defective in quality. If a dangerous defect in a chattel is discovered before it causes any personal injury or damage to property, because the danger is now known and the chattel cannot safely be used unless
B the defect is repaired, the defect becomes merely a defect in quality. The chattel is either capable of repair at economic cost or it is worthless and must be scrapped. In either case the loss sustained by the owner or hirer of the chattel is purely economic. It is recoverable against any party who owes the loser a relevant contractual duty. But it is not recoverable in tort in the absence of a special relationship of proximity imposing on the tortfeasor a duty of care to safeguard the plaintiff from economic loss. There is no such special relationship between the manufacturer of a chattel and a remote owner or hirer.

I believe that these principles are equally applicable to buildings. If a builder erects a structure containing a latent defect which renders it dangerous to persons or property, he will be liable in tort for injury to persons or damage to property resulting from that dangerous defect. But if the defect becomes apparent before any injury or damage has been caused, the loss sustained by
C the building owner is purely economic. If the defect can be repaired at economic cost, that is the measure of the loss. If the building cannot be repaired, it may have to be abandoned as unfit for occupation and therefore valueless. These economic losses are recoverable if they flow from breach of a relevant contractual duty, but, here again, in the absence of a special relationship of proximity they are not recoverable in tort. The only qualification I would make to this is that, if a building stands so close to the boundary of the building owner's land that after discovery of the dangerous defect it remains a potential source of injury to persons or property on neighbouring land or on the highway, the building owner ought, in principle, to be entitled to recover in tort from the negligent builder the cost of obviating the danger, whether by repair or by demolition, so far as that cost is necessarily incurred in order to protect himself from potential liability to third parties.

D The burden of Mr Green's submissions, relying upon *Murphy*, was to the effect that in the instant case the court was concerned with what had become a patent defect in the flight of steps because, prior to the accident, the plaintiff was well aware of the conditions prevailing and the potential danger by reason of the absence of the hand rail and adequate lighting. The only remedy available to the plaintiff, so submitted Mr Green, was to put right the defects himself and charge the local authority with the expense involved.

For my part I do not read the speeches in *Murphy* as supporting Mr Green's propositions. A weekly tenant is in an entirely different position from an owner of a house defectively constructed who discovers the defect thereby rendering it no longer a latent defect. In
E my judgment, the scope of *Murphy* was defined by Lord Keith at p 463H where he said:

The question is whether the appellant council owed the respondent a duty to take reasonable care to safeguard him against the particular kind of damage which he has in fact suffered, which was not injury to person or health nor damage to anything other than the defective house itself.

The point was emphasised in the speech of Lord Oliver, relied upon by Mr Lloyd Williams, at p 487H:

Anyone, whether he be a professional builder or a do-it-yourself enthusiast, who builds or alters a semi-permanent structure must be taken to contemplate that at some time in the future it will, whether by purchase, gift or inheritance, come to be occupied by another person and that if it is defectively built or altered it may fall down and injure that person or his property or may put him
F in a position in which, if he wishes to occupy it safely or comfortably, he will have to expend money on rectifying the defect. The case of physical injury to the owner or his licensees or his or their property presents no difficulty. He who was responsible for the defect — and it will be convenient to refer to him compendiously as "the builder" — is, by the reasonable foreseeability of that injury, in a proximate "neighbour" relationship with the injured person on ordinary *Donoghue* v *Stevenson* principles. But when no such injury has occurred and when the defect has been discovered and is therefore no longer latent, whence arises that relationship of proximity required to fix him with responsibility for putting right the defect?

Rimmer's case was not even cited in *Murphy*, for in my view it had nothing whatever to do with the problem posed in *Murphy*. Likewise, I am satisfied that their lordships did not refer to *Rimmer* because it had no relevance.

I am firmly of the opinion that the House of Lords in *Murphy* did G
not overrule *Rimmer* by implication or otherwise and, in my view, *Rimmer* remains an authority binding upon this court.

On the issue of primary liability, therefore, I have come to the conclusion that the recorder's judgment cannot be faulted.

Mr Green's final submission was that the learned recorder erred in his apportionment of liability. The pleaded allegation against the plaintiff in the defence was effectively that the plaintiff failed to look where he was walking. In argument it was suggested that he should have carried a torch or should have left the light on in his kitchen to create illumination. I regard the latter as unrealistic. Quite apart from the reluctance of this court to interfere with the exercise of the judge's discretion upon matters of apportionment I am abundantly H
satisfied that the culpability of the defendants far outweighs the momentary carelessness of the plaintiff found by the learned recorder. I see no error of principle in the way in which the recorder arrived at his apportionment and I would not disturb it. I would dismiss this appeal.

Agreeing, **Leggatt LJ** said: In 1982 the defendant council had a house built in a village near Pontypool and let it to the plaintiff. In 1984 on returning to his house at night the plaintiff fell down the steps leading from the road to his house. The recorder held that "at night they do present a foreseeable risk of danger of possible stumbling or falling in the absence of direct lighting and/or absence of a handrail". He found the defendants liable for the plaintiff's accident J
and also assessed the plaintiff's blame for it at 25%. Against those decisions the defendants appeal.

By application of the principle in *Donoghue* v *Stevenson* [1932] AC 562 a landlord who is responsible for the design and construction of the house let is under a legal duty to take reasonable care that the house is free from defect likely to cause injury to any person whom he ought reasonably to have in contemplation as likely to be affected by such defect. But the duty arises only where there is no reasonable opportunity of intermediate examination. The very reason why expense incurred in putting right such a defect was held by the House of Lords in *Murphy* v *Brentwood District Council* [1991] 1 AC 398 not to be recoverable was that it had ceased to be latent. At p 464D Lord Keith of Kinkel said: K

. . . an essential feature of the species of liability in negligence established by *Donoghue* v *Stevenson* was that the carelessly manufactured product should be intended to reach the injured consumer in the same state as that in which it was put up with no reasonable prospect of intermediate examination . . . It is the latency of the defect which constitutes the mischief. There may be room for disputation as to whether the likelihood of intermediate examination and consequent actual discovery of the defect has the effect of negativing a duty of care or of breaking the chain of causation (compare *Farr* v *Butters Brothers & Co* [1932] 2 KB 606 with *Denny* v *Supplies & Transport Co Ltd* [1950] 2 KB 374) . . . In the case of a building, it is right to accept that a careless builder is liable, on the principle of *Donoghue* v *Stevenson*, where a latent defect results in physical injury to anyone, whether owner, occupier, visitor or passer-by, or to the property of any such person.

At p 475E Lord Bridge of Harwich, after considering the liability L
under *Donoghue* v *Stevenson* of a manufacturer of chattels, said:

I believe that these principles are equally applicable to buildings. If a builder erects a structure containing a latent defect which renders it dangerous to persons or property, he will be liable in tort for injury to persons or damage to property resulting from that dangerous defect. But if the defect becomes apparent before any injury or damage has been caused, the loss sustained by the building owner is purely economic.

Similarly, Lord Oliver of Aylmerton said at p 488A:

The case of physical injury to the owner or his licensees or his or their property presents no difficulty. He who was responsible for the defect . . . is, by the reasonable foreseeability of that injury, in a proximate "neighbour" relationship with the injured person on ordinary *Donoghue* v *Stevenson* M
principles.

Later on the same page Lord Oliver observed in the context of economic loss that "once the danger ceases to be latent there never could be any liability", adding that "The injury will not now ever occur unless the plaintiff causes it to do so by causing a danger of which he is aware . . .".

In *Department of the Environment* v *Thomas Bates & Son Ltd* [1991] 1 AC 499* Lord Keith of Kinkel summarised (at p 519F) thus the conclusion of the House of Lords in *Murphy's* case:

It was the unanimous view that, while the builder would be liable under the principle of *Donoghue* v *Stevenson* . . . in the event of the defect, before it had

*Editor's note: Also reported at [1990] 2 EGLR 154; [1990] 46 EG 115.

A been discovered, causing physical injury to persons or damage to property other than the building itself, there was no sound basis in principle for holding him liable for the pure economic loss suffered by a purchaser who discovered the defect . . .

In reaching this conclusion the House had departed from *Anns* v *Merton London Borough Council* [1978] AC 728 and overruled *Dutton* v *Bognor Regis Urban District Council* [1972] 1 QB 373.

This court referred to those latter cases in *Rimmer* v *Liverpool City Council* [1985] QB 1, but only for the proposition that "a landlord of an unfurnished house may be liable to his tenant for defects rendering it dangerous if he has constructed them himself": see p 13F. *Rimmer's* case concerned personal injury sustained by a tenant falling on to a thin glass panel which was part of an inside wall of a flat
B designed and built by the defendant local authority. This court concluded at p 13H that the defendants owed the plaintiff:

not as tenant but, like his wife or his child, as a person who might reasonably be expected to be affected by the provision of the glass panel in the flat, a duty to take such care as was reasonable in all the circumstances to see that he was reasonably safe from personal injury caused by the glass panel.

Without referring to the principle that liability under *Donoghue* v *Stevenson* would be negatived by knowledge of the defect or reasonable opportunity of examination of it the court nevertheless applied *Denny* v *Supplies & Transport Co Ltd (supra)*, remarking that the defendants were not exonerated from liability in that case by the plaintiff's knowledge of the dangerous process there involved if
C "there was no practical alternative" to using it. In *Slater* v *Clay Cross Co Ltd* [1956] 2 QB 264 Denning LJ had said at p 271:

Even in cases which concern the condition of premises . . . knowledge of the danger is only a bar where the party is free to act on it, so that his injury can be said to be due solely to his own fault . . . Where knowledge of the danger is not such as to render the accident solely the fault of the injured party, then it is not a bar to the action but only a ground for reducing the damages.

Of that statement, Lord Somervell said in *A C Billings & Sons Ltd* v *Riden* [1958] AC 240 at p 266:

I think the words "free to act" in the context may be ambiguous . . . There may be many cases where a plaintiff is "free" in one sense to avoid the risk altogether but where it would be reasonable to run the risk. When the plaintiff has full or partial knowledge of the danger the question must always be: Was
D the injury, in all the circumstances, including the plaintiff's knowledge, due solely to his own negligence or was it due solely to the negligence of the defendant or was it due to the negligence of each?

This court in *Rimmer's* case [at p 14] therefore took the law to be that:

. . . an opportunity for inspection of a dangerous defect, even if successfully taken by A who is injured by it, will not destroy his proximity to B who created the danger, or exonerate B from liability to A, unless A was free to remove or avoid the danger in the sense that it was reasonable to expect him to do so, and unreasonable for him to run the risk of being injured by the danger. It was not reasonable or practical for the plaintiff to leave the flat or to alter the glass panel. He remained in law the council's neighbour, although he had complained that the glass was too thin.

E In this case the recorder held the defendants liable as designers of a "foreseeably dangerous feature." That without more would not have been enough. But here, as in *Rimmer's* case, it was not reasonable to expect the plaintiff to remove or avoid the danger, nor was it unreasonable for him to run the risk of being injured by the danger. It was not reasonable or practical for the plaintiff to leave the house or to provide a hand rail or fixed lighting. On the other hand, it was practical for him to take greater care for his own safety by steadying himself on the wall or putting his foot down firmly on each successive step, even if not by equipping himself with a torch.

Despite the fact that the recorder reached his conclusion without expressly considering whether the plaintiff had any practical
F alternative but to put up with the danger constituted at night by steps which were neither lighted nor safeguarded by a hand rail, I agree with his decisions that the defendants were liable for the plaintiff's accident but that in all the circumstances he was 25% contributorily negligent. It follows that I, too, would dismiss the appeal.

Also agreeing, **Sir Donald Nicholls V-C** said: I, too, would dismiss this appeal, for the reasons given by Russell and Leggatt LJJ. I add some observations only on the contention that the decision of this court in *Rimmer* v *Liverpool City Council* [1985] QB 1 must be taken to have been overruled by the decision of the House of Lords in *Murphy* v *Brentwood District Council* [1991] 1 AC 398. *Rimmer* seems not to have been cited in the House of Lords but, so it was submitted to us, the decision in *Rimmer* cannot stand with the reasoning applied in *Murphy*. Particular reference was placed on G
passages in the speeches of Lord Keith of Kinkel and Lord Bridge of Harwich in *Murphy*. Lord Keith said at p 464:

There may be room for disputation as to whether the likelihood of intermediate examination and consequent actual discovery of the defect has the effect of negativing a duty of care or of breaking the chain of causation (compare *Farr* v *Butters Brothers & Co* [1932] 2 KB 606 with *Denny* v *Supplies & Transport Co Ltd* [1950] 2 KB 374). But there can be no doubt that, whatever the rationale, a person who is injured through consuming or using a product of the defective nature of which he is well aware has no remedy against the manufacturer. In the case of a building, it is right to accept that a careless builder is liable, on the principle of *Donoghue* v *Stevenson*, where a latent defect results in physical injury to anyone, whether owner, occupier, visitor or passer-by, or to the property of any such person. But that principle H
is not apt to bring home liability towards an occupier who knows the full extent of the defect yet continues to occupy the building.

Lord Bridge said at p 475:

If a dangerous defect in a chattel is discovered before it causes any personal injury or damage to property, because the danger is known and the chattel cannot safely be used unless the defect is repaired, the defect becomes merely a defect in quality.

Before us the appellants' argument was that in *Rimmer*, as in the present case, the person injured knew of the dangerous defect which subsequently caused his personal injuries. In *Rimmer* the plaintiff considered the glass panel was a danger because it was too thin and he had complained about this to the defendant local authority. In the present case the plaintiff knew of the lighting position and of the J
absence of the hand rail. Accordingly, it was argued, in accordance with the observations of Lord Keith and Lord Bridge the claim by the plaintiff in *Rimmer*, and the claim in the present action, must fail.

I cannot accept this. In *Murphy* the House of Lords was concerned to consider the principles applicable to a claim to recover the cost of making good a defective product, there a building, which had not yet caused any physical injury to persons or damage to other property. Their lordships explained why the *Donoghue* v *Stevenson* principle did not embrace such a claim. When the defect was discovered, the goods or building ceased to be dangerous, although they remained defective. The loss sustained by the owner, for example, in making good the defect, was economic loss. Thus, their lordships were concerned with examining the interrelation between the *Donoghue* v K
Stevenson principle and a claim by the owner of goods or a building to recover purely economic loss. That examination did not call for a consideration of whether there were any circumstances in which, under *Donoghue* v *Stevenson*, there might be liability for injuries sustained even though the plaintiff was aware of the existence of the hazard. In *Murphy* a statement of the broad principle was sufficient.

That is the context of the passages relied on in the speeches of Lord Keith and Lord Bridge. Given that context, I cannot read those speeches as an expression of view that in no circumstances can a person recover compensation for personal injuries sustained by using defective goods or a defective building once he has become aware of the existence of the defect. The general principle is, indeed, that such L
a person cannot recover compensation, because in the ordinary way his knowledge of the existence of the dangerous defect, at any rate in the case of goods, will suffice to enable him to avoid the danger. If he finds there is a decomposed snail in his ginger beer he will not drink it. He does not use underwear which he knows contains a mischievous chemical. Thus in *Grant* v *Australian Knitting Mills Ltd* [1936] AC 85 at p 105, Lord Wright said:

The principle of *Donoghue's* case can only be applied where the defect is hidden and unknown to the consumer, otherwise the directness of cause and effect is absent: the man who consumes or uses a thing which he knows to be noxious cannot complain in respect of whatever mischief follows, because it follows from his own conscious volition in choosing to incur the risk or certainty of mischance. M

But knowledge of the existence of a danger does not always enable a person to avoid the danger. In simple cases it does. In other cases, especially where buildings are concerned, it would be absurdly unrealistic to suggest that a person can always take steps to avoid a danger once he knows of its existence and that if he does not do so he is the author of his own misfortune. Here, as elsewhere, the law seeks to be realistic. Hence the established principle, referred to by this court in *Rimmer* (at p 14), that knowledge or opportunity for inspection *per se*, and without regard to any consequences they may have in the circumstances, cannot be conclusive against the plaintiff. Knowledge, or opportunity for inspection, does not by itself always negative the duty of care or break the chain of causation. Whether it

A does so depends on all the circumstances. It will do so only when it is reasonable to expect the plaintiff to remove or avoid the danger and unreasonable for him to run the risk of being injured by the danger.

None of this impinges on the rationale underlying the decision in *Murphy*. The rationale of *Murphy* was that the *Donoghue* v *Stevenson* principle is not designed to compensate for losses arising from defects in quality as such; it is designed to compensate for losses arising from physical injuries to persons or damage to other property B flowing from the use of a dangerous product when the user, usually through lack of knowledge of the danger, was unable to take steps to avoid the danger. *Rimmer* is consistent with this. In my view, it remains good law.

Appeal dismissed with costs; legal aid taxation of respondent's costs; application for leave to appeal to the House of Lords refused.

For further cases on this subject see p 43

LANDLORD AND TENANT

BUSINESS TENANCIES

Court of Appeal

November 25 1991

(Before Lord DONALDSON of LYMINGTON MR and NOLAN LJ)

PETER GODDARD & SONS LTD v HOUNSLOW LONDON BOROUGH COUNCIL

Landlord and Tenant Act 1954 — Whether landlord entitled to oppose tenant's application for a new tenancy on ground *(f)* of section 30 to that Act — Whether sufficient evidence of intention — Appeal by tenant dismissed

The appellant tenant had held a lease of a plot of land — Upon the expiration of that lease the respondent landlord opposed the grant of a new tenancy on the ground that it proposed to develop the site through a company called Tablelodge Ltd with whom it had a building agreement — Outline planning permission had been obtained and upon the building agreement becoming unconditional Tablelodge Ltd would pay the respondent £400,000 by way of a premium and would then be granted a lease of the site to carry out development — On a preliminary issue, Judge Edwards (sitting in Brentford County Court) held that the respondent landlord was entitled to oppose the appellant's application on the grounds set out in para *(f)* of section 30 to the Act — The appellant tenant appealed contending that plans and specifications were not sufficiently advanced, there might be financial problems with Tablelodge Ltd and there were occupiers of the site claiming to be subtenants from whom possession might not be obtained without some difficulty

***Held*: The appeal was dismissed — Although there might have been problems in bringing forward detailed specifications and drawings without having possession of the site in order that the surveys might be undertaken, the prospective developers had spent £70,000 on the project, which suggested that this was not a case of hope but one of intention — On the need for financing of the development by Tablelodge Ltd, it was a question of taking a view as to what view the lending bank itself would be likely to take when the appropriate date was reached — Although it was not directly relevant, the court was told that the judge's view that there would be no problem in getting rid of the occupiers claiming subtenancies had been found to be correct in the event — These matters were essentially a question of fact and degree for the trial judge — It was essentially a matter for the judge who tried the case to assess the conviction which the evidence carried — There were no grounds for interfering**

No cases are referred to in this report.

This was an appeal by Peter Goddard & Sons Ltd, who had made an application under Part II of the Landlord and Tenant Act 1954 for a new tenancy of premises known as The Yard, The Ham, Brentford, Middlesex, from a decision of Judge Edwards (sitting in Brentford County Court) on a preliminary issue as to whether the respondent landlords were entitled to oppose the application on the grounds set out in para *(f)* of section 30 of the 1954 Act.

Neil Mendoza (instructed by Mendoza's) appeared for the appellant; and Colin Ishmael (instructed by the solicitor to Hounslow London Borough Council) represented the respondents.

Giving judgment, **LORD DONALDSON OF LYMINGTON MR** said: This is an appeal from a judgment of Judge Edwards, sitting in Brentford County Court, on a preliminary issue under the Landlord and Tenant Act 1954. The issue was whether the respondent landlords were entitled to oppose the tenant's application for a new tenancy of premises known as The Yard, The Ham, Brentford, Middlesex, on the grounds set out in para *(f)* of section 30 of the Act. That section provides as follows:

> (1) The grounds on which a landlord may oppose an application under subsection (1) of section twenty-four of this Act are such of the following grounds as may be stated in the landlord's notice under section twenty-five of this Act or, as the case may be, under subsection (6) of section twenty-six thereof, that is to say:
>
> . . . (f) that on the termination of the current tenancy the landlord intends to demolish or reconstruct the premises comprised in the holding or a substantial part of those premises or to carry out substantial work of construction on the holding or part thereof and that he could not reasonably do so without obtaining possession of the holding;.

I turn to the appellant's skeleton argument, which very helpfully sets the background in the following three paragraphs:

> 1. The tenant was the lessee of a plot of land in Brentford, Middx, which the landlords, the local authority, wished to develop. The development was not to be done by the local authority themselves, but by a company called Tablelodge Ltd. For the purposes of the development the local authority entered into a building agreement with Tablelodge Ltd. The building agreement is at pp 69-90 of the bundle.
> 2. In essence, within 60 days of the building agreement becoming "unconditional" Tablelodge Ltd would pay to the local authority the sum of £400,000 by way of premium; Tablelodge Ltd would be granted a lease of the land to be developed and Tablelodge Ltd would then carry out the development works.
> 3. The overall development site, and that covered by the planning permission that had been obtained (pp 64-68 of the bundle), concerned 2 parcels of land, known as "site A" and "site B". Site B is the area of land demised to the appellants. Immediately next to it is site A which was demised to a Mr Cutbush.

The skeleton agrument later states, quite accurately, of course, that it was common ground that the matters which the landlord had to prove to the court were accurately set out in *Hill and Redman's Law of Landlord and Tenant* (18th ed) at B 322, which were as follows:

> To summarise, a landlord who is seeking to establish his case under ground *(f)* would be wise to ensure:
>
> (1) that, if a corporation, there is a recent and appropriately worded resolution of the corporation,
>
> (2) that any requisite consents and approvals, such as planning permission or listed building consent, have been obtained,
>
> (3) that plans and specifications have been obtained,
>
> (4) that either a building contract has been entered into for the proposed works or at least that evidence is available that a contract will be entered into in good time,
>
> (5) that the necessary finance for the project is or can be made available,
>
> (6) that the project will not be held up by some impediment such as a tenant of another part of the building who cannot be removed.

In the proceedings before the learned judge Mr Mendoza, appearing for the tenant, said that no issue arose on the *Hill and Redman* points other than in respect of 3, 5 and 6. It should, of course, be said, and Mr Mendoza did not argue to the contrary, that, helpful though these paragraphs are, they do not form part of the statute. They are merely the distilled wisdom of a large number of

A cases where it has been found that these are the points to which attention should be directed. But the issue is and always will remain whether, upon the termination of the current tenancy, the landlord intends to demolish or reconstruct the premises and, when the statute refers to "on the termination of the current tenancy", of course it does not mean at the instant at which the tenancy comes to an end, but within a reasonable time thereafter, so that the work of reconstruction or demolition can be considered to be a continuum with the termination of the tenancy.

Taking the three points which were in issue, it was said that the plans and specifications were not sufficiently advanced. What happened was that outline planning permission had been obtained, but there were difficulties in going further because this was a riverside
B site with various underground works holding up the river bank, and it was said that there would be problems in bringing forward detailed specifications and detailed drawings without having possession of the site in order that surveys might be undertaken. However that may be, the prospective developers had in fact spent some £70,000 on this project, which does not, of itself, suggest that this was a case of hope rather than intention.

The second point that was taken was that there might be financial problems. Tablelodge apparently, like many businesses, carried on its business largely with bank finance and at the time of the hearing the bank's agreement to advance money to it on overdraft was about to be reviewed. There was no firm evidence that it would be renewed.
C It was a question of taking a view as to what view the bank itself would be likely to take when the appropriate date was reached.

As to the third point, it was said that, although Mr Cutbush, as tenant of site A, could be evicted without difficulty, there were two other occupiers who were claiming to be subtenants, although their D
claims were in dispute. It is not, I think, directly relevant, but we were in fact told by Mr Mendoza that the judge's view that there would be no problem in getting rid of those occupiers has been found to be correct in the event.

The judge, who heard all the evidence and was able to form a much better view of the true intentions of the landlord than we can, looking purely at the matter on paper, summarised the points and said this as to the question of plans and specifications:

> . . . it seems to me that much of that work could be undertaken straight away. The developers could not reasonably have been expected to go further under this heading than they have already done.

That was a question of fact and degree for him, as it seems to me, and E
I can see no grounds for disagreeing. He then said:

> As far as 5 is concerned, I am quite satisfied that the £400,000 facility [which would be needed when the contract became unconditional] will be extended beyond February [the review date] and that finance for the building works will be obtained from the bank, or the end users or from other sources.

As far as the third point was concerned, he said that he was quite satisfied that it would be possible to get rid of the occupiers, as indeed has been proved to be the case.

This was essentially a question of fact and degree. It was essentially a matter for the judge who tried the case to assess the conviction which the evidence carried. He was convinced of the statutory requirements and, for my part, I can see no grounds for interfering. Accordingly, I would dismiss the appeal. F

Nolan LJ agreed and did not add anything.

Appeal dismissed with costs.

For further cases on this subject see p 95

LANDLORD AND TENANT

RENT REVIEW

Chancery Division

July 25 1991

(Before Judge Paul Baker QC, sitting as a judge of the High Court)

BUYNEAT LTD v TANAP INVESTMENTS (UK) LTD AND ANOTHER
SAME v SAME

Landlord and tenant — Rent review — Construction of clause providing that reviewed rent of headleases was to be a proportion of the rents, service charges and other payments under occupational subleases

The Westmoreland Place Shopping Centre, opposite Bromley South Railway Station, consisting of a main office block, 23 shop units and other components, such as parking centres, was divided into three sites, which were subject to headleases and subleases — There were differences between the exact arrangements of the different sites but the general idea was to provide machinery to ensure that the subleases would have up-to-date rents of which the headleases would take one-third — The rent revision clauses in the leases provided for the review intervals of 25 years — The net rack-rental value was defined as the total of the gross rents, service charges and other payments payable to the lessees by the occupying tenants under the subleases — The total of these payments was subject to a deduction on account of increases in the lessees' expenditure on repairs, maintenance and insurance and the provision of services and management

(1) whether in the computation of rent payable to the headlessees account should be taken of rent reviews under the subleases pending on the headlease review date or whether regard should be had only to the passing rents; (2) whether the service charges should be taken to be the amounts ultimately payable for the preceding service-charge year after the necessary adjustments had been made or whether they should be the provisional amounts payable in advance; (3) whether the lessees could rely on estimates of expenditure where records had been lost or destroyed or whether they had to provide strict proof; (4) whether, in cases where such a deduction fell to be made, the amount was deductible only from the service-charge components of the revised rent or whether it was deductible from the totality of the revised rents

***Held*: Judgment was given in favour of the defendant lessee's submitted construction on issues (1), (3) and (4)**

(1) Regard should be had only to the passing rent — (2) Service charges should be taken to be the amounts ultimately payable for the preceding service-charge year, after necessary adjustments — (3) Estimates of expenditure can be relied on where estimate lost — (4) The deduction fell to be made from the totality of the revised rents

No cases are referred to in this report

These were two related actions in which the court's determination was sought on the construction and application of rent review clauses in the headleases relating to the large development of shops and offices in the Westmoreland Place Shopping Centre, Bromley. The plaintiff was Buyneat Ltd in both actions and the defendants in the first were Tanap Investments (UK) Ltd and British Telecommunications plc and in the second Tanap Investments (UK) Ltd.

David Neuberger QC and Robert Farber (instructed by Clifford Chance) appeared on behalf of the plaintiff; Terence Cullen QC and Alastair H Norris (instructed by Bischoff & Co) represented the defendants.

Giving judgment, **JUDGE PAUL BAKER QC** said: I have before me two related actions concerning the rent review clauses in the headleases of a large development of shops and offices in Bromley. Owing to negotiations and events which have occurred since the actions began, the issues originally pleaded have narrowed down to four points of construction of the review clauses in the leases. I can state them briefly, prefacing them with the explanation that the reviewed rent under the headleases is a proportion of the rents, service charges and other payments payable under the occupational subleases.

The four points are: first, whether one takes account of rent reviews under the subleases pending at the headlease review date, as the plaintiff landlord says, or whether one simply takes account of the passing rents at the review dates, as the defendant lessees say; second, whether the service charges are the amount ultimately payable for the preceding service-charge year after adjustment, as the plaintiff says, or whether it is confined to the amount paid and payable in advance for the years in which the review date falls, as the defendants say; third, where a deduction falls to be made on account of an increase in the cost of services, is it dependent on the continued existence of records of expenditure, as the plaintiff says, or can estimates be adduced as evidence if the records have been lost or destroyed, as the defendants say; and, fourth, where such a deduction falls to be made, is it deductible only from the service-charge component of the revised rent under the headleases, as the plaintiff says, or is it deductible from the totality of the revised rent, as the defendants say. That states the issues.

I now turn to the facts. The Westmoreland Place Shopping Centre lies opposite Bromley South Railway Station. It consists of a multi-storey office block and adjacent shopping precinct on two levels. It was constructed in the 1960s around an open square or piazza. The final layout plan, it seems, dates from 1963 and shows the site divided into three sections known as sites 1, 2 and 3 respectively. As developed, there was the main office block and some 23 shop units and other components such as parking spaces. On March 12 1965 two leases were granted of site 1, which included the main office block, one in possession and the other reversionary. By the former Bromley Borough Council, the landowners, demised the site to PIC (Bromley) Ltd, the developer, for 20 years from June 1 1964 at an annual rent of £17,264 payable quarterly in arrears on the usual quarter days.

By clause 2(24), the lessee covenanted not to grant on first letting any lease to extend beyond June 1 1989. That was a date beyond the term granted by this lease and foreshadows the contemporaneous reversionary lease. By that lease the site was demised for 85 years from June 1 1984 initially at the same rent but subject to revision from June 1 1989 in terms to which I shall have to return. By clause 2(24) of that lease, the lessee covenanted not to grant on first letting any lease to extend beyond June 1 1989; so far it is identical with the covenant in the other lease. The clause goes on:

> and generally arrange for general renewal of all leases or tenancies on or within five years prior to the dates upon which the annual ground rent falls to be revised, in accordance with clause 1(1) hereof and as far as practicable in

accordance with good estate management, grant such renewal leases for terms of not less than 14 years.

By clause 28, the lessee covenanted to give the corporation notice together with a copy of, *inter alia,* all documents whereby the demised premises or any part should be underlet. The other lease contained an identical covenant.

I now come to the headlease of sites 2 and 3. It is between the same parties as the lease of site 1, but it is dated over five years later, that is September 15 1969. However, from a reading of clause 6 of the reversionary lease, its grant would appear to have been regulated by an agreement of November 26 1962, hence the material negotiations apparently date from the early 1960s, a factor which may have some relevance when we come to consider the review clauses. By this lease, sites 2 and 3 were demised for 100 years from June 1 1969 at the yearly rent of £32,736 payable quarterly in arrears on the usual quarter days, subject to revision in identical terms to those contained in the lease of site 1. The aggregate initial rent of the two leases is thus £50,000 and both leases run until June 1 2069.

The clause 2(24) regulating underlettings is in slightly different terms. The lessee is not to grant on first letting any lease to extend beyond December 28 1990, as distinct from June 1 1989. This may be a slip. I note that the date for the commencement of the term had to be changed after engrossment, but I think that nothing turns on this. The rest of the clause as to the renewal of the underleases generally is identical, as is the clause requiring notices of underlettings to be given to the lessor.

I now return to the rent revision clause in the reversionary lease of site 1. As I have said, that and the lease of sites 2 and 3 are identical, *mutatis mutandis.* After providing for the initial rent of £17,264 and its payment on the usual quarter days, the *reddendum* continues:

> Provided that, (1) the total annual ground rent payable by the lessee to the corporation shall be revised from 1st day of June 1989, the 1st day of June 2014, the 1st day of June 2039 and 1st day of June 2064.

Pausing here, one notices that the review periods are very long by present-day standards, that is 25 years. The present dispute arises from the revision due on June 1 1989. Having identified the dates from which the rent is to be revised, the clause declares that:

> The ground rent shall represent one third of the then net rack rental value of the demised premises (ascertained as hereinafter mentioned).

Mr Neuberger, for the plaintiff lessor, places some emphasis on the word "value", but the mode of ascertainment to which I shall come does not depend on any means of valuation. It further demonstrates its early unsophisticated form, as does the proviso which now follows.

> Provided always that the yearly ground rent payable to the corporation throughout the term of this lease shall not be less than £17,264.

This upwards-only clause is fully effective only in the 1989 revision.

We now come to the all-important definition of the net rack-rental value of the demised premises. It

> . . . means the total of the gross rents, service charges and other payments payable to the lessee by the occupying tenants under subleases and tenancies granted by the lessee.

So the general idea is simple. According to clause 2(24), all the underleases have to be renewed on or within five years prior to June 1 1989, hence they will have up-to-date rents of which the lessor takes a one-third share. However, that leaves several questions. What if the leases are not renewable, but the rents in them are reviewable in the way that has become common in recent years? What is comprised within service charges and other payments and why should the ground landlords share in them in so far as they represent a recoupment of expenditure incurred by the headleases? Some light is thrown on these questions by the remainder of the rent revision clause. It goes on to declare that the total of the gross rents, service charges and other payments is

> subject to adjustment in respect of:
>
> (a) any increase in the Lessee's average annual expenditure during the last ten years of the relevant twenty-five year period over the Lessee's average expenditure during the first fifteen years of the relevant twenty-five year period on repairs, maintenance, insurance and provision of services and management (but no adjustment shall be made in respect of sinking fund ground rent or property tax) . . .

The paragraph concludes by declaring that the first 25-year period is deemed to have commenced on June 1 1964. It is agreed that the purpose of this is to allow a deduction from the net rental value or some element in it, where an increase could be demonstrated, but its precise meaning and application is not wholly clear and has given rise to the third and fourth questions before me.

By way of introduction, I can make some general remarks about the paragraph. In the first place it makes clear that the gross rents, service charges and other payments in the definition are indeed to include payments by the underlessees by way of reimbursement for the sums expended by the lessee on repairs, maintenance, insurance and the provision of services. Second, it shows a direct relationship between gross rents and service charges in the definition. This can be demonstrated by considering repairs. Under the headlease, the lessee is to repair the demised premises (clause 2(3)). That is not an obligation that can be passed on or wholly passed on to the many occupying sublessees directly by way of repairing covenant. The financial burden can only be passed on either by setting the rent of the subleases at an appropriate level or by including it in a service charge. Similar considerations apply to the lessee's obligation to cleanse, maintain and light the shopping precinct (clause 2(9)).

Following from this, it seems to me that any adjustment required by para (a) should be made by deduction from the total of the gross rents, service charges and other payments and not from the service-charge element alone as the plaintiff landlord suggests. There is no necessary and exclusive connection between that element and the expenditure referred to in para (a).

Looking at para (a) itself, one is required to discover whether there is any increase in the average annual expenditure on the listed items during the last 10 years of the 25-year period as compared with the first 15 years. The increase, when discovered, is presumably deducted, although the paragraph does not in terms say so. The saving for sinking fund and ground rent would seem to be an abundance of caution. A sinking fund is not an expenditure but a saving against eventual expenditure and the ground rent does not change throughout the 25-year period. Be that as it may, the difficulty which has arisen with regard to the application of this paragraph is an absence of records of expenditure for the first six years of the period and for one half-yearly period in the 13th year. It is said by the landlord that the lessee cannot take advantage of this paragraph unless he can prove the actual expenditure. While it is in general understandable that records of this nature are not kept for decades, in this case it was clear from the outset that they were going to be required. Though the present lessee has come on the scene only recently, it is said that it took the lease with its eyes open and, if it did not or could not procure the necessary records from its predecessors, it must take the consequences.

I cannot take that strict view of the absence of the records. The lease contains no provision that accounts are to be kept and made available for inspection. The sole obligation set out later in the clause is to provide such information as the landlord may reasonably require in respect of the expenditure. I take that as meaning such relevant information as the lessee may have. If, say, the records of one-half year had been accidentally destroyed, but all the rest had been preserved, would that vitiate the whole exercise? Surely not. But that is the conclusion to which the landlord's submissions would lead. If there is evidence, say, of the works done, but the financial records have been lost, I could see no objection to expert valuation evidence from a building or quantity surveyor putting a price on those works. The expert evidence proffered in this case employs a much broader brush. Mr David Large [ARICS], the defendant's surveyor, deposes that efforts were made to procure records from the former lessee's surveyors, but unavailingly. All seems to have been done that reasonably could be done by the lessee.

The first available figure for the seventh year was £5,479.81. Mr Large has estimated £5,000 for each of the preceding years. He explained it in para 11 of his statement:

> I took as my starting point the earliest available actual figure, reduced it by 10 per cent and applied it to all missing periods. As I acknowledged in my letter of June 16 this process made no allowance for inflation nor did it allow for the fact that expenditure in early years may well have been lower because of the limited maintenance necessary to what one can assume to have been new buildings.

I would also comment that not all the buildings may have been completed by 1965. The lease of sites 2 and 3 was not granted until 1969, so the cost of servicing of what was there in 1965 would presumably be less. Accordingly, I would accept Mr Large's methods.

Any further inquiries as to the precise state of the buildings and what in fact was done is likely to result in a reduction of Mr Large's estimate. That would be to the advantage of the lessee, as it would add to the increase referred to in para (a).

I now come to the remaining two issues, which were of considerable difficulty and arise from the definition of the net rack-rent itself. To deal with them, I shall have to look at some further provisions of the review clause, examine the overall structure of the review procedure and set out the material facts as to the subleases. We have just seen that the total of the gross rent, service charges and other payments was subject to adjustment in respect of increases in average annual expenditure on repairs and so on. It was also subject to the three further adjustments. Para (b) states:

> Premiums, surrenders or other considerations, paid rendered payable or receivable for or under the subleases or tenancies which said premises . . .

That is an obvious slip for "premiums". It is correctly stated in the other lease.

> . . . which said premiums surrenders or other considerations shall be equated over the period of the sublease or tenancy in respect of which they were paid accepted or given.

The purpose of this provision is obvious. The interest lies in the simple but crude way in which the adjustment is made: a division of the premiums by the number of years in the term. If the premium had not been paid, presumably the rent would have been higher than the adjusted rent to reflect the deferred payments. The third adjustment reads as follows:

> any part of the demised premises occupied by the Lessee or by its associated or subsidiary companies at less than its full rack rental value or any vacant or unlet part of the demised premises PROVIDED that for the purpose of this clause such parts of the demised premises shall be deemed to be let for the period during which they are let at a reduced rent vacant or unlet at the full rack rent or average full rack rent at which they were last previously let (or if not previously let at a full rack rent at their gross value for rating purposes).

This again provides a simple way of bringing subleases not let at arm's length and voids into account. It is done by reference to historic rents or rateable value. There is no question there of valuation.

The fourth adjustment (d) states:

> rents which are due to vary during the currency of a sublease or tenancy in which case the rent shall be taken to be the average annual rental throughout the terms of the sublease or tenancy.

This provides for the case of a rent subject to fixed increases throughout the term. When it is recalled that the headlease contains a covenant requiring subleases to be renewed or granted on or shortly before the review date, this provision could not have been applied to rents which were to be fixed on some future review.

The overall structure of the rent review provisions displays the following features. First, there is no element of valuation. The review takes the form of calculations from present or historic facts. Second, by modern standards the review periods are very long. If the negotiations for the grant of the headleases took place before 1962, as seems likely, it was before the present widespread system of frequent reviews at market valuation became apparent. Third, the lessee's covenant required the renewal of all leases at or shortly before the review date. They were to be for substantial terms if possible. The purpose of this, as I see it, is to ensure that the maximum rent is likely to be paid. If the rent is fixed throughout the term, the initial rent is likely to be higher than the rent for a short term. Alternatively, the rent might be set at fixed levels throughout the term. Yet again, the lease might command a premium. These possibilities are all catered for. What is not contemplated is subleases subject to a rent review determined by valuation. The covenant requires all subleases to terminate before the review date and to be renewed.

Fourth, the draftsman also, as it seems to me, contemplated that the service charge and other payments referred to in the definition should be of some fixed sum, although possibly subject to later variation. The service charges there referred to must be those payable under the renewed leases, whereas the adjustment under para (a) relates to expenditure during the currency of the old leases. In effect, the lessee is getting an allowance in the subsequent period for his excess expenditure in the earlier. It is very rough and ready.

Although, for reasons I have given, the headleases required the grant of new or renewed subleases at determined rent for substantial terms, commencing on or within five years before the renew date, the actual subleases, or some of them, did not comply with this requirement. In particular, they contained review provisions of the now familiar kind; that is for an open market rent to be assessed by arbitration, if not agreed. In the case of three of them, the arbitration procedure had not been completed or the new rent had not been agreed on June 1 1989, although when the procedure was completed or the rent agreed, the new rent would be payable from June 1 or some date prior to it.

Thus the question has arisen whether for the purpose of determining the net rack-rental value under the headleases, one takes the rent actually being paid on June 1 or whether one takes into account the review rent as and when it becomes known.

I must briefly refer to the three subleases. The most significant is an underlease made on November 21 1979 by PIC (Bromley) Ltd to the Post Office, now British Telecom, for 50 years and one month from June 1 1964. The rent for the first 25 years, that is to say to May 31 1989, was £128,150 payable quarterly in arrears on the usual quarter days. From June 1 and for the rest of the term, that is to say 25 years and a month, the rent was:

> a rent equal to a fair yearly rent for the demised premises as at the last quarter day but one before the end of the last preceding rent period the amount of such fair yearly rent to be agreed between the Lessor and the Lessee at least one year before the end of the first period of twenty five years or in the absence of agreement to be determined by an arbitrator to be nominated by the President for the time being of the Royal Institution of Chartered Surveyors on the application of the Lessor before but not more than two quarters before the end of the first period of twenty five years . . .

The parties could not agree and consequently an arbitrator was appointed. The proceedings leading to the appointment were set in train in the middle of May 1989 and it was not until December 11 1989 that the arbitrator made his award. It was in the sum of £1,057,000.

By clause 6 of the underlease, British Telecom had the right to terminate the lease on June 1 1991 if the reviewed rent was greater than £128,158, as it was. It exercised that right and quitted the premises on that date; that is to say June 1 1991. Under the terms of the break clause, the new rent had to be paid from June 1 1989 to June 1 1991.

The next is an underlease made on May 10 1965 to Barclays Bank for 50 years and one month from June 1 1964. The rent for an initial period of 16 years to 1960 was £7,500. The rents for the following period of nine years 1980 to 1989 and then the periods of 18 years and seven years and one month were to be determined by the same formula as in the Post Office lease. In 1980 the rent was agreed at £28,000, which was the rent being paid on June 1 1989. At that time the parties had not agreed the rent for the following period. On June 1 1989 it was the subject of arbitration. The arbitrator's award was £75,500, which was recorded by a memorandum dated August 15 1990. Unlike the Post Office underlease, rents under this underlease were payable quarterly in advance; also there was no break clause.

The third underlease was made on September 21 1977 to Derek John Weatherley for a term from September 30 1976 to September 28 1999, paying the rent specified in the fourth schedule quarterly in advance on the usual quarter days. The fourth schedule provides for a rent of £2,500 pa, up to and including September 28 1981, and for the residue of the term it provides for:

> . . . a reviewed rent determined and calculated in accordance with clause 7 hereto or, if the rent is not reviewed in accordance therewith, the rent payable prior to the last previous date of review (whether such rent is itself a reviewed rent or not).

Clause 7 is a slip for clause 8. Clause 8(1) says:

> On the 29th September 1981 1st June 1989 and 29th September 1994 (hereinafter referred to as the dates of review) the Landlord shall have the right to review the yearly rent then payable hereunder . . .

Subclause (4) shows that the procedure for review is set off by a notice by the landlord given not earlier than 18 months before the date for review but may be given after as well as before the date for review. In default of agreement, the rent is to be determined by an arbitration.

By subclause (5), if the reviewed rent is not determined prior to the date of the review,

> then in respect of the period of time (hereinafter called "the interval") beginning with the said date of review and ending on the quarter day immediately following the date on which such determination shall have been made as aforesaid the rent payable hereunder shall continue to be paid at the same rate as prior to the date of review PROVIDED that at the expiration of the interval there shall be due by way of additional rent payable by the Tenant to the Landlord on demand a sum of money equal to the amount whereby the reviewed rent shall exceed the rent prior to the date of renewal, but duly apportioned in respect of the interval together with interest thereon on the amount outstanding during the interval.

The reviewed rent for this underlease was agreed on September 15 1990 at £9,000 pa and a memorandum to that effect was endorsed on a lease on December 7 1990.

A I now return to the question whether the rents of the subleases, as settled subsequently to June 1 1989, are to be taken into account. Mr Neuberger, for the landlord, says that one must not shut one's eyes to the commercial purpose of the review clauses in the headleases. The provisions for review in the subleases on June 1 1989 demonstrate that they were entered into with an eye to the headleases. The parties must have envisaged that reviews might not take effect until subsequently. If one takes one's stand at June 1 1989 and asks what is the rent then payable, the answer would be the pre-existing rent subject to review. Mr Cullen, for the lessee, says that one has to find the total of the gross rents, service charges and other payments payable on one specified day, June 1 1989; that is to say, the rental rates per annum payable on that day. It was not contemplated that B one would wait for some unspecified date but that on June 1 1989 all the data would be available and hence the new rent would be payable and normally could be paid on the following quarter day, June 24.

In my judgment, the submissions of the lessee are correct. I have already indicated reasons for saying that the headleases contemplated that the initial underleases would be terminated before the review date and replaced by others, which were so structured that the rent or rents payable under them would be known from the outset. The hallmark of the rent revision provision in the headleases is the attempt to attain precision and predictability over long periods of 25 years. Very little is left to tests where more than one view is possible. A question might arise under para (c) as to whether lettings C to subsidiary or associated companies are at less than the full rack-rental value, but unlikely. Such a rent would probably clearly be below the market value. A question might arise with regard to good estate management under clause 2(24). All else is made to depend on arm's length negotiations between the headlessees and the occupying subtenants with nothing left to be determined by some outside body. Yet leases containing review clauses requiring arbitration in default of agreement have become commonplace.

It would be consistent with clause 2(24), as I see it, for an underlease to be granted or renewed four years before June 1 1989 for 25 years, with five-yearly reviews of the common type. The initial rent would almost certainly be lower than the subsequent rents. It might even be artificially low, yet it will be the rent on June 1 1989. It D will not be possible to apply para (d) where the subsequent rents cannot be known until they are respectively agreed or determined.

Another gap is the absence of any consideration of the effect of Part II of the Landlord and Tenant Act 1954. If the subleases have all to be terminated before June 1 1989, some of them may be subject to proceedings under that Act on June 1 1989. The reference to not less than 14 years, more than the court can award, demonstrates an absence of any thought given to the Act.

The actual underleases with which I am particularly concerned stray even further from the discernible rent revision scheme of the headleases. They would all seem to be in breach of clause 2(24). They are all granted for terms extending beyond June 1 1989. So far from E arranging for a general renewal of the Post Office underlease on or within five years prior to June 1 1989 for a substantial term, that underlease gives the underlessee the right to break it, if the rent is increased. In the case of that underlease and that of Barclays Bank, the underlessee may defer going to arbitration until immediately before June 1. Furthermore, in the third case the procedure can be deferred until after the review date. It is true that either expressly or by implication the rent when determined is back-dated to on or before June 1 1989, but that is not what the rent revision scheme contemplated.

Faced with such failure to contemplate the cases which have occurred and which might have been foreseen, the court can do one of two things: it can apply the revision provisions strictly or it can F attempt to extend them to cover the cases which have arisen. That of course has its attractions. Though the underleases are in breach of clause 2(24), an attempt has been made to secure the revision of the rent on June 1 1989. Further, the landlord is powerless to influence the negotiations. Nevertheless, I have to reject it. It is no function of the court to remake contracts. In so far as the grant of the underleases is in breach of the covenants in the headleases, the landlord has or did have remedies. Moreover, as regards the Post Office underlease, I do not think that the court could simply accept the subsequent rent without taking some account of the break clause which has been exercised. But, in my judgment, the court should not interfere at all.

The final question concerns the computation of the service-charge element in the revised rent. Is it the amount of the charge ultimately payable in respect of the last service-charge year ending before the G review date, as the plaintiff says, or is it the amount payable in advance for the year in which the review date falls, as the defendants say? I have already said that it appears to have been contemplated that the service charge in the underleases should have been at some fixed sum, in which case the defendants' contention would have been correct. However, the underleases actually granted took the form of the variable charge as is common. I have seen only the underleases to which I have already been referred. In the Post Office underlease, the underlessee covenanted to pay: (a) the due proportion of the insurance premium on the quarter day following its payment by the underlessor; (b) a due proportion of the costs incurred by the underlessor in cleansing and lighting the forecourt to be paid by equal H quarterly payments on the usual quarter days; and (c) a due proportion of the costs incurred by the underlessor in complying with its repairing covenant. There are no provisions as to the mode of payment of the third element.

The underlease to Barclays Bank is similar to that of the Post Office underlease in respect of the insurance and cleansing charges, but contains no provision for the reimbursement of the cost of repairs.

The third underlease contains more elaborate provisions in relation to the service charge. A proportion of the insurance premium is reserved as rent. The tenant covenants to pay on demand the service charge as defined in clause 4(1). That provides for payment of a fair contribution in respect of anticipated or actual J payments or costs of a list of items including repairs and maintenance of the forecourt. The year runs to March 31. There is a provisional sum payable quarterly which is subject to adjustment after the end of March, upwards or downwards, according to the actual cost. If the tenant has been undercharged, the excess is payable on demand; if he has been overcharged, the excess is credited against the net quarterly payment of rent.

It would appear that many of the underleases contain similar provisions. At all events, the provisional sums, where payable during the year which includes June 1 1989, amount to £22,260 in all. The total of the amounts payable in respect of the previous year to March 31 1989 amounted to £119,750.07. I think there has to be some small adjustment to that, but I will leave that for the moment. That of K course includes sums payable under those underleases which do not allow for a provisional sum to be paid in advance.

In my judgment, the latter figure is the correct one for the following reasons. (a) To take the provisional figures would not be the true service charge. It does not include some of the charges where no advance payment is provided for. It is liable to adjustment when the true cost of the services for the year April 1 1989 to March 31 1990 became known. (b) As the service charge for the year in which the review date falls cannot be known for a long time afterwards, I consider that the charge for the preceding year can properly be regarded as then payable. For example, the insurance premium is paid only once a year. It is not payable on June 1 1989. Thus one is L compelled to look to the past payment, similarly as to the other charges. To read "payable" as "are" or "have been payable" seems consistent in this case with the purpose of the review provision. I note the provision for bringing voids into account based on past facts. On the other hand, to read it as "will be payable" would be to introduce an unwarranted element of uncertainty into the operation of the provision, on which I have already commented in relation to the gross rents.

My conclusion is that on the first question I shall declare that it is not permissible to take account of the reviews pending on June 1 1989. The part of the pleadings which seems to relate to that would appear to be the 1990 action, para 5(1) of the statement of claim. It is not otherwise specifically dealt with. M

The second question, the one with which I have just been dealing, about the amount of service charges payable, would seem to come under the statement of claim in the first action, para 11. I think that I should make a declaration along the lines that the amount of the service charge payable within the meaning of the aforesaid provision is the amount of service charges ultimately payable in respect of the service-charge year ending immediately before the relevant review date, taking into account any balancing charge or credit made after the end of that year.

Under the third question, determining the average annual expenditure for the purposes of para (a) of clause 1(1) of the lease, I shall declare that the lessees can rely on estimates where records are

A not available. I think that that is covered by the statement of claim, section 12(1), although of course the declaration would be in the opposite sense in that case.

In relation to the fourth question, I shall declare that deductions are permitted to be made from the total of the gross rents, service charges and other payments and are not limited to any specific element of that. I think that that is covered by the statement of claim 12(2), the declaration of course being in the opposite sense to the declaration there sought.

The judge made the appropriate declarations and ordered the plaintiff lessors to pay three-quarters of the costs.

B

Chancery Division

October 10 1991

(Before Mr E C EVANS-LOMBE QC, sitting as a deputy judge of the Division)

WGTC NOMINEES LTD v HAMILTON INSTALLATIONS LTD

C **Construction of rent review clause — Provision for additional rent payable by tenant to the landlord from each relevant review date until the next relevant review date — Additional rent to be such a sum as is equal to 50% in the increase in the under-rents payable (or deemed to be payable) on each relevant date in excess of the apportioned basic annual rent or the aggregate of the relevant parts as set out in the lease — Relevant date defined — At first rent review the increase of rent was agreed and resulted in an increase to a level of rent representing half the under-rents actually passing**

Demised premises consisting of six separate warehousing units at Caxton Road, Bedford — Plaintiff landlords D **contended that rent review clause to be construed so that at each relevant date there was added to the part of the basic rent apportioned to each of the underlet units half the difference between the figure to which the relevant under-rent had been increased and the apportioned basic annual rent — The defendant tenants contended that the important comparison was between the under-rents before and after review and to which the 50% was to be applied, this figure to be added to the basic rent which would be exactly half the rent passing under the relevant underlease**

***Held*: If it had been the intention of the parties to make the** E **rent half the passing under-rents, it would have been extremely simple so to provide — The construction of the clause supported by the defendants does not work once one gets to the second and later rent reviews — The construction contended for by the plaintiffs does little, if any, violence to the words used by the rent review clause, while by contrast the defendants' construction requires either additional words to be read into the section or a special meaning to be given to the words "the increase in the under rents"**

No cases are referred to in this report.

F By an originating summons the plaintiffs, WGTC Nominees Ltd, the owners of a reversion to a lease, sought a declaration as to the proper construction of the rent review clause in a lease of warehouse premises at Caxton Road, Bedford. The defendant tenants were Hamilton Installations Ltd.

John Furber (instructed by Nabarro Nathanson) appeared for the plaintiffs; Michael Rich QC (instructed by Wood Nash Kimber) represented the defendants.

Giving judgment, **Mr E C Evans-Lombe QC** said: This case concerns the construction of provisions as to the payment of rent under a lease dated April 23 1976. The plaintiffs are the successors in title of the original lessors. The defendants are the original lessees.

By clause 1 of the lease, the demise was for a period of 99 years from April 23 1976 of premises described in the first schedule. These G consist of six separate sets of premises used for warehousing [at Caxton Road, Bedford], which are referred to in the proceedings as units 1 to 6.

The lease results from a sale and lease-back transaction between the defendants and the plaintiffs' predecessors in title. Prior to the transaction, the defendants had let units 2 to 5 by a lease dated February 20 1976 between them and J Lyons & Co Ltd. Clause 1 of the lease is expressed to be subject to and with the benefit of that lease, referred to as the "Lyons lease". It seems that at the date of the lease units 1 and 6 were unlet.

Further, clause 1 of the lease contains provisions for payment of rent as follows: H

> Yielding and paying therefor unto the landlord yearly during the said term and so in proportion for any less period than a year (1) the basic annual rent of £37,550.38; (2) the additional rent calculated in accordance with the provisions of the second schedule hereto.

There then follow provisions for the time for payment of the rent and also for the payment of additional rent to cover the premiums for insuring the premises and, by clause 2(1)(b), to pay a management charge representing 2½% of each quarterly payment of rent.

The dispute between the parties arises from the proper construction of the second schedule, which contains the provisions defining the additional rent payable under clause 1(2). The operative paragraph of those provisions is para 2, which reads:

> The additional rent payable by the Tenant to the Landlord from each relevant J date until the next relevant date shall be such a sum as is equal to 50% of the increase in the under rents payable or deemed to be payable on each relevant date in excess of the apportioned basic annual rent or the aggregate of the relevant parts as set out in para 1(4) of this schedule, or the additional rent payable during the period immediately preceding the appropriate relevant date, whichever shall be the greater.

It is accepted that I am not concerned with the words "or the additional rent payable" to the end of the clause, since these are provisions designed to ensure that any alterations to the rent payable under the lease shall be upwards only.

A relevant date, for the purposes of that paragraph, is defined in para 1(1) of the schedule as capable of taking place in one of four different circumstances, namely: K

1. By subpara (a), the date of the commencement of the term created by the lease.
2. By subpara (b), any date upon which rents become payable under any underlease granted in pursuance of the lease after the date of its commencement.
3. The date of any rent review of any underlease.
4. By subpara (d), the expiration of five years from the last relevant date in respect of the whole or any appropriate part of the premises' demise, subject to an exception, the details of which are not material to this judgment.

For the purposes of para 2, the under-rents are defined in subpara (a) of para 1 as being: L

> The full and open market yearly rent of any part of the premises sublet

and contains provisions for a deemed rent to deal with the circumstance which existed in relation to units 1 and 6 at the commencement of the lease, where any part of the premises let had not been sublet at the date of the lease or (which does not seem to have occurred) where any premises have been sublet at less than a full open market rent.

Para 1(4) of the schedule, "for the avoidance of doubt", apportions the basic annual rent of £37,550.38 as between those units 2 to 5 covered by the Lyons lease in the sum of £24,545.50 and units 1 and 6 in the sum of £6,502.44 for each of them.

It seems that notwithstanding the taking place in the interim of events capable of constituting "a relevant date", within the meaning M of para 1(1), in the intervening period from the granting of the lease, the first increase in the rent payable under the lease resulted from a review of all the rents payable under the underleases on May 3 1979, at a time when the landlord was still the original lessor, the Provident Mutual Life Assurance Association.

The increase of rent was agreed and resulted in an increase to a level of rent representing half the under-rents actually passing. The calculation of such increased rent is set out in two letters of May 17 1979 and November 1979 written by the Provident Mutual's assistant surveyor. It is accepted, in particular on behalf of the defendants, that the method of calculation shown in these letters bears no identifiable relation to the actual words used by the second schedule,

A although the result is in accordance with the defendants' contentions as to how the schedule falls to be construed.

The next relevant date occurred as a result of a review of the under-rents taking place on November 3 1984 and it is because the parties are unable to agree the additional rent payable as a result of the consequent relevant date that these proceedings have been brought. It appears that since the dispute has arisen the defendants have been paying rent calculated in accordance with their construction of the lease.

The plaintiffs contend that para 2 is to be construed so that at such relevant date there is added to the part of the basic annual rent apportioned to each of the underlet units, under para 1(4) of the
B second schedule, half the difference between the figure to which the relevant under-rent has been increased and the apportioned basic annual rent. The plaintiffs contend that this is the natural meaning of the words used by para 2 which, omitting irrelevant passages, reads:

> The additional rent payable by the Tenant to the Landlord from each relevant date until the next relevant date shall be such a sum as is equal to 50% of the increase in the under rents payable on each relevant date in excess of the apportioned basic annual rent. . .

It is accepted that the words "50% of the increase in the under rents" are possibly confusing, since those words lead one to look to a comparison between different levels of under-rent and not between the level of under-rent as increased and the apportioned basic annual rent. Mr Furber, for the plaintiff landlords, conceded that the
C drafting would have been more felicitous had it read "50% of the increased under-rents payable on each relevant date". None the less, it was his contention that the meaning of the paragraph was tolerably plain, namely an intention to compare the level of under-rents with the apportioned basic annual rent.

For the defendants, Mr Rich drew particular attention to the words "50% of the increase in the under-rents payable". He contended that this demonstrates an intention that the important comparison is to be drawn between levels of under-rent. He took as an example the first review of the premises subject to the Lyons lease in 1979. Prior to that review, the under-rent payable was £49,091. After the review, it was £68,230. Applying the provisions of para 2 to the difference between
D those two figures, namely £19,139, by dividing that figure in two, throws up, so Mr Rich contends, an additional rent of £9,569.50, which when added to the basic rent of £24,545.50 throws up a new rent payable in relation to units 2 to 5 of £34,115. This is exactly half the rent passing under the relevant underlease and this method of calculation will always throw up a figure which is half such rent passing. It was Mr Rich's contention that this was, as a result, E
consistent with such indications as there are in the lease of the intentions of the lessor and the lessee, in particular the fact that the basic annual rent fixed by the lease in respect of units 2 to 5 is exactly half the under-rent payable under the Lyons lease.

The defendants' contentions are subject to two objections. The first is that if it had been the intention of the parties to the lease to make the rent half the passing under-rents, it would have been extremely simple so to provide and not necessary to draw a complicated provision like para 2 to carry such intention into effect.

The second objection is that without reading words into the paragraph or giving a special meaning to some of them, it does not work once one gets to the second and later reviews. If the comparison F
is to be taken between levels of under-rent between the last two relevant dates, then on the review in 1984 the difference, of which 50% becomes the additional rent, is the difference between the level resulting from the 1979 review and that resulting from the 1984 review. This is plainly not what the parties intended. Mr Rich's contention was that the words of the paragraph indicate an intention to compare levels of under-rent. To give effect to that intention, the words "increase in the under-rents" must be read as meaning the increase resulting from the original under-rent ruling at the commencement of the lease or the appropriate deemed under-rent in pursuance of para 1(2) and the rent resulting from the most recent review.

I am unable to accept this contention. In agreement with the G
submissions of the plaintiffs, it seems to me that the construction contended for by the plaintiffs does little, if any, violence to the words used by para 2 of the schedule. The words used are not as clear as they might have been but their effect is tolerably plain and it is the effect for which the plaintiffs contend.

By contrast, in my judgment, the defendants' construction requires either additional words to be read into the section or a special meaning to be given to the words "the increase in the under-rents", which neither the rest of the paragraph nor the other provisions of the schedule or of the lease justifies. In those circumstances, it is not open to me to take into account, as Mr Rich contended I could, any extrinsic fact, shown by the statement of facts agreed between the H
parties to which I might have been able to have regard had I taken the view that para 2 was ambiguous.

For these reasons, it seems to me that the plaintiffs' claim succeeds and I invite counsel to address me on the form of order which I should make.

For further cases on this subject see p 119

NEGLIGENCE

Official Referees' Business

October 28 1991

(Before Judge Peter BOWSHER QC)

HEATLEY AND ANOTHER v WILLIAM H BROWN LTD

Negligence — Structural survey — Whether chartered surveyor negligent — Assessment of damages

Prior to their purchase of 72 High Causeway, Whittlesea, Cambridgeshire, the plaintiffs instructed the defendant firm to carry out a standard structural survey of the property—The defendants' conditions of engagement made reference to the limitations of the report where parts of the property were unexposed or inaccessible — The plaintiffs believed they could rely on the defendants' report without providing access to the roof voids for inspection — The defendants' report noted that parts of the structure had fallen into disrepair but concluded that the structure was in a reasonable condition for its age; the report recommended certain work to be carried out — Following the plaintiffs' purchase, the defendants' surveyor carried out a further survey with access to the roof void — In the event there were substantial defects in the property and by September 1990 the plaintiffs moved out as they saw no alternative to total demolition of the property and the purchase of another house

***Held*: Judgment for the plaintiffs — The statement in the surveyor's report that "the structure of the property was in a reasonable condition for its age" was not correct — The defendants' surveyor failed to gather all the relevant matters together in the report and draw the obvious conclusion, which was that the house was not a good buy at any price other than the site value — There were serious bulges in some of the walls — The defendants' surveyor ought to have made further examination or put suspicions as to the cause in his report and recommended further investigation — The defendants were negligently in breach of contract in relation to both the original report and the subsequent investigation — By reason of the later investigation the plaintiffs were encouraged to stay in the property and spend more money on it and their damages were increased — There was no liability in the tort of negligence — Damages assessed on the difference between the value of the property as it was described in the surveyor's report and its value as it should have been described, namely £50,000; additional damages awarded for distress and inconvenience arising out of the physical consequences of the defendants' breach**

The following cases are referred to in this report.

Hill v *Debenham, Tewson & Chinnocks* [1958] EGD 328; (1958) 171 EG 835
Philips v *Ward* [1956] 1 WLR 471; [1956] 1 All ER 874, CA
Tai Hing Cotton Mill Ltd v *Liu Chong Hing Bank Ltd* [1986] AC 80; [1985] 3 WLR 317; [1985] 2 All ER 947; [1985] 2 Lloyd's Rep 313, PC
Watts v *Morrow* [1991] 2 EGLR 152; [1991] 43 EG 121

This was a claim by the plaintiffs, Neil Alexander Heatley and Patricia Margaret Heatley, for damages against the defendant, William H Brown Ltd, arising out of a structural survey report and advice concerning 72 High Causeway, Whittlesea, Cambridgeshire.

Michael Parroy QC (instructed by Shepherd Harris & Co, of Enfield) appeared for the plaintiffs; and James Guthrie (instructed by Roythorne & Co, of Spalding) represented the defendants.

Giving judgment, **Judge Peter Bowsher QC** said: The defendants are surveyors and valuers who are sued for alleged breach of contract and negligence in and about surveying and reporting upon a house known as 72 High Causeway, Whittlesea, Cambridgeshire ("the property").

The plaintiffs are husband and wife. On July 30 1988, they became interested in buying the property for occupation by themselves and their three children, who were then aged 4, 2 and 1. About six miles from Peterborough, which has good rail links to London and York where the plaintiffs' relations lived, the property was well situated for the plaintiffs. Moreover, the plaintiffs were attracted by the character of the property and by the size of the accommodation, which was much more generous than their existing home.

The property is L-shaped. A two-storey building (the main building) fronts on to the street and abuts the pavement. The front of the main building faces west and the left-hand end faces north. The main building fills the whole of the frontage to the street, apart from a narrow passageway between the south gable wall and the neighbouring property. From the northern end of the east face of the main building there lies an extension whose main face is to the south, forming a courtyard behind the main building. The rear extension is in two parts: a two-storey part with a dormer roof (the "dormer extension") and then a single-storey part (the "single-storey extension"). The whole property is old and the dormer extension is probably about 200 years old. The building is at the front of a very long narrow site, which apparently used to form a part of a market garden. On each of two sides of the site is a high brick wall about 100 years old. The garden contains various outhouses. The property has considerable charm. It is in a quiet road, surrounded by a number of very attractive properties, in a conservation area, within a short walk of Whittlesea town centre.

The property was advertised for sale at £89,500. It had three living rooms, five bedrooms and a kitchen and bathroom. The only wc and the kitchen and bathroom were in the single-storey extension. There was no garage and no access to the quarter-acre garden for a garage.

When the plaintiffs first visited the property, they were told by the vendors, Mr and Mrs Searle, that the roof of the dormer extension had recently been repaired: they added that there was every possibility that the plaintiffs might obtain a grant for reroofing this extension. It appeared from evidence at the trial that the local authority had in fact refused to make a grant for that purpose on the ground that the general condition of the dormer extension was such that it was not worth spending money on the roof alone. The first plaintiff noticed that there were no access points to the three roof spaces (above the main building, the dormer extension and the single-storey extension) and he asked whether Mr Searle had any objection to entry being made to allow inspection of the roof voids. Mr Searle said that he had no objection.

Mr Searle told the plaintiffs that the asking price was a reduction from the original price of £95,000 because there was no central heating. The plaintiffs offered to buy for £89,500 subject to contract and survey and subject to the plaintiffs' being able to get a suitable mortgage. The plaintiffs calculated that they expected to net about £50,000 from the sale of their existing house and needed to borrow about a further £42,000 to cover the purchase price plus fees relating

A to purchase. While they could have borrowed more than that sum, in the event they borrowed only £32,500.

On August 15 1988, Mr Ian Matthew [FRICS], a valuation surveyor, acting on behalf of the Abbey National Building Society, inspected the property. On August 16 1988, he reported to the building society as follows:

> This old house has been extended and altered from time to time. Amongst the many defects we observed rising dampness, damp penetration to exterior walls and pantile roof, weak and defective ceilings, woodworm and old electric wiring. Some settlement has occurred in the past and although we did not observe significant recent movement the applicants should seek the advice of a structural engineer regarding the stability of the 4½ inch brick walls. In the rear garden is a large willow tree and roots from this are probably affecting the
> B drains and brick outbuildings. . . . We do not want to make a retention in view of the margin of security. However we understand that the applicants are obtaining a full structural survey report. In our opinion this is essential prior to purchase and a full survey might well uncover defects not mentioned in our brief general observations. . . . Some old houses in Whittlesea are affected by settlement.

The pantile roof to which Mr Matthew referred was the roof of the dormer extension. The 4½-in brick walls were walls of the single-storey extension. I was reminded of Mr Matthew's observation that some old houses in Whittlesea are affected by settlement, during a view on October 14 1991 when I observed that brickwork between first- and ground-floor windows in a number of old properties showed signs of displacement and rebuilding. When Mr Matthew
C gave evidence, it became clear that by his advice that a construction engineer should be instructed he meant that there should be a structural survey by either a building surveyor or a structural engineer.

On behalf of the defendants, the point was strongly and properly made that it did not occur to Mr Matthew that this property might be worth only its site value or that the plaintiffs should be advised not to buy, and he was aware of his potential liability to a purchaser as well as to his clients, the building society. But against that it was pointed out that Mr Matthew was surveying for a different purpose than the purpose required by the plaintiffs; he was not a structural surveyor, he advised a structural survey, he himself relied on the defendants' structural survey, and he said that if the exposure of the building
D society had been greater he would have advised against giving the loan.

Mr Matthew's report did not come to the attention of the plaintiffs before they contracted to buy the property. They had instructed their own surveyor before they knew of Mr Matthew's recommendation that there should be a full structural survey, although they knew that Mr Matthew had inspected the property on August 15 because the vendor told them so and they anticipated that he would advise a structural survey.

From his experience as a housebuyer, the first plaintiff knew that, because of the age of the property, the building society would require a structural survey in addition to the usual building society valuation,
E but I am entirely satisfied that that was not the only or main reason why the plaintiffs obtained a structural survey. After inquiries, the first plaintiff contacted Mr Michael E Stanton [FRICS] of the defendant company.

Mr Stanton is in charge of the defendants' offices in Cambridgeshire, varying between two and three in number. Starting as an apprenticed carpenter and joiner in the early 1960s, he later became a chartered surveyor. At the time of receiving instructions from the plaintiffs he was an Associate and now is a Fellow of the Royal Institution of Chartered Surveyors. From 1968 to January 1986, Mr Stanton worked for local authorities. His work included responsibility for such things as dangerous structure notices. From 1986 to January 1987, Mr Stanton worked for the Property Services
F Agency.

On August 16 1988, the first plaintiff telephoned Mr Stanton's office and as a result Mr Stanton rang him back. Mr Stanton took particulars of the property and made it plain that he would not proceed until the plaintiffs had sent to him a signed copy of the defendants' printed conditions together with a cheque for £667 (including VAT) for his fees. The same day, Mr Stanton sent the plaintiffs a copy of the defendants' "Conditions of engagement for a standard structural survey". Those conditions were typed on three and a half A4 pages. No point was taken under section 2 of the Unfair Contract Terms Act 1977.

The defendants' conditions included the following terms:

> 1. The purpose for which the survey is required shall be as agreed between the client and the surveyor. The surveyor will advise the client as to his opinion G of the state of repair and condition of the property specified by the client in accordance with the scope, assumptions and conditions of engagement relating to the report commissioned by the client.
>
> 2. In making the report the following assumptions will be made:
>
> . . .
>
> (c) That the property and its value are unaffected by any matters which would be revealed by a local search and replies to the usual enquiries
>
> . . .
>
> 3. The report will be provided for the stated purpose and for the sole use of the named client. It will be confidential to the client and his professional advisers. The surveyor accepts responsibility to the client alone that the report will be prepared with the skill, care and diligence reasonably to be expected of a Chartered Surveyor but accepts no responsibility whatsoever to any person other than the client himself. Any such person relies on the report at his own H risk.
>
> . . .
>
> 4. It is not practical to inspect those parts of the structure which are covered, unexposed or inaccessible. However, within those constraints it is our policy to give as clear a picture as possible on the overall condition of the property. Unless prior arrangements have been made for opening up and the removal of furniture, fitted floor coverings etc, the *exceptions* may include:
>
> . . .
>
> (b) Unexposed wall areas and inaccessible voids including cavities.
>
> (c) Inaccessible sub-floors, joists and covered floor surfaces;
>
> . . .
>
> (h) An opinion on the condition of parts of the property not inspected should not be construed as making an implied representation nor statement about such parts. J
>
> 5. The paragraphs in the heading "purpose and scope of report" further detail the extent and limitations of the inspection.
>
> PURPOSE AND SCOPE OF REPORT
>
> . . .
>
> 2. . . . Inspection will, therefore exclude both the roof space if there is no readily accessible hatch, joists and wall plate details where covered by loft insulation and outer surfaces of the roof if they cannot readily be seen. . . .
>
> (i) Externally, all roof areas will be inspected as closely as practicable from ground level. . . . The interior of accessible roof voids will be inspected in detail and inaccessible connecting voids will be inspected to the extent practicable.
>
> (ii) . . . The surface of all floors not covered with fixed coverings will be inspected as far as practicable and loose boards lifted where possible . . . Where there are fixed coverings these will not normally be lifted but the K surveyor will lift loose areas sufficiently to identify the nature of the finish beneath.
>
> . . .
>
> (ix) Where repair costs are given they are for guidance purposes only and should not be construed as a quotation nor estimate and should be substantiated prior to purchase by proper competitive quotations or estimates.

The first plaintiff signed those conditions on August 17 and returned them with a cheque for the defendants' fees.

Also on August 17, the first plaintiff rang Mr Stanton again.

Either in their first telephone conversation on August 16 or in their second conversation on August 17, the first plaintiff and Mr Stanton L discussed the absence of access to the roof voids. They both agree that roof voids were discussed in terms which made it plain to Mr Stanton that there were no existing means of access to the roof voids. Beyond that, the parties disagreed about this conversation, but Mr Stanton's recollection was vague and in cross-examination consisted mainly of a reconstruction of what he would have said. For the most part, I prefer the account of this telephone conversation given by the first plaintiff. When the first plaintiff raised the question of lack of access to the roof voids, Mr Stanton said that he would not himself make any access holes, the plaintiffs would have to arrange that through a builder. I accept from the first plaintiff that Mr Stanton warned him that should he not proceed with the purchase he would have to pay for the holes being made good. With this in mind, the first plaintiff asked M Mr Stanton if he could get an adequate impression of the roof's condition from the outside and he replied that he should be able to do so. Mr Stanton claimed that he added that you cannot say that roof timbers are sound or unsound as a definite statement without going into the roof void: I do not accept that he expressed himself as clearly as that. In any event, the plaintiff did not arrange for access to the roof voids and Mr Stanton did not advise him before or after the inspection of the property that he should definitely arrange for access to the roofs before purchasing the property. Mr Stanton's notes regarding this property begin with the words "Requested to inspect roof voids".

Not unreasonably, the first plaintiff decided that as Mr Stanton

A was not advising him to have the access holes made for him prior to purchase, the plaintiffs need not incur unnecessary costs.

In his written report, Mr Stanton wrote: "We strongly recommend that proper access hatches be formed into each roof space to facilitate maintenance inspections in the roof voids". In cross-examination, he said that the exterior of the roof did not lead him to advise that access should be gained to the roof voids before purchase.

In advance of discussing the report, I should say at this point that in the context of this conversation it was reasonable for the plaintiffs to think that they could rely on the report for the purpose of deciding whether to purchase without providing access to the roof voids for inspection by Mr Stanton.

Mr Stanton accepted that he was specifically asked at some point B by the first plaintiff to give an opinion on the price of the property and also to give some guidance as to the repair costs of the defects he found.

It was submitted on behalf of the defendants, having regard to the decision of the Privy Council in *Tai Hing Cotton Mill Ltd* v *Liu Chong Hing Bank Ltd* [1986] AC 80, that in view of the detailed terms of the contract there is no room for any liability in tort and that the liability of the defendants lay in contract only. I accept that submission.

The covering letter sent with those conditions was addressed to "Mr Heatley". It was submitted on behalf of the defendants that the "named client" referred to in clause 3 of the conditions of C engagement was Mr Heatley alone. Counsel's written submission contained the following:

> The contract is with the first plaintiff only. The correct analysis of the position is that he contracted also as agent for the second plaintiff, she being, jointly with him, purchaser of the property. In this situation she has the benefit of the contract and is also bound by its terms. . . . The practical effect in the present case is only that as a matter of law . . . damages are only recoverable by the first plaintiff, although general damages may include compensation for the additional effect on him of his wife and family's physical inconvenience and discomfort.

I accept that the correct analysis of the position is that the first plaintiff contracted also as agent for the second plaintiff, she being, jointly with him, purchaser of the property and that she has the D benefit of the contract and is also bound by its terms: but I reject the remainder of the submission. Since the first plaintiff contracted as agent for the second plaintiff as well as on his own behalf, both of them are for all purposes parties to the contract and both are entitled to damages for its breach (if any). In the circumstances of this case, the reference to "the named client" must be taken as a reference to both plaintiffs notwithstanding that only one of them is named in the covering letter. This finding may not have any effect on the total amount of any award of damages, but the form of the judgment may affect their rights as between each other and may have implications as regards the Inland Revenue and also as regards legal aid assessments.

The duty accepted by the defendants in condition 3 is related to E "the stated purpose", which relates back to condition 1, "The purpose for which the survey is required shall be as agreed between the client and the surveyor". It is plain from the first plaintiff's written statement (which stood as his examination in chief) and from the defendants' telephone message book that the first plaintiff asked and paid for a full structural survey and that he disclosed that the property was 18th century. As I have already indicated, the first plaintiff showed his concern about the roof and made it plain that he expected that in a property of this age some repairs would be required and he wanted to have an indication of their cost at the same time as having the information about the structure which would be revealed by a full structural survey. While it was not expressed concisely in a single sentence, it must have been plain to the defendants and agreed F that the purpose of the survey was that the plaintiffs should be informed and advised whether the state of the structure was such as to justify the expenditure of whatever was required in the way of necessary repairs and consequently whether it was worth buying the property and at what price.

It was submitted on behalf of the defendants that the purpose of the survey was simply to satisfy the building society and that the plaintiffs did not themselves rely on it. I reject that submission. It is inconsistent with the conversations between the first plaintiff and the defendants before the inspection of the property and it is also inconsistent with the care taken by the plaintiffs (established in cross-examination of the first plaintiff) to read and study the report when they received it.

Para 4(ix) of the "Purpose and Scope of Report" attached to the G conditions of engagement should be read in the light of the agreed purpose of the report. The statement that repair costs are "for guidance purposes only" must mean, in the context, for guidance in achieving the purpose of advising whether the state of the structure was such as to justify the expenditure of whatever was required in the costs of necessary repairs.

Mr Stanton inspected the property on August 18. By August 19, the first plaintiff was already pressing for Mr Stanton's report because he wanted to speak to the vendor and to have the report with him when he did so. Mr Stanton told the first plaintiff that it was his practice not to discuss his findings until the client had his report. The report was dated August 22, but it was not ready until August 24 1988, when it was collected by the plaintiffs, together with one copy, H from the staff in Mr Stanton's office. Although the plaintiffs wanted to meet Mr Stanton, he was not available when they called at his office, and they did not in fact meet him before they signed the contract to purchase.

The plaintiffs had persuaded themselves, or had been persuaded by the vendors, that they had to make a decision whether or not to go ahead with the purchase on that day, August 24. They had been told by the vendor that there were other people interested in buying and they thought that by comparison with other properties they had viewed this property had real character. Another consideration was that the completion date for the sale of their existing home was September 16 1988. J

The plaintiffs read the report with care, although they were not skilled in building. The original document was produced in evidence. That document is marked by a number of ticks and underlinings in pencil which, on the balance of probabilities, I find were made by the first plaintiff. The first plaintiff also identified some brief comments, also in pencil, as having been written by him. It is plain that the plaintiffs did indeed read the whole document with some care. The first plaintiff said, and I accept, that the plaintiffs based their decision to go ahead with the purchase mainly on the introduction to the report and on its conclusion and on the budget figures for remedial works given at the conclusion of the report.

I also accept the evidence of the first plaintiff that after reading the report he was able to speak to Mr Stanton on the telephone to discuss K it and he was assured by Mr Stanton that subject to the remedial works listed in the report the property was "OK". Mr Stanton did not in that conversation warn of any major structural defects.

The introduction and the conclusion of the report relied on by the plaintiffs were in the following terms:

> 3.0 STRUCTURAL CONDITION AND REPAIR
> We confirm that we found the property to be in a reasonable condition for its age and we discovered no serious structural defects or other major short comings that would lead us to advise against proceeding with the proposed acquisition as planned. It is apparent, however, that certain parts of the structure have been allowed to fall into disrepair over the years. Consequently, there are now several defects and items of disrepair which have to be remedied and these should be carefully considered and borne in mind in L any prepurchase negotiations. It is appropriate therefore to report upon the various structural elements under specific headings as follows:
>
> . . .
>
> CONCLUSION
> 17.00 This is an attractive property with some character set in a very convenient location. We generally found the structure of the building to be in a reasonable condition for its age, although there are some defects and items of disrepair requiring attention which have been highlighted in the body of this report. We strongly recommend that the suggested works and investigations are undertaken as the consequences of delayed maintenance in an old property of this type are invariably more serious than in a building of modern construction. We would suggest the following budgets be set aside for remedial work to the various parts of the property but would state that these M may be varied when you seek quotations from contractors depending on their workload and the extent of the work being carried out.
> Damp proof course and associated replastering works: £3,000.
> Brickwork repair; repointing; cement render north elevation: £2,500.
> New electrical installation: £1,500.
> De-infestment treatment to softwood timbers: £500.
> To replace roof and floor repair: £1,000.
> Valuation — Assuming there are no onerous restrictions which we are aware of [*sic*], we are of the opinion that the current market value in its present condition would be in the region of £84,000 for freehold with vacant possession upon completion.
> Subject to the purchaser being fully aware of the foregoing we can see no reason why the proposed purchase should not proceed as planned. However, we would point out that whilst our survey was as detailed as circumstances

permitted, we did not examine components or other parts of the structure which were covered, unexposed or otherwise inaccessible and we are unable to comment upon the condition thereof.

On August 24 1988, with the aid of Mr Stanton's report, and in particular by reference to the figures given for guidance as to the likely cost of remedial works totalling £8,500, the plaintiffs persuaded Mr and Mrs Searle to reduce the purchase price by £7,000 to £82,500 (that is, £2,000 below Mr Stanton's valuation) and on that basis and relying on the report given by the defendants, the plaintiffs told Mr and Mrs Searle that they were going ahead with the purchase.

On August 25, the building society made an offer of a loan which the plaintiffs accepted. That offer was conditional upon a structural survey report to be obtained by the plaintiffs.

The plaintiffs sent a copy of the defendants' report to the building society but, possibly owing to a prolonged postal strike, the building society did not receive it, so the plaintiffs sent their only remaining copy, without taking a further copy for themselves. The plaintiffs were then without a copy of the report until October 8 1988 when, at their request, they received a further two copies of the report from the defendants.

Mr Matthew was satisfied by Mr Stanton's report, in view of the comparatively small amount of the loan, and the building society approved the loan.

On September 8, the contract of sale was exchanged stating September 16 as the completion date and on September 16 1988 the plaintiffs moved into the property.

Before entering into the contract the plaintiffs did not get any estimates to substantiate the figures given in the defendants' report.

Detailed investigation of the property has since shown that much more extensive repair work is required to the property than is detailed in the body of the report given by the defendants, and the cost of repairs would on any footing be far greater than the cost of repairs put forward by the defendants as guide. There is a strong body of opinion, which is disputed by the defendants, that the property was worth only its site value and that the only sensible thing to do was to knock the whole property down and rebuild. I have to resolve the dispute as to what is required by way of repair. There are even greater disputes as to what ought to have been discovered and reported on by the defendants in 1988 within the restrictions then imposed on them and within the limitations stated in their report.

The burden of proof rests upon the plaintiffs to prove breach of contract on the part of the defendants with all that means for the professional reputation of the defendants. Did they fall below the standard to be expected of the ordinary careful surveyor? It is not enough to show that the defendants gave the plaintiffs wrong advice. It is not enough to show that where there was room for two views they took the wrong view. I must be careful not to be wise after the event: I must not look at the premises as they now are or see what has in fact happened: rather I must put myself back in the position which [Mr Stanton] found himself in in [1988], and judge the matter by the evidence as to the conditions which then existed. The plaintiffs must prove, in order to succeed, an absence or a lack of skill on the part of the defendants: see *Hill* v *Debenham Tewson & Chinnocks* (1958) 171 EG 835 *per* His Honour W K Carter QC. As regards the conditions existing at the time of the original inspection, it is not suggested that the structure has become any worse, but some defects have become more apparent either because of opening up of the building or because attention has been drawn to them.

Matters developed quickly after the plaintiffs moved into the property.

On September 23 1988, the second plaintiff's sister and Mr Anthony Lister [ARICS] (now that sister's husband) called in for a short visit in the evening. Mr Lister is a quantity surveyor. He expressed disquiet about the property and asked if he could return to look at the property in daylight, which he did on September 26.

Mr Lister gave evidence. Objection was rightly taken to his giving any opinion evidence (no leave having been given for him to give expert evidence), but the evidence of what he saw is important as contemporaneous evidence of what was there to be seen. Mr Lister was concerned about damp and the possibility of dry rot. No real complaint is made about this: the defendants had warned about damp and in fact there was no dry rot. On examination of the front wall he found that many areas of the render to the front wall were hollow. He also saw that certain floors and walls of the dormer extension and the single-storey extension were distorted and overloaded. He also saw a large willow tree in the garden and queried whether it had affected the drains. The plaintiffs had already found that the drain from the wc blocked regularly and that the drain outside the utility room was blocked by roots, and there were also roots in the floor of the utility room.

As a result of the concern expressed by Mr Lister, the first plaintiff asked Mr Stanton to visit the property again. He agreed to do so and a fee of £30 per hour was agreed. As a result, Mr Stanton visited again on September 30 1988. This was the first time that he met the plaintiffs.

Counsel for the plaintiffs submitted that the defendants were at this time under a continuing duty under the original contract. I accept that submission. Counsel also submitted that arising out of this visit and the advice given in relation to it the defendants were under a duty in tort. I disagree. There was plainly a contractual agreement in relation to this visit and, in my judgment, that agreement was impliedly subject to all the previously agreed conditions. There was no room for any duty in tort.

Among the requests made by the first plaintiff to Mr Stanton was a request that he should inspect the roof voids through access holes which were made by the first plaintiff. Mr Stanton agreed to do that and did so on his visit on September 30.

Following that visit, Mr Stanton sent to the plaintiffs a letter dated October 5 1988. There was some dispute over whether it was intended that this letter should be shown to a builder to enable him to undertake work on the property. I find that the first plaintiff did ask for the letter for that purpose and that it was intended by the defendants to be used for that purpose, otherwise a number of dimensions and other details would not have been included. The principal significance of this letter is that it shows that after inspecting the roof voids and noting that in the roof of the main building certain ceiling joists, ceiling binders, and purlins were overspanned and overstressed, the defendants did not withdraw any of the previous recommendations for repair and added further recommendations for repair in the roof of the main building and of the single-storey extension. This would have some significance in the assessment of damages.

On November 12 1988, a Mr Trevor Storey, a carpenter and joiner, gave an estimate for doing work to timbers in the roof. He did the work in November or December and was paid. Mr Storey had refused to do one of the things recommended in the letter of October 5, namely, to anchor to the new timber the triangular area of brickwork at the top of the gable of the north gable wall. Mr Storey said that the brickwork was visibly from the inside in such a poor state that if he drilled into it the wall would fall down. The defendants' response was that the letter began with a requirement that it should be read in conjunction with the defendants' original report and that report contained a recommendation that that brickwork should be taken down and rebuilt: accordingly, the plaintiffs ought to have realised that the intention was that the wall should be rebuilt before being tied back to the new timber. No doubt that is right but little turns on it.

On January 17 1989, at the plaintiffs' request, Mr Stanton visited the property again. By then the first plaintiff had done a certain amount of work on the property which had the effect of opening up a number of apparent defects which were shown to Mr Stanton.

In May 1989, Mr Manning of the district health department of the Fenland District Council inspected the property in connection with an application for a grant for work on the property. He sent a list of defects to the plaintiffs, but no evidence was called to prove the defects on that list.

The plaintiffs became so perturbed that Mr Heatley personally instructed Mr Robin Hall [FRICS], of Snow & Astill, of Leicester, to make a further survey and report, Mr Hall reported on August 21 1989. In that report he summarised his conclusions as follows:

> In conclusion we would say that had we seen the property prior to your purchase we would have strongly advised against proceeding with the purchase.
>
> . . . we are unable to let you have a valuation of the property as it is so far outside our normal location of work . . .
>
> We would however say that we consider that it would be difficult to substantiate any value at all other than site value as a prospective purchaser is unlikely under any circumstances to buy the property in its present condition.

On December 1 1989, the plaintiffs' solicitors wrote a letter before action to the defendants. In that letter, the plaintiffs' solicitors referred to certain advice which had been given by some structural engineers named Travers Morgan Two Ltd. No evidence has been called from that company and no more is known of their involvement

A save that some of their drawings were used as annexures to a notice from the local authority in March 1990. No inference can be drawn from the failure of the plaintiffs to call evidence from that company or from another firm of surveyors instructed on behalf of the plaintiffs.

The writ in this action was served on the defendants on December 8 1989.

On December 18 1989, the first plaintiff went to see the local authority and was referred to Mr Manning, the engineering services manager with responsibility for dangerous structure enforcement procedures. According to Mr Manning's notes, the first plaintiff invited the initiation of the dangerous structure procedure against B himself "to pressurise his solicitor to get the vendor to do something". I have heard no evidence of any complaints addressed to the vendor. It is possible that Mr Manning misunderstood the first plaintiff and that the first plaintiff in fact mentioned pressure on the solicitor to get the defendants to do something.

Mr Manning visited the property on December 19 1989. He noted that the first-floor timbers to the dormer extension had sagged excessively and the first-floor walls of the dormer extension lean outwards, that the gable of the north wall leans in, that there was a pronounced sag in the roof facing the road above the more northerly window and a less pronounced sag above the southerly window, and there was an outward bulge in the west wall of the main building between a ground-floor bay window and the bedroom window at the C northern end. He noted that the floor at the north-west end of the main building (that is the floor in the bedroom behind the bulge to which I have referred) was excessively springy and from the inside the wall in the south-west bedroom bulged 40 mm outwards. By taking off a skirting board he was able to ascertain that the front wall in the north-west bedroom had bulged outwards and was only just sitting on the brickwork. He was also able to see that there was an oblique crack in an internal wall between a box room and the north-west bedroom: that latter crack would probably not have been visible at the time of Mr Stanton's first inspection.

Having visited the property, I comment at this point that the photographs of the property do not give an adequate impression of the roof of the main building as seen from the ground. I agree with Mr D Manning's description of the roof.

In a letter of December 12 1989 to the plaintiffs, Mr Manning confirmed those findings. Mr Manning also gave evidence about those observations. At the plaintiffs' request, Mr Manning inspected again with a consulting engineer instructed by the council and by letter dated February 8 1990 told the plaintiffs that the consultant confirmed his views. Under threat of a dangerous structure notice, the plaintiffs had bracing inserted in the front wall to make it secure and they ceased to use the dormer extension, it being agreed that the dormer extension should be demolished, though it has not yet been demolished. On May 21 1990, the plaintiffs were given planning permission to demolish the dormer extension and the single-storey E extension and to rebuild.

In September 1990, the plaintiffs with their children moved into a flat in Whittlesea. In their evidence they stated that they saw no alternative to total demolition of the property and the purchase of another house. At present they cannot afford to demolish and rebuild the property (if that were practicable) or to buy another house. They are legally aided.

In the light of all the evidence now available, the decision of the plaintiffs to move from the property and to buy another house is entirely reasonable. Moreover, in the light of all that evidence, it is clear that the advice given by the defendants which I have quoted was wrong. It does not, of course, follow that it was necessarily given in breach of contract.

F I turn to consider the report given by the defendants dated August 22 1988. The report should, of course, be read as a whole and it is plain that the plaintiffs did indeed read all of it with care. Many, though not all, of the defects existing in the property were mentioned in the report. On behalf of the plaintiffs some of the statements in the body of the report are criticised and further criticisms are made that important matters were omitted. As to omissions, it is said that some matters were noticed by Mr Stanton and wrongly disregarded when it came to writing the report and others which ought to have been noticed were not noticed. But the central criticism, which I find is well founded, is that Mr Stanton failed to gather all the relevant matters together in his mind and draw the obvious conclusion, which was that the house was not a good buy at any price other than the site value. It may be that a skilled surveyor who read the report as a whole might G doubt its conclusions, which were directly contrary to the obvious conclusion, but there was no reason why the plaintiffs or any other ordinary house purchaser should do so.

Para 3.00 of the report was misleading in the statement "we discovered no serious structural defects or other major shortcomings that would lead us to advise against proceeding with the proposed acquisition as planned". That statement went to the heart of the agreed purpose of the report. As Mr Stanton acknowledged in cross-examination, he did see serious structural defects, and on the evidence it appears that the serious structural defects which he saw, even without others which he ought to have seen but did not, should have led the defendants to advise against proceeding with the H purchase. Turning to the conclusion on p16 of the report, it was not correct that "the structure of the building [was] in a reasonable condition for its age" and the suggestion that "the following budgets be set aside for remedial work", totalling £8,500, was very misleading indeed.

Without referring to every part of the body of the report, I shall mention some parts of it.

Mr Stanton examined the roof of the main building with the aid of binoculars from the outside. He reported: "Apart from a slight sag in both the front and rear slopes this roof remains true and even with no signs externally to indicate any deterioration, apart from the slight sagging of the supporting timber framework". Later, having commented that there was no access to the roof voids, he wrote: "We J strongly recommend that proper access hatches be formed into each roof space to facilitate maintenance inspections in the roof voids". By the reference to "maintenance inspections" the defendants indicated no need for access hatches to be made to the roof voids before purchase. Mr Stanton continued:

It is difficult to assess the condition of the timbers in the roof voids without actually visually inspecting them, however, the external roof slopes on the main roof seem to indicate some deflection of the common rafters particularly in the centre of the roof towards the eaves. This may indicate that some strengthening of the purlins is required. From within the first floor roofs [*sic*: rooms] it can clearly be seen that the ceilings undulate considerably. One can assume that the undulations are caused by the deflection of the timber joists. This could indicate damaged ceiling joists through say beetle K infestation or rot. On the other hand it may be that the timbers were rough hewn and were twisted as indeed branches often are.

To say that there was a slight sag in the front and rear slopes of the roof was to make a serious understatement. Mr Hall in his first report wrote: "there is evidence of considerable movement in the underlying structure, the roof lines being very uneven". Mr Manning referred to a "pronounced sag" at the north end of the front of the roof and a "less pronounced sag" at the other end. Mr Malcolm York [BSc CEng, MIStruct E, MICE], a structural engineer from John Burrow & Partners, who inspected in April and June 1991, wrote:

Externally the front, west, slope has significant sagging easily visible from ground level, particularly to the north end where deflection estimated to be over 50mm (2in). To the south end sag less but still noticeable. Rear slope L partly covered by dormer to north, again sags but less severe.

The sagging was indeed easily visible from the road on my view.

Mr Stanton put forward two possible explanations, beetle infestation or rough-hewn timbers. In cross-examination he admitted that there was a third possibility which he should have considered, namely that the roof had been overloaded when a new roof covering had been put on (as Mr Stanton knew) about 15 years earlier. Mr Stanton said that a roofer does not put a heavier load on a roof without strengthening the timbers and for that reason it did not cross his mind that the roof had been deflected by an over-heavy covering. He assumed that any obviously defective timbers would have been repaired 15 years before and that if the new roof covering was heavier, appropriate strengthening works would have been M done. It cannot be right for a surveyor to assume that building works have been done properly. It is his job to check.

Inspection of the roof from the outside showed a substantial deformity in the roof which should have led any reasonably careful surveyor to advise that the property should not be bought without an inspection of the roof void. A proper inspection of the roof void would have revealed, as it did to Mr York (but not to Mr Stanton when he looked in the roof void after purchase), that the roof of the main building needed to be totally replaced.

With regard to the walls of the property, the defendants reported:

7.00 *MAIN WALLS*

The property is constructed with mainly 9" solid brick load bearing perimeter walls. The walls to the kitchen/bathroom [in the single storey extension] however, appear to be in mainly 4.5" brickwork with some timber shiplap boarding over cladding to the north face.

Part of the south gable end wall to the main part of the building and the east elevation of the main building together with the south elevation of the rear wing are covered with some sort of creeper. We do not know what kind but it makes the inspection of the wall surfaces very difficult. The majority of the walls are generally in a sound upright condition and as far as we can say, not having inspected the foundations, they are structurally sound.

It emerged that the creeper had very little limiting effect on Mr Stanton's inspection. There was no creeper on the north and west faces of the main building. A contemporaneous photograph and inspection of the suckers from the creeper remaining on the wall showed that the creeper on the south wall had been limited to a small part at the top of the gable. The east wall of the main building had been substantially shielded from inspection, but that wall was not in fact substantially defective. The south wall of the dormer extension was so bad that despite the creeper the defects could be seen from the sagging window frame, from a door opening in the upper part of the wall, and also from the inside of the wall.

It was quite wrong to say that the majority of the walls were in an upright condition. They were seen by Mr Stanton not to be upright for the most part and, at the very least, he had no or inadequate ground for describing them as being in sound condition. In cross-examination Mr Stanton agreed that the following walls were not upright: the north, south and west walls of the main building and the north and south walls of the dormer extension. He also agreed that the east wall of the dormer extension is unsound. I do not accept Mr Stanton's assertion that the walls were acceptable in terms of the age of the building. However old the building may be, the walls are either upright or not, though in the case of an old building it may well be proper to say that the walls are not upright but in view of their age this is acceptable because they have been checked and found to be sound.

The north gable wall to the main building was correctly described by Mr Stanton as constructed in 9in brickwork from the ground to the first floor and in 4.5in brickwork from there upwards. The triangle at the top of the gable is severely distorted, as can be seen clearly from ground level. The brickwork is dished inwards, going in in the middle of the triangle and back out again at the top, though it does not come back out sufficiently to extend out from the vertical centre line of the wall. Mr Stanton misdescribed this section of the wall as leaning outwards, though it would be more accurate to describe it as leaning inwards, an obviously more dangerous state. With regard to this brickwork, Mr Stanton advised that "some amount of rebuilding will be required". He also recommended that this wall should be rendered. He added that "this work should be carried out after first repairing the defective and distorted brickwork at the top of the wall". The amount of rebuilding of brickwork required was not made clear. In fact, certainly the whole of the top of the gable needed to be taken down and rebuilt. In evidence, Mr Stanton tried to minimise the need for rebuilding by suggesting that the erection of a light-weight timber wall which he had advised on the inside of that wall to provide thermal insulation would also provide support for that wall. That suggestion was not supported by any expert evidence and it is plainly unfounded. When it was put to him in cross-examination that the report does not say that the wall is unstable and does not convey any sense of urgency and does not state the area of brickwork to be rebuilt, Mr Stanton replied: "I take the point that it was not as explicit as it could have been". He accepted that there was a serious structural defect in the north gable. He also accepted that the figure of £2,500 given at the end of the report would have to cover the cement render of the whole of the north gable apart from new brickwork at the top and would also have to include repair and rebuilding of brickwork both on the north elevation and elsewhere.

Of the brickwork to the south wall of the dormer extension, Mr Stanton wrote:

The south side wall to the rear two storey wing is also covered in creeper. We were able to see that the brickwork is distorted but it is very difficult to determine the condition of the brickwork beneath the creeper, one could reasonably expect to have areas of pointing and maybe isolated areas of defective brickwork. Although the wall is distorted it does not appear to be structurally unstable. The brickwork over the top of the dining room window opening shows sign of sagging. This does not appear to reflect in any damage to the brickwork above it and we would suggest it is not current or progressive.

Although it is understandable that the creeper prevented Mr Stanton from giving an opinion about the pointing or the appearance of the bricks, he was able to, and did, see the sagging over the window and the distortion in the brickwork. He said that he felt that he understood what was happening despite the creeper. He correctly advised that the roof had to be replaced. He also noted that the floor of the first-floor room was sagging very badly despite support from an upright timber in the centre of the ground-floor room.

Mr Stanton advised that the roof and the floor to the first floor of the dormer extension together with a part of the brickwork to the south wall should be removed and replaced. All the experts, including Mr Rodney Auckland [FRICS], a building surveyor called on behalf of the defendants, agreed that if that were done as Mr Stanton had advised, the south wall of the dormer extension would fall out, possibly bringing with it other parts of the building. In his budget costings at the end of the defendants' report, Mr Stanton had allowed £1,000 for replacing the roof and this floor in the dormer extension and repairing. He accepted that those costings made no allowance for the support of the building while the work was done, nor did the costings allow for the fees of an architect or engineer to design a new dormer roof which would not (as the present one does) spread out from the base of the dormer, pushing out the walls. Mr Auckland costed the work of taking off the roof and removing the floor and rebuilding at £5,321. That costing included doing this work on the single-storey extension, but even so, Mr Stanton was right to admit that on those figures he was several 100% out in his costings for the dormer extension and that his estimate of £1,000 for the roof and floor of the dormer extension was basically and fundamentally inaccurate.

The general view of the experts was that the only sensible thing to do with the dormer extension was to demolish it and rebuild. That view does not depend on inspection of roof voids or anything else which was not open to inspection by Mr Stanton. Moreover, in his manuscript notes made during his inspection, Mr Stanton wrote: "This rear wing is old and probably should be demolished". He said that this referred to the dormer extension, and from the context that appears to be the case. He said that when he sat down with the notes to write the report it occurred to him that the dormer extension might be renovated. It was not explained what were the reasons for moving from the view in favour of demolition of the dormer extension to partial taking down and repair. In the light of his note that the dormer extension "probably should be demolished" Mr Stanton admitted that para 3.00 of his report was totally misleading in saying "we discovered no serious structural defects". Counsel for the defendants did not remove the effect of this admission by pointing out that para 3.00 in fact said: "we discovered no serious structural defects . . . that would lead us to advise against proceeding with the proposed acquisition as planned".

The front wall was much considered in evidence before me. As one faces the front wall from the road, the front door is at the left-hand (south) end, and a wide bow window is on the left. Above are two bedroom windows. The whole wall is covered in Tyrolean render. Mr Stanton in his report wrote of this wall:

7.01 The front (west elevation) has a cement render and pebble dash finish and has recently been redecorated. The bottom of the cement render terminates in a bell mouth with three courses of bricks below the bell mouth forming a plinth which has been painted with bitumen. The rendered surfaces generally appear to be satisfactory, however, there are hollow areas of render at either end of the elevation adjacent to the front door and bow window. This we would suggest is due to moisture penetration through the ends and could also occur at higher level.

In his manuscript notes, he wrote:

Area of hollow render south end adjacent to front door
Also north end bow window
may also be higher up either end
could *not* see any sign of cracking
The wall is generally plumb apart from odd bulge.

The words "could also occur at higher level" in the report and the words "may also be higher up either end" in the notes strongly suggest that Mr Stanton did not test the wall above his reach when standing on the pavement, either by standing on a ladder or by leaning out of a bedroom window. His oral evidence did not alter that impression. There was no explanation for failing to confirm whether on this fairly small area of wall there was further hollow render as he suspected might be the case.

There were two bulges on this wall, one over the bay window on the left and the other below the right-hand bedroom window. The bulge on the right is less prominent than the bulge on the left. There was some debate as to whether the ordinary competent surveyor could be

A expected to have seen either or both bulges in August 1988. It would be normal to look along the face of the wall to check that it was vertical and even. It was difficult to do this from the north because at the north end there was a downpipe from the gutter obscuring vision. But from the south it was perfectly possible to get a good view of the frontage and to check the straightness of the surface aided by the straight edge of the downpipe. It was also possible to make other checks from the bedroom windows. But quite apart from what could be expected of an ordinary competent surveyor, the fact is that Mr Stanton did see a bulge and noted it in his notes. In oral evidence he said that as far as he knew he did not see the bulge above the bay window but he did see a bulge at the opposite end of the wall. He said: B "The bulge I referred to was in the context of render which I did not see to be of any consequence". That appears to be his reason for not referring to it in the report.

Bulges in the wall below the bedroom windows might have been due to lack of adhesion in the render, which would not be very important, or due to bulges in the brickwork, which would be another matter. In view of the tendency in the locality for brickwork below bedroom windows to be defective, and in view of other observable defects in this property, it was particularly important to establish whether there were any bulges below the windows and as far as possible to establish their cause.

I am by no means satisfied that the bulge which Mr Stanton admits seeing was on the right rather than on the left, there being no help to C be derived from his notes on the precise position. Since he regarded the bulge as of no significance, it is the less likely that he will have remembered its position independently of any record. But since Mr Stanton did see a bulge he should have looked to see whether there were any other bulges and tried to establish their cause or causes. If he had tapped the bulge he saw, he would have discovered that it was not hollow for the most part. If he had then looked inside, he would have discovered, if he was looking for it (though perhaps not if he was not specifically looking for it), under the right hand window a corresponding bulge outwards in the inside of the wall, 40mm in depth (which Mr Manning saw), indicating a bulge in the totality of the brick skin. If he had found the bulge on the left side and then found that the render was not hollow and then examined the inside D wall, he would have discovered, as Mr Hall did, that there was no corresponding bulge on the inside and that there was ground for suspecting that the inside skin of the wall had separated from the outside wall. As the vendors were still in occupation, Mr Stanton might well not have been given permission by them to take off a skirting board to follow up this suspicion, but he did not ask for permission. If he had been refused permission to make this further examination, he should have mentioned his suspicion in his report and recommended further investigation. Further investigation would have revealed, as it did to Mr Hall, very serious defects in the wall. Mr Stanton's cross-examination again suggests a lack of suspicion due to an assumption that good practice will have been E followed in construction. When it was put to him that the bulge in the wall could be due to the inner and outer leaves of the 9-in wall separating, he replied that that could be so, but it is normal practice to build a 9-in wall so that the two leaves are interlocking.

In the body of the report Mr Stanton wrote of the south gable of the main building:

> 7.06 The south gable end wall to the main property is constructed in 225mm (9in) brickwork. The ground to first floor level of brickwork is much older than that from first floor to roof level. The lower level of brickwork shows signs of distortion and past movement, however, none of the movement appears to be current or progressive and the wall would appear to be structurally sound. Creeper is growing along the east corner and along the top of the verges making it impossible to visually inspect these areas of brickwork. F Some isolated areas of repointing will be required.

Objection was taken on behalf of the plaintiffs to the description of this wall as "structurally sound" in view of the other comments, but otherwise this paragraph, taken on its own, is not criticised. Objection is made that the defendants failed to take the state of this wall into account with all the other defects and assess the property as a whole to consider in particular what was happening to the main building.

Mr Hall in his second report dated May 16 1990, based upon his original inspection and also on a further inspection in April 1990, expressed the following conclusions:

> It is my opinion based on my discussions with Mr Heatley concerning the then condition of the property that had I inspected it on or about the 18 August 1988 my conclusions would have been very much the same as set out in that report. G
>
> In my opinion Mr Stanton should have been aware of the severe deformation in the north, west and southern elevations of the main house and in particular the southern elevation of the rear wing and if this had been the case the only advice that he could have given would have been to advise strongly against proceeding with the purchase.
>
> The recommendations with regard to "maintenance inspection" of the roof structure in my opinion were unsound in that they did not give a sense of urgency about the matter or indeed require interior inspection prior to purchase nor was the description of what must then have been a structurally deformed main roof correct.
>
> In my opinion the property as it stands at the present time and almost certainly as it stood on the 22 August 1988 is in considerable danger of collapse until adequately shored up and that collapse if it once occurs could well be H progressive particularly with regard to the northern gable of the main house and roof structure and certainly with regard to the two storey section of the rear wing.
>
> I consider as previously set out in this report that other items were not adequately covered but the foregoing items are of major importance and go to the root of the report. I cannot see that under any circumstances I would have recommended a purchase of this property to a client.

Mr Hall did have the advantage over Mr Stanton that he was inspecting a property which belonged to his client, and he was able to make more detailed investigation, and he was investigating a property known by him to be suspect. All of that was extensively considered during the trial and I am satisfied that Mr Hall has properly made allowances for those considerations. The evidence J which I have heard supports Mr Hall's conclusions and bears out the facts on which they are based.

Mr York expressed himself in even stronger terms, save that he concluded that the defendants should *either* have recommended further investigation by a structural engineer *or* have advised against purchase, depending on the purchase price. In the light of Mr Hall's evidence, I find it difficult to accept that Mr Stanton ought to have recommended further investigation by a structural engineer. But Mr Stanton certainly ought to have advised at the very least that the purchase should not go ahead unless further investigations were made of the structure either by himself as a building surveyor or by a structural engineer.

I have to add that the case brought by the plaintiffs was K substantially strengthened by admissions made by Mr Stanton in cross-examination and also by some of the evidence quite properly given by Mr Auckland, who was called as an expert on behalf of the defendants.

On the whole of the evidence, I find, regretfully but without any doubt, that the defendants by the delivery of the report dated August 22 1988 were negligently in breach of contract and that the plaintiffs in reliance on a negligent report bought a property which they should have been warned against and suffered damage. The defendants were also in breach of contract by their conduct in and about the second inspection on September 30 1988 and delivery of the letter of October 5 1988. At that point the defendants had access to the roof voids and L the principal restriction on their previous inspection was removed. Having the benefit of being able to reconsider all their previous advice, and having the further advantage of being able to reconsider that advice in the light of their examination of the roof voids, the defendants ought then to have advised the plaintiffs that their previous advice was wrong, that the property was worth nothing like what the plaintiffs had paid for it, and that the plaintiffs would be in some danger if they continued living in the property, particularly if they continued to use the dormer extension. By not giving that advice, and by advising repairs in the letter of October 5 1988, the defendants were in breach of their continuing duty under the original contract and were in breach of the fresh contract made in September 1988. By reason of those breaches, the plaintiffs were encouraged to M stay in the property and to spend more money on it and their damages were increased.

I turn to assess that damage.

The principal head of damages should be assessed as what is generally known as the "difference in value". How that difference should be computed is not agreed.

In *Watts* v *Morrow* (CA) unreported July 30 1991*, it was taken as common ground that the diminution in value rule is more accurately expressed as the difference between the price paid and the value in its true description at least where no point is taken that the plaintiff

---

*Editor's note: Reported at [1991] 2 EGLR 152.

chose to pay above the market value. In *Philips* v *Ward* [1956] 1 WLR 471, Denning LJ at p 473 said:

> The proper measure of damages is therefore the difference between the value in its assumed good condition and the value in the bad condition which should have been reported to the client.

In the present case, the property should not be taken in any "assumed good condition" because no good condition was assumed: the defendants did report the property as having defects valued by the defendants, as a guide, at £8,500. The defendants have in this action sought to rely on para 4(ix) of that part of their conditions headed "Purpose and scope of report" to suggest that that guide figure should be ignored for all practical purposes if not substantiated by competitive estimates or quotations. But I do not think it right to construe that paragraph in that way either in favour of the defendants on liability or against them on assessment of damages. The figures amounting to £8,500 were put forward to be relied on as a guide and were relied on as a guide, though obviously not as a guarantee of what would be quoted. There was a contractual requirement that the figures should have a reasonable degree of accuracy in the right area. As a matter of caution, the plaintiffs added 10% to the guide figures before acting on them for budget purposes. It should not be forgotten that variation from guide figures might in the event be expected to be up or down.

In the circumstances of this case, it would be more appropriate to adopt the measure approved by Morris LJ in *Philips* v *Ward* at p 475, namely the difference between the value of the property as it was described in the defendants' report and its value as it should have been described.

The value of the property as it was described in the defendants' report was established by the plaintiffs' bargaining with the vendors in reliance on the report and obtaining the reduced purchase price, namely £82,500, although Mr Smith valued the property as described at £85,000. The value of the property as described in the report is therefore in this case, but not in all cases, the same as the price paid. The main head of damage is therefore the difference between £82,500 and the true value of the property.

I accept the opinion of both Mr Hall and Mr David Smith [FRICS], of Colleys, a local valuer, that the true value of the property is its site value. In their Further and Better Particulars of Statement of Claim served on March 1 1991, the plaintiffs put the site value at £35,000 less the costs of clearance £5,000, making a net site value of £30,000. In his evidence, Mr Smith said that the site value in 1988 was £32,500 less £2,500 for clearing the site, making again a net site value of £30,000. Mr Auckland did not quarrel with the pleaded site value of £35,000 but put the cost of demolition at £3,000, though he qualified this by saying that, in view of the desirability of the site in relation to the town centre, it is more than likely that a purchaser would be likely to pay the £35,000 in full without deduction for costs of clearance. In resolving that conflict, I think it right that I should take the plaintiffs' pleaded site value £35,000. I think that Mr Auckland was optimistic in thinking that a purchaser would not make any deduction for site clearance, and for that element I deduct the smallest figure given, £2,500, being the figure given by Mr Hall. The net site value is therefore £32,500.

The diminution in value is therefore £82,500 less £32,500, amounting to £50,000.

The defendants have denied that the site should be valued on the basis that the building on it is worthless, but if one assesses the diminution in value by reference to the cost of repairing the existing house, one would reach a much larger figure for damages.

The plaintiffs also claim damages for distress and inconvenience. In *Watts* v *Morrow* at p 35 of the transcript it was held that "in the case of the ordinary surveyor's contract, damages are only recoverable for distress caused by physical consequences of the breach of contract" and any award should be moderate.

The plaintiffs complain that the physical consequences of the defendants' breach of contract were that the plaintiffs lived with their young children from September 1988 until September 1990 in premises which became in September 1989 wholly unsuitable for family living because of restrictions imposed as a result of warnings of danger. The reason for not moving out before September 1990 was because of the difficulty of facing up to paying both mortgage repayments and rent. In September 1989 the plaintiffs were warned not to use the dormer extension because of the danger of collapse. The only hot water, bath and lavatory were in the single-storey extension, which could only be approached under cover through the dormer extension. The plaintiffs were thus in the position of having in effect only an outside bath and lavatory. The ordinary inconveniences arising out of this physical circumstance were increased when the children were ill and not fit to be taken outside to the bath or lavatory. The plaintiffs bought bathroom fittings to convert a bedroom into a bathroom but were told not to do this in view of apprehended danger from unstable structure above that bedroom. Instead, the plaintiffs bought a portaloo, which was placed in the inner hall. During winter months, the children were strip washed in the lounge using hot water from kettles. The lack of proper toilet facilities was a source of acute embarrassment when there were visitors. The health of the second plaintiff suffered and she consulted her doctor concerning nausea and head pains which largely subsided when the family moved into a flat. The relationships between members of the family became strained. With the winter of 1990 approaching, the second plaintiff insisted that they must move out before winter, and the first plaintiff's brother insisted that they should now accept financial help to pay the rent on top of the mortgage. In the flat the three children share one bedroom and there is no garden.

This distress and inconvenience began a year after the original breaches of contract. It was foreseeable that, no warnings of danger having been given, the plaintiffs would buy the property, settle down in it, and then be subjected to great strain and inconvenience when they found themselves financially trapped in a property in which it was difficult to use essential amenities because of danger of which they were warned by others.

On a very moderate assessment of the damage to the plaintiffs, I assess the damages for the first plaintiff at £1,500 and for the second plaintiff at £3,000.

Other damages also are claimed, being legal and moving expenses, cost of repairs, cost of investigating the premises and cost of alternative accommodation.

There has been some discussion of the recoverability of some of these heads of damage. In *Philips* v *Ward* at p 478, Romer LJ said:

> It may well be that if, on learning of the real condition of the house, he had decided to leave and resell, he would have been entitled to recover from the defendant, in addition to the £4,000, his costs and expenses of moving out and of the resale. As, however, he elected to stay, after all the facts had become known to him, this point does not arise.

In *Watts* v *Morrow* at p 22, Ralph Gibson LJ said:

> In the absence of any point on mitigation, the buyer will recover the diminution in value together with costs and expenses thrown away in moving in and out and of resale: see Romer LJ in *Philips*. I will not here try to state the nature or extent of any additional recoverable items of damage.

At p 29 Ralph Gibson LJ also said:

> It is, I think, clear law that where a claimant is caused to enter into a transaction in consequence of negligent advice, as in the case of a surveyor employed under the ordinary contract, the claimant may be entitled to all the losses incurred as a result of entering into the transaction where he would not have entered into the transaction if properly advised and the losses are caused by entry into the transaction and by extrication from it.

At p 24, Ralph Gibson LJ added:

> We do not have to decide whether, if the plaintiff has to rent other accommodation during the carrying out of repairs, such costs will be recoverable in the absence of any contractual warranty as to the existence of defects requiring repairs, and I would reserve my decision upon it.

At p 41, Bingham LJ said:

> I would be willing to accept, as Romer LJ did in *Philips* v *Ward* that if, on learning of the true state of the house, the plaintiffs had at once moved out and sold, they might well have been able to recover the costs thrown away in addition to the diminution in value.

Bingham LJ's use of the words "at once" in that citation must be read in the light of the considerable financial resources of the plaintiffs in *Watts* v *Morrow*, which are to be contrasted with the financial circumstances of the plaintiffs in the present case. The first plaintiff is a laboratory technician training to be a factory inspector. The second plaintiff works as a nurse when her family commitments allow her to do so.

It has not been argued that the plaintiffs are not entitled to compensation for the damage suffered during the extensive period during which the plaintiffs have been the reluctant owners of this house since September 1989, first in terms of damages for inconvenience and distress while they were in the property and then for rent of alternative accommodation. Both forms of damage are the natural and probable result of the plaintiffs' buying the property

A in reliance on the defendants' report. For this purpose, I discount the period up to September 1989 when the plaintiffs were suffering some lesser inconvenience while doing repairs to the property which they expected to undertake in the light of the defendants' report. The claim for rent for alternative accommodation includes a claim for six months after judgment. No evidence was given that that was a reasonable time to take in finding and buying alternative accommodation. I therefore approach the matter on the basis of my experience. Six months seems to me to be excessive and I substitute a period of two months.

In the circumstances of this case, the plaintiffs are entitled to damages for moving into and extricating themselves from 72 High Causeway and the contrary was not argued. It must be borne in mind B that the plaintiffs were intending to move from their former house in any event, so that if they had not bought 72 High Causeway, they would, in any event, have incurred legal costs of sale of their old house and of the purchase of the new house and would also have incurred moving costs to the new house. The legal fees on purchase of 72 High Causeway were proved in the sum of £1,070.75: it was suggested that that sum included the legal fees on sale of the old house, but the first plaintiff said that was not so and I accept that. No figure was proved as the legal cost of sale of 72 High Causeway so far as I can find, but if in fact it can be shown that a figure does appear in the evidence before me, that sum would be recoverable. The plaintiffs incurred no removal costs in moving into 72 High C Causeway. A figure has been given for the cost of removal into the plaintiffs' next home, but they would have been involved in one removal in any event, so I do not find that expense recoverable.

Building works were undertaken at 72 High Causeway partly in reliance on the report of August 22 1988, partly in reliance on the letter of October 5 1988, and partly to make the front wall safe in response to the requirements of the local authority. In addition, the plaintiffs bought some materials and tools for work to be done by the first plaintiff and his friends. These costs were not in general challenged, but it was suggested on behalf of the defendants and accepted on behalf of the plaintiffs that there is some residual value in some of the tools, and also in the bathroom which are still on the premises unused: in this regard I was invited to make a broad D assessment without any evidence, which I shall do.

A claim was made for damages for the cost of investigating the premises in the sum of £557.35 for the first report of Snow & Astill (Mr Hall) in August 1989. For the defendants it was submitted that the cost of the report should be treated as part of the costs of the action. It appears that Mr Hall was instructed direct by the first plaintiff and not by his solicitors, but there was no enquiry as to whether that instruction was in fact given on the advice of the solicitor and for the purpose of obtaining legal advice for the purpose of contemplated legal proceedings. The matter can best be tested by reference to the plaintiffs' list of documents served on April 2 1991. In that list, no report of Snow & Astill appears in Schedule 1 Part 1, E and it must therefore be included among the documents for which privilege is claimed by a formula set out in Schedule 1 Part 2. I therefore hold that the cost of that report should be included in the costs of the action and not be the subject of a claim for damages.

The special damages are therefore as follows:

(a) Costs of moving into 72 High Causeway	
Legal fees on purchase	£1,070.75
Furniture moving — cost	nil
(b) Building works undertaken to no 72	
Trevor Storey, March Roofing, Goakes, Firebrace	1,676.64
Hire and purchase by plaintiffs of materials and equipment (less an allowance for usable tools)	1,000.00
Loss on bathroom fixtures bought by plaintiffs, say	250.00
F (c) Cost of alternative accommodation (flat in Whittlesea)	
Removal cost to flat	86.25
Rent September 1990 to October 1991	3,575.00
Continuing for a further, say, 2 months	550.00
(d) Miscellaneous agreed expenses	
Travel, stationery, lost pay	428.00
Total of (a), (b), (c) and (d)	£8,636.74

Subject to any correction on points of detail, I therefore propose to give judgment as follows:

For both plaintiffs jointly:	
£50,000 + £8,636.74	£58,636.74
For the first plaintiff	£1,500.00
For the second plaintiff	£3,000.00

To those figures should be added interest. For the defendants it is G argued that the appropriate rate should be the special account rate: the rates are now set out on p 40 of the *Supreme Court Practice 1991*. That does not seem to me to be right. The largest part of the damages relates to the plaintiffs' investment in a commercial market, the housing market, and interest should be more nearly equated to a commercial rate rather than to any supposed comparison to damages in personal injury actions. I award interest at the judgment rate, which for most of the period was 15% from September 16 1988 to the date of judgment on £50,000. In considering interest on the figure of £8,636.64 it is necessary, first, to deduct £550 for liability for rent which has not yet been incurred, leaving £8,086.64. That balance was paid out in varying amounts between late 1988 and October 1991. It H would do broad justice if interest were calculated at 15% on the sum of £8,086 from October 28 1990 to judgment.

Interest on the damages for distress and inconvenience should be paid at the rate of 2% from October 28 1990.

Judgment for the plaintiffs.

Chancery Division

December 10 1991

(Before Sir Donald NICHOLLS V-C) J

GRAN GELATO LTD v RICHCLIFF (GROUP) LTD AND OTHERS

Misrepresentation — Negligence — Representation relating to grant of lease — Whether lessee entitled to damages under the Misrepresentation Act 1967 — Whether lessor and its solicitors owed a duty of care to the lessee when answering preliminary enquiries

Pursuant to a contract dated May 30 1984 the first defendant, Richcliff (Group) Ltd, granted the plaintiff an underlease dated June 26 1984 of the basement and ground K **floor of nos 7 and 9 Montpelier Street, London SW7 — The first defendant held the premises under two headleases both of which contained redevelopment break clauses, exercisable by 12 months' notice expiring on or after June 1989 — Prior to the contract, the first defendant's solicitors, the second defendants, Gershon Young Finer & Green, supplied answers to "Enquiries before Lease" on May 10 1984 — In answer to an enquiry as to whether there were any covenants, restrictions or rights affecting the superior leasehold titles which would impose a greater obligation on the tenants than contained in the draft lease or in any way inhibit the** L **enjoyment of the property by the tenant, the second defendants stated "not to the Lessor's knowledge" — Pursuant to break clause options in the headleases, the superior landlord gave 12 months' notice in November 1988 — The plaintiff asserted that the reply to the enquiry constituted a representation that there was no provision for early determination, the representation was intended to induce the plaintiff to take the underlease and in reliance on that representation it did so — The plaintiff asserted that the representation was false and made negligently and claimed damages under the Misrepresentation Act 1967 against the first defendant — The plaintiff also asserted negligence** M **against both defendants contending, *inter alia*, that the second defendants owed a duty to a buyer when answering preliminary enquiries**

***Held*: Judgment for the plaintiff against the first defendant — Claim against second defendants dismissed — Given the existence of the break clauses in the headleases, the answer to the inquiry was a misrepresentation — On the evidence the plaintiff was induced to make the contract by that representation and had a good cause of action for damages against the first defendant under section 2(1) of the Misrepresentation Act 1967**

The second defendants did not owe a duty of care to the plaintiffs — Although there was no problem about foreseeability or a close and direct relationship as explained by the House of Lords in *Caparo Industries Ltd* v *Dickman* and that the second defendants must have foreseen that the plaintiff might suffer financial loss if the answers to the enquiries were incorrect, in normal conveyancing transactions solicitors who are acting for a seller do not in general owe to the would-be buyer a duty of care when answering enquiries before contract or the like — In the context of a contract for the sale of an interest in land, the buyer seeks information from the seller and answers are given by the solicitors on behalf of the seller; the law provides the buyer with a remedy against the seller — The seller himself owes a duty of care to the buyer — There was no good reason to impose a duty of care when the agent already owes to his principal a duty which covers the same ground and the principal is responsible to the third party for his agent's shortcomings

In relation to a defence of contributory negligence, section 1 of the Law Reform (Contributory Negligence) Act 1945 applies to a claim by a plaintiff for damage under the Misrepresentation Act 1967 — On the evidence it would not be just or equitable to make any reduction in the plaintiff's damages — Carelessness in not making other enquiries provides no answer to a claim when the plaintiff has done that which the representor intended he should do — The measure of damages for the misrepresentation and for the breach of the duty of care were in the present case the same — By reason of the period of occupation of the plaintiff it was entitled to recover by way of damages only one-half of the premium of £30,000 paid for the underlease, one-half of the legal costs of acquisition and other abortive legal costs in all totalling £25,304

The following cases are referred to in this report.

Al-Kandari v *J R Brown & Co* [1988] QB 665; [1988] 2 WLR 671; [1988] 1 All ER 833, CA
Allied Finance & Investment Ltd v *Haddow* [1983] 1 NZLR 22
Caparo Industries plc v *Dickman* [1990] 2 AC 605; [1990] 2 WLR 358; [1990] 1 All ER 568, HL
Cemp Properties (UK) Ltd v *Dentsply Research & Development Corporation* [1989] 2 EGLR 205; [1989] 37 EG 133
Esso Petroleum Co Ltd v *Mardon* [1976] QB 801; [1976] 2 WLR 583; [1976] 2 All ER 5; [1976] 2 Lloyd's Rep 305, CA
Forsikringsaktieselskapet Vesta v *Butcher* [1989] 1 AC 852, CA & HL
Hedley Byrne & Co Ltd v *Heller & Partners Ltd* [1964] AC 465; [1963] 3 WLR 101; [1963] 2 All ER 575; [1963] 1 Lloyd's Rep 485, HL
Nocton v *Lord Ashburton* [1914] AC 932
Redgrave v *Hurd* (1881) 20 Ch D 1; [1881-5] All ER Rep 77; 57 LJ Ch 113; 45 LT 185; 30 WR 251, CA
Resolute Maritime Inc v *Nippon Kaiji Kyokai* [1983] 1 WLR 857; [1983] 2 All ER 1; [1983] 1 Lloyd's Rep 431
Reynell v *Sprye* (1852) 1 De GM&G 660
Rowe v *Turner Hopkins & Partners* [1980] 2 NZLR 550
Smith v *Eric S Bush (a firm)* [1990] 1 AC 831; [1989] 2 WLR 790; [1989] 2 All ER 514; [1989] 1 EGLR 169; [1989] 17 EG 68 & 18 EG 99, HL

This was a claim for damages alleging breach of a duty of care against the first defendant, Richcliff (Group) Ltd, and against the second defendants, Gershon Young Finer & Green, the first defendant's solicitors, and a claim for damages under the Misrepresentation Act 1967 against the first defendant alone in relation to enquiries before contract concerning an underlease dated June 26 1984 of premises at 7 and 9 Montpelier Street, London SW7. The first defendant held two headleases of the premises from the third defendant, Eagle Star plc.

Christopher Pymont (instructed by Mackenzie Mills) appeared for the plaintiff; Ralph Wynne-Griffiths (instructed by Freeman Pollard) represented the first defendant; Dirik Jackson (instructed by Ince & Co) represented the second defendants; the third defendant did not appear and was not represented.

Giving judgment, **Sir Donald Nicholls V-C** said: This case arises out of the circumstances in which the plaintiff, Gran Gelato Ltd (which I shall refer to as "Gran Gelato"), acquired from the first defendants, Richcliff (Group) Ltd ("Richcliff"), an underlease dated June 26 1984 of the basement and ground floor of 7 and 9 Montpelier Street, London SW7. The underlease was expressed to be for a term of almost 10 years, that is, until March 24 1994. The underlease was carved out of two headleases. The headlease relating to no 7 was due to expire on June 24 1995 and the headlease relating to no 9 was for a term expiring on September 29 1997. Unknown to Gran Gelato and its solicitors, both headleases contained redevelopment break clauses, exercisable by 12 months' notice expiring on or after June 1989. Thus, if the headlessor, Eagle Star Insurance Co Ltd, exercised the break clauses at the first opportunity, the term granted to Gran Gelato by the underlease would be reduced by about one-half to five years. In the event, Eagle Star did exercise the break clause options, by giving 12 months' notice in November 1988.

In the course of the negotiations and steps leading up to the grant of the underlease, Gran Gelato's solicitors, Mackenzie Mills, sent "Enquiries before Lease" in the usual way to Richcliff's then solicitors, Gershon Young Finer & Green. They are the second defendants in the action. Those enquiries, together with the answers supplied by Gershon Young on May 10 1984, included the following:

Q. 1(C) If the landlord does not own the freehold, please supply a copy of the Superior Lease or leases and deduce title thereto and confirm that an unqualified receipt for the latest rent due under the superior lease will be produced on completion?
A. 1(C) The Superior Title is not to be deduced.
. . .
Q. 3(A) Are there any covenants, restrictions or rights affecting the freehold or superior leasehold titles which would impose a greater obligation on the Tenant than contained in the present draft lease or in any way inhibit the enjoyment of the property by the Tenant in accordance with the terms of the present draft lease?
A. 3(A) Not to the Lessor's knowledge.

In this action Gran Gelato claims that the answer to enquiry 3(A) constituted a representation that there was no provision for early termination in the two headleases; that the representation was intended to induce Gran Gelato to take the underlease; and that in reliance on that representation Gran Gelato did so. Gran Gelato asserts that the representation was false and made negligently. Gran Gelato claims damages in negligence against both defendants and also against Richcliff damages under the Misrepresentation Act 1967. Thus, this action raises, among other matters, the question whether a solicitor acting for a seller owes a duty of care to a buyer when answering preliminary enquiries.

The history

Shortly stated, the background and history of the transactions were this. In 1983 Mr David Landau conceived the idea of selling top-quality Italian ice cream in London. The idea was that the ice cream would be made by an Italian chef with particular expertise in this field. The *laboratorio* in which the manufacture took place would be visible to the public, who could see that the ingredients used, such as fresh milk and fresh eggs, were all far removed from any image of back-street manufacture of ice cream sold in mobile vans. The product would be displayed and sold in the shop in the familiar style of an Italian gelateria. The fittings would be to a high, indeed spectacular, standard of design. It was hoped that the brand, for which the name Gran Gelato was adopted, would become well known and that the ice cream would be sold at other outlets in London.

Mr Landau, with three Italian colleagues, Mr Max Rabino, since deceased, Mrs Ilaria Borletti and Mr Marco Rocca, decided to proceed with this as a joint venture. They were directly or indirectly the shareholders of Gran Gelato, which was a shelf company acquired for the purpose in 1984. They were to provide the necessary finance, in varying proportions. Mr Landau, who lived in England, looked at premises in Brompton Road and in Covent Garden. In February 1984 he heard that the ground floor and basement of 7 and 9 Montpelier Street were available. This property was then being used as a delicatessen shop. The situation, in Knightsbridge, was ideal, as was the layout, with two adjacent shop fronts. Mr Landau initially met Mr Trapani, and subsequently Mr Tesai, from whom he understood that a 10-year lease of the two shops was available. Mr Tesai seems to have been the moving spirit behind Richcliff. On April 9 1984 Mr Landau and Mr Tesai orally agreed terms in principle: a rental of £26,000 pa and premium of £30,000. Mr Tesai told Mr Landau that his company's lease of no 7 was until June 1995 and its

A lease of no 9 was until September 1997. Nothing was said about break clauses. Mr Tesai insisted on payment of a premium of £30,000, ostensibly for fixtures and fittings and goodwill, although none of these had any value to Gran Gelato.

Mr Landau informed Gran Gelato's solicitor, Mr David Mills, who was also the secretary of Gran Gelato, of the gist of the arrangements he had negotiated. The usual conveyancing steps proceeded, in the course of which the enquiries before contract already mentioned were asked and answered. Gran Gelato was very anxious to proceed quickly, so as to get the necessary shop-fitting work done in time to open before the summer season was too far advanced. Contracts were exchanged on May 30, and Gran Gelato went into possession at once. The contract provided that Richcliff
B was not required to deduce any title to the grant of the underlease. Gran Gelato agreed to pay Richcliff's solicitors' costs. Completion took place on June 26 by which date Eagle Star had given its consent in principle. It was not until much later, on November 29 1985, that Eagle Star's formal licence to sublet and carry out alterations was granted. Mr Landau entered into both the underlease and the licence as a guarantor of Gran Gelato's obligations. The underlease was excluded from the protection afforded to business tenancies by Part II of the Landlord and Tenant Act 1954 pursuant to an order made by West London County Court.

The shop-fitting works were carried out to an arresting, award-winning design by Miss Dinah Casson. The cost was of the order of
C £100,000. The ice cream manufacturing equipment, costing about another £100,000, was imported from Italy. The business opened in early August 1984. Unfortunately, the summer of 1984 was cold and wet. So was summer 1985. The business did not prosper. In September 1985 Gran Gelato decided to cut back. The services of the Italian chef were dispensed with. The company decided to make the ice cream elsewhere in cheaper property and to dispose of no 9. Difficulties arose in obtaining the necessary consents for the assignment of part only of the property comprised in the underlease and for the relaxation of the existing user restriction. The underlease confined the use of nos 7 and 9 to the retail sale of high-quality ice cream and the ancillary making of such ice cream. Mr Tesai insisted on being paid £4,000 for this licence. It was not until February 27
D 1987 that a formal deed was entered into by Richcliff. By then a proposed sale of no 9 had fallen through.

1986 was another poor summer and the business made heavy financial losses. In March 1987 or thereabouts the four principals behind Gran Gelato decided that the underlease should be sold. Hillier Parker, estate agents, were instructed accordingly. In September Gran Gelato quit the property and ceased trading with accumulated losses of over £600,000. On August 11 Hillier Parker wrote to Mr Landau and told him that a problem had arisen with a potential purchaser. His solicitors had discovered that the headleases contained a rebuilding break clause. Thus, although the underlease did not contain such a clause, if the headlessor operated this clause,
E the underlease would necessarily end. That sale then fell through and further marketing efforts were unsuccessful.

Gran Gelato ceased paying rent. As already mentioned, break clause notices were served by Eagle Star in November 1988. Eagle Star recovered possession in or about October 1989. In this action Richcliff counterclaims for unpaid rent from June 1987 up to Michaelmas 1989, totalling £50,420, plus contractual interest.

Misrepresentation

Richcliff denies that the answer to enquiry 3(A) contained any misrepresentation. I cannot accept this. The enquiry asked, among other things, whether there were any "rights affecting the . . . superior leasehold titles which would . . . in any way inhibit the
F enjoyment of the property by the Tenant in accordance with the terms of the present draft lease". Whatever be the precise limits to the scope of this question, I am in no doubt that a break clause in the headlease is within those limits. If the headlessor exercised its rights under such a clause, the tenant would have to quit the property before the expiration of the term expressed to be granted by the underlease. The tenant would cease to be able to enjoy the property in accordance with the terms of its lease.

Mr Wynne-Griffiths, counsel for the first defendant, contended that the existence of the break clause did not inhibit the enjoyment of the property in accordance with the terms of the underlease. The tenant's enjoyment of the property would be inhibited only if the headlessor exercised its rights. Hence, he argued, until the break option was exercised the legal position was that there was no right G
affecting the headleases which inhibited Gran Gelato's enjoyment of nos 7 and 9 in accordance with the terms of its underlease. In my view, this is not a tenable interpretation of enquiry 3(A). The existence of a right in the headlease, whether a user restriction or whatever, will affect what the undertenant actually does in the property only if and to the extent that the headlessor chooses to exercise its right. But the enquiry cannot sensibly be regarded as directly only at instances where the headlessor has already taken steps to enforce, or is presently in a position to enforce, its rights. The potential undertenant is concerned with what the headlessor is or may become entitled to do during the currency of the proposed underlease. The word "rights" in question 3(A) is apt to embrace such an entitlement. H

In my view, therefore, given the existence of the break clauses in the headleases, the answer to enquiry 3(A) was a misrepresentation. Let me hasten to add that nobody has suggested that there is any question of fraud. A mistake was made in Gershon Young's offices. It is to Mr Young's credit that he has never sought to suggest otherwise.

Reliance

Clearly, the answers given by Gershon Young on behalf of Richcliff were intended to be relied on and acted on by Gran Gelato and its solicitors. That is the whole purpose for which the enquiries were made.

The next issue is whether Gran Gelato did in fact rely on the answer
to enquiry 3(A). In particular, would Gran Gelato have proceeded J
with the transaction even if Gershon Young's answer had disclosed the existence of the break clauses in the headleases? The evidence of Mr Landau and Mr Rocca was that Gran Gelato would not have gone ahead. Mrs Borletti did not give evidence.

This is a difficult question. It is not a question which arose for consideration at the time. Had it done so, there were several features which might have prompted Gran Gelato to proceed even though there was a risk that the underlease would be determined after five years; Mr Landau was very enthusiastic about the project; he was convinced that the property was ideal; and he hoped that the capital outlay would be recovered over a period of five years. Further, by May 10 preliminary arrangements for the new business had already
got under way and Gran Gelato had entered into certain K
commitments: the chef had been engaged; Miss Casson had prepared her splendid design; the availability of the ice cream equipment in Italy had been checked; a shop-fitting contractor had been approached; some professional fees, although only in modest sums, had been incurred; it was too late to look elsewhere if trading was to begin in time to catch any worthwhile part of the 1984 summer season; and there was a serious risk that, if the project did not proceed in 1984, it would not proceed at all. Still further, Mr Landau was not temperamentally averse to taking risks. He went ahead and installed a new shopfront without having obtained planning permission, confident in his own judgment (later vindicated by a
successful appeal to the minister) that permission would not be L
withheld for a design of such quality. Finally, there was no certainty that redevelopment would take place: no planning application had been made by 1984.

I consider that the other principals would have accepted Mr Landau's advice on the point had it arisen in May 1984. But had the point arisen then, I do not believe the answer would have been as cut-and-dried as Mr Landau's evidence suggested. I believe he would have been very reluctant indeed to abandon the project. Nevertheless, in my view, even Mr Landau would have felt that a certainty of five years was insufficient to justify a capital outlay of some £250,000. I accept his view that, had he known the true position
about the headleases on or about May 10 1984, he would not have M
proceeded to acquire the underlease of 7 and 9 Montpelier Street.

Accordingly, subject to proof of loss, Gran Gelato has established a good cause of action for damages against Richcliff under section 2(1) of the Misrepresentation Act 1967.

Duty of care

Gran Gelato's claim against Gershon Young depends upon the solicitors themselves owing directly to Gran Gelato a duty to take reasonable care when answering the preliminary enquiries on behalf of their client Richcliff. That Richcliff itself owed such a duty of care is common ground. Indeed, in the light of authorities such as *Esso Petroleum Co Ltd* v *Mardon* [1976] QB 801 the contrary could not be seriously argued. I was told that the existence of a claim directly

A against the solicitors may have practical importance in this case because it is questionable whether Richcliff is in a position to satisfy any substantial judgment debt.

The approach now to be adopted by the court when considering issues relating to the existence of a duty of care in the context of negligent misrepresentation is set out in the decision of the House of Lords in *Caparo Industries plc* v *Dickman* [1990] 2 AC 605. For there to be a duty of care there must be a foreseeability of damage and a close and direct relationship which has come to bear the label of "proximity". In addition, to adopt the phraseology of Lord Bridge of Harwich (at p 618), the situation must be one in which the court considers it "fair, just and reasonable" that the law should impose a
B duty of a given scope upon the one party for the benefit of the other.

Here there is no problem about foreseeability or a close and direct relationship. Gershon Young intended, or must be taken to have intended, that Gran Gelato and its solicitors should rely on the accuracy of the answers to the enquiries and that they should do so in connection with this particular transaction. Gershon Young foresaw, or are to be taken to have foreseen, that Gran Gelato might suffer financial loss if the answers were incorrect. Certainly, in so far as the enquiries related to matters of title, as did enquiry 3(A), Gran Gelato and its solicitors would be relying on Gershon Young to apply their legal expertise as solicitors when answering the enquiries. Thus far, all the indications point towards it being just and reasonable to impose on Gershon Young a duty of care in favour of Gran Gelato.
C Indeed, all the factors which lead to the conclusion that a duty of care was owed by Richcliff exist also in the case of Gershon Young. The only material difference is that in making the representations Gershon Young were acting not as principals but as agents on behalf of Richcliff. They gave their answers as Richcliff's solicitors, for and on behalf of Richcliff. Does this make any difference?

By itself, it does not. It is now established that the fact that the person making the representation was acting for a known principal does not necessarily negative the existence of a duty of care owed by him to the representee. The mortgagee's valuer may owe a duty of care to the mortgagor: *Smith* v *Eric S Bush* [1990] 1 AC 831*. Further, in *Resolute Maritime Inc* v *Nippon Kaiji Kyokai* [1983] 1 WLR 857 at p 861, Mustill J held that no cause of action in damages
D lies under the Act of 1967 against a negligent agent acting within the scope of his authority. One of the strands in his reasoning was that the Act does not need to be interpreted as having this effect, because there was no gap here which required to be filled. In such a case the representee can sue the agent at common law, in accordance with *Hedley Byrne* principles.

That was in the context of agents generally. In the particular context of enquiries before contract in a normal conveyancing transaction, Morritt J expressed a different view in *Cemp Properties (UK) Ltd* v *Dentsply Research & Development Corporation* [1989] 2 EGLR 205† at p 207. He observed that it would be absurd if the solicitor for one party to the transaction owed a duty of care to
E another party as well as to his own client.

In my view, in normal conveyancing transactions solicitors who are acting for a seller do not in general owe to the would-be buyer a duty of care when answering enquiries before contract or the like. In reaching the conclusion that the law should not generally import a duty of care in such circumstances, three factors have weighed with me. The first lies in the context in which such representations are made. The context is a contract for the sale of an interest in land. The buyer is formally seeking information from the seller about the land and his title to it. The answers given by the solicitor are given on behalf of the seller. The buyer relies upon those answers given on behalf of the seller; although the confidence of the buyer and his solicitors in the reliability of the answers may be increased when they
F see the answers have been given by a solicitor in the ordinary way. They will expect the seller's solicitor, as a professional acting on behalf of his client, to have got the answers right. I venture to think that in these circumstances one would expect to find that the law provides the buyer with a remedy against the seller if the answers were given without due care. I am far from persuaded that the fair and reasonable reaction to these facts is that there ought also to be a remedy against the other party's solicitor personally.

Second, what one finds is that the buyer indeed does have a remedy against the seller in respect of any misrepresentation in the answers given on his behalf. As already noted, the seller himself owes a duty G of care to the buyer. When, as is usual, the answers are given by the seller's solicitor, the seller will be as much liable for any carelessness of his solicitor as he would be for his own personal carelessness. He will be so liable, because in the ordinary way the solicitor has implied authority from the seller to answer on his behalf the traditional enquiries before contract made on behalf of the buyer. In providing the answers the solicitor is acting within the scope of his authority. Some of the enquiries will raise questions of fact. Others will raise legal conveyancing points which the client cannot answer himself. The client leaves all these matters to the solicitor to handle for him, after seeking instructions where appropriate from the client on any particular points on which the client may be expected to have relevant information. Thus, the purchaser to whom incorrect answers are H given is not without a remedy even if the fault was that of the seller's solicitor and not the seller himself. Whoever was at fault, the buyer has a remedy for damages at common law against the seller. (This, I interpose, is to be contrasted with a case such as *Smith* v *Bush*. There the mortgagor would have been without remedy if he did not have one against the valuer personally or his employer.)

Third, at the forefront of his submissions, Mr Jackson, for the second defendants, presented an argument that to impose a duty of care on solicitors would be to expose them to conflicting duties, with one duty owed to their clients and another different duty owed to the buyer. I am not persuaded that this would be so. The duty to the buyer would be to take reasonable care to see that the answers provided J were accurate. That duty would march hand-in-hand with a duty to the same effect owed by the solicitor to his own client. There would be no conflict. Nevertheless, and although I am not impressed by this argument based on conflict, it does seem to me that in the field of negligent misrepresentation caution should be exercised before the law takes the step of concluding, in any particular context, that an agent acting within the scope of his authority on behalf of a known principal, himself owes to third parties a duty of care independent of the duty of care he owes to his principal. There will be cases where it is fair, just and reasonable that there should be such a duty. But, in general, in a case where the principal himself owes a duty of care to the third party, the existence of a further duty of care, owed by the agent to the third party, is not necessary for the reasonable protection K of the latter. Good reason, therefore, should exist before the law imposes a duty when the agent already owes to his principal a duty which covers the same ground and the principal is responsible to the third party for his agent's shortcomings. I do not think there is good reason for such a duty in normal conveyancing transactions.

I add this. I appreciate that one consequence of this conclusion is that the buyer may be left without an effective remedy if the seller becomes insolvent. I do not think that is sufficient reason for adding on to the solicitor-client relationship a duty owed directly by the solicitor to the non-client in normal conveyancing transactions. That those with whom one deals may become insolvent is an ordinary risk of everyday life. L

Caveats

I must emphasise two points. First, there will be special cases where the general rule does not apply and a duty of care will be owed by solicitors to a buyer. A good illustration is the New Zealand decision of *Allied Finance & Investment Ltd* v *Haddow* [1983] 1 NZLR 22. There solicitors acting for a borrower certified to an intending mortgagee that specified documents had been duly executed and were fully binding on their client and that there were no other charges on the boat which was the intended security. The solicitors were held to owe a duty of care to the lender in connection with the giving of that certificate. On any reasonable appraisal of that arrangement, the solicitors must be taken to have assumed personal responsibility to the mortgagee for the accuracy of their certificate. That is a case in M which, to adapt the language used by Lord Donaldson MR in *Al-Kandari* v *J R Brown & Co* [1988] QB 665 at p 672, in the context of a solicitor acting in adversarial litigation, the solicitors had stepped outside their role as solicitors for their client and had accepted a direct responsibility to the lender.

Second, I must emphasise that nothing I have said detracts in any way from the duties owed by a solicitor to his own client when answering enquiries before contract. Nor does it relieve him from full financial responsibility for any carelessness on his part. If by his carelessness he exposes his principal to a claim by a buyer for negligent misrepresentation, he will be liable to indemnify his client on well-established principles.

*Editor's note: Also reported at [1989] 1 EGLR 169.
†Editor's note: Also reported at [1989] 37 EG 133.

A *Not a special case*

Mr Pymont, for the plaintiff, submitted that this was a special case: enquiry 3(A) related to legal rights, the answer was provided by Gershon Young without any disclaimer of liability, there was an express refusal to deduce title, there were practical difficulties in Gran Gelato's going direct to Eagle Star to obtain information about the headleases, in correspondence the underlease was said to follow the headleases closely, Gershon Young knew or should have known that Gran Gelato was likely to rely on what Gershon Young said and Gran Gelato paid Gershon Young's bill.

Suffice it to say, I cannot see in these features, taken separately or altogether, anything which would make this a special case to which the general rule should not apply. In particular, I cannot think that B payment by the buyer of the solicitor's bill as such, as distinct from the bill being met out of the purchase price paid by the buyer to the seller, can determine whether or not the seller's solicitors owe a duty of care to the buyer. My conclusion, therefore, is that Gran Gelato's claim against Gershon Young fails.

Contributory negligence

Richcliff has advanced a defence of contributory negligence. Clearly, this is available as a defence to the claim against Richcliff for damages for breach of the common law duty of care; but is it available to the claim against Richcliff under section 2(1) of the Misrepresentation Act 1967? In other words, does section 1 of the C Law Reform (Contributory Negligence) Act 1945 apply to a claim by a plaintiff for damages under the Misrepresentation Act 1967? So far as is material, section 1 (1) of the 1945 Act provides:

> Where any person suffers damage as the result partly of his own fault and partly of the fault of any other person or persons, a claim in respect of that damage shall not be defeated by reason of the fault of the person suffering the damage, but the damages recoverable in respect thereof shall be reduced to such extent as the court thinks just and equitable having regard to the claimant's share in the responsibility for the damage.

Fault is defined in section 4 as "negligence, breach of statutory duty or other act or omission which gives rise to a liability in tort or would, apart from this Act, give rise to the defence of contributory negligence". Section 2(1) of the 1967 Act provides:

> D Where a person has entered into a contract after a misrepresentation has been made to him by another party thereto and as a result thereof he has suffered loss, then, if the person making the misrepresentation would be liable to damages in respect thereof had the misrepresentation been made fraudulently, that person shall be so liable notwithstanding that the misrepresentation was not made fraudulently, unless he proves that he had reasonable ground to believe and did believe up to the time the contract was made that the facts represented were true.

Thus, in short, liability under the Misrepresentation Act 1967 is essentially founded on negligence, in the sense that the defendant, the representor, did not have reasonable grounds to believe that the facts represented were true. (Of course, if he did not believe the facts represented were true he will be liable for fraud.) This being so, it E would be very odd if the defence of contributory negligence were not available to a claim under the Act. It would be very odd if contributory negligence were available as a defence to a claim for damages based on a breach of a duty to take care in and about the making of a particular representation but not available to a claim for damages under the Act in respect of the same representation.

In my view, the answer to this point is provided by the decision of the Court of Appeal in *Forsikringsaktieselskapet Vesta* v *Butcher* [1989] 1 AC 852. There the court held that the 1945 Act applies to a case where there is a claim for damages for negligence at common law even if, in addition, there is a claim in contract to the same effect. O'Connor LJ (at p 866) adopted the view expressed by Prichard J in *Rowe* v *Turner Hopkins & Partners* [1980] 2 NZLR 550 at p 556, F regarding the equivalent section of the New Zealand legislation:

> I therefore conclude, in the absence of any clear authority to the contrary, that the first limb of the definition of s2 determines the meaning of the word "fault" as it relates to the plaintiff's cause of action: that accordingly, the Contributory Negligence Act cannot apply unless the cause of action is founded on some act or omission on the part of the defendant which gives rise to liability in tort: that if the defendant's conduct meets that criterion, the Act can apply — whether or not the same conduct is also actionable in contract.

Neil LJ (at p 875) agreed with this approach and I do not read Sir Roger Ormrod's judgment as differing on this point.

In the present case the conduct of which Gran Gelato complains founds a cause of action both in negligence at common law and under the 1967 Act. As already noted, under the 1967 Act liability is essentially founded on negligence. By parity of reasoning with the G conclusion in *Forsikringsaktieselskapet Vesta* v *Butcher* regarding concurrent claims in negligence in tort and contract, the 1945 Act applies in the present case where there are concurrent claims against Richcliff in negligence in tort and under the 1967 Act.

Having decided that the 1945 Act applies, I turn to consider whether there should be any apportionment in this case. Mr Wynne-Griffiths submitted that Gran Gelato and its solicitors were at fault in proceeding without first seeing the headleases, especially since Gran Gelato and Mr Landau were required to enter into covenants with Eagle Star to observe and perform the covenants (save as to rent) contained in the headleases so far as they affected the ground floor and basement of nos 7 and 9. H

In my view, it would not be just or equitable to make any reduction in Gran Gelato's damages. The essential feature of the present case is that Gran Gelato's claim, both at common law and under the 1967 Act, is based on misrepresentation. Richcliff intended, or is to be taken to have intended, that Gran Gelato should act in reliance on the accuracy of the answers provided by Gershon Young. Gran Gelato did so act. In those circumstances it would need to be a very special case before carelessness by Gran Gelato, the representee, would make it just and equitable to reduce the damages payable to compensate Gran Gelato for loss suffered by it in consequence of doing the very thing which, in making the representation, Richcliff intended should happen, *viz* that Gran Gelato should rely on the representation. In principle, carelessness in not making other J enquiries provides no answer to a claim when the plaintiff has done that which the representor intended he should do. This is a well-established principle when misrepresentation, whether innocent or fraudulent, is being relied on as a ground for seeking recision of the ensuing contract. In *Reynell* v *Sprye* (1852) 1 De GM & G 660 at p 710, Lord Cranworth LJ said:

> No man can complain that another has too implicitly relied on what he himself has stated.

In *Redgrave* v *Hurd* (1881) 20 ChD 1 at p 14, one of the leading authorities on this subject, Sir George Jessel MR, said:

> Another instance with which we are familiar is where a vendor makes a false statement as to the contents of a lease, as, for instance, that it contains no K covenant preventing the carrying on of the trade which the purchaser is known by the vendor to be desirous of carrying on upon the property. Although the lease itself ought to be produced at the sale, or might have been open to the inspection of the purchaser long previously to the sale, it has been repeatedly held that the vendor cannot be allowed to say, "You are not entitled to give credit to my statement." It is not sufficient, therefore, to say that the purchaser had the opportunity of investigating the real state of the case, but did not avail himself of that opportunity.

Likewise, in *Nocton* v *Lord Ashburton* [1914] AC 932 at p 962, Lord Dunedin said:

> No one is entitled to make a statement which on the face of it conveys a false impression and then excuse himself on the ground that the person to whom he made it had available the means of correction. L

In my view, the like approach is applicable when considering what is Gran Gelato's share in the responsibility for the damage flowing from the misrepresentation. It may be surprising that Mackenzie Mills did not advise their clients against proceeding without seeing the headleases containing the covenants their clients were undertaking to perform. But that has no relevance to the loss suffered by Gran Gelato in consequence of its reliance on the representation in answer 3(A).

Damages

Stated broadly, the measure of damages payable under the Misrepresentation Act 1967 is that sum of money which will place the plaintiff in the position he would have been in if the representation M had not been made. Had the representation not been made in the present case, Gran Gelato would not have gone ahead with the Montpelier Street transaction. Copies of the headleases were not being produced for inspection. It is inconceivable that Gran Gelato's solicitors would have proceeded without receiving a clear answer to enquiry 3(A). An accurate answer would have disclosed the existence of the break clauses.

As to the claim for damages for breach of the duty of care in answering the preliminary enquiries, the measure of damages is that sum of money which will place the plaintiff in the position it would have been in if reasonable care had been exercised. Had due care been exercised by Gershon Young when giving the answers, Gran Gelato

A would have learned of the break clauses and, hence, would not have gone ahead. In the present case, therefore, the two measures are the same.

Gran Gelato did proceed. It paid a premium of £30,000 to acquire the underlease, it incurred rental obligations, it incurred expenses and it laid out considerable sums of money in fitting out the property. *Prima facie* the fair measure of its loss is the total of all these outgoings, less the value of what Gran Gelato acquired for all this expenditure.

Gran Gelato's claims can be classified under four headings.

(1) *Premium of £30,000 and rental payment.* The parties' contentions, and also their valuation evidence, were remarkably B disparate. Gershon Young's valuation evidence was to the effect that even without a break clause Gran Gelato substantially overpaid. The market rent of the property in 1984 was £18,000 a year and without a break clause up to £2,000 more; and no premium, let alone one of £30,000, was justified. Richcliff adduced no evidence, but its case was that the market rental in 1984 was £26,000 even with a break clause and that the rental might have increased above this figure in September 1987 when the rent fell to be reviewed. As to the £30,000, this sum was paid for fixtures, fittings and goodwill, which Gran Gelato duly received. Gran Gelato's valuation evidence was that with a break clause the agreed rental should be discounted by 15 to 20% and that in August 1987 the property had no market value because by then redevelopment proposals were afoot and were a matter of public C knowledge.

As I see it, the position is this. The premium was effectively a lump sum which Gran Gelato had to agree to pay in order to get the underlease. In agreeing to pay £30,000 and an annual rental of £26,000, Gran Gelato, acting without professional advice, was paying a very full price, but these figures were not so high that they should not be taken fully into account in assessing damages. They were not absurd figures. With a break clause, the appropriate figures would have been smaller. The underlease acquired was less valuable. But it does not follow from this that, in assessing damages, Gran Gelato should have to give credit and accept responsibility for a rental, albeit reduced, for the whole period of the underlease. It would not be right to apply the *prima facie* rule in this case, because D what happened was that when the misrepresentation came to light in August 1987 it was too late for Gran Gelato to dispose of the property. Gran Gelato found that, because of the break clauses, it now had an unsaleable asset. Indeed, if anything, the underlease was a liability, with a negative value. Thus, when the misrepresentation was discovered Gran Gelato found itself saddled with a rental obligation for a property from which henceforth (*viz* after September 1987) it received no benefit and which it could not turn to financial advantage. In my view, Gran Gelato is entitled to be reimbursed in respect of this rental obligation. It was a liability which flowed directly from the misrepresentation.

But, on that footing, the matter stands quite differently for the E period up to September 1987. In respect of that period it is fair that Gran Gelato should give credit for a rental and, moreover, that should be at the agreed rate. The misrepresentation did not affect the quality of Gran Gelato's enjoyment of the property while in occupation.

As to the premium, Gran Gelato is not entitled to repayment of the whole £30,000. Gran Gelato paid this sum in order to obtain a 10-year underlease. In fact, Gran Gelato acquired an underlease subject to Eagle Star's rights under the headlease break clauses. When the misrepresentation came to light, the underlease was not saleable because of the existence of the break clauses. By then Gran Gelato had enjoyed the property for three years. In my view, the fair course is to require Gran Gelato to give credit for an apportioned part of the F premium; Gran Gelato received some benefit in return for the premium and this it should bring into account. There was no evidence before me directed as to apportionment, so I must do the best I can by the light of nature. A straight-line apportionment, whereby Gran Gelato would have to credit for only three-tenths of the premium, would be unduly favourable to Gran Gelato, since a higher proportion of the premium ought to be attributed to the earlier years of the term. I consider a fair apportionment is that Gran Gelato should give credit for one-half of the premium.

(2) *Legal costs of acquisition.* Gran Gelato paid legal fees and G disbursements totalling £6,308 in connection with the acquisition of the underlease. The figure is made up of stamp duty (£520), Gershon Young's fees and disbursements which Gran Gelato had agreed to pay (£2,388), and Mackenzie Mills' fees and disbursements (£3,400). In arriving at the latter figure I have deducted those parts of the bills which Mr Mills agreed were attributable to other matters. These expenses would not have been incurred if Gran Gelato had not acquired the Montpelier Street underleases. On the other hand, Gran Gelato derived some benefit from this expenditure because it traded for three years from the property. Painting with a broad brush, I consider Gran Gelato should recover one-half of this expenditure, but not more.

(3) *Shopfitting costs.* Gran Gelato spent £99,030 in respect of the H design and installation of shopfittings at the property and subsequent works of alteration. Mr Pymont submitted that all these costs are recoverable. None of this expenditure would have been incurred if the misrepresentations had not been made. In my view, that would be to spread the net of the damages claim far too wide. This argument proves too much. If sound, it would mean that Gran Gelato could equally well recover all its trading losses, because they would not have been incurred if the Montpelier Street premises had not been acquired and, hence, the business had not started. That cannot be right. In my view, none of the shopfitting costs are recoverable. These works were carried out to enable Gran Gelato to trade and they were used by Gran Gelato while it traded. They ceased to be of any J value to Gran Gelato when it stopped trading. But Gran Gelato stopped trading for reasons wholly unconnected with the misrepresentation. Gran Gelato's loss here flowed solely from the commercial failure of its business. There was no causal connection between the misrepresentation and this loss.

(4) *The deed of variation and other legal costs.* Gran Gelato paid £4,000 for Richcliff's licence for change of use of no 9 and paid legal expenses totalling £3,150 in connection with that deed and with abortive attempts to sublet: Gershon Young's fees (£1,350), Eagle Star's solicitors' fees (£50), and Mackenzie Mills' fees (£1,750). These expenses were all incurred before the misrepresentation came to light. Gran Gelato derived no benefit from this expenditure. Any possibility of deriving benefit from the deed of variation ceased when K the misrepresentation came to light. Nor is it likely that Gran Gelato would ever have derived any benefit from the other legal expenditure: the existence of the break clauses presumably would have stymied the proposed assignments or sublettings ultimately, even though this was not appreciated in the course of the abortive negotiations.

(5) *Conclusion on damages.* Gran Gelato is therefore entitled to recover the following sums by way of damages:

	£
One-half of the premium	15,000
One-half of the legal costs of acquisition	3,154
The deed of variation licence fee and other legal costs	7,150
	£25,304

L In addition, Gran Gelato is entitled to recover and set-off against and thereby almost wholly extinguish Richcliff's counterclaim for rent sums totalling £48,420 in respect of unpaid rent after September 1987. Richcliff is entitled to recover £2,000, being the unpaid balance of rent for the quarter ending September 1987.

Footnotes

I mention two final points. First, in its defence Richcliff raised a point to the effect that by consenting to the grant of the underlease which contained no break clause Eagle Star was estopped as against Gran Gelato from relying on the break clauses in the headleases. Eagle Star was then joined as a defendant in the action and this issue was decided, as preliminary issue, in favour of Eagle Star in June M 1989. Eagle Star did not participate in the trial of the action.

Second, there are on foot claims by Richcliff against Gershon Young for an indemnity and contribution. The parties agreed, as indeed seemed sensible, that the determination of these claims should await judgment in the action. So, these claims remain to be resolved.

Judgment for the plaintiff against the first defendant for £25,304. Judgment for the first defendant on its counterclaim for £2,000. Plaintiff's claim against the second defendant dismissed.

For further cases on this subject see p 163

REAL PROPERTY AND CONVEYANCING

A

Court of Appeal

(Before Lord Justice PURCHAS, Lord Justice RALPH GIBSON and Lord Justice TAYLOR)

December 11 1991

SWANSEA CITY COUNCIL v GLASS (EXECUTOR)

Limitation Act 1980 — Claim by local authority for expenditure — A landowner failing to comply with compulsory repair notices — Whether proceedings issued to recover expenditure statute-barred — Appeal by council
B **dismissed**

The defendant (as executor of the will of the late Mr H Glass) had control of 96 King Edward Road, Swansea, a dwelling-house let to a tenant — Following the tenant's complaint about the condition of the property, the council issued two repair notices under section 9(1)(*b*) of the Housing Act 1957 requiring the defendant to put the dwelling-house in proper repair — The respondent did not appeal and failed to comply with the notices — The council did the necessary work and completed the works required by the first notice on September 7 1983 — The council served a demand for the
C **expenses in respect of the works required by the notices, respectively on May 31 1984 and April 25 1985 — The defendant did not pay and the council issued a summons seeking payment on May 30 1990 — The summons was issued more than six years after completion of the works but less than six years from the service of the demands — On a preliminary point, Judge Michael Evans QC decided that the council's cause of action accrued when the works were completed — The council appealed**

***Held*: The appeal was dismissed — It is implicit in sections 10 and 11 of the Housing Act 1957 that a demand for expenses must be served by the local authority before proceedings can**
D **be taken to recover them — A cause of action may well accrue before, for procedural reasons, the plaintiff can bring proceedings — A "cause of action" is every fact which it would be necessary for a plaintiff to prove, if traversed, in order to support his right to the judgment of the court: see *Coburn* v *Colledge* — The council had to prove (1) that the repair notice was served on the person having control of the house to execute works; (2) that he failed within the time specified to execute them; (3) that the local authority have themselves carried out the works specified in the notice; and (4) that they have incurred expenses in so doing —**
E **Accordingly, time runs from the accrual of the cause of action, ie when these four elements are complete — The requirement to serve a demand is a procedural condition precedent to bringing proceedings; it is not part of the cause of action**

The following cases are referred to in this report.

Central Electricity Generating Board v *Halifax* [1963] AC 785
Coburn v *Colledge* (1897) 1 QB 702
Read v *Brown* (1888) 22 QBD 128
Sevcon Ltd v *Lucas CAV Ltd* [1986] 1 WLR 462; [1986] 2 All ER 104, HL

This was an appeal by Swansea City Council against a decision of Judge Michael Evans QC given in Swansea County Court on February 6 1991 on a preliminary point of law in a claim against the defendant, Mr Martin Glass, to recover expenditure incurred by the council in carrying out works the subject of repair notices under section 9 of the Housing Act 1987. F

Kenneth Thomas (instructed by the solicitor to Swansea City Council) appeared for the appellants; Robert Kirk (instructed by Peter Williams & Co, of Swansea) represented the respondent.

Giving judgment, **TAYLOR LJ** said: This is an appeal by Swansea City Council against a decision of Judge Michael Evans QC, given in Swansea County Court on February 6 1991. The learned judge had to consider a preliminary point concerning the application of the Limitation Act 1980 to the appellants' claim. He decided in favour of the respondent defendant, Mr Martin Glass, but granted leave to appeal. G

The facts were not in dispute. The respondent (as executor of Mr H Glass) was the person in control of 96 King Edward Road, Swansea, for the purposes of the Housing Act 1957. The house was let to a tenant. In 1983 the tenant complained to the appellant council about the condition of the property. As a result, the appellants issued two notices under section 9(1)(*b*) of the 1957 Act requiring the respondent to put the house into proper repair. The first notice was served on June 1 1983 and the second on October 28 of the same year. The respondent did not appeal and failed to comply with the notices. H
Accordingly, the appellant council did the necessary work themselves. They completed the works required by the first notice on September 7 1983. There is some uncertainty as to when they completed the works required by the second notice but the precise date is immaterial to the present appeal.

The appellants served upon the respondent written demands for the expenses they had incurred in respect of the works required under the notices, respectively on May 31 1984 and April 25 1985. The respondent did not pay. There was then a long delay for which no explanation has been given. Finally, on May 30 1990 the appellant council issued a summons in the county court seeking payment of both amounts. The summons was thus issued more than six years J
after completion of the works but less than six years from the service of the demands.

The local authority were empowered to do the work themselves by section 10(1) of the Act, which provided that if the person in control of the house did not comply with a notice to execute work then, after specified periods of time, "the local authority may themselves do the work required to be done by the notice . . .". The entitlement to recover their expenses for doing the work derived, at the relevant time, from section 10(3) of the Act, which provided, so far as is relevant, as follows:

> Any expenses incurred by the local authority under this section, together with interest from the date when a demand for the expenses is served until payment, may . . . be recovered by them, by action or summarily as a civil debt, from K
> the person having control of the house . . .

The relevant limitation period was prescribed in section 9 of the Limitation Act 1980 as follows:

> (1) An action to recover any sum recoverable by virtue of any enactment shall not be brought after the expiration of six years from the date on which the cause of action accrued.

The preliminary point to be decided by the learned judge was: "when did the local authority's cause of action accrue?" The respondent contended it accrued when the works were completed, in which event the action is statute-barred. The appellants contend the

A correct date is when the demand for payment was served or when it became operative.

It is convenient first to consider an argument on behalf of the appellants that the relevant statutory provisions are those contained in the Housing Act 1985. Mr Thomas took this point before the learned judge, who rejected it. He has repeated the argument before us. The 1957 Act was clearly in force at the time of the notices, the works and the demands. It was repealed by the Housing (Consequential Provisions) Act 1985, which came into force on April 1 1986, as did the Housing Act 1985, which is a consolidating Act. Section 2 of the Housing (Consequential Provisions) Act 1985 provides, so far as is relevant, as follows:

B (1) The re-enactment of provisions in the consolidating Acts, and the consequent repeal of those provisions by this Act, does not affect the continuity of the law.

(2) Anything done . . . or having effect as done, under a provision reproduced in the consolidating Acts has effect as if done under the corresponding provision of the consolidating Acts.

Section 5(1) of the same Act provides that Schedule 3 has effect with respect to transitional matters. The relevant provisions of Schedule 3 are as follows:

1. — (1) The general rule is that the provisions of the consolidating Acts apply, in accordance with section 2 of this Act (continuity of the law), to matters arising before the commencement of those Acts as to matters arising after that commencement.

C . . .

(4) The general rule does not apply so far as a provision of the consolidating Acts gives effect to an amendment (in pursuance of a Recommendation of the Law Commission and, in some cases, the Scottish Law Commission).

The provisions which replace those contained in section 10(3) of the 1957 Act are to be found in section 193 of, and Schedule 10 to, the Housing Act 1985. In particular, if one compares the provisions relating to recovery of expenses in section 10(3) of the 1957 Act with those in Schedule 10 to the 1985 Act it is clear that the whole scheme has been rewritten and amended in accordance with recommendations in the Law Commission Report No 144 Cmnd 9515 dated May 1985. In those circumstances, the learned judge held D that the provisions in the 1985 legislation did not apply in the present case, which continued to be governed by the 1957 Act. In my judgment, he was right. I should add that the outcome would have been no different even if the 1985 Act had applied, since Mr Thomas concedes that the time when the course of action arises must be the same under both Acts.

Approaching the crucial issue in this case by reference to the 1957 Act, Mr Thomas put forward two propositions: (1) although the Act does not expressly say so, it is implicit in sections 10 and 11 that a demand for expenses must be served by the local authority before proceedings can be taken to recover them; (2) service of the demand is not merely a procedural requirement but an essential ingredient in the cause of action.

E In my judgment, his proposition (1) is correct. It is supported by reference to section 10(4), section 11(1) and section 37(1) of the Act, which provide as follows:

10. — . . . (4) In all summary proceedings by the local authority for the recovery of any such expenses, the time within which the proceedings may be taken shall be reckoned from the date of the service of the demand or, if an appeal is made against that demand, from the date on which the demand becomes operative.

. . .

11. — (1) Any person aggrieved by —

(*a*) a notice under the foregoing provisions of this Part of this Act requiring the execution of works,

(*b*) a demand for the recovery of expenses incurred by a local authority in executing works specified in any such notice,

F (*c*) an order made by a local authority with respect to any such expenses,

may, within twenty-one days of the service of the notice, demand or order, appeal to the county court within the jurisdiction of which the premises to which the notice, demand or order relates are situate, and no proceedings shall be taken by the local authority to enforce any notice, demand or order in relation to which an appeal is brought before the appeal has been finally determined.

. . .

37. — (1) Any notice, demand or order against which an appeal might be brought to a county court under this Part of this Act shall, if no such appeal is brought, become operative on the expiration of twenty-one days from the date of the service of the notice, demand or order, and shall be final and conclusive as to any matters which could have been raised on such an appeal, and any such notice, demand or order against which an appeal is brought shall, if and so far as it is confirmed by a county court judge, or the Court of Appeal, become operative as from the date of the final determination of the appeal. G

Clearly section 10(4) assumes a demand will be served before summary proceedings are taken. There seems no good reason why the same should not apply before county court proceedings are brought. Moreover, under sections 11 and 37, a person aggrieved by a demand has a right of appeal against it and proceedings cannot be brought until any appeal has been finally determined. It would be surprising if the right of appeal depended on whether the local authority chose to serve a demand before taking proceedings. It is of interest that the Law Commission, at p 33 of their report (*supra*), were of the view that service of a demand was implicit in section 10 of the 1957 Act. On their recommendation, the obligation to serve a H demand is imposed expressly in the 1985 Act.

However, given that a demand must be served before proceedings could be taken, it does not follow that Mr Thomas' second proposition is correct.

The classic definition of a cause of action is that originally propounded by Lord Esher in *Read* v *Brown* (1888) 22 QBD 128 and repeated by him in *Coburn* v *Colledge* (1897) 1 QB 702. He said at p 706:

The definition of "cause of action" which I gave in *Read* v *Brown* has been cited. I there said that it is "every fact which it would be necessary for the plaintiff to prove, if traversed, in order to support his right to the judgment of the Court." The language I used obviously means this: the plaintiff in order to J make out a cause of action must assert certain facts which, if traversed, he would be put to prove.

Coburn was a solicitor who did some work for the defendant Colledge in 1889. He sent a bill of costs in 1889 which did not reach the defendant in Australia until 1891. Action was commenced in 1896. Section 37 of the Solicitors Act 1843 provided that no solicitor could commence an action for fees until one month after he delivered to his client a bill of fees. The question was whether delivery of the bill amounted to an ingredient of the cause of action or whether the cause of action arose when the work was done and delivery of a bill was only a procedural requirement. Lord Esher said at p 705:

Before any enactment existed with regard to actions by solicitors for their costs, a solicitor stood in the same position as any other person who has done K work for another at his request, and could sue as soon as the work which he was retained to do was finished, without having delivered any signed bill of costs or waiting for any time after the delivery of such a bill. Then to what extent does the statute alter the right of the solicitor in such a case, and does the alteration made by it affect or alter the cause of action? It takes away, no doubt, the right of the solicitor to bring an action directly the work is done, but it does not take away his right to payment, for it, which is the cause of action. The Statute of Limitations itself does not affect the right to payment, but only affects the procedure for enforcing it in the event of dispute or refusal to pay. Similarly, I think, s 37 of the Solicitors Act, 1843, deals, not with the right of the solicitor, but with the procedure to enforce that right. It does not provide that no solicitor shall have any cause of action in respect of his costs or any right to be paid till the expiration of a month from his delivering a signed bill of costs, but merely that he shall not commence or maintain any action for the L recovery of fees, charges, or disbursements until then. It assumes that he has a right to be paid the fees, charges, and disbursements, but provides that he shall not bring an action to enforce that right until certain preliminary requirements have been satisfied. If the solicitor has any other mode of enforcing his right than by action, the section does not seem to interfere with it. For instance, if he has money of the client in his hands not entrusted to him for any specific purpose, there is nothing in the section to prevent his retaining the amount due to him out of that money.

Coburn v *Colledge* was a case arising from a contract for services and it is well established at common law that the cause of action in such cases accrues when the work is complete. In the present case the cause of action arises under statute and Mr Thomas argues that different considerations apply. M

However, in *Central Electricity Generating Board* v *Halifax Corporation* [1963] AC 785, where a similar issue arose, the cause of action derived from statute. The Electricity Act 1947 transferred the assets of electricity undertakings to electricity boards and provided that the property held by local authorities in their capacity as authorised undertakers should, on vesting day (April 1 1948), vest in the relevant board. The question whether certain moneys had been held by the respondent in the relevant capacity was referred under section 15 of the Act to the minister, who decided on September 18 1958 that it was. The appellants sought to recover the money. A limitation point was taken. The House of Lords held that the cause of action accrued at vesting date, even though the appellants could not

A have proved the moneys were held by the respondents as authorised undertakers until the minister so decided. At p 801 Lord Reid said:

> No new right or liability came into existence at [the date of the minister's decision]. It is quite clear, and it is now admitted by the appellants, that the effect of the Minister's decision was merely to prove that this sum had belonged to the appellants ever since the vesting date. It created no new right of property or chose in action: it merely enabled a pre-existing right to be enforced.

Again, in *Sevcon Ltd* v *Lucas CAV Ltd* [1986] 1 WLR 462, the House of Lords had to consider section 13(4) of the Patents Act 1949, which provided:

> After the date of the publication of a complete specification and until the sealing of a patent in respect thereof, the applicant shall have the like
> B privileges and rights as if a patent for the invention had been sealed on the date of the publication of the complete specification: Provided that an applicant shall not be entitled to institute any proceedings for infringement until the patent has been sealed.

It was held that an applicant for a patent has all the rights which a holder has from the moment the complete specification of the patent is published. Accordingly, his cause of action for an infringement before the date of sealing accrues at the date of the infringement but cannot be enforced until the procedural requirement of sealing is met. Lord Mackay said at p 467E:

> If he were to institute proceedings for infringement before the patent for the invention was sealed, the procedural requirement of the proviso would not be
> C satisfied but a statement of claim could not be struck out as disclosing no cause of action although it might be liable to be struck out as an abuse of the process of the court.

Although not on all fours with the present case, these decisions show that a cause of action may well accrue before, for procedural reasons, the plaintiff can bring proceedings. Where the cause of action arises from statute, the question as to what is merely procedural and what is an inherent element in the cause of action is one of construction.

In the present case, the learned judge carefully analysed the provisions of section 10(3) and (4) and, in my judgment, reached the correct conclusion as to when the cause of action arose. Section 10(3) gives the right to recover "any expenses incurred by the local
D authority under this section". The local authority must therefore prove: (1) that notice in accordance with section 10(1) was served on the person having control of the house to execute works; (2) that he failed within the time specified to execute them; (3) that the local authority have themselves carried out the works specified in the notice; and (4) that they have incurred expenses in so doing. A statement of claim alleging those matters could not, in my view, be struck out as disclosing no cause of action.

The phrase "together with interest from the date when a demand for the expenses is served until payment" in section 10(3) shows that interest runs only from the date of the demand and (by implication) not from the date when the cause of action arose.

Section 10(4) provides expressly that where the local authority opt E
to take summary proceedings to recover their expenses, the limitation period runs from the date of service of the demand or, if there is an appeal, the date when the demand becomes operative. Again, by implication, since no such provision is applied to proceedings in the High Court or county court, time in those proceedings does not run from the date when the demand is served or becomes operative. It will run from the accrual of the cause of action which, *ex hypothesi*, is a different time.

The rationale of the distinction between summary and other proceedings probably lies in the respective limitation periods. In summary proceedings the period is six months. If time were to run from the accrual of the cause of action, ie when the expenses were F
incurred, summary proceedings might often be statute-barred before they could be brought, especially where there was an appeal against the demand. In other proceedings, however, the limitation period of six years gives, or should give, the local authority ample time to sue even after an appeal against their demands. In my judgment, the expression, special to section 10(4), that time runs from service of the demand or when it becomes operative is intended to distinguish summary proceedings from other proceedings. *Inclusio unius, exclusio alterius*. In other proceedings, time runs from the accrual of the cause of action, ie when the four elements identified above are complete. Thus, I conclude that the requirement to serve a demand is a procedural condition precedent to bringing proceedings. It is not part of the cause of action. G

I am fortified in this view by consideration of what could result if the appellants were right. Upon their argument, the local authority could delay service of a demand indefinitely. Then, having served their demand long after the works were complete, they would have a further six years in which to take proceedings in the High Court or the county court. In *Coburn* v *Colledge* (*supra*) Lopes LJ said at p 709:

> There is nothing in the section, so far as I can see, inconsistent with the view that the cause of action arises when the work is completed. It was urged that, if this construction were adopted, a solicitor would have a shorter time during which he may abstain from bringing his action for work done than the rest of Her Majesty's subjects. That may be so; but on the other hand, if the plaintiff's contention is correct, the solicitor may abstain from delivering his H
> bill for twenty years, and then at the end of that time he may deliver it and sue after the expiration of a month from its delivery. It seems to me that that would be a very anomalous and inconvenient result.

For the reasons given above, I am of the opinion the learned judge's answer to the question posed was correct and I would dismiss this appeal.

RALPH GIBSON and PURCHAS LJJ agreed and did not add anything.

Appeal dismissed with costs.

For further cases on this subject see p 185